Fabric and Mineral Analysis of Soils

Fabric and Mineral Analysis of Soils

ROY BREWER

Division of Soils
Commonwealth Scientific and Industrial Research Organisation (Australia)

John Wiley & Sons, Inc., New York · London · Sydney

Library of Congress Catalog Card Number: 64-23828

Printed in the United States of America

Preface

This book is intended primarily for pedologists and for the training of pedologists—that is, scientists whose main interests are soil morphology and soil genesis. The overall purpose is to set out a system of description and classification of the phenomena observed in soils and the uses of mineral analysis of the sand and silt fractions in the general context of studies of soil genesis. The prime objectives are to facilitate precise descriptions of soil materials to any desired level of detail and to quantify measurements of degree of weathering within profiles and the effects of soil-forming processes. While the literature published in journals is increasing rapidly, there is no textbook at present which brings all these aspects together with these objectives.

Many avenues of investigation of pedogenesis have already been explored, such as total chemical analysis, variations of ratios of elements with depth, experimental work in columns in the laboratory, and so on. Researchers engaged in such work are, in fact, pedologists, even though their basic training and immediate interest may be in some specific discipline, but interpretation of the data of such work in pedogenetic terms depends on knowledge of the kinds of materials to which the studies appertain and their arrangement in soils. It behoves all researchers who are studying the genesis and properties of soils, therefore, to understand as much as possible the details of the morphology and the constitution of the materials they are studying. It is hoped that the contents of this book will contribute to the better understanding of the nature of soils.

The contents are set out in a number of sections. The introduction traces the events that have led to a realization of the need for more detailed morphological data and classification of soil features. Part I sets out the concepts of soil used to develop the themes of mineral and fabric analysis, and points out the analogy of pedology, pedogenesis, and pedography to petrology, petrogenesis, and petrography. Part II deals with the application of mineral analysis to the study of soils. Certain subjects, such as size and shape analysis and mineral weathering, are introduced for completeness but are not dealt with exhaustively because of the availability of detailed treatises in other texts. Part III deals with structure and fabric analysis of soils. Chapter 7 gives the concepts of structure and fabric, the general nature of soil materials and their constituents, and a broad outline of the system of description according to the levels of organization; this is a most important chapter for the understanding of what follows. Part IV deals with description, classification, and interpretation of pedological features and the various levels of structure and fabric. Many of the characteristics described are observed best in thin section on a petrological microscope, so the whole book has a bias toward this technique. It should always be borne in mind, however, that the subject matter is structure and fabric, and any technique is admissible that will provide pertinent data. In fact the significant characteristics vary in magnitude from ultramicroscopic to megascopic. Part V is a general discussion of the role fabric and mineral analyses can be expected to play, and are already playing, in the study of soils.

Several previously published papers have been used almost in their entirety, with some alterations necessitated by more recent data. These are:

Brewer, R. (1960). The petrographic approach to the study of soils. *Trans. 7th Intern. Cong. Soil Sci.*, Madison.

Brewer, R. (1960). Cutans: their definition, recognition, and classification. *Jour. Soil Sci.* **11,** 280–292.

Brewer, R., and J. R. Sleeman (1960). Soil structure and fabric: their definition and description. *Jour. Soil Sci.* **11,** 172–185.

Brewer, R., and J. R. Sleeman (1963). Pedotubules: their definition, classification, and interpretation. *Jour. Soil Sci.* **14,** 156–66.

Brewer, R., and J. R. Sleeman (1964). Glaebules: their definition, classification, and interpretation. *Jour. Soil Sci.* **15**, 66–78.

I wish to acknowledge permission to use this material. A number of the figures have been taken from other publications, and I am grateful to the various publishers for permission to reproduce them; acknowledgment is made in the description of the appropriate figures.

I am indebted to Mr. J. R. Sleeman, Dr. R. J. Hunter, and Dr. D. S. McIntyre of the Division of Soils, Commonwealth Scientific and Industrial Research Organization, Australia, for criticism and help. In particular, Mr. Sleeman was co-author of several papers on which Parts III and IV are based, and his constructive criticism of the manuscript has been invaluable. Thanks are also due to Dr. W. D. Keller, Professor of Geology, Columbia University, Missouri, who supplied the thin sections of sedimentary rocks, which are described; to Dr. G. H. Taylor of the Division of Coal Research, C.S.I.R.O., Australia, for the electron micrograph (Fig. 121); and to Dr. J. Lovering of the School of Advanced Studies, Australian National University for the electron microprobe trace of a ferri-argillan (Fig. 124).

R. Brewer

June 1, 1964
Canberra, Australia

Contents

Introduction

As soil science has developed, the systematic description of soil profiles in the field and in the laboratory has become progressively more detailed. In the early days the prime objective was to distinguish different kinds of profiles on the number and arrangement of recognizable horizons, which were described very broadly in terms of color and texture. These descriptions were amplified later to include "structure," consistence, and the occurrence of detectable amounts of relatively soluble materials, such as carbonates and sulfates. Still more recently such morphological features as "clay skins" have been described (U.S.D.A., 1960). Concomitant with this development there has been an increasing demand for chemical, physical, and mineralogical analyses.

This progression in complexity of description points to the fact that many soil materials are highly heterogeneous, in contrast to earlier concepts of relative homogeneity of single soil horizons. Morphological studies by microscopic methods have emphasized this heterogeneity and revealed many significant expressions of it that hitherto received no attention. They have also shown that the physical constitution of soil materials has many similarities to sedimentary rocks; the problems of description of soils are similar to those for sediments, and the techniques used in the study of sediments, particularly petrographic techniques, are proving just as useful in the study of soils.

As early as 1931 Kubiena was examining soils with a petrological microscope. In 1933 Fry, who had the title of "soil petrographer," published a bulletin on petrographic methods for soils laboratories that dealt with the separation and recognition of minerals in soils. In 1937 Kubiena published a fairly satisfactory method of preparation of thin sections of soil materials, and in 1938 he published a book dealing essentially with fabric analysis of soils. With more recent techniques satisfactory thin sections of most soil materials can be made with relative ease. Simultaneously methods for separation and examination of minerals in soils have been improved greatly, especially in regard to freeing the mineral grains (dispersion) and cleaning them. As a result of these advances petrographic study of soils has developed rapidly throughout the world during the past decade.

In spite of this, no systematic scheme for the examination of soils by petrographic methods exists. Several significant profile studies based on quantitative estimation of the minerals separated from various horizons in soil profiles have been made, and numerous papers have been published on identification of the material from which soils have developed, using mineralogical methods. This is the application of the techniques of sedimentary petrology to soil materials, and it has been developed to a relatively satisfactory level. On the other hand, the level of fabric analysis of soil materials is far below that of the fabric analysis of rocks, both in regard to completeness and system. Until recently most fabric studies of soil materials were aimed at recording specific phenomena or producing a limited amount of additional evidence for a hypothesis that depends mainly on other data. Undoubtedly this is primarily because soil materials have important characteristics to which the descriptive terms and system of fabric analysis used for rocks cannot be applied directly; further development of fabric analysis of soil materials depends on enumeration of the properties which should be described, development of a system of description, and definition of specific descriptive terms.

The object of this work is to set out such a system of description for soil materials and to propose a system of mineral analysis for specific studies of the origin and development of soils. Such a system naturally involves a number of concepts and definitions. A search for suitable existing concepts has emphasized that there is considerable confusion in published works on petrology and pedology. Various interpretations of many concepts and definitions overlap considerably, and occasionally the same term is used for different concepts, even within the one science. An attempt has been made, therefore, to develop a system specifically for soils but allied as closely as possible to concepts used in petrology because of the close relationship between the nature of rocks and soil materials.

There is a school of thought among soil scientists strongly opposed to the introduction and definition of new terms and concepts. Unsystematic proliferation of names is certainly to be avoided, especially in the older sciences where the existing vocabulary specific to that science is adequate for normal investigation. Pedology, however, is not yet in the happy position of having an adequate vocabulary. Indeed pedology has very few terms specific to the science; many terms have been taken from other sciences or from English dictionaries and redefined so that their meanings in pedology are not quite what they are in normal English usage or in the science from which they were taken; the meanings have been broadened or narrowed or given a new slant. On the other hand, no terms have been proposed for many quite specific pedological phenomena, and so these must be described in considerable detail if they are to be recorded; all that is needed, is a single, simple, precisely defined term.

These deficiencies in nomenclature militate against development of pedology as a science. Many observers record only those features which they have been trained to recognize and for which adequate terms exist. Certainly it is most inconvenient and inefficient to repeat long descriptions for features or arrangements that are sufficiently specific to have a concise name; such descriptions are tedious to read and difficult to interpret. Through experience, we find that there are few greater difficulties in pedology than the simple interchange of morphological data between workers; misunderstandings are unavoidable unless comparisons can be drawn between soil materials and profiles with which all the workers concerned are familiar. Even then many of the details of a soil material are ignored or cannot be adequately conveyed from one worker to another, if either is unfamiliar with the particular soil material under discussion.

No apology is made, therefore, for the relatively large number of new terms introduced in this work or for the reassessment and redefinition of a large number of terms presently in use. It is not expected that all these terms and definitions will be accepted without question, but the objective of this work will have been achieved if the suggestions embodied in it lead to the introduction and use of a sufficient number of properly defined terms and concepts to facilitate precise description of soil materials and recording of the many details of soil materials which are given too little importance at present.

This work is not intended to be a complete review of past work. The principles involved in some aspects of pedology and sedimentary petrology, and the results of certain investigations, are reviewed to some extent but only as far as is necessary to establish the principles involved in pedographic studies, to underline the difficulties involved in certain aspects of such studies, or to reach certain conclusions.

PART I

SOIL AND PEDOGRAPHY

1

Soil and Pedography

To avoid any possible misunderstanding it is appropriate to discuss at the outset the definition and nature of soils and the concept and scope of pedography, which together form the basis of the approach to the study of soils used in this book.

1.1 SOIL

Jenny (1941) pointed out that there is no generally accepted definition of *soil* and that it is practically impossible to define soil in such a way as to satisfy all soil workers. This is due to the wide divergence of interests concerned with studies involving soil—from those interested primarily in agriculture to those concerned with rock and mineral weathering. This breadth of interest is reflected in the variety of definitions proposed (e.g., Hilgard, 1914; U.S.D.A., 1951; Crocker, 1959; Joffe 1936; de Sigmond, 1938; Kubiena 1953). Although he did not define soil Jenny (loc. cit.) emphasized what is probably its most distinctive characteristic when he stated that "all soils are anisotropic"; that is, the sequence of soil properties differs profoundly along a line extending from the surface of the soil toward the center of the earth from that along lines parallel to the surface. This anisotropy is basically the result of weathering of hard rocks or sedimentary bodies and of reorganization, translocation, and concentration of the more mobile constituents due to the effects of the environment. Soil, therefore, can reasonably be defined in the following way, which is appropriate for pedographic studies:

Soil is the collection of natural bodies formed by alteration of sedimentary and/or igneous bodies due to exposure at the earth's surface and having an anisotropic arrangement of properties along an axis normal to the earth's surface.

This is not so much a definition as a concept since specific properties and variations in properties with depth are not listed. For a precise definition of soil such characteristics as are due to "exposure at the earth's surface" must be identified and listed.

1.2 SOIL PROFILE

Most soil workers have recognized the importance of the vertical anisotropy of soil properties in their definitions of a *soil profile,* and there is much less divergence in concept between these definitions than between those of soil. The point of greatest disagreement is the depth which should be regarded as the lower limit of the soil profile. In most cases this is set in reference to the "parent material." There is general agreement that the soil profile is the vertical succession of genetic horizons or layers, but Hall (1945) limited the depth "down to the raw parent material," Robinson (1949) included the "upper part of the undifferentiated parent material," and the *U.S.D.A. Soil Survey Manual* (1951) included "the parent material or other layers beneath the solum that influence the genesis and behavior of the soil." Apart from the disagreement as to the inclusion or otherwise of the parent material and specific underlying horizons, the application of these definitions is unsatisfactory because of the difficulty in recognizing the "parent material." Inherent in all discussions of parent material is the implication that geological weathering can be distinguished from pedological weathering. For example, the *U.S.D.A. Soil Survey Manual* (1951) defined parent material as ". . . the unconsolidated mass from which the solum develops. It includes the *C*-horizon and other materials above the *C* from which the solum developed. The unconsolidated material directly beneath the solum is called *C* or parent material only if the evidence suggests that at least a part of the solum is developed from material of the same kind. The term *parent rock* is used for rock from which the parent material was formed by weathering."

It is virtually impossible to recognize the parent material of a sedentary soil. One extreme example of such soils is where the boundary between highly altered soil material and solid rock is sharp; the other extreme is where there is a very gradual transition from the upper soil horizons to solid rock. In the former parent material is considered to be absent (*U.S.D.A. Soil Survey Manual,* 1951, p. 147) although the "parent rock" is present, while in the latter there is a great depth of parent material whose characteristics change gradually with depth (Brewer 1955b). On the other hand, alluvial, colluvial, and aeolian deposits are considered as soil parent material even when they are completely unaltered since deposition. These are, in fact, sedimentary deposits just as much as are shales or sandstones, and they are more analogous to parent rocks than parent materials as defined by the *U.S.D.A. Soil Survey Manual* (1951).

De Sigmond's (1938) concept of a soil profile extending down to the parent rock is more logical. If the term "parent material" is to be used it should be synonymous with parent rock for sedentary soils, and then it could be used also, as at present, for alluvial, colluvial, and aeolian deposits, which are, in fact, rocks according to the geological concept. The soil profile can then be defined as follows:

> *Soil profile* is the complete vertical succession of genetic horizons down to the unaltered parent material or parent rock.

1.3 SOLUM

The position regarding the definition of the *solum* is similar to that for the soil profile. All agree (e.g., Hall, 1945; Robinson, 1949; *U.S.D.A. Soil Survey Manual,* 1951) that the solum includes the *A* and *B* horizons but not the parent material. Here again the major difficulties are recognizing the parent material and distinguishing it from the lower horizons of the solum. If the definition of soil profile just proposed is accepted, and parent material is accepted as synonymous with parent rock, then the solum does not include the parent material or the less significantly modified parent material immediately overlying it, thus some other lower limit must be set for it. The following, rather broad definition is proposed:

> *Solum* is the upper horizons of the soil profile, which in a normal soil consists of the *A* and *B* horizons, or those horizons representing the more drastically modified parent material—that is, the soil profile to the depth where the *fabric* of the parent material can be recognized.

The defintion of fabric is dealt with in some detail later (Section 7.1).

1.4 SOIL MATERIAL

The techniques discussed later are not applied directly to complete soil profiles or sola; they are most suited to description of samples of materials whose characteristics are comparatively uniform. It is convenient, therefore, to introduce the term *soil material.* Hall (1945) made this distinction between soil as a material and soil as an individual. He stated that the soil individual is the soil profile, which is the ultimate object of study and unit of classification. He recognized, however, that "much of the preliminary study of the soil must have to do with soil as material; just as the materials of the human body must be studied in order to understand the anatomy, physiology, and psychology of human beings." This is an exact parallel to the approach necessary for the study of soils. The study of a soil individual, or soil profile, is an integration of the data obtained from the study of a series of soil materials which form the various horizons in the profile.

Because of the variable nature of the characteristics to be studied, a soil material must be defined in such a way that certain characteristics may vary considerably while others—those being studied at the moment—are kept constant within rather strict limits. Certain characteristics, such as a series of shear planes, may extend over a wide area and to a relatively great depth in the soil, thus to describe the form, distribution, and relationships of such planes, the unit of study must be large. On the other hand, certain characteristics may be so small that they can be described only under the microscope, and the unit of study is very small indeed. To cover these contingencies a soil material is defined as follows:

> *Soil material* is the unit of study; it is that unit in which the characteristics being studied are relatively constant, and it will vary in size with the kind and extent of development of those characteristics.

These, then, are the basic concepts of soil within which fabric and mineral studies can be made. Preliminary detailed study of the soil is made on soil materials; data obtained from such preliminary studies can be integrated into a study of individual soil profiles, and this data, in turn, can be integrated into a study of soil or the relationships of individual profiles with each other.

1.5 THE CONSTITUENTS OF SOIL MATERIALS

Like rocks, most soil materials consist essentially of various shapes and sizes of mineral grains occurring in various kinds of arrangements, which result in the occurrence of voids of particular shapes, sizes, and arrangements. Soils are usually formed by weathering and transformation of solid rock *in situ* or weathering and/or transformation of transported materials derived ultimately from solid rocks. The basic constituents of soil materials, therefore, are the primary rock-forming minerals, their products of weathering, authigenic minerals, such complex grains as rock fragments, and materials of organic origin such as humus, phytoliths (Smithson, 1956a, 1956b, 1958, 1959; Baker and Leeper, 1958), diatom skeletons, and sponge spicules (Brewer, 1955a; Smithson, 1959). Kubiena (1938) pointed out that these constituents can be divided into two broad groups on the basis of their physical and physicochemical properties: "fabric skeleton" and "fabric plasma." The former is relatively stable in its physical and chemical properties and in its position in the soil profile; rather drastic conditions and/or a long time are necessary to significantly change either its properties (by weathering) or its position. The latter group is relatively unstable because of the ease with which it can be translocated through the soil profile and because its physical and chemical properties can be modified quite readily by relatively small changes in conditions and environment. Each of these groups is composed of diverse constituents, but it is the activity of the unstable group that causes many of the significant features and properties of soil materials.

The difference in stability or activity between the constituents of the two groups is due to three factors: particle size, particle shape, and composition. Soil materials are usually composed of grains with a relatively complete size distribution from the extremely small to something at least 2 mm in diameter. The smaller size fractions in this range exhibit colloidal properties and can be translocated and reorganized within the soil profile by movement in colloidal suspension. Although almost any of the constituents can behave as colloids if their particle size is small enough, certain of them are particularly prone to do so, particularly the layer-lattice types of clay minerals which usually have a platy or rod shape. These minerals, because of their shape, are also prone to reorganization by processes in addition to translocation in suspension—for example, reorientation due to pressure. Other constituents, such as calcite, gypsum, and halite, may be unstable because of relatively high solubility and can be readily translocated through the soil profile in solution. While these characteristics contribute to instability and activity, coarseness of particle size and insolubility contribute to stability.

Since Kubiena (1938) did not define his terms concisely, they are defined here retaining his original concepts; the adjective "fabric" seems unnecessary since the terms refer simply to the components of the soil materials.

> *Skeleton grains* of a soil material are individual grains which are relatively stable and not readily translocated, concentrated or reorganized by soil-forming processes; they include mineral grains and resistant siliceous and organic bodies larger than colloidal size (Brewer and Sleeman, 1960).

Complex grains are not considered as skeleton grains but as pedological features which will be dealt with later (Section 7.5:B). Skeleton grains are relatively immobile except for rather extreme processes, such as washing down cracks, but they are capable of weathering to form plasma.

> *Plasma* of a soil material is that part which is capable of being or has been moved, reorganized, and/or concentrated by the processes of soil formation. It includes all the material, mineral or organic, of colloidal size and relatively soluble material which is not bound up in the skeleton grains (Brewer and Sleeman, 1960).

The material need not be in a suitable state for translocation or concentration at present to be termed plasma: colloid-size clay minerals that occur as aggregates stabilized by iron oxides are still part of the soil plasma, as are the iron oxides, since their characteristics are such that under certain conditions they are capable of translocation, reorientation, and concentration.

Besides these two groups of constituents a sample of a soil material often contains living and dead plant roots and various species of soil fauna and flora. These are not considered as part of the soil material in fabric and mineral analysis; the soil material is regarded simply as the medium for their activities, although its pedography may be affected by these activities. Similarly, recognizable undecomposed organic remains are not considered; these materials have characteristics so different from those of the mineral constituents that they require a different method of treatment not within the scope of fabric and mineral analysis. Humified organic matter, however, can be considered; humification releases mobile organic constituents (which are part of the plasma) and resistant siliceous and organic bodies (which can be included in the skeleton).

Soil materials, therefore, are composed of mineral grains and other constituents that can be identified mineralogically and/or chemically; have characteristics of size, shape, color, etc.; and exhibit various kinds of arrangements with regard to each other. Further, the various arrangements result in the occurrence of voids of various sizes and shapes and with various arrangements. Considering these characteristics and the scope of pedography as defined (Section 1.7), it is apparent that the characteristics involved in fabric and mineral analyses are the size, shape, and arrangement of the constituents and associated voids, and identification of the constituents.

1.6 ORGANIZATION IN SOIL MATERIALS

Soil-forming processes cause reorganization of the constituents in soil materials, as compared with the organization in the parent material or parent rock, by weathering of minerals and by translocation and rearrangement of the more mobile constituents. Such processes may cause fractionation, translocation, and reorganization of the plasma, resulting in formation of concentrations of relatively pure constituents on the walls of voids and/or cementing of parts of the soil material by some fraction of the plasma; these features may have various internal arrangements of the constituents. Other processes lead to the development of recognizable composite grains or aggregates which may be packed together in particular arrangements compatible with the size and shape of the aggregates. Soil faunal activity may produce fecal pellets and burrows which may be empty or filled with soil material or droppings. Plant root activity may cause particular patterns of channels. Differential forces due to shrinking and swelling or other processes may cause rearrangement of some of the constituents, especially the clay mineral fraction of the plasma. In addition, certain features may be inherited from the parent rock or parent material, such as rock fragments, or as a result of sedimentary processes during deposition of a sedimentary parent material.

Many soil materials, therefore, have an extremely complex organization as a result of various combinations of these features. Pedographic studies involve the systematic description of all these manifestations of organization. Fabric and mineral analyses deal with identification of the single grains; recognition of the various kinds of composite units or features; description of the size, shape, and arrangement of the single grains in both the basic soil material and within the different kinds of composite units; and the size, shape, and arrangement of the composite units.

1.7 PEDOGRAPHY

The aspects of soil studies discussed in this work can be conveniently grouped under the heading of *pedography.* We can argue that pedology, or "the science of soils" (*Shorter Oxford Dictionary,* 1956), is logically a branch of the broader science of geology, or "the science which treats of the earth in general" (*Shorter Oxford Dictionary,* 1956), since soils are indeed a part of the surface of the earth, having their origin basically in the materials derived from rocks and owing their characteristics to physical, chemical, and biological alteration of these materials. Pedology then assumes the stature of petrology as a science, and the term pedography is proposed as the equivalent of petrography within the science of pedology. Petrology and pedology are encompassed in the broad science of geology; petrology consists of petrography (description) and petrogenesis (the study of origins of rocks), while pedology consists of pedography (description) and pedogenesis (the study of the origins of soils). Bryan (1952) spoke of the "pedographic" description of soils in this sense.

Although the scope of petrography as defined by various authors (e.g., Rice, 1954; Krumbein and Pettijohn, 1938; Milner, 1962) varies considerably, all agree that it primarily concerns *description* of rocks; the difference in scope depends on interpretation of the scope of description, which ranges from purely optical description to description that includes all possible analytical data. Most textbooks on petrography, however, deal primarily with structure and fabric analysis and mineralogy in the sense of granulometric analysis, particle shape, and mineral identification, usually with subsidiary sections on analyses of other kinds and on description of color. A suitable concept for the study of soils is contained in the following definition of pedography:

> *Pedography* is the systematic description of soils based on observations in the field, on hand specimens and thin sections, and on data from other techniques on the size, shape, arrangement, and identification of the constituents.

One of the objectives of pedology is elucidation of the origin and genesis of soil materials and soil profiles. Such objectives cannot be achieved by confining investigations to soil materials or even soil profiles. The origins of parent materials of soils can be studied only with the aid of exhaustive studies of rocks, and the modifications of parent materials and parent rocks due to soil-forming processes can be assessed only if the original characteristics of the parent materials and parent rocks are known. At this point the study of soil overlaps the study of rocks, and the pedographer cannot afford to ignore the petrographic characteristics of the hard rocks, which may be the parent rocks of the soil or may have been the source material for sedimentary bodies that become the parent material of the soil.

PART II

MINERAL ANALYSIS

2

Size and Shape

Although the principles and mathematical treatment of size and shape analyses have been dealt with exhaustively by a number of workers in sedimentary petrology, they are summarized here because size and shape are an integral part of both mineral and structure analysis of soil materials. This chapter deals only with the general principles and methods usually applied to mineral grains. There are other aspects, particularly of shape, which apply more specifically to features in soil materials other than mineral grains; these are discussed in the appropriate chapters on such features.

The aim of size and shape analyses is to express quantitatively the frequency distribution of particular shape and size ranges as part of the description of the material and as a basis for interpretation of origin and/or mode of formation. In sedimentary petrology the principal aim has been to make estimates for the individual or composite grains which make up the sediment, excluding cementing materials. This is justified by the particular objective, which is to calculate the degree of sorting, kurtosis, and skewness, and the mean and modal values as indicators of the origin, mode of transportation, and conditions of deposition of the sediments. Such objectives are also legitimate in the study of soils, particularly soil genesis, because there is a basic need for characterization of the parent material and for recognition of changes in geological nature or mode of deposition of the parent material within a single soil profile.

2.1 CONCEPTS OF SIZE AND SHAPE

A. Size

Wadell (1932) discussed the concept of size and rejected all definitions except that of volume, which can be expressed as his *true nominal diameter* (i.e., the diameter of a sphere which has the same actual volume as the particle).

B. Shape

Shape analyses are based on the concepts of sphericity and roundness originated by Wadell (1932, 1933, 1935). *Sphericity* concerns the overall form of the particle irrespective of the sharpness of edges and corners; it is a measure of the degree of conformity of the shape to that of a sphere. *Roundness* concerns only the sharpness of corners irrespective of the form or sphericity.

Sphericity is calculated according to the equation:

$$\psi = \frac{s}{S}$$

where ψ = the true sphericity,
s = the surface area of a sphere of the same volume as that of the particle, and
S = the actual surface area of the particle.

Because of the difficulty of measuring these surface areas (s and S) the equation was modified to a simpler form, which gives a close approximation to true sphericity:

$$\phi = \frac{dc}{Dc}$$

where ϕ = the shape value,
dc = the diameter of a circle equal in area to the area obtained by planimeter measurement of the projection of the grain when it lies on its largest face, and
Dc = the diameter of the smallest circle circumscribing the projection.

Zingg's (1935) shape classification, similar in concept to sphericity, was based on the ratios of the lengths of the principal axes of particles; if a is the longest axis, c the shortest, and b intermediate, and all three are mutually at right angles, then Zingg classified particle shape

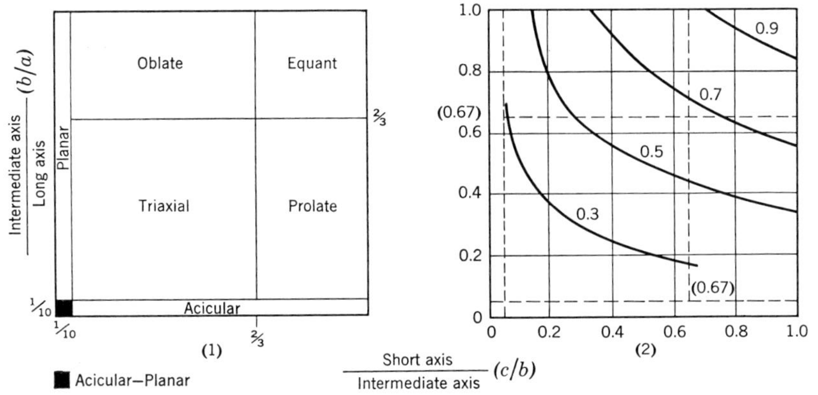

(A)

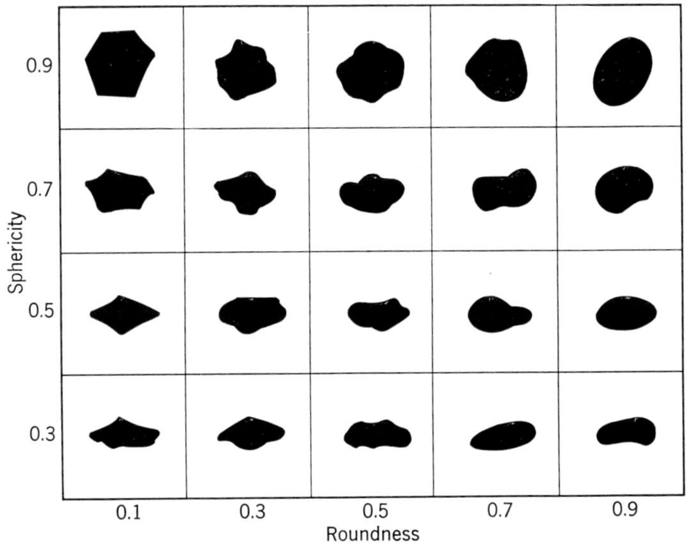

(B)

Fig. 1. Sphericity and roundness classes. *A*(1) Shape classes based on ratios of the principal axes (modified from Krumbein, 1941). *A*(2) shows the relation of sphericity to the shape classes. *B* Chart for visual estimation of sphericity and roundness. (From *Stratigraphy and Sedimentation* by W. C. Krumbein and L. L. Sloss, 1955.) *Courtesy W. H. Freeman & Co., California.*

TABLE 2-1. Zingg's (1935) Shape Classes

Class I	Disc	$b/a > \frac{2}{3}$	$c/b < \frac{2}{3}$
Class II	Spheroid	$b/a > \frac{2}{3}$	$c/b > \frac{2}{3}$
Class III	Blade	$b/a < \frac{2}{3}$	$c/b < \frac{2}{3}$
Class IV	Rod	$b/a < \frac{2}{3}$	$c/b > \frac{2}{3}$

and acicular-planar); these are shown in Fig. 1*A*. Sneed and Folk (1958) proposed a larger number of classes; their sphericity-form scale is shown in Fig. 2. Roundness scales have been proposed by Pettijohn (1949) (Table 2-2) and Krumbein and Sloss (1955) (Fig. 1*B*).

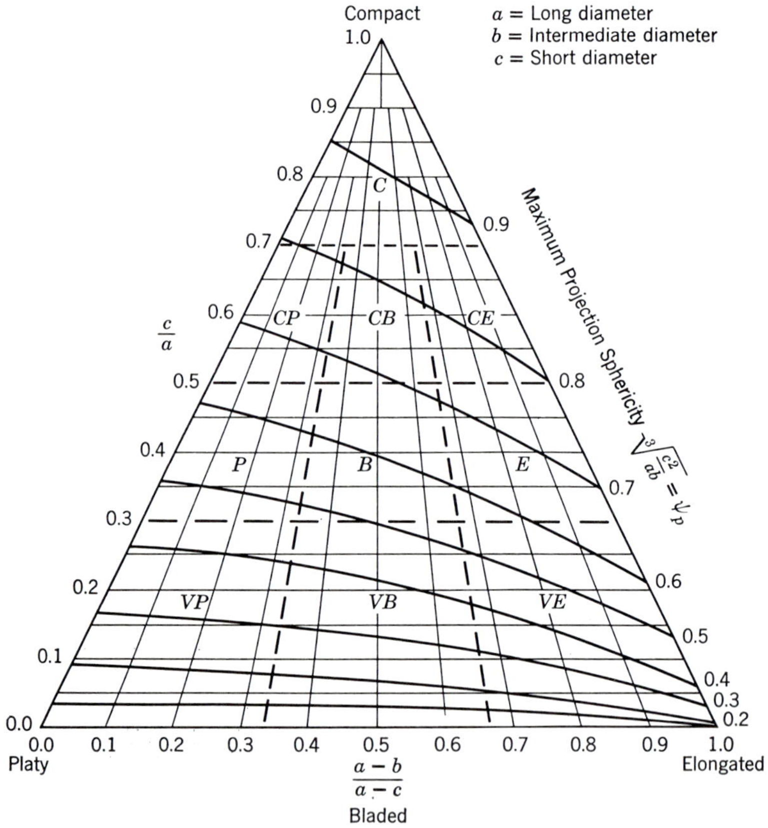

Fig. 2. Sphericity-form scale (after Sneed and Folk, 1958). *Courtesy Jour. Geol.* Copyright (1958), The University of Chicago.

TABLE 2-2. Pettijohn's (1957) Roundness Classes

Class Name	Roundness Value	Description
Angular	0–0.15	Strongly developed faces with sharp edges and corners; secondary corners* are numerous
Subangular	0.15–0.25	Strongly developed faces with somewhat rounded edges and corners; secondary corners are numerous
Subrounded	0.25–0.40	The edges and corners are rounded and the area of flat faces is comparatively small; secondary corners are much rounded and reduced in number (5–10)
Rounded	0.40–0.60	Flat faces are practically absent; all edges and corners are rather broad curves, and there may be broad re-entrant angles; secondary corners have disappeared
Well rounded	0.60–1.00	There are no flat faces; the entire surface consists of broad curves

* Secondary corners are "the many minor convexities seen in the grain profile. Primary corners are the principal interfacial edges and are few in number (3–5)" (Pettijohn, 1957).

Powers (1953) criticized the former on the grounds that there were insufficient classes at the lower end; he suggested a modified geometric scale with the class intervals and nomenclature set out in Table 2-3. Folk (1955) pointed out that a *rho scale* could be set up similar to

TABLE 2-3. Roundness Grades

	Class Limits		
Grade Term	Pettijohn (1957)	Powers (1953)	Rho Scale* (Folk, 1955)
Very angular		0.12–0.17	0.00–1.00
Angular	0–0.15	0.17–0.25	1.00–2.00
Subangular	0.15–0.25	0.25–0.35	2.00–3.00
Subrounded	0.25–0.40	0.35–0.49	3.00–4.00
Rounded	0.40–0.60	0.49–0.70	4.00–5.00
Well rounded	0.60–1.00	0.70–1.00	5.00–6.00

* Based on Powers' (1953) class limits

Krumbein's (1941) phi scale for size intervals. For roundness he based his approach on Powers' (1953) logarithmic class intervals. A comparison of Pettijohn's, Powers' and Folk's scales is shown in Table 2-3.

2.3 METHODS OF ANALYSIS

Having selected an appropriate grade scale there are a number of methods which can be used to perform the analysis; the choice of grade scale and of the particular method depends principally on the size or shape range of the particles in the material being analyzed.

A. Sieving

The most common method used for size analysis of large particles is sieving; the principles and theory of sieving have been discussed by Krumbein and Pettijohn (1938) and in some detail by Herdan (1960). Inaccuracies are due to the fact that sieves sort according to shape as well as size (the intermediate and short diameters of a particle determine whether it passes a sieve or not), incomplete sieving, and lack of uniformity in apertures of the sieves particularly in the smaller sizes. Sieving is effective only down to a lower size limit of approximately 50 μ, but it has the advantages of simplicity and ease of handling large samples so that sampling errors are reduced to a minimum. In general the method gives satisfactory results.

B. Methods Based on Stokes' Law

These methods and their modifications are used for size analysis of small particles; they include elutriation, sedimentation, centrifugation, pipette and hydrometer methods, and the bottom-end withdrawal tube. Inaccuracies occur in all these methods because the particles in soil materials are not usually spheres, are not always smooth, may have too great a settling velocity in the coarser ranges, and do not have a uniform density (Krumbein and Pettijohn, 1938). Sedimentation methods suffer further inaccuracies due to the difficulty of stirring without creating turbulence and the impossibility of achieving complete separations by decantation. In elutriation in water there is danger of flocculation in the finer grades caused by the dilution of electrolytes necessary for dispersion, and inaccuracies may be introduced due to lack of uniformity of the rising current throughout the cross section of the tube; the lower limit for elutriation is 10 μ equivalent diameter (Krumbein and Pettijohn, 1938). The pipette (Robinson, 1922; Piper, 1942), hydrometer (Bouyoucos, 1927, 1934, 1937; Day, 1950, 1953), and bottom-end withdrawal tube (U.S. Engineers, 1943, 1953) methods partially overcome some of these difficulties. The last method is effective down to 4, or even 2, μ equivalent diameter, but for smaller size grades centrifugation is necessary because of the long settling times and the effects of Brownian movement in grades smaller than about 1 μ; Tanner and Jackson (1947) have discussed the principles of this method. The upper limit of equivalent diameter for these methods, using water as the fluid, is approximately 50 to 100 μ except that the bottom-end withdrawal tube can cope with particles up to an equivalent diameter of 1000 μ, or even larger if the length of the tube is increased. In spite of these theoretical difficulties the degree of agreement between expected theoretical values and actual settling velocities of particles from sediments and soils is very good except for the coarser grades (Andreason and Lungberg, 1929).

C. Direct Measurement

This method can be used for size analysis of large and small particles. The volume of individual large particles can be estimated by displacement of a liquid, allotted to the grades in the selected grade scale, and converted to weights within each grade class. Small particles can be analyzed microscopically in thin section or grain mounts. Krumbein and Pettijohn (1938) examined the mathematical theory underlying this approach, which has been based on a number of different measurements, such as the intermediate dimension, the area of the section of the grain expressed as the diameter of a circle having the same area [Wadell's (1935) "nominal sectional diameter"], and the longest apparent diameter (Friedman, 1958). Friedman claimed that by traversing a section at predetermined intervals, so that more than 75 percent of the section was included between the two outside traverses, and then applying an overall regression graph and a correlation graph, his method could reproduce the cumulative frequency curve obtained by sieving. He measured a minimum of 500 grains, but it is probable that a larger number should be measured in less well-sorted materials (Section 3.3:A). Van der Plas (1962) examined data of granulometric analyses obtained by line and point count techniques, and he concluded that results for the same sand were so different that they were obviously biased by the technique used; in line counts, larger particles were overestimated in proportion to their diameter (twofold for a twofold increase in diameter), and in point counts in proportion to their area (fourfold for a twofold increase in diameter). As a remedy he advocated "ribbon sampling," that is, "a

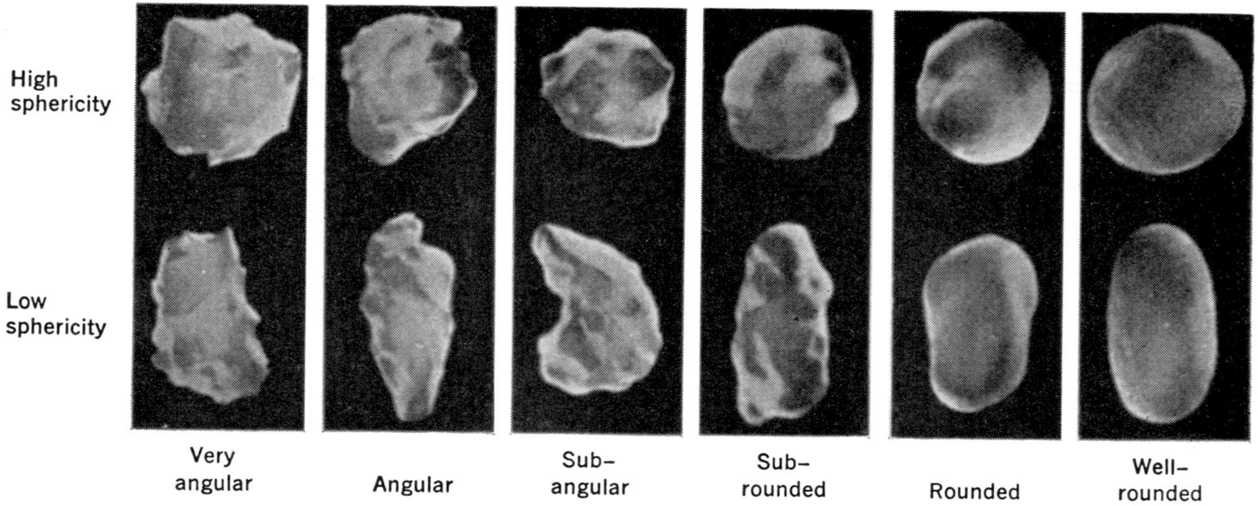

Fig. 3. Chart for visual estimation of roundness (after Powers, 1953). *Courtesy Jour. Sed. Petrol.*

central section of the preparation containing a sufficient number of intersections was selected; then an arbitrary starting point was chosen and fixed by readings of the two perpendicular scales of the mechanical stage. Now the preparation was very slowly moved . . . in a direction exactly perpendicular to the scale of the eyepiece micrometer. All fat globules (particles) whose centres crossed this scale between two fixed points (e.g., the 20th and 30th mark on the scale) were classified" (van Kreveld, 1942). The ribbons should be laid out systematically across the section and should be at least as wide as the largest diameter of the largest visible grain; areas are the best measurements to record. Inaccuracies in such methods are due to errors in measuring; it is often difficult to identify accurately the margins of grains, especially in thin sections, and such errors are greatest in the smallest grades, so the method should not be used for particles less than 20 μ in diameter. Other inaccuracies in grain mounts are due to preferred orientation of the particles in mounting and, in thin section, to the fact that the observed surfaces must always be equal to or less than the maximum possible cross sectional areas of the grains.

Direct measurement and application of the appropriate equations (Section 2.1:B) or charts are the only accurate methods of analysis of sphericity and roundness. Zingg's (1935) shape- (sphericity) class limits (Table 2.1) are fixed simply by limits of the ratios of the principal axes, as are the limits of the modified classes shown in Fig. 1. The classes established by Sneed and Folk (1958) are shown in Fig. 2 in relation to their maximum projection sphericity and the ratios of the principal axes.

D. Visual Estimation

Analyses of sphericity and roundness are often made visually. Pettijohn (1957) supplemented his roundness classes with detailed descriptions so that the particles can be classified visually (Table 2-2). Krumbein and Sloss (1955) presented a chart for visual estimation of both sphericity and roundness (Fig. 1*B*), and Powers (1953) published a chart for visual estimation of roundness classes based on actual photographs of grains (Fig. 3).

2.4 PRESENTATION OF DATA

Presentation of the data of size and shape analyses should be such that it presents at a glance as much information as is possible and facilitates interpretation and analysis of the data.

A. *Histograms*

For size analyses, the diameter of the grade classes or a function of the diameters, such as their logs, is the independent variable, and frequency is the dependent variable. Common practice is to plot diameters on the horizontal axis and frequency on the vertical axis, and also to plot the coarser to finer grades from left to right along the horizontal axis. A vertical rectangle of appropriate *area* is then drawn for each size class. The scale used for the diameters may be arithmetic or logarithmic. With a logarithmic horizontal scale, the width for each geometric grade class on the X-axis is equal, and the height of each rectangle is proportional to the frequency in that grade. If an arithmetic scale is used the width for each grade along the horizontal axis decreases successively, and care must be taken to insure that the area of each rectangle represents the frequency in that grade.

For shape analyses sphericity and roundness are usually plotted separately along the horizontal axis on an arithmetic scale with intervals equivalent to 0.1 units. The vertical scale is usually one of percentage by number, and measurements are usually expressed for each size grade because roundness and sphericity may be functions of size. The horizontal scale can be converted to a geometric scale by using Folk's (1955) rho scale.

B. *Frequency Polygons*

These polygons are constructed by plotting the frequency corresponding to a given grade class on a line midway between the class limits and connecting the resulting points successively with straight lines. This line is brought to the zero point at the centers of the classes just beyond the two limiting classes in the analysis. The horizontal scale may be arithmetic or geometric, as in histograms.

C. *Frequency Curves*

These smooth curves show the variation of the dependent variable as a continuous function of the independent variable. An approximation of such curves can be produced by joining the midpoints of the top sides of the histogram rectangles by a smooth curve which is brought to the zero point at the limits of the two limiting histogram rectangles; the narrower the range of each class the more accurately will this curve approximate the true frequency curve. Krumbein and Pettijohn (1938) set out a graphical method of deriving the frequency curve accurately from the cumulative curve.

D. Cumulative Curves

For size analyses these curves are constructed by plotting ordinates which represent the total amount of material larger than a given diameter. An ordinate is erected at the lower limit of the coarsest grade equal in height to the frequency in that class. Another ordinate is drawn at the end of the next grade equal in height to the sum of the frequencies in the first two grades, etc. A smooth curve is drawn through the points so obtained; this is the cumulative curve representing the continuous distribution of sizes. The horizontal scale may be arithmetic or logarithmic. Similar curves can be drawn for sphericity and roundness analyses.

E. Descriptive Terms

When accurate, detailed analyses are not required we can use descriptive terms. For size analysis simple terms for *relative size* (a rough estimate of the degree of sorting) and *absolute size* (actual size of the particles) may meet the needs; a list of such terms is set out in Table 2-4. The descriptive terms used by various authors for shape

TABLE 2-4. Descriptive Terms for Size (after Brewer and Sleeman, 1960)

(1) Relative Size

Class Name	Description
Very even	There is a single strong peak in the frequency distribution curve
Even	There is a single distinct peak in the frequency distribution curve
Uneven	There is a single slight peak in the frequency distribution curve
Very uneven	The distribution is irregular

(2) Absolute Size

Class Name	Class Limits
Extremely fine	<0.005 mm
Very fine	0.02–0.005 mm
Fine	0.1–0.02 mm
Medium	0.5–0.1 mm
Coarse	2–0.5 mm
Very coarse	10–2 mm
Extremely coarse	>10 mm–state actual size

analyses and their general relationships to each other are set out in Table 2-3.

2.5 ANALYSIS OF DATA

Treatment of the data of size and shape analyses depends on the information required. Frequency distributions can be characterized by a number of well-established mathematical treatments; cumulative curves and frequency distribution curves are the most useful for this purpose. Full mathematical treatment of the data is beyond the scope of this work, but there are certain aspects of the mathematical analysis which may be mentioned with advantage. Of particular interest in the study of soils are the origin and mode of transport of sedimentary parent materials. One method of studying such materials is by size and shape analyses of the discrete mineral grains with special reference to the degree of sorting, skewness, and kurtosis of the distributions.

A. *Sorting Coefficients*

These are single values used to express the degree of sorting, which is essentially a measure of the evenness of grain size or spread of the size frequency distribution. One of the earliest sorting coefficients proposed for size frequency distributions was the "geometric quartile deviation" (Trask, 1932), which is derived from the equation:

$$QDg = \sqrt{Q_1/Q_3} \qquad (1)$$

where Q_1 = the larger quartile, and
Q_3 = the smaller quartile.

The quartiles are obtained from the cumulative curves and lie on either side of the median; they are the diameter values where the 75 and 25 percent lines intersect the cumulative curve, and the median is the diameter value where the 50 percent line cuts the cumulative curve. Since QDg increases geometrically it cannot be used for direct comparison of different materials, but *log* QDg can since it forms an arithmetic series. Taking logs of equation (1):

$$\text{Log } QDg = \frac{\log Q_1 - \log Q_3}{2} \qquad (2)$$

and a material that has *log* QDg twice the value of another material has twice the spread of grain sizes. On the basis of experience Trask (1932) suggested that sediments with QDg less than 2.5 were well sorted, those between 2.5 and 4.5 were normally sorted, and those greater than 4.5 were poorly sorted.

This sorting coefficient has been criticized on two main counts (Friedman, 1962): that it does not take sufficient account of the geometry of the frequency distribution curves and that the suggested limits for degree of sorting are unreal in the light of wider experience. Most of the sorting coefficients proposed in later years are based on the premise that the grain-size distribution of many sediments approaches lognormality; the arguments are based on the degree of deviation of the curves from lognormality, and the parameters selected are based on the phi scale. The modifications of Trask's coefficient are designed to take into consideration a larger section of the curve and to place more emphasis on both extremities.

Inman (1952) modified Trask's equation to:

$$\sigma_\phi = \tfrac{1}{2}(\phi_{84} - \phi_{16}) \tag{3}$$

on the basis that the sixteenth and eighty-fourth percentiles are one standard deviation on either side of the mean in a normal size frequency distribution; Folk (1955) proposed this same coefficient for roundness sorting. Folk and Ward (1957) added a second term to form the equation:

$$\sigma_\phi = \tfrac{1}{4}(\phi_{84} - \phi_{16}) + \tfrac{1}{6.6}(\phi_{95} - \phi_5) \tag{4}$$

Friedman (1962) proposed direct use of standard deviation calculated from the equation:

$$\sigma_1 = \frac{\sqrt{\Sigma f(m_\phi - \bar{x}_\phi)^2}}{100} \tag{5}$$

where Σ = "the sum of,"
f = frequency or abundance of the different grain size grades present in the sediment,
m_ϕ = the midpoint of each grain size grade in ϕ values, and
$\bar{x}_\phi$ = the mean in ϕ values.

McCammon (1962) considered the efficiency of some of the graphic measures in assessing a sorting coefficient and proposed further modifications. He suggested that

$$\sigma_\phi = \tfrac{1}{5.4}(\phi_{85} + \phi_{95} - \phi_5 - \phi_{15}) \tag{6}$$

was as efficient as the Folk and Ward coefficient (eq. 4), and that

$$\sigma_\phi = \tfrac{1}{9.1}(\phi_{70} + \phi_{80} + \phi_{90} + \phi_{97} - \phi_3 - \phi_{10} - \phi_{20} - \phi_{30}) \tag{7}$$

TABLE 2-5. Approximate Correlation of Sorting Classes*

Trask (1932) $\sqrt{Q_1/Q_3}$ Interval	Schneiderhoehn (1953) $\sqrt{Q_1/Q_3}$ Interval	Fuechtbauer (1959) $\sqrt{Q_1/Q_3}$ Interval	Folk (1954) $\sqrt{Q_1/Q_3}$ Interval	Folk and Ward (1957) $\sigma_1 = \frac{\phi_{84} - \phi_{16}}{4} + \frac{\phi_{95} - \phi_5}{6.6}$	Friedman (1962) $\sigma_\phi = \sqrt{\frac{\Sigma f(m_\phi - \bar{x}_\phi)^2}{100}}$
1.0	1.0–1.23	1.0–1.23	1.0–1.165	<0.35	<0.35
			1.165–1.26	0.35–0.50	0.35–0.50
	1.23–1.32	1.23	1.26	0.50	0.50–0.80
		to		to	
	1.32–1.41	1.41		1.00	
			to		0.80
				1.00	
	1.41–1.51	1.41			
to		to			to
	1.51–1.74	1.74	1.60		
			1.60		
	1.74–1.87	1.74			1.40
		to		to	
	1.87–2.00	2.00			1.40
			to		
	>2.00	>2.00			
2.5					to
2.5–4.5					
			2.63		
				2.00	2.00
>4.5				>2.00	>2.00

* Each author named the sorting classes within these ranges; the differences in range reflect each author's concept of the degree of sorting and the importance of the limits of the ranges.

was significantly more efficient; other coefficients, such as Trask's (eq. 2) and Inman's (eq. 3), were considerably less efficient. Sharp and Fan (1963) proposed a new sorting index (S_i) derived from the equation:

$$S_i = 100 + k \sum_{i=1}^{n} f_i \log_{10} f_i$$

where S_i = sorting index expressed in percentage,

k = a constant depending on the scale specified $= \dfrac{100}{\log_{10} N}$

(N = total number of allowed classes),

f_i = the fraction observed in the ith class, and

Σ = "the sum of."

Alternatively, $$S_\phi = 100 - \sum_{i=1}^{n} U_i$$

where $U_i = -kf_i \log_{10} f_i$.

Tables for U_i for the scale range —10 to 15 ϕ are given. S_i has a maximum of 100 percent when all particles lie within a single grade class and a minimum of zero percent when all grade classes contain equal masses of particles.

Schneiderhoehn (1953), Folk (1954), and Fuechtbauer (1959) proposed separate modifications of Trask's (1932) class ranges of degree of sorting, while Folk and Ward (1957) suggested classes for their graphic measure, and Friedman (1962) for his moment measure of standard deviation; these are shown in Table 2-5. Folk (1955) pointed out that at present there is insufficient data on the significance of coefficients of roundness sorting to set up reliable classes; he proposed the classes shown in Table 2-6 based on his σ_ρ derived from equation 3.

TABLE 2-6. Classes of Roundness Sorting (after Folk, 1955)

Class Name	Class Limits (σ_ρ*)
Very good roundness sorting	<0.7
Good roundness sorting	0.7–0.9
Poor roundness sorting	1.1–1.3
Very poor roundness sorting	>1.3

* $\sigma_\rho = \frac{1}{2}(\rho_{84} - \rho_{16})$

B. Skewness

The deviation of the frequency distribution curve from the symmetry of a normal curve is described by skewness. Inman (1952) suggested two measures of skewness:

$$\alpha_\phi = \frac{\phi_{84} + \phi_{16} - 2\phi_{50}}{\phi_{84} - \phi_{16}}$$

to measure the deviation from symmetry of the central part of the distribution, and

$$\alpha_{2\phi} = \frac{\phi_{95} + \phi_5 - 2\phi_{50}}{\phi_{84} - \phi_{16}}$$

for the extremes of the distribution. Folk and Ward (1957) discussed the methods of assessing skewness and proposed a formula for an *inclusive graphic skewness* (Sk_i):

$$Sk_i = \frac{\phi_{16} + \phi_{84} - 2\phi_{50}}{2(\phi_{84} - \phi_{16})} + \frac{\phi_5 + \phi_{95} - 2\phi_{50}}{2(\phi_{95} - \phi_5)}$$

which they claimed is a better measure of overall skewness and which is independent of sorting, as Inman's measures were not, so that perfectly symmetrical curves have an Sk_i value of 0.00, and the limits of Sk_i are -1.00 and $+1.00$. Friedman (1962) presented a general formula for calculating skewness, the third moment about the mean ($\alpha_{3\phi}$):

$$\alpha_{3\phi} = \tfrac{1}{100} \cdot \sigma_\phi^{-3} \Sigma f(m_\phi - \bar{x}_\phi)^3$$

For this measure, also, normal curves have $\alpha_{3\phi} = 0$.

C. Kurtosis

Kurtosis is a measure of the peakedness of a curve. Folk and Ward (1957) proposed an equation to calculate their *graphic kurtosis* (K_g):

$$K_g = \frac{\phi_{95} - \phi_5}{2.44(\phi_{75} - \phi_{25})}$$

such that K_g is 1.00 for normal curves; larger values indicate excessive peakedness, while smaller values indicate excessive flatness. For plotting graphs and for statistical analysis, the distribution of K_g can be approximately normalized by using:

$$K_g' = \frac{K_g}{K_g + 1}$$

Friedman (1962) presented a general formula for calculating kurtosis, the fourth moment about the mean ($\alpha_{4\phi}$):

$$\alpha_{4\phi} = \tfrac{1}{100} \cdot \sigma_{\phi}^{-4} \Sigma f(m_{\phi} - \bar{x}_{\phi})^4$$

$\alpha_{4\phi}$ is 3 for a normal curve; lower values indicate excessive flatness and higher values excessive peakedness.

D. Mean

Folk and Ward (1957) discussed the assessment of mean size and pointed out that Inman's (1952) measure:

$$\tfrac{1}{2}(\phi_{16} + \phi_{84})$$

is adequate only for nearly normal curves. For skewed and bimodal distributions they proposed a measure (M_z) calculated from:

$$M_z = \tfrac{1}{3}(\phi_{16} + \phi_{50} + \phi_{84})$$

2.6 INTERPRETATION

Frequency distribution curves, coefficients of sorting, skewness and kurtosis, and mean values are simply factual statements of these characteristics, but they are not interpretations of the origin or cause of the characteristics. The basis of interpretation in sedimentary petrology is correlation of particular degrees of sorting, kurtosis, and skewness with sediments of known origin by the accumulation of a vast amount of corroborative data from analyses of a great number of samples.

A. Size Analysis

The basic assumption for interpretation of size analysis must be that the source material could supply a wide range of sizes so that the observed characteristics of the frequency distribution are due to the processes of transportation and sedimentation and are not inherited from the source material. Given such a source material the data available for sediments suggest that the uniformity and effectiveness of the process of transportation is reflected in the degree of size sorting, and the kind of process *may* be reflected in modal size, kurtosis, and skewness. More important for interpretation of the process of transportation is the areal pattern of variation of the coefficients of the size distribution curves in relation to the location of the source material.

In interpretation of size analyses of the particles in soil materials which compose a soil profile, the effects of the soil-forming processes must be considered. A common effect is a progressive reduction in grain size of the mineral grains which composed the parent material or parent rock; a progressive decrease in size of all the mineral species from the lower to the upper soil horizons has been demonstrated in a yellow podzolic soil formed on granodiorite (Brewer, 1955*b*), and it is apparent, even from field observations, that many soil materials formed from relatively coarse-grained rocks contain a high percentage of clay and silt size material. Thin sections of soil materials also show that primary mineral grains break down physically during soil formation; the micas commonly exfoliate and fracture along cleavages and across them as the flakes become thin (Fig. 127); quartz and other minerals may break down physically along planes of weakness due to strain, cleavage, or inclusions (Raeside, 1959). Carried to extremes such processes may reduce the majority of mineral grains to the finer fractions, that is, significantly decrease the size range even to the extent of producing a modal size and coefficient of sorting comparable with some loesses or alluvial deposits. Kurtosis, skewness, and other such measures may distinguish this kind of sorting due to weathering from other kinds due to various modes of transport, but the possibility of making such distinctions has not been investigated systematically.

In comparison with the large amount of detailed data available on size analysis of sediments, that available for soil materials is infinitesimal. Standard practice in pedology has been to make size analyses according to a grade scale whose classes are very much broader than the $\phi/4$ intervals used in modern practice in sedimentology. Indeed, many analyses of soil materials are made only on the "fine earth," all the material coarser than 2 mm diameter being discarded. Such analyses are quite unsatisfactory for mathematical treatment and interpretation, and this is one reason why so little is known of the effects of soil formation on degree of sorting.

A major obstacle in attempting studies of the effects of soil formation is the difficulty of being certain of the uniformity of the parent material throughout a soil profile and of its derivation from the underlying *in situ* rock. Recently Butler (1958, 1959) and his coworkers (Walker, 1959, 1962; Churchward, 1959, 1961; van Dijk, 1958, 1959) and Jessup (1960) have shown by field correlations that many soil profiles in Australia, which were formerly thought of as single profiles on uniform parent materials, are actually a series of more or less truncated profiles overlying each other. Other workers have reached similar conclusions for various reasons (e.g., Nikiforoff, Humbert, and

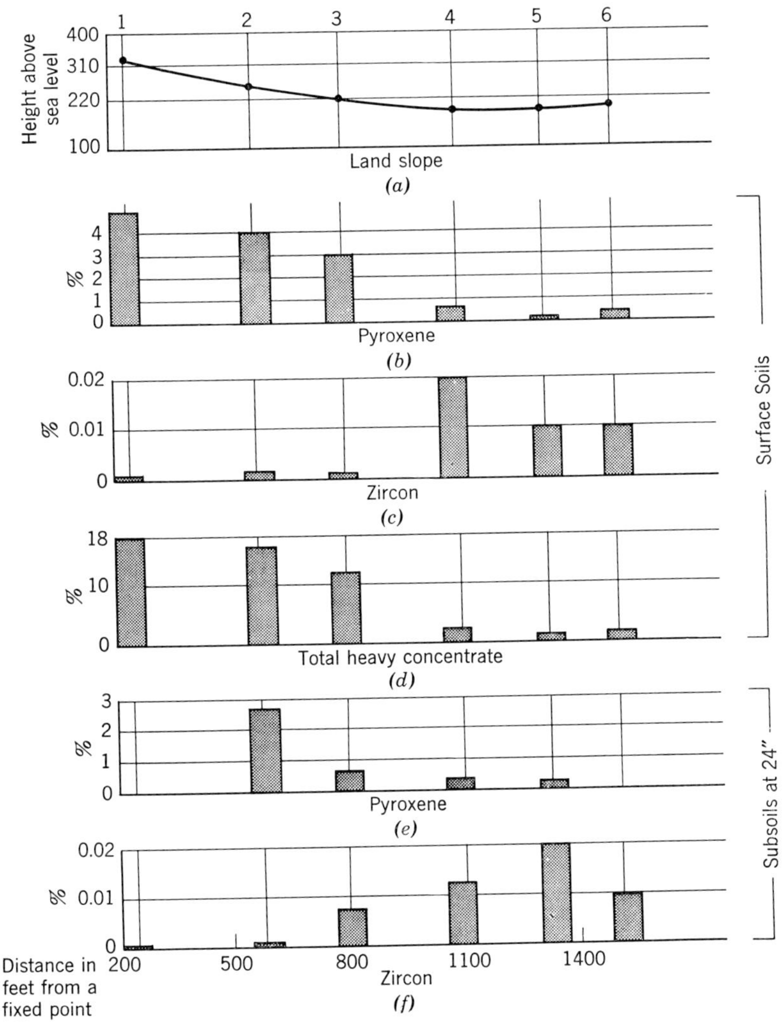

Fig. 4. Distribution of pyroxene, zircon, and total heavy concentrate in relation to topography at Prospect Hill, N.S.W. (after Brewer, 1948). (*a*) Land slope. (*b*) Pyroxene in the surface soils. (*c*) Zircon in the surface soils. (*d*) Total heavy concentrate in the surface soils. (*e*) Pyroxene in the subsoil at 24 in. (*f*) Zircon in the subsoil in 24 in. *Courtesy Roy. Soc. N.S.W.*

Cady, 1948; Nicholls, 1936; Ruhe, 1956). Similar relationships have been demonstrated by laboratory data, for example, in the following cases:

(1) A study at Prospect Hill, New South Wales (Brewer, 1948), showed that there had been considerable movement of material downslope on a hill-side; the soil profiles consisted of mixtures of materials, the degree of mixing diminishing with depth in each profile and with distance downslope (Fig. 4).

(2) The *A* horizons of a strongly differentiated yellow podzolic soil profile were shown mineralogically to be a later deposit of material that had been transported only a short distance (Brewer, 1955*b*).

(3) Fabric studies indicated the formation of a younger profile in material overlying an older profile (Brewer, 1956).

(4) Zircon distribution and zircon-tourmaline ratios with depth in four profiles specifically selected as formed *in situ* on granodiorite provided strong evidence of layering of materials in all cases (Brewer, unpubl. data).

In spite of these difficulties a number of detailed studies based on size analysis of mineral grains have been made in relation to the origin of transported parent materials. Examples of these are the studies made in Europe and the United States on the origin, distribution, and characteristics of loess sheets. More recently, Butler and Hutton (1956) have demonstrated the occurrence of a sheet of aeolian clay material (parna) in the Riverine Plain of New South Wales by size analysis of the accompanying mineral grains and the variation in its size distribution characteristics laterally from the source. In this study the samples were taken as deeply in the profiles as was possible without contamination from the underlying materials so as to avoid changes in grain size due to soil formation. Interpretation was based on size analyses of the mineral grains of diameter greater than 2 μ and less than 2 mm; Trask's (1932) criteria indicated significant sorting in all samples, and Doeglas' (1946) method of plotting the data gave a curve that he claims is characteristic of loess, flood plain, and estuarine deposits.

B. Shape Analysis

As with size analysis, in interpreting data on roundness and sphericity of pebbles and grains in terms of degree and kind of transport we must assume that the observed characteristics are not inherited from the source material and are not due to the processes of soil formation. Even in these circumstances the accumulated data on which interpretation is based are not as convincing as for size analysis.

In general, sphericity and roundness are believed to increase with distance of transportation (Krumbein and Sloss, 1955), and Pettijohn (1957) stated that "secondary corners quickly disappear during abrasion, whereas primary corners remain and are discernible, though rounded, in much abraded grains." The results of a number of studies of naturally abraded particles, however, are contradictory in regard to the effects of transport on sphericity. Some indicate increasing sphericity with transport (Grogan, 1945; Wentworth, 1922), others indicate little or no change (Krumbein, 1941), while still others indicate decreasing sphericity with distance of transport (Russell and Taylor, 1937; Pettijohn and Lundahl, 1943). Rayleigh (1942, 1944) pointed out that the final shape rock materials will assume on continued wear has not yet been determined. Carroll, Brewer, and Harley (1950) suggested that "it is probably only homogeneous materials, fine and even grained, nonlaminated materials, or materials from thick even beds among sedimentary rocks" that will tend to become spherical; observations on pebbles from a number of situations indicate that lamination, bedding, and jointing are important factors in determining the final shape. Sneed and Folk (1958) concluded from their studies that sphericity and form are complex functions of rock type, size, and distance of transport, and they suggested that only nearly isotropic rocks should be used as a measure of the effect of transport on form and sphericity. They also showed that there is a large difference in sphericity and form between large and small pebbles.

It is well established that larger particles wear more rapidly than smaller ones. Thiel (1940) demonstrated this experimentally, while Kuenen's (1958, 1959, 1960) experiments on abrasion of mineral grains by wind and water confirmed it and showed that original shape has a considerable effect on final shape, but anisotropic crystallography does not. Of particular interest is Kuenen's conclusion that "the mechanical abrasion of sand-size particles by rivers is so slight as to have little significance even for coarse sand and even after the cumulative effect of a succession of sedimentary cycles" (Kuenen, 1960).

No studies have been made on the modifications of sphericity and roundness that can be achieved during soil formation. It is probable that the effects are small, except in weathering of primary minerals to clay minerals, when change of shape to platy or rod shapes may occur, although Sorby (1880) and Tricart (1958) both noted rounding by solution during weathering. Changes may be affected also by physical breakdown of minerals along cleavages and zones of strain. Breeze

(1960) presented evidence that secondary outgrowths on quartz grains during soil formation may change the grain shape, although this is probably negligible where statistical counts of grains are made for shape analysis. In spite of these difficulties some information concerning origin and transport of parent materials can be derived under favorable circumstances from shape analyses; the most suitable materials are those containing relatively large particles and pebbles.

3

Mineral Estimations

Perhaps the greatest obvious difference between soils and sediments is the development of horizons, or layers, within soil profiles caused by soil-forming processes. Such horizons can occur even in profiles formed on a relatively uniform, unconsolidated, transported sediment or on a relatively homogeneous hard rock mass, whether of sedimentary or igneous origin. In some profiles the horizons merge into each other without any apparent break, and so there is perhaps little possibility of mistaking the soil horizons for sedimentary beds. In others, however, the boundaries between certain soil horizons, particularly that between *A* and *B* horizons, may be very sharp, and the differences between the soil materials of the two horizons may be quite marked; such horizons may be mistaken for separate sedimentary beds. This possibility is equally applicable in reverse; pedologists have come to expect soil profiles with *A* and *B* horizons (and other successive layers) with contrasting properties and relatively sharp boundaries between them, and many such arrangements are accepted as soil profiles without question. The work of Butler (1959), Walker (1962), Churchward (1961), and van Dijk (1958), however, has demonstrated the common occurrence of superimposed sedimentary beds in southeastern Australia which are difficult to distinguish from soil horizons, and there are many reports of similar occurrences from other parts of the world. Recognition of these sedimentary layers in the field is often difficult because they may be old buried soil profiles or truncated remnants of profiles.

The first essential, therefore, in studies of soil formation, is to differentiate soil horizons in single soil profiles from super-imposed beds of transported sediments, whether these beds are represented by old soil profiles, parts of profiles, or unaltered sediments. Pedologists have generally assumed that this can be done by studying the variation in properties of soil materials with depth in the section. It can be done, probably with more certainty, by analysis of variations in mineralogy with depth in the profile. Such studies have the same basis as (and come within the scope of) studies in provenance and variations in sedimentary beds.

Having established that a soil profile has been formed throughout its depth from a relatively uniform parent rock, changes in the profile due to soil formation can be studied by other kinds of mineral analysis. Such studies include the kind and degree of weathering of minerals in the soil materials at various depths in the profile, progressive changes in mineral composition of the soil materials with depth in the profile, and increases or decreases in volume and weight of the various horizons in the profile. Such investigations are beyond the normal field of study of both the sedimentary and igneous petrographer, but they are an integral part of mineral analysis of soil materials and soil profiles.

Mineral estimations of soil materials are not necessarily confined to entities which would be strictly defined as mineral grains; all the constituents that can be recognized must be considered in the analysis (e.g., complex grains and phytoliths), if each mineral species is to be estimated as a percentage of the whole soil material.

Such mineral analyses involve identification of the mineral species and varieties present and estimation of their relative proportions.

3.1 PRINCIPLES OF IDENTIFICATION

Little need be said here about the principles of identification of the constituents of soil materials. Numerous excellent textbooks are available for the systematic determination of mineral species by optical methods in thin sections (e.g., Ford, 1947; Winchell, 1951) and on separated grains and mixtures of grains (e.g., Milner, 1962). Standards are also available for identification by X-ray analysis and other methods, such as chemical analysis. Broadly there are two chemical techniques which are useful: complete analyses (Washington, 1930; Shapiro and Brannock, 1956) and spot tests (Feigl, 1954) with which may be included the various staining tests for felspars (Gabriel and Cox, 1929; Hawkins and Graham, 1950; Graham, 1955; Bailey and Stevens, 1960), carbonates (Fairbanks, 1925; Walger, 1962; Warne, 1962), various clay mineral species (Grim, 1953; Shukevich, 1954), etc.

Practically all soil materials contain significant amounts of material that cannot be identified with an optical microscope, such as the very fine grained and/or intimately mixed mineral fractions of the plasma, the smallest grades of skeleton grains, and humified organic matter. These constituents can be identified by X-ray diffraction, electron diffraction, D.T.A., infrared absorption, electron microscopy, and/or chemical analysis. If they are present in high proportions, identification of the coarser mineral grains may be difficult in thin section because of the restricted number and variety of sections of any particular mineral species present; indeed the rarer species may not be represented at all in any one section.

3.2 PRINCIPLES OF CONCENTRATION

Studies in sedimentary petrology have shown that the greatest variety of mineral species in sediments and transported parent materials of soils are those of relatively high specific gravity. Although quartz and the various kinds of felspars are the commonest rock-forming minerals, they are so ubiquitous that their usefulness depends on interpretation of morphological varieties (e.g., Krynine, 1950) and the diagnostic value of kinds and arrangements of inclusions. The most useful diagnostic minerals in soil materials are relatively resistant to weathering and so can persist for a long time; such minerals usually have a relatively high specific gravity, but they usually occur in relatively small amounts in sedimentary layers so that it is necessary to concentrate them for quantitative analyses.

A. Size Separations

Before attempting more specific methods of concentration the sample is usually divided into a number of grades according to the principles discussed earlier (Section 2.3). This in itself tends to partially concentrate groups of mineral species. The clay minerals are concentrated in the finer grades although individual grains of clay minerals, such as kaolinite, and stable aggregates of others, such as nontronite, may occur in the coarser grades. In a sediment that has some degree of size sorting, the mineral grains of higher specific gravity are concentrated in the grades smaller than the mode; in practice the "heavy minerals" (S.G. $>$ 2.83) tend to be concentrated in the grade one degree smaller than the mode. Particular mineral species may also concentrate in certain grades because of their original size since many igneous rocks tend to have a relatively even grain size for specific minerals, and those which are resistant to weathering and physical breakdown tend to retain the evenness of grain size.

B. Specific Gravity Separations

Semiquantitative specific gravity separations can be made on large samples by panning or with the vibrating table (Milner, 1962); more accurate separations are made with liquids of known specific gravity. The efficiency of the method depends on the difference in specific gravity between the liquid and the mineral grains and the concentration of the mineral grains per unit volume of the liquid and per unit area of the surface of the liquid; the latter factor is more significant than the former, except when particularly shallow vessels of very large diameter are used so that the ratio of volume of liquid to surface area of liquid is low. The maximum desirable concentration of grains also depends on the size of the grains; coarse grains can be separated without centrifuging and with quite high concentrations, while fine grains require centrifuging and much lower concentrations; there is also danger of flocculation with very fine grains.

C. Magnetic Separations

Magnetic, electromagnetic, electrostatic, and dielectric methods of separation have all been used successfully. Magnetic and electromagnetic methods are the most useful; practically pure concentrates of particular minerals can often be obtained with suitable equipment. Electrostatic and dieletric methods are cumbersome by comparison and are of restricted usefulness; these are described by Milner (1962).

Separation by magnetic methods depends on variations in magnetic strength of the mineral species. The purity of the separation, therefore, depends on the difference in magnetic strength between the mineral species being separated and the accuracy and range of operation of the instrument used to produce a magnetic field. Strongly magnetic minerals, such as magnetite, can be concentrated by a permanent magnet of the bar or horseshoe type, rather less magnetic minerals by a horseshoe magnet with adjustable pole pieces, and a wide range of weakly magnetic minerals by an electromagnet in which the strength of the magnetic field can be controlled by varying the current and/or the gap between the poles. A simple bench type illustrated by Milner (1962) is very similar in operation to the horseshoe magnet with adjustable poles.

More specific separations can be made by using one of the commercial electromagnetic separators, such as the Cooke or Frantz isodynamic magnetic separators. These instruments consist essentially of a vibrator table fitted between two elongated poles of a powerful electromagnet. The vibrator table is fitted with a hopper at the top, to take the sample, and two catchers at the bottom end; it is divided by a ridge for approximately the last two-thirds of its length. The whole instrument can be tilted so that the vibrator table can be set for slope along its length thus controlling the rate of movement of grains down the table; the rate of feed is controlled by adjusting the hopper. The mineral species separated are selected by controlling the strength of the magnetic field (by controlling the current) and the side tilt of the vibrator table; the more magnetic fraction is attracted to the high side of the table and is separated from the less magnetic fraction by the ridge dividing the table. The efficiency of separation depends largely on the rate of flow of the grains down the vibrator table, and care should be taken to insure that this is not too fast; fast rates of flow may also lead to overflowing from the vibrator table. As strongly magnetic minerals adhere to the poles of the electromagnet, even at a setting for the weakest possible field, they should be separated first either with a horseshoe magnet or a special vertical table and feed attachment. Further separations are made by removing the more strongly magnetic minerals before the less magnetic.

Usually the sample must be passed through the electromagnetic separator several times to achieve a complete single separation; the finer the grain size and the faster the rate of flow, the larger the number of times the process must be repeated for complete separation.

3.3 PRINCIPLES OF ESTIMATION

Quantitative estimation of the various mineral species in a concentrate involves counting and/or measurement of individual grains; this, of course, applies to optical not to X-ray techniques. The results of grain counts are usually expressed as percentage by number, but for many calculations based on these data and for some kinds of interpretations, the results must be expressed as percentage by volume or weight. To do this the specific gravity of the mineral species is measured, or assumed from published tables, and either the estimations are made on grades with restricted limits, so that number can be translated to volume without introducing undue errors, or the volume of each grain is measured directly or indirectly. Certain procedures have been established for counting grains and measuring their volume.

A. Accuracy of Grain Counts

The accuracy of grain counts has been considered by a number of authors (e.g., Carroll, 1938, 1957; Krumbein and Pettijohn, 1938; Nicholls, 1936), most of whom based their conclusions on considerations put forward by Dryden (1931, 1935). There is, however, some confusion of thought regarding the bases of these estimates of accuracy.

Suppose that a hypothetical sediment contains mineral species 1, 2, . . . , r in amounts $y_1, y_2, \ldots, y_r$ expressed as percentages (so that $\Sigma y_i = 100$). The chance that a given grain in the sediment will belong to a species i is $y_i/100$. Assume, then, that each individual mineral species follows a binomial distribution and that, for the sample sizes normally considered, the binomial distribution approximates a normal distribution. Then, in a sample containing n grains drawn from the above sediment, the expected percentage of mineral species i occurring will be y_i with a standard deviation σ_i, given by:

$$\sigma_i = \sqrt{y_i(100 - y_i)/n} \tag{1}$$

σ_i may be converted to a probable error:

$$p_i = k\sigma_i \tag{2}$$

by a choice of the constant k which fixes the confidence limits of the estimate. Both σ_i and p_i are expressed as *percentages* of the *total number of grains* in the sample. For $k = 0.6745$ (50 percent confidence limits) the expression

$$\text{Percentage of mineral species } i \text{ present} = (y_i \pm p_i)\%$$

means that in 50 percent of samples (of n grains) drawn from the sediment an actual count of the mineral species i would give a value lying between the limits

$$(y_i - p_i)\% \quad \text{and} \quad (y_i + p_i)\%$$

As it stands this is of very little use since what is required is an estimate of the likelihood that the estimate of percentage is close to the true percentage. Fortunately, provided the sample is fairly large (and $n = 300$ grains is certainly large enough), it can be shown that the converse interpretation of the above statement also holds (Simon, 1941); that is, a count of the mineral species i in a sample containing a total of n grains yields a value y_i with a probable error of

$$p_i = k\sqrt{y_i(100 - y_i)/n} \tag{3}$$

and the statement

$$\text{Percentage of mineral species } i \text{ present} = (y_i \pm p_i)\%$$

means that the *actual* percentage of mineral species i present in the original sediment lies between the limits

$$(y_i - p_i)\% \quad \text{and} \quad (y_i + p_i)\%$$

in 50 percent of the cases (where $k = 0.6745$). For $k = 1.96$ the statement holds in 95 percent of cases.

Equation 1 shows that the probable error (p_i) is inversely proportional to $\sqrt{n}$, so that four times as large a sample is required to reduce the error by half. This "absolute" error is also higher for the more abundant species.

This error may now be considered in relation to the actual percentage of mineral species i present. A *probable percentage error* (P_i) can be defined, so that

$$P_i = \frac{p_i}{y_i} \times 100 = 100k\sqrt{\frac{1}{n}\left(\frac{100}{y_i} - 1\right)} \tag{4}$$

P_i decreases with increase in n in the same way as p_i, but it is also lower for the more abundant species. From equation 4,

$$n = \left(\frac{100}{y_i} - 1\right)\left(\frac{100k}{P_i}\right)^2 \tag{5}$$

or, alternatively,

$$n = p(1 - p)\left(\frac{100 \cdot k}{P \cdot a}\right)^2 \tag{6}$$

where n = total number of grains counted,
p = mineral species as a fraction of the sample,
k = constant for the confidence interval (50% = 0.6745; 80% = 1.28; 95% = 1.96),
P = probable percentage error, and
a = true frequency of the mineral species.

or

$$n = \left(\frac{1 - p}{p}\right)\left(\frac{100 \cdot k}{P}\right)^2 \tag{7}$$

assuming that $a \doteqdot p$.

Dryden (1931) pointed out that beyond a certain limit the number of grains that must be counted to significantly decrease the probable percentage error, at any confidence level, increases very rapidly; it also increases rapidly with increase in confidence interval.

Table 3-1 shows values of n calculated from equation 7 for various values of p and P, and for three confidence intervals ($50\%:k = 0.6745$, $80\%:k = 1.28$, $95\%:k = 1.96$). This table thus provides an estimate of the number of grains that must be counted in the sample so that the percentage of a given mineral species present (y_i) may be estimated to the desired degree of precision (P_i) at three confidence intervals.

In deciding the number of grains that should be counted to estimate the percentage of a particular mineral species in a sample, a decision has to be made concerning the confidence interval and probable percentage error that are acceptable. The usual acceptable confidence interval for comparable work in other disciplines is 95 percent ($k = 1.96$), and the acceptable probable percentage error (P_i) should

TABLE 3-1. Relationship between Number of Grains Counted, Probable Percentage Error, Confidence Interval, and Percentage of the Mineral Species in the Sample*

Mineral Species as a Fraction of Sample	Probable Percentage Error = 5%			Probable Percentage Error = 10%		
	Confidence Interval, 50%	Confidence Interval, 80%	Confidence Interval, 95%	Confidence Interval, 50%	Confidence Interval, 80%	Confidence Interval, 95%
0.005	36,019	130,345	305,863	9055	32,636	76,416
0.01	16,119	64,848	152,163	4504	16,236	38,016
0.02	8,869	32,095	75,313	2230	8,036	18,816
0.05	*3,439*	12,445	29,203	865	3,116	7,296
0.10	*1,629*	5,895	15,370	406	*1,476*	*3,456*
0.15	1,029	3,712	8,709	258	929	2,176
0.20	724	2,620	6,148	182	656	*1,536*
0.25	543	1,965	4,611	136	492	1,152
0.30	422	*1,528*	*3,586*	106	383	896
0.40	271	983	2,306	68	246	576
0.50	181	655	*1,537*	46	164	384
0.60	121	437	1,025	30	109	256
0.70	78	281	659	20	70	165
0.80	45	164	384	11	41	96
0.90	20	73	171	5	18	43

* Calculated from equation (7), Section 3.3:A.

be no greater than the variability of the material being studied and/or the errors introduced by techniques of preparation of the samples; the latter, of course, depend on the percentage of the mineral species being estimated in the sample, the probable percentage error increasing with decreasing proportion of the mineral species in the sample.

Carroll (1938) suggested that a probable percentage error (P_i) of 5 percent is all that is warranted in studies of sediments and soil materials because of the known variability within sediments and because of errors inherent in sampling and preparation of the sample. The ideal situation, therefore, would be a probable percentage error of 5 percent at a confidence interval of 95 percent. This ideal is unattainable for mineral species which comprise a small proportion of the sample because of practical considerations of time involved in counting very large numbers of grains; Table 3-1 shows that over 150,000 grains must be counted if the mineral species is 1 percent of the sample. Therefore, some loss of accuracy and/or confidence level must be accepted.

As a balance between accuracy and time involved in making counts, it has been generally accepted that counts should be restricted to between 300 and 1000 grains per sample (e.g., Krumbein and Pettijohn, 1938; Carroll, 1938, 1957; Milner, 1962; Haseman and Marshall, 1945); this is based on a confidence interval of 50% and a probable percentage error that varies according to the percentage of the mineral species in the sample. Table 3-1 shows that, for a count of 300 grains to be adequate at the desirable confidence interval of 95% and probable percentage error of 5%, the mineral species must comprise more than 80 percent of the sample; even at a confidence interval of 50% and a probable percentage error of 5%, the mineral species must be approximately 40% or more of the sample, or more than 13% of the sample for a probable percentage error of 10%. For a count of 1000 grains to be adequate at a confidence interval of 95% and probable percentage error of 5%, the mineral species must be 60% or more of the sample; at a confidence interval of 50% it must be 15% or more, and for a probable percentage error of 10% it must be 5% or more of the sample. Counts of between 300 and 1000 grains per sample, therefore, are apparently inadequate, except for the more abundant mineral species, if the desirable degree of accuracy and precision is required.

Variability of samples is more likely to affect estimates of low mineral percentages rather than high ones, so low percentages are less useful for interpretative work unless the mineral species is a specific "marker" mineral, in which case simply presence or absence is usually diagnostic. It is reasonable, therefore, for normal work, to increase cording to the limits of the size fraction and, to some extent, the shapes of the grains.

Friedman's (1958) method of deriving sieve-size distribution from measurements in thin sections (Section 2.3:C) can also be adapted to estimating percentages by volume. Instead of grouping the grains into classes according to the apparent long diameter, the measurements for each mineral species are added together and reduced to a percentage of the total. This percentage can be used as a measure of percentage by volume. The technique is again subject to considerations of number of grains measured similar to those discussed for the Rosiwal method.

Chayes' (1956) point-count method can be used simply by counting the number of points occurring within the outline of the grains of each mineral species, summing them, and reducing them to a percentage of the total.

Inaccuracies inherent in these methods are due to inaccuracies in measurement of the individual lengths of the minerals along the lines of traverse (or of deciding whether points of the grid which touch the boundaries of the grains should be counted or not) and to mounting methods, for example, grains mounted in an oil tend to take up a preferred orientation. Inaccuracies of measurement are greatest for the smallest grain sizes, while those due to preferred orientation are greatest for such physically anisotropic minerals as the micas. This error can be reduced by embedding the grains in a cold polymerizing plastic, making separate estimations on two sections cut at right angles to each other, and averaging the results. The results approximate the true volume much more closely than either one individually. There are two difficulties in this approach. First, the sample must be large enough to allow sections to be cut in two directions at right angles to each other; this is not always possible where a large number of particle-size, specific gravity, and magnetic separations have been made so that the number of grains in each fraction is small. Second, and less important, the grains in the mounts of small particle-size grades may overlap because of their small size compared with the thickness of the section (25 μ). This may cause difficulty in measurement, but it can be overcome to some extent by grinding the section thinner; with care such mounts can be reduced to about 15 μ thickness.

D. Weight Estimations

Estimations of percentage by weight can be made by separating pure concentrates of the various mineral species. This has the advantage that no approximate assumptions concerning (or measurements of) specific gravity of the mineral species, their size or shape, or the accuracy of counting methods are necessary. Errors due to subsampling or measurement of too small a number of grains are virtually eliminated also because most methods of concentration of mineral species, such as specific gravity and magnetic separations, can cope with relatively large samples. On the other hand, the method suffers from inaccuracies inherent in the methods of separation, such as the difficulty of making complete separations and unavoidable small losses of material. These errors, however, are likely to be insignificant compared with sampling errors and variations in composition of the material being analyzed. The major disadvantage of the method is its tediousness. It is a time-consuming task to prepare clean separates of individual mineral species from a sample containing 8 to 15 mineral species, as many samples from sediments and soil materials do. For this reason the method is little used except in special circumstances (e.g., when the weight percentage of a single mineral species is required).

The alternative to making pure separations is to estimate percentage by volume and convert to percentage by weight using specific gravity measurements of the mineral species. Errors of measurement of percentage by volume and specific gravity can be reduced by using separates that have narrow limits of size and specific gravity and include as small a number of mineral species as possible; magnetic separations are very useful for this purpose. Specific gravity determinations made on handpicked separates of pure minerals from the actual sample are preferable to published data because many minerals have a wide range of specific gravity.

Percentage by volume can be converted to percentage by weight according to the equation:

$$W_a = \frac{100 V_a G_a}{\sum_{i=1}^{n} V_i G_i}$$

where W_a = percentage by weight of mineral species a,
V_a = percentage by volume of mineral species a,
G_a = specific gravity of mineral species a,
V_i = percentage by volume of all the mineral species taken one at a time,
G_i = specific gravity of all the mineral species taken one at a time, and
Σ = "the sum of."

Percentage by weight of each mineral species in the whole sample of sediment or soil material can be calculated by the equation:

$$P_a = \frac{\Sigma W_a W_s}{W_t}$$

where P_a = percentage by weight of mineral species a in the whole material,
W_a = percentage by weight of mineral species a in each separate taken one at a time,
W_s = weight of each separate corresponding to the values W_a, and
W_t = total weight of the material = ΣW_s.

3.4 PRESENTATION OF DATA

The four methods most generally used to present the data of mineral estimations are histograms, profile diagrams, frequency number tables, and simple tabulation of estimated percentages. Oertel (1961) proposed a further method which he termed "common measure," which should prove particularly useful in special cases.

A. *Histograms*

Histograms of mineral estimations are constructed simply by marking off, along the horizontal X-axis, equal divisions to represent each mineral species and erecting vertical rectangles whose heights represent the estimated percentage of each species. Such diagrams can be constructed for estimations in single size fractions, in the whole material, or in any required specified fraction of it. Quick visual comparisons of the mineralogical composition of materals can be made by maintaining a constant order for the mineral species and arranging the histograms under each other. Modifications of this method can be used to show variation in percentage of certain minerals with some other factor, such as variation in mineral composition with position in the landscape down a slope (Brewer, 1948, and Fig. 4).

B. *Profile Diagrams*

Profile diagrams are similar to histograms, except they are designed to show mineral frequencies in relation to stratigraphic horizons. The horizontal X-axis is marked off in intervals representing fixed vertical intervals in the stratigraphic series being studied, and verticals are erected at the points along the X-axis that represent the stratigraphical position of each sample. Heights are marked off on the verticals to represent the frequency of a single mineral species in each of the samples studied, and these points are joined by straight lines so that the result is a line profile representing the distribution of a single mineral species throughout the stratigraphic sequence. This is repeated for each mineral species, and the profiles are arranged one under the other. Other data, such as unconformities, can be represented on the diagrams. This device can be used effectively for soil profiles.

C. Frequency Number Tables

This method simply consists of lists of tables of integers or letters representing ranges of frequencies. Variations between the proposed tables are mainly in the limits of frequency ranges. Krumbein and Pettijohn (1938) favor a table put out by the Burmah Oil Company (Evans, Hayman, and Majeed, 1933) which is set out in Table 3-2. The frequency numbers in this table are, in a general way, logarithms of the frequency percentage determined by counting; thus the lower frequency values are emphasized. It is pertinent to note that a true frequency of three-quarters of one percent will be estimated as between one-half and one percent with a confidence interval of 80 percent when

TABLE 3-2. Frequency Number Scale (after Evans, Hayman, and Majeed, 1933)

Frequency	Approximate Percentage
8+	90–100%
8	75–89%
8−	60–74%
7+	45–59%
7	35–44%
7−	28–34%
6+	23–27%
6	18–22%
6−	13–17%
5	7–13%
4	4–6%
3	2–3%
2	1–2%
1	$\frac{1}{2}$–1%
1*	1 grain only

approximately 1800 grains are counted. The frequency number 1 (½ to 1 percent), therefore, has significance if this number of grains is counted. A count of 1500 grains, which was suggested for normal work (Section 3.3:A), would give a level of precision slightly lower than this.

Frequency number tables offer little avantage over the simple statement of estimated percentages of mineral species. Although they make comparison of data easier in a visual sense, they tend to hide the inherent inaccuracies in the methods used. For example, a true frequency of a mineral species of 2.2 percent will be estimated as between 1.7 and 2.7 percent at the 80 percent confidence interval by counting 1500 grains. This means that sometimes it will be represented by the frequency number 2 and sometimes by 3 in Table 3-2. A true frequency of 2.5 percent, however, will be represented by frequency number 3 at the same level of precision.

D. Tabulation

This method is simply a listing of the mineral species by name across the sheet, the sample numbers vertically down the left hand margin, and the percentages in the appropriate positions. Various comparisons can be made easier by judicious arrangement of the samples. Tabulation completely eliminates such misconceptions regarding accuracy that may arise from the use of frequency number tables, provided a statement is made concerning the number of grains counted so that the probable percentage error at any required confidence level can be calculated. Without such a statement tabulation can be even more misleading than frequency numbers by the bald statement of such low percentages as one-half and one-quarter.

E. Common Measure

Oertel's (1961) presentation is obtained by reducing each profile, and the measurements made for each sample, to a common measure. A corresponding horizon is chosen in each profile and given a depth value of unity, irrespective of the actual depth of the solum to this horizon. All samples from all the horizons of each profile are added and meaned, and the results for each horizon are expressed as a decimal of this mean value. By plotting the common measure of depth against the common measure of the characteristic analyzed, the plots for like profiles all fall on a single curve. This data is eminently suitable for statistical analysis.

3.5 ANALYSIS OF DATA

Perhaps the most important application of mineral analyses is in the study of the genesis of soil profiles as reflected by changes in mineralogy with depth in the profile due to weathering and leaching, which may involve complete removal of some of the constituents from the soil profile and/or translocation of certain constituents from one horizon to another. The first essentials in such studies are recognition of the parent material from which the soil profile was developed and proof that the parent material was uniform throughout the profile; this is achieved by comparative, quantitative, mineral analyses. It is often of interest, also, to identify the parent material of soil profiles, especially in areas of transported parent materials where profiles may be developed in stratified sediments, and older profiles may be buried by a more recent sedimentary deposit in which younger profiles have developed.

A. General Considerations

Quantitative mineral estimations facilitate comparison of separate soil materials within the one soil profile or between soil profiles and of soil materials with their presumed parent material. The kind of mineral analyses required and the treatment of the data in such studies are the same as those required for study of the origin and variability of sediments, that is, the principles of analysis of the data are the same as those used in studies of provenance. These principles have been firmly established by the sedimentary petrologists who have used the methods for a long while; many detailed studies in provenance have proved the value of the method.

Certain detailed studies in provenance may demand complete quantitative mineral analysis of the materials, for example, where the objectives are to identify the sources of a mixed transported material and to estimate the proportions of the various kinds of source materials present in it. This involves counting individual grains in a large number of restricted size classes from each sample of material. For many studies, however, the amount of detail required in the analyses can be greatly reduced. If the study is aimed only at estimation of uniformity of the parent material of a soil profile, complete analyses are not strictly necessary, although the results become increasingly dependable with increasing completeness of data. Justification for limiting the number of size classes studied lies partly in logical deduction from the known properties and processes of formation of sediments and partly in the experience and great volume of data accumulated by sedimentary petrologists which indicate the validity of the method.

For direct correlation of a number of different materials, Rubey (1933) suggested examination of two grade classes from each material: the same actual grain size fraction, which eliminates errors due to abrasion and irregular size distribution in the source material, and the same grain size fraction relative to the size distribution curve,* which eliminates errors due to sorting. This suggestion was based on his observations that certain minerals were increasingly abundant in increasingly finer-grained samples of the same origin. Haseman and Marshall (1945) concluded that the origin of a soil (the origin of the parent material) is reflected in the kinds and amounts of heavy minerals present and that a qualitative and roughly quantitative determination of the heavy minerals is usually sufficient to show what this origin is, provided the kind and relative abundance of heavy minerals in the possible source materials have been determined. They suggested that only one fraction (the same size fraction) need be used in making heavy mineral comparisons between soils and parent materials. Marshall and Jeffries (1946) summarized the bases of interpretation quite succinctly when they pointed out:

(1) Qualitative and quantitative differences in the suite of the most resistant heavy minerals are characteristic of differences in geological materials.

(2) Qualitative similarity of the suite of heavy minerals but quantitative differences in mineral ratio are characteristic of differences in conditions of deposition (Cf. Section 3.5:C).

More recently, Briggs, McCulloch, and Moser (1962) pointed out that the shape of particles can have a large effect on particle size distribution of mineral species in sediments. They stated that "variation in shapes of heavy minerals can produce about the same differences in size frequency distribution as the expected variations in specific gravity. . . . If minerals are to be used to discriminate sedimentary environments, or to establish consanguinity of heavy minerals suites, their hydraulic shapes should be accurately measured, and the effect of shape on the mineral distribution evaluated."

* The choice of this size fraction should be guided by the observation that the fraction one degree smaller than the mode of the distribution curve is most likely to contain the greatest abundance and variety of diagnostic minerals (the "heavy" minerals).

B. Coefficient of Determination

Dryden (1935) considered the possibility of expressing the "likeness" of mineral assemblages to each other by means of a single figure rather than the subjective method of comparing tabulated columns of data or graphical presentations of such data; he developed the *coefficient of determination.* This is derived from the *coefficient of correlation* got from the equation:

$$r = \frac{\Sigma(XY) - n(M_x M_y)}{\sqrt{[\Sigma(X^2) - nM_x{}^2] \cdot [\Sigma(Y^2) - nM_y{}^2]}}$$

where r = coefficient of correlation,
Σ = "the sum of,"
n = the number of pairs of percentages used, that is, the number of mineral species,
X = any percentage from sample A,
Y = the corresponding percentage from sample B, that is, the percentage of the same mineral species as that to which X refers,
M_x = the mean of the X values, and
M_y = the mean of the Y values.

Then,

$$r^2 = \text{the coefficient of determination.}$$

The answer for r is a decimal, but its value for different pairs of samples cannot be compared directly. For example, $r = 0.7$ and $r = 0.35$ does not mean that the first pair are twice as alike as the second pair. On the other hand, r^2 is a simple percentage as an expression of the proportion of elements common to the two samples or suites, and so $r^2 = 0.48$ and $r^2 = 0.16$ mean that the first pair contains three times the number of common elements as the second pair.

Dryden made several remarks concerning the derivation and characteristics of this coefficient of determination:

(1) Usually r^2 is between 0.5 and 1 because of the small number of mineral species involved; he is referring here particularly to estimations of the heavy minerals.

(2) If no species are common, $r^2 = 0$. If all are common, r^2 depends on the frequency of each species.

(3) If a mineral species is absent its percentage as zero must be used to calculate r and r^2.

(4) Mineral varieties are treated as species.

(5) Since the probable error of r is $\frac{0.6745(1 - r^2)}{\sqrt{n}}$, for a confidence interval of 50 percent, the reliability increases as the square root of the number of observations.

(6) The absolute value of r^2 means little as yet because of lack of data on its significance.

Just as the basis of interpretation of the data of mineral analysis depends on logic and accumulated data, so does the use of the coefficient of determination. One of the bases of correlation by heavy mineral analyses is the assumption that the assemblages originated in the same way at the same time, and any factors which bear on the proof or disproof of this assumption should be considered; for example, geographical proximity and stratigraphic evidence may enhance or disprove the assumption.

C. *Application to Soil Profiles*

Due consideration must be given to the nature of soil materials in applying mineral estimations to studies of uniformity and origin of parent materials. Many heavy minerals are not particularly resistant to weathering. In fact, Haseman and Marshall (1945) cite hornblende as a particularly good mineral to use in comparing severity of weathering in different horizons of a soil profile and of soils formed under different climatic conditions. Other heavy minerals weather at different rates to that of hornblende, and the rates of weathering of all species vary with conditions. Consequently the proportions and total amounts of the various mineral species may vary considerably with depth in profiles formed on uniform parent materials. This situation has been demonstrated adequately by a number of workers (e.g., Haseman and Marshall, 1945; Brewer, 1955*b*). Weathering can also modify the size distribution of the heavy mineral species (Brewer, 1955*b*). This also is effective to different degrees on different mineral species, so it can cause differences in mineral ratios within certain size fractions in the various horizons of a profile formed from a uniform parent material.

There are two ways of dealing with this problem:

(1) One is to compare the series of analyses at different depths in the profile and plot percentages of the various weatherable mineral species against depth. The shape of this curve indicates, but does not prove, whether the differences in amounts of minerals are due to

weathering or to differences in parent materials; a smooth continuous curve indicates progressive weathering, while relatively sharp breaks in the curve indicate differences in parent material—that is, either differences in conditions of deposition or differences in geological materials if the breaks are associated with the appearance of different mineral species (Section 4.1). This method must be used with caution since sharp interhorizonal boundaries, such as that between the *A* and *B* horizons in planosolic soils, may cause a break in the curve even though the profile is developed from a uniform parent material. Ruhe (1956) used the variation in ratio of resistant-to-weatherable minerals (zircon plus tourmaline to pyroxenes plus amphiboles, and felspars to quartz) with depth in the profile for this purpose.

(2) The second and perhaps more dependable method is to base the comparison mainly on the more resistant mineral species, such as zircon and tourmaline. Although these will show an increase in total amount in horizons where weathering has reduced the amount of the less resistant minerals, the ratio of amount of zircon to amount of tourmaline in the whole material, or in specific size fractions, should be much more constant with depth if the parent material is uniform. Variations in this ratio can be plotted against depth in the profile, and they should give a smooth, almost flat curve for a uniform parent material. The use of such a ratio of resistant minerals largely eliminates the difficulty involved in profiles with sharp interhorizonal boundaries which coincide with sharp differences in degree of weathering; the ratio of two resistant minerals should not change as significantly across such boundaries as the amount of weatherable mineral species.

At this point stress should be laid on the importance of recognizing varietal differences within mineral species in the analysis of heavy minerals. Brammall (1928) stressed the importance of varietal differences in studies of sedimentary rocks, and Dryden (1935) stated that varieties of mineral species should be considered in deriving his coefficient of determination. In practice, mineral varieties are as significant and diagnostic as mineral species. Both zircon and tourmaline are good examples of mineral species within which a wide range of varieties may be recognized. A single igneous rock often contains a single variety of one or more mineral species, and the source of certain materials can often be established on the varieties of mineral species, and a possible source eliminated on the same basis. Lack of observation of varietal differences may cause misinterpretation, as when the source of a sediment containing two varieties of zircon is assumed to be solely an igneous rock that contained only one of the varieties. Varietal differences are also important when using ratios, such as zircon to tourmaline, to establish uniformity of parent material. Consistent ratios of this kind are significant only if the varieties are also consistent. Indeed this method can be applied where, say, tourmaline is absent by estimating the ratios of zircon varieties to each other.

It is worth noting, too, that several workers, (e.g., Carroll, 1937; Jeffries and White, 1937) have suggested that, besides the actual mineral analysis, the total percentage of heavy minerals is a useful characteristic for correlation of parent materials. This is useful only as additional corroborative data in conjunction with the quantitative mineral analysis.

Although this discussion has been confined to data from mineral estimations of the coarser size fractions, the same principles can be applied to analyses of the clay-size fraction (less than 2 μ equivalent diameter). Many soil materials contain relatively high proportions of clay-size mineral grains, but use of estimations of clay-size minerals for studies of provenance and of uniformity of parent material throughout soil profiles is complicated because new clay mineral species can be formed during sedimentation and by weathering during soil formation; detailed mineralogical studies of soil profiles have shown that clay-size minerals are sometimes inherited from the parent material, while other times they are completely new species formed by weathering (e.g., Droste, 1959; Jackson, 1959; Harrison and Murray, 1957; Mitchell, 1962). Studies of origin and uniformity of sediments and parent materials, therefore, are generally more reliable when based on size fractions greater than 2 μ and, more especially, on the "heavy mineral" fractions because they contain the greatest variety of mineral species, especially in sediments, and the species that are most likely to be diagnostic for particular igneous rocks and sedimentary beds. Complete analyses of all the size fractions are desirable in studies of weathering and soil formation.

4

Calculations of Soil Formation

The techniques of quantitative mineral analysis can be used to estimate gains and losses of weight, thickness (or volume), and particular constituents in the various soil horizons due to soil formation. The calculations are generally based on quantitative estimations of the constituents of the soil materials of the horizons in relation to the amount of some substance considered to be essentially stable to the soil-forming processes which have been operative in the particular profile. Apart from sampling and analytical errors, the accuracy of the calculations depends on the validity of basic assumptions concerning uniformity of the parent material, selection of the true parent material, and the properties of the stable constituent. It is appropriate, therefore, to examine the principles involved in reducing these errors to a minimum.

4.1 THE PARENT MATERIAL

The first essential characteristic of the parent material, for calculations relating to profile development, is that it should be uniform throughout the depth of the profile, or at least sufficiently so that errors due to variations in it will be less than those due to sampling, pretreatment, and estimation of the constituents. Profiles formed on parent material with this requirement are relatively rare. Barshad (1955) pointed out that most studies based on index minerals (the stable constituents) to that date had been made on profiles formed on stratified parent materials, or those in which the parent material had "gained clay through migration." There is, in fact, considerable difficulty in finding a profile developed on a uniform parent material.

Recently a number of profiles specifically selected as appearing to be formed *in situ* from relatively uniform parent rocks were found to be formed either on stratified parent materials or on rocks with significantly variable mineralogy (Brewer, unpubl. data). It is important, therefore, to be certain that the parent material is sufficiently uniform for the purpose.

The second requirement for the parent material is that it should be the true parent material unaltered in any way by the soil-forming processes. Such a statement may seem obvious and trite, but Barshad (1955) found it necessary to state that "changes due to soil development can be evaluated with the greatest certainty in profiles in which the clay distribution with depth indicates that the assumed parent material, or what appears to be the preserved parent material, has gained no clay through migration and that any clay found in the parent material was there when soil formation began." Others (e.g., Haseman and Marshall, 1945; Brewer, 1955*b*) have shown that profiles studied by such quantitative estimations should be sampled to a sufficient depth to avoid selecting as the parent material one which has been altered by the soil-forming processes; the calculations used in both these studies would give varying results according to the depth selected as representing the parent material. Since the soil-forming processes responsible for the development of the yellow podzolic soil on granodiorite (Brewer, 1955*b*) were still operative at depths well over 14 ft, the unweathered granodiorite was selected as the parent material. This is in contrast to the suggestion that in such profiles, in which the soil material below the *B* horizon grades into rock, the decomposed rock material could be assumed to represent the parent material (Barshad, 1955). This presents the question: What stage of decomposition of the rock should be accepted as parent material in such a case?

These difficulties in selecting the parent material, even for soil profiles developed on a uniform material, arise largely from a current concept of specific differences between "geologic weathering" and "pedologic weathering" (Jenny, 1941). Certain weathering processes may be geologic, rather than pedologic, in the sense that they take place at depth under the influence of such phenomena as deep ground water. This, however, is not a valid argument against basing the calculations on unweathered rock as the parent material. The object of the calculations is to estimate the changes due to soil formation; "geologic weathering" may be a stage in development of the soil profile, and it should not be ignored (Whiteside, 1953). For a full appreciation of the changes involved in soil formation, and to avoid confusion due to subjective selection of material at some stage of decomposition as the parent material, the calculations should be based on comparison with the parent rock. Soil profiles formed on unconsolidated, transported parent materials, however, may extend to the full depth of a sedimentary layer, or even across the boundary between two layers, so it is not always possible to select the parent material completely unaltered by soil-forming processes. If quantitative estimations of soil formation are to be attempted on such profiles, the deepest horizon of a uniform material should be selected to represent the parent material and its degree of alteration assessed by comparison of its structure, fabric, and mineralogy with that of horizons higher in the profile.

Uniformity of the parent material throughout a soil profile can be tested by the principles of mineral analysis used in the study of provenance (Section 3.5). The soundest basis for selection of the true parent material is that it should show little or no variation with depth within itself; the percentage of the stable constituent and the results of certain of the calculations should be essentially constant with depth within the parent material, and fabric analysis should show no evidence of changes with depth.

On the basis of estimation of a stable constituent, the percentage of each mineral species that has been weathered in the formation of each soil horizon can be plotted against depth in the profile (depth plotted along the X-axis). In the simplest case the resultant curve is smooth and gradually approaches the X-axis, touching it at a very low angle at the depth at which the parent material is unaltered by soil formation. The curve for the stable constituent as a percentage of the whole soil material, plotted against depth in the profile, is similar in form but never meets the X-axis; it becomes parallel to the X-axis at a value of y equal to the percentage of the stable constituent in the parent material. The greater the number of such curves that can be drawn the more certain is the diagnosis since the percentage weathered varies for each mineral species according to its relative resistance to weathering under the specific conditions of formation of the soil profile. The calculations, therefore, should be made for a number of mineral species, especially those which are least resistant to weathering, and so give the most sensitive measure of weathering of the parent material.

The shape and smoothness of these ideal curves can be used both to select the true parent material and as an indication of uniformity of parent material throughout the profile; in general, smooth curves without inflections indicate uniformity. Considerable care must be exercised in their application, however, because certain pedological processes cause translocation and deposition of constituents within the profile with formation of particular horizons that may have sharp or diffuse boundaries. Other processes, such as severe weathering *in situ* and removal of constituents by lateral leaching, can cause sharp *A–B* horizon boundaries. Such sharp horizon differentiation tends to cause sharp inflections in the curves for the percentages weathered of mineral species and the percentage of the stable constituent, even in profiles formed on uniform parent materials. If the curves are not entirely smooth, therefore, the inflections and maxima and minima must be interpreted in light of the characteristics of the profile; the inflections simply indicate an additional problem in the study of the development of the profile.

The technique of using these depth functions to select the true parent material is dependable even if the "stable constituent" weathers to some extent, provided its rate of removal is significantly less than that of the mineral species for which the calculations are made and provided it does not accumulate anywhere in the profile. Under these conditions the curves for the degree of weathering of the mineral species and for the percentage of the stable constituent in the soil material have the same general form but a lower slope than when the stable constituent is quite immobile. This is an important consideration because there is evidence that even the most stable minerals, such as zircon and tourmaline, do weather to some degree under certain climatic conditions (Section 4.2).

The application of these principles can be illustrated by the following examples:

(1) Consider a sedentary soil profile, developed *in situ* on hard rock, in which the soil material passes imperceptibly into rock by a gradual decrease in degree of weathering with depth. Figure 5 shows an example of the application of the criteria for selecting the true parent material in such a case for a yellow podzolic soil formed on granodiorite (Brewer, 1955*b*). The curves for the weathering of felspar and biotite with depth based on hard rock as parent material [Fig. 5*a* (1) and (2)] show every indication of meeting the *X*-axis at a very low angle at the depth at which hard rock occurs (approximately 60 to 70 ft). Thus, although the profile was sampled only to a depth of 14 ft, 6 in., the curves indicate a progressive change in characteristics of the material down to the hard rock; therefore, the true parent material is the hard rock within which there is no change with depth. The curve for weathering of felspar in the same profile

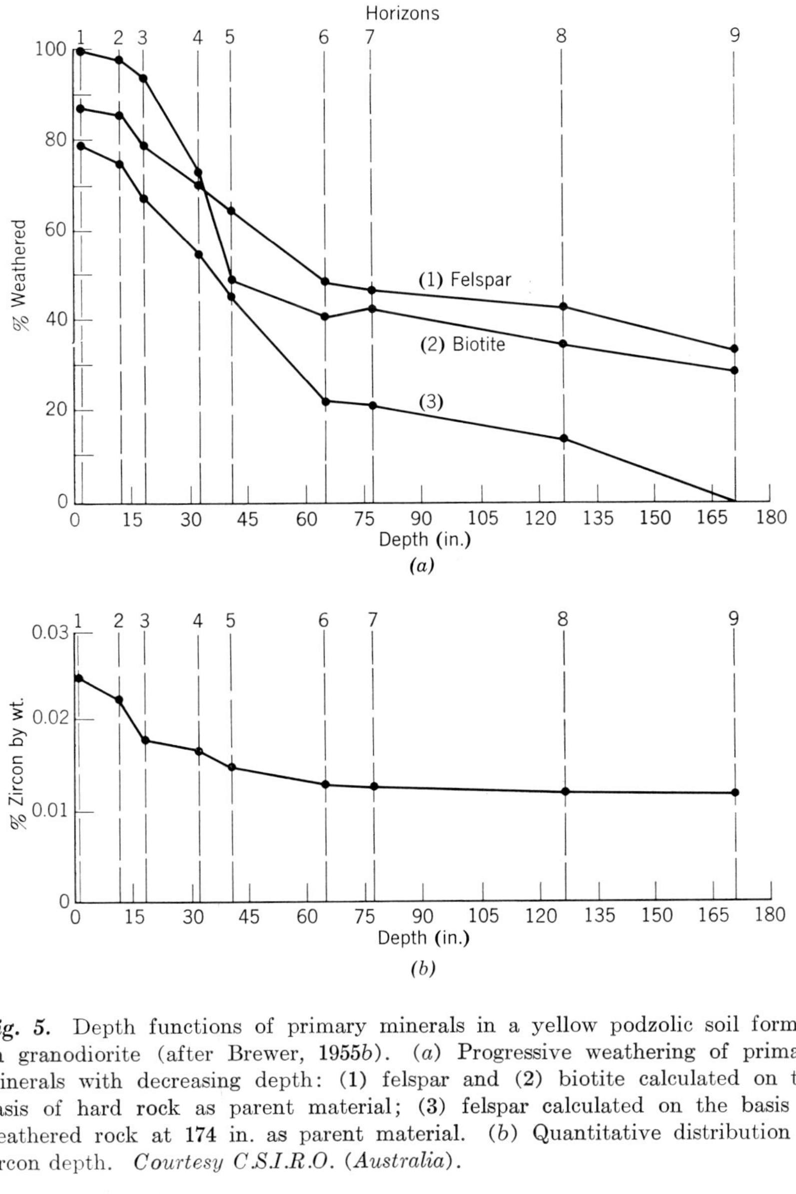

Fig. 5. Depth functions of primary minerals in a yellow podzolic soil formed on granodiorite (after Brewer, 1955*b*). (*a*) Progressive weathering of primary minerals with decreasing depth: (1) felspar and (2) biotite calculated on the basis of hard rock as parent material; (3) felspar calculated on the basis of weathered rock at 174 in. as parent material. (*b*) Quantitative distribution of zircon depth. *Courtesy C.S.I.R.O. (Australia).*

based on weathered rock at 14 ft, 3 in. as parent material [Fig. 5*a*(3)] meets the X-axis at a fairly large angle; it has a relatively steep slope at the greatest depth, which demonstrates lack of constancy above and below 14 ft, 3 in. Calculations based on the material at 14 ft, 3 in. as parent material, therefore, are inaccurate; the inaccuracies would be greater if the material selected were higher in the profile, and less if it were lower in the profile. The curve for the stable constituent (zircon) as a percentage of the whole soil material (Fig. 5*b*) also indicates that hard rock is the correct choice as parent material. The curve is still at a finite angle to the X-axis at a depth of 14 ft, 3 in., which indicates lack of constancy of the material above and below this depth, but it is tending to parallelism with the X-axis at a value of $y = 0.011$ percent—the percentage of zircon in the hard rock. The inflection at a depth of 12 in. in all these curves was shown by mineral analysis to be a depositional break and not due solely to soil-forming processes.

(2) Haseman and Marshall (1945) made calculations of soil formation on a Grundy silt-loam profile formed on loess. They used particle-size distribution and mineral analysis to establish uniformity of parent material to a depth of 69 in. The curve for the stable constituent (zircon) as a percentage of the whole soil material, plotted against depth (Fig. 6*a*) has a minimum at 27 in., a flattening between 59 and 66 in., and a continuous rise from 66 to 111 in. (taking the mean depths of the samples for each horizon). These data suggest either that the true, unaltered parent material had not been reached by 111 in., and the inflections are due to soil-forming processes, or that there are three sedimentary layers with their boundaries marked by the inflections. Mineral and particle-size analyses, however, showed that beyond 69 in. the loess was contaminated with till, that the material was mineralogically identical above and below the inflection at 27 in., and that there was little quantitative difference in the layers from 63 to 69 in. The 63- to 69-in. layer, therefore, was confidently selected as the parent material. The plot of the degree of weathering of the heavy minerals with depth (Fig. 6*b*) supports this choice since it flattens and becomes virtually parallel to the X-axis from 49 to 66 in. If the soil-forming processes had affected the whole depth of the sedimentary bed (loess in this case), the curve would not have become parallel to the X-axis within the bed of uniform material, the unaltered, true parent material could not have been sampled, and any calculations of soil formation must have been qualified and interpreted in the light of this knowledge.

(3) Consider next a sedentary profile developed on *in situ* hard

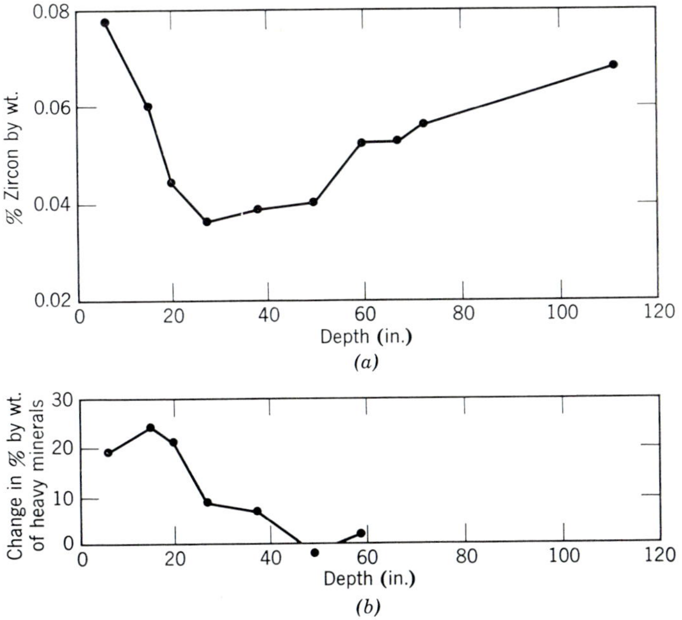

Fig. 6. Depth functions of primary minerals in the Grundy silt loam profile [calculated from Haseman and Marshall's (1945) data]. (*a*) Quantitative distribution of zircon with depth. (*b*) Change in percent by weight of heavy minerals with depth.

rock, in which there is an abrupt change from soil material to hard rock; such profiles have been observed in basaltic areas in Australia. The curve for the percentage weathered of each mineral species against depth in the profile, presuming the hard rock to be the parent material, again should tend to become parallel to and meet the X-axis, but it will be cut short so that it will not meet the X-axis; there will be a more or less sharp break in the curve depending on the sharpness of the transition from soil material to hard rock, and the curve will not have reached parallelism with the X-axis. This is a special case of a sudden change of slope in the curve due to soil formation, and it may be emphasized by accumulation of constituents at the transition from soil to rock. A hypothetical case is set out in Figure 7, which shows the kind of curve to be expected for the stable constituent as a percentage of the soil material plotted against depth. The data from the yellow podzolic soil (Brewer, 1955*b*) is used assuming there is a sharp break to hard rock at 66 in. Applying the principle for recog-

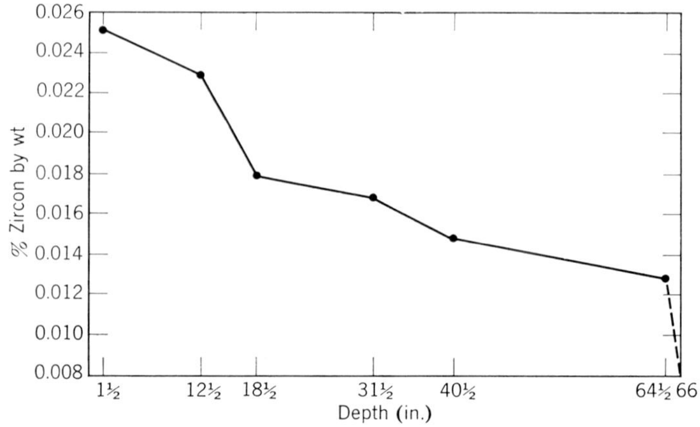

Fig. 7. Quantitative distribution of zircon with depth in a yellow podzolic soil formed on granodiorite assuming hard rock at 66 in. [calculated from Brewer's (1955*b*) data].

nizing the true parent material by the slope of such curves, the solid rock must be selected. The break in the curve, however, destroys its usefulness as evidence that the profile has been formed from the underlying rock, and this must be established by the standard mineralogical methods for proving uniformity of parent material. On the other hand, the steep slope of the curve for the deepest horizons in the soil profile demonstrates that this material cannot be accepted as the true parent material.

(4) Figure 8 shows plots of the index constituents (stable minerals), expressed as a percentage of the whole soil material, against depth for three profiles formed on sedimentary beds (Barshad, 1955); the values were calculated from his results. According to the principles for recognition of the parent material there is no certainty that the true, unaltered parent material has been properly selected for the profile of Sheridan clay loam (Fig. 8*a*) since there is a very significant difference in the percentage of the stable constituent (quartz plus albite) between the horizons centered on depths of 27 and 33 in. and an inflection in the curve at 21 in. Without further mineralogical proof of uniformity of parent material this inflection may indicate the boundary between two sedimentary layers. If the parent material is uniform it has not been sampled to sufficient depth. The curve for the stable constituent (the fine sand fraction) for the Adelanto loamy sand profile (Fig. 8*b*) is difficult to interpret without additional data. If the

true parent material has been selected the curve should have become parallel to the X-axis. The slope from 23 to 52 in., therefore, is due to experimental error or else the true parent material has not been properly selected; mineral analysis may decide this question. The curve for the percentage of the stable constituent (tourmaline) for the Bethel silt-loam profile (Fig. 8*c*) indicates that the 24- to 31-in. layer is the true parent material since it differs very little from the overlying horizon. This horizon, however, is reported as a B_3 horizon

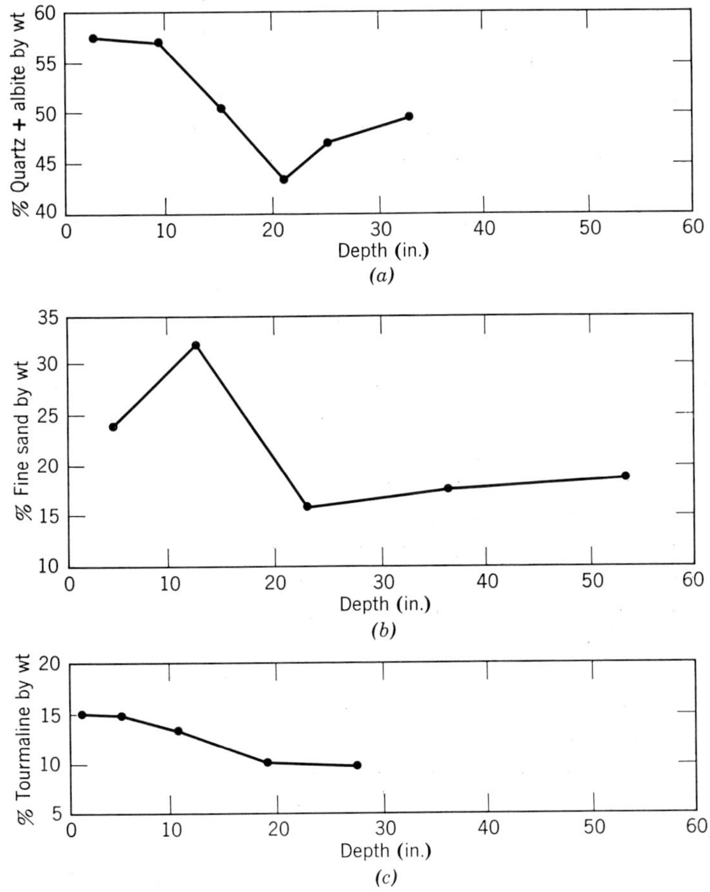

Fig. 8. Quantitative distribution of the stable constituent with depth in three profiles [calculated from Barshad's (1955) data]. (*a*) Quartz plus albite in the Sheridan clay loam profile. (*b*) Fine sand fraction in the Adelanto loamy sand profile. (*c*) Tourmaline in the Bethel silt loam profile.

containing illuviated clay (Barshad, 1955). The flattening of the curve, therefore, must be due to soil-forming processes, and this fact would be revealed and the true parent material identified by sampling at greater depths, provided the sedimentary body were uniform to great enough depths. The micromorphology of the B_3 horizon, as revealed in thin sections, would provide evidence concerning the presence of illuviated clay.

4.2 CHOICE OF THE STABLE CONSTITUENT

The problem is to choose the most stable and immobile constituent so that the least possible error will be made in calculations of soil development. This is unresolved at present except to the extent that the range of constituents proposed is relatively limited. Certain workers (e.g., Merrill, 1921; Muir, 1951; Nikiforoff and Drosdoff, 1943) favor chemical estimations of silica and/or alumina, while others (e.g., Haseman and Marshall, 1945; Brewer, 1955*b*; Khan, 1959) favor chemical and/or mineralogical estimations of relatively stable minerals. Joffe (1936) favored chemical estimations but suggested that the constituents used should be chosen according to the climate; he suggested titanium for the podzol zone and silica for the chernozem zone.

If chemical estimations of such compounds as silica and alumina are used, great care must be taken in applying Joffe's principle of selecting the most stable and immobile compound in the soil being studied. This assumes that the climate has remained uniform, whereas many soil profiles have undergone relatively drastic changes so that the present climate and soil morphology may not be dependable guides for selection of the appropriate compound. All the compounds that have been used are known to be mobile under certain conditions, and in some profiles they are all mobile at once. Khan's (1959) work on soils formed on limestone, using zircon as the stable constituent, showed that silica, alumina, and titania were all lost from the upper soil horizons as compared with the parent rock. These data are apparently quite dependable since the losses of silica, alumina, and titania are underestimates if zirconium is also mobile. There are no dependable criteria at present, therefore, for recognizing soil profiles that have formed under uniform conditions where one particular constituent can be expected to be a reliable stable constituent on which to base calculations of profile development.

Use of the relatively stable mineral species as a basis for the calculations also entails some difficulties since no minerals are absolutely immobile and stable (Polynov, 1944). Zircon and quartz have been used most frequently because of their apparent relative stability under a wide range of conditions (Section 5.1). There is considerable evidence, however, that both quartz and zircon do weather to some extent under certain conditions. Quartz grains may break down physically due to strain or inclusions (Humbert and Marshall, 1943; Raeside, 1959) or go into solution (Englehardt, 1937); physical breakdown and weathering of quartz relative to zircon during soil formation have been demonstrated under some conditions (Brewer, 1955*b*). Similarly zircon may break down physically (Raeside, *loc. cit.*) and weather gradually by solution (Rankama and Sahama, 1950; Marel, 1949; Wild, 1961*b*), and zirconium may accumulate within the soil profile, especially in accumulation zones of iron oxides (Brewer, *loc. cit.*); there is evidence that zircon weathers most rapidly under the alkaline conditions of laterite formation (Carroll, 1953). A further difficulty is that authigenic quartz (Hoehne, 1954; Smithson, 1956*b*; Breeze, 1960) and zircon (Butterfield, 1938) have both been reported.

The degree of weathering of zircon, however, especially of the coarser grains, is usually relatively slight even in the case of laterite formation quoted by Carroll (1953). Laterite formation is apparently an extremely severe process, and perhaps no mineral will be stable enough under such conditions to allow sufficiently reliable calculations of soil formation to be made. The most dependable procedure at present, therefore, is to base the calculations on estimations of stable minerals, especially zircon. Zircon and quartz have the added advantage of widespread occurrence; a large variety of igneous rocks, particularly the more basic types, may not contain either of them, but in terms of area in most landscapes they are almost ubiquitous since they occur in practically all sediments and in most common igneous rocks. Quartz has an advantage over zircon in that it is generally more abundant, and so quantitative estimation is more accurate.

Minerals other than quartz and zircon may be suitable as the stable constituent in some instances, but care must be taken to check their suitability in each individual profile. Variations in the lists of mineral stabilities set out by different authors indicate that minerals weather at different rates under different conditions (Section 5.1), and they vary in the ease with which they break down physically into smaller grains. Some species, such as rutile and anatase, are formed rather readily as authigenic minerals. The most dependable stable mineral after zircon is probably tourmaline, but some of the metamorphic minerals, such as kyanite and sillimanite, may be suitable in some circumstances. Tourmaline, like zircon, has been shown to weather slightly under certain conditions (Druif, 1937), and authigenic tourmaline has been reported in sediments (Stow, 1932).

4.3 ESTIMATION OF THE STABLE CONSTITUENT

There are two methods of estimation of the stable mineral: chemical analysis of the whole material or of specific separates (size, specific gravity, and/or magnetic separates), and grain counts in specific separates. Chemical analysis can only be used when some element is specific for the stable mineral, for example, boron for tourmaline and zirconium for zircon in materials containing no other boron or zirconium minerals. It cannot be used for estimation of quartz because the majority of minerals in soil materials and sediments are silicates, and all forms of secondary silica are difficult, if not impossible, to distinguish chemically from primary quartz.

Chemical analyses of the whole material for such minerals as zircon and tourmaline are likely to cause inaccuracies in calculations based on them because of translocation and redistribution of the specific element released by weathering, such as occurred in a yellow podzolic soil (Brewer, 1955*b*). Inaccuracies due to the occurrence of more than one mineral containing the specific element are least likely for such a mineral as zircon which is usually the only source of zirconium, especially in igneous rocks (Poldervaart, 1956). Certain relatively rare minerals, however, contain a high proportion of zirconium, for example, rinkite, zirconalite, and zirkelite which may have up to 53 percent zirconium dioxide (Semenov, 1959). These inaccuracies can be eliminated, or assessed, by analyzing specific separates that contain concentrates of the particular stable mineral with other species that do not contain the specific element. Grain counts have two advantages: every grain counted is identified, thus eliminating errors due to other mineral species containing the specific element, and each grain counts as one, thus eliminating errors due to slight losses by solution. Chemical analysis has the advantage of ease of conversion of percentage by weight of zirconium to percentage by weight of zircon, whereas grain counts require measurements of specific gravity for conversion.

The accuracy of both methods of estimation can be enhanced by careful selection of the separates analyzed. The rate of weathering of mineral grains is directly related to surface area of the grains, that is, inversely related to size of grains. The coarser grains of zircon (larger than 10 μ equivalent diameter) in the majority of soils show little or no evidence of weathering or solution; many of the grains are euhedral with sharp corners and edges. This was so even in the yellow podzolic soil which showed evidence of considerable release of zirconium (Brewer, 1955*b*) which, therefore, must be from weathering of the very small grains of zircon occurring as inclusions in other minerals. Choice of the size classes to be analyzed should be governed also by the size distribution of the selected stable mineral in the parent rock or parent material; the range of size classes used should include the modal size for greatest accuracy. Accessory minerals such as zircon often occur in two generations in igneous rocks—as discrete grains and as inclusions in other minerals—and each of these generations may be relatively uniform in grain size; the size range selected should include the modal size of the coarser of these two generations. In sediments zircon tends to concentrate in the class one size smaller than the mode of the grain-size distribution.

4.4 THE CONSTANT VOLUME HYPOTHESIS

The constant volume hypothesis suggests that calculations of soil formation should be made on the assumption that there is no change in volume during weathering of the parent rock and formation of the soil profile; unit volume of parent rock forms the same volume of soil material in all horizons of the profile.

A. Basis of the Hypothesis

Wild (1961*a*, *b*) proposed the constant volume hypothesis because his analytical data from twelve soil profiles formed on granite suggested that a high proportion of the zircon in the parent rock had weathered during soil formation and that the zirconium was removed from the profiles. This conclusion was based on two observations:

(1) If zircon were completely stable, the ratio of zirconium in the parent rock to that in the soil horizons could be explained only by a large increase in volume from parent rock to soil material. In profile 229 the parent rock had 0.013 percent zirconium, and the soil horizon at 28 to 32 in. had 0.014 percent, while the bulk densities were 2.7 and 1.7, respectively. "Originally, therefore, 100 g rock occupied 37 cc, and now 93 g soil occupies 55 cc, an increase of 50%" (Wild, 1961*a*).

(2) In the field, quartz veins were observed running continuously and undisturbed from the weathered, but apparently undisturbed, parent rock into the more weathered soil horizons (Wild, 1961*b*). This suggests little or no change in volume for these horizons, certainly not

of the order of 50 percent as was calculated to be necessary in observation (1).

Oertel (1961), in a study of six red-brown earth profiles from South Australia, also proposed that there had been no change in volume during soil formation, at least below the top of the *B* horizons. This conclusion was based on analysis of his data using gallium as the stable constituent. He pointed out, however, that the results could be explained by illuviation of clay size material, but he rejected this hypothesis in favor of the constant volume hypothesis because of the necessity for a very nice balance between weathering and illuviation for this explanation to hold. An equally delicately balanced process is necessary to preserve constant volume throughout the *B* horizons during weathering and other processes of soil formation, particularly since the degree of weathering would vary with depth, and red-brown earths usually have a horizon with an accumulation of carbonates.

A closer examination of Wild's data throws some doubt on the general applicability of the constant volume hypothesis to the profiles he studied:

(1) The criteria for evaluating uniformity of parent rock (Section 4.1) cast considerable doubt on the suitability of many of the profiles for quantitative studies without additional evidence. Few of the curves for zirconium plotted against depth (Fig. 9) are smooth

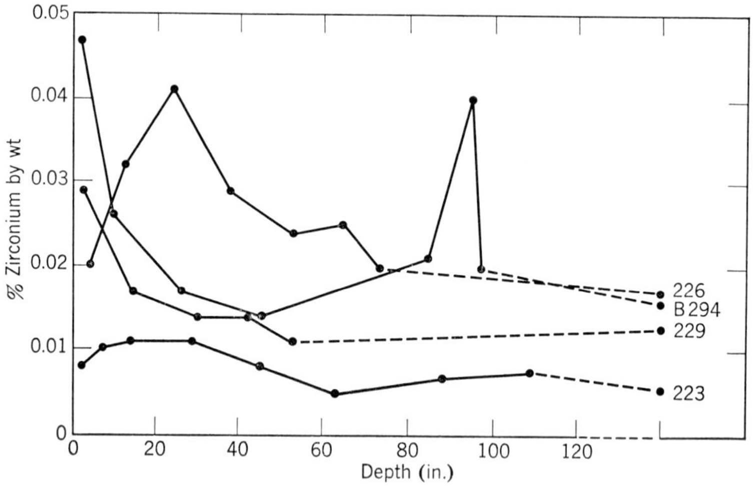

Fig. 9. Quantitative distribution of zircon with depth in four profiles formed on granodiorite [calculated from Wild's (1961*b*) data].

enough to indicate uniformity of parent material throughout the profile, and no other data was offered. Profile B294 is of particular interest in that the sampled rock did underlie the profile, and yet the curve for zirconium provides strong evidence that the parent material was anything but uniform. For other profiles the presumed parent rock was sampled from adjacent outcrops or floaters and is less reliable on field evidence as the true parent material. The simple fact that a profile is apparently formed on granite is not sufficient to assume neighboring outcrops, or even floaters, within the profile to have been the uniform parent material. This is exemplified by one red earth on granite from Batlow (the location of Wild's profile 231), which had a sudden change in zircon-tourmaline ratio from 2 to 20 within the solum, and another which had a sudden change in mica type from biotite to muscovite within the solum (Brewer, unpubl. data).

(2) Six of the twelve profiles have a higher percentage of zirconium in the parent rock than in the deepest soil horizon sampled, while the bulk density of the rock is considerably higher than that of the soil material. This indicates that zircon weathered faster than the average for the rest of the mineral species in the rock. Zircon is generally regarded as one of the most stable minerals (Section 5.1), and, although it may weather faster than some other species under some conditions, it is unlikely to do so under the wide range of environments in which these profiles occur. Data for a yellow podzolic soil on granodiorite (Brewer, 1955*b*) did not show either this anomaly or anomalous volume estimations discussed under (1) for profile 229, even though it occurred in the same general environment on the Southern Tablelands of New South Wales.

(3) Although the field observations concerning the undisturbed nature of quartz veins appear to be legitimate for the lower horizons of some of the profiles, it is difficult to see their applicability to surface horizons.

(4) Since the limits of accuracy of estimations of zirconium by spectrographic methods are ±13 percent, considerably different results can be obtained from the calculations (Wild, 1961*b*) if the analyses are close to the limits of analytical error. In profile 229 the rock could have had 0.011 percent zirconium and the soil horizon 0.016 percent, which would mean practically no change in volume; 100 g of rock which occupied 37 cc would give 63 g of soil which occupies 37 cc. Certainly this is stretching the probability of analytical error to the limit, but smaller errors would considerably reduce the apparent loss of zirconium from this horizon from the estimated 33 percent.

B. Test of General Applicability

In profiles for which the constant volume hypothesis is true, the skeleton grains will gradually become smaller and eventually disappear as weathering progresses, and so the distance between them will remain constant or progressively increase; that is, the number of skeleton grains per unit volume will remain constant or decrease. The hypothesis may not be valid even for profiles in which this requirement is fulfilled because of the effects of relative rates of weathering and removal of other constituents and of mineralogically different skeleton grains.

A preliminary attempt has been made to test the general applicability of the constant volume hypothesis by studying two profiles (P106 and P105) formed on alluvium in which the skeleton grains were dominantly quartz with some felspar. Particle size distribution of the skeleton grains was measured as weight (or number) per unit volume. The results are shown in Figure 10. For P106 the weight of skeleton grains per unit volume decreased progressively with depth in the profile. As a parent material composed entirely of skeleton grains weathers, clay size material is produced and the number of skeleton grains decreases, so the skeleton grains become gradually further apart; that is, there are fewer per unit volume of soil material. This trend will continue until the clay size material is being destroyed or translocated at a faster rate than the weathering of the skeleton grains; at this point collapse of the soil material and continued weathering will bring the skeleton grains closer together. In P106, therefore, if the parent material was uniform, the results indicate progressively decreasing removal of material with depth and collapse of the soil material in the upper horizons to bring the skeleton grains closer together, that is, a significant decrease in volume.

Alternatively the lower horizons may have received illuviated material and expanded considerably, with some collapse in the upper horizons. The change in numbers of grains per unit volume with depth could hardly be due to sedimentation, which would be more likely to give the opposite trend with a greater proportion of fine-grained material with decreasing depth.

Assuming that the material at 47 in. in P106 is the unaltered parent material, that there has been no weathering of skeleton grains, and that the increase in grains per unit volume is due solely to collapse, then the material at 3 in. has collapsed to approximately 0.42 times the original volume (the ratio of grains per unit volume = $^{47}\!/_{111}$ = 0.42); the other horizons have collapsed proportionately (Fig. 10*c*).

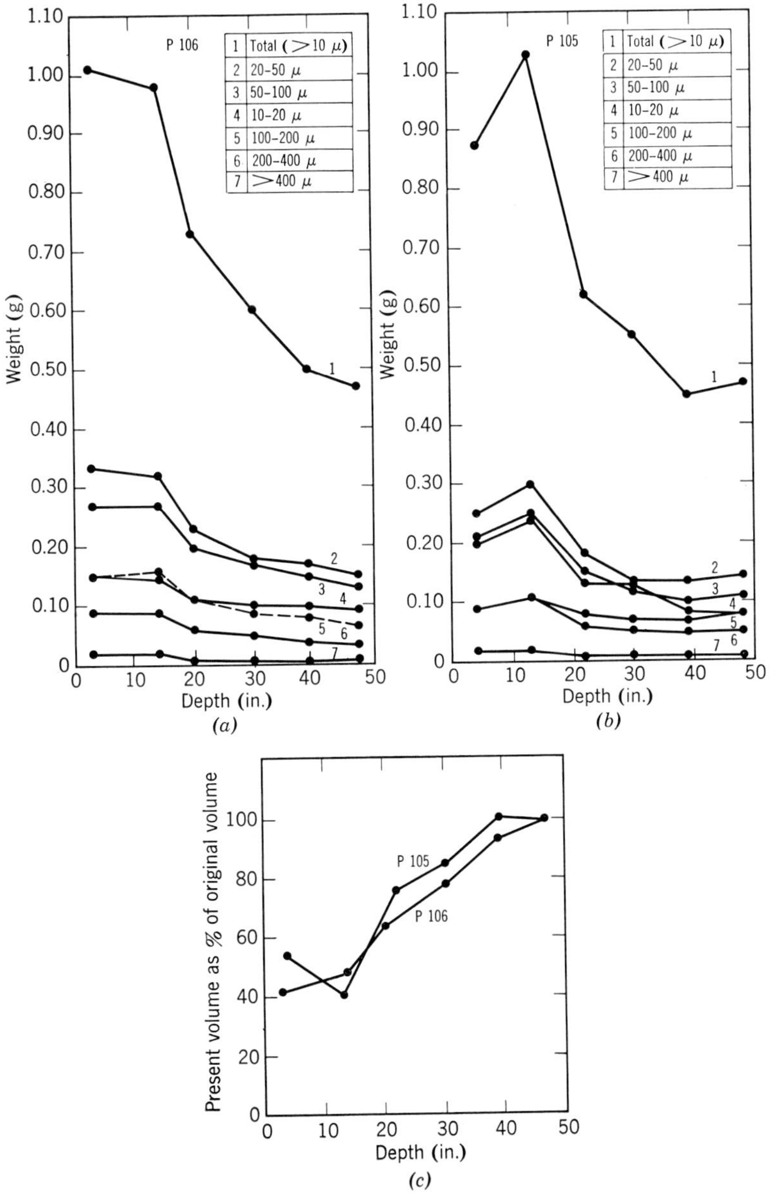

Fig. 10. Depth functions of skeleton grains (> 10 μ) and volume for two profiles formed on alluvium. (*a*) Change in total weight of skeleton grains and weight of each size fraction per unit volume for profile P106. (*b*) Similar data for profile P105. (*c*) Present volume of each horizon as a percentage of the original volume of parent material for each profile.

If the true parent material had fewer skeleton grains per unit volume than the material at 47 in. (as appears likely from the curves, which have not reached parallelism with the *X*-axis at 47 in.), or if the skeleton grains have weathered, the upper horizons have collapsed more than estimated. If the lower horizons have expanded, the collapse of the upper horizons has been overestimated. A similar analysis can be applied to P105 except that the uppermost horizon does not fall on the smooth curve (Fig. 10*b*, *c*); field evidence strongly suggests that this is an overlie unrelated to the main soil profile.

The constant volume hypothesis, therefore, can be applied only in special circumstances where other evidence, especially field evidence, strongly supports the assumption. The hypothesis is particularly suspect for any horizons above the "*C* horizon." On the other hand, the results for profiles P105 and P106 cannot be applied generally because many profiles show a progressive decrease in number of grains per unit volume, at least from the parent material to the top of the *B* horizon; calculations of changes in volume in such profiles can only be based on estimations of a stable constituent, so the argument becomes rather circuitous. Although doubt has been cast on the suitability of the profiles studied by Wild (1961*b*) as regards uniformity of parent material, the weight of evidence is that zircon has weathered. This is in agreement with other observations (as stated by Wild), and it strengthens the conclusion reached earlier (Section 4.3) that calculations should be based on estimations in selected classes.

4.5 CALCULATIONS OF SOIL FORMATION

Changes that occurred during development of a soil profile from parent material can be estimated quantitatively if the parent material is sufficiently uniform and a suitable basis for the calculations can be established. Calculations based on the constant volume hypothesis and on a stable constituent are independent of each other; the former can be used to calculate losses of constituents (Wild, 1961*b*) and the latter to calculate changes in volume, weight, and thickness of each soil horizon and of the whole profile, and gain or loss of specific constituents. Merrill (1921) first proposed an equation for calculation of losses of materials from rocks during weathering based on estimation of a stable constituent; this equation has been used by Khan (1959). Several modifications and extensions of the method of calculation have been made since (e.g., Nikiforoff and Drosdoff, 1943; Haseman and Marshall, 1945; Barshad, 1955; Brewer, 1955*b*). The basic calculations concern changes in volume and weight during development of the soil horizons and profile from the parent material. Nikiforoff and Drosdoff (1943) coined three terms in relation to these basic calculations:

Gram equivalent. The amount of a modified material in any soil horizon that represents a residue or modification of a definite amount (100 g) of the original material.

Parent material quotient (Q). One-hundredth part of the gram equivalent, that is, the ratio of the precentage of the stable constituent in the parent material to the percentage in the soil horizon.

Volume factor (v). The ratio of the thickness of the soil horizon to the thickness of the parent material that went to form it.

A. Change in Volume, Weight, and Thickness

These calculations are based on a stable constituent:

(*a*) **Volume.** The relationship of the volume of a present-day soil horizon (V_s) and the volume of parent material from which it was derived (V_p) is given by the equation:

$$\frac{V_s D_s R_s}{100} = \frac{V_p D_p R_p}{100} \tag{1}$$

where D_s = bulk density of the present-day soil horizon,
R_s = percentage by weight of the stable constituent in the present-day soil horizon,
D_p = bulk density of the parent material, and
R_p = percentage by weight of the stable constituent in the parent material.

From equation 1,

$$V_s = V_p \frac{D_p R_p}{R_s D_s} = \frac{K V_p}{R_s D_s} = Q V_p \frac{D_p}{D_s} \tag{2}$$

or

$$\frac{V_s}{V_p} = v = \frac{K}{R_s D_s} = Q \frac{D_p}{D_s} \tag{3}$$

where v = the volume factor,
K = a constant for the parent material = $R_p D_p$, and

Q = the parent material quotient = $\frac{R_p}{R_s}$.

Equation 3 gives the ratio of the volume of the soil horizon to the volume of the parent material from which it was formed. Alternatively,

$$V_p - V_s = V_p - \frac{KV_p}{R_sD_s} = V_p(1 - v) \qquad (4)$$

gives the change in volume during soil formation.

(*b*) ***Weight.*** The weight of the present-day soil horizon (W_s) derived from volume V_p or weight V_pD_p of parent material is given by the equation:

$$W_s = V_sD_s$$

$$= \frac{KV_p}{R_sD_s} \cdot \frac{D_s}{1} \qquad \text{(from equation 2)}$$

$$= \frac{KV_p}{R_s} = QV_pD_p \qquad (5)$$

and the change in weight during soil formation is

$$W_p - W_s = V_pD_p - \frac{KV_p}{R_s} = V_pD_p(1 - Q) \qquad (6)$$

(*c*) ***Thickness.*** The relationship between thickness of the present-day soil horizon (T_s) and the original thickness of the parent material from which it was derived (T_p) is given by the equation:

$$T_s = \frac{R_pD_pT_p}{R_sD_s}$$

$$= \frac{KT_p}{R_sD_s} = Q\frac{D_pT_p}{D_s} \qquad (7)$$

and change in thickness is $T_p - T_s$. At first glance equation **7** appears erroneous since it states that

$$\frac{T_s}{T_p} = \frac{R_pD_p}{R_sD_s}$$

$$= \frac{V_s}{V_p} \qquad \text{(from equation 3)} \qquad (8)$$

whereas, mathematically, if V_p is a cube,

$$\frac{T_s}{T_p} = \left(\frac{V_s}{V_p}\right)^{1/3} \qquad (9)$$

Equation 7 can be justified, however, by consideration of the processes involved in volume changes during soil formation. If the volume V_p of parent material changes during soil formation it does so in three dimensions according to equation 9. This changes the thickness and also the dimensions of the cross section. The latter, however, is manifested by adjustments in the volume occupied by voids, particularly cracks. This causes a change in bulk density of the present-day soil horizon, so equations 7 and 9 are reconciled by adjustments in D_s; that is, the cross-sectional area remains constant. It is important, therefore, that sufficiently large samples should be considered in estimating bulk density of soil materials to take account of crack patterns; these crack patterns are widely spaced in some soil materials.

B. Gain or Loss of Constituents

(a) Based on a Stable Constituent. The weight in grams of constituent X (X_s) in volume V_s of the present-day soil horizon is given by the equation:

$$X_s = \frac{V_s D_s P_x}{100} \tag{10}$$

where P_x = the precentage by weight of constituent X in the present-day soil horizon. The weight in grams of constituent X (X_p) in the volume of parent rock (V_p) from which the volume of the present-day soil horizon (V^s) was derived is given by:

$$X_p = \frac{V_p D_p P_x'}{100}$$

$$= \frac{V_s R_s D_s}{R_p D_p} \cdot \frac{D_p P_x'}{100} \quad \text{(substituting for } V_p \text{ from equation 1)}$$

$$= \frac{V_s R_s D_s}{R_p} \cdot \frac{P_x'}{100} = \frac{V_s D_s}{Q} \cdot \frac{P_x'}{100} \tag{11}$$

where P_x' = the percentage by weight of constituent X in the parent material. The weight in grams of constituent X (X_g) lost or gained during formation of the present-day horizon is given by:

$$X_g = X_p - X_s$$

$$= \frac{V_s R_s D_s P_x'}{100 R_p} - \frac{V_s D_s P_x}{100}$$

$$= \frac{V_s D_s}{100}\left(\frac{R_s P_x'}{R_p} - P_x\right) = \frac{V_s D_s}{100}\left(\frac{P_x'}{Q} - P_x\right) \tag{12}$$

and the percentage gain or loss of constituent X (X_{gp}) is given by:

$$X_{gp} = \frac{(V_sR_sD_sP_x'/100R_p) - (V_sD_sP_x/100)}{V_sR_sD_sP_x'/100R_p} \cdot \frac{100}{1}$$

$$= 100\left(\frac{R_sP_x' - R_pP_x}{R_sP_x'}\right) = 100\left(1 - Q\frac{P_x}{P_x'}\right) \qquad (13)$$

(b) Based on Constant Volume (after Wild 1961*b*). The weight in grams of constituent X (X_g) lost or gained during formation of the present-day horizon from volume V of parent material is given by:

$$X_g = \frac{VD_pP_x'}{100} - \frac{VD_sP_x}{100} \qquad (14)$$

where V = volume in cc of soil formed from the same volume of parent material,
D_p = bulk density of the parent material,
D_s = bulk density of the soil material,
P_x' = percentage by weight of constituent X in the parent material, and
P_x = percentage by weight of constituent X in the present-day soil horizon.

The percentage gain or loss of constituent X (X_{gp}) is given by:

$$X_{gp} = 100\left(\frac{P_x'D_p - P_xD_s}{P_x'D_p}\right)$$

$$= 100\left(1 - \frac{P_xD_s}{P_x'D_p}\right) \qquad (15)$$

4.6 INTERPRETATION OF THE CALCULATIONS

Calculations of soil development can be used in two ways to interpret the formation of soil profiles: to reconstruct the original profile and as a basis for interpretation of formation and migration of particular constituents, particularly clay size material.

If the calculations are based on a stable constituent and on oven-dried material or some other suitable basis, the volume and weight of parent material that has gone to form each of the present-day horizons, and the whole profile, can be estimated. Losses and gains of any of the mineral constituents, size fractions and chemical elements can be determined (e.g., Haseman and Marshall, 1945; Brewer, 1955*b*; Khan, 1959) whether the calculations are based on a stable constituent or on the assumption of constant volume.

Beyond the factual statement of losses and gains the data from the calculations must be interpreted with caution. They represent total losses and gains in the transformation of parent material to the soil materials of particular horizons, and losses of a single constituent cannot be interpreted in terms of translocation of material out of the horizon, nor can gains be interpreted as additions from outside the horizon or from weathering of minerals or transformations of other materials, unless additional evidence is available. Such additional data can be obtained from structure and fabric analyses (Part III), which may indicate translocation of materials or lack of it, and from more complete calculations concerning soil development. For example, the possibility of a gain in iron oxide being due to weathering of minerals can be assessed by calculating the loss, or amount of weathering, of the iron-bearing minerals in the transformation of the parent material to the material of the soil horizon; see equation 12. Knowing the composition of the mineral species the amount of iron oxide released by weathering can be calculated. This, of course, does not show the ultimate fate of the constituents released; elements released by weathering of primary minerals may occur in the soil material as oxides, carbonates, sulfates, etc., in such secondary minerals as the layer-lattice clay minerals, or they may be redistributed through the soil profile or completely removed from it. The more complete the calculations the more dependable are the interpretations of the data, and other data, such as micromorphology, are invaluable for interpretation.

Barshad (1955, 1959) suggested a method for calculating the formation and translocation of clay size material. He considered a column of soil of unit cross-sectional area, for a soil profile in which the parent material is uniform and unaltered parent material is preserved, and calculated the weight of clay size material formed in the whole profile (C_f) and the weight of nonclay material weathered in the whole profile (NC_l), that is, the gain in grams of clay size material and the loss in grams of nonclay material; these are obtained from equation 12. The amount of clay size material formed in the horizon (C_{fa} for horizon a) was then calculated from the equation:

$$C_{fa} = \frac{C_f}{NC_l} \cdot \frac{NC_{la}}{1} \tag{16}$$

where C_{fa} = the amount of clay size material actually formed by weathering of NC_{la} nonclay, and

NC_{la} = the loss in grams of nonclay from the horizon (also derived from equation 12).

The weight of parent material ($W_p = V_p D_p$) that went to form each horizon can be obtained from equation 5, from which can be calculated the weight of clay size material originally present in each horizon (C_0), and the weight of clay size material now present in each soil horizon (C_a for horizon a) can be measured. Then the amount of translocation of clay size material into and out of each horizon (C_{lma} for horizon a is given by the equation

$$C_{lma} = C_{fa} + C_0 - C_a \tag{17}$$

The term C_f/NC_l in equation 16 infers that the total clay size material formed can be apportioned to the various horizons. This assumes that the conditions of weathering were uniform throughout the profile, so that the nonclay minerals weathered at the same relative rate in all horizons to form the same clay size minerals, and that, if clay size material weathered with loss of the weathering products in solution, it did so equally in all horizons. Different mineral species, however, weather at different rates and lose different proportions by weight in weathering to form the same clay mineral species (Brewer, 1955*b*). Further, the conditions of weathering may vary considerably with depth in the profile so that a single nonclay mineral species may form different clay minerals in different horizons with consequent different losses in weight. These factors would cause differences in the ratio of the weight of clay minerals formed to the weight of nonclay minerals weathered from horizon to horizon.

Certainly we would expect the rate of weathering of the clay size material originally present in the parent material to vary with depth. Barshad's (1955) interpretation of these calculations also assumes that none of the clay size material formed migrates beyond the profile either vertically or laterally; this may be so, but only in profiles for which other evidence suggests that the parent material is unaltered by weathering or migration of clay size material.

Another factor that may contribute to inaccuracies in interpretation of these calculations is that certain clay size materials may occur originally or become segregated into coarser individuals, such as aggregates of layer-lattice minerals or nodules of carbonates. Adjustments must be made in the calculations for such occurrences.

A rather more precise method of estimating formation of clay size material can be based on calculations of the percentage loss by weight of the primary minerals in formation of the clay minerals, the amount of each of the nonclay mineral species lost from each horizon (Brewer, 1955*b*), and chemical analysis of the nonclay minerals, which allow calculation of the precise amounts of silica, alumina, potash, etc., released by weathering. Clay mineral analysis and chemical analysis of the clay size fraction give the proportions in which these counstituents are retained in that fraction, and so the percentage loss by weight in formation of these clay minerals from the nonclay minerals can be calculated. Such calculations give the *maximum possible* amount of clay size material of the kind present that could have been formed by weathering of the amount of nonclay material lost, assuming no loss of constituents that could have combined. These results for the whole profile and for each horizon can be compared with estimates of the amounts of clay size material actually formed in the whole profile and in each horizon, that is, the actual gains in clay size material. The difference between the maximum possible amounts of clay size material that could be formed and the actual gain is a measure of the migration of constituents.

Interpretation of these results in terms of translocation of clay size material is impossible without other evidence, especially fabric analysis. If such evidence indicates no migration of clay size material the difference between the maximum possible amounts of clay size material that could be formed by weathering and the actual gain is a measure of losses of constituents, which may have some correlation with the degree of leaching of the profile during the period of formation of the clay minerals. If the evidence indicates migration of clay size material within the depth of the profile, the migration of constituents must be apportioned between losses in solution and clay migration; this can be done only by estimation of the amount of movement of clay in thin section. If clay minerals originally present in the parent material have weathered in the profile the situation is impossible for calculations of clay migration.

Barshad (1955) also set out a series of calculations concerning formation and migration of clay size material for profiles in which the "parent material" has received clay size material by migration. It has already been argued that calculations of soil development cannot be considered reliable unless the true, unaltered parent material is preserved or its original characteristics can be assessed. On these criteria the calculations proposed by Barshad, which are similar to those already set out except that additional assumptions are made or implied, cannot be considered reliable. More reliable calculations would result from attempting to separate or estimate quantitatively the migrated clay in the parent material; this can be done by dissection under a stereomicroscope or by measurements in thin sections. Even so it must still be shown that the isolated parent material is the true unaltered parent material according to the criteria set out earlier (Section 4.1).

5

Mineral Stability and Weathering

Since soil formation is largely the weathering of minerals to form new mineral species with possible losses of constituents from the profile, the stability of minerals, the conditions under which they weather, and the kind and conditions of formation of weathering products are all of considerable importance in understanding soil genesis. Mineral analysis of soils provides the opportunity for assessing stability and weathering, provided the original, true parent material can be identified and the conditions of weathering within the profile specified. As a corollary the kind and amount of weathering of the minerals of the parent material can be used as a measure of the "degree of weathering" of soil profiles.

Subdivision of the mineral constituents of soil materials into primary and secondary minerals is quite arbitrary insofar as many may be either primary in the sense that they were present in the parent material before soil development commenced, or secondary in the sense that they were formed by weathering of constituent minerals of the parent material. It is convenient, however, to consider the minerals that are normally formed in soil materials by weathering of pre-existing minerals separately from those which normally occur as the constituents of igneous and metamorphic rocks.

5.1 STABILITY OF PRIMARY MINERALS

The stability, or resistance to weathering, of a mineral species depends on a large number of factors, which include hardness, cleavage, coefficient of expansion, original cracks in the crystal, and solubility under a specific environment (Haseman and Marshall, 1945). In certain instances other factors, such as grain size or their specific surface, may become dominant. Some of these characteristics are inherent in the mineral species, while others, such as original cracks and grain size, may be regarded as accidental. Accidental characteristics are quite variable and are usually ignored in considering the relative stability of mineral species in broad terms. Several workers have classified the primary minerals according to their stability on this basis.

Pettijohn (1941) examined the frequency of occurrence of a number of mineral species in sedimentary rocks of increasing age and proposed a sequence which is a measure of their relative *persistence* (Table 5-1). This sequence has been generally interpreted as a stability table, the more stable minerals appearing first in the sequence and the degree of stability decreasing from anatase to olivine. This interpretation is valid in a general way because Pettijohn's method of arriving at the sequence is equivalent to considering the average reaction of the mineral species to the great variety of environments in which they may have occurred during geologic history. His table of persistence, therefore, can be regarded as a statistical estimation of the relative persistence, or stability to weathering, of the mineral species in an average environment. Tables of stability of minerals to weathering proposed by other workers do not have the mineral species arranged in precisely the same order. A number of these are shown in Table 5-1, which also attempts to correlate them with Pettijohn's sequence.

Discrepancies between the various tables refer principally to the accessory and ferromagnesian minerals. Pettijohn (1941) showed that the persistence of hornblende is greater than sillimanite and much greater than olivine, whereas Weyl (1952) grouped hornblende with olivine and augite and considered sillimanite much more stable than hornblende. Marel's (personal communication; see Jackson and Sherman, 1953) sequence differs from Pettijohn's especially in the relative ratings of the micas (muscovite and biotite). Of particular interest is the rating of quartz as more resistant than zircon and tourmaline, especially in relation to the choice of a stable constituent on which to base calculations of soil formation (Section 4.2). Pettijohn did not consider quartz, but a study of weathering in a yellow podzolic soil demonstrated that quartz was much less stable than zircon under the conditions of formation of that profile (Brewer, 1955*b*). Graham's table (personal communication; see Jackson and Sherman, 1953) also has discrepancies from Pettijohn's table, for example, in the rating of

TABLE 5-1. Comparative Stability of Mineral Species according to Various

Pettijohn (1941)	Weyl (1952)	Marel (Jackson and Sherman, 1953)	Graham (Jackson and Sherman, 1953)
Anatase (−3)*			Quartz
Muscovite (−2)	Rutile	Tourmaline	Muscovite
Rutile (−1)	Zircon	Rutile	Rutile
	1 Tourmaline†	Staurolite	Zircon
Zircon (1)	Sphene	Zircon	Tourmaline
Tourmaline (2)	Magnetite		1 Ilmenite
Monazite (3)			Andalusite
Garnet (4)		Garnet	Kyanite
Biotite (5)		Muscovite	Sphene
Apatite (6)	Kyanite		Magnetite
Ilmenite (7)	2 Andalusite		
Magnetite (8)	Sillimanite		2 Staurolite
Staurolite (9)			
Kyanite (10)			Biotite
Epidote (11)	Epidote	Epidote	Epidote
	3		
Hornblende (12)	Garnet	Amphibole	3 Garnet
Andalusite (13)			Augite
Topaz (14)			Hornblende
Sphene (15)			
Zoisite (16)			
Augite (17)	Augite	Augite	
Sillimanite (18)	4 Hornblende	Biotite	
Hypersthene (19)	Olivine	Hypersthene	
Diopside (20)			
Actinolite (21)			Apatite
			4
Olivine (22)		Olivine	Olivine

* Negative numbers indicate relative greater abundance in older sediments
† Stability of minerals expressed as groups without differentiation within
‡ Stability index numbers taking garnet as 1

Authors (Stability Decreasing from Top to Bottom of the Tables)

Fieldes and Swindale (1954)	Smithson (1941)	Dryden and Dryden (1946)
5 Quartz		
	Zircon	
Felspars	Rutile	
4		Zircon (100)‡
Acid volcanic glass	Tourmaline	Tourmaline (80)
	1	
Muscovite		Sillimanite (40)
3		
Biotite		
	Apatite	Monazite (40)
	2 Monazite	Chloritoid (20)
	Garnet	
	3 Staurolite	
Zeolites	Kyanite	Kyanite (7)
2		
Basic volcanic glass		
		Hornblende (5)
Augite		Staurolite (3)
Hornblende		
1 Hypersthene	4 Mafic minerals	Garnet (1)
Olivine		Hypersthene (1)

than in younger ones
groups

apatite, and the stability rating of quartz is high, being roughly equivalent to zircon and tourmaline. Smithson's (1941) table agrees reasonably well with Pettijohn's, the only slight difference being the higher rating of apatite relative to monazite and garnet by Smithson. In contrast Dryden and Dryden's (1946) table bears relatively little relation to Pettijohn's except for the high rating of zircon and tourmaline. Fieldes and Swindale's (1954) table does not contain a wide enough range of mineral species to make a proper comparison, but the high rating of felspars compared with the micas is worth noting.

There does seem to be general agreement, however, as to the relative stability of the felspars and quartz and little disagreement concerning the more common ferromagnesian minerals. Goldich's (1938) stability sequence (Table 5-2) summarizes general opinion quite well. This sequence is similar to the Bowen (1922) sequence for temperature of crystallization (Jackson and Sherman, 1953), and this has been interpreted as indicating that the stability of a mineral species increases with decreasing difference between the temperature of formation and that of the present environment. The sequence is also one of acidity in the geological sense—stability increasing with acidity (proportion of silica in the crystal lattice). The mafic minerals are arranged in this table in the same order as in Pettijohn's table of persistence.

Many and possibly all of the discrepancies between the stability tables can be explained in terms of variations in chemical and physical characteristics of the mineral species studied in each specific case and the environment of weathering from which the data were obtained.

Plagioclase felspars vary widely in their stability to weathering

TABLE 5-2. Mineral Stability Table* (after Goldich, 1938)

Mafic Minerals		Felspathic Minerals
Olivine		Calcic plagioclase
Augite		Calcic-alkalic plagioclase
Hornblende		Alkali-calcic plagioclase
Biotite		Alkalic plagioclase
	Potassium felspar	
	Muscovite	
	Quartz	

* Stability increasing from top to bottom

according to their composition (Goldich, 1938; Graham, 1950), albite being the most stable and very much more stable than anorthite. A number of other minerals form isomorphous series comparable to that of the plagioclases, and the stability to weathering of the various members of such series can vary widely; for example, dark green hornblende was considerably less stable to weathering than the grey-green variety in the podzol soil profiles of Michigan (Matelski and Turk, 1947). The accidental characteristics also may be important factors leading to apparently contradictory conclusions on the stability of a mineral species to weathering; for example, sillimanite can occur as relatively large grains or as masses of small fibers; the latter are likely to break down physically to produce a large specific surface which materially decreases its stability to weathering. Strain and inclusions are also likely to increase physical breakdown (Raeside, 1959), and twinned crystals of any mineral species are likely to weather more rapidly than untwinned crystals.

Most important perhaps is the specific environment of weathering. There is considerable agreement between different workers where there is evidence that the environment of weathering of the mineral species was much the same in each case; for example, Cady (1940), Richard and Chandler (1943), Matelski and Turk (1947), and Marel (1949) all agreed that hornblende weathered rapidly in podzol soil profiles; the first two also agreed that hypersthene weathered rapidly; Cady, Matelski and Turk, and Marel noted weathering of garnet. This close agreement is undoubtedly due to the fact that the micro-environments within these podzol profiles were similar. On the other hand, Marel found that zircon was less stable to weathering than quartz and even felspar, while hornblende was much less stable than any of these; Brewer (1955*b*) found that quartz and felspar were both much less stable than zircon, and felspar was only slightly more stable than hornblende. These contradictory results are probably due to the studies being made on two entirely different kinds of soil profiles whose micro-environments, which are primarily responsible for weathering of the minerals, were quite different.

Similarly the small amount of data available on the two minerals apatite and hypersthene emphasize the effect of the micro-environment on stability of mineral species. Pettijohn rated apatite as a relatively persistent mineral species, while Graham rated it as one of the least stable (Table 5-1). Pettijohn's assessment averaged the effect of environment, while Graham's assessment was based on the stability of the mineral under specific conditions of low pH; these two contradictory ratings are to be expected when it is realized that apatite is much more stable in an environment of high pH than low pH. On the other hand, both Pettijohn and Marel rate hypersthene as highly unstable, Pettijohn on the basis of its persistence in sediments, and Marel on the basis of its stability under the conditions of low pH obtaining in podzol soil profiles. It was also reported to disappear completely in the formation of neutral-to-alkaline black earths from hypersthene andesites (Brewer and Butler, 1953). The evidence, therefore, is that hypersthene is quite unstable under a wide range of micro-environments.

Pettijohn's (1941) table of persistence, therefore, is the most reliable general guide to the relative stability of mineral species to weathering because the method used to compile it tends to average the various factors that can give anomalous or contradictory results in specific cases. The discrepancies between the various tables of stability, however, point to the need for estimations of the relative stability of mineral species under the different conditions of weathering in soil profiles. If such estimations are to be reliable and have any real significance, they must be based on studies of profiles that can be shown to be formed on uniform parent materials. Only in recent years has it been realized that many profiles are either formed in layered sediments or actually consist of superimposed profiles. Since Pettijohn's table of persistence is derived from a broad, essentially statistical survey, any data that deviates significantly from it should be closely examined from two aspects, that is, whether the apparent deviation is due to lack of uniformity of parent material with depth in the profile, or to some specific characteristic of the micro-environment such as pH level or permeability.

5.2 STABILITY OF SECONDARY MINERALS

Jackson and Sherman (1953) set out a weathering sequence for clay size minerals (less than 2 μ equivalent diameter), which is a modification of an earlier proposal (Hseung and Jackson, 1952), consisting of thirteen stages (Table 5-3). It is proposed essentially as a stability table in which stability increases from the first-mentioned mineral species to the last. It lists mineral species that are normally recognized as secondary formed from primary minerals during soil formation, such as gypsum, calcite, the layer-lattice minerals (except the micas), and the oxides and hydroxides of iron, aluminium, and titanium. Gypsum and calcite are listed as the least stable of these because of their relatively high solubility, and the layer-lattice minerals and oxides and hydroxides are arranged in what amounts to an

TABLE 5-3. Weathering Sequence for Clay Size Minerals (after Jackson and Sherman, 1953)

Stability Increasing from (1) to (13).

(1)	Gypsum (also halite, sodium nitrate, ammonium chloride, etc.)
(2)	Calcite (also dolomite, aragonite, apatite, etc.)
(3)	Olivine-hornblende (also pyroxenes, diopside, etc.)
(4)	Biotite (also glauconite, magnesium chlorite,* antigorite, nontronite, etc.)
(5)	Albite (also anorthite, stilbite, microcline, orthoclase, etc.)
(6)	Quartz (also cristobalite, etc.)
(7)	Muscovite (also 10Å zones of sericite, illite, etc.)
(8)	Interstratified 2:1 layer silicates and vermiculite (including partially expanded hydrous micas, randomly stratified 2:1 layer silicates with no basal spacings, and regularly stratified 2:1 layer silicates)
(9)	Montmorillonite (also beidellite, saponite, etc.)
(10)	Kaolinite (also halloysite, etc.)
(11)	Gibbsite (also boehmite, allophane, etc.)
(12)	Hematite (also goethite, limonite, etc.)
(13)	Anatase (also zircon, rutile, ilmenite, leucoxene, corundum, etc.)

* Additional evidence (Jeffires, Rolfe, and Kunze, 1953; Caillere, Henin, and Mering, 1947) indicates that chlorite may occur at any stage from (4) to (9) according to it composition.

order of simplicity; that is, the more simple the structure of the mineral the more stable it is. Interstratified minerals are considered to be less stable than the montmorillonite series, which is less stable than kaolinite, which, in turn, is less stable than the oxides and hydroxides of iron and aluminium.

The table also contains minerals that are normally regarded as primary but do occur as clay size grains in soil materials. These are arranged in the same order in Jackson and Sherman's table (Table 5-3) as they are in Pettijohn's table (Table 5-1), that is, in decreasing order of stability:

> (13) anatase (zircon, rutile, ilmenite, etc.), (7) muscovite, (6) quartz, (5) albite, (4) biotite, (3) olivine-hornblende (pyroxenes, diopside, etc.).

Through these mineral species the order of stability of secondary minerals can be compared with that of primary minerals; this should be legitimate since the effects of grain size are largely eliminated by comparison of the stability of minerals in a particular size range. The comparison is only relatively broad because a number of the primary minerals are grouped together in Table 5-3, whereas Pettijohn observed significant differences in persistence between each of these mineral species. A number of these minerals, too, may be authigenic in sediments, such as zircon (Butterfield, 1938), felspars (Gruner and Thiel, 1937; Tester and Atwater, 1934), quartz (Hoehne, 1954), and anatase (Marshall and Jeffries, 1946), while quartz has been reported as secondary in soil materials (Breeze, 1960). It is interesting to note, however, that all the layer-lattice minerals, oxides, and hydroxides are listed as being more stable than muscovite and less stable than anatase.

5.3 WEATHERING OF MINERALS

The details of mechanisms of weathering of individual mineral species to form secondary minerals are beyond the scope of this work, but it is pertinent to consider weathering of minerals in broad terms and the kind of secondary mineral that can be expected from the various mineral species under different conditions of soil formation. Stephen (1959) and Keller (1955) have reviewed research on weathering of rocks and minerals and some theories of the processes involved.

Early workers were forced to rely on the data of bulk changes in chemical composition of rock and the products of weathering to follow the processes of weathering, but more recent techniques have made it possible to follow the weathering of individual grains. The petrological microscope and electron microscope make it possible to record the arrangement of secondary minerals with regard to the weathering grain, while X-ray techniques, differential thermal analysis, and electron microscopy make it possible to identify the smallest grains of the weathering products. An obvious feature of the results of such studies is the wide variety of secondary minerals that has been reported as the weathering products of a single primary mineral species. The following examples illustrate this variability.

A. Micas

The greatest amount of detailed work has been done on weathering of the micas. Walker (1949) studied the weathering of biotite in a soil profile from Glen Buchot, Aberdeenshire, that was formed on a hypersthene gabbro. Color of the biotite changed progressively with increasing weathering from fresh black biotite, through dark brown, glistening yellow, and dull buff to dull brownish flakes. The color change was paralleled by decrease in refractive index and definition of the interference figure, increase in optic axial angle, decrease in specific gravity, and disappearance of pleochroism. These changes are due to oxidation of ferrous iron to ferric iron, followed by loss of iron, magnesium, potassium, and sodium and gain in water; the biotite weathered first to mixed-layer biotite-vermiculite and then to vermiculite with coatings and penetrations of goethite; there were no indications of chlorite. Comparable changes were reported by Denison, Fry, and Gile (1929). In a second study of some soils from northeastern Scotland, Walker (1950) reported vermiculite as the end product of weathering of a biotite-like mica in well-drained profiles on acid and intermediate rocks, and montmorillonite in poorly drained profiles on basic or ultrabasic rocks. Similar results were reported by McAleese and Mitchell (1958), and MacEwen (1954) reported a trioctahedral montmorillonoid occurring as pseudomorphs after biotite in a soil from Aberdeenshire.

On the other hand, a number of workers have reported chlorite as a weathering produce of biotite: as an intermediate between biotite and the end product vermiculite (Stephen, 1952*a*), as the end product (Humbert and Marshall, 1943), and as an intermediate in the weathering sequence from mica, through mica intermediate, hydrated mica, and chlorite, to kaolinite (Jeffries, Rolfe, and Kunze, 1953). Biotite has also been reported as weathering directly to gibbsite and iron oxides (Alexander, Hendricks, and Faust, 1941) and to material amorphous to X-rays (Neuwirth, 1954).

It is of interest to note that biotite has been converted to vermiculite under relatively natural conditions in the laboratory by growing four successive crops of wheat in sand cultures containing biotite (Moreland, Lawton, and Uehara, 1956). It was concluded that biotite is unlikely to persist in the clay fraction of soil materials which are subjected to intense biological activity.

B. Hornblende and Felspars

Hornblende has been reported as weathering to chlorite in a brown earth formed from a basic appinite (Stephen, 1952*b*), in a profile formed on granite (Humbert and Marshall, 1943), and in association with a little vermiculite in immature soils from Devon and Cornwall on a variety of rocks (Butler, 1953); in the last, impeded drainage caused alteration of chlorite to a montmorillonoid. It has also been observed as weathering directly to gibbsite, which may be resilicified to kaolinite, in soils of the Southern Piedmont Plateau of the United States (Alexander, Hendricks, and Faust, 1941).

A similar variation in data has been obtained for the felspars. In the Southern Piedmont Plateau they have been reported as weathering in the same manner as hornblende, that is, directly to gibbsite, which may be resilicified to kaolinite (Alexander, Hendricks, and Faust, 1941). Cady (1950), working in the same area, observed felspar weathering to gibbsite, allophane, and possibly kaolinite in a profile on diorite and to halloysite in a profile on metagabbro; Neuwirth (1954) also reported felspar weathering to halloysite. Haldane and Brewer (unpubl. data) observed labradorite weathering directly to gibbsite with a little kaolinite in an alpine humus soil formed on granodiorite. It has also been reported as weathering to illite (Stephen, 1952*a*), kaolinite (Ellis, 1952; Claisse, 1953), sericite and saussurite (Humbert and Marshall, 1943), and through allophane to hydrated halloysite (Aomine and Wada, 1962).

C. General

Fieldes and Swindale (1954) attempted to rationalize the weathering of silicate minerals. They considered that the minerals in each of their stability groups follow specific courses of weathering that are modified according to the micro-environment in the following way:

Group 1 (*Olivine, Augite, Hypersthene, Hornblende*). Amorphous hydrous oxides of titanium, iron, aluminium, and silicon are formed, and subsequently crystallize to anatase, goethite, hematite, gibbsite, and boehmite, or kaolinite is formed by resilication.

Group 2 (*Basic Volcanic Glass, Zeolites*). Amorphous hydrous oxides of iron, aluminium, and silicon are formed and may age to form allophane or gibbsite, or kaolinite or montmorillonite by resilication, depending on the micro-environment; kaolinite and montmorillonite can be converted from one to the other depending on the influence of hydrogen and calcium ions.

Group 3 (*Biotite, Muscovite*). Illite is formed under the influence of hydrogen ions and may be reconverted to micas in the presence of potassium ions. Illite weathers to montmorillonite if calcium ions dominate the micro-environment or to vermiculite if hydrogen ions are dominant. Vermiculite may form montmorillonite under the influence of calcium ions, and vermiculite and montmorillonite may form kaolinite under the influence of hydrogen ions.

Group 4 (Felspars, Acid Volcanic Glass). Amorphous hydrous oxides of aluminium and silicon are formed and may age to form allophane or be resilicified to kaolinite.
Group 5 (Quartz). Amorphous silicic acid is formed, and ages to form "chalcedonite," which may form secondary quartz.

These proposed weathering sequences do not agree with many observations of other workers, and they are fundamentally different from the views of Jackson and Sherman (1953) who regard their table of clay size minerals as a weathering sequence, the kind of clay mineral formed progressing from (1) to (13) (Section 5.2) with increasing severity of weathering.

5.4 FACTORS AFFECTING WEATHERING

A. Lattice Structure

In general the micas weather to form layer-lattice minerals. This is to be expected because they already possess such a lattice structure, and a number of workers (e.g., Walker 1949; Fieldes and Swindale, 1954) have shown that this is not destroyed by release of a high proportion of their constituents. As a result, only particular constituents are released from micas to contribute to the formation of other secondary minerals, except under conditions where layer-lattice type secondary minerals are themselves unstable; the constituents released are mainly potassium from muscovite and magnesium, potassium, and iron oxidized to the ferric form from biotite (Denison, Fry, and Gile, 1929). Under extreme conditions the layer-lattice structure is broken down (Alexander, Hendricks, and Faust, 1941; Neuwirth, 1954; and Section 5.3:A).

In contrast to the micas the structures of most of the other primary rock-forming minerals do not resemble those of the common secondary minerals so closely. Fieldes and Swindale (1954) proposed that the first stage of weathering of these minerals is destruction of the crystal lattice and release of amorphous hydrous oxides which may subsequently age to form crystalline oxides or be resilicified to form layer-lattice silicates (Section 5.3:C). It should not be inferred from this hypothesis, however, that destruction of the crystal lattice is catastrophic so that all the constituents of an entire mineral grain are suddenly released with complete freedom to interact; if such destruction of the crystal lattice does occur it may well be on an ultramicroscopic scale. More recent work suggests that the felspars need not break down completely to amorphous hydrous oxides as a stage in the formation of layer-lattice structures. De Vore (1959) suggested that single chains with the felspar Al-Si ratio can be removed as units from the (100)–(010) surfaces of ordered felspars, and "without further cation transformations, the (100)–(010) derived chains can polymerize directly into Al-Si ordered mica sheets. A simple transformation in these chains permits the kaolinite sheet structure to form." He also suggested that "the mica-like surface chemistries of the felspar" may act as "nucleation sites for the micaceous minerals." Similar studies on other primary mineral species, such as the ferromagnesian minerals, may show that they can behave in a similar manner.

B. *Micro-environment*

The wide variety of secondary minerals reported from the weathering of primary minerals can be reconciled provided due weight is given to the effects of the micro-environment. The secondary minerals produced during soil formation depend largely on the constituents released, the opportunity for their interaction, and the suitability of the micro-environment for formation of particular secondary mineral species. In most soil materials a relatively wide variety of minerals is being weathered simultaneously so that, within limits, the secondary mineral formed is not strictly limited by the absence of some necessary constituent. The secondary mineral species formed in soils, therefore, are controlled essentially by the effects of leaching, the chemical micro-environment, and the accessibility, or opportunity to interact, of constituents, whether they are formed as pseudomorphs or in the body of the soil material.

(*a*) **Leaching.** Crompton (1960) discussed the importance of the ratio of rate of weathering to rate of leaching in controlling the kind of secondary mineral formed. Free drainage tends to remove the products of weathering (the hydrous oxides or lattice groups first released), while impeded drainage holds them within the soil material thus increasing the opportunity for interaction. Also, since the chemical micro-environment is controlled by the constituents present, it is ultimately controlled by the drainage status of the profile. The balance between rate of weathering and rate of leaching, therefore, can account for the formation of a wide range of secondary mineral species from many of the primary species, from allophanes and crystalline oxides to montmorillonoids and vermiculites; impeded drainage conditions tend to form the more complex layer-lattice silicates due to retention of a larger variety of the constituents required for their formation, while free drainage tends to form the simpler secondary minerals according to the relative retention of various constituents. Numerous examples of this effect appear in the literature (e.g., Mitchell, 1962; Bolter, 1961; Povoas and Macedo, 1959; White, Bailey, and Anderson, 1960).

(*b*) ***Chemical Micro-Environment.*** With free drainage the kind of constituents retained in the profile, from which the secondary minerals are formed, and the kind leached out depend on the chemical micro-environment. Swindale and Jackson (1956) proposed that two separate processes, which can act together but independently, control the relative rate of removal of the constituents:

> *Soluviation.* The process of solution of the minerals and eluviation of the decomposition products depending only on the presence of moving water; since aluminium and ferric iron are less soluble than the other constituents of minerals, they tend to accumulate as a result of pure soluviation.
>
> *Cheluviation.* The process in which the minerals decompose, and the weathering products are eluviated in the presence of chelating agents; the ions are removed according to their susceptibility to chelation, and so the process removes iron and aluminium faster than silica.

The weathering of felspar in rhyolite was quoted as an example of the effects of these two processes. With increasing soluviation alone the felspar formed 2:1 layer-lattice silicates, kaolinite-halloysite, and gibbsite, with liberation of increasing amounts of free iron oxides. By superimposing cheluviation the kaolinite-halloysite and gibbsite decomposed, the free iron oxide was removed, and the surface horizons of the soil profile became enriched in primary silica.

Both pH and temperature have a marked effect on solubility. In particular pH affects the relative rates of removal of constituents by leaching and the kind of secondary mineral formed. Keller (1958) pointed out that most aluminium silicate minerals have an abrasion pH of 10 or more, while natural hydrolyzing systems often have a pH of 7.0 to 9.5 at which silica is much more soluble than alumina, so alumina tends to accumulate. At higher pH values both silica and alumina are more mobile, while at lower values both have a low solubility and the more complex clay minerals are formed, the type depending on the amounts and proportions of the cations present. Other workers (e.g., Correns, 1962; Gritsaenko, Berkhin, and Rudnitskaya,

1956) have shown that kaolinite tends to form at low pH and montmorillonite at higher pH. The pH of the micro-environment is controlled by the drainage status (leaching) and the nature of the parent rock. This is exemplified by data obtained on the soil clays of northeast Scotland (Macaulay Institute, 1948) which showed that high basicity of parent rocks and low leaching tend to cause formation of montmorillonite, while acidity of parent rocks and high leaching cause formation of kaolinite.

The constituents released from one mineral species may have a considerable effect on the weathering of other minerals and on the kind of secondary mineral formed. The micro-environment can be changed by formation of relatively stable secondary minerals, such as epidote, which effectively remove calcium from the micro-environment (Humbert and Marshall, 1943). Stephen (1952*a*) reported similar circumstances where calcium and silica released by weathering of biotite to vermiculite resulted in the formation of sphene, epidote, and calcite.

(*c*) ***Accessibility.*** The details of the first breakdown of the crystal lattice of primary minerals to release hydrous oxides or lattice groups are important with regard to the opportunity for these constituents to react to form secondary minerals. In one extreme they may be released completely with complete freedom of interaction except as modified by rate of leaching. In the other extreme, since weathering is known to occur most readily along fractures, cleavages, and twin planes, ultramicroscopic decomposition may occur in these situations, and there may be no opportunity for entrance of a thoroughly mixed soil solution or suspension; the secondary minerals formed then are controlled by the constituents released by a single mineral species and by the micro-environment in the situations where weathering is taking place.

This concept of the effects of leaching, micro-environment, and accessibility explains the formation of a wide variety of secondary minerals, either as pseudomorphs after primary minerals or unassociated with the grains supplying the constituents in situations in the profile where suitable constituents are available and the micro-environment is favorable. It also explains the many variations in kind of secondary mineral reported from weathering of similar mineral species, the formation of discrete grains of such authigenic minerals as epidote, anatase, and rutile, and the occurrence of such materials as chalcedony filling voids in the soil material.

5.5 CORRELATION OF WEATHERING WITH ENVIRONMENT

The problem of assessing the effectiveness of the environment in soil formation has been considered from other points of view than mineral analysis. Jenny (1941) proposed that soil formation is the result of the effects of climate, drainage, parent material, topography, and organisms, all of which affect the micro-environment within soil materials and so affect the intensity of weathering of minerals. Various attempts have been made to correlate soil characteristics with these factors. In particular, the effectiveness of temperature and leaching has been recognized by Prescott (1949), who attempted to correlate the occurrence of great soil groups with measurements of rainfall and evaporation: his P/E^m, where P is the average annual rainfall, E the average annual evaporation from a free water surface, and m a value varying from 0.67 to 0.80 with a probable mean of 0.73. This is a modification of the Meyer ratio P/Sd where P is the precipitation and Sd the saturation deficit. Later Hallsworth (1952) developed a number of equations in an attempt to explain the distribution of various soil groups on basalt in the Richmond-Tweed region of New South Wales. These equations are a more direct attempt to assess the effectiveness of weathering and leaching within soil profiles. He first proposed an equation for evaporation:

$$Ep = K_1 \frac{VPt(100 - RH)}{100}$$

where Ep = the evaporative power,
K_1 = a constant,
VPt = the vapor pressure at temperature t°C, and
RH = the relative humidity.

Then he calculated the water passing through the soil profile from the equation:

$$Wp = K_2 \frac{P}{Ep}$$

where Wp = the water passing through the profile,
K_2 = a constant, and
P = rainfall.

Then P/Ep is analogous to the Meyer ratio P/Sd and the *leaching potential of the climate.* Since higher temperatures increase hydrolysis of minerals, and so increase the supply of bases, he proposed the equation:

$$Rb = K_3 Dwt$$

where Rb = the rate of base release,
K_3 = a constant, and
Dwt = the dissociation of water at t°C, i.e. the mean annual temperature.

From these equations he derived the acidification potential of the soil, which is a balance between the water passing through the profile and the bases released by weathering, from the equation:

$$Ap = \frac{K_2(P/Ep)}{K_3 Dwt}$$

$$= K\left(\frac{P}{Ep}\right) \Big/ Dwt$$

where Ap = the acidification potential, and
K = a constant.

Then, $\dfrac{P/Ep}{Dwt}$ = the *acidification index.*

To date, this mathematical evaluation of the environment has not been correlated with the kind of weathering products of minerals over a wide enough range of soils and parent materials to assess its value.

5.6 EVALUATION OF WEATHERING

The pedologist is particularly interested in evaluation of what has been loosely referred to as the "degree of weathering" in soil profiles. Several attempts based on mineral analysis have been made to calculate quantitative measures of the degree of weathering of soil materials and soil profiles so that independent materials and profiles can be compared. The problem, however, is not nearly so simple as it might seem at first sight. Data from mineralogical analysis of primary and secondary minerals in soil profiles show that there are a number of aspects of the problem that must be considered if reliable comparative figures are to be obtained. The two most important of these are: a factual measurement of the amount of weathering of the constituent minerals, and an estimation of the severity of the microenvironment in regard to weathering. The first aspect is by far the simpler in that it is a factual measurement that does not depend on any other knowledge. The second aspect is much more complex since it depends on a knowledge of the relative stability of the mineral species and the effects of environment and grain size on weathering.

It is proposed, therefore, that the best method of evaluating the degree of weathering of soil materials available at the present state of knowledge is to use two separate indexes each of which has its own particular function. These can be termed *absolute weathering* and *intensity of weathering*.

A. Absolute Weathering

Absolute weathering can be measured simply by the amounts of the various original mineral species in the parent material that have been weathered, that is, the loss of each mineral species during soil formation. By appropriate techniques these measurements can be applied to any size fraction, including the clay size fraction; the more size fractions considered the more reliable will be the results. Since primary mineral grains may decrease significantly in size during weathering, estimations made on a single size fraction may be misleading because larger grains may break down physically with increasing weathering, thus materially increasing their proportions in the finer size fractions.

These estimations should be quoted as absolute amounts of the original mineral species weathered rather than as percentages of the mineral species weathered. For example, a siliceous sand may have very small amounts of minerals which are more easily weathered than quartz, and soil development may completely remove all of these minerals leaving the quartz virtually unweathered. If the estimations are quoted as absolute amounts, this material is only slightly weathered. If they are quoted as percentages, certain mineral species are completely weathered, but only because they were present in very small amounts.

These calculations are tedious and can be made only for soil materials whose parent material can be identified, that is, on soil profiles formed on uniform parent material in which the unaltered parent material is preserved or can be assessed with reasonable accuracy (Section 4.1). Approximate estimates can be obtained by restricting the number of size fractions examined, but the results become less dependable the narrower the size range examined.

Alternatively absolute weathering can be rated according to specific mineral ratios. Ruhe (1956) proposed two weathering ratios, which are related to the aforementioned calculations, for approximate assessments of the comparative degree of weathering of soil materials. They are based on quantitative estimation of the mineral species present in the very fine sand fraction (0.62–0.125 mm equivalent diameter); they are derived from the formulas:

$$Wrh = \frac{\text{Zircon} + \text{Tourmaline}}{\text{Amphiboles} + \text{Pyroxenes}}$$

and

$$Wrl = \frac{\text{Quartz}}{\text{Felspars}}$$

where Wrh is the weathering ratio based on heavy minerals and Wrl the weathering ratio based on light minerals. Each mineral species is expressed as a percentage by number of the size fraction studied. In expressing the ratios the denominator is given the value 1.00, and the numerator is expressed as a decimal. These ratios were chosen on the basis that zircon and tourmaline are generally held to be the most stable of the common "heavy minerals" and quartz the most stable of the "light minerals," while amphiboles, pyroxenes, and felspars are considered to weather at moderate rates. The indexes, therefore, are generalized quantitative estimates of weathering based on stable constituents.

As measures of weathering, these ratios are applicable only to single profiles on uniform parent materials or separate profiles on common parent material with the same mineral ratios and the same size frequency distribution of the mineral species used to calculate the ratios. For example, a siliceous sediment that contains virtually no amphiboles, pyroxenes, or felspars, but large amounts of quartz and considerable zircon and tourmaline, has a high value for both Wrh and Wrl even with no soil formation, and a small actual amount of weathering greatly increases the values of the ratios; a parent material with high proportions of amphiboles, pyroxenes, and felspars and small amounts of quartz, zircon, and tourmaline must lose large amounts of minerals by weathering to give a comparable increase in value of Wrh and Wrl. It could be argued that Wrh and Wrl values for a soil material are an expression of the overall degree of weathering of the material, but they give no information concerning the amount of weathering during soil formation, especially from a sedimentary parent rock. The *difference* between Wrh for the soil material of each of the soil horizons and for the parent material, and the *difference* between Wrl for those materials, however, could be used in most cases for comparison of the absolute weathering between different profiles, even when the parent materials differ.

The variation in grain size distribution in sediments and igneous rocks and the fact that the grain size of most mineral species decreases progressively with increasing weathering (Brewer, 1955*b*; Raeside, 1959) suggest that estimations of *Wrh* and *Wrl* based on a single, limited size range are not sufficient for comparison of independent soil materials; they should be based on estimations of the total proportions of the mineral species involved, that is, estimations of all the size fractions. Even then the values of *Wrh* and *Wrl* depend on the *relative* rates of weathering of amphiboles, pyroxenes, and felspars so that neither *Wrh* nor *Wrl* is satisfactory without the other. More serious is the fact that the values may be influenced by the relative rates of weathering of other mineral species not considered in the ratios, which may be present in some parent materials but not in others. If the ratios are based on estimations of a wide size range, as they must to be effective, relatively little more work would be required to estimate the absolute amounts of the minerals weathered.

Other weathering indexes proposed depend simply on estimation of the percentage of weatherable minerals present in each soil horizon, for example, the percentage by weight of heavy minerals in a given size class (Cady, 1940). This approach has the same disadvantages as *Wrh* and *Wrl,* and even the difference between the indexes for the parent material and soil horizons is not very useful because the relative rates of weathering of mineral species differ under different conditions; weathering may increase the proportion of heavy minerals in some soils and decrease it in others.

Other workers have suggested that the index of weathering should be based on the percentage weathered of particular mineral species, such as hornblende (Haseman and Marshall, 1945), felspars (Graham, 1950), and the micas (Rolfe and Jeffries, 1952).

B. Intensity of Weathering

Intensity of weathering is a measure of the amount of weathering of primary minerals of known relative resistance to weathering (Section 5.1) and of the effectiveness of leaching as expressed by the kind of secondary mineral formed.

Jackson and Sherman (1953) proposed a table of the clay size minerals found in soil materials (Section 5.2), which they designate as a "mineral weathering sequence." Their discussion suggests that the order of the minerals in this table is intended to be one of increasing weathering of soil materials; that is, as the "degree of weathering" increases the kind of weathering product found in the profile changes down the table from weathering stage (1) gypsum to (13) anatase.

This proposal is based on the authors' assessment of the "weathering intensity factors," which apparently depends on correlation of the various mineral species quoted in the table with profiles that occur in various climatic zones and on different kinds of parent materials. Nevertheless the weathering stage of the clay size minerals in soil profiles has been used directly as an assessment of intensity of weathering.

Although there is some basis for this proposal as a broad approximation, there are considerable data available that show it should not be accepted without question in all cases; for example, the clay size fraction in an alpine humus soil profile was dominated by gibbsite and hydrobiotite (Brewer and Haldane, unpubl. data), the gibbsite being derived from felspar and the hydrobiotite from biotite; very little of either of these primary minerals had been weathered, the total clay size fraction being approximately 11 percent of the whole soil material. Hydrobiotite appears at stage (8) in Jackson and Sherman's table (Table 5–3) and indicates a moderately weathered soil material, while gibbsite is at stage (11) and indicates a strongly weathered soil material. The occurrence of these two minerals together as the dominant clay size minerals can be explained quite adequately by considering the effect of leaching in this highly permeable soil material which occurs at an elevation of approximately 6000 feet under a 60-in. rainfall in a subalpine to alpine environment. In fact there has been comparatively little weathering in terms of the amount of primary minerals weathered, but the felspar has gone directly to gibbsite because of the rapid, effective, differential leaching of all the other constituents released by weathering of both the felspar and the biotite; the biotite itself has been only weakly weathered, both in regard to the amount weathered and the degree to which it has been altered.

The "intense" weathering that takes place in tropical and subtropical environments is also probably a function of leaching. Certainly the high temperature and moisture in these areas would make for rapid weathering in terms of the amount of primary minerals weathered, but the common occurrence of particular clay size minerals in the great majority of profiles, irrespective of the kind of parent rock (Merwe and Heysteck, 1952), can logically be attributed to the effects of leaching. In other words, leaching differentially removes all the constituents released by weathering other than those which form the clay size minerals of such soils, that is, kaolinite, gibbsite, goethite, hematite, etc., and these are the minerals that occur in the more stable groups in Jackson and Sherman's (1953) table—groups (10) to (13). The nature of these clay size minerals indicates the effectiveness of leaching in removing a great bulk of material. Jackson and Sherman's table, therefore, is more appropriate as a table of minerals arranged in regard to the effectiveness of leaching during weathering rather than one of stability.

Some of these difficulties were overcome by using the *weathering mean* as a measure of the degree of weathering of soil materials (Jackson and Sherman, 1953). They stated that "the results of chemical weathering are measured both by the residual minerals (residue mechanism) and the solids formed from solution (synthesis mechanism). In soils unsaturated with bases the stage of weathering can be measured by the clay size minerals alone, whereas in supersaturated soils the calcite and gypsum must be included in the calculation in addition to clay size minerals (Hseung and Jackson, 1952)." Their weathering mean is derived from the equation:

$$m = \frac{\Sigma(pS)}{\Sigma(p)}$$

where m = the weathering mean,
p = the percentage of a mineral in the soil,
S = the weathering stage of the mineral to which p refers, according to the weathering sequence table of Jackson and Sherman (1953), and
Σ = "the sum of."

The weathering mean is usually derived entirely from analyses of the clay size fraction, but Jackson and Sherman suggested that it can be applied also to the silt fraction where it has been weathered appreciably. They pointed out that the weathering mean is sensitive to changes in the layer-silicate weathering stage, according to their table of stability of clay size minerals, whereas such ratios as the silica–sesquioxide ratio are not (Marshall, 1935). It takes account of the proportions of clay size mineral species, but not the absolute amounts, nor does it take account of the amount of the various clay size minerals present originally in the parent material, for example, in sedimentary bodies whether they are hard rocks or unconsolidated beds.

These objections can be overcome to different degrees by various modifications in calculating the weathering mean. Some allowance for the proportion of primary minerals weathered can be made simply by multiplying the weathering mean (m) by the percentage of clay size minerals expressed as a fraction, that is,

$$m_1 = m\frac{x}{100}$$

where m_1 = the new weathering mean,
m = Jackson and Sherman's weathering mean, and
x = percentage of clay size minerals in the soil material.

A similar allowance can be made by including in the estimate of the weathering mean the primary minerals still present in the soil material:

$$m_2 = \frac{\Sigma(pS)}{\Sigma(p)}$$

where p = percentage of all the minerals, both primary and secondary, in the soil material taken one at a time, and
S = the weathering stage of the mineral to which p refers, derived entirely from Jackson and Sherman's (1953) table, or from that table for secondary minerals and from Pettijohn's (1941) table for primary minerals.*

Neither m_1 nor m_2 take account of any clay size material present originally in the parent material. This can be done by calculating the difference between m_2 for the soil material and the parent rock; thus,

$$m_3 = m_{2s} - m_{2p}$$

where m_{2s} = m_2 for the soil material, and
m_{2p} = m_2 for the parent material.

Alternatively two separate indexes (m_p and m_s) can be used:

$$m_p = m_{pp} - m_{ps}$$

and

$$m_s = m_{ss} - m_{sp}$$

where m_{pp} = weathering mean $\left(\frac{\Sigma(pS)}{\Sigma(p)}\right)$ of the primary minerals for the parent material, using Pettijohn's (1941) table of persistence,
m_{ps} = weathering mean of the primary minerals for the soil material,
m_{ss} = weathering mean of the secondary minerals for the soil material, using Jackson and Sherman's (1953) table, and
m_{sp} = weathering mean of the secondary minerals for the parent material.

* By comparison of the stability tables of Pettijohn (1941), Graham (personal communication quoted by Jackson and Sherman, 1953) and Jackson and Sherman (1953), it seems reasonable for the purposes of these calculations to rate quartz, albite, and microcline as stage (3), orthoclase as (9), intermediate plagioclase as (13), and basic plagioclase as (21) in Pettijohn's sequence.

The indexes, m_3 and m_p and m_s combined, are the most complete measures of intensity of weathering because they take into account the amount of secondary clay size minerals formed, the different accepted degrees of stability of the various primary and secondary minerals, and the effectiveness of leaching by considering the kind of secondary clay size minerals formed. They are deficient in that they do not take account of the absolute amounts of the various mineral species weathered.

Table 5-4 shows the rating of intensity of weathering for four different kinds of soil profiles according to the various indexes:

(*a*) an alpine humus soil formed on granodiorite and having 11 percent clay size minerals in the B_2 horizon consisting dominantly of gibbsite with some hydrobiotite;

(*b*) a yellow podzolic soil formed on granodiorite and having 42 percent clay size minerals in the B_2 horizon consisting of approximately equal amounts of hydrobiotite, illite, and kaolinite with negligible amounts of felspar and goethite;

(*c*) a red-brown earth formed on parna and having 74 percent clay size minerals in the B_2 horizon consisting of approximately 60 percent illite and 40 percent kaolinite with negligible quartz; the parna parent material has the same mineralogical composition but less (59 percent) clay size material. Particle size and mineral analysis of this profile indicate that weathering is negligible and that profile differentiation is due solely to reorganization of the constituents, especially illuviation;

(*d*) a young, weakly differentiated windblown sand having 4 percent clay size minerals in the B_2 horizon consisting of approximately 60 per-

TABLE 5-4. Variations of Jackson and Sherman's (1953) Weathering Mean for a Series of Soil Materials

Weathering Mean	Alpine Humus Soil (a)	Yellow Podzolic Soil (b)	Red-Brown Earth (c)	Windblown Sand (d)	Maximum Possible Values
m	9.6	8.3	8.2	8.2	13.0
m_1	1.1	3.5	6.1	0.3	13.0
m_2	5.3	6.5	7.8	6.1	13.0
m_3	0.5	1.8	0.4	0.1	10.0*

* Developed from rocks very rich in minerals of stage (3) (Jackson and Sherman, 1953), that is, olivine, hornblende, pyroxenes, etc.

cent illite and 40 percent kaolinite; the parent material has the same mineralogical composition and approximately 1 percent clay size minerals.

The differences in weathering between these four profiles is so great that they form a suitable series on which to assess the efficiency of the various indexes in rating soil materials according to intensity of weathering. In broad terms the B_2 horizon of the yellow podzolic soil (*b*) is much more strongly weathered than the alpine humus soil material (*a*), and the red-brown earth on parna (*c*) and the windblown sand (*d*) are essentially unweathered; that is, the apparent order of decreasing intensity of weathering is:

$$(b) > (a) > (c) = (d)$$

The results for m indicate that

$$(a) > (b) = (c) = (d)$$

The results for m_1 indicate that

$$(c) > (b) > (a) > (d)$$

The results for m_2 indicate that

$$(c) > (b) > (d) > (a)$$

and they show relatively little difference between all four materials. The results for m_3 put all the soil materials in the order indicated by other data; that is,

$$(b) > (a) > (c) > (d)$$

They also indicate that (*a*) is very weakly weathered and (*d*) is practically unweathered; the index for (*c*) is too high [close to that of (*a*)] because of the presence of illuviated clay, which could be assessed by fabric analysis in thin section and subtracted before making the calculations.

6

Classification of Parent Rocks

There is little doubt that the nature of the parent rock has a considerable influence on the kind of soil profile developed (Jenny, 1941) and on the course of weathering of primary and secondary minerals in most climatic zones, except perhaps the humid tropics where the intensity of weathering apparently can have an overriding effect if the soils are old enough. The literature contains many references to these effects; for example, many of the major soil boundaries in southeastern Australia coincide with geological boundaries between different rock types (Brewer, 1954), especially if unconsolidated alluvial, colluvial, and aeolian sediments are considered as rocks in accordance with Rice's (1954) definition, and the effects of parent material on the clay mineral species which occur in the various great soil groups of the United States have been recognized (Jackson, 1959). Jackson recognized the importance of such characteristics as permeability of the parent rock in determining the course of weathering and soil formation, while he and others (e.g., Keller, 1958) recognized the importance of the supply of particular chemical elements for formation of specific secondary minerals. In spite of this realization of the effects of the detailed characteristics of parent materials, few attempts have been made to classify them except in terms of broad lithological groups for igneous rocks and consolidated sediments and groupings according to mode of origin for unconsolidated sediments; "granite," "weathered granite," "loess," and "alluvium" are common names used to describe the parent rocks of soil profiles.

One reason for the lack of attention given to parent rocks is the wide acceptance of the zonal concept of soil distribution, as outlined by Smith (1949), which lays particular emphasis on climate as a soil forming factor. A corollary to this concept is that the normal zonal soil of an area should be predictable from a study of the climate of the area. As a result, a number of workers have attempted to correlate soil distribution with a climatic index, for example, Hallsworth's (1952) acidification index (Section 5.5) and Prescott's (1949) P/E^m ($m = 0.75$), while the effects of parent rocks have been largely neglected even though it has been recognized that such climatic indexes are meaningful only when the other soil-forming factors are relatively constant (e.g., Stephens, 1947; Butler, 1950; Crocker, 1952; Brewer and Butler, 1953).

A more important reason for the lack of attention to classification of parent rocks is the widely accepted view that the parent material is not the rock but the weathered rock or, as the *U.S.D.A. Soil Survey Manual* (1951) puts it, "the unconsolidated mass from which the solum develops." The considerations put forward earlier (Section 4.1) show the errors involved in such an assumption and the practical impossibility of consistently recognizing, and hence classifying, such materials. On the other hand, recognition of the parent rock is usually an achievable objective, and equating parent rock to parent material presents a basis for classification.

Whiteside (1953) pointed out that the term "parent rock" for unconsolidated materials is not incongruous. Rice (1954) defined rock as "the material that forms the essential part of the earth's crust and includes loose, incoherent masses such as a bed of sand, gravel, clay, or volcanic ash as well as the very firm, hard, and solid masses of granite, sandstone, limestone, etc. Most rocks are aggregates of one or more minerals, but some are composed of glassy matter or a mixture of glass and minerals." Such a definition includes loess, parna (Butler, 1956; Butler and Hutton, 1956), and other unconsolidated sediments. In fact, Whiteside asserts that this definition includes soils as rocks and that the first problem of the pedologist is to differentiate soils from other rocks. If the definition of soil proposed earlier (Section 1.1) is accepted, soil is rock (in the widest sense of Rice's definition) that has undergone some degree of alteration due specifically to soil-forming processes which operate because of exposure of the rock at the earth's surface; they are formed from *parent rocks.*

These arguments lead to the conclusion that practically all soils are sedentary in the sense that they were formed *in situ* on materials, whether unconsolidated or hard, that conform to the definition of rock. The few possible exceptions to this are bodily movements of soils or parts of soil profiles in land slips and the lesser movements involved in some kinds of soil creep. Apart from these exceptions the concepts of rock and soil establish a foundation on which can be built a classification of the parent rocks of soils. There will still be instances where it is difficult to decide precisely where the boundary occurs between soil and parent rock, especially where there is a gradual transition in the lower soil horizons from soil to unconsolidated rock (Section 4.1). Of course, the material recognized as the parent rock is not the actual material from which the soil profile was developed, but the term "parent" is meant to indicate that the material is representative of the kind of material from which the soil was developed.

It is not to be expected that the geological classification of rocks should be an ideal one for the classification of parent rocks of soils since, as Whiteside (1953) pointed out, the "tests used by pedologists and geologists in evaluating a rock classification" are different. The objective of any classification is to group together like individuals; the same individuals can be classified in different ways according to the characteristics used to assess likeness. Within any particular "descriptive" science the aim usually is to group together individuals that are morphologically similar. The most useful morphological classification, within the particular science concerned with detailed study of the individuals, is one that also groups the individuals according to their genesis; mineralogy and texture are very effective morphological characteristics to achieve this objective for rocks. There have been many such attempts by petrologists to reach a satisfactory classification of rocks, from the geological point of view, on the basis of chemistry, mineralogy, and texture. None of these, as reviewed by Johannsen (1939), is satisfactory as a classification of parent rocks of soils because the elements are grouped according to their occurrence in the rock-forming minerals or in arbitrary groups of "normative" minerals and not according to their effects on soil formation; for example, ferrous iron is usually grouped with magnesium because these elements are commonly associated in ferromagnesian minerals, whereas their effects on soil formation are quite distinct. In addition, some of the important characteristics of rocks as soil parent materials are not considered.

When a classification of such a group of individuals is required for a specific purpose, it may be found that quite different characteristics are more suitable as a basis than those used within the science concerned with the complete, detailed study of the individuals; the characteristics must be chosen according to their importance for the specific purpose. This is the situation with regard to the classification of rocks as the parent rocks of soils. Those characteristics must be chosen that are the most important in their effects on the development of soil profiles without reference to their importance in a classification of rocks according to their genesis. Pedologists are not concerned primarily with the age or geological relationships of rocks as such but with those physical and chemical properties which affect soil formation.

6.1 MODIFICATION OF THE GEOLOGICAL CLASSIFICATION

Whiteside (1953) developed an arrangement of rock types based primarily on geologically defined types, which are treated in two broad groups: (1) dominantly igneous rocks, and (2) dominantly sedimentary rocks. Within each group the various rock types are arranged according to their mineralogical composition, porosity, orientation of the

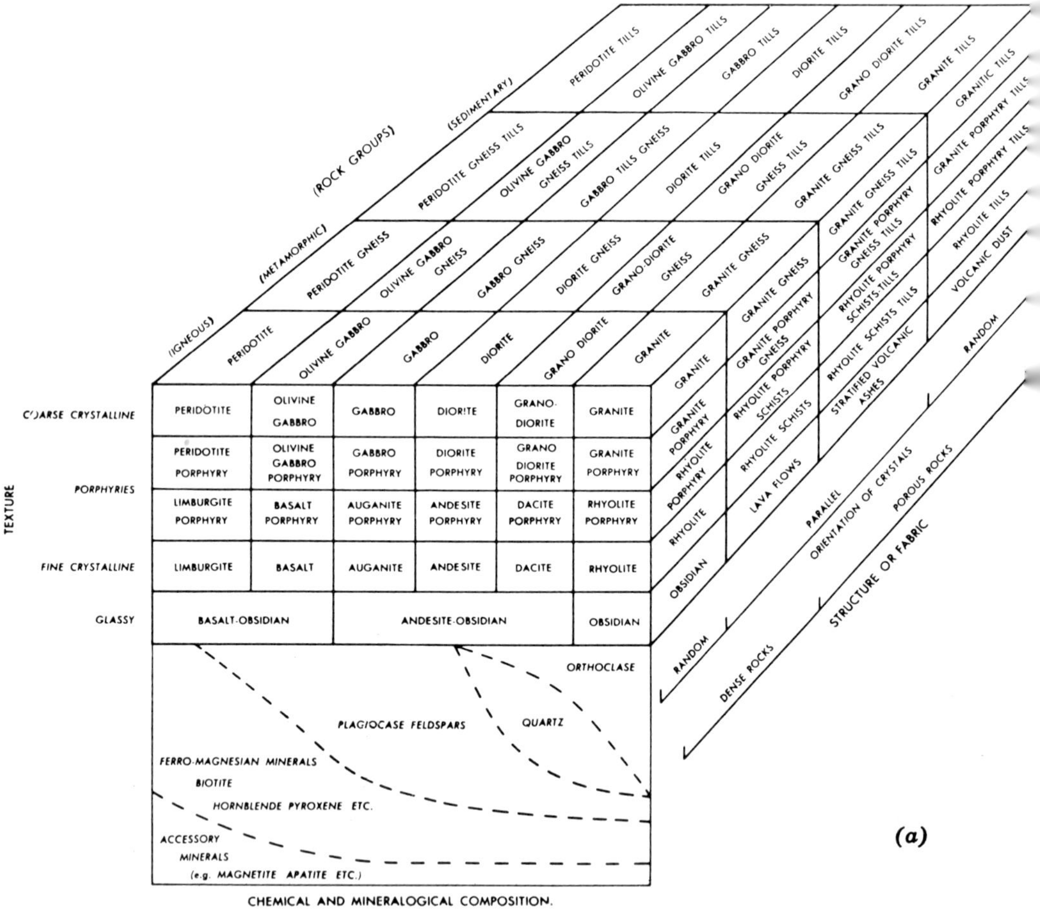

(a)

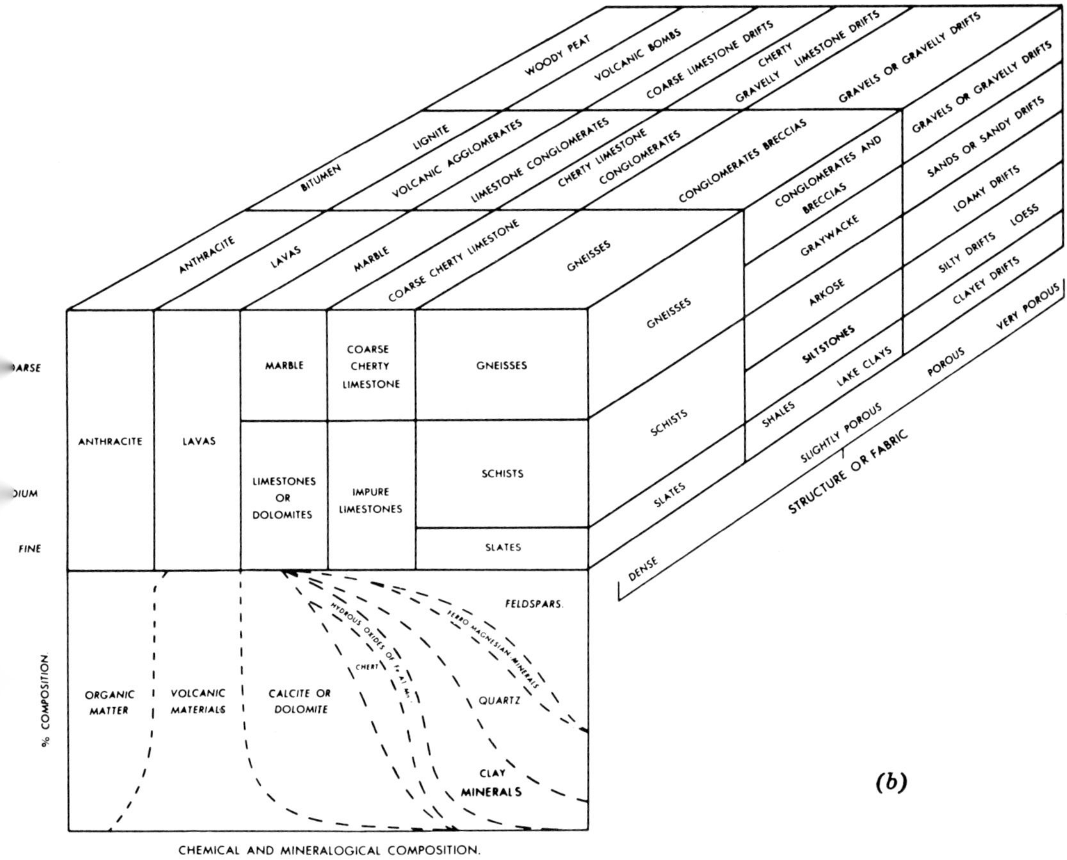

Fig. 11. Texture, mineralogical composition, and structure of rocks as a basis for classification as parent rocks of soil profiles (after Whiteside, 1953). (*a*) "Igneous" rocks. (*b*) "Sedimentary" rocks. *Courtesy Soil Sci. Soc. Am.*

crystals, and grain size—the characteristics he considered to be of paramount importance in soil formation. His diagrams (Fig. 11) show at a glance the general characteristics of a large number of rock types. This is not a true classification of parent rocks, nor yet a simple modification of the geological classification of rocks since the arrangement cuts across many of the most important geological classes, even the distinction between sedimentary and igneous rocks. It is, however, a useful classification for subdividing parent rocks into broad classes that have some meaning in soil formation.

Care must be exercised in using this classification, particularly as regards the estimation of mineralogical composition. Whiteside recognized the variability in mineralogical composition within rock types and the continuous nature of the variation between rock types. Such variations within a rock type can be very wide indeed, especially in the rather broad groupings used. Geologists are interested primarily in a genetically sound classification, and variations in mineralogy at the level of classification used by Whiteside are of less significance to the geologist than they would be in their effect on soil formation; for example, Johannsen (1939) pointed out the very wide mineralogical variation within the group of granites as classified by competent geologists (Fig. 12). The mineralogical composition of parent rocks, therefore, must be estimated quantitatively if Whiteside's arrangement is to be of significant use and if some of the advantages claimed for the system are to be realized. It is not sufficient to classify a parent rock as "granite" if conclusions are to be drawn concerning the probable status of the parent rock or the soil profile developed from it in

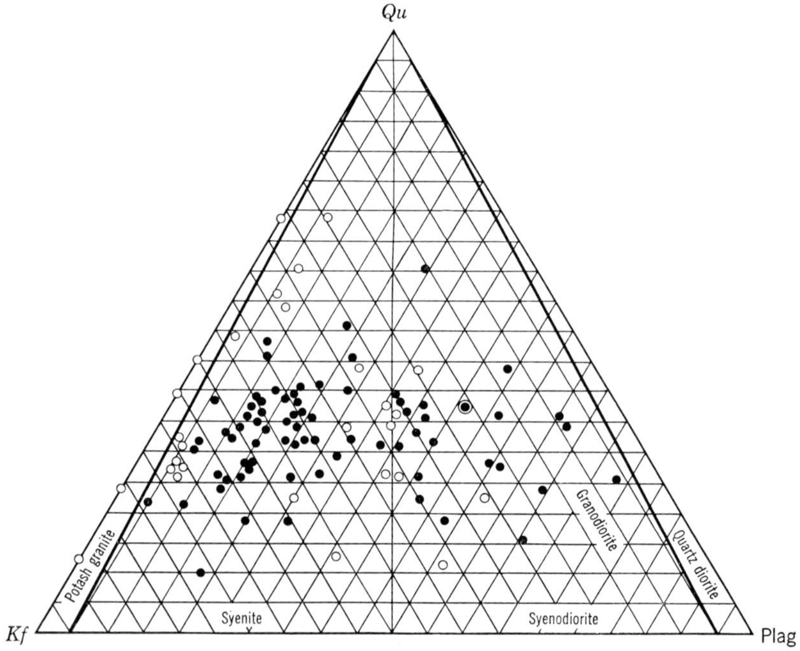

Fig. 12. Classification and mineralogical composition of 109 "granites" (after Johannsen, 1939). *Courtesy Chicago University Press.* Copyright (1939) by The University of Chicago.

regard to minor elements. The kind of mineral species and varieties present, their proportions, and whether they do carry the expected minor elements must be ascertained. Different species and varieties of ferromagnesian minerals may carry different minor elements, and the same species and varieties from different igneous bodies and complexes may carry different proportions of the minor elements which are compatible with their crystal structure.

6.2 PEDOLOGICAL CLASSIFICATION

Classification of parent rocks of soils should be based on those characteristics that have a significant effect on the properties used to distinguish different soils from each other. For classification of soils according to their level of plant nutrition, parent rocks could be classified simply on their content of the major and minor elements important for plant nutrition. A pedological classification, however, should be based on those characteristics for which there is evidence of significance in the genesis of different kinds of soil profiles. Such a classification was proposed by Brewer (1954).

A. *The Basis of the Classification*

The classification is based on the assumption that there are five major considerations involved in the effect of parent rocks on soil development. These are:

(1) The proportion of clay size minerals, particularly the proportion of true clay minerals, that can be formed from the parent rock by weathering *in situ*.

(2) The maximum amount of alkali and alkaline-earth metals that can be released from the parent rock by weathering of its constituent minerals. These are the major cations concerned in base exchange, and their effect on the chemical and physical condition of the exchange complex is considerable. They are also the major ions that, in association with such anions as sulfate, carbonate, chloride, etc., affect the pH value of the environment; in large amounts they inhibit profile differentiation. Excess amounts of calcium and magnesium usually occur as carbonates and sulfates with consequent relatively high pH values that tend to inhibit weathering and indicate low rates of leaching, as well as forming relatively stable calcium- and magnesium-saturated clays, which are flocculated by relatively low proportions of

electrolytes. Excess amounts of sodium usually occur as chlorides, which again indicate low levels of leaching, and the excess electrolyte tends to flocculate the clay size fraction. In lower proportions, however, and in the absence of high proportions of calcium and magnesium, a sodium-saturated clay may be formed that is mobile except in the presence of relatively high amounts of electrolytes. It is necessary in classifying parent rocks, therefore, to consider the major alkali and alkaline-earth metals separately.

(3) The maximum amount of iron (ferrous and ferric) that can be released from the parent rock by weathering of its constituent minerals. This element can have a considerable effect on the physical and chemical environment of the soil profile, especially on the physical condition of the exchange complex. Its presence in large amounts tends to inhibit particle size differentiation in the profile by its flocculating and stabilizing effect on the clay size particles.

(4) The ease of release of all these elements by weathering.

(5) The permeability of the rock, that is, the ease of removal of the products of weathering by leaching. The importance of this factor in controlling the kind of clay size secondary minerals formed by weathering has been discussed earlier [see Section 5,4:B(*a*)].

At first glance rocks apparently can be classified according to silica content alone since the less silica a rock contains the more alkali and alkaline-earth metals and iron and aluminium it may contain, and vice versa. Examination of a large number of complete chemical analyses, however, showed that such a classification is not satisfactory because a rock that is low in silica may be high in iron and aluminium and low in alkali and alkaline-earth metals, or vice versa.

Classification according to the relative ease of release by weathering of the elements most significant in soil formation presents several difficulties. Rate of release of an element depends on the mineral species in which it occurs, grain size of those minerals, permeability of the rock, and interaction with other constituents; coarse-grained, more fractured rocks are more permeable and more readily leached than fine-grained types, so they are effectively less basic, and the percolating solutions that are largely responsible for weathering are likely to be less saturated with solutes. Also, coarser grains are less readily weathered because of the formation of a protective coating of weathered material on the mineral surfaces and because proportionately less surface area was exposed originally.

For igneous rocks, the basis of petrological research is the association of mineral groups, grain size, and, to some extent, porosity with mode of formation of the rocks (Mohr, 1944). Petrologists have long classified igneous rocks as volcanic, hypabyssal, and plutonic, and sedimentary rocks as psammites and pelites, while rocks altered by heat or pressure are classified as metamorphic. Because of the association of mineral groups and grain size with mode of formation, a modification of this grouping of rocks into broad classes will overcome some of the difficulties of evaluating the relative ease of release of the major elements, especially in view of Mohr's statement that "practically all the mineral components of the rocks poor in silicic acid weather faster than those of the rocks rich in silicic acid." This, of course, is the main basis of Whiteside's (1953) classification.

B. *The Limits of the Classes*

The classes are defined by the following limits (the limiting values used for the various subgroups based on the amounts of iron, calcium, magnesium, sodium, and potassium were selected after examination of a very large number of complete chemical analyses for a wide range of rock types):

(*a*) The first division is made into ten arbitrary groups based on the proportion of *kaolinite* that the rock can form by weathering *in situ* if all the aluminium released by weathering is retained as kaolinite while excess bases and silica are removed by leaching. Each of the groups has a range of 10 percent of kaolinite:

1. 0–10%	6. 50–60%
2. 10–20%	7. 60–70%
3. 20–30%	8. 70–80%
4. 30–40%	9. 80–90%
5. 40–50%	10. 90–100%

The selection of kaolinite as the basis for this subdivision is quite arbitrary. If the clay minerals occurring in the soil material are identified, calculations can be made concerning the maximum proportions of these minerals that can be formed by weathering, that is, the maximum proportions of these minerals that could occur in the resulting soil material. It is necessary, however, to select some standard mineral to compare different parent rocks that may occur under a wide range of environments, thus forming a wide range of clay size minerals.

(*b*) Each of the ten major groups is then subdivided into nine subgroups according to the number of equivalents of ferrous plus ferric iron in 100 g of the parent rock:

1. less than 0.01 equivalents
2. 0.01–0.02 equivalents
3. 0.02–0.05 equivalents
4. 0.05–0.10 equivalents
5. 0.10–0.20 equivalents
6. 0.20–0.30 equivalents
7. 0.03–0.40 equivalents
8. 0.40–0.50 equivalents
9. more than 0.50 equivalents

No distinction is made between ferrous and ferric iron in the parent rock because the form that the iron will take after release by weathering depends not so much on its form in the rock as on the conditions of weathering, the micro-environment in the soil material, and the kind of secondary minerals formed.

(*c*) Similar subdivisions are made progressively in four further stages, using the same range of equivalents for nine subgroups in each stage as was used for iron, according to the number of equivalents in 100 g of parent rock of: (1) calcium, (2) magnesium, (3) sodium, and (4) potassium. Each of these elements must be assessed separately because their individual effects on the chemical and physical characteristics of the soil material, and so on the course of soil formation, are rather specific for each element; the details of their effects are not completely understood.

(*d*) Further subdivision is then made into ten groupings on the basis of rock types and mode of formation. The following symbols and groupings are proposed:

Va: Efflata volcanic rocks; fragmental and usually porous, for example, tuffs, volcanic ash.

Vb: Effusive volcanic rocks; fine grained lavas, for example, basalt, rhyolite.

H: Hypabyssal rocks; medium grained or porphyritic with fine-grained groundmass, for example, porphyry.

P: Plutonic rocks; coarse grained, for example, granite, gabbro.

Ma: Coarse-grained metamorphic rocks, for example, schist, gneiss.

Mb: Fine-grained metamorphic rocks, for example, slate, phyllite.

S: Psammites; coarser grained arenaceous sediments, for example, sandstones.

Pe: Pelites; finer-grained argillaceous sediments, for example, shales.

W: Unconsolidated alluvial and colluvial materials.

A: Unconsolidated aeolian material.

This part of the classification together with the chemical groupings gives a general indication of the mineral composition of the parent rocks. This is important since if sesquioxides, alkalies, or alkaline earths, for example, are present in significant amounts in sedimentary rocks, they are usually in very different forms than in igneous rocks. Similarly such elements as iron and potassium are held in a different manner in micas than in pyroxenes and amphiboles.

(*e*) The final subdivision in the classification is into three broad groups on the basis of fracturing and jointing:

- U: Massive, where joint planes are absent or very widely spaced, as in most granites.
- F: Fractured, where the system of joint planes would materially affect permeability, but the rock is still relatively solid, as in the younger sedimentary rocks, such as the Permian.
- S: Shattered, where the joint planes are close together and have numerous directions, as sometimes seen in the older Ordovician and Silurian sediments, and in fault shatter zones.

These broad groupings give a general assessment of the permeability of the rocks on a broad scale, as distinct from porosity due to grain size and packing of the grains.

The subdivisions in all these stages of classification are rather narrow and give rise to a vast number of possible groupings, probably a much larger number than can be handled conveniently. However, the system was presented as a concept of a method of classification rather than a complete system. The numerous possible groupings have the virtue of allowing a narrow definition of any particular parent rock so that the significance of the various ranges of values of the characteristics suggested can be assessed; it should be possible, with further experience, to widen some of these ranges and so arrive at a complete classification system. On the other hand the effects of many of the chemical constituents of rock on soil formation are not fully understood at the present time and so other groupings may become necessary, even groupings based on additional elements.

C. *Classification Formulas*

The classification of each rock is written as a formula from left to right in the order of the steps just set out, with the classes according to the chemical composition appearing in the center of the formula between two colons. Table 6-1 sets out the classification formulas for a limited number of rock types; for example, 10:77842:VbU is an olivine basalt. The formula classifies this as a fine-grained volcanic

TABLE 6-1. Classification Formulas for Selected Rock Types Based on Actual Chemical Analyses (after Brewer, 1954)

Rock Type	Percent Kaolinite (Calculated)	Rock Group	Classification Formula
Olivine basalt	99	Vb	10:77842:VbU
Basalt	99	Vb	10:77933:VbU
Hornblende granite	81	P	9:97732:PU
Biotite granite	33	P	3:55554:PU
Biotite granite	55	P	6:32253:PU
Quartz-orthoclase porphyry	46	H	5:21112:HU
Hornblende—pyroxene schist	99	Ma	10:68722:MaF
Albite—muscovite gneiss	44	Ma	5:11143:MaF
Average igneous rock*	60	?	6:66554:?
Average psammite*	12	S	2:35423:S
Average pelite*	39	P	4:65534:P

* Derived from data by Clarke (1924)

rock with little or no fracturing, a high clay-forming potential, high iron, calcium, and magnesium, and relatively low sodium and potassium. All these characteristics indicate the probable formation of a well flocculated fine-grained soil material resistant to horizon differentiation. The table also brings out the need for a close examination if the full significance of the effects of parent rocks on soil formation is to be appreciated, for example, the visual description of "gneiss" or "schist" is insufficient; the representatives of these two rock types in the table are obviously very different as parent rocks, the schist being rather like the basalts except in grain size and fracturing. The three granites are quite different from each other, the hornblende granite being somewhat similar to the schist. In particular the clay-forming potential of the three granites is very different. With regard to the sedimentary rocks, although only average formulas are given in Table 6-1, it is worth pointing out that the amounts of iron, calcium, and magnesium, in particular, can vary widely in these rocks as can the clay-forming potential.

D. Determination of the Classification Formulas

The classification formulas for parent rocks are best determined by a combination of a partial chemical analysis and a petrographic analysis. Total iron, aluminium, calcium, magnesium, sodium, and potassium are estimated chemically and the resistant minerals* petrographically. The amount of aluminium is converted to kaolinite, which is then reduced to a percentage:

$$\frac{\text{Kaolinite}}{\text{Kaolinite} + \text{Resistant Minerals}} \times \frac{100}{1}$$

The suffix indicating the permeability of the rock due to jointing and fracturing is determined by field examination. In the absence of chemical analyses an approximation to the actual classification formula can be obtained by petrographic analysis alone. This is done by petrographic estimation of the percentage of all the major minerals in the rock and then a series of calculations for iron, aluminium, sodium, calcium, potassium, and magnesium based on correlation of the chemical composition of the minerals with optical properties. Kennedy's (1947) graphs for correlation of optical properties with chemical composition for several groups of minerals are helpful in this technique. The accuracy of this technique is increased greatly by the estimation of the alkalies and alkaline earths by means of the flame photometer.

* The resistant minerals are those normally regarded as persistant under normal weathering conditions, that is, Pettijohn's (1941) "Persistence 4" and less, in addition to quartz.

PART III

STRUCTURE
AND FABRIC ANALYSIS

7

Concepts of Structure and Fabric

The concepts of *structure* and *fabric* vary in both geology and pedology, even to the extent that the same terms have been defined in different ways, and different terms have been used for the same concept. It is necessary, therefore, to define the terms relating to structure and fabric as they are used in the following chapters and to relate these definitions to the concepts used by other authors. The relationships of the proposed concepts to those current in pedology and petrology are set out in Table 7-1. Most of the concepts in use at present are similar inasmuch as they deal with some aspects of the physical constitution of the materials being described, that is, with all or some of the aspects of size, shape, and arrangement of recognizable entities or solid particles.

The recognizable solid particles in soil materials include single mineral grains and a wide variety of composite particles of varying degrees of complexity. To keep descriptions of soil materials which contain such complex features to a standard of simplicity that can be handled conveniently, it is necessary to introduce a number of concepts that facilitate study and description of any part of the complex material without detailed reference to any other part. Two such basic concepts are the *units of organization* and the *levels of organization.*

7-1 SOIL FABRIC

Current definitions of fabric used in geology (Rice, 1954; Pettijohn, 1957) and pedology (Kubiena, 1938, 1953) are similar in that they deal primarily with *arrangement* of the constituents. Differences between the geological definitions relate to whether shape of constituents

TABLE 7-1. Relationships of Proposed Concepts to Current Concepts (after Brewer and Sleeman, 1960)

		Proposed Pedological Usage					Current Pedological Usage					Geological Usage						
Characteristics	Soil components*	Field Grading	Pedality	Fabric	Structure	Texture	Field Texture (U.S.D.A. manual)	Fabric (Kubiena)	Structure (Zakharov)	Structure (Baver)	Structure (U.S.D.A. manual)	Grade	Packing	Fabric (sedimentary)	Fabric (igneous)	Structure† (sedimentary)	Texture (sedimentary)	Texture (igneous)
Crystallinity	Skeleton grains					x												x
	Plasma					x												x
Size	Skeleton grains	x			x	x	x					x				x	x	x
	Peds		x		x	x			x		x							
	Pedomogical features				x	x												
	Plasn a grains	x			x	x	x					x				x	x	x
	Voids				x	x												
Shape	Skeleton grains				x	x									x	x	x	x
	Peds		x		x	x			x		x							
	Pedological features				x	x												
	Plasma grains				x	x									x	x	x	x
	Voids				x	x												
Arrangement	Skeleton grains			x	x	x		x		x			x	x	x	x	x	x
	Peds		x	x	x	x		x		x	x							
	Pedological features			x	x	x		x										
	Plasma			x	x	x		x		x			x	x	x	x	x	x
	Voids			x	x	x												

* For geological usage substitute rock components, namely, solid elements and voids
† Restricted usage, e.g., oolitic structure, equivalent to sedimentary features

should be included in fabric descriptions; Kubiena's concept deals with arrangement only. Jongerius (1957), in his studies of the structure of some soils of the Netherlands, concluded that a positive treatment of voids was desirable; this is in contrast to early concepts when voids were considered only as a complementary result of the arrangement and packing of solid particles requiring no specific mention. Soil fabric, then, is defined as follows:

> *Soil Fabric.* The physical constitution of a soil material as expressed by the spatial arrangement of the solid particles and associated voids.

Fabric is not only concerned with the spatial arrangement of simple, discrete grains and the associated voids but also with that of compound particles, which are so common in many soil materials, and the voids between the compound particles.

7.2 SOIL STRUCTURE

The petrological concepts of structure (e.g., Krumbein and Sloss, 1955; Pettijohn, 1957) deal with regional features of sedimentary rocks or specific arrangements (the sedimentary features of Shrock, 1948), so they are generally unsuitable for pedographic studies. The definitions of structure in pedology include three broad variations that overlap in their treatment of the physical characteristics of soil materials and reflect the interests and disciplines of the workers responsible for the definitions. Baver (1948), a soil physicist, defined structure as "the arrangement of the soil particles," or, in more detail, "the arrangement of sand, silt, and clay, and of secondary particles (aggregates or structural elements) into a particular structural pattern"; size and shape of the particles were not considered. This concept is almost identical with that of fabric as defined above. Zakharov's (1927) concept of structure is one that would satisfy the agronomist or agriculturalist; he defined it as "the very fragments or clods into which the soil breaks up." This concept considers only the size and shape of compound particles and ignores arrangement. The field pedologist's concept is expressed in the *U.S.D.A. Soil Survey Manual* (1951), which defined structure as "the aggregation of primary soil particles into compound particles, or clusters of primary particles, which are separated from adjoining aggregates by surfaces of weakness." This definition considers the size, shape, and arrangement of Baver's secondary particles but ignores primary particles.

These three variations encompass most of the shades of meaning implied by definitions put forward by various pedologists from time to time. Butler (1955) stated that Zakharov's concept is applicable to the study of soils in the field, as is that of the *U.S.D.A. Soil Survey Manual;* Kubiena (1953) used structure in a similar way. Jongerius (1957) used it in a similar way to Baver, except that he also considered the arrangement of voids. These variations are due to the different interests of the various workers and the sensitivity of their techniques, and they can be combined in such a way that each defines a part of structure, that is, all or some of the characteristics of size, shape, and arrangement of primary and/or secondary particles. The following definition, therefore, should be acceptable to workers in all fields of pedology:

> *Soil Structure.* The physical constitution of a soil material as expressed by the size, shape, and arrangement of the solid particles and voids, including both the primary particles to form compound particles and the compound particles themselves; fabric is the element of structure which deals with arrangement.

Philosophically this concept of structure is the most satisfying of those proposed. It is applicable to practically any naturally occurring or fabricated material and at any level of organization in these materials. In soil materials it can be applied satisfactorily to compound particles, to primary particles (single grains), and to even more specific levels, such as the structure of individual minerals (size, shape, and arrangement of atoms) and even the structure of the atoms (size, shape, and arrangement of electrons, protons, neutrons).

There are several points concerning the nature of structure and fabric that should be emphasized:

(1) The definitions place no limitations on the scale of development of phenomena, so observations at all possible magnifications must be considered. As the magnification at which observations are made is increased greater detail of arrangement (fabric) can be described, so descriptions of the same material may vary considerably according to the magnification used.

(2) Fabric and structure do not take account of the frequency of occurrence of the constituents of a soil material; they are not measurable quantities, but they do include descriptions of size and shape of the constituents and an estimate of the variability of these characteristics. Such descriptions may indicate certain genetic processes, such as sorting, but not the absolute frequency of occurrence; the descriptions are chiefly qualitative. Low (1954) pointed out that measurements such as permeability are not measurements of structure; they are measurements of properties determined by the structure.

(3) Fabric and structure are not actually concerned with mineralogical or chemical identification of the constituents of the soil material or the color of the constituents. Variations in color, however, may reflect variations in mineralogy, which reflect variations in concentration and/or arrangement of certain constituents (fabric), so color and mineralogy are used in classifying and naming certain structural features as a matter of convenience.

7.3 PEDALITY

Great stress has always been laid on description of the size, shape, and arrangement of compound particles, particularly in field studies. This aspect is a useful subdivison of structure, and it is convenient to have a separate name for it; the term *pedality* is proposed:

> *Pedality.* The physical constitution of a soil material as expressed by the size, shape, and arrangement of peds.*

Thus pedality is a part of structure and satisfies the needs shown by the concepts put forward by Zakharov (1927), Butler (1955), and the *U.S.D.A. Soil Survey Manual* (1951).

7.4 SOIL TEXTURE AND FIELD GRADING

Since analogies have been drawn between the proposed definitions of soil fabric and soil structure and the concepts of fabric and structure as used in petrology, it is pertinent to consider the relationships of the concepts of *texture* in petrology and pedology and the relationship of texture to structure and fabric.

Texture is defined in igneous petrology as "the appearance, megascopic or microscopic, seen on a smooth surface of a homogeneous rock or mineral aggregate, due to the degree of crystallization (crystallinity), the size of the crystals (granularity), and the shapes and interrelations of the crystals or other constituents (fabric)" (Rice, 1954; after Holmes 1928). This definition can be summarized as the crystallinity, size, shape, and arrangement of the constituents, so it is structure, as defined for soil materials, with the addition of crystallinity. In sedimentary petrology texture is defined as "the size, shape, and arrangement (packing and fabric) of the component particles" (Pettijohn, 1957); this definition is synonymous with that of soil structure given earlier. In pedology, texture has been used with various shades of meaning: as an estimation of the percentage of clay size particles, the

* The nature of peds is dealt with later (Section 7.5:A).

complete particle size distribution, the effects of structure and consistence on the feel of a soil material of particular particle size distribution, and other allied properties (see Robinson, 1949; *U.S.D.A. Soil Survey Manual,* 1951; Leeper, 1952). Such concepts are incompatible with those used in petrology, which come closer to normal English usage of the term.

Although texture, as used in soil science, is ill defined, it is a strongly established concept that is widely used as one of the major characteristics of soil materials recorded in the field description of soil profiles. Even so a strong case exists for redefining the term in line with the petrological meaning and for substituting another term for the present pedological concept. The following terms and definitions are proposed:

> *Soil Texture.* The physical constitution of a soil material as expressed by its structure, and by the degree of crystallization (crystallinity) of the solid particles; fabric is a part of structure, which is a part of texture in this context.
>
> *Field Grading.* The assessment of the particle size distribution by the method of moistening and manipulation of the soil material in the field.

These definitions should satisfy the needs of the field pedologist and of workers engaged in pedographic studies.

This concept of texture is not as essential as a basis for pedographic work as are the concepts of soil structure and soil fabric; the method can be applied within the framework of structure and fabric, and crystallinity can be described as a separate characteristic. It would be most convenient, however, to be able to use the term *texture* in relation to pedography, especially as it is an accepted petrographic term.

7.5 THE UNITS OF ORGANIZATION

The organization of a soil material is expressed by the development of three units of organization: *peds, pedological features,* and *s-matrixes,* each of which can be treated separately. The first step in structure analysis of soil materials is recognition and description of these units.

A. Peds

Peds are the units used for description of "field structure" in the *U.S.D.A. Soil Survey Manual* (1951). Although not defined specifically peds are referred to as "compound particles, or clusters of primary particles, which are separated from adjoining aggregates by surfaces of weakness. The exteriors of some aggregates have thin, often dark-colored, surface films which perhaps help to keep them apart. Other aggregates have surfaces and interiors of like color, and the forces holding the aggregates together appear to be wholly internal." It is stated also that "an individual natural soil aggregate is called a *ped,* in contrast to (1) a *clod,* caused by disturbance, such as plowing or digging, that molds the soil to a transient mass that slakes with repeated wetting and drying, (2) a *fragment* caused by rupture of the soil mass across natural surfaces of weakness, or (3) a *concretion* caused by local concentrations of compounds that irreversibly cement the soil grains together" (U.S.D.A., 1951). The salient points of these statements are that a ped is an individual natural soil aggregate separated from adjoining aggregates by surfaces of weakness, which may or may not be recognizable by their appearance.

Fig. 13. Macropeds of similar size and shape with offset packing. Monolith (after Sleeman, 1963). *Courtesy Aust. J. Soil Res.* ¾×.

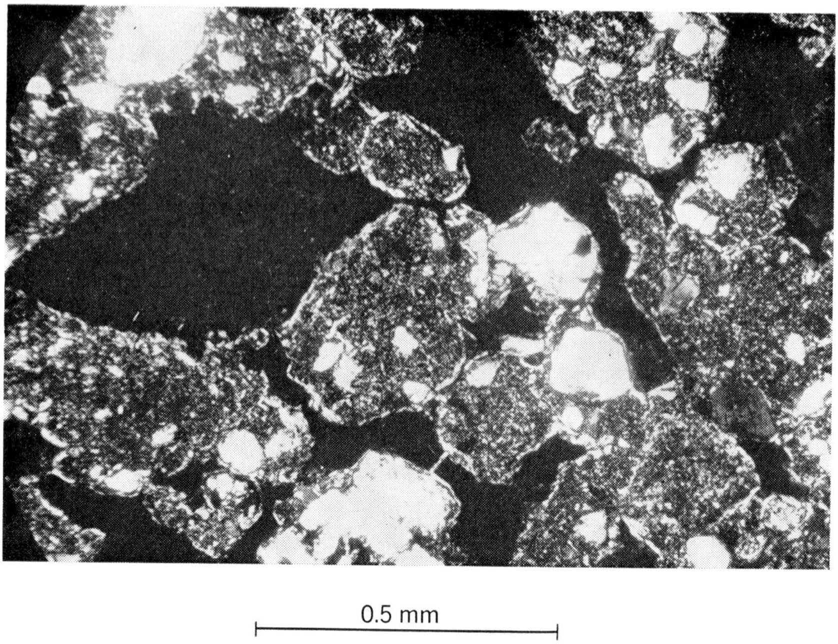

Fig. 14. Micropeds of similar size and shape. Thin section under crossed polarizers. The thin white borders to the peds are ped argillans with strong, continuous orientation. The black areas are compound packing voids due to the packing of the micropeds. 80×.

This concept has arisen from observations made in the field that are conditioned by the methods used. The main criterion of pedality of a soil material, in this sense, is the presence or absence of a population of units of similar size and shape after application of a force to a mass of the soil material. If a population of such units results, the material is *pedal* (consists of peds), if not it is *apedal* (Butler, 1955); the occurrence of surfaces of weakness is inferred from the occurrence of a population of such units since they cannot always be recognized by their appearance. This basis for recognition is inferred in the statement: "breakage into fragments larger than the soil grains without some orderly shape and size should not be confused with soil structure" (*U.S.D.A. Soil Survey Manual,* 1951). This same criterion was used by Butler (1955). Inherent in this method is the assumption that surfaces of weakness always occur in a regular pattern throughout the soil material, thus producing units of similar size and shape. There is some basis for assuming that cracking on drying will occur in a uniform pattern throughout homogeneous, fine-grained materials, but review of the literature suggests very strongly that crack patterns are highly irregular in more heterogeneous materials, as many natural soil materials are (Sleeman, 1963).

In contrast to this deductive method of recognizing surfaces of weakness and peds, natural surfaces in soil materials can be distinguished from fracture surfaces (Sleeman, *loc. cit.*) by their characteristics, as compared with those of a purposely produced fracture surface, in the field, on hand specimens, and in thin sections. The criteria used are the pattern of voids in the soil material and observable characteristics of the surfaces including smoothness, luster, occurrence of coatings, presence or absence of vesicles and shrinkage cracks, and the relationships of the skeleton grains to the surface; most of these characteristics are those associated with one or another of the various kinds of cutans (Chapter 10), which may be so thin that they may be recognized most readily in reflected light. True peds, then, are entirely enclosed by

Fig. 15. Cuneate macropeds of dissimilar size separated by inclined sets of joint planes. Monolith (after Sleeman, 1963). *Courtesy Aust. J. Soil Res.* ¾×.

Fig. 16. Macropeds of very dissimilar size and shape separated by craze planes. Monolith (after Sleeman, 1963). *Courtesy Aust. J. Soil Res.* ¾×.

surfaces of weakness, or natural surfaces, which are recognizable by these characteristics, and a ped is defined as follows:

> "A *ped* is an individual natural soil aggregate consisting of a cluster of primary particles, and separated from adjoining peds by surfaces of weakness which are recognizable as natural voids or by the occurrence of cutans (Brewer, 1960*b*)" (Sleeman, 1963).

This is an objective definition that enables recognition of peds by the characteristics of their surfaces without inferences drawn from the occurrence or otherwise of populations of units of relatively uniform size and/or shape.

This concept of a ped was derived from examination of a number of soils from western New South Wales, which showed that production of populations of units of similar size and shape by applying forces to a soil material is not a dependable criterion for the recognition of peds. In some soil materials true peds with recognizable surfaces of weakness, or natural surfaces, occurred in relatively wide ranges of size and shape. Other soil materials were essentially apedal, but the distribution of the megascopic voids, especially planar voids, predisposed the soil material to break into units of similar size and shape; examination of the surfaces of such units, however, showed that, while some were natural surfaces, others were fracture surfaces. Other soil materials, of course, did break down to form populations of true peds of similar size and shape in which all the surfaces of the peds were natural surfaces or surfaces of weakness. A number of pedal and apedal soil materials are depicted in Figs. 13–18. Figures 13 and 14 are of soil materials with true peds of similar size and shape, Fig. 13 showing macropeds and Fig. 14 micropeds. Figures 15 and 16 are of soil materials with true peds of dissimilar size and shape; in Fig. 15 the peds are due to the intersection of sets of subparallel planar voids, while in Fig. 16 the voids have a very irregular form. Figure 17 is an apedal material that breaks into fragments of similar size and shape when a force is applied, because of the distribution of planar voids. Figure 18 shows an apedal soil material that breaks into fragments of

Fig. 17. Apedal soil material with a set of joint planes. Monolith (after Sleeman, 1963). *Courtesy Aust. J. Soil Res.* ¾×.

Fig. 18. Apedal soil material. Thin section under crossed polarizers. The soil material is very friable because of the arrangement of strongly interconnected vughs (black areas). 30×.

similar size and shape because of the distribution of irregular, interconnected vughs (Section 9.5:B).

B. Pedological Features

Fractionation and reorganization of the mobile, active plasma by soil-forming processes may result in formation of coatings of various fractions of the plasma on the walls of voids, cementing of the constituents in specific locations in the soil material to form such recognizable entities as nodules and concretions, changes in arrangement of the constituents in specific locations in the soil material, and other manifestations. Such features are directly analogous to specific arrangements or organizations observed in sedimentary rocks that Shrock (1948) referred to as *sedimentary features,* such as cone-in-cone structure, oolitic structure, animal tracks and trails, and concretions. Much of the complexity of pedographic description can be handled

Fig. 19. Plasma concentrations. Hand specimen. The dark, cracked feature filling the left half of the figure is a thick coating of clay minerals with organic staining on a vertical joint plane (joint-plane organo-argillan). $1\frac{3}{8}\times$.

relatively simply by introducing the concept of *pedological features,* which can be described in terms of their size, shape, and internal organization without reference to the rest of the soil material, and whose distribution and arrangement can be described in relation to the organization of the constituents of the rest of the soil material. These units are defined in the following way:

Pedological Features. Recognizable units within a soil material which are distinguishable from the enclosing material for any reason, such as origin (deposition as an entity), differences in concentration of some fraction of the plasma, or differences in arrangement of the constituents (fabric).

This definition is not intended to include peds, but it does include features inherited from the parent rock or formed by the processes of deposition of transported parent material.

Fig. 20. Plasma concentration. Hand specimen. The black areas are thin discontinuous coatings of manganese oxides on planar voids (plane mangans). $1\frac{3}{8}\times$.

Pedological features can be divided into three groups according to their pattern of occurrence with regard to peds: *intrapedal,* which occur entirely within peds, *interpedal,* which occur between peds, and *transpedal,* which traverse the soil material without specific relationships with the peds. They can also be divided into two broad classes according to their origin: *orthic features* and *inherited features;* inherited features may be *in situ* or transported (Brewer and Sleeman, 1964). Each of these groups can be subdivided into three groups as follows:

(1) *Orthic pedological features* are formed *in situ* in the soil material by soil-forming processes; three main groups have been recognized so far:

Plasma Concentrations. Concentrations of any of the fractions of the plasma in various parts of the soil material due to soil forma-

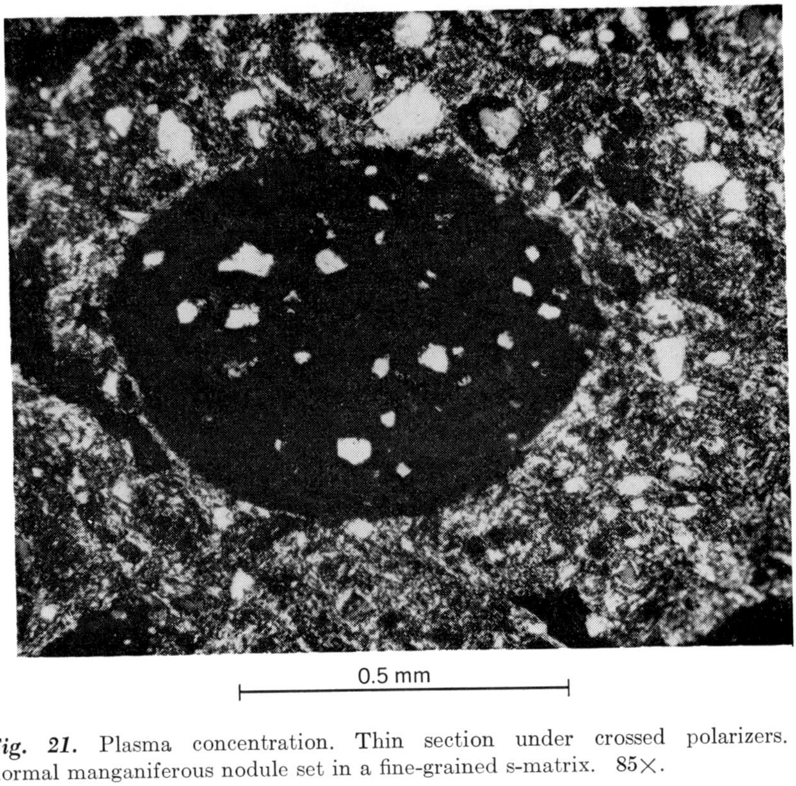

Fig. 21. Plasma concentration. Thin section under crossed polarizers. A normal manganiferous nodule set in a fine-grained s-matrix. 85×.

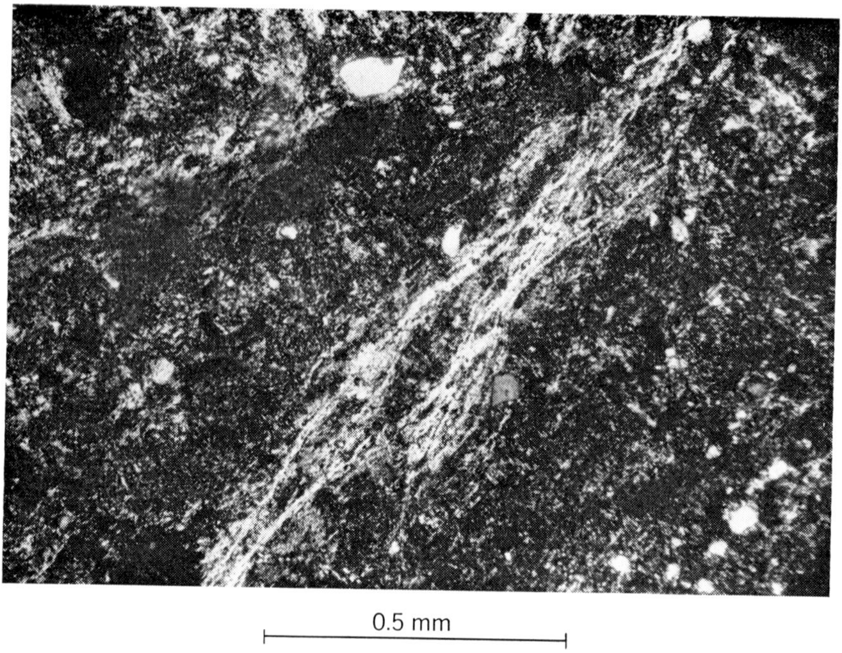

Fig. 22. Plasma separation. Thin section under crossed polarizers. The diagonal striated zone is a plasma separation having a different arrangement of the plasma grains than in the surrounding material. The whole is an s-matrix with masepic plasmic fabric. 80×.

tion. Examples of these are carbonate nodules, iron oxide nodules, and clay mineral coatings.

Plasma Separations. Features characterized by a significant change in the arrangement of the constituents rather than a change in concentration of some fraction of the plasma. An example of this is the change in orientation of the clay mineral fraction near the surface of slickensides. Plasma separations are not three-dimensional *entities,* but their internal organization can be described and related to that of the rest of the soil material.

Fossil Formations. Preserved features resulting from biological activity such as burrows and root channels. These may or may not be filled with various materials.

(2) *Inherited pedological features* are relicts of the parent rock or parent material or of underlying materials; three groups have been recognized so far:

Fig. 23. Fossil formation. Hand specimen (after Brewer and Sleeman, 1963). The curved, generally horizontal feature is a filled faunal burrow (curved, porous, weakly adhesive isotubule) embedded in an s-matrix. *Courtesy J. Soil Sci.* 1¾×.

Lithorelicts. Features derived from the parent rock; usually recognizable by their rock structure and fabric, for example, "floaters" of parent rock.

*Pedorelicts.** Features formed by erosion, transport, and deposition of nodules of an older soil material or pedological features from it, or by preservation of some part of a previously existing soil horizon within a newly formed horizon (Brewer and Sleeman, 1964).

Sedimentary Relicts. Features formed during deposition of a transported soil parent material, for example, the clay galls of Pettijohn (1957).

* Pedorelicts, as defined by Brewer and Sleeman (1960), were confined to *pedological features* formed in older soil materials. Subsequent observations have revealed the occurrence of transported inherited nodules of soil material and *in situ* inherited nodules of *B* horizon material in the A_2 horizons of strongly differentiated profiles; both of these types of nodules are properly pedorelicts.

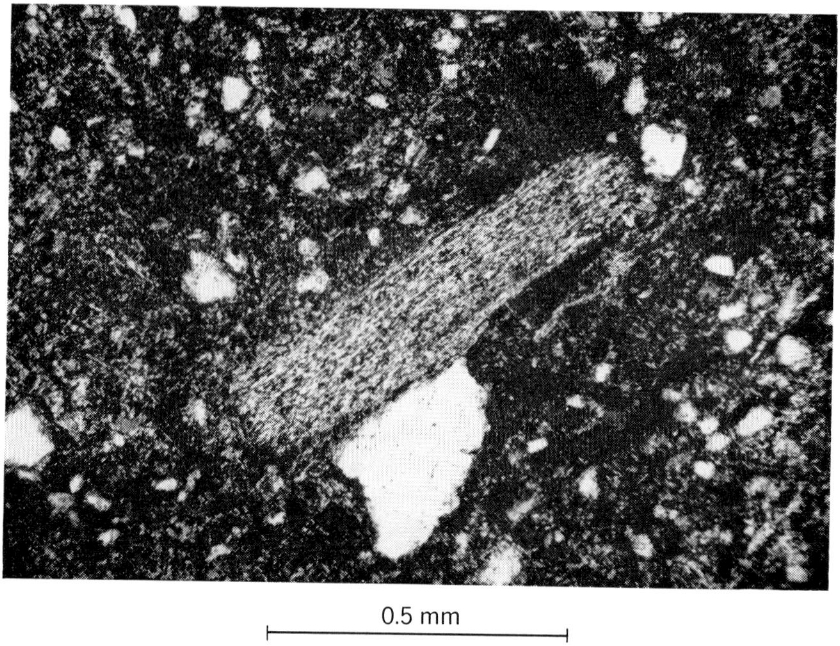

Fig. 24. Lithorelict. Thin section under crossed polarizers. The central elongated feature with striated birefringence is a nodule of shale set in an s-matrix with flecked birefringence. 80×.

Examples of each of these classes are shown in Figs. 19–26. Specific types of pedological features are shown in the figures in Chapters 10–14.

The two broad classes of pedological features, and the three subdivisions within each, are based on mode of origin rather than morphological features, so classification of any observed feature into one or other of these groups depends partly on interpretation. Within each group specific features with characteristic form and internal organization can be described and named. This aspect of the morphological classification of pedological features is dealt with later (Chapters 10–13).

C. S-matrix

In an earlier paper on soil structure and fabric (Brewer and Sleeman, 1960) the concept of "matrix" or "groundmass" was not used specifically but was inferred in the concepts of basic structure and fabric. Later it was found convenient to use some specific term for the material in which pedological features are usually embedded within peds or in unaggregated soil materials; the term *matrix* was proposed for this material (Brewer 1960*a*). Both "groundmass" and "matrix" have been used in geology for parts of rocks similar to this part of a soil material, but the geological definitions infer some differences in concept.

In petrology, groundmass is "the relatively fine crystalline or glassy portion of a porphyritic rock as contrasted with its phenocrysts" (Rice, 1954; after Fay, 1920), and "porphyritic" is now used only for igneous rocks. Although soil materials are somewhat analogous to porphyritic rocks in that they often consist of a mass of fine plasma with embedded skeleton grains, they usually have a continuous distribution of grain sizes; that is, there are some representatives of all grain sizes from the largest to the smallest present. "Groundmass," therefore, is not a suitable term to apply to soil materials.

Matrix is not defined nearly so specifically in general geology (Rice, 1954); it is simply a material that encloses other recognizable features or materials. In sedimentary petrology, however, "matrix" has a more specific meaning. Sedimentary rocks are considered to consist of a framework of mineral grains with finer-grained material packed between the members of the framework; this finer material is termed the "matrix" (Pettijohn, 1957). The grain size distribution of this matrix is defined only in relationship to the size of the members of the framework, and its use in pedography is seriously limited because of the continuous particle size distribution in the majority of soil materials; the morphological distinction between framework members and matrix is not clear although most sedimentary petrologists would accept an upper size limit of 30 μ for the matrix (Spencer, 1963).

"Matrix," therefore, is preferable to the more specifically defined geological term "groundmass," but to avoid confusion with "matrix" as used in sedimentary petrology *s-matrix* is proposed (i.e., matrix including skeleton grains). It can be defined as follows:

> *S-matrix* of a soil material is the material within the simplest (primary) peds, or composing apedal soil materials, in which the pedological features occur; it consists of the plasma, skeleton grains, and voids that do not occur in pedological features other than plasma separations.

Plasma separations are treated as part of the s-matrix because they are not recognizable as three dimensional entities (Chapter 14). This

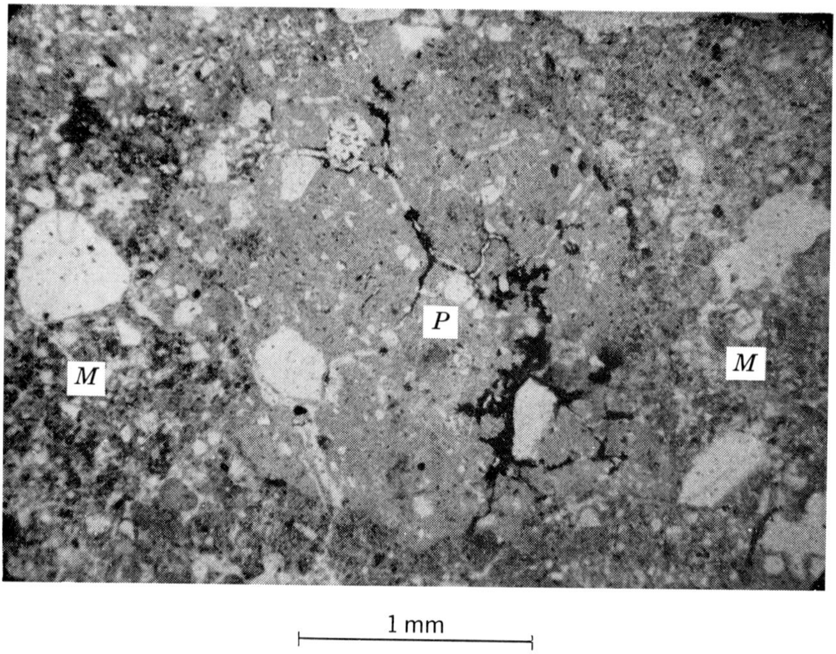

(a)

Fig. 25. Pedorelict. (*a*) Thin section in plain light. (*b*) Thin section under crossed polarizers. A large, rounded nodule (*P*) of soil material set in a different s-matrix (*M*). Note the black manganiferous coatings on planar voids

is analogous to the use of matrix and groundmass in petrology where variations in arrangement, such as flow structures, are considered as part of the organization of the matrix or groundmass. It should be noted that the concept of *s-matrix* can be applied to sedimentary rocks for that material in which features such as nodules, oolites and concretions are enclosed, and the sedimentary petrological concept of *matrix* can be used in pedography. Also, just as a soil material has an s-matrix in which the pedological features occur, pedological features themselves have an s-matrix in which other smaller features may occur. This aspect is treated in more detail when considering the levels of organization (Section 7.6).

The general concept of an s-matrix is illustrated in Figs. 21–26; numerous other examples are shown in the figures in Chapters 10–13, and the details of a number of specific types in Figs. 99–112 in Chapter 14.

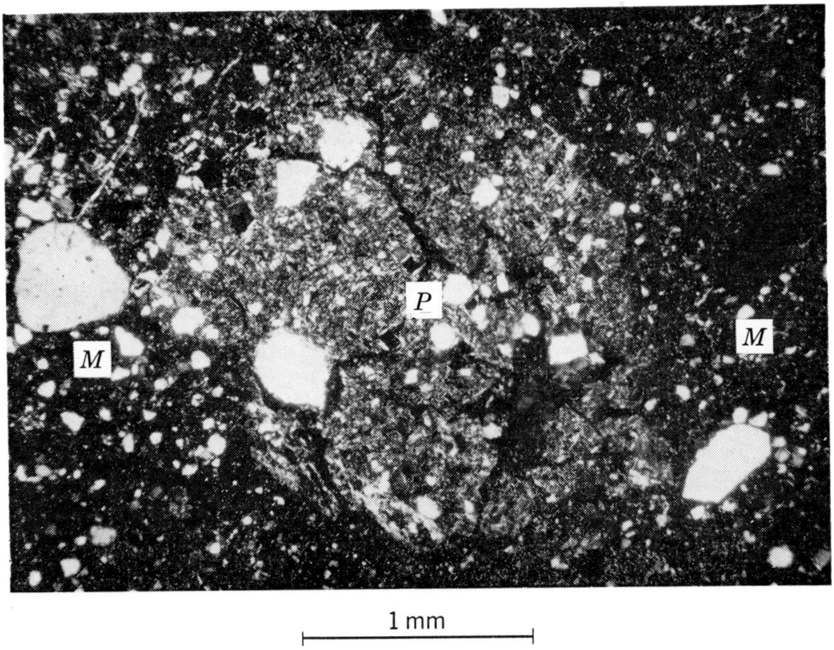

(b)

(plane mangans) of the s-matrix of the pedorelict which are absent from the s-matrix of the enclosing soil material. (*a*) 30×; (*b*) 30×.

7.6 LEVELS OF ORGANIZATION

To treat a soil material according to the levels of organization developed in it, a standard starting point or unit of description must be established. Soil materials may be apedal; that is, there are no peds, and the whole soil material is the basic unit of description. Alternatively they may consist of simple peds packed together in a simple manner so that no larger units are recognizable, or of simple peds packed together in such a manner as to form larger recognizable units which, in turn, may be packed in such a way as to form even larger recognizable units. The levels of organization due to the arrangement of peds, therefore, can be based conveniently on the size, shape, and arrangement of the simplest peds. Also, the s-matrix of soil materials with such complex arrangements of peds is, by definition, the material within the simplest peds exclusive of pedological features other than plasma separations. These simple peds, therefore,

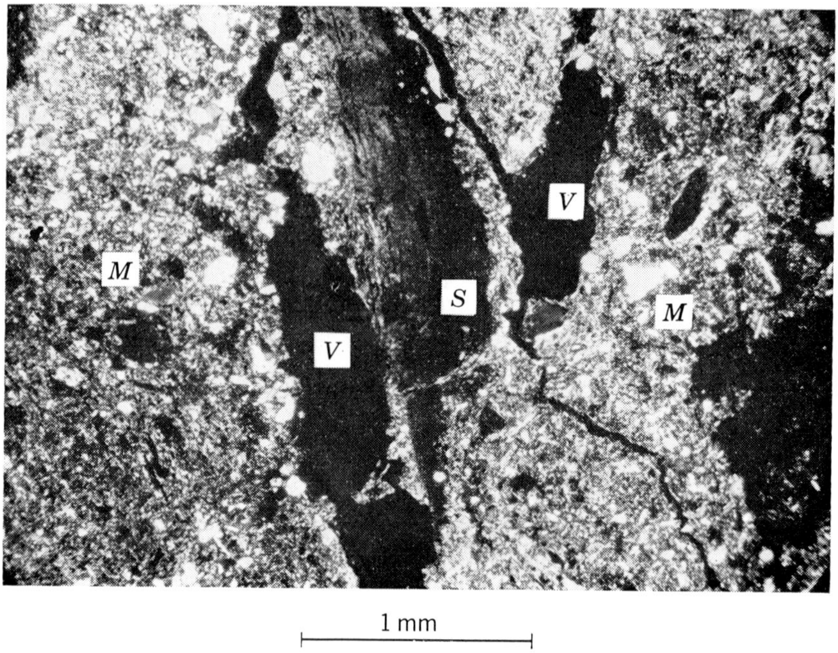

Fig. 26. Sedimentary relict. Thin section under crossed polarizers. A laminated strongly oriented clay flake or papule (*S*) set in an s-matrix (*M*). Black areas (*V*) are voids. 30×.

are convenient basic units for pedographic descriptions of pedal soil materials; they are termed *primary* peds and are defined as follows:

> *Primary Peds.* The simplest peds occurring in a soil material; they cannot be divided into smaller peds, but they may be packed together to form compound peds of a higher level of organization.

According to previously accepted concepts (*U.S.D.A. Soil Survey Manual,* 1951) primary peds would be the smallest recognizable peds, but the simplest peds in a soil material may vary considerably in size and shape (Sleeman, 1963), so the definition given here is rather more appropriate than an earlier one (Brewer and Sleeman, 1960), which referred to the smallest instead of the simplest ped.

Having established the primary ped and/or apedal soil material as the basic units of description, structure analyses can be made at a number of levels of organization, either independently of each other or in a logical progression. Kubiena (1938) applied this principle in splitting fabrics into "elementary fabric" and "fabrics of higher order," but the principle has far wider application than this. The following treatment and definitions are somewhat generalized without statements of the specific characteristics, especially as regards arrangement, which should be described. It is apparent from the definitions of the levels of structure, and from the nature of soil materials, that structure analysis is concerned with units with very different properties: plasma, skeleton grains, peds, pedological features, and

TABLE 7-2. The Levels of Structure

Structure Level	Unit of Organization	Components Described
Plasmic structure	S-matrix* (including plasma separations)	Plasma†
Basic structure	S-matrix* (including plasma separations)	Plasma,† skeleton grains, voids
Elementary structure	Primary ped	Plasma,‡ skeleton grains, voids (s-matrix), specific intrapedal pedological features as units
	Host pedological feature	Plasma,‡ skeleton grains, voids (s-matrix), specific included pedological features as units
Primary structure	Primary ped	Plasma,‡ skeleton grains, voids (s-matrix), all intrapedal pedological features as units
	Host pedological feature	Plasma,‡ skeleton grains, voids (s-matrix), all included pedological features as units
Secondary structure	Secondary ped	Primary peds as units, interpedal voids, interpedal pedological features
	Host pedological feature	Plasma, skeleton grains, voids, included pedological features and secondary included pedological features as units
Tertiary structure	Tertiary ped	Secondary peds as units, interpedal voids, interpedal pedological features

* S-matrix of soil material or pedological feature

† Excluding pedological features other than plasma separations

‡ Plasma of the s-matrix

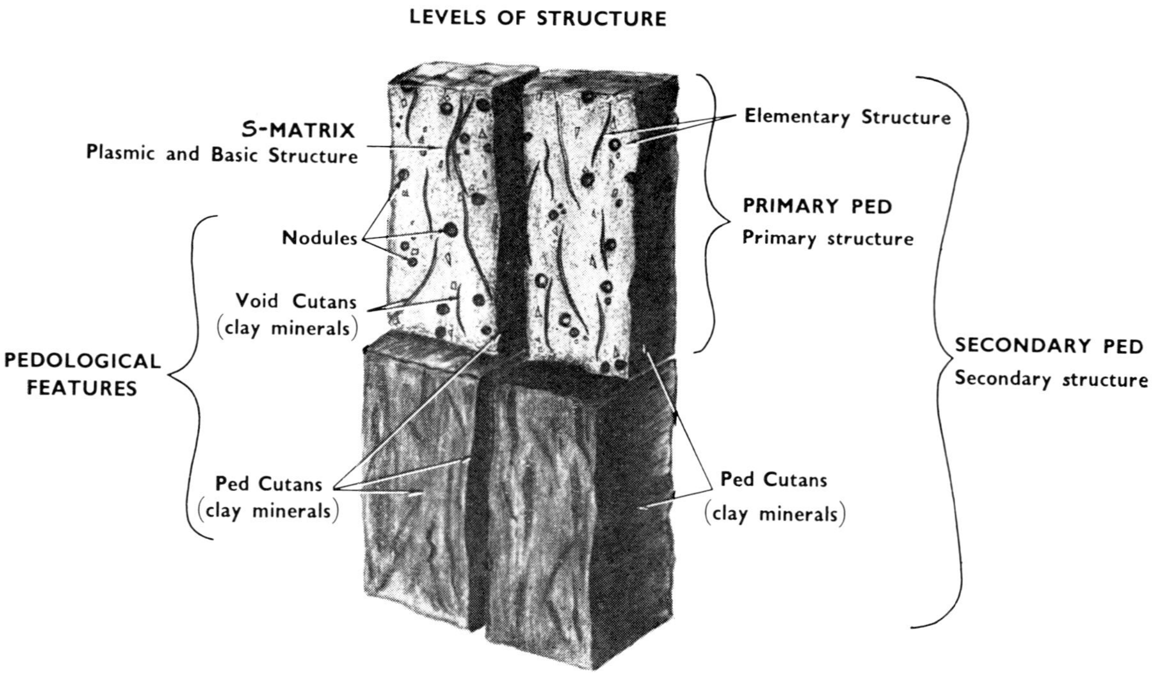

Fig. 27. Illustrations of the levels of structure in a complex, pedal soil material. *Courtesy Trans. Seventh Intl. Cong. Soil Sci.*

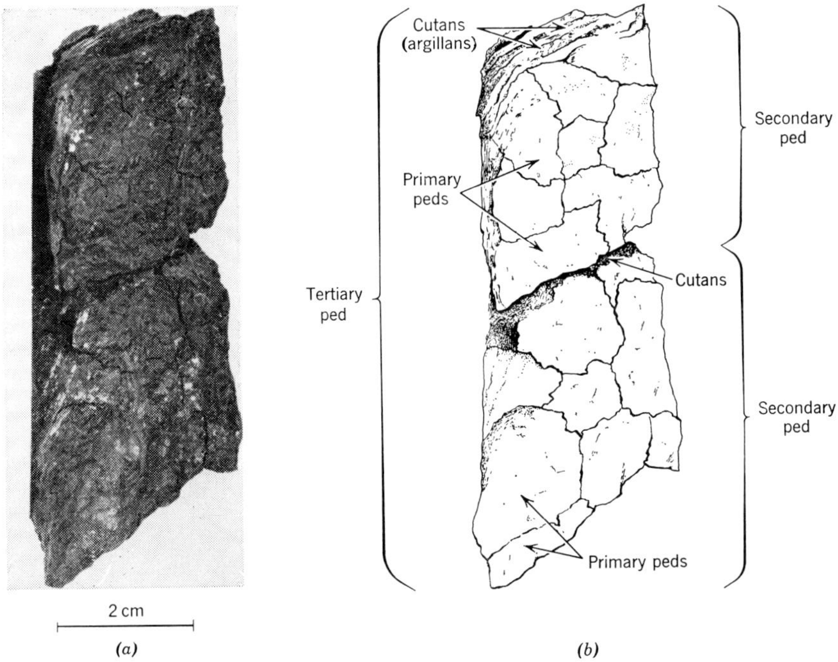

Fig. 28. Structure of a compound ped. (*a*) Hand specimen. (*b*) Illustration. ⅞×.

voids. Each of these features must be described in terms of their size, shape, and arrangement, but different principles govern the methods used to describe each of them because of their different characteristics; a full treatment of such characteristics is given elsewhere (Chapters 9–14).

The defined levels of structure, the units of organization in which they occur, and the components to be described are summarized in Table 7-2; the relationships of the structure levels are illustrated in Figs. 27–30. The definitions deal only with structure, but there is an equivalent fabric level for each structure level, fabric being the aspect of structure that deals with arrangement.

A. *Organization within the Basic Unit*

The levels of organization that occur within the basic unit (the primary ped or an apedal soil material) are expressed by:

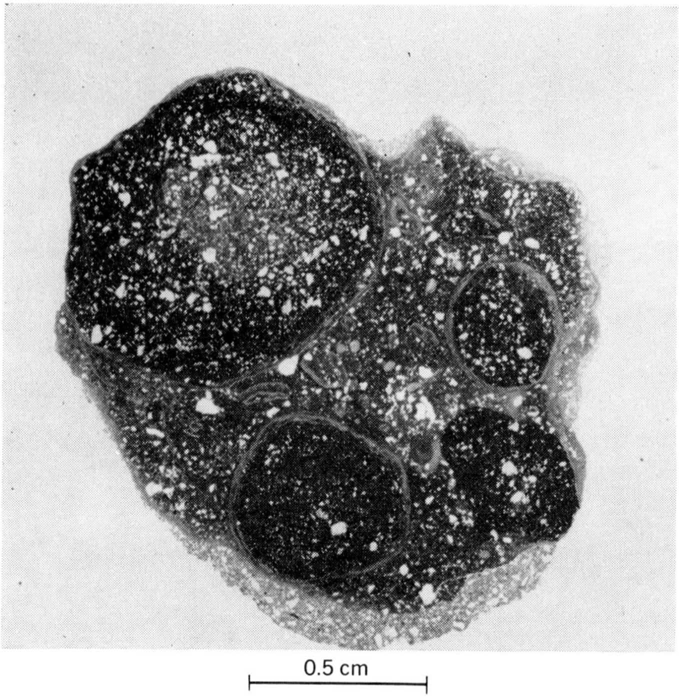

Fig. 29. Compound pedological feature (sesquioxidic nodule). Thin section in plain light. The large feature is a sesquioxidic nodule (host pedological feature). The dark, almost circular features are included sesquioxidic nodules. The thin zones bordering the included nodules are coatings of clay minerals (included argillans). 4¾×.

(1) the organization within the s-matrix, which can be considered as a whole (plasma, skeleton grains, and voids) or as separate parts so that the organization of the plasma can be treated independently (Brewer, 1960*a*);

(2) the kind, size, shape, and arrangement of the pedological features included in the s-matrix.

The s-matrix can be considered as the basic material of the soil material containing expressions of rather specific processes or relicts of other materials which are represented by the included pedological features. It is appropriate, therefore, to refer to the organization within the s-matrix as the *basic structure:*

Basic Structure. The structure of the s-matrix, that is, the size, shape, and arrangement of simple grains (plasma and skeleton grains) and voids in primary peds or apedal soil material, excluding pedological features other than plasma separations.

The organization of the plasma of the s-matrix can be referred to as the *plasmic structure:*

Plasmic Structure. The structure of the plasma of the s-matrix, that is, the size, shape, and arrangement of the plasma grains and associated simple packing voids (Chapter 14).

The next level of organization within the basic unit is that which includes the pedological features associated with the s-matrix; that is, it is the complete internal structure of the basic unit; it is termed the *primary structure:*

Primary Structure. The structure within an apedal soil material or within the primary peds in a pedal soil material; it is an integra-

Fig. 30. Internal fabric of part of a compound pedological feature (sesquioxidic nodule). Thin section under crossed polarizers. The plasma of the host feature (*M*) is virtually isotropic. The included features (*A*) are coatings of strongly oriented clay minerals (argillans). The large anisotropic grains (*S*) are the skeleton grains of the host feature. 30×.

tion of the size, shape, and arrangement of all the pedological features enclosed in the s-matrix and the basic structure, or structure of the s-matrix.

A level of organization between primary and basic structure has proved useful as a convenience in description of complex soil materials; it is termed the *elementary structure:*

Elementary Structure. A simplified level of primary structure; it is an integration of a *characteristic* size, shape, and arrangement of *specific* pedological features and the basic structure, or structure of the s-matrix.

In primary and elementary structure the detail of classification of the pedological features can be varied according to the purpose of the description. At one extreme the pedological features can be treated simply as entities whose size, shape, and arrangement are described without any reference to their internal structure. At the other extreme they can be classified in great detail according to their internal fabric and composition (Part IV), and then the size, shape, and arrangement of specific, restricted groups of pedological features can be described as part of primary and elementary structure.

B. Organization between Peds

Pedal soil materials can exhibit additional levels of organization due to the occurrence of peds and interpedal pedological features. These levels are expressed by: (1) the size, shape, and arrangement of the primary peds and interpedal voids, (2) the size, shape, and arrangement of compound peds and their associated interpedal voids, and (3) the relationships (size, shape, and arrangement) of the interpedal pedological features with regard to the peds, whether they are primary or compound peds.

Levels of organization beyond the primary level, therefore, deal with recognizable peds, interpedal pedological features, and voids. Few soil materials are likely to exhibit more than two such levels, the *secondary and tertiary structure:*

Secondary Structure. The size, shape, and arrangement of the *primary* peds, their interpedal voids, and associated interpedal pedological features in a soil material (Fig. 28).

Tertiary Structure. The size, shape, and arrangement of the *secondary* peds of a soil material (compound peds resulting from the packing of primary peds), their interpedal voids and associated interpedal pedological features (Fig. 28).

Higher levels of organization due to the presence of even more complex peds are theoretically possible depending on the feasibility of recognizing these more complex units.

It is noteworthy that these higher levels of organization are analogous to the basic and primary levels if the peds of the appropriate level are considered as individuals analogous to skeleton grains at the basic level, the interpedal voids analogous to intrapedal voids, and all the plasma occurring as pedological features. The only analogue absent is that of the plasma of the s-matrix. Some soil materials are dominantly apedal but contain occasional isolated peds which then are analogous to pedological features set in an s-matrix of apedal material.

C. Organization within Pedological Features

The levels of organization within a pedological feature are analogous to those within a primary ped. A pedological feature is usually first recognized as an entity, such as a nodule or concretion, and levels of organization are manifested in the internal organization of such units. Many pedological features have relatively uniform internal fabrics, so their internal organization is analogous to that of primary peds which have no pedological features. Others, however, have quite complex organizations, due to the occurrence of included pedological features (Brewer, 1960*a*) and exhibit several levels of organization. This arrangement is analogous to the occurrence of intrapedal pedological features in a soil material in that the smaller features can be considered as enclosed in the s-matrix of the larger feature, just as the latter is enclosed in the s-matrix of the soil material (Figs. 29 and 30). The larger feature is referred to as the *host pedological feature* and the smaller ones as the *included pedological features.* This kind of arrangement may be exhibited at an even higher level of complexity in that an included pedological feature may act as the host for even smaller features; these are referred to as *secondary host pedological features* and *secondary included pedological features,* respectively. Included pedological features are often similar in shape and internal structure to the host feature, but the host feature may contain more than one kind of included feature.

The organization of such complex pedological features can be described in a manner analogous to the treatment of the basic units of soil materials. Each feature, whether host or included, has an s-matrix and so a *plasmic* and a *basic structure.* If there are included pedological features the *primary structure* of the larger feature (the host) is an integration of the kind, size, shape, and arrangement of the included features and the basic structure of the host feature. If the included features are simple types they have only plasmic and basic structure; if they are secondary hosts to secondary included features they exhibit primary structure as well as plasmic and basic structure, and the original or *primary host* feature exhibits a *secondary structure,* which is an integration of the basic structure of the secondary included features, the size, shape, and arrangement of the secondary included features, the basic structure of the secondary host features, the size, shape, and arrangement of the secondary host features, and the basic structure of the primary host feature. Examples of *compound pedological features* are shown in Figs. 29 and 30.

D. Complete Organization

The complete organization in a soil material, or the *complete structure,* is the size, shape, and arrangement of all the pedological features in relation to the structure at all levels of organization; it is an integration of the size, shape, and arrangement of *transpedal* pedological features and the structure at all levels of organization, including the levels within pedological features.

8

Principles of Fabric Analysis

Billings (1942) defined petrofabric analysis as "the study of the internal structure of rocks and its relations to the movements involved in their formation and deformation." In petrology, fabric analyses are concerned primarily with the distribution and orientation of the constituents of the rocks, whether they be pebbles or mineral grains in sediments or mineral grains in igneous rocks, and whether their relationships are measured according to the physical dimensions or crystallography of the individuals.

In pedology, fabric analysis involves both distribution and orientation patterns at all levels of organization. It is the most complex part of the description of structure because it involves not only the arrangement of the constituents (plasma, skeleton grains, and voids) with regard to each other, but also the arrangement of individuals with regard to each other within a group of like individuals (*basic* arrangement), and between groups of unlike individuals (*related* arrangement), and the arrangement of individuals with regard to *specific reference features* (*referred* arrangement); specific reference features are features that can be specified in any particular soil material, such as the surfaces or shapes of peds and pedological features, the walls of voids, interhorizonal contacts of the soil profile, the surface of the soil, the horizontal and vertical, etc. The reference features are different for different soil materials, and each one is used for the description of fabric only if there is a positive correlation between it and the arrangement of the entities being described.

This complexity inherent in fabric analysis of soil materials can be treated adequately by considering each phase separately. There are two separate aspects of arrangement (fabric) that apply to all the individuals (mineral grains, voids, and compound units) in a soil material at all levels of organization: *distribution* and *orientation patterns.* The treatment of distribution patterns is relatively simple; most of the complexity of fabric analysis is associated with the description of orientation patterns.

8.1 DISTRIBUTION PATTERN

The treatment of distribution patterns entails little more than recognition, description, and classification of the individuals whose distribution pattern is to be described and definition of adequate terms for observed patterns. As far as possible such terms should be general ones so that they can be applied to the patterns of any kind of individuals recognizable in soil materials, that is, plasma and skeleton grains, peds, pedological features, and voids.

A. *Basic Distribution Pattern*

This is the distribution pattern of like individuals with regard to each other; like individuals may be distributed randomly throughout the soil material at approximately equal or randomly varying distances, or they may occur in characteristic arrangements (basic distribution patterns), for example, in groups in which the distance between individuals within a group is less than the distance between groups, in bands in which the distance between individuals within a band is less than the distance between bands, or in radial or concentric arrangements.

B. *Referred Distribution Pattern*

This is the distribution pattern of like individuals with regard to a specific reference feature. Any of the basic distribution patterns, except that in which the individuals are randomly distributed, may exhibit a characteristic arrangement (referred distribution pattern) in relation to one or more prominent directions or surfaces within the soil material or to the surface of the soil, the horizontal, or the vertical. The distribution pattern of like individuals also has a relationship to the occurrence of specific phenomena, especially peds, and so individuals and basic distribution patterns can be described as being within specific entities (*intra-*, e.g., intrapedal), between specific entities (*inter-*, e.g., interpedal), or independent of and cutting across specific entities (*trans-*, e.g., transpedal).

C. Related Distribution Pattern

This is the distribution pattern of like individuals with regard to the distribution of groups of individuals of a different kind; it can usually be inferred from the referred distribution patterns of the two groups of individuals exhibiting the relationships. The characteristics described are very similar to those described in referred distribution pattern except that the reference feature is the basic distribution pattern of another group of individuals, so few, if any, additional terms are required to describe related distribution patterns, except for the distribution of plasma grains in relation to skeleton grains in basic fabric [Section 8.3: A(*c*)].

The nature of these distribution patterns is such that a quantitative treatment is hardly possible except for certain aspects that can be treated in a partially quantitative manner. If a group of like entities exhibits a basic distribution pattern characterized by some obvious pattern of grouping, the intensity of this grouping can be expressed by the ratio of the average distance between individuals within a group to the average distance between groups. Certain referred distribution patterns lend themselves to similar measurements, for example, entities arranged parallel to a specific reference feature, such as the surface of a ped, and at a distance from the surface that is a fixed proportion of some dimension of the ped.

8.2 ORIENTATION PATTERN

Theoretically the orientation of all the individuals in soil materials can be described and measured as the proportion by number oriented within certain limits. In practice, however, such measurements can be made only if the techniques used enable direct observation of the individuals. With magnifications that can be achieved with the petrological microscope, orientation analyses can be made and orientation diagrams drawn only for individuals with a lower size limit of approximately 5 μ diameter or, under ideal conditions, of approximately 2 μ diameter; this excludes all clay size individuals. With electron microscopic techniques orientation patterns of even the smallest individuals in soil materials can be described directly, but, failing this, the orientation pattern must be inferred from other characteristics, such as extinction phenomena under crossed nicols on the petrological microscope. Little or nothing, however, can be said of the orientation pattern of isotropic or opaque clay size plasma grains by optical techniques.

The orientation of megascopic and microscopic individuals* is described by measurement of the directions of the principal axes of the individuals. The axes used for measurement may be one or more of the principal crystallographic axes for mineral grains, or one or more of the principal axes of shape for any individual. For the latter the long axis, as used for determination of shape, is usually used, and its azimuth and angle of dip are measured. For entities that are flattened (disc- or plate-shaped), however, the longest axis may not indicate preferred orientation, whereas parallelism of the largest faces of the individuals may; the azimuth and angle of dip of the pole of the largest face can be used; this is the requirement for micas. Other cases may be more complex and may require measurement of more than a single axis or plane, for example, individuals that have both a prominent long axis and a large flat face; the orientation of such particles is completely stated only by measurement of the azimuth and angle of dip of both the long axis and the pole of the flat face (the normal to the maximum projection area).

Interpretation of extinction phenomena of the plasma in terms of preferred orientation depends on knowing the optical properties of the mineral species involved. Under crossed nicols on the petrological microscope, anisotropic grains are extinguished or illuminated when they are arranged in particular directions with regard to the vibration directions of the polarizer and analyzer. Uniform extinction or illumination, therefore, indicates a high degree of preferred optical orientation. The majority of the mineral species in the plasma exhibiting preferred orientation are the true clay minerals, especially those with a layer-lattice structure and those with an elongated form. These mineral species have a number of optical and physical properties in common that enable a considerable amount of information concerning their preferred orientation patterns to be obtained from the optical properties of the plasma.

The clay minerals with a layer-lattice structure generally have a micaceous, or platy, form, and the crystallographic and optical axes are related to this physical form in a similar way to the relationships in the common micas (Fig. 31; Stephen, 1960). The two long physical axes lie in the 001 crystallographic plane, which is parallel to the cleavage plane. This plane also contains the two optical directions that represent the vibration directions with almost equal refractive

* Megascopic and microscopic individuals include skeleton grains, inherited complex grains (pebbles, etc.), peds, pedological features, and voids that can be seen and measured megascopically or microscopically.

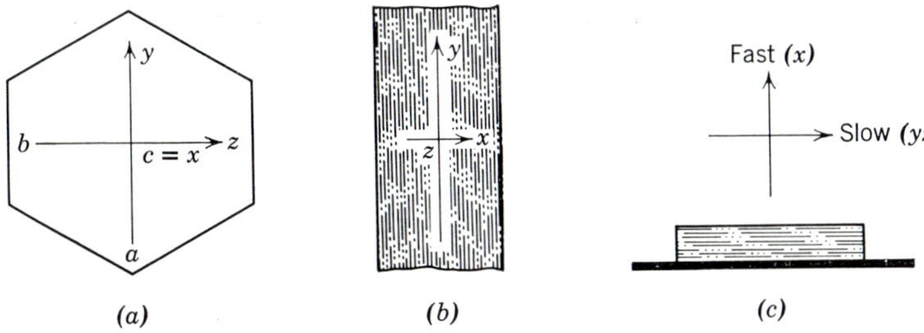

Fig. 31. Optical and crystallographic relationships of the micas (after Stephen, 1960). *Courtesy Sci. Prog.*

indexes. The shortest physical axis is perpendicular to this plane; it is essentially parallel to the *C*-crystallographic axis and to the vibration direction with the smallest refractive index in optically negative species and the largest in optically positive species. These minerals extinguish under crossed nicols when the cleavage traces, that is, the traces of the edges of the platy form, are parallel to the vibration directions of the polarizer or analyzer. The extinction positions, therefore, indicate that the vibration direction of either the polarizer or analyzer is in the plane of the flat face and, therefore, that these faces are parallel to each other. Also, because of the relationships of the physical axes to the crystallographic and optical axes, the minerals have a positive or negative elongation, according to the optical sign, measured in the direction of the cleavage traces. It is possible, therefore, to decide whether the orientation at a particular position of extinction is parallel to the vibration direction of the polarizer or analyzer simply by the normal procedure for determining the vibration directions of the fast and slow rays; this determination is greatly simplified because two of the refractive indexes are usually almost equal in these minerals. The majority of the layer-lattice minerals occurring in soil materials are optically negative.

A. Basic Orientation Pattern

This is the orientation pattern of like individuals with regard to each other. For individuals that can be observed directly it can be analyzed by the normal methods of petrofabric or orientation analysis (Pettijohn, 1957) based on statistical analysis of the directions of the principal axes of the individuals. For anisotropic plasma grains it must be inferred from extinction phenomena in thin section.

Since individual plasma grains are usually less than 2 μ equivalent diameter, and a thin section is approximately 25 to 30 μ thick, *any* anisotropism in the plasma in thin section must be due either to anisotropic crystals of certain fractions of the plasma larger than 2 μ or to some degree of optical orientation of the smaller anisotropic grains, that is, preferred orientation of the individual grains with regard to each other; relatively large crystals of plasma can usually be identified and distinguished from anisotropy due to preferred orientation. Plasma grains in soil materials, in fact, do commonly occur as compound anisotropic units that may be so small that they are recognizable only at high magnifications. Also, because of their small dimensions compared with the thickness of a thin section, they may have rather indistinct boundaries due to overlapping and superposition of several units. Recognition of these units makes it possible to describe the *basic orientation* of the plasma at two levels: the *degree* of basic orientation and the *pattern* of basic orientation.

The *degree of basic orientation* in s-matrixes in which plasma aggregates can be recognized is a measure of three factors: the size of the individual anisotropic units, the degree of preferred orientation of individual grains within the units, and the ratio of the area of the plasma in thin section occupied by recognizable anisotropic units to the area occupied by apparently isotropic material. Where plasma aggregates cannot be recognized it is simply a measure of the degree of parallel orientation of the individual plasma grains. Additional information concerning the degree of parallel orientation of the individual grains within the compound units can be obtained by observing the extinction phenomena within the units. If the preferred orientation of the grains is strong and linear the unit extinguishes suddenly as a whole as the microscope stage is rotated; all the grains extinguish in the one position. If, however, the grains have a strong preferred orientation that follows a curve, the high degree of parallel orientation of adjoining grains and the gradual change in direction of orientation across the unit are expressed by strong dark extinction bands that move in an orderly manner across the section of the unit as the microscope stage is rotated. The maximum darkness of these bands expresses the degree of parallel orientation of adjoining grains, and their width expresses the rate of change of direction of orientation with distance across the section of the unit.

The *pattern of basic orientation* is a measure of the degree of preferred orientation of the compound anisotropic units. If these units are oriented with regard to each other the plasma is more strongly illuminated in particular positions under crossed nicols. This usually causes a striated appearance, and the density and degree of parallelism of the striae is a measure of the degree of preferred orientation of the units. This varies from a distinct band of plasma which virtually extinguishes as a unit, or in a manner similar to a curved sheet of mica cut parallel to the *C*-axis, to faint, thin, broken, elongated patches of plasma which are illuminated and extinguished at approximately the same positions as the microscope stage is rotated (Section 14.1).

In practice the apparent degree of anisotropy of the fine-grained plasma, as seen in thin section, varies with the magnification used for the observations. At any particular magnification the plasma of the s-matrix often consists partly of anisotropic units and partly of apparently isotropic material, but at other magnifications the ratio of the area of plasma occupied by anisotropic units to that occupied by apparently isotropic material may be different, and the degree of anisotropy within units may appear different. Both these factors affect assessment of the degree and pattern of basic orientation. It is important, therefore, to state both the characteristics of the extinction phenomena used to differentiate the various degrees and patterns of basic orientation and the magnifications at which the observations were made. The method is essentially comparative rather than absolute as can be achieved by direct measurement.

B. Referred Orientation Pattern

This is the orientation pattern of like individuals with regard to a specific reference feature. Measuring azimuth and angle of dip of axes and poles of faces of observable individuals is measuring their referred orientation pattern with regard to true north and the horizontal. Many of the reference features within soil materials, however, are not straight lines or flat faces, so measuring the orientation of individuals with reference to compass bearings and the horizontal may indicate little or no preferred orientation whereas referring it to the orientation and conformation of the specific reference feature may show a high degree of preferred orientation.

Because plasma commonly occurs as compound units of grains as well as single grains, it can exhibit a referred orientation pattern by preferred orientation of the individual grains and/or of the anisotropic aggregates with regard to a specific reference feature. The treatment for assessment of referred orientation is similar to that for basic orientation in that extinction phenomena must again be used to assess degree of parallelism. If the reference feature is linear or planar a well-developed referred orientation pattern is manifested by the plasma extinguishing as a unit, but if the reference feature is curved detailed observations of the extinction behavior of the plasma with reference to very small sections of the reference feature are necssary. If the mineral species have been identified, even in broad terms, the actual optical referred orientation of individual grains can be measured by determination of elongation, and even the physical referred orientation of individual grains can be inferred from the relationships of the physical and optical axes (Fig. 31).

Three common manifestations of referred orientation of the clay minerals in soil materials are illustrated in Fig. 32, assuming the clay mineral to be a layer-lattice type with micaceous form:

(1) A unit composed of strongly oriented clay mineral grains along a flat face of a skeleton grain (Fig. 32*a*). In Fig. 32*a*(1) the parts of the unit at extinction are those adjacent to faces of the skeleton grain that are parallel to the vibration directions of the polarizer and analyzer. The individual grains within the unit, therefore, are oriented either parallel to or at right angles to the faces of the skeleton grain. In Fig. 32*a*(2) the grain has been rotated through 45 degrees in which position the elongation can be determined, that is, whether the plasma grains are oriented with their platy form parallel or normal to the faces of the skeleton grain.

(2) A coating of clay mineral grains on the walls of a void with a circular cross section. Such an arrangement is indicated by the occurrence of an extinction cross in the oriented material, which remains virtually stationary as the stage is rotated [Fig. 32*b*(1)]. The grains at 45 degrees between the arms of the cross are in a suitable position for determination of elongation, and insertion of the appropriate plate gives an additive effect in two opposite quadrants and a subtractive effect in the other two quadrants [Fig. 32*b*(2)]. This shows whether the plasma grains are oriented with their platy form parallel or normal to the walls of the void.

(3) A more difficult situation where the reference feature is neither linear nor regular in conformation so that the extinction lines that indicate preferred orientation of individual grains move in a rather complex manner across the section as the stage is rotated [Fig. 32*c*(1)]. The best that can be done is to choose the position that gives the simplest extinction pattern, rotate through 45 degrees, and determine the elongation [Fig. 32*c*(2)]. By this means referred orientation can usually be determined even in the most difficult cases.

The universal stage may be helpful in determining referred orientation patterns of plasma grains where the extinction pattern is par-

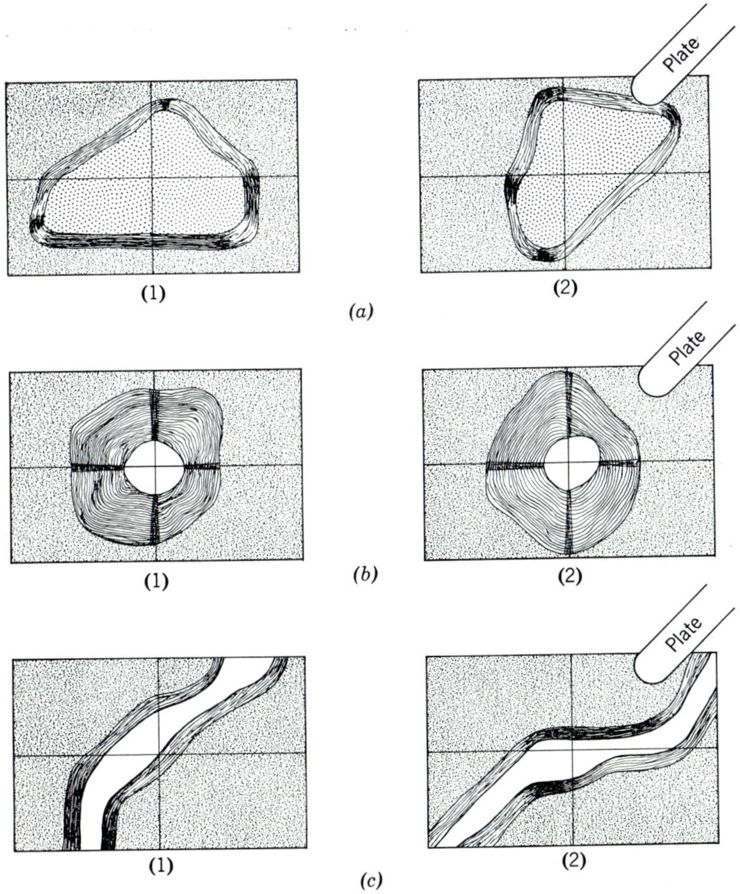

Fig. 32. Common referred arrangements of clay minerals in soil materials. (*a*) With reference to the surfaces of a skeleton grain. (*b*) With reference to the walls of a channel. (*c*) With reference to the walls of an irregular planar void. The three positions on the right, at 45 degrees to those on the left, are suitable for determining whether the individual clay mineral grains are parallel or normal to the reference features.

ticularly complex or where the section is close to the 001 crystallographic plane so that extinction is indistinct.

C. Related Orientation Pattern

This is the orientation pattern of like individuals with regard to the orientation of groups of individuals of a different kind. If more than one group of like individuals have a basic preferred orientation, the related orientation pattern describes the relationships between the patterns of the two groups of individuals; this is a special case of referred orientation where the reference feature is the orientation pattern of another group of individuals. Large individuals, plasma grains, and anisotropic aggregates of plasma grains can exhibit related orientation patterns with regard to the orientation patterns of other individuals, such as skeleton grains, pedological features, and voids.

8.3 PRESENTATION OF DATA

The data of the two aspects of fabric analysis (distribution patterns and orientation patterns) are applicable to all the recognizable individuals in soil materials at all levels of organization, and the same terms and criteria for description can be used for each group.

A. Distribution Pattern

The following terms have been found satisfactory for the distribution patterns observed to date; others undoubtedly will be found necessary as experience grows and new patterns are observed.

(*a*) Basic Distribution

Random. The individuals are distributed randomly throughout the soil material; there is no recognizable specific pattern of distribution.

Clustered. The individuals are concentrated in clusters or groups.

Banded. The individuals are concentrated in bands or sheets (Fig. 38).

Radial. The individuals are grouped along radiating lines.

Concentric. The individuals are grouped along approximately concentric lines or surfaces, or a single approximately circular or elliptical line, or spherical or ellipsoidal surface.

All these basic distribution patterns can be characterized to some extent by measurement. The distance between entities can be quoted in random patterns, or within the groups in clustered, banded, radial, and concentric patterns. In the last four, the distances between groups, and the ratio of this distance to the distance between entities within each group, can be quoted. These measurements, in conjunction with the size of the individuals, provide a fairly quantitative description of the basic distribution pattern.

(*b*) *Referred Distribution*

Unrelated. The pattern of distribution is unrelated to any reference feature.

Normal. Individuals, or groups of them, are arranged perpendicular to the specific reference feature.

Parallel. Individuals, or groups of them, are arranged parallel to the specific reference feature (Figs. 38 and 40).

Inclined. Individuals, or groups of them, are arranged at roughly a constant angle to the specific reference feature; this angle should be measured and stated. This term replaces "transverse" (Brewer and Sleeman, 1960), which has the connotation of extending across the length.

*Cutanic.** The individuals are associated with natural surfaces in the soil materials.

Subcutanic. Individuals, or groups of them, are arranged parallel and close to adjoining natural surfaces. This is a special case of a parallel referred pattern.

Noncutanic. The individuals have no relationships to natural surfaces.

The referred distribution pattern may also be described in relation to the occurrence of peds or pedological features by using the prefixes *intra* (within), *inter* (between), and *trans* (across). Certain individuals may be intrapedal, intranodular, intraconcretionary, etc. where they occur within peds, nodules, or concretions, respectively. Similarly they may be interpedal where they occur between peds, or transpedal where they occur independently of the peds, a single entity extending through and between peds.

(*c*) *Related Distribution.* Although related distribution patterns can be inferred from the referred distribution patterns for the two groups of related individuals, it is convenient to have specific names for specific arrangements in the basic fabric. The terms porphyritic, agglomeratic, intertextic, and granular have been proposed (Brewer, 1960*a*; following Kubiena, 1938). The first two of these terms are inappropriate in that they are used in geology in a rather different sense; they are replaced here by *porphyroskelic* and *agglomeroplasmic.* The four terms are defined as follows:

* Cutanic is a coined term whose derivation and use are discussed in the treatment of cutans, which are a group of pedological features (Chapter 10).

Porphyroskelic. The plasma occurs as a dense groundmass in which skeleton grains are set after the manner of phenocrysts in a porphyritic rock.

Agglomeroplasmic. The plasma occurs as loose or incomplete fillings in the intergranular spaces between skeleton grains.

Intertextic. The skeleton grains are linked by intergranular braces or are embedded in a porous groundmass (matrix in the sedimentary petrological sense).

Granular. There is no plasma, or all the plasma occurs as pedological features.

Additional terms will be necessary for other specific arrangements as more data are obtained.

B. Orientation Pattern

Theoretically the orientation of an individual can be measured with regard to some reference provided the individual is anisotropic in some characteristic; that is, some standard property, such as the *C*-crystallographic axis, a flat face, or a long axis, is recognizable. Such data can be presented graphically. Where direct methods are not possible, or such detailed analyses are considered unnecessary, orientation patterns can be presented by descriptive terms.

(*a*) Graphical Presentation. The most effective method of presenting the direct measurement of orientation is by plotting the data on *polar coordinate paper.* Each individual is considered as being placed at the center of a sphere. The direction of the property used to measure orientation (e.g., the long axis) is produced to intersect the surface of the sphere in the southern hemisphere, and the latitude and longitude of this point is plotted on the polar map. Polar coordinate paper allows these points to be plotted directly from measurements of azimuth and dip. Clustering of points, or lack of it, on this *polar diagram* indicates the degree of orientation of the individuals. Contours of the density of the points can be drawn from the polar diagram as an aid in visualizing preferred orientation patterns. Plotting azimuth and dip of some characteristic of like individuals gives a polar diagram that expresses the basic orientation pattern of the individuals.

Referred and related orientation patterns can be represented by plotting the orientation of the reference feature or of several groups of different individuals on one polar diagram. If the individuals have a strongly developed preferred orientation with regard to a curved reference feature, such as the walls of a void, the wall of the void can be considered as a flat horizontal sheet, irrespective of its shape and position, so that its pole plots as the center of the polar diagram. Then the poles of the individual grains are plotted in relation to the surface of the void, and the concentration of these around the center of the diagram indicates the degree of preferred orientation with reference to the walls of the void (the referred orientation pattern) and the basic orientation pattern in the sense of the degree of orientation of individual grains with regard to their immediate neighbors.

Alternatively, preferred orientation can be recorded graphically by setting up orientation classes, whose limits are fixed intervals of azimuth and/or dip measured in degrees, and plotting the frequency distribution within these classes as frequency distribution curves, cumulative curves, or histograms.

(b) Descriptive Terms. Often a descriptive term is used to distinguish classes of preferred orientation where detailed measurements of individuals are not possible or are considered unnecessary. The *basic orientation pattern* of individuals that can be observed and measured directly can be classified as:

Strongly Oriented. More than 60 percent of the individuals are oriented with their principal axes within 30 degrees of each other.

Moderately Oriented. Between 40 and 60 percent of the individuals are oriented with their principal axes within 30 degrees of each other.

Weakly Oriented. Between 20 and 40 percent of the individuals are oriented with their principal axes within 30 degrees of each other.

Unoriented. There is no preferred orientation.

While aggregates of plasma grains are frequently large enough to measure directly, the orientation of individual clay size particles can be stated for optically anisotropic grains only on the basis of their extinction phenomena at various magnifications; the most useful magnifications are approximately 30×, 80×, 200×, 400×, 800×, and 1200×. The descriptions are made at each of these magnifications, and the increasing observable degree of orientation with increasing magnification is noted as a comparative measure of orientation. The *degree of basic orientation* of plasma grains is assessed by quoting the magnification used (which is related to size of the plasma aggregates, if present) and the proportions of the plasma present that can be placed in the following classes:

Strongly Oriented. At the magnification used, the plasma aggregates* are continuously birefringent in thin section; that is, under crossed nicols dark extinction lines run across the aggregates, or they extinguish as a unit.

Moderately Oriented. At the magnification used, the plasma aggregates have rather indistinct boundaries and an incomplete, "mottled" extinction reminiscent of the extinction of some micas.

Weakly Oriented. At the magnification used the plasma aggregates exhibit weak birefringence and a wavy, undulose extinction.

Unresolved.† At the magnification used the plasma aggregates are anisotropic but too small to observe in detail; the mass has a flecked appearance under crossed nicols.

Unoriented. At the magnification used the plasma is isotropic because of the low degree of orientation.

Indeterminate. The plasma is isotropic because of its crystallographic nature or opacity.

The qualifying adjectives "strongly," "moderately," and "weakly" should not be read too literally at the present state of knowledge. The manifestations of extinction phenomena that these terms describe are presumed, from first principles, to be due to differing *degrees* of preferred orientation; such a deduction is not yet proven conclusively.

Besides the preferred orientation of individual plasma grains the aggregates of plasma exhibit various *patterns of orientation* which are also inferred from observations of extinction phenomena. To facilitate comparison of different materials it is important again to quote the magnification at which the observations are made. The following classes have been recognized:

Continuous Orientation. Individual aggregates are not observable at any magnification achievable with the petrological microscope except as expressed by parallel layers of plasma. Under crossed nicols the mass of plasma exhibits extinction lines or extinguishes a unit. This kind of orientation pattern is most com-

* Where plasma aggregates are not observable the definitions refer to the whole of the plasma of the s-matrix; such s-matrixes usually do not have degrees of orientation higher than the weakly oriented.

† The term "unresolved" replaces "weakly oriented" as used by Brewer and Sleeman (1960), since the definition refers not to the degree of orientation of individual plasma grains within a compound unit, but rather to the fact that the compound units are too small to resolve at the magnification used; the degree of basic orientation is unresolved. Weakly oriented is more appropriately used in the sense defined here.

mon in those pedological features consisting of a single unit of strongly oriented plasma grains.

Striated Orientation. The plasma aggregates are physically and/or optically anisotropic and exhibit a linear or curved linear arrangement. The orientation pattern is *strongly striated* when more than 60 percent of the material exhibits the linear arrangement, *moderately striated* when 20 to 60 percent exhibits the linear arrangement, and *weakly striated* when less than 20 percent exhibits the linear arrangement.

Flecked Orientation. The plasma aggregates are arranged randomly.

Referred orientation patterns can be described in terms similar to those used for referred distribution patterns provided the use of the terms is made clear; they can be classified as follows:

Unrelated (Fig. 51). There is no constant relationship between the orientation pattern and any specific reference feature.

Normal (Fig. 50). The direction of preferred orientation is perpendicular to the specific reference feature.

Parallel (Fig. 48). The direction of preferred orientation is parallel to the specific reference feature.

Inclined. The direction of preferred orientation is at a constant angle to the specific reference feature.

The terms *cutanic* and *subcutanic* can also be applied to referred orientation patterns, the former for patterns of individuals associated with, and oriented parallel to, natural surfaces, and the latter for patterns of individuals in the s-matrix adjoining, and oriented parallel to, natural surfaces.

Related orientation patterns are virtually classified by description of the referred or basic orientation patterns of the two groups of individuals whose patterns are related. It may be useful, however, to describe related orientation patterns positively, and this can be done by using the same terms as for referred orientation, that is, unrelated, normal, parallel, and inclined.

C. Sharpness of Boundaries

Much of the description of soil fabric depends on recognition and description of pedological features and of their arrangement with regard to other constituents and voids. An essential part of this description is assessment of the sharpness of their boundaries. Theoretically a detailed description of fabric (distribution and orientation patterns) within the features should describe sharpness of boundaries, but in practice it is convenient to describe this characteristic directly in the following terms:

Very Diffuse. There is no observable boundary but merely a waxing and waning effect of the material or fabric by which the pedological feature is recognized.

Diffuse. The boundary consists of a transition over a distance greater than one-quarter of the shortest dimension of the feature.

Rather Diffuse. The boundary consists of a transition over a distance between one-quarter and one-tenth of the shortest dimension of the feature.

Rather Sharp. The boundary consists of a transition over a distance less than one-tenth of the shortest dimension of the feature.

Sharp. The boundary is quite sharp.

8.4 ANALYSIS OF DATA

The data of fabric analysis of soil materials do not lend themselves to mathematical treatment and interpretation to the same extent as do size and shape analyses. When the individuals involved in a particular fabric are large enough for measurements to be made, distribution patterns can be characterized by certain measurements discussed earlier, such as distance between individuals and groups of individuals. Similarly, if the orientation of the individuals can be presented as an orientation diagram or a cumulative curve, such values as the *arithmetic mean* of the measurements made on all the individuals and the *median value* (50 percent value) can be obtained and used as an estimate of the direction of preferred orientation, in regard to a specific reference feature (direction of referred orientation), in regard to the arithmetic mean or median value for another group of individuals (direction of related orientation), or in regard to true north and the horizontal (a special case of referred orientation). Walker (1959) used such measurements to interpret the origin of the parent materials of some soils on the south coast of New South Wales. The treatment and analysis of data he used were similar to those used in sedimentary petrology, and he was able to elucidate the kind of transport of the material by the orientation of physically anisotropic pebbles in the sedimentary deposit. Most of the important aspects of fabric, however, are usually exhibited by the plasma, and even such restricted mathematical treatment is not possible except by using special techniques for measuring the orientation of the minute grains of plasma. Analysis of fabric data, therefore, is often restricted to interpretation of extinction phenomena and assessment of the size of the plasma aggregates; either the magnification at which the observations were made or the actual size of the aggregates of oriented clay minerals can be recorded. In fact a size analysis of such aggregates can be made.

On a broader scale, interpretations must necessarily be based on observing the consistent mutual occurrence of two or more specific expressions of fabric; this indicates that the phenomena are related in some way, either as cause and effect or as due to similar sets of conditions. The larger the number of such observations, the more dependable is the interpretation. The concepts of levels of organization and levels of structure simplify such interpretations of fabric data by focusing attention on various aspects of fabric in some specific order. For example, interpretation of the formation of pedological features (one of the units of organization) is of particular importance in interpretation of the processes and history of formation of soil materials, and so of soil profiles. Such features can be treated as entities whose distribution and orientation patterns can be observed and described as part of the fabric analysis of the soil material. One or both of these patterns may correlate with some other characteristic of the soil material or soil profile.

Many such correlations have been made in the field description of soil profiles, which is equivalent to the megascopic phase of structure and fabric analysis. A common example is the occurrence of a layer of ferruginous nodules or concretions at the base of an *A* horizon immediately above an impermeable *B* horizon in many planosolic profiles. These nodules exhibit a banded distribution pattern parallel to the upper surface of the impermeable horizon. This relationship is so common that it is legitimate to presume some causal relationship between the two. Field observations have also shown that many such profiles develop a seasonal or intermittent perched watertable on top of the impermeable horizon, suggesting that the nodules are formed under the accepted conditions for formation of such features, that is, an intermittent, fluctuating watertable, and that the direct cause of the watertable is the impermeable *B* horizon. This method of deduction from observation of consistently related features of fabric and environment is most important in interpretation of soil fabric.

In addition to treating pedological features as entities, their internal fabric can be examined with a view to interpreting the actual processes of their formation. For example, the ferruginous nodules or concretions just discussed are formed because of a fluctuating watertable, but variations in their internal fabric may indicate differences in the actual processes of concentration of the iron oxides. Little is known of the processes of formation of many of the pedological features, and the problems and hypotheses of their formation will be discussed later (Chapters 10–14).

Pedological features have been chosen here as an example of the kind of reasoning and correlations of fabric needed to understand the processes of their formation and of soil formation. Similar arguments apply to the study of the other units of organization (peds and s-matrixes) and to the study of voids. Fabric analysis at each level of organization (plasmic, basic, elementary, primary, secondary, and tertiary) can be studied in the same way with the objective of correlating particular phenomena with other phenomena or known sets of conditions. A system of interpretation of fabric phenomena can be built up in this way, and it can be continually checked against further observations.

Many of the hypotheses of formation of specific fabric phenomena are amenable to testing by experimentation in the laboratory. Several such experiments have been made. Bloomfield (1953*a,b,c,* 1956*a,b*) demonstrated that certain processes could be responsible for the formation of the illuvial horizons of podzol profiles. Although he did not make fabric analyses of either his experimentally produced horizons or natural podzol horizons, they are megascopically the same, which is equivalent to saying that their megascopic structure and fabric are the same. Similarly Thorp, Strong, and Gamble (1957) demonstrated that clay size material moves in suspension under certain conditions, so the possibility of formation of clay skins on the walls of voids due to deposition from suspension has been demonstrated. Again Thorp did not make fabric analyses. Brewer and Haldane (1957) went further in their experiments on the behavior of suspensions of clay minerals when passed through columns of sand. They demonstrated not only that clay minerals can be deposited from suspension but that the internal fabric of the deposited material was identical with that of similar material observed in soil materials and that the distribution and orientation patterns of the grains of plasma in relation to the skeleton grains were identical with some patterns observed in soil materials.

These, then, are the principles on which must be built the interpretation of fabric analyses. So few analyses have been made to date that hypotheses must necessarily be tentative until much corroborative evidence in the way of observation and experimentation is obtained. Some of these aspects of interpretation of fabric analysis are discussed in more detail in the section on the classification and interpretation of the units of organization (Part IV).

PART IV

CLASSIFICATION OF THE UNITS AND LEVELS OF ORGANIZATION

9

Voids

9.1 VOIDS AS INDIVIDUALS

All voids are interconnected in any material which consists of packing of individuals, so soil materials can be regarded as having a single void of intricate shape which varies considerably in its dimensions throughout its extent; this emphasizes the difficulty of description and classification of voids. The shape of the space which results from the regular packing of spheres is known (see Pettijohn, 1957), and simple packing of individual grains in soil materials would give similar characteristics with differences due mainly to divergence of the grains from a spherical shape and the presence of a range of sizes. The size, shape, and arrangement of the voids in many soil materials, however, show that factors additional to simple packing have influenced their formation, for example, the quite large prolate, acicular, and planar voids common in fine-grained soil materials.

In many soil materials a significant proportion of the void space consists of voids of comparable size interconnected by a population of very much smaller voids, so the larger ones can be treated as entities, for example, large prolate or equant voids in the s-matrix connected by the very small voids which result from the packing of plasma and skeleton grains. Other soil materials may have systems of interconnected planar or acicular voids which meet at relatively sharp angles so that again they can be treated as entities even though the actual extent of a single individual may be a matter of judgement, especially when they are curved or have rather irregular walls. Even the interconnected voids that result from packing of individual grains can be treated as entities if the convention is adopted that the significant constrictions where the walls of the solid individuals come closest together are the boundaries of individual voids, which are then connected through relatively narrow "necks."

If this concept of voids as entities, which are interconnected with each other either through voids of dissimilar size and shape, through narrow necks, or through intersection with voids of similar size and shape, is adopted, they can be described and classified in terms of their size, shape, and arrangement.

9.2 SIZE

Accepting the concept of voids as individuals, their size frequency distribution theoretically can be measured, and the data treated in the same way as size analyses of mineral grains (Sections 2.4 and 2.5). In practice, the procedure is complicated by the difficulty of measuring the true volume of voids, which is the only true measure of size (Wadell, 1932) and because most soil materials change volume with moisture content with accompanying change in void size.

A. *Methods of Size Analysis of Voids*

The size frequency distribution of voids in soil materials, or the "pore size distribution," is usually measured by methods based on the theory of the retention of water in voids of different sizes. An undisturbed specimen is wetted under conditions that prevent its breakdown or disruption, and then it is drained at increasing moisture tensions with measurement of the volume of water removed at each tension. The moisture characteristic (or moisture release) curve can be plotted, and a pore size (void size) frequency distribution determined for soil materials whose volume remains constant with change in moisture (Marshall, 1959, p. 12). The radius of voids containing the air-water interface at a given moisture suction is defined in terms of a capillary model by the expression for curvature of the surface in a capillary tube ($S = \frac{2\sigma}{r}$, where S is the soil moisture suction, σ the surface tension of water, and r the radius of curvature of the air-water interface which, for a circular capillary, is very closely that of the capillary). The radius of a void is given as the equivalent circular radius or "effective" radius (Childs and Collis-George, 1950; Childs, 1957) in much the same way as particle size is defined as the equivalent spherical particle when measured by sedimentation rates. The amount of water released by an increment of soil moisture tension gives the volume of voids in a known "effective" void size range. For soils whose volume changes with moisture content, pore size distribution is undefined, and no constant void size distribution exists. Measurement of void size distribution of such soils can be made with nonpolar liquids at air-dry moisture content (Quirk and Panabokke, 1962), but little of this work has been done.

These calculations of void size have no relationship with Wadell's (1932) true nominal diameter, since the tension or pressure required to drain a void of certain size is affected by its shape, being controlled to some extent by its shortest dimension; progressively higher tensions or pressures are required to drain voids of the same true nominal diameter as the shape changes from spherical (or equant) to prolate, to acicular, to planar. Also, because of considerations of shape, successive tensions often drain successively larger proportions of the same void.

The alternative is to estimate void size frequency distribution by direct measurement of individuals. In practice it is virtually impossible to measure the true volume; many voids in soil materials have such irregular forms and intricate patterns of distribution that only the dimensions of cross sections can be measured with any degree of accuracy. The ideal of size classification according to true volume, therefore, is unattainable, and some other characteristic of size must be used. The most practical method is to measure the area of individuals in section (Anderson and Binnie, 1961) and attempt to correlate these measurements with volume in the same way as is done for mineral grains (Section 2.3:C). Even this method has serious practical difficulties because of the large range of sizes and shapes of voids; it is most practical for voids with diameters from 20 μ to about 2 mm, which can be seen and measured in thin section or on polished blocks, because samples of appropriate size can be studied at any moisture content by freeze-drying techniques before impregnation. Although voids are rarely circular capillaries a close correlation has been found between distributions determined by microscopic and suction methods (Swanson and Peterson, 1942) in soils in which a void size distribution can be determined from the moisture characteristics, that is, sands and sandy soils.

Alternatively, void size can be measured according to the length of one of the principal axes. This method is much less satisfying than measurement of cross-sectional area or actual volume, but it has the advantage of relative simplicity in practice. Physical measurements of the properties of soil materials, especially of water movement and "pore-size distribution," indicate that the shortest dimension is the most significant. This measurement, in conjunction with a description of shape, provides a reasonable classification according to size since shape descriptions provide certain limits for the ratios of the lengths of the principal axes. It is proposed, therefore, that measurements of voids for classification according to size should be standardized on the length of the shortest dimension.

B. Classification according to Size

Jongerius (1957) classified voids into three groups, according to size, which he apparently measured as the length of one of the principal axes. The basis of his groupings is assessment of the role of the size classes in soil water—plant relationships. He considered that *micropores* (diameter less than 30 μ) are "important as a reservoir for the plants and for the microorganisms in the soils," *mesopores* (diameter between 30 and 100 μ) are "important for intense renewal of air in the elements and serving to transport and distribute water" (by "elements" Jongerius meant soil units or peds), and *macropores* (diameter greater than 100 μ) "enable air to penetrate the soil rapidly and deeply" and enable "the rapid withdrawal of large quantities of water." The choice of the limits of size classes is quite arbitrary unless some basic criterion such as this, or the standard phi scale for mineral grains, is used; at the present state of knowledge the classes used by Jongerius appear to have some significance both in soil genesis and for plant growth. Johnson et al. (1960) classified

TABLE 9-1. Size Classification of Voids

Class Name	Subclass	Class Limits
Macrovoids		>75 μ
	Very fine macrovoids	75–1000 μ
	Fine macrovoids	1000–2000 μ
	Medium macrovoids	2000–5000 μ
	Coarse macrovoids	>5000 μ
Mesovoids		30–75 μ
Microvoids		5–30 μ
Ultramicrovoids		<5 μ
Cryptovoids*		<0.1 μ

* The water in cryptovoids is generally accepted as being unavailable to plants

voids into five size classes for field description of soil materials. Their smallest class was less than 75 μ, this limiting value being chosen because of an apparent relationship between percolation rate and porosity as measured by a tension of 40 cm, which corresponds to a theoretical pore diameter of 75 μ; the other four classes dealt with voids from 75 μ diameter up to greater than 5 mm diameter. These two classifications can be combined and another size class added to cope with the very small voids comprising a high proportion of the pore space in many soil materials (Katchinski, 1956) and which can be observed only by sensitive techniques such as electron microscopy; the new proposal is set out in Table 9-1.

9.3 SHAPE

The shape of individual voids can be described in terms of the lengths of the principal axes (sphericity) and roundness (Section 2.1:B). In practice, such shape classes are not always specific or even sufficiently adequate to insure a reasonable conception of the shapes which occur, because they do not take account of the details of variations in *conformation* and *smoothness* of the walls. The walls of voids in soil materials are rarely perfectly smooth, and the degree of roughness or irregularity becomes more obvious as the magnification at which the observations are made is increased. Due allowance must be made for this in dealing with conformation and smoothness, both of which must be considered as relative characteristics.

A. *Conformation*

Conformation has to do with the grosser characteristics of the walls, that is, the overall shape of the walls other than irregularities due to projection of skeleton grains, pedological features, and plasma aggregates into the void. Many voids in soil materials have highly irregular walls; voids that would be described as equant or prolate, according to the ratios of their principal axes, may have short curved arms (or strongly re-entrant angles), may be star-shaped in three dimensions, or may have any other shape with highly irregular or contorted walls. Such shapes are described inadequately by the lengths of the principal axes, and roundness measures are difficult and not completely satisfactory.

Beckmann (1962) proposed the Lappungsquotient (*Lq*) as a

measure of the divergence of the conformation of the walls of voids (and peds or other features) from a circular shape. It is a simple ratio:

$$Lq = \frac{U_g}{U_f}$$

where U_g = circumference of the unit, and
U_f = the circumference of a circle of the same area.

The measurements are made by projecting large thin sections (6 × 8 cm) on a specially adapted table projector, LEITZ TP200 (Kubiena, Beckmann, and Geyger, 1961; Geyger, 1962) or by photogrammetric methods (Kubiena, Beckmann, and Geyger, 1961); macroscopic voids can be measured directly. Beckmann proposed classes with limits of values for U_g/U_f; these are shown in Table 9-2. *Lq* can be used, only in relation to the shape classes based on the lengths of the principal axes since an acicular void with regular walls has a high *Lq,* perhaps as high as an extremely lobate equant void; in such instances Beckmann's class names are misleading. This objection can be overcome to some extent by applying *Lq* measures only to sections through the void that are approximately equidimensional ($c/a > 2/3$, where c is the shortest diameter and a the longest).

A second aspect of conformation is the overall degree of curvature of voids which are pronouncedly unequidimensional, such as planar and acicular types.

For rapid description of the structure of soil materials descriptive terms that place voids in broad classes according to the conformation of their walls are desirable. The following terms are defined to convey some idea of the deviations from ideal shapes; they are to be

TABLE 9-2. Shape Classes Based on Lappungsquotient (after Beckmann, 1962)

Class Name	Lappungsquotient (U_g/U_f)
Sharp edged	1.0–1.2
Very weakly lobate	1.2–1.5
Weakly lobate	1.5–2.0
Moderately lobate	2.0–2.5
Strongly lobate	2.5–3.0
Very strongly lobate	3.0–3.5
Extremely lobate	>3.5

used as qualifying terms for the shape classes based on the lengths of the principal axes:

Curved (Fig. 46). This term is applied particularly to acicular and planar voids that deviate significantly from a straight line or flat plane in the direction of their long axes. *Bi-curved* is used when the deviation is in more than one plane. The terms may be applied to any physically anisotropic forms.

Regular. This term is used when there are virtually no re-entrant or acute angles between faces or segments of the walls.

Irregular (Fig. 34). This term is applied particularly to equant and prolate voids whose walls have a significantly irregular conformation. *Mammillated* (Fig. 36) is used for the specific case where the walls consist of rounded, interfering, spheroidal surfaces. The terms may be applied to other forms, especially the less physically anisotropic.

B. Smoothness

Smoothness has to do with the minute details of the surface that result from the arrangement of plasma and skeleton grains. Walls of voids that consist of the s-matrix in its normal random state of packing are not smooth on a microscale; skeleton grains and random arrangements of plasma and skeleton grains project into the void. Similarly the walls of planar voids that are caused solely by cracking due to shrinkage have a certain roughness from the tearing apart of the soil material. On the other hand, the walls of many of the larger voids in soil materials are so smooth that some other process must have been superimposed on the normal random packing of plasma and skeleton grains or simple cracking due to shrinkage upon drying. From observations to date the following simple classification according to smoothness of the walls seems sufficient:

Ortho (Figs. 34 and 40). A prefix for voids whose walls appear morphologically to be due to the unaltered, normal, random packing of plasma and skeleton grains.

Meta (Figs. 35–38). A prefix for voids whose walls appear morphologically to be significantly smoother than would result from the normal random packing of plasma and skeleton grains. The usual types are referred to simply as being *smoothed,* while the term *slickensided* is used in the same sense as in geology for a subclass of metavoids in which the walls show evidence of differential movement under pressure by the occurrence of striations and/or elongated crests and depressions on the plane surfaces.

There is sometimes considerable difficulty in recognizing the smoothed surfaces of metavoids because of modifications to the surfaces by processes subsequent to formation of the voids. Illuviation or diffusion cutans (Chapter 10) on the surfaces cause a smooth appearance and mask the original appearance of the walls, as also do fillings in acicular, branched voids (Chapter 11).

9.4 ARRANGEMENT

Voids can also be classified according to certain aspects of their distribution and orientation patterns; that is, voids of similar size, shape, conformation, and smoothness may be classified differently according to their occurrence in specific situations in the soil material (Section 8.5); the number of patterns which appear to be significant is rather limited. The first separation is into three broad groups according to distribution in relation to peds:

Intrapedal. Voids occurring within the s-matrix of peds or apedal soil materials.

Interpedal. Voids occurring between peds.

Transpedal. Voids traversing the soil material without any specific relationships to the occurrence of peds; they usually extend beyond the limits of a single ped.

They can be classified further according to their distribution and orientation patterns into groups with *random, banded,* and *clustered* distribution patterns, *parallel* orientation patterns, and specific *branching* patterns (single, dendroid, anastomosing, and trellised; Section 9.5:D).

Intrapedal voids that result from packing of plasma and skeleton grains necessarily exhibit a random distribution pattern, the details of which are determined by the shape, packing, and particle size distribution of the individual grains. Those which are much larger than would result from normal random packing of plasma and skeleton grains usually have a random basic distribution, but they may exhibit a banded pattern that can usually be referred to the horizontal or to ped faces; occasionally they occur in a clustered basic distribution pattern, both the voids within clusters and the clusters themselves having a random distribution. The distribution of interpedal voids can be referred to the size, shape, and arrangement of peds, so their distribution pattern is usually random within a single soil material whether the peds are accommodated or not; banded distributions may occur when the soil profile as a whole is considered, for example, a banded distribution of horizontal planar voids may occur in one horizon of a profile (Fig. 40). Transpedal voids are usually acicular metavoids that traverse the soil material in various branching patterns; they usually have a random basic distribution pattern but can be subdivided into more specific groups on the details of the branching patterns. Any of the voids that are physically anisotropic can exhibit preferred orientation (parallel basic orientation pattern).

9.5 MORPHOLOGICAL CLASSIFICATION OF VOIDS

A number of morphological characteristics are consistently associated with particular groups of voids, so it is appropriate to define specific names for such groups. This does not preclude the possibility of describing individual voids in detail in terms of size, shape (conformation and smoothness of the walls), distribution, and orientation patterns. The specific names defined in the following constitute a morphological classification of voids; the characteristics used in their classification are set out in Table 9-3.

A. *Packing Voids*

These voids are due to *random* packing of individuals (either single grains or such compound units as peds). They are usually equant to prolate, irregular orthovoids, or smoothed metavoids; they have a random orientation and distribution and are strongly interconnected by relatively narrow necks. This group of voids is subdivided into:

(*a*) ***Simple Packing Voids*** (Fig. 33). These voids are due to random packing of single grains. Since the grains in a soil material are never all spheres their packing usually does not even approximate any of the six possible regular packing systems illustrated by Pettijohn (1957). Simple packing voids, then, are irregular orthovoids approximately equant to prolate in overall dimensions as measured between adjoining grains, and strongly interconnected. They are usually too small to observe even microscopically except in relatively coarse sands; electron microscopic techniques are necessary to make a study of them so that further subdivisions, particularly according to shape, could be made.

TABLE 9-3. Morphological Classification of Voids

Void Type	Distribution Pattern	Orientation Pattern	Shape (ratios of axes)	Conformation (of walls)	Smoothness (of walls)
Packing voids:	Between solid individuals	Random	Usually equant to prolate	Irregular	Orthovoids or metavoids
(i) Simple	Between single grains				Orthovoids
(ii) Compound	Between unaccommodated peds (interpedal)				Orthovoids or metavoids (smoothed)
Vughs*	Random	Random	Usually equant to prolate	Irregular to mammillated	Orthovoids or metavoids (smoothed)
Vesicles	Usually random, but sometimes a specific referred distribution	Random or parallel	Usually equant to prolate, or arcuate	A specific form	Usually metavoids (smoothed)
Chambers	Random	Random	Usually equant to prolate	Usually a specific form	Usually metavoids (smoothed)
Joint planes	Random or interpedal	Parallel to subparallel sets with specific referred orientation	Planar	Regular	Usually metavoids (smoothed; sometimes slickensided)
Skew planes	Random	Random	Planar	Regular or irregular	Orthovoids or metavoids (smoothed; sometimes slickensided)
Craze planes	Interpedal (between accommodating equant to prolate peds) or random	Random	Planar	Irregular due to compound nature; small individual planes may be regular	Orthovoids or metavoids (smoothed)
Channels	Branching patterns		Usually acicular; sometimes prolate	Regular with specific cross-sectional shapes	Metavoids (smoothed)

* When vughs are very numerous they may intersect, thus forming a network of large voids; these are *interconnected vughs*.

Fig. 33. Simple packing voids. Thin section under crossed polarizers. The black areas between skeleton grains are voids. Plasma is virtually absent. 30×.

(***b***) ***Compound Packing Voids*** (Fig. 14). These voids result from packing of compound individuals, such as peds, which *do not accommodate each other.* Compound packing voids are comparable to the interconnected voids that result from random packing of spheres, but considerable variations are allowed in the preciseness of packing and the shape of the individuals. They are unoriented, have a random distribution, and are usually equant to prolate, irregular orthovoids or smoothed metavoids; they are strongly interconnected.

B. Vughs (Figs. 18 and 34–37)

Vughs are relatively large voids, other than packing voids, usually irregular and not normally interconnected with other voids of comparable size; at the magnifications at which they are recognized they appear as discrete entities. They may be orthovughs (Fig. 34) or smoothed metavughs (Fig. 35), are usually unoriented, and have a random distribution pattern. Rice (1954) gave several definitions of "vugh" as used in geology, one of which is simply "small cavities in the rock" qualified by a suggested mechanism of formation (after Cox, Dake, and Muilenberg, 1921); such voids have the same morphological characteristics as those used here for the definition of vughs in soil materials. These are the "normal voids" of Brewer (1960*b*).

Vughs are normally equant to prolate but may be any shape, according to the ratios of their axes, except planar. Many vughs are so irregular in shape, with numerous curves, angles, and contortions in the conformation of their walls, that classification according to the ratios of the axes is inappropriate. If they are relatively regular in form, the shape names can be used as adjectives, for example, *equant, prolate,* and *acicular vughs.* If they are highly irregular they can be classified as *irregular vughs* (Fig. 34) when the walls have no distinctive conformation but consist principally of relatively sharp protuberances, and as *mammillated vughs* (Fig. 36) when the conformation of the walls consists of rounded, interfering, spheroidal surfaces; irregular vughs usually have normal walls, and mammillated vughs

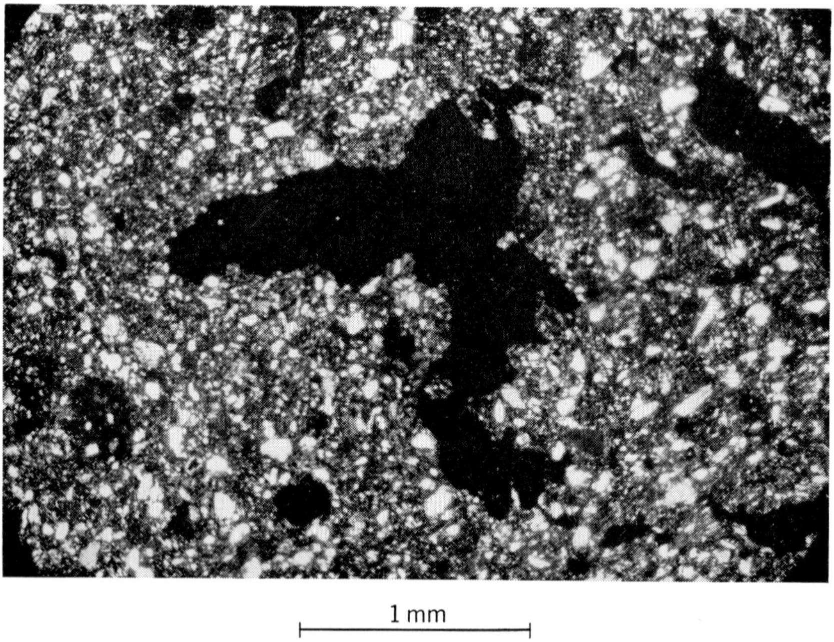

Fig. 34. Irregular orthovugh. Thin section under crossed polarizers. 30×.

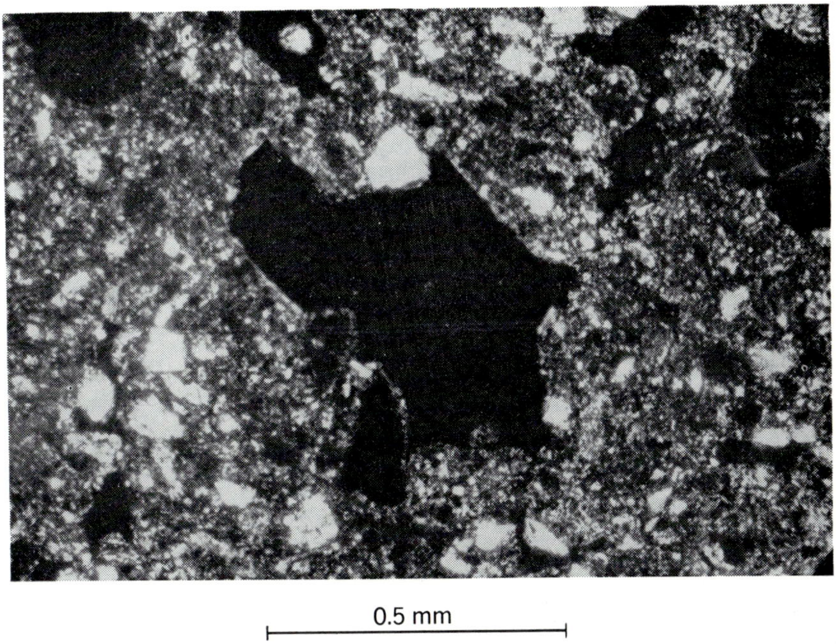

Fig. 35. Irregular smoothed metavugh. Thin section under crossed polarizers. 80×.

have smoothed walls. Vughs may be so irregular and numerous that they intersect each other so that the soil material is ramified by numerous, interconnected, highly irregular, relatively large voids; these may be referred to as *interconnected vughs* (Figs. 18 and 37).

C. Vesicles (Fig. 38)

Vesicles differ from vughs principally in that their walls consist of smooth, simple curves; that is, they are smoothed (metavoids) and regular. Rice's (1954) definition of vesicle (after Fay, 1920) has a genetic context, "being formed by bubbles of gas or steam," while that of Grabau and Shimer (1909–1910) is simply a "bladder-like vessel; a cell." Thus vesicles are recognized in geology from the morphological characteristics used here for the definition of vesicles in soil materials.

Vesicles of several different shapes have been observed in soil materials. Since their walls have a regular conformation most of them can be classified simply by qualifying the term vesicle with the appropriate adjective for the shape class according to the ratios of the principal axes, for example, equant, prolate, and oblate vesicles; some vesicles are arcuate. Other specialized shapes requiring specific names can be expected.

There is insufficient data as yet to specify particular distribution and orientation patterns for vesicles, but at least three patterns have been observed. In soils exhibiting thin surface seals on "scalded" patches that carry little or no vegetation, equant vesicles commonly occur in a thin band (banded distribution) parallel to the sealed surface and at a very shallow depth below it (of the order of 1 to 3 cm). The "platy structure" in the *A* horizons of some strongly differentiated profiles is due to a parallel arrangement of triaxial vesicles; the vesicles are oriented with the planes of their two longer principal axes parallel to the surface of the soil so that the application of a slight pressure causes rupture into plate-like fragments parallel to the sur-

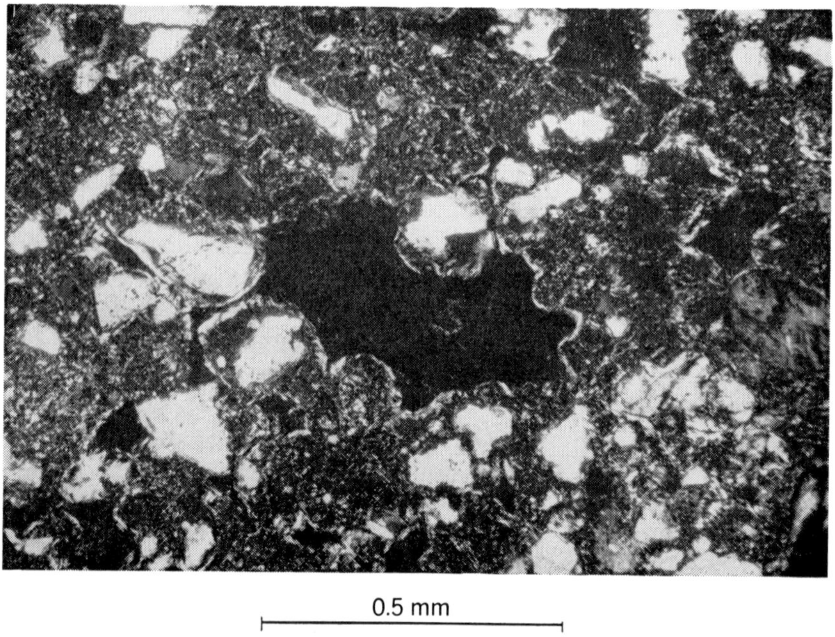

Fig. 36. Mammillated metavugh. Thin section under crossed polarizers. The thin white border to the vugh is a vugh argillan with strong continuous orientation. 80×.

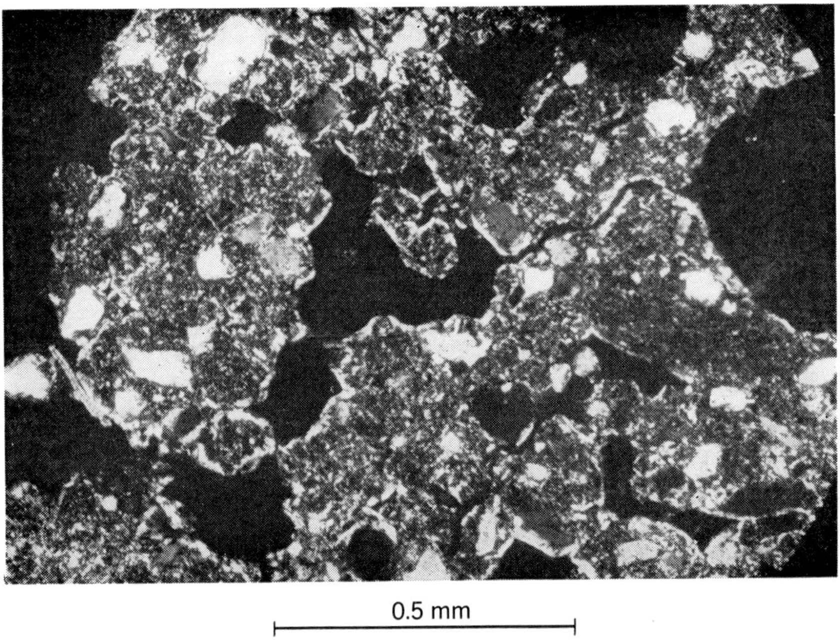

Fig. 37. Interconnected mammillated and smoothed metavughs. Thin section under crossed polarizers. The thin white borders to the voids are vugh argillans with strong continuous orientation. 80×.

face. The "subangular blocky structure" of a number of soil materials, including some of the Australian chocolate soils, is due to a random distribution and orientation of small arcuate vesicles.

D. Channels (Figs. 39 and 46)

Channels are voids that are significantly larger than those which would result from normal packing of single grains, and have a generally cylindrical shape. They commonly have smoothed walls, regular conformation, and a relatively uniform cross-sectional size and shape over significant proportions of their length. They may be transpedal or intrapedal.

Channels can be subdivided according to their cross-sectional shape and pattern of branching. The three common cross sections are *circular, elliptical,* and *arched;* the last consists of one relatively flat, planar surface compounded with an arc of a circle. Four broad patterns of branching have been observed and described (Brewer and Sleeman, 1963):

Single. The channels occur as single, discrete entities without branches, usually with a downward trend.

Dendroid. The channels branch "after the manner of a tree" (Rice, 1954), each branch having a downward trend but not necessarily vertical; there is no rejoining of branches.

Anastomosing. The channels interconnect by branching and rejoining of the branches to produce an irregular network.

Trellised. The channels branch and rejoin to form a regular network characterized by relatively long horizontal-to-subhorizontal channels interconnected by shorter vertical channels.

Since channels are cylindrical they are usually acicular, but not always. Single channels can be almost any shape since the axis perpendicular to the plane of the typical, regular cross section of the

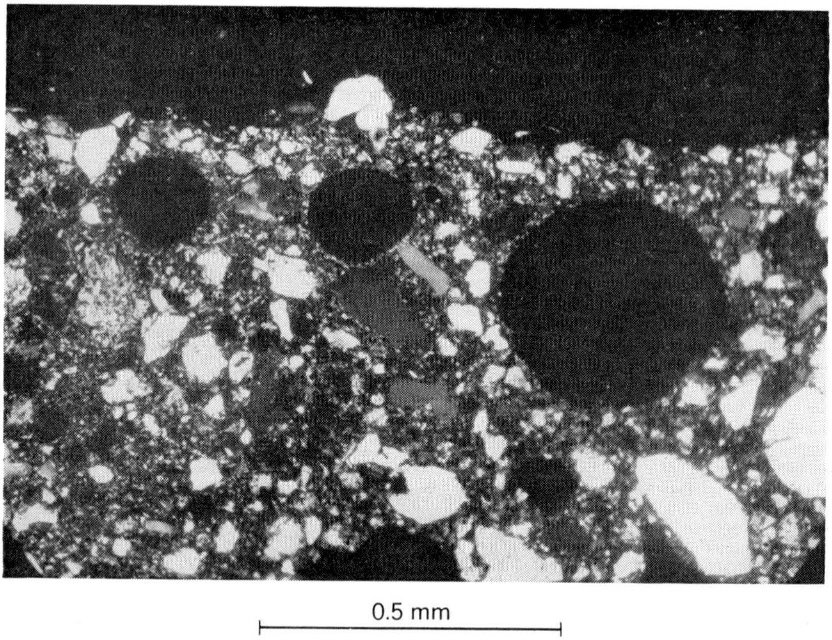

Fig. 38. Vesicles. Thin section under crossed polarizers. The nearly circular features are vesicles associated with a "scalded" surface. The vesicles have a banded distribution pattern parallel to the surface. 80×.

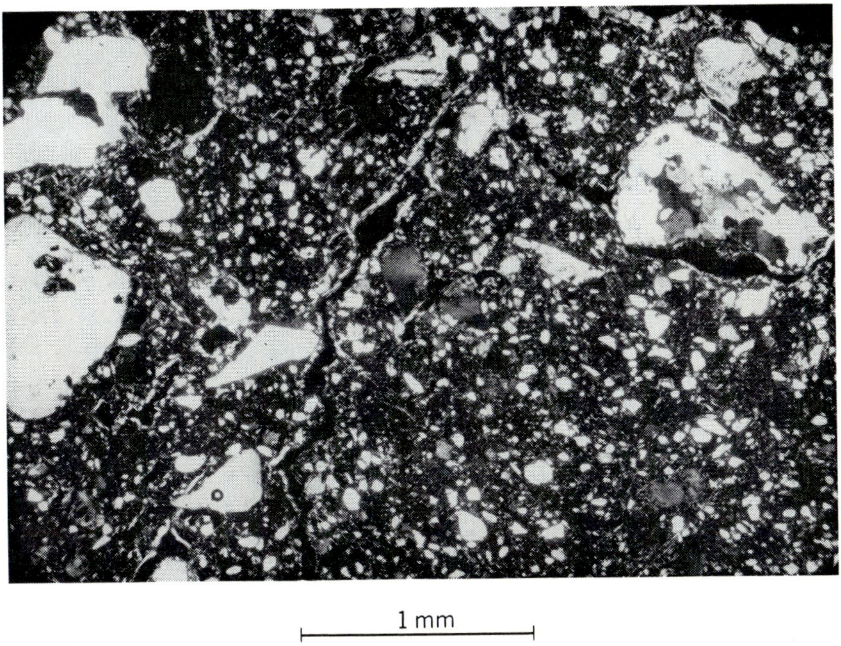

Fig. 39. Channels. Thin section under crossed polarizers. The elongated black areas are a bifurcated channel. The thin white borders to the channels are channel argillans with strong continuous orientation. 30×.

cylinder can have any ratio with the axes within that plane. It is unusual, however, for this axis to be less than 3/2 times either of the other two principal axes, so they are rarely other than prolate or acicular. Single channels, therefore, are distinguished from vesicles on shape characteristics, while branching patterns are diagnostic for the more complex types.

E. Chambers

Chambers differ from vughs in that their walls are regular and smoothed (metavoids) and from both vesicles and vughs in that they are interconnected through channels and usually have a characteristic shape. They can be subdivided according to their shape characteristics other than the ratios of the principal axes; *subspherical* and *arched* types have been observed. They are the voids that lead to the formation of chamber pedotubules (Chapter 11).

F. Planes

These are simply voids that are planar according to the ratios of their principal axes; by virtue of their shape and extent they constitute an obvious deviation from the normal packing of single plasma and skeleton grains in a soil material.

Brewer (1960*b*) subdivided planes into joint planes, slip planes, and contraction planes. Sleeman (1963) re-examined this classification in the light of his detailed study of the characteristics and distribution of planar voids in a number of soil materials. He pointed out that most planar voids are formed by contraction of the soil material, so he renamed "contraction" planes as skew planes. Also, both joint planes and skew planes can exhibit evidence of differential movement. It is proposed, therefore, that planes should be subdivided into two main groups which can be subdivided further according to the characteristics of their surfaces (conformation and smoothness):

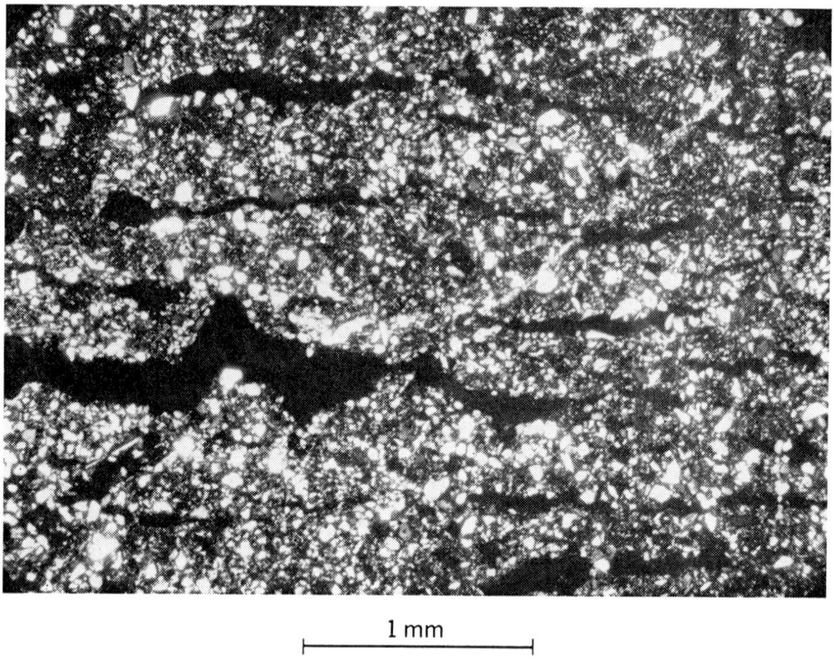

Fig. 40. Irregular ortho-joint planes. Thin section under crossed polarizers. The planes have a banded distribution pattern parallel to the surface of the soil. 30×.

Joint planes (Figs. 15, 17, and 40). Planar voids that traverse the soil material in some fairly regular pattern, such as parallel or subparallel sets;* they may occur as a series* of more than one of these sets. Their surfaces are usually smoothed, quite commonly slickensided, but they may be orthovoids.

Skew planes (Fig. 41). Planar voids that traverse the soil material in an irregular manner, having no specific basic distribution or orientation pattern between individuals; they are usually orthovoids, but they may be smoothed or slickensided metavoids.

A third kind of void best related to planes is common in many soil materials in which peds accommodate each other. Such peds may belong to the polyhedral, polysphedral, or some spheral form (Chapter 15), and the interpedal voids between them can have a wide variety of conformations but are all related to the planar form. Such voids can be thought of as a number of very short, intercon-

* The terms *sets* and *series* follow geological usage where one group of subparallel joint planes is termed a set while the presence of two or more sets arranged with different orientations are termed a series.

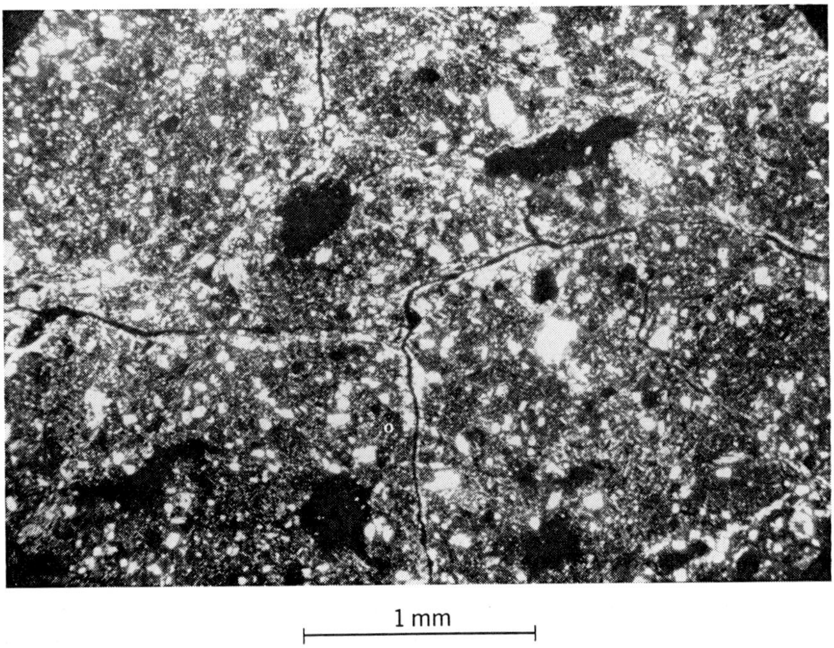

Fig. 41. Meta-skew planes. Thin section under crossed polarizers. 30×.

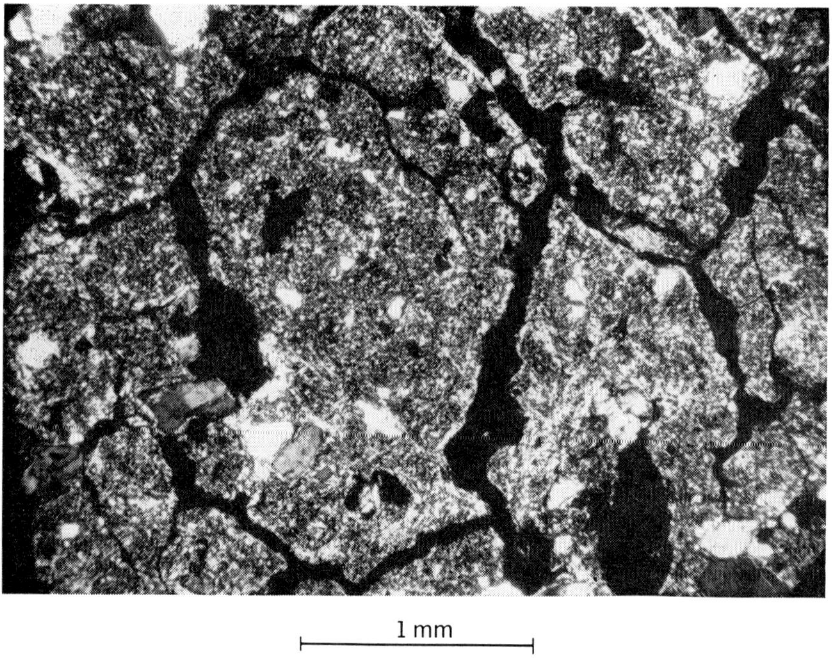

Fig. 42. Craze planes. Thin section under crossed polarizers. 30×.

nected planes, each plane being the size and shape of a pair of accommodating ped faces, or as a single planar void with a complex conformation of the walls. Similar short planar voids with random distribution and orientation occur in some apedal soil materials that have a fine system of cracking on drying. Such voids have been termed:

Craze planes (Figs. 16 and 42). These are essentially planar voids with a highly complex conformation of the walls due to the interconnection of numerous short flat and/or curved planes.

9.6 INTERPRETATION OF VOIDS

Voids in soil materials can be divided broadly into three genetic groups: those which result simply from the random packing of individuals, those which are much larger than those resulting from simple systems of packing, and those which result from some additional process superimposed on simple packing. The problem is to differentiate these three classes on the available data (i.e., on morphology) and to assess the processes by which the more complex types are formed.

A. Packing Voids

If the means are available to observe *simple packing voids* at magnifications appropriate for the grain size of the material, it is relatively easy to recognize the arrangement of single grains to form intergranular voids whose walls are the surfaces of the mineral grains. The size, shape, and arrangement of such voids depend on the size, shape, and system of packing of the grains.

Elucidation of the genesis of *compound packing voids* is not so straightforward. The individuals, whose size, shape, and system of packing result in the occurrence of the voids are compound units, particularly unaccommodated peds, and the origin of these units must be explained to explain the occurrence of the interpedal voids. Virtually nothing is known of the formation of unaccommodated peds except that some examples undoubtedly originate as faunal (especially earthworm) excreta. Such excreta are generally subspherical to spheroidal or ellipsoidal, and they often occur in a botryoidal form because of their system of packing. In other instances, unaccommodated peds are apparently due to a system of cracking caused by shrinking and swelling strains; they are bounded by a number of plane or curved faces; that is, they are polyhedric, polyspheric, or polysphedric forms (Chapter 15).

B. Vughs and Vesicles

Orthovughs have walls that are indistinguishable in smoothness from what would be expected as a result of normal packing of plasma and skeleton grains. The question, then, is why such large voids occur. They usually occur in soil materials with a wide particle size distribution, including clay size particles composed in part of layer-lattice types. These clay size particles have colloidal properties, including the ability to flocculate and to adhere to each other and, often, to adhere to the coarser mineral grains. It is feasible, therefore, that floccules of clay size materials could be formed under certain conditions, and these may adhere quite strongly to the associated skeleton grains. Under such conditions a sediment may consist not of normally packed single grains but of a complex packing of compound particles; the nature of this packing is conditioned by the degree of adhesion and attraction between particles, floccules, and skeleton grains. Such a process may result in the formation of *orthovughs.* Orthovughs may also be formed by differential weathering of mineral skeleton grains and complete removal of the weathering products by leaching.

Mammillated vughs have smoothed walls and mammillated conformation. These features are difficult to explain by either of the hypotheses proposed for the formation of orthovughs. They occur most commonly in soil materials with strong evidence of extensive faunal activity, especially of earthworms; obvious faunal excreta are often abundant, commonly in the form of "welded" excreta, which consist of clusters of coalescing fecal pellets with a mammillated external form very similar in conformation and smoothness to the walls of mammillated vughs. It is probable, therefore, that they are the direct result of faunal activity; they often exhibit other features characteristic of faunal activity, such as darkening of the soil material for a short distance inward from the surface of the walls associated with increased denseness of the soil material, which may be due to close packing of the constituent grains caused by pressure.

The very regular conformation and pronounced smoothness of the walls of *vesicles* indicate a more specific mode of formation than by mere accidents of flocculation and packing. Voids with similar characteristics in rocks have been attributed mainly to expansion and movement of gas bubbles. A similar explanation may be feasible in some soil materials, as for the vesicles often occurring in a horizontal zone immediately underlying surface seals in "scalded" areas (Fig. 38); entrapped air could produce such subspherical vesicles, and this hypothesis is enhanced where two or more such vesicles which overlie each other in the vertical plane are joined by a narrow neck. Lapham (1932) considered that vesicles near the surface of a sierozem were formed by air entrapped by rainfall during dry, dusty periods. Springer (1958) verified this hypothesis in the laboratory but pointed out that the process was operative only in soil materials with certain textural and structural properties characteristic of certain kinds of desert soils; the vesicular structure was unstable and transitory. Air bubbles entrapped during deposition of sedimentary material may be responsible for some of the clustered patterns of vesicles in the subsoils of some profiles developed on sediments; in these, the vesicular material often occurs sporadically as masses associated with denser material; vesicles formed in this way may be physically anisotropic, especially prolate, oblate, and triaxial, possibly due to compaction during further burial by sedimentation subsequent to their formation. The vesicles that form in the crusts of takyrs on drying after heavy rainfall have been attributed to liberation of carbon dioxide in the formation of carbonates from bicarbonates under conditions where the temperature of the soil solution is high; carbon dioxide liberated by biological activity is considered to be important also (Paletskaya, Lavrov, and Kogan, 1958).

There is no completely satisfactory explanation at present, however, for some of the physically anisotropic vesicles, such as the triaxial and oblate types in *A* horizons of some differentiated soil profiles, which are usually oriented with their long axes parallel to the surface of the soil. They may be due to alternate wetting and drying since this process is known to produce horizontal fissures in some surface soil materials (Czeratzki and Frese, 1958), but the specific conditions are not known. Alternatively they may be due to lenses of ice formed during freezing of the surface soil. Small arcuate vesicles observed in some subsoils are probably formed by some specific, systematic process of cracking, and experimental production of arcuate voids by alternate wetting and drying of a soil clod is suggestive in this regard (Czeratzki and Frese, *loc. cit.*).

C. Channels and Chambers

On the basis of size, cross-sectional shape, and kind of branching pattern, the most logical explanation of the formation of channels is by the activity of fauna or plant root systems or possibly by certain geological processes (Brewer and Sleeman, 1963; Chapter 11). *Single branching patterns* at depth in the soil profile apparently originate as earthworm burrows; *anastomosing branching* is most likely to originate as ant burrows; *trellised branching* is likely to be caused by termites; and *dendroid branching patterns* probably originate from decay of plant root systems, which may be modified by subsequent earthworm activity. Single branching patterns near the surface may be due to the activity of a wide variety of fauna, especially if the channels do not extend to any great depth, and, in special instances, they may be due to escape of entrapped air (Springer, 1958) or carbon dioxide (Paletskaya, Lavrov, and Kogan, 1958).

In pedotubules (Chapter 11), the surface characteristics of the walls of the voids are often masked by the infilling material, so they cannot be used in interpretation of the origin of the voids. In unfilled channels, however, such characteristics may be useful. Plant root systems often leave some evidence of their previous existence in the form of decayed organic material, which may diffuse into the outer margins of the walls of the channels causing a dark, diffuse stain; and they may exert considerable pressures on the walls, which may cause slight striations, denser packing near the surfaces of the walls as compared with the normal soil material, and even a slight degree of striated orientation of the plasma close to the walls. Excreta in channels may suggest their origin as faunal burrows, although fauna may use channels formed originally in other ways. Earthworms may cause smoothing, discoloration, cementing, and dense packing in the walls by forcing excreta into the small packing voids (Jacot, 1936; Kevan, 1955). Termites, too, commonly coat the walls of their galleries with a dark stain (Ratcliffe and Greaves, 1940; Hill, 1942), and the galleries are elliptical in cross section.

The formation of *chambers* has also been attributed to the activity of fauna (Brewer and Sleeman, 1963). They often exhibit the characteristics of smoothness, dense packing, and dark staining in walls that are attributable to faunal activity. They may originate as chambers at the end of earthworm burrows (ovoid shapes) or as chambers connected by ant or termite galleries, both of which are usually arched; that is, they have relatively flat floors and domed roofs.

D. Planes

Planes in soil materials originate primarily through shrinking and swelling during wetting and drying; shrinkage by itself produces tensions within the soil material by a decrease in volume of the plasma fraction, and this produces systems of cracks, or planes. This is in agreement with conclusions reached by geologists who have studied jointing in igneous and sedimentary rocks; such joint planes have been attributed to stresses and tensions due either to earth movements in sedimentary rocks or to stresses from crystallization in igneous rocks such as columnar basalts.

Sleeman (1963) discussed the origin of planes in soil materials and made the following points:

(1) Since up to 50 percent of the total shrinkage in soil materials often occurs at moisture contents greater than that corresponding to pF 4.2, large planes would originate only under very dry conditions (Stirk, 1954).

(2) Once formed, planes tend to open in the same place during successive drying cycles, so they are relatively permanent (Robinson, 1949; Czeratski and Frese, 1958; Gotschalk, 1957).

(3) The distribution and orientation patterns of planes depend

on the kind and uniformity of the soil material, the presence of inherited cleavages, and the pattern of wetting and drying. Heterogeneity of material and irregular drying both tend to form discontinuous, irregularly arranged planes. Soil materials are generally quite heterogeneous, and they dry most irregularly at the surface and less so with depth in the profile.

(4) The more plastic the material, the greater is the internal cohesion and the more widely spaced are the planes and the smoother their surfaces (Wolkewitz, 1958; Czeratzki and Frese, 1958). There is some confusion here since the experiments conducted by these workers show that calcium-saturated clays form wider-spaced planes with smoother surfaces than do either hydrogen-saturated or sodium-saturated clays. On the other hand, Baver (1948) asserted that sodium-saturated clays are more plastic, based on Atterberg numbers, than calcium-saturated clays. Added humic acid had the same effect as sodium saturation, and high iron oxides had the same effect as calcium saturation.

(5) Homogeneous clay soil materials on drying form a crack pattern (pattern of planes) that is hexagonal in the horizontal plane (Zonneveld, 1956; Winterkorn and Tschebotarioff, 1947).

(6) Higher clay size content and higher exchange capacity are correlated with smoother surfaces.

These points suggest the mode of origin of joint planes and skew planes. The occurrence of sets and series of *joint planes* suggests that conditions were such that a relatively regular system of cracking was produced; Sleeman (1963) also pointed out that the surfaces of joint planes are usually relatively smooth. The characteristics that aid in development of such systems are high clay size content and homogeneity of the material, calcium saturation, and uniform regular drying. Slickensided planes usually occur as sets and series of subparallel planes; they are usually joint planes whose surfaces have been modified by differential movement during alternate wetting and drying. White and Bonestall (1960) suggested that sets of slickensided joint planes at angles of 20 to 30 degrees to the horizontal in gilgai soils may be formed by stresses produced during wetting from the surface downward, plugging of vertical cracks by the expanding wet clay size soil material, and, finally, shearing in a regular pattern. They suggested that the curved planes of Krishna and Perumal's (1948) lenticular pedality are due to a similar process.

Sleeman (1963) also suggested that inherited planes, such as bedding planes, may persist in soil materials, especially in soil profiles developed from sedimentary materials. He explained some of the joint planes observed in such soils as inherited bedding planes from aeolian cross-bedded sediments (Twenhofel, 1950) and others as inherited bedding planes in fluviatile deposits.

Skew planes do not occur in regular patterns and usually have rougher surfaces than joint planes; they would be produced under conditions of higher heterogeneity of the material, sodium saturation, and/or irregular drying.

Craze planes may originate in a similar way to skew planes if it is assumed that such irregular patterns of planes of small extent can be produced by shrinking under conditions of highly irregular drying and/or in very heterogeneous materials. Many such planes have relatively smooth surfaces which are not suggestive of high sodium saturation, but high humic acid content may be a contributing factor since such planes often occur in chernozem-like soil profiles.

10

Pedological Features I: Cutans

One of the more common pedological features reported (e.g., Allen 1932; Peterson, 1937; Johnson and Peterson, 1941; Frei and Cline, 1949; McCaleb, 1954; Altemuller, 1956; Kubiena, 1938, 1956*a*; Karpachevskii, 1960; Dalrymple, 1962) consists of plasma concentrations or plasma separations associated with natural surfaces within the soil material, that is, associated with the faces of peds and skeleton grains and the walls of voids; these are the situations where the greatest reorganization of plasma is to be expected. Some of these phenomena have been recognized as coatings of clay minerals, and they have been referred to as "clay skins" or "tonhautchen"; both these terms have a strongly genetic inference indicating that the surfaces of peds or the walls of voids have been coated with "clay" by deposition after illuviation in suspension.

These phenomena were originally recognized (in reflected light at magnifications up to 60×) by their waxy luster and/or irregular shape, "like the appearance of tallow which has run down the side of a candle and hardened" (Kubiena, 1938). A closer study of phenomena identified as "clay skins" has shown that there are many variations in the kind and arrangement of the material that cannot be determined by reflected light characteristics. Surface luster is particularly sensitive to changes in organization of material, so a difference in luster between the broken interior of a ped and the ped surface may be due to a change in organization (fabric) only. The term "clay skins" and "tonhautchens," therefore, are inappropriate as group names for all these phenomena because they mean "coatings of clay," and such an identification cannot always be made by the methods used. In addition, "clay" is a broad, inconcisely defined term that generally to soil workers means any material whose particle size is less than 2 μ equivalent diameter, but to some workers it means particular mineral species irrespective of size. Similar accumulations of materials on the faces of primary mineral grains and the walls of voids in sedimentary rocks have not been given specific names but are referred to generally as "cementing materials."

Because of these considerations a new term, *cutan** (plural form *cutans;* adjectival form *cutanic*), has been coined to include all the variations in phenomena associated with natural surfaces in a soil material:

> *Cutan.* A modification of the texture, structure, or fabric at natural surfaces in soil materials due to concentration of particular soil constituents or *in situ* modification of the plasma; cutans can be composed of any of the component substances of the soil material.

By definition cutans are pedological features; those formed by deposition or diffusion of plasmic material are *plasma concentrations,* while those formed by *in situ* modification of the plasma are *plasma separations* (Section 7.5:B). Buol and Hole's (1961) definition of a clay skin as "the assemblage of optically oriented clay (<0.002 mm), with included coarser particles, formed on the walls of interstices in the soil and exhibiting abrupt internal and external boundaries" is based on microscopic characteristics in thin section; apparently any material of less than 2 μ equivalent diameter is accepted as "clay." Such materials may not be recognizable as "optically oriented" by microscopic methods, for example, opaque forms of manganese and iron oxides; anisotropic clay minerals may occur as very weakly oriented "skins," and skins of clay size materials, such as manganese oxides, may be optically unoriented. The definition, therefore, is of a particular group of cutans composed dominantly of optically oriented clay minerals.

Although the definition is strongly genetic, cutans are recognized by their location (distribution pattern) as evidenced by a change in concentration, texture, structure, or fabric and by their shape, which conforms to the shape of the natural surfaces with which they are

* "Cutan" is derived from the Latin "cutis" meaning "a coating, rind, or surface of a thing" (Gepp and Haigh, 1928); the ending has been changed for convenience in usage, so that the term can be defined without fear of confusion due to translations of the original Latin.

associated. The definition has been made broad so that cutans can be recognized in the field and classified to the extent that field data will allow; for example, all the phenomena that have been identified as "clay skins" are cutans irrespective of their method of formation.

A number of varieties of cutans are shown in Figs. 43–59. These are referred to later as examples of particular characteristics of cutans.

10.1 DISTINGUISHING CHARACTERISTICS

Cutans are recognized by their referred distribution pattern where the specific reference features are natural surfaces within the soil material, especially the surfaces of peds, skeleton grains, voids, and even other pedological features that occur as individuals, such as nodules. They are subdivided according to the characteristics of the surfaces with which they are associated and their internal structure, texture, or fabric. A further subdivision can be made on their composition,

Fig. 43. Simple free grain argillans with strong continuous orientation. Thin section under crossed polarizers. The thin white borders to the skeleton grains are the argillans. Most of the black areas are simple packing voids since all the plasma occurs as argillans. 80×.

but shape characteristics are of no use because cutans are not discrete, three-dimensional entities. Their shape is controlled by the characteristics of the surface with which they are associated, and extent and thickness are the only significant measures of size. The most useful sequence for considering these differentiating characteristics is:

(1) the characteristics of the surfaces affected,
(2) the mineralogical nature of the cutanic material,
(3) internal fabric, and
(4) additional characteristics.

By using these characteristics, cutans can be classified and the various kinds named and qualified by descriptive adjectives.

A. Characteristics of the Surfaces Affected

Cutans are classified at the highest level according to the kind of surface with which they are associated. The following sub-groups

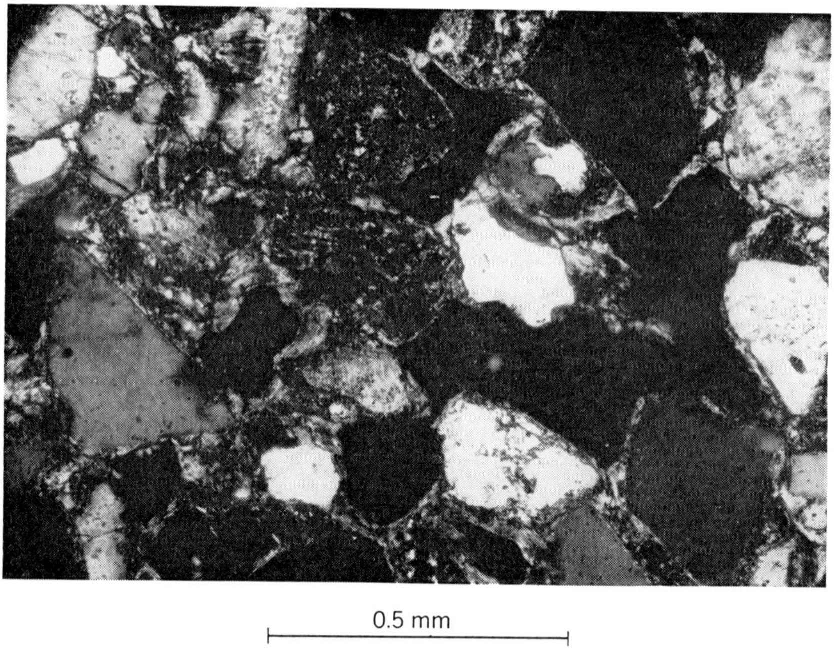

Fig. 44. Simple free grain argillans with moderate orientation. Thin section under crossed polarizers. The mottled borders to the skeleton grains are the argillans. 80×.

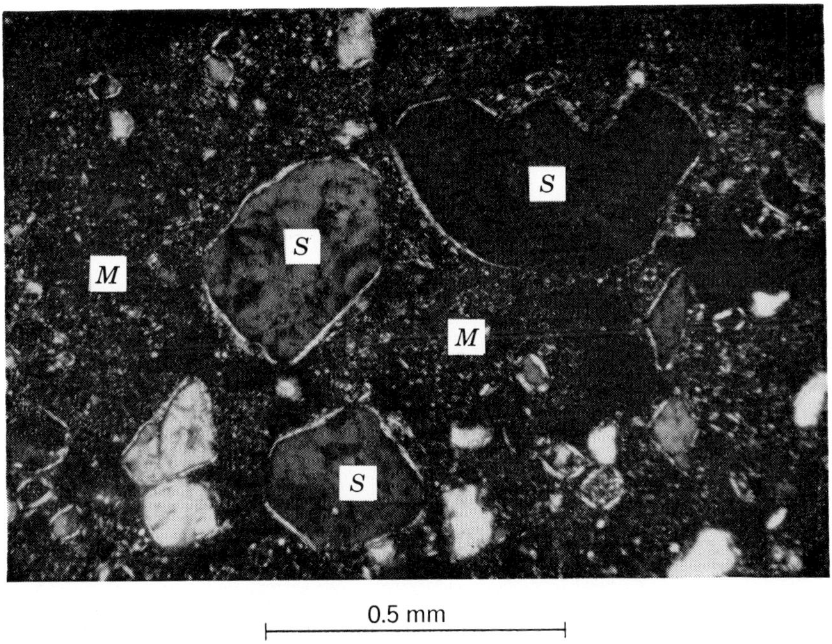

Fig. 45. Simple embedded grain argillans with strong continuous orientation. Thin section under crossed polarizers. The thin white borders to the larger skeleton grains (*S*), in the s-matrix (*M*), are the argillans. 80×.

follow those previously proposed (Brewer, 1960*b*) with some modifications due to the more complete classification of voids (Chapter 9).

(*a*) ***Grain Cutans.*** Cutans associated with the surfaces of skeleton grains or other discrete units, such as nodules, concretions, etc. There are two types:

Free grain cutans (Figs. 43 and 44) occur on the surfaces of grains which form the walls of voids. Such an arrangement results from normal packing of grains in a loose sand. The loosely packed grains usually adhere to their neighbors at the points of contact because of the binding effect of the cutanic material. Although these cutans can be considered as being on the walls of simple packing voids, it is more appropriate to refer them to the surfaces of the grains.

Embedded grain cutans (Fig. 45) occur on the surfaces of grains embedded in relatively densely packed plasma; displacement of the grains reveals cutanic material either adhering to the grains or on the surfaces of the impression left by its removal. The cutans are not visibly (megascopically or microscopically) connected to voids which could act as conducting channels.

(b) Ped Cutans (Fig. 14). Cutans associated with the surfaces of peds. They are subdivided into primary, secondary, and tertiary types according to the order of the peds with which they are associated (primary, secondary, tertiary peds).

(c) Channel Cutans (Figs. 39 and 46). Channel cutans are associated with the walls of channels whether these be of biological origin (fossil formations of worm channels, etc.) or not.

(d) Plane Cutans. Cutans associated with the walls of planar voids other than those between peds. They are divided into joint-plane cutans (Fig. 19), skew-plane cutans (Figs. 47, 49, 51–53, 55, and 58), and craze-plane cutans. Joint planes and craze planes can occur as interpedal voids, but for the classification of cutans the term *ped*

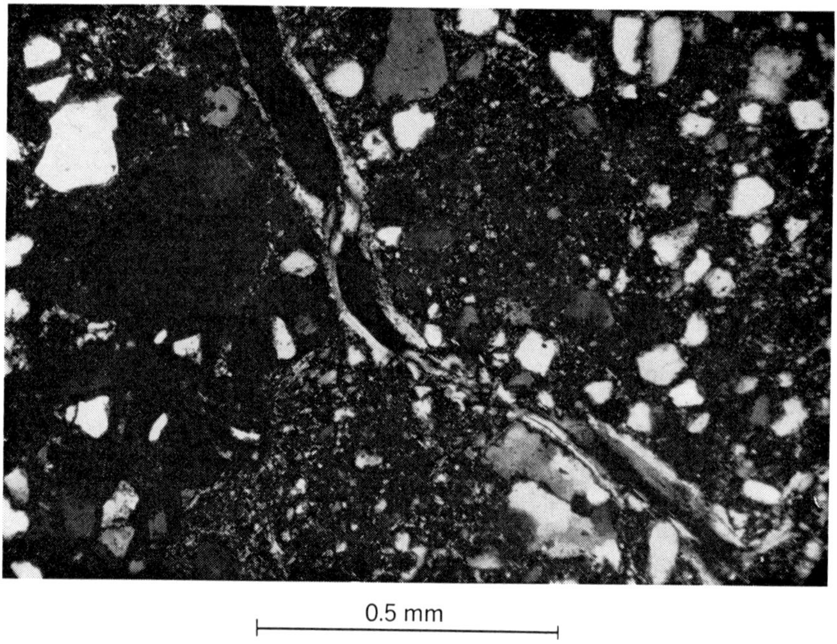

Fig. 46. Simple channel argillan with strong continuous orientation. Thin section under crossed polarizers. The diagonal black area is a curved acicular void (channel). The white border is the argillan. Fine black extinction lines are visible in the argillan. 80×.

Fig. 47. Simple skew-plane sesquan (ferran). Thin section in plain light. The curved diagonal black area is the sesquan. 135×.

cutans is more specific and descriptive than joint-plane and craze-plane cutans where these planes delineate peds. For these reasons the concepts of ped cutans and plane cutans, as just defined, are retained.

(*e*) ***Normal Void Cutans.*** Normal void cutans are associated with the walls of the normal equant, triaxial, and prolate voids within primary peds or in apedal soil materials. On the basis of the morphological classification of voids they can be subdivided into *vugh cutans* (Figs. 48 and 57), *vesicle cutans,* and *chamber cutans.*

B. Mineralogical Nature of the Cutanic Material

Although the mineralogy of the cutanic material is not a part of structure and fabric descriptions, it is a convenient and significant characteristic for classification of cutans, especially in regard to interpretation of their genetic significance. The usual techniques of

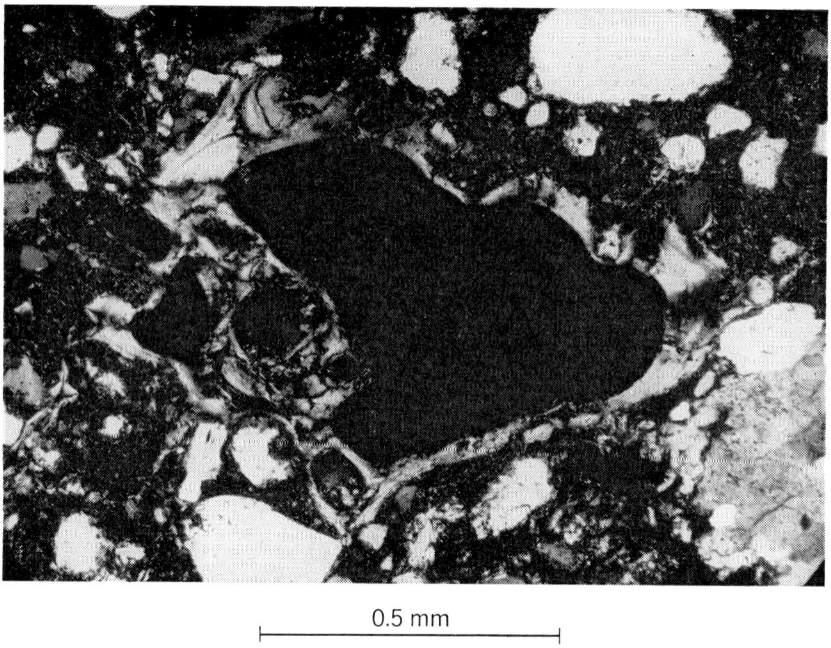

Fig. 48. Simple vugh argillan with strong continuous parallel orientation. Thin section under crossed polarizers. The central black area is a vugh. The rather broad white zone surrounding the vugh, and showing black extinction lines, is the argillan. 80×.

identification are optical in thin section on the petrological microscope and chemical on material isolated by dissection of the soil material. It should be possible by these means to identify the composition of the cutanic material and even to estimate the proportions of the constituents quantitatively. Such work, however, is tedious and time consuming, and little has been done as yet. The following classification, therefore, places cutans in rather broad groups according to their chief constituent, or rather the constituent that is most obvious with the techniques used. Further groups, and subdivisions of them, will be necessary as experience grows and more precise techniques are applied. The names are derived simply by adding the ending *ans* to an abbreviation of the dominant cutanic material.

(*a*) ***Argillans*** (Figs. 43–46, 48, 49, and 54–56). Argillans are composed dominantly of clay minerals. The term "clay minerals" is used advisedly in contrast to the general term "clay." It refers to the layer-lattice minerals, the attapulgites, sepiolites, and palygorskites, the chlorites, and the allophane types; crystalline oxides, carbonates, and sulfates are excluded from the clay mineral group for the purpose of classifying cutans. In reflected light, thin argillans have a smooth, glazed surface, while thick ones usually have an uneven, ropy appearance with waxy luster. A more detailed subdivision is possible by identification of the kind of clay mineral and measurement of the size and shape of the individual clay mineral grains. Although cutans composed of pure clay minerals are rare they do occur, usually as white, creamy, or grey coatings; cutans of relatively pure palygorskite have been reported from New South Wales (Beattie and Haldane, 1958), and such types may be given a more specific name, *palygorskans* (Fig. 49). More commonly the clay minerals are mixed with some contaminant and can be subdivided according to this on the basis of color and a few simple spot tests:

Ferri-argillans consist of a mixture of clay minerals and iron oxides

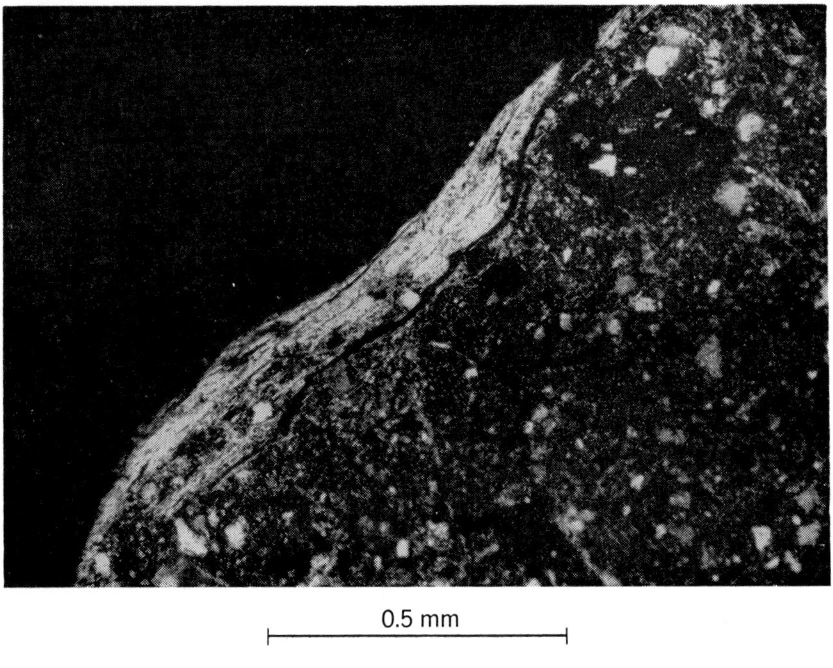

Fig. 49. Simple skew-plane palygorskan with moderate orientation. Thin section under crossed polarizers. The rather thick, diagonal, white zone on the left is a moderately oriented skew-plane palygorskan.

or hydroxides colored in shades of yellow, red, green, or blue depending on the degree of hydration and oxidation of the iron oxides or hydroxides. Ignition produces a bright red color, and comparison of this with the original color gives some idea of the degree of hydration and oxidation of the iron compounds within a single cutan.

Organo-argillans (Fig. 19) consist of clay minerals stained by organic compounds which usually produce a grey to dark grey color, although organic matter may be colorless, yellow, or even red. The degree of charring on heating gives some indication of the amount of organic matter, and ignition reveals the presence or absence of iron compounds.

(*b*) *Sesquans* (Figs. 47 and 50). These consist of sesquioxides or hydroxides. In reflected light the material may have a glazed appearance but no waxy luster. If iron oxides or hydroxides are present

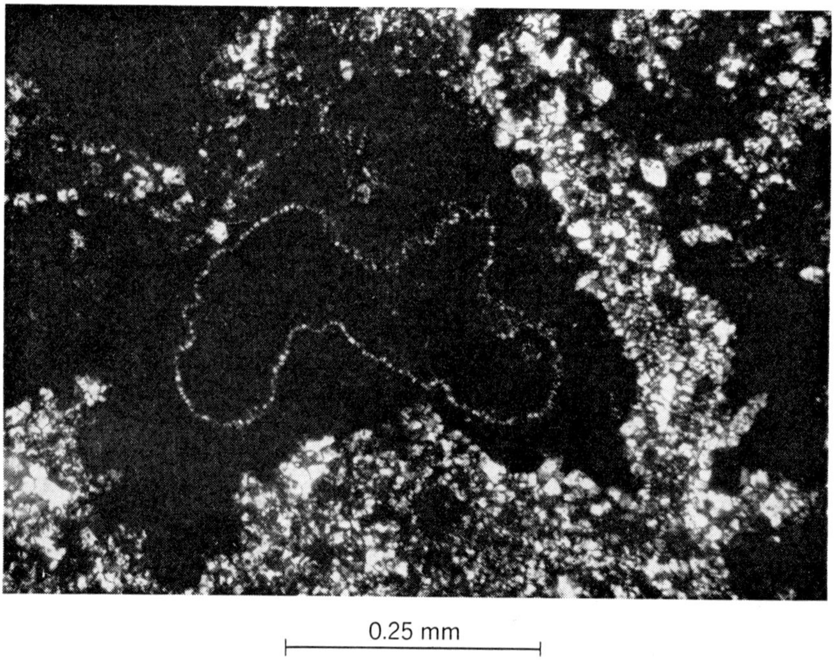

Fig. 50. Simple vugh gibbsan with normal orientation. Thin section under crossed polarizers. The central black area is a vugh. The thin, broken, white line is the gibbsan. The black area outside the gibbsan is opaque sesquioxides. Note that the individual gibbsite grains are oriented normal to the walls of the vugh. 135×.

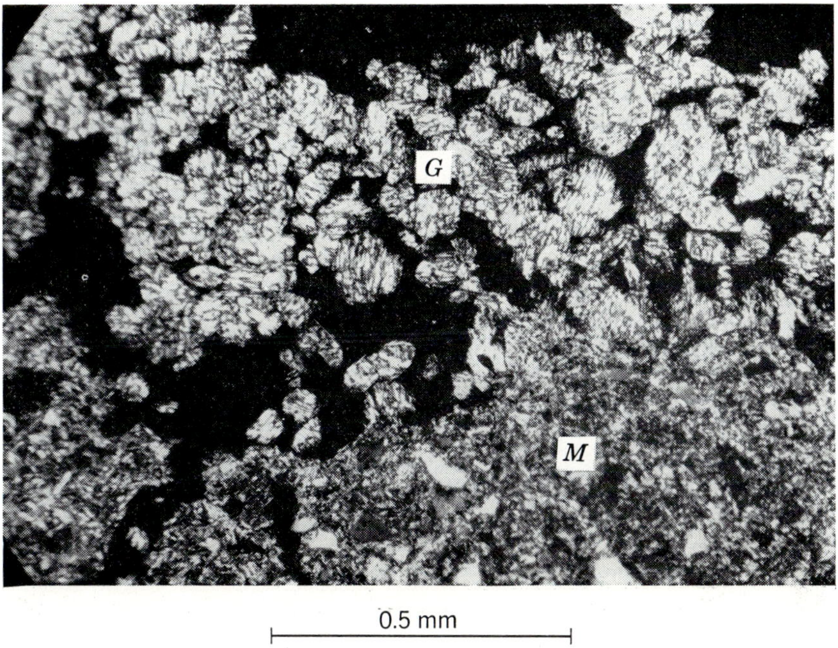

Fig. 51. Simple soluan; skew-plane gypsan (*G*). Thin section under crossed polarizers. *M* is the s-matrix. 80×.

it is red, yellow, blue, or green in reflected light according to the degree of hydration and oxidation; ignition produces the typical bright red color of ferric oxide, and solution by Jeffries' method (1946) or the dithionite method (Aguilera and Jackson, 1953) leaves little or no residue. In thin section cutans of iron oxides or hydroxides (ferrans, Fig. 47) are translucent to opaque and commonly isotropic because of the very fine grain size. If the mineral can be identified they can be classified as hematans, goethans, etc. Aluminium oxides and hydroxides are often identifiable in pure forms, for example, gibbsite (gibbsans, Fig. 50), but are difficult to identify when they are associated with highly colored iron minerals in sesquans; X-ray analysis of dissected cutanic material may be necessary for their identification.

(*c*) ***Mangans*** (Figs. 20 and 25). Mangans consist of manganese oxides or hydroxides. The material may have a glazed appearance and is very dark brown to black in reflected light; it is usually opaque in thin section. The periodate method (Piper, 1942) will identify the presence of manganese quantitatively.

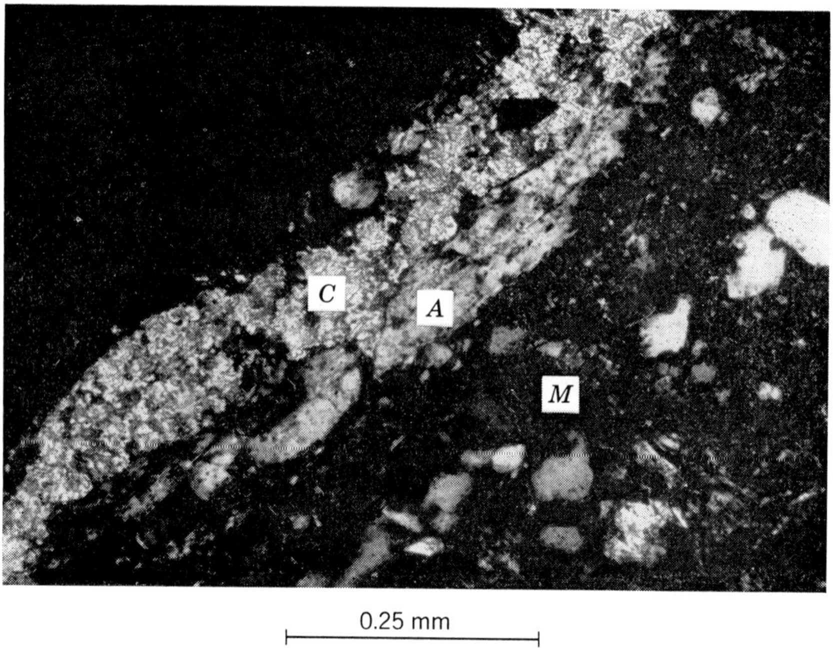

Fig. 52. Compound skew-plane argillan-calcitan. Thin section under crossed polarizers. The calcitan (*C*) overlies the argillan (*A*). *M* is the s-matrix. 135×.

(*d*) ***Soluans*** (Figs. 51 and 52). These consist of crystalline salts, such as carbonates, sulfates, and chlorides of calcium, magnesium, and sodium. The material occurs as a fine powdery efflorescence or as masses of coarser crystals; it is usually white or colorless with yellow and grey tints. It may have a glassy or dull earthy luster according to the degree of crystallization. The various salts can be identified by separation, solution and microtests, and flame tests; coarser crystals can be identified optically. If the cutanic material is identified more specific names may be applied, for example, *gypsans* (gypsum, Fig. 51), *calcans* [carbonate: *calcitans* (calcite, Fig. 52)], and *halans* (halite).

(*e*) ***Silans*** (Fig. 53). Silans consist of silica in its various forms; the most common forms are silt or clay size quartz and poorly crystalline chalcedony. It is important genetically to distinguish primary from secondary silica, but this is difficult in practice. Staining tests for amorphous silica have been used (Kubiena, 1938) but are generally unreliable. Infrared absorption (Fieldes et al., 1956) has also been used. Cutans of silt and clay size quartz look generally like soluans in reflected light, but they are usually distinguishable from the latter on optical properties and hardness. Cutans of chalcedony are usually readily recognizable, if they are reasonably thick, because of their vitreous luster in reflected light and hardness. The former may be termed *quartzans* and the latter *chalcedans* (Fig. 53). Overgrowths on quartz grains in some soil materials from Australia have been recently identified optically in thin section (Breese, 1960). These are, in fact, special cases of *embedded grain quartzans.*

(*f*) ***Skeletans.*** These consist of skeleton grains adhering to the cutanic surface. In reflected light they are usually similar in appearance to quartzans of which they are, in fact, a variety differing only in the mineralogical composition of the grains. The presence of other minerals suggests that the quartz is primary rather than secondary.

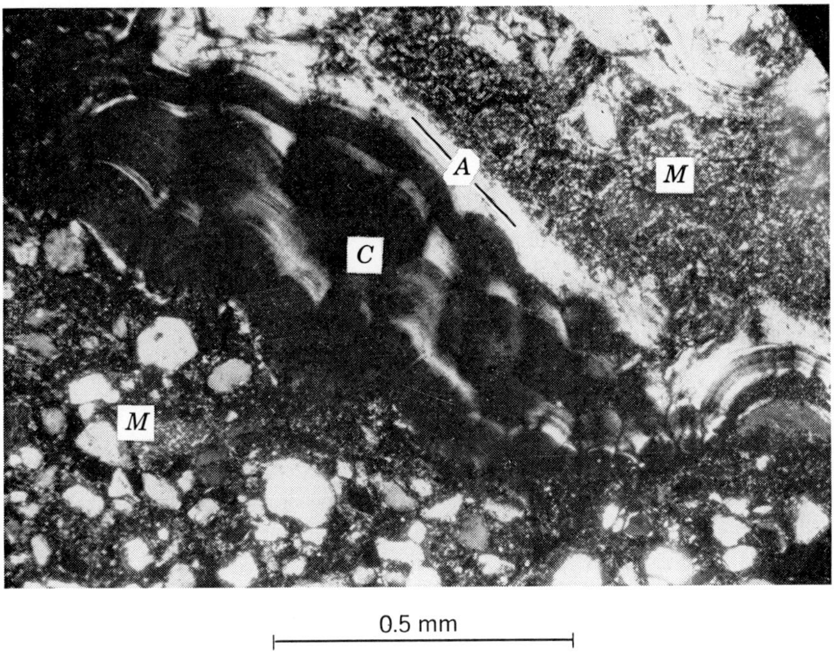

Fig. 53. Compound skew-plane argillan-chalcedan. Thin section under crossed polarizers. The chalcedan (*C*) overlies the argillan (*A*). *M* is the s-matrix. 80×.

An important characteristic of skeletans is their mineralogical composition compared with that of the skeleton grains of the s-matrix of the soil materials. Skeletans formed in channels are a morphological intergrade to granotubules (Section 11.1).

C. Internal Fabric

Further subdivisions of cutans can be made on the physical arrangement (fabric) of the cutanic material. These characteristics are described under the following headings in the terms set out earlier in the descriptive system for structure and fabric (Section 8.3):

(1) Sharpness of boundary refers to the boundary between the cutanic material and the normal soil material since the other boundary is sharp, either because it is within a void or because it is the boundary with a solid entity, such as a skeleton grain or nodule.

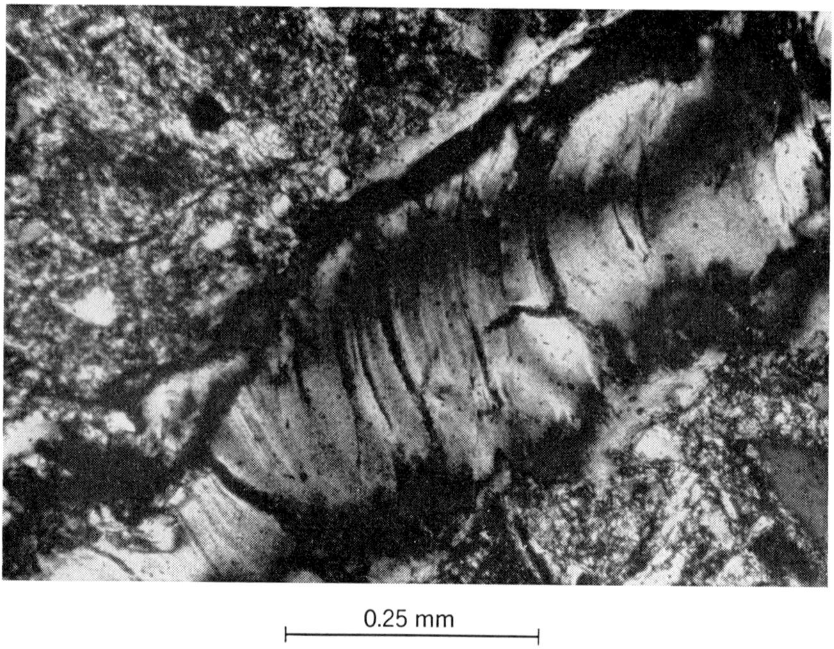

Fig. 54. Elongated section through the edge of a simple channel argillan with strong continuous orientation. Thin section under crossed polarizers. Note the lamination in the argillan, across its length, and the dark extinction lines almost parallel to its length. 135×.

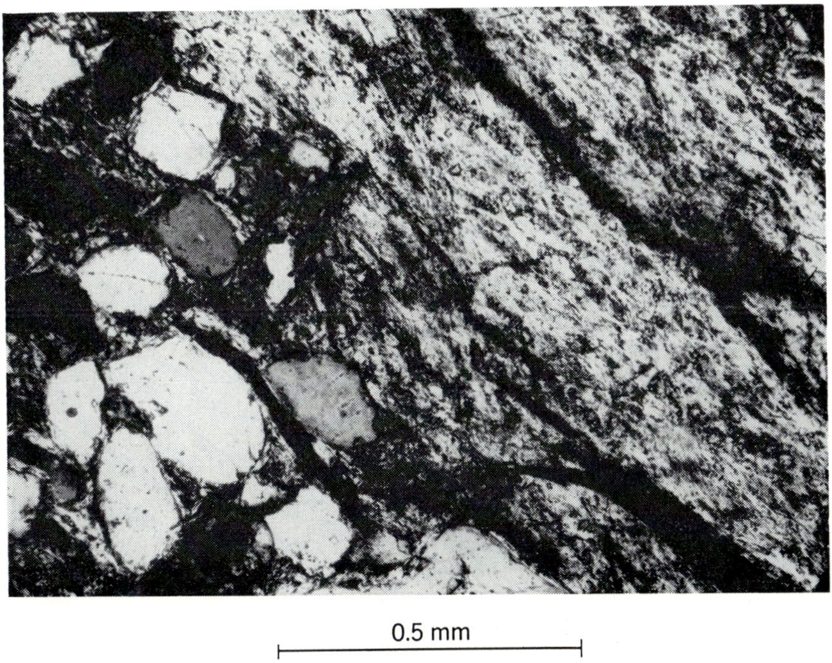

Fig. 55. Simple skew-plane argillan with striated orientation. Thin section under crossed polarizers. Note the lack of distinct extinction lines in the argillan. 80×.

(2) The basic fabric, in particular the basic orientation (degree and pattern), which is effectively the plasmic fabric of many cutans since they are often composed entirely of plasma, and the referred orientation with reference to the cutanic surface with which the cutan is associated. Figures 54–56 show three kinds of basic orientation (strong continuous, striated, and flecked orientation) in argillans composed of intimate mixtures of illite and kaolinite.

D. *Additional Characteristics*

For the interpretation of soil genesis, as opposed to identification and structure descriptions, two other characteristics may be important and should be described. These are the *degree of separation* and the *degree of adhesion:*

(1) Degree of separation deals with the degree of contrast in fabric or concentration of material between the cutanic and noncutanic soil

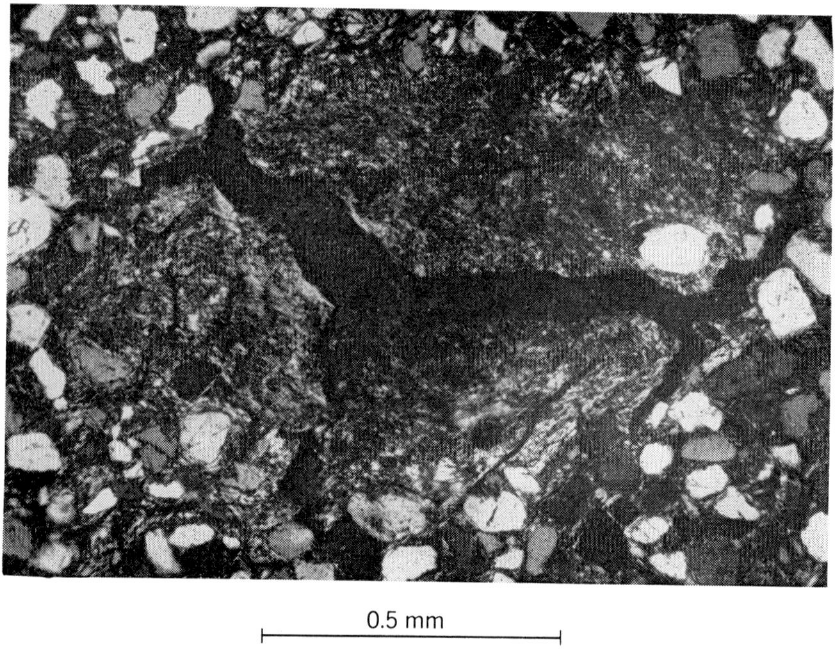

Fig. 56. Section through the filled end of a simple channel argillan with flecked orientation. Thin section under crossed polarizers. Note the lack of any directional extinction pattern in the argillan. 80×.

material. It is recorded as *slightly separated* where there is just an observable difference, *weakly separated* where there is a definite difference, *moderately separated* where there is a significant difference, and *strongly separated* where the characteristics of the cutanic material are entirely those due to the characteristics of the concentrated component of the plasma, or where there is a large decrease in the proportion of skeleton grains in the cutanic material.

(2) Degree of adhesion refers to the strength of adhesion between the cutanic material and the cutanic surfaces. It is estimated by the amount of cutanic material adhering to the cutanic surface after gentle crushing. It is recorded as *discrete** when virtually none of the surfaces are coated with cutanic material after gentle crushing, *weakly adhesive* when less than 30 percent of the surfaces are coated

* "Discrete" replaces "free" (Brewer, 1960*b*) because of the danger of confusion with "free" as referring to the mode of occurrence of the group of free grain cutans.

after crushing, *moderately adhesive* when 30 to 60 percent of the surfaces are coated, and *strongly adhesive* when more than 60 percent of the surfaces are coated after gentle crushing.

10.2 CLASSIFICATION AND NOMENCLATURE

Although insufficient data are available to be precise about the genetic interpretation of all the various kinds of cutans, the task can be made somewhat simpler by classifying them into broad groups on their morphological characteristics before superimposing an interpretive classification. The first subdivision can be made on the basis of the complexity of the cutanic material into *simple* and *compound cutans:*

Simple Cutans. Composed of a single mineralogical and/or chemical substance or of a uniformly intimate mixture; there is no

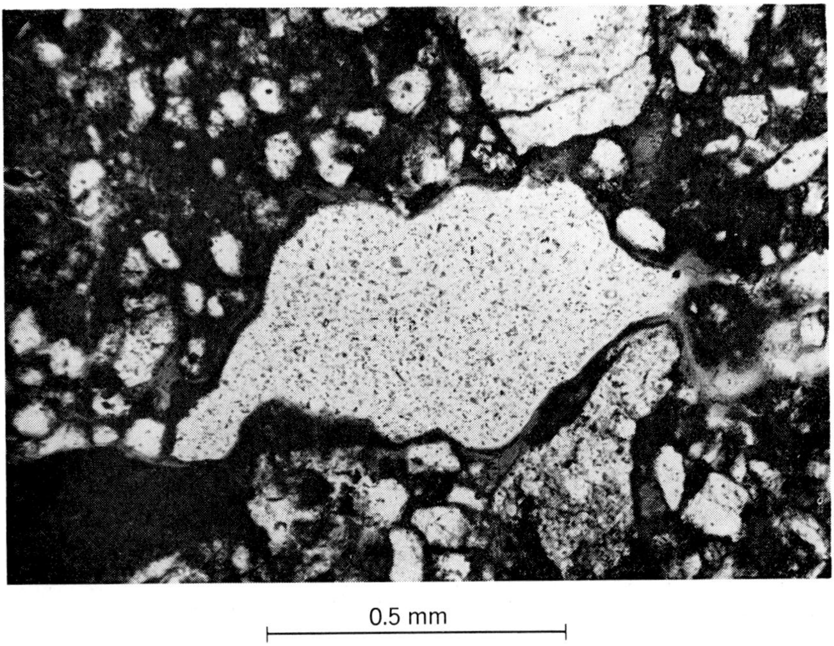

Fig. 57. Compound vugh argillan-mangan. Thin section in plain light. The central white area is a vugh with a thin, black mangan overlying a thin, paler argillan. 80×.

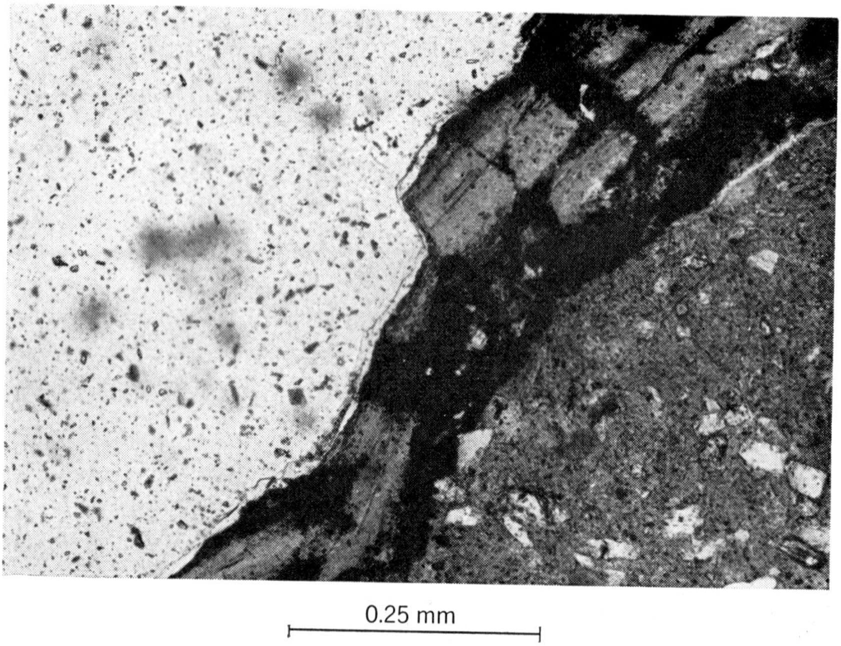

(a)

Fig. 58. Compound skew-plane mangan-argillan. (*a*) Thin section in plain light.

detectable layering effect in composition or fabric (Figs. 43–51 and 54–56).

Compound Cutans. Composed of alternate layers of mineralogically and/or chemically different substances or of different fabrics; more than one process may be involved in their formation (Figs. 52, 53, and 57–59).

These two broad groups can be subdivided on the details of their morphological characteristics. Since, at the present state of knowledge, it is not always possible to classify any particular observed cutan completely, no morphological categories have been set up, but cutans are described by means of qualifying adjectives and phrases used in a standard sequence. The operative qualifying terms, in their order of importance, are as follows:

(1) kind of surface affected,
(2) simple or compound cutan,
(3) kind of cutanic material,

0.25 mm

(b)

(*b*) Thin section under crossed polarizers. The black manganese oxides underlie and break through the thick paler argillan. 135×.

(4) orientation pattern, or internal fabric,
(5) sharpness of boundary,
(6) degree of separation, and
(7) degree of adhesion.

These characteristics can be used individually or collectively according to the evidence available. If (1) and (2) are used they should precede the noun; the other terms are best used after the noun in the sequence set out. The descriptive classification can be simplified slightly by using the terms proposed for cutans composed of various fractions of the plasma (Section 10.1), for example, "joint-plane argillan with strong, continuous, parallel orientation, sharp boundary, strong separation, and weak adhesion." These names based on composition of the cutan are particularly useful in the description of compound cutans; the names can be compounded systematically so that the name appearing first is that closest to the cutanic surface;

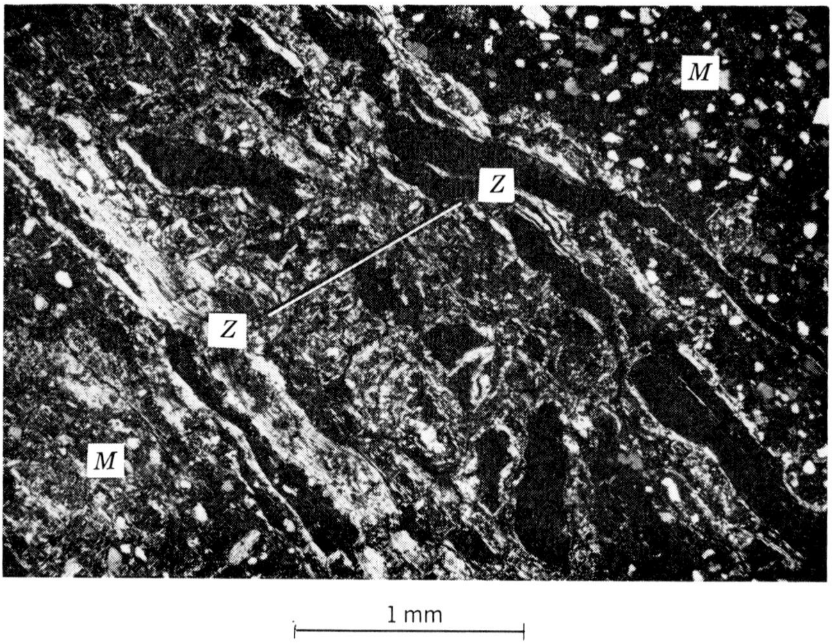

Fig. 59. Section through a thick complex zone (*Z*) of vugh and plane argillans and chalcedans. Thin section under crossed polarizers. *M* is the s-matrix. 30×.

for example, the compound cutan composed of a layer of clay minerals and a layer of calcite shown in Fig. 52 can be classified as an argillan-calcitan. An outline of the morphological classification of simple cutans is set out in Table 10.1.

In addition to this morphological classification cutans can be subdivided into four major groups on the basis of interpretation of the process of formation:

Illuviation Cutans. Formed by movement of the cutanic material in solution or suspension and subsequent deposition.

Diffusion Cutans. Concentrations at a surface due to diffusion; they are usually associated with concentrations within the s-matrix of the soil material that reach a maximum at the cutanic surface (Section 13.2).

Stress Cutans. *In situ* modifications of the plasma due to differential forces such as shearing; they are not true coatings. Most stress cutans are simple types.

TABLE 10-1. Morphological Classification of Simple Cutans

Cutan Type	Surface Affected	Composition*†	Other Characteristics†
1. Grain cutan	Surfaces of skeleton grains and glaebules	Argillans (ferri-, organo-argillans; palygorskans, kaolinans, etc.)	Internal fabric (sharpness of boundary and internal orientation pattern—especially of argillans)
(1) Free grain	Surfaces of loosely packed grains		
(2) Embedded grain	Surfaces of grains embedded in densely packed plasma		
2. Ped cutan	Surfaces of peds (natural surfaces)	Sesquans (ferrans, aluminans, geothans, etc.)	
3. Plane cutan	Surfaces of planar voids		
(1) Joint-plane	Surfaces of joint planes		
(2) Skew-plane	Surfaces of skew planes	Mangans	Degree of separation
(3) Craze-plane	Surfaces of craze planes		
		Soluans (gypsans, calcitans, halans, etc.)	Degree of adhesion
4. Channel cutan	Surfaces of channels		
5. Normal void cutan	Surfaces of "normal" voids		
(1) Vugh	Surfaces of vughs	Silans (quartzans, chalcedans)	
(2) Vesicle	Surfaces of vesicles		
(3) Chamber	Surfaces of chambers	Compound cutans	

* *Stress cutans* have the same composition as the s-matrix of the soil material. *Skeletans* usually occur on the walls of conducting voids, that is, joint planes and channels.

† Applicable to all types classified according to the surface affected.

Complex Cutans. Formed by a combination of more than one of the above processes.

This interpretive classification can be superimposed on the morphological classification if sufficient evidence is available. If it is used, the qualifying adjective immediately precedes that describing the kind of surface affected; for example, a cutan may be classified as an "illuviation, joint-plane argillan, and so on." Such statements, whether they are complete or not, provide the interpretation and a statement of the observed evidence on which the interpretation is based (see Table 10-2).

In addition to this classification within the group of cutans they can be classified in the broader sphere of soil materials. Most cutans are *in situ* pedological features; illuviation and diffusion cutans are plasma concentrations, and stress cutans are plasma separations at a surface.

10.3 CODE FOR CUTANS

Since the complete classification of cutans requires a rather extended statement of their morphological characteristics and their genetic interpretation, it is convenient to use abbreviations or symbols, especially when the description of the cutans is only a part of an even more protracted description of a soil material. The following code has been developed for this purpose. The symbols can be used in the order in which they are set out or the various characteristics can be grouped in a convenient way as shown in the examples at the end of the list of symbols.

A. *Symbols*

(*a*) Genetic type of cutan.
- C: genesis unspecified.
  - C^1: illuviation cutan.
  - C^2: diffusion cutan.
  - C^3: stress cutan.

(*b*) Kind of cutanic surface.
- A: surface of a ped.
  - A^1: surface of a primary ped.
  - A^2: surface of a secondary ped.
  - A^3: surface of a tertiary ped.
- G: surface of a solid entity other than a ped, such as a skeleton grain.

G^1: surface of a free grain.
G^2: surface of an embedded grain.
C: surface of a channel.
P: surface of a planar void.
P^1: surface of a joint plane.
P^2: surface of a skew plane.
P^3: surface of a craze plane.
V: surface of a normal void.
V^1: surface of a vugh.
V^2: surface of a vesicle.
V^3: surface of a chamber.

(*c*) Nature of the cutanic material.
p: unpigmented, or pure, clay minerals.
f: clay minerals pigmented by, or mixed with, iron oxides or hydroxides.
f^1 : red.
f^2 : yellow.
f^3 : shades of blue or green.
h: clay minerals pigmented with organic matter.
r: sesquioxides.
r^1 : red iron oxides or hydroxides.
r^2 : yellow iron oxides or hydroxides.
r^3 : iron oxides or hydroxides in shades of blue or green.
r^4 : aluminium oxides or hydroxides.
m: manganese oxides.
s: soluble salts.
s^1 : carbonates (dominantly calcite).
s^2 : sulfates (dominantly gypsum).
s^3 : chlorides (dominantly halite).
q: silica.
q^1: quartz.
q^2: chalcedony.
sk: skeleton grains.

(*d*) Internal fabric.

Orientation pattern:

O: domains are not observable; the cutan as a whole has strong, continuous orientation (continuous birefringence).

M: domains are not observable; the cutan as a whole has moderate orientation (micaceous birefringence).

W: domains are not observable; the cutan as a whole has weak orientation (weak birefringence).

U: domains are not observable; the cutan is isotropic because of the low degree of orientation.

D: domains are not observable; the cutan is isotropic because of opacity of its plasma.

I: domains are not observable; the cutan is isotropic because of the crystallographic nature of its plasma.

F: anisotropic domains are arranged randomly (flecked extinction).

S: anisotropic domains have a preferred parallel arrangement (striated extinction).

Crystallization of cutanic materials other than clay minerals:

X: crystalline; birefringent except for cubic minerals such as halite.

X^1: cryptocrystalline.

X^2: opaque.

Referred orientation with regard to the cutanic surface:

R: preferred orientation is parallel to the cutanic surface.

R^1: normal to the cutanic surface.

R^2: inclined to the cutanic surface.

(*e*) Nature of the boundary with the s-matrix.

/: sharp.

//: rather sharp.

///: rather diffuse.

////: diffuse.

(*f*) Degree of separation.

D: strongly separated.

D^1: moderately separated.

D^2: weakly separated.

(*g*) Degree of adhesion.

H: strongly adhesive.

H^1: moderately adhesive.

H^2: weakly adhesive.

H^3: discrete.

B. Writing the Code

In writing the code the symbols are arranged in the following order: genetic type, kind of cutanic surface, nature of the boundary with the s-matrix, nature of the cutanic material, internal fabric (the symbol referring to the orientation of the plasma grains with regard to each other precedes that for the referred orientation with regard to the cutanic surfaces), degree of separation, and degree of adhesion. The last two symbols are dissociated from the preceding ones by a horizontal dash (–). Thus, the code,

$$C^1P^1/f^1OR\text{–}DH^3$$

is a red, illuviation, joint-plane ferri-argillan with strong, continuous, parallel orientation and a sharp boundary, and it is discrete.

This method of writing the code is also suitable for recording the detailed characteristics of compound cutans by using the convention of stating the mineralogical nature of the material closest to the cutanic surface first. A compound cutan that consists of a layer of clay minerals against the cutanic surface, overlain by a later of calcite (Fig. 52), can be coded as:

$$C^1P^2/f^2OR/s^1X\text{–}DH^2$$

This is interpreted as a compound, depositional, skew-plane cutan composed of crystalline calcite and yellow iron-impregnated clay minerals with strong, continuous, parallel orientation, with a sharp boundary both between the clay minerals and the s-matrix and between the clay minerals and the calcite, the cutan as a whole being strongly separated and weakly adhesive. Alternatively it is a ferri-argillan-calcitan with all the other qualifications. This example illustrates the conventions involved; the kinds of cutanic material are listed from left to right according to the order of their occurrence from the cutanic surface, the symbols for sharpness of boundary refer to the cutanic layer immediately following each symbol, the symbols for fabric of the cutan refer to the cutanic material immediately preceding them, and the symbols for degree of separation and degree of adhesion refer to the compound cutan as a whole.

10.4 INTERPRETATION OF CUTANS

If the cutans present in a soil material can be classified as simple or compound and, further, into one of the four groups of the interpretive classification, a great amount of information concerning the processes of formation of the soil material has been achieved. This can be amplified by identification of the cutanic material and measurement and comparison of the amount of cutanic material compared with noncutanic material in the soil material. The greatest need, then, is for a list of characteristics for classifying cutans at the two highest levels. This is the interpretive stage of the study, and such interpretations must necessarily be largely tentative at present; more definite and dependable interpretations will be possible as experience grows, particularly if the various kinds of cutans can be reproduced under controlled conditions in the laboratory. Certain facts are known, however, and some other inferences seem legitimate.

There is usually little difficulty in classifying cutans as simple or

TABLE 10-2. Diagnostic Characteristics of Cutans (after Brewer, 1960*b*)

Characteristic	Kind of Cutan			
	Illuviation		Diffusion	Stress
Kind of material	Those which form suspensions	Those which form solutions	Those which form solutions	Those which readily undergo reorientation
Sharpness of boundary	Sharp	Sharp	Diffuse (to sharp)*	Diffuse (to sharp)
Orientation pattern	Strong continuous (to weak, flecked or unoriented)	Indeterminate, or specific orientation patterns of recognizable crystals	Indeterminate, or specific orientation patterns of recognizable crystals	Moderate, striated (to strong continuous) with associated striated subcutanic plasma
Distribution pattern	Associated with conducting voids	Associated with conducting voids	Associated with voids of any size or shape	Associated with planar voids or embedded grain surfaces
Degree of separation	Strongly separated	Strongly separated	Strongly separated	Weakly to moderately separated and associated with weakly separated to unseparated subcutanic material with striated orientation

* The characteristics in parentheses may occur but are less useful diagnostically than those not in parentheses.

compound on evidence from thin sections. Simple cutans indicate relatively uniform conditions during their formation, whereas compound cutans indicate one or more changes in conditions during their formation. The real difficulty in interpretation of cutans is to distinguish between illuviation, diffusion, and stress cutans. An examination of the characteristics of cutans, however, gives some logical basis for their interpretation. The diagnostic characteristics are set out in Table 10-2.

A. Kind of Cutanic Material

The cutanic materials can be divided into those which form suspensions and readily undergo reorientation, such as the clay minerals, those which form solutions or suspensions but are not readily reoriented, such as the oxides and hydroxides of iron and secondary silica, those which form solutions, such as hydroxides, oxides, and soluble salts, and those which form suspensions only with difficulty and are not prone to take up preferred orientations, such as skeleton grains. The first group can form deposition or stress cutans but are unlikely to form diffusion cutans, the second and third groups can form deposition or diffusion cutans but are unlikely to form stress cutans, and the fourth group can only form depositional cutans under rather drastic conditions.

B. Internal Fabric

The fabric of the cutanic material can be considered in two separate phases:

(*a*) ***Sharpness of Boundary.*** In general the sharper the boundary between the cutanic material and noncutanic material the more definite is the diagnosis as an illuviation cutan. Similarly the more diffuse the boundary the more definite is the diagnosis as a diffusion or stress cutan. It is possible, however, for all these types of cutan to have sharp or diffuse boundaries.

(*b*) ***Orientation Pattern.*** Strong orientation of the plasma grains is indicative of illuviation clay mineral cutans; simple, illuviation, free grain cutans composed dominantly of clay minerals have been produced in the laboratory and always had strong, continuous orientation (Brewer and Haldane, 1957). In certain circumstances, however, illuviation argillans have moderate or weak, striated or flecked orientation (Figs. 55 and 56), and illuviation cutans of other materials are unoriented, opaque, optically isotropic, or crystallized (Figs. 51 and 52). Stress cutans are often very thin and the orientation at the surface difficult to see, but typically they have associated with them a striated orientation of the subcutanic plasma (Figs. 103–105). Embedded grain stress cutans have been attributed to rotation of skeleton grains under pressure (Lafeber, 1962). Diffusion cutans typically are crystallized or have an indeterminate orientation pattern.

C. Distribution Pattern

Illuviation cutans are always associated with conducting voids unless they have been disturbed and disrupted by subsequent mass movement or differential movement of the soil material. Nothing is known at present concerning the size of void through which particles of particular size and shape will pass in suspension, but experience suggests that illuviation argillans are usually associated with voids that can be seen under the microscope at magnifications up to 800×. Illuviation cutans of soluble materials may be associated with much smaller voids. Diffusion cutans can be associated with voids of any size. Stress cutans are usually associated with planar voids or surfaces of skeleton grains and pedological features embedded in densely packed plasma; that is, there may be no visible voids.

D. Additional Characteristics

(a) Degree of Separation. Illuviation cutans are typically strongly separated since they consist entirely of the deposited material and so contrast strongly with the undifferentiated materials of the soil mass. Similarly diffusion cutans are often strongly separated, but they may be weakly separated. Stress cutans may have any of the degrees of separation; usually they are weakly to moderately separated, and the associated subcutanic material with striated orientation is weakly separated to unseparated.

(b) Degree of Adhesion. The significance of the degree of adhesion in interpretation of the genesis of cutans is not yet understood. It is probable that variations in this property are due to variations in properties of the cutanic material and the plasma of the soil material, so it may have no interpretive value in regard to the formation of cutans.

It is apparent from this analysis of the interpretive value of the characteristics of cutans that no one characteristic is completely diagnostic for the process of formation (illuviation, diffusion, or stress) in all cases. In some instances a single characteristic is strongly diagnostic, such as strong continuous orientation for illuviation cutans composed of clay minerals and striated orientation in the subcutanic plasma for stress cutans. In many instances, however, all possible properties must be considered, and sometimes a definite diagnosis is not possible. Some of the troublesome diagnoses may be verified with more experience and the use of other characteristics of the soil material, such as bulk density and porosity, especially by comparison with soil materials from other parts of the soil profile.

11

Pedological Features II: Pedotubules

A number of pedological features with the morphology of natural voids filled with soil material and approximately tubular external form have been observed in a wide variety of soils and attributed by pedologists to infilling of root channels and animal tunnels and burrows (e.g., Hugie and Passey, 1963; Vermaat and Bentley, 1953). The term *krotovina* has been defined as irregular tubular streaks within one horizon of material transported from another horizon due to the filling of tunnels made by burrowing animals (U.S.D.A., 1951); this definition is genetic involving both morphology and interpretation. While this interpretation is undoubtedly correct in many instances, other occurrences of tubular shaped pedological features with similar morphology cannot be attributed definitely at present to such a process; their origin is unexplained. Such a genetic terminology depends on there being sufficient data available for such interpretations to be made. A purely morphological classification and terminology is desirable so that such features can be placed in a relatively stable system independent of interpretation; their classification in a genetic, interpretive system may change as additional interpretive data become available.

In geology "tubule" was originally used for "a small tube or pipe" (Rice, 1954; after Grabau and Shimer, 1909–1910), and later for "an irregular hollow, twig-like, calcareous concretion characteristic of loess . . ." (Rice, 1954; after Fay, 1920). Fay's "tubules" are specific types of the original definition and are akin to Kubiena's

(1938) crystal tubes, which are included in a separate group of pedological features (Chapter 13). Grabau and Shimer's definition refers simply to a shape that is comparable to the shape of this group of pedological features. Tubes and pipes, however, may be unfilled or filled with a wide variety of materials (see Rice, 1954), so neither definition of "tubules" can be applied directly to this group of pedological features. Therefore the term *pedotubule* is proposed (plural

Fig. 60. Discrete pedotubule. Hand specimen. An ortho-isotubule. The apparent difference in color between the isotubule and the enclosing soil material is due to a reddish cutan on the walls of the isotubule. 1⅝×.

form, *pedotubules;* adjectival form, *tubulic**). It is defined as follows:

> *Pedotubule* (Figs 23 and 60). A pedological feature consisting of soil material (skeleton grains or skeleton grains plus plasma, as distinct from concentrations of fractions of the plasma) and having a tubular external form, either single tubes or branching systems of tubes; its external boundaries are relatively sharp. Tubular form, in this context, means that the feature as a unit, or its impression in the enclosing soil material, has a relatively uniform cross-sectional size and shape, most commonly circular or elliptical; that is, the impression of the pedotubule conforms to the definition of channels (Section 9.5:D).

* "Tubulic" is used instead of "tubular" to avoid confusion since the latter is an accepted term for shape; the prefix *pedo* is unnecessary in the adjectival form since the filling in any tubular feature may be referred to as tubulic.

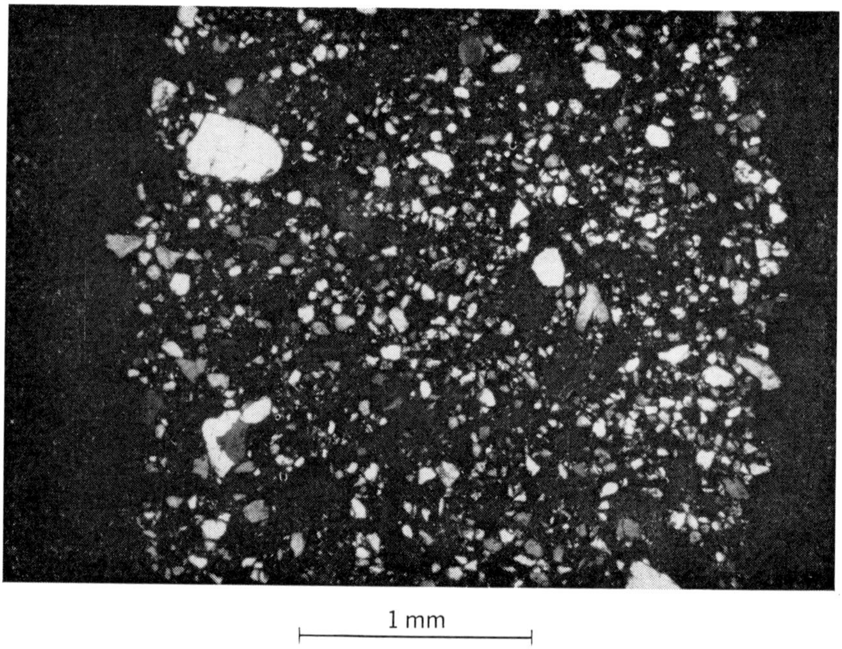

(a)

Fig. 61. Meta-granotubule. (*a*) Longitudinal section. Thin section under crossed polarizers. (*b*) Soil material enclosing the granotubule. Thin section under crossed polarizers. Note the difference in fabric, principally in the presence

Pedotubules, by this definition, are distinguished from similarly shaped crystallized material (the crystal tubes of Kubiena, 1938) and illuviation cutans by their internal fabric, which is usually comparable to that of a soil material (at the level of basic or primary fabric) as compared with accumulations of relatively pure fractions of the soil constituents. In addition, as crystallized material and illuviation cutans are formed by crystallization or deposition from suspension on the walls of voids, a remnant of a central channel usually remains, whereas such remnants are rare in pedotubules.

11.1 DISTINGUISHING CHARACTERISTICS

Pedotubules can be subdivided into more specific groups on the characteristics by which they are recognized, that is, by:

(1) the internal fabric,
(2) the composition of the plasma,

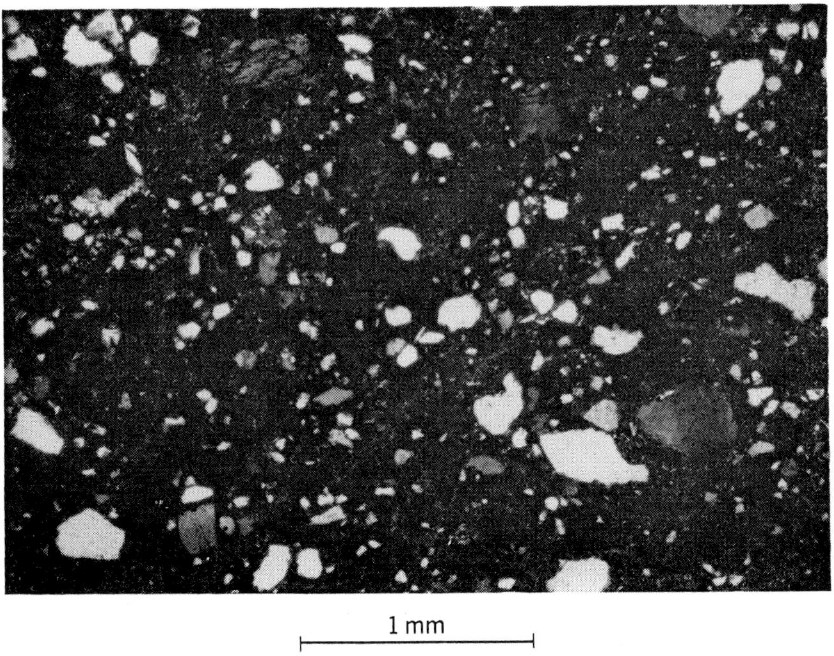

(b)

of plasma in the soil material, compared with that of the granotubule. (*a*) 30×; (*b*) 30×.

Fig. 62. Curved, moderately adhesive aggrotubule. Hand specimen (after Brewer and Sleeman, 1963). The smooth areas on the surface are adhering soil material. 5×.

(3) their external form, including cross-sectional size and shape and the pattern of distribution or branching pattern, and
(4) their distinctness in the sense of ease of separation from the enclosing soil material.

A. Internal Fabric

The internal basic fabric of pedotubules and their complete fabric and structure are quite variable. Pedotubules have been observed that consist of dense complex soil material with a basic fabric akin to some normal densely packed soil materials with anisotropic plasma but virtually no plasma separations. Others exhibit preferred orientation of the anisotropic plasma and/or voids and/or skeleton grains with regard to the external form. Others consist of loosely packed aggregates, and some of these can be identified with considerable certainty as faunal excreta. Others, again, consist of normally packed skeleton grains with virtually no plasma. All these varieties can contain included pedological features, and so they can exhibit a complex primary fabric. Pedotubules, therefore, can be grouped conveniently according to their basic fabric into a number of classes as follows, and these can be subdivided on the basis of smaller differences in basic fabric:

Granotubules (Fig. 61). Pedotubules composed essentially of skeleton grains without plasma, or all the plasma occurs as pedological features; the basic fabric is essentially granular fabric.

Aggrotubules (Figs. 62, 63, and 70). Pedotubules composed of skeleton grains and plasma which occur essentially as recognizable aggregates within which there is no directional arrangement with regard to the external form; individual skeleton grains may also occur; the basic fabric approaches agglomeroplasmic fabric in some cases.

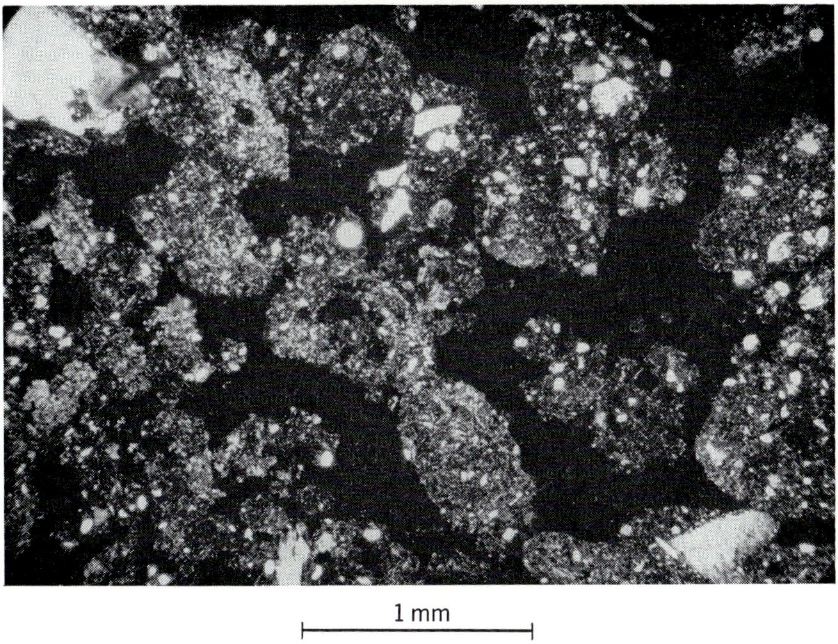

Fig. 63. Internal fabric of an aggrotubule. Thin section under crossed polarizers. Note the ellipsoidal shape of the aggregates suggesting their origin as fecal pellets. 30×.

Three variations from the normal have been recognized within aggrotubules:

Variety 1. The proportion of plasma is very low. These grade into granotubules.

Variety 2. The amount of filling is very small and consists essentially of skeleton grains and small aggregates clinging to the

Fig. 64. Discrete, segmented ortho-isotubule. Hand specimen (after Brewer and Sleeman, 1963). 2¼×.

walls of the void; such arrangements approach intergrades to cutans.

Variety 3. The aggregates are more distinct, have spheroidal or ellipsoidal shapes, and occur as groups with a mammillated or botryoidal external form; in more densely packed types the voids have a mammillated form.

Isotubules (Figs. 23 and 64–66). Pedotubules composed of skeleton grains and plasma that are not organized into recognizable aggregates and within which the basic fabric shows no directional arrangement with regard to the external form; the basic fabric is essentially porphyroskelic.

Theoretically a wide range of basic fabrics is possible within the group of isotubules, and, undoubtedly, it will be necessary to name subgroups on this basis when a wider range of variation has been observed.

Striotubules (Figs. 67 and 68). Pedotubules composed of skeleton grains and plasma that are not organized into recognizable aggregates but exhibit a basic fabric with a directional arrangement related to the external form. The directional arrangement is semi-ellipsoidal with the walls of the pedotubule approximately tangential to the semi-ellipsoid.

The fabric within striotubules is expressed by varying degrees of striated orientation pattern of the plasma (plasma separations) and/or parallelism of elongated voids and even skeleton grains. Three degrees of development have been recognized according to the development of striated orientation: *weakly developed* (plasma separations are developed parallel to, but only close to, the walls), *moderately developed* (in longitudinal section the constituents exhibit a semi-elliptical arrangement and/or orientation, and the two traces of the walls of the pedotubule are tangential to the semi-ellipses), and *strongly developed* (the semi-ellipsoidal pattern is strongly expressed; even the voids and/or skeleton grains exhibit the typical semi-elliptical arrangement in longitudinal section). Another expression of striotubules occurs in materials in which parallel orientation of the plasma does not occur. These have been termed *cleavage striotubules* which are characterized by a quite closely spaced "parting" in the form of the typical semi-ellipsoidal arrangement.

All the pedotubules that consist of relatively dense materials can exhibit a saucer-shaped parting, which is distinct from that shown by cleavage striotubules in that it is much more widely spaced; it occurs

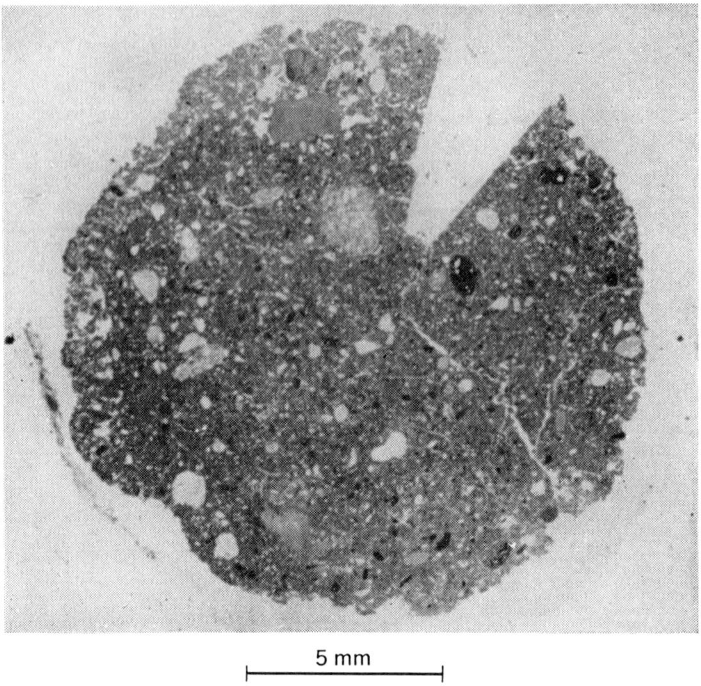

Fig. 65. Ortho-isotubule. (*a*) Transverse section (damaged) of the isotubule shown in Fig. 64. Thin section in plain light. Note circular cross-sectional

particularly in striotubules and isotubules. The adjective *segmented* is proposed to describe this feature; segmented isotubules (Fig. 64) are intergrades to striotubules which lack plasma separations.

Each of these groups of pedotubules can be subdivided according to variations in fabric, especially the occurrence and distribution of included pedological features. Of particular importance are the occurrence, nature, and distribution of cutans which occur as coatings on the walls of the pedotubules (Fig. 69) and/or on the voids within them (Fig. 70). Interpretation of these cutans and other pedological features within pedotubules provide additional evidence of the history of development of the soil material. At present these features are simply recorded together with their size, shape, and arrangement, and no specific names for the many observed variations have been proposed. Pedotubules containing such included pedological features are termed *compound pedotubules;* they are compound pedological features.

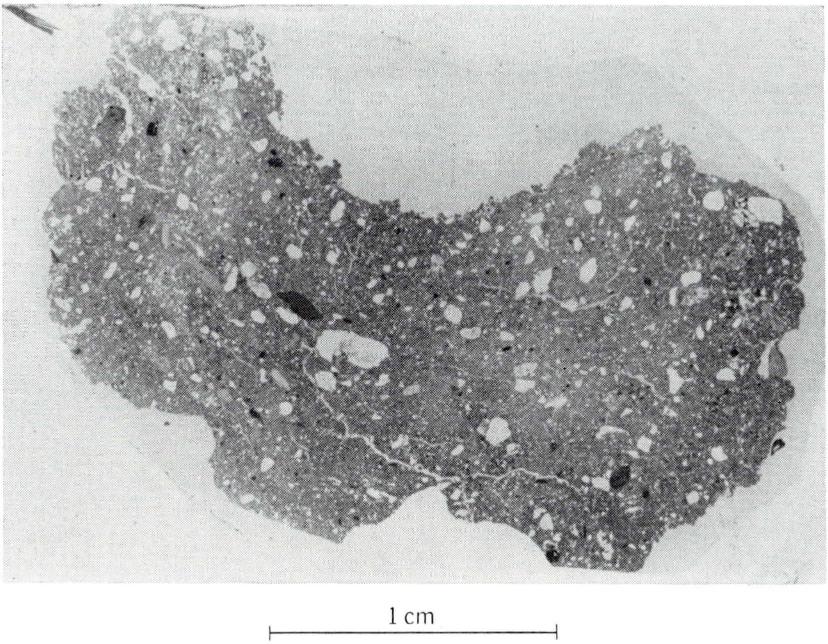

(*b*)

shape. (*b*) Section of part of the mould left after removal of the isotubule. Thin section in plain light. (*a*) 5×; (*b*) 4×.

B. Composition of the Plasma

Superimposed on the various kinds of basic fabric there can be apparent accumulations of various fractions of the plasma within pedotubules, either as pedological features (plasma concentrations) or as material intimately dispersed throughout the s-matrix of the pedotubule. Accumulations occurring as pedological features present no difficulties in treatment; they can be described in terms of their nature (composition and internal fabric), size, shape, and arrangement, and these descriptions can be used to characterize pedotubules. The treatment of apparent accumulations of fractions of the plasma, which occur essentially as part of the s-matrix, however, requires some discussion.

Since the apparent accumulation is some fraction of the plasma that probably occurs to some degree in the soil materials of all horizons of the soil profile, it can be recognized only by comparison with

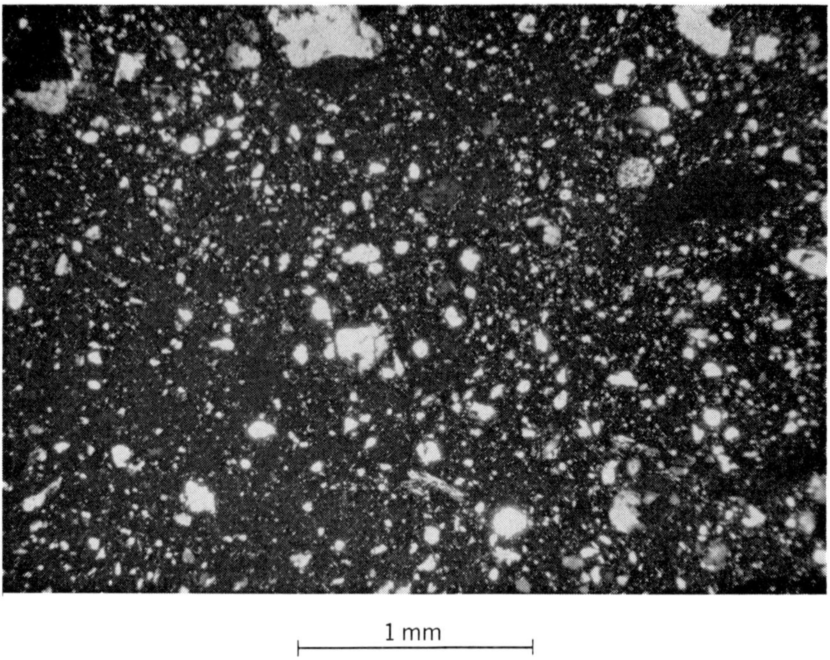

(a)

Fig. 66. Ortho-isotubule. (*a*) Internal fabric of ortho-isotubule shown in Fig. 65*a*. Thin section under crossed polarizers. 30×. (*b*) Internal fabric of the enclosing soil material shown in Fig. 65*b*. Thin section under crossed polarizers.

these soil materials. The question is whether the comparison should be with the soil material enclosing the pedotubule or with that from which the tubulic material is assumed to be derived; a pedotubule in the *B* horizon, but composed of material from the A_1 horizon, will have an apparent accumulation of organic matter as compared with the enclosing soil material but not as compared with the soil material of the A_1 horizon. Since the classification of pedotubules is intended to depend as little as possible on interpretation, the comparison is made with the enclosing soil material. A further reason for this is that such comparisons are useful in interpretation of the source of the tubulic material. Whether observed apparent accumulations of this kind are due to concentration within the pedotubule subsequent to its formation, or to the original nature of the tubulic material, is a matter of interpretation.

These accumulations of certain fractions of the plasma as compared

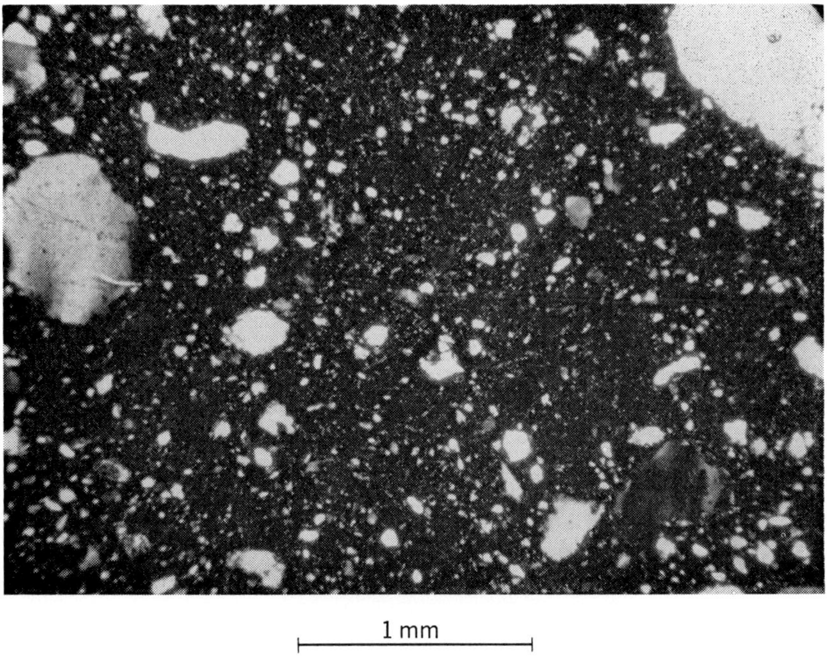

(b)

Note the similarity in fabric between the isotubule and the enclosing soil material. 30×.

with the enclosing soil material can be estimated quantitatively by comparative analyses. Usually, however, they can be recognized qualitatively on optical data and by the simple tests used for identifying the composition of cutans composed of intimately mixed constituents (Chapter 10). On these bases the following relative accumulations have been observed:

(*a*) ***Organic.*** Usually the material is notably darker in shades of grey or brown as compared with the enclosing soil material. Charring occurs on heating. *Organic-deficient* types may occur where the enclosing material is notably darker than the tubulic material and exhibits charring on heating.

(*b*) ***Sesquioxidic.*** The tubulic material has a notably stronger color than the enclosing soil material in shades of red or yellow, and ignition produces the typical bright red color of ferric oxide. Comparative quantitative estimation of free iron oxides can be made chemically.

Fig. 67. Dissected, discrete striotubule. Thin section in plain light. 6×.

Qualitatively it is convenient to distinguish pedotubules that are redder or yellower than the enclosing soil material; ideally crystallinity and mineralogy of the sesquioxides should be determined. Aluminium oxides and hydroxides are difficult to identify optically because they are usually associated with highly colored iron minerals; chemical or X-ray methods are necessary.

(*c*) ***Siliceous.*** The tubulic material has an accumulation of *secondary* silica as compared with the enclosing soil material. Secondary silica

Fig. 68. Longitudinal section of a strongly adhesive para-striotubule. Thin section under crossed polarizers. 33×.

is difficult to recognize when it occurs as thin layers of cementing material around grains. Some pedotubules, however, are quite highly silicified and can be recognized as such in thin section or even hand specimen.

Pedotubules relatively rich in other fractions of the plasma, such as manganese oxides, calcite, and gypsum, may also occur but have not been observed as yet. These would be described as manganiferous, calcareous, and gypseous, respectively. Types rich in iron sulfites, sulfides, and oxides (jarosite, natrajarosite, cerphosiderite, pyrite) have been reported from acid, poorly drained soils (Clark, Gabin, and Sprout, 1961). Types deficient in certain constituents as compared with the enclosing soil material are described as calcite-deficient (Fig. 70), etc.

C. External Form

Pedotubules commonly occur in a variety of branching patterns as well as single tubes. There are, therefore, two aspects of their ex-

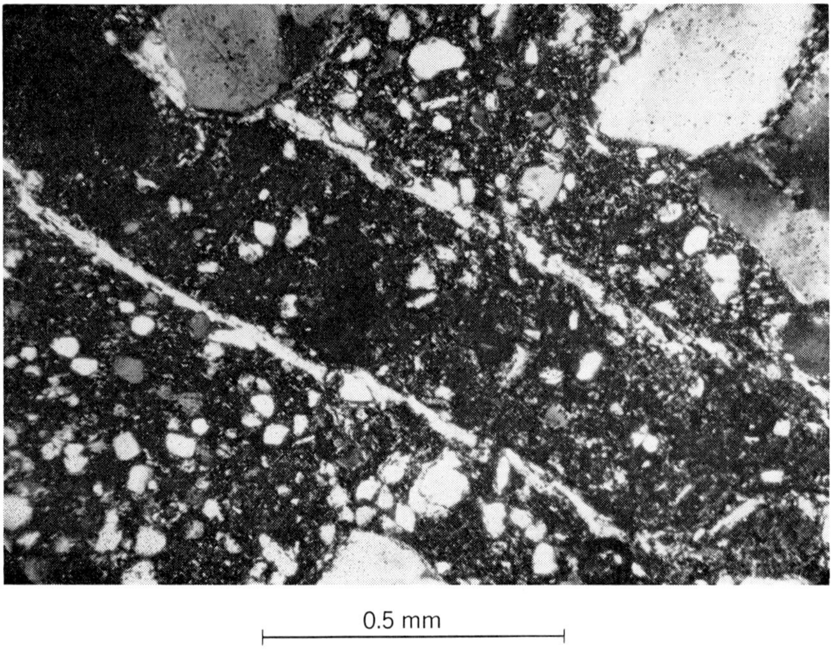

Fig. 69. Longitudinal section of an isotubule. Thin section under crossed polarizers. The central, diagonal feature is the isotubule. The white borders on the walls are an argillan with strong continuous orientation. Note that the argillan does not penetrate the s-matrix of the isotubule. 80×.

ternal form: cross-sectional size and shape, and branching pattern or pattern of distribution. Most of the pedotubules that have been observed are megascopic features, hence their cross-sectional form can be measured and described readily. Description of the branching pattern requires careful dissection, either in the field or in undisturbed monoliths in the laboratory under a stereoscopic microscope.

(*a*) ***Cross-sectional Shape.*** All the pedotubules observed so far can be referred to as *circular* (Fig. 65*a*) or *elliptical*. Many, however, have been interpreted as fillings in faunal burrows, and, if this is correct, they may have the form of any of these burrows. One common form attributed to ants (Wheeler, 1960) has a relatively flat floor and a domed roof; this form may be referred to as *arched*.

(*b*) ***Cross-sectional Size.*** The diameters of pedotubules are usually between 3 and 10 mm, but very much larger types have been observed. If their interpretation as fillings of faunal burrows is correct they may have a wide size range; size is best described simply by measurement of the shortest axis, or more specifically both axes of the cross section.

(*c*) ***Branching Pattern.*** There can be little doubt that the branching pattern is an important characteristic regarding interpretation of the origin and genesis of pedotubules. On present data three broad branching patterns (single, dendroid, anastomosing), and a fourth expression of shape related to branching patterns (chambers), have been observed. A fifth branching pattern (trellised) may be ex-

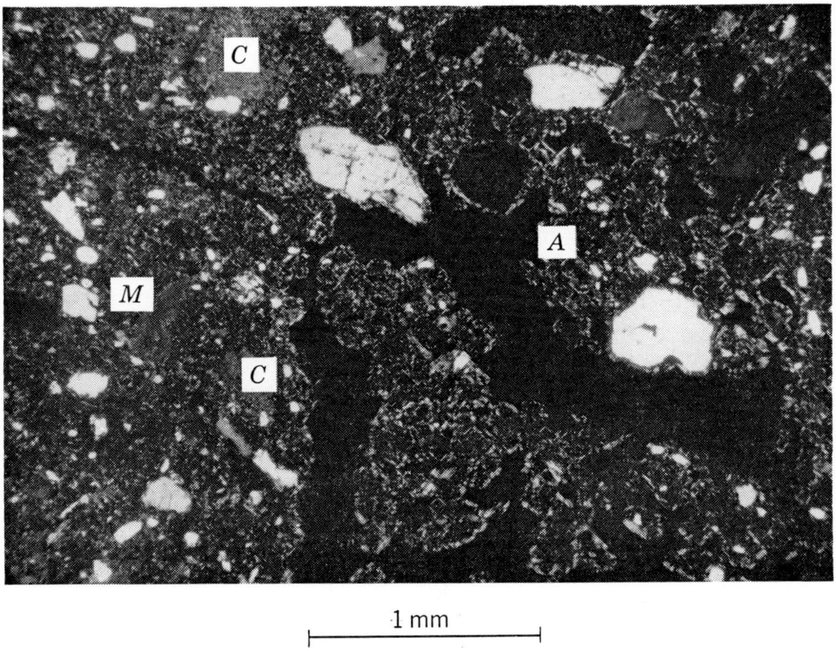

Fig. 70. Section showing the junction between a meta-aggrotubule (*A*) and the enclosing soil material (*M*). Thin section under crossed polarizers. The thin white borders to the aggregates within the aggrotubule are argillans with strong continuous orientation. Note that the enclosing soil material contains calcite (*C*) whereas the aggrotubule does not (a calcite-deficient meta-aggrotubule). 30×.

pected if termite burrows should be filled to form pedotubules; colonies have been observed where some galleries are filled (Ratcliffe and Greaves, 1940). These patterns, which are identical with those observed in channels (Section 9.5:D), are used as adjectives, for example, *single, dendroid, anastomosing,* and *trellised* pedotubules. *Chamber* pedotubules are classified with pedotubules because of their association with tubular connecting features and the nature of the tubulic material; they are defined as follows:

Chamber Pedotubules. Equant to prolate features connected by channels which are much smaller in diameter than the chambers themselves; the interconnecting voids are often empty.

Other broad groups may be found with wider experience, and undoubtedly the defined groups will be subdivided as more detailed data becomes available. It should be pointed out that pedotubules are rarely straight tubes; usually they have curves and kinks along their length rather reminiscent of the various shapes seen in the smaller limbs and twigs of trees.

D. Distinctness

Pedotubules have been defined as having sharp to rather sharp external boundaries, so distinctness refers primarily to ease of separation, as a unit, from the enclosing soil material. Because of the sharpness of the external boundaries pedotubules may be expected to be readily separated from the enclosing soil material. This is so in a large proportion of cases, but all degrees of adhesion have been observed. The best method of describing these differences is by attempting to separate the pedotubule by applying a shearing force tangential to its walls and estimating the proportion of the walls still with adhering soil material. The following terms and limits are analogous to those used for cutans:

Discrete (Figs. 60 and 64). The pedotubule is virtually free from the enclosing soil material and falls out with the application of the slightest manual* force; virtually no soil material adheres to the walls.

Weakly Adhesive (Fig. 23). The pedotubule is relatively free from the enclosing soil material and can be separated with the application of a moderate manual shearing force after which soil

* Between the thumb and forefinger.

material adheres to less than 30 percent of the surface of the pedotubule.

Moderately Adhesive (Fig. 62). The enclosing soil material adheres quite strongly to the pedotubule, which can be separated by the application of a moderate manual shearing force after which soil material adheres to between 30 and 60 percent of the surface of the pedotubule.

Strongly Adhesive (Fig. 68). The enclosing soil material adheres very strongly to the pedotubule which can be separated only by dissection or by application of a maximum manual shearing force, which still leaves soil material adhering to more than 60 percent of the surface of the pedotubule.

11.2 CLASSIFICATION AND INTERPRETATION

An outline of the morphological classification of pedotubules is set out in Table 11-1. Specific names have been used only at the highest levels of classification—granotubules, aggrotubules, isotubules, striotubules—and classification at the lower levels depends on the use of qualifying terms. There are two reasons for this: first, there are insufficient data yet to justify naming specific kinds of pedotubules with consistent characteristics at lower levels of classification; second, qualifying terms are convenient in that only those need be used for which there is sufficient evidence. If the qualifying terms are written in a systematic sequence, from the lowest categories of classification to the highest, the level of classification and the specific subgroup become apparent in the name, for example, "discrete, circular, single isotubule," and "strongly adhesive, elliptical, dendroid, sesquioxidic striotubule." Complete description of the pedotubule, however, still requires description of the detail of its basic and primary fabric. If such fabrics have been previously described and named, the pedotubule can be qualified for this characteristic, for example, "discrete, circular, single isotubule with Kempan* fabric" (Section 14.5).

Besides this morphological classification, a genetic classification can be set up on the basis of interpretation of the morphology of pedotubules in terms of their origin. Since a generally accepted interpretation of many pedotubules is that they originate as faunal burrows and tunnels which are later filled with soil material from various horizons of the soil profile [e.g., "krotovinas" (U.S.D.A., 1951), "cicada krotovinas" (Hugie and Passey, 1963), and the channels in

* Kempan being a specific name for a particular basic fabric.

TABLE 11-1. Morphological Classification of Pedotubules

Pedotubule Type	Internal Fabric	Sharpness of External Boundary	Degree of Adhesion	Comparison of Tubulic Material with Enclosing Soil Material		Other Characteristics
Granotubule	Single grains, randomly packed, without plasma, or all the plasma occurs as pedological features	Sharp	Variable; isotubules are usually discrete or weakly adhesive	*Rich* or *deficient* in some fraction of the plasma as compared with enclosing soil material	*Orthotubules:* materials comparable in fabric and composition	Cross-sectional size
Aggrotubule	Dominantly recognizable aggregates randomly packed and consisting of skeleton grains and plasma				*Metatubule:* tubulic material unlike enclosing soil material in fabric and/or composition but like the material in some horizon of the profile	Cross-sectional shape
Isotubule	Randomly packed skeleton grains and plasma; aggregates cannot be recognized					Branching pattern: single, dendroid, anastomosing, trellised, chambers connected by channels
Striotubule	Skeleton grains and plasma packed so that the basic fabric has a directional arrangement related to the external form of the pedotubule				*Paratubule:* tubulic material unlike that of any horizon in fabric and/or composition	

Ceylon laterites (Vermaat and Bentley, 1955)], it is pertinent to examine the characteristics of pedotubules in the light of this hypothesis.

A. Distinctness

Pedotubules have sharp external boundaries with the enclosing soil material, and many are discrete. These two characteristics together suggest that pedotubules either originate as channels in the soil material, which were later filled with other soil material, or else they have undergone a process of case hardening similar to that observed in many discrete concretions and nodules (Pettijohn, 1957). Many discrete types, however, have little or no demonstrable accumulation of particular fractions of the plasma, and evidence of case hardening is absent. It is logical, therefore, to interpret such types as having originated as channels. Less distinct types cannot be interpreted in this way on these characteristics; nor can it be said categorically that they did not originate in this way.

B. External Form

With experience the insect responsible for burrows observed in the soil can be identified, especially by the size and shape of the burrows (Bryson, 1939). Similar remarks apply to other fauna which burrow or tunnel in soil materials; Jacot (1940) described the size of burrow and depth of working for a wide range of fauna. Until more detail is known of the size, shape, and patterns of tunnels and burrows of soil fauna, however, positive interpretations of the origin of pedotubules cannot be made on the character of the branching pattern alone. Certain logical conclusions can be drawn from the patterns already observed, and these will serve as working hypotheses based on the evidence available at present.

There is a wide range of animals which burrow to a limited extent in the surface 1 to 2 ft, while the more extensive burrowers, such as earthworms, ants, and termites, frequently operate to depths of 15 ft or more. Within the surface 15 in., earthworm burrows may be irregular, but with increasing depth the burrow systems become simpler, are seldom branched, and tend to be near vertical with the lower ends frequently enlarged into ovoid chambers. The burrows vary in diameter from 0.5 to 12.0 mm, and the walls are smooth, cemented, and often dark colored with ejected soil pressed into the interstices (Jacot, 1936; Kevan, 1955; Barley, 1959). Excavations by ants are usually widely spaced but are occasionally anastomosing; they may be divided into chambers and galleries. The chambers have a vaulted roof and flat floor (arched) and vary in size up to 100 cm long and 30 cm broad and high; the galleries are generally circular in cross section. The walls are generally pitted and lack staining. Thus, while some ants do construct vertical galleries, these can be distinguished from earthworm burrows on cross-sectional shape, presence of chambers, surface characteristics, and a more irregular, meandering branching pattern (Jacot, 1936; Wheeler, 1960). Termite burrow systems are similar to those of ants in having chambers linked by galleries, but the galleries are mainly horizontal with occasional vertical linking shafts (trellised pattern); the cross section is elliptical and the walls are coated with a dark stain (Ratcliffe and Greaves, 1940; Hill, 1942).

It is probable, therefore, that pedotubules with specific branching patterns originate as faunal burrows. On the other hand, dendroid patterns probably originate from the decay of plant root systems; these may be modified in part by earthworms, which feed on organic matter and so tend to follow the root pattern. Unfortunately the characteristics of the walls often cannot be used effectively as diagnostic characteristics for their origin, even though it is known that certain fauna, and sometimes plant roots, impress rather specific characteristics on them. This is because the processes of infilling and packing usually destroy most of the evidence or make interpretation uncertain.

In soils formed on transported parent materials, attempted correlations of the external form with the fauna responsible for originating the voids is complicated by the fact that the void pattern, and even the pedotubules, may have been formed during pauses in the sedimentary processes of deposition of the parent material. The voids may have been due to fauna that no longer live in the environment of the present soil materials; subsequent sedimentation may bury voids and pedotubules originally formed at the surface (e.g., fossil wormtracks and trails on bedding planes). Similar considerations may apply to very old soil profiles.

All the interpretations of external form presented here suggest that their origin is due to faunal or plant activity, but other processes, such as weathering and solution, can produce channels in soils and especially in rocks; Blumel (1962) reported tubular shaped concentrations of iron hydroxides (sesquioxidic pedotubules) that were apparently formed by concentration of the iron hydroxides in waterlogged horizons of groundwater gley soils. A wide range of tubular features, either filled or unfilled, has been observed in rocks, and the accepted origins for some of these are (Shrock, 1948):

(1) solution structures, including solution of stromatolites,
(2) ascending water currents in sandy materials producing spring pits and cylindrical structures,
(3) gas bubbles ascending through sediments producing pit-and-mound structure and through lavas producing pipe and spike vesicles and amygdules,
(4) lightning striking loose dry sand or soil producing fulgurites,
(5) extrusion of lava through an opening in the hardened crust of a flow to form linear ridges with hollow centres (squeeze-ups), and
(6) molds of organic remains that have since been removed.

Such features may be preserved during weathering of rocks to form soils and so would properly be classified as pedotubules, since their morphology conforms to the definition.

C. Basic Fabric and Composition

The internal fabric of pedotubules can provide evidence of the origin of both the tubular form and the tubulic material. The granular fabric of granotubules suggests washing in of surface material or material from a highly eluviated horizon. The spheroidal and ellipsoidal aggregates that fill some aggrotubules are typical of faunal excreta, and this interpretation is strengthened if they have other characteristics of known excreta forms; earthworm excreta frequently have a diffuse dark outer border of humus, are commonly 2 to 5 mm in diameter (Kubiena, 1938), do not contain skeleton grains greater than 2 mm diameter (Barley, 1959), often have an outer layer of material of fine grain size, and commonly form botryoidal or mammillated clusters. Unfortunately little is known of the morphological characteristics of the excreta of many species of soil fauna, and it is possible that old earthworm tunnels may be filled with excreta of other soil fauna. In other aggrotubules the loose tubulic material may be due to washing in of material, especially when the aggregates do not have the smooth, curved surfaces or other specific characteristics of faunal excreta.

The more densely packed isotubules and striotubules contain no evidence of faunal activity in the form of excreta. In particular the fabrics of some striotubules indicate rather unusual pressures to form the striated pattern of orientation. Even if these arrangements are attributed to faunal activity the difference between processes which cause simple dense packing in isotubules and the parallel arrangement in striotubules is unexplained.

The source of the tubulic material may also be indicated by its fabric and the plasma fractions accumulated. In some pedotubules the tubulic material is indistinguishable from the enclosing soil material, while in others it is very like the soil material in some other horizon, usually an overlying horizon; the latter is particularly obvious when the tubulic material of a pedotubule in the *B* horizon is dark and relatively highly organic, similar to the material of the A_1 horizon. Disturbance, movement, and packing, or concentration of particular fractions of the plasma, may modify the plasmic, basic, and primary fabrics of the tubulic material. The fabric features most likely to be modified during packing are the voids, orientation pattern of plasma separations in striotubules, and plasma concentrations. The origin of the tubulic material, however, can often be assessed with some certainty by comparing the plasmic fabric, the nature and distribution of the skeleton grains, and the included pedological features of the pedotubule with those of the enclosing material, especially in conjunction with evidence of the composition of the plasma of the s-matrix. When such interpretations can be made pedotubules can be classified into three broad genetic groups:

Orthotubules (Figs. 60 and 66). The tubulic material has been derived from the soil material of the horizon in which they occur.

Metatubules (Figs. 61 and 70). The tubulic material has been derived from the soil material of another horizon from that in which they occur, usually an overlying horizon.

Paratubules (Fig. 68). The tubulic material is quite unlike that of the soil materials of any of the horizons in the profile.

D. Included Pedological Features

Pedological features included in, and associated with, pedotubules provide evidence of the history of development of soil materials. Included pedological features, such as concretions, nodules, and cutans, may be derived from the soil material that supplied the tubulic material, or they may have formed in the pedotubule. Evidence for these two origins can be obtained by comparison of the tubulic material with the soil material from which it was derived. Cutans may be associated with the contact between the pedotubule and the enclosing soil material (Fig. 69). Their type and mineralogical nature provide information concerning the conditions and processes of soil formation immediately preceding the infilling, presuming the cutans were formed at this stage. Cutans may also form after infilling when the pedotubule has assumed the form of a relatively cohesive tube. If this has been so, and if the cutans are illuviation types, then the cutans probably will have penetrated the infilling material so that included cutans will be connected with those on the walls of the pedotubule (Fig. 70). Knowledge of the processes of formation and the distribution pattern of the included pedological features, therefore, may provide evidence for the time of their formation relative to the time of formation of the pedotubules in which they occur.

12

Pedological Features III: Glaebules

A number of features that occur characteristically as units embedded in a matrix have been described in both pedology and sedimentary geology. Pettijohn (1957) named one group of such features "accretionary structures," which he defined simply as "segregations of the rarer constituents of the rock"; they are formed by accretion, that is, "the process by which inorganic bodies grow larger, by addition of fresh particles to the outside" (Rice, 1954; after Fay, 1920). Kubiena (1938) termed similar features in soil materials "amygdali" when they are microscopic in size and formed in voids by secretion, that is, by deposition in voids from solution (Rice, 1954; after Holmes, 1928), "concretions" when they are "macroscopic amygdali," and "invasion amygdali" when they are formed in the body of the soil material by accretion. All these terms have a strongly genetic connotation and infer formation by accretion or secretion of particular constituents *in situ*. On the other hand the *U.S.D.A. Soil Survey Manual* (1951, p. 23) classified all these features as "concretions," which are defined as "hardened local concentrations of certain chemical compounds that form indurated grains or nodules of various sizes, shapes, and colours." This morphological definition can be criticized on the use of "concretion" and "nodule"; Pettijohn (1957) proposed specific definitions for both terms, true concretions being "concentrically laminated structures" and nodules "structureless." To use "concretion" in a broad sense is to destroy a term with a specific meaning for features with a specific structure.

This confusion in terminology means that many of the pedological

Fig. 71. Discrete, normal, baritic nodule. Hand specimen. 3¼×.

features that occur embedded in the s-matrix of soil materials cannot be categorized with any precision. It is preferable, therefore, that a morphological classification be set up and subsequent interpretation based on morphology, mode of occurrence, experimental data, and any other available evidence. The term *glaebule** (plural form, *glaebules;* adjectival form, *glaebular*) has been coined, therefore, to include a broad group of these features. It is defined as follows:

> *Glaebule.* A three dimensional unit within the s-matrix of the soil material, and usually approximately prolate to equant in shape; its morphology (especially size, shape, and/or internal fabric) is incompatible with its present occurrence being within a single void in the present soil material. It is recognized as a unit either because of a greater concentration of some constituent and/

* The term *glaebule* is derived from the Latin *glaebula,* a small clod or lump of earth, a small piece, a bit of anything (Gepp and Haigh, 1928).

or a difference in fabric compared with the enclosing soil material, or because it has a distinct boundary with the enclosing soil material (Figs. 21 and 71).

Glaebules include Pettijohn's (1957) nodules, concretions, septaria, and related forms but not his spherulites and other regular crystal growths whose internal fabric suggest formation and present occurrence in a single large void. They also include Kubiena's (1938) invasion amygdali but not his amygdalus types formed in voids or his crystal growths. Besides these, additional types are necessary to accommodate all the glaebules in soil materials. Bryan (1952) classified all "rounded lumps" in soil materials as "nodules" but preserved the specific meaning of "concretion," so his nodules include glaebules; he did not consider some of the features which are included in glaebules.

12.1 DISTINGUISHING CHARACTERISTICS

Within the broad group of glaebules individuals can be classified on the following characteristics, which are arranged in what is considered to be their most useful order:

(1) the internal fabric,
(2) the mineralogical nature of the relatively concentrated glaebular material,
(3) their distinctness as a unit, which includes sharpness of the external boundary and ease of separation of the glaebule, as a unit, from the enclosing soil material, and
(4) their shape.

In selecting the order of these characteristics, consideration has been given to what are considered the important characteristics of the features described by Pettijohn (1957) and Kubiena (1938) which belong to glaebules, as well as the characteristics of glaebules in soil materials.

A. *Internal Fabric*

Pedological features can exhibit levels of organization comparable to those of soil materials (Section 7.6:C); a single glaebule (the host feature) may enclose smaller pedological features (the included features) such as cutans and other glaebules. The fabric of the host feature is the *basic fabric* of this complex glaebule, and classification of glaebules at this highest level is made on the characteristics of this basic fabric independent of the included pedological features. Such complex glaebules are referred to simply as *compound* types (Figs. 29 and 30) if the distribution and/or orientation patterns of the included pedological features do not give the glaebule as a whole a specific fabric that places it in a specific class.

The basic fabrics observed in glaebules from soil materials can be grouped into the following classes:

(*a*) ***Undifferentiated Fabric*** (Figs. 72–75). There is no specific plasmic orientation pattern or distribution of the various constituents or voids with regard to the shape of the glaebule. The "structureless" structure of Pettijohn (1957) is an undifferentiated fabric.

(*b*) ***Concentric Fabric*** (Figs. 76 and 77). The constituents are arranged in zones which are approximately parallel to each other and usually to the surfaces of the glaebule. *Convolute fabric* is a variant of concentric fabric in which the fabric features are arranged in parallel but highly contorted zones. Concentric fabric can be subdivided further on the degree and kind of orientation of the material within the laminae, for example, the degree and direction of preferred orientation of the clay mineral fraction of the plasma in relation to the shape of the glaebule and the configuration of the laminae.

(*c*) ***Lamellar Fabric*** (Figs. 26 and 82). The constituents are arranged in parallel planar zones. The laminae may be somewhat curved but do not conform to the shape of a three-dimensional body. Lamellar fabric can be subdivided on the degree and kind of orientation of the constituents within the laminae.

(*d*) ***Continuous Fabric*** (Fig. 81). The material is anisotropic, and the glaebule as a whole has the extinction characteristics of strong, continuous orientation (Section 8.2); laminae as such are not visible, but a micaceous cleavage may cause a laminated appearance.

(*e*) ***Void Patterns.*** Several basic fabrics of glaebules are characterized by the arrangement of relatively large voids:

Central void (Figs. 79 and 80): there is a single, large, central void.

Concentric voids (Fig. 78): there is a concentric pattern of voids.

Radial voids (Fig. 78): there is a radial pattern of voids.

B. Mineralogical Nature of the Glaebular Material

Many glaebules are *relative* accumulations of fractions of the plasma, which are also present in the enclosing soil material; thus, even though a glaebule may be separated easily by dissection, qualitative identification of its constituents is not sufficient to identify the accumulated fraction of the plasma. Identification is also complicated because glaebules usually contain a wide variety of constituents of which only a limited number, often only one, has been concentrated. It is necessary, therefore, to compare quantitative estimations of the mineralogical composition of the glaebule with that of the enclosing material to identify the accumulated constituent. In many instances, however, a broad classification of glaebules on the basis of the mineralogy of the glaebular material can be made on evidence from hand specimens, in thin sections, and from a few simple spot tests as were used for cutans and pedotubules (Chapters 10 and 11). The following separations can usually be made on the basis of such data, and the classification can be made more specific if the accumulated constituents are identified as specific minerals.

(*a*) ***Sesquioxidic*** (Figs. 72, 73, 76, 77, 83, and 84). The glaebules are translucent to opaque and of variable anisotropy in thin section depending on the proportion, kind and grain size of the iron compounds. They are usually red or yellow in reflected light depending on the degree of hydration; the contrast in color between reds and yellows in soil materials, however, is not always due to differences in iron content, but it does indicate differences in mineralogy. The degree of accumulation can be measured only by comparative quantitative analyses. Jeffries' (1946) and the dithionite (Aguilera and Jackson, 1953) methods for removing iron oxides, however, have little effect on many sesquioxidic glaebules. Aluminium oxides or hydroxides are difficult to identify optically because they are usually associated with the colored iron minerals; relatively pure glaebules of pale aluminium oxides do occur.

(*b*) ***Manganiferous*** (Fig. 21). Glaebules rich in manganese oxides are usually very dark brown to black in reflected light and opaque in thin section, thus contrasting with the enclosing soil material. Doubtful cases can be checked by the periodate method (Piper, 1942), and a comparative estimation can often be made by comparing the color of the solution resulting from equal weights of the glaebule and the enclosing soil material after making up to a standard volume.

(*c*) ***Carbonate*** (Figs. 78–80). The carbonate minerals, which are usually dominantly calcite, occur as masses of small white or colorless crystals, sometimes with yellow or grey tints; the luster is glassy or dull earthy depending on the degree of crystallization. Calcite is readily estimated quantitatively by chemical methods, but optical methods are usually sufficient to recognize calcareous glaebules.

Glaebules of siderite have also been reported (Yamasaki and Yoshizawa, 1961). Carbonate glaebules can be subdivided into calcitic, dolomitic, sideritic, etc.

(*d*) ***Sulfate*** (Fig. 71). The sulfate minerals occur in a similar form to the carbonates; they can usually be recognized optically, provided the crystals are large enough, and estimated quantitatively by chemical methods. Two varieties have been reported: gypseous types, which are quite similar in appearance to calcitic glaebules, and baritic types (Beattie and Haldane, 1958), which are white to pale yellowish and chalky in hand specimen (Fig. 71).

(*e*) ***Siliceous.*** Secondary silica is difficult to identify, especially if it occurs as thin coatings. In glaebules, however, it may cause very significant hardening that can be recognized in hand specimen. Glaebules of opal-like silica have been reported (Kovda, Zimovets, and Amchislavskaya, 1958).

(*f*) ***Argillaceous*** (Figs. 81 and 82). Many glaebules are equivalent to Kubiena's (1938) invasion amygdali and so include the clay minerals of the s-matrix of the soil material. Accumulations of clay minerals can be determined in these types only by comparative quantitative estimations between the glaebule and the enclosing soil material; argillans are recognized as included pedological features and are not diagnostic for argillaceous glaebules. Glaebules composed of relatively pure clay minerals do occur, however, and they are readily recognized as such by their optical properties.

C. Distinctness

There is usually a close correlation between the two aspects of distinctness of glaebules, that is, sharpness of external boundary and ease of removal as a unit from the enclosing soil material. Previous authors (e.g., Pettijohn, 1957; Kubiena, 1938; *U.S.D.A. Soil Survey Manual,* 1951) defined their glaebular types as having sharp external boundaries and thus being readily separable as a unit. Many glaebules in soil materials, however, do not have sharp boundaries and cannot be readily separated. The sharpness of the external boundary can be described as *sharp* (Figs. 21, 83 and 84), *rather sharp, rather diffuse, diffuse* (Fig. 73), and very *diffuse* (Section 8.3:C). The ease of removal of the glaebule as a unit from the enclosing soil material can be described in the terms used for pedotubules (Chapter 11) as *discrete* (Fig. 71), *weakly adhesive, moderately adhesive,* and *strongly adhesive* (Fig. 73).

D. Shape

The shape of glaebules can be described in terms of sphericity and roundness (Section 2.1). Many glaebules are prolate to equant and rounded, their surfaces consisting of a series of smooth curves; some, however, are angular to subangular. Sphericity and roundness alone do not describe in detail the surface conformation, which may be characteristic in some types. It is useful, therefore, to define terms for specific shapes. Many suitable terms are available in mineralogy, and to avoid proliferation of names the following list has been selected largely from Read (1948) with some additions:

Amygdaloidal. Almond-shaped; the surface consists of smooth curves.

Bladed. In forms shaped like a knife-blade or lath.

Botryoidal. Consisting of spheroidal aggregations somewhat resembling a bunch of grapes; the surface consists of smooth curves.

Convolute. Grotesquely contorted shape; the surface consists of smooth curves.

Ellipsoidal. Shaped like an ellipsoid; the surface is smooth.

Lenticular. With the form of flattened balls or pellets; shaped like a lens or lentil; the surface is smooth.

Lamellar. In the form of plates which are much larger in two dimensions than in the third.

Mammillated. Displaying large, mutually interfering spheroidal surfaces; the surface consists of smooth curves.

Reniform. Kidney-shaped, the rounded surfaces resembling those of kidneys.

Spherical. Shaped like a sphere; the surface is smooth.

Tuberose. Having very irregular rounded surfaces, often giving rise to gnarled, root-like shapes.

Irregular. Irregular shapes with relatively flat surfaces and sharp angles at the intersections of the surfaces.

Other specific shapes may be encountered with further experience and require additional descriptive names.

12.2 CLASSIFICATION OF GLAEBULES

It is desirable that the groupings of glaebules should conform, as far as possible, to those already in use, having regard to the characteristics of the glaebules found in soil materials. The following classification follows that of Pettijohn (1957) with additions for the types he did not consider; it is set out in Table 12-1.

TABLE 12-1. Morphological Classification of Glaebules

Glaebule Type	Internal Fabric	Shape	External Boundary	Composition	Degree of Adhesion	Specific Shape Names†
Normal nodules	Undifferentiated*	Usually prolate to equant and well rounded	Sharp to rather sharp	Any fraction of the plasma	Discrete or weakly adhesive	Amygdaloidal Bladed Botryoidal Convolute Ellipsoidal Lenticular Lamellar Mammillated Reniform Spherical Tuberose Irregular
Normal concretions	Concentric (or convolute)					
Septaria	Undifferentiated except for series of radiating and concentric cracks			Usually calcareous		
Contraction spheroids	Undifferentiated except for a series of concentric cracks					
Pedodes	Undifferentiated except for a large central void					
Papules	Continuous or laminated			Dominantly clay minerals (argillaceous)	?	
Glaebular halos	Undifferentiated	Indeterminate	Diffuse		Strongly adhesive	
Irregular nodules and concretions	As for normal types	Irregular shape with flat planar surfaces meeting at sharp angles	Sharp to rather sharp	Any fraction of the plasma	Any degree of adhesion	
Diffuse nodules and concretions	As for normal types	As for normal or irregular types	Rather diffuse to diffuse		Moderately to strongly adhesive	

* Undifferentiated fabric includes normal rock and soil fabrics
† Applicable to all glaebule types

A. *Nodules* (Figs. 71–75)

Pettijohn (1957) defined nodules as "irregular, tuberous bodies. . . . Their surface is warty or knobby . . . under the microscope the nodules are structureless"; structureless in this context means undifferentiated fabric. Rice (1954; after Fay, 1920) defined nodules on shape characteristics alone: "a small roundish lump of some mineral or earth, as a nodule of ironstone"; this shape definition is less specific than that of Pettijohn. Considering these definitions a group of glaebules from soil materials can be termed nodules and defined as follows:

> *Nodules.* Glaebules with an undifferentiated internal fabric; in this context undifferentiated fabric includes recognizable rock and soil fabrics.

More detailed categorization of nodules involves the use of adjectives for the shape and distinctness classes discussed earlier. To reduce

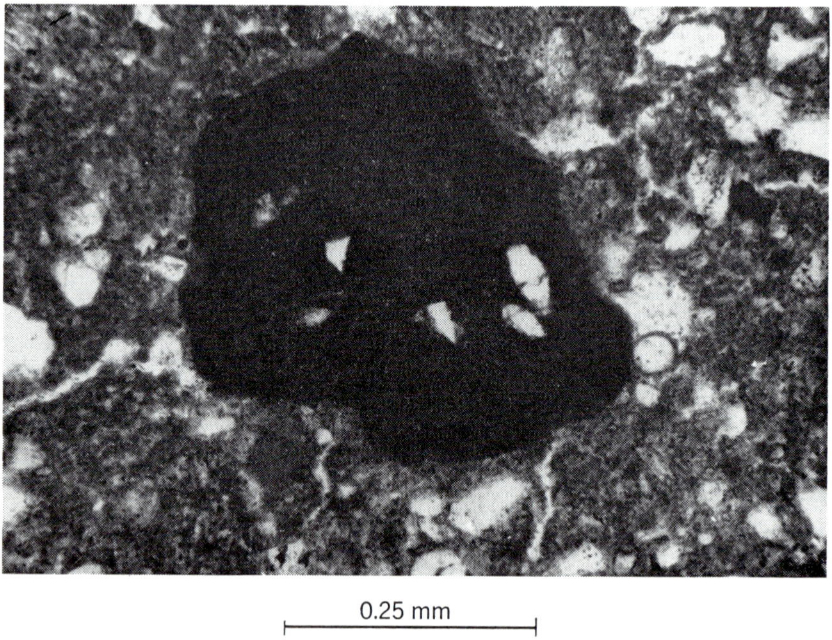

Fig. 72. Irregular sesquioxidic nodule (undifferentiated fabric). Thin section in plain light. Note the thin, rather pale glaebular halo. Alternatively, a weakly developed concretion, the halo constituting a concentric fabric. 135×.

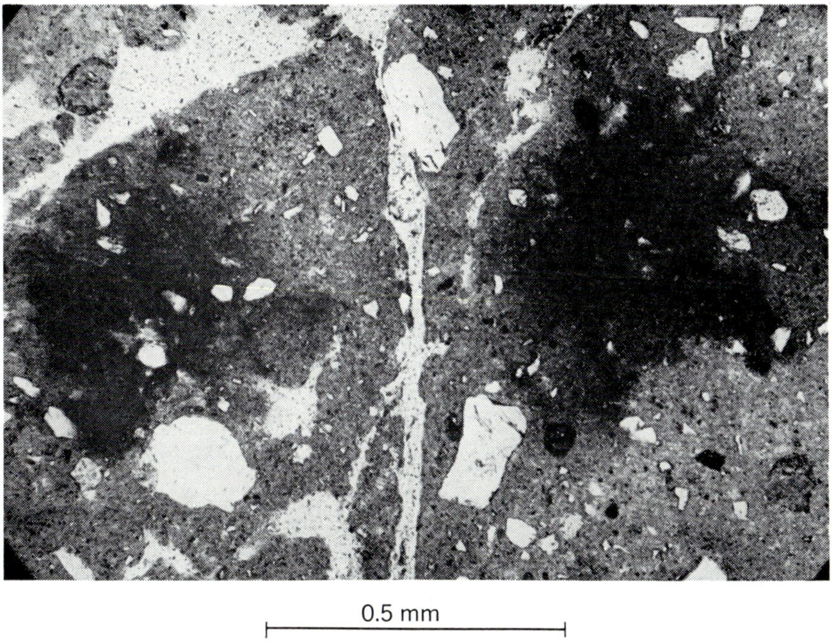

Fig. 73. Strongly adhesive, diffuse, irregular, sesquioxidic nodules (undifferentiated fabric). Thin section in plain light (after Brewer and Sleeman, 1964). 80×.

this rather lengthy nomenclature, the term *normal nodule* can be applied to the commonly occurring types with sharp to rather sharp external boundaries whose surfaces consist of relatively smooth curves.

B. Concretions (Figs. 76, 77, and 84)

Rice (1954; after Fay, 1920) and Kubiena (1938) defined concretions on characteristics of shape. The *U.S.D.A. Soil Survey Manual* (1951) defined them on hardness and distinctness. Other definitions (e.g., Krumbein and Sloss, 1955; Scott, 1943) convey the concept simply of inorganic aggregations. All these definitions are very broad, whereas Pettijohn (1957) stated that "true concretions are concentrically laminated structures, normally subspherical, though commonly highly irregular." In fabric analysis of soil materials it is necessary to distinguish glaebules with various internal fabrics, and the term "concretion" is the most appropriate for concentric arrangements. Some glaebules, however, consist of parallel laminae with a

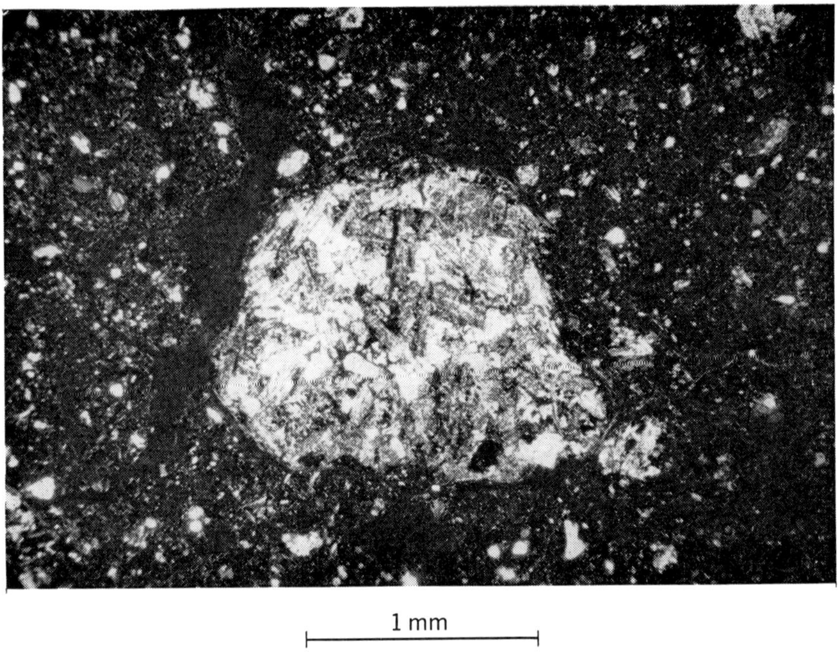

Fig. 74. Irregular nodule with rock fabric (lithorelict). Thin section under crossed polarizers. 30×.

convolute arrangement related to concentric fabric, and these, too, should properly be classified as concretions, which can be defined as follows:

> *Concretions.* Glaebules with a generally concentric fabric about a center which may be a point, a line, or a plane.

Concretions can be categorized further according to shape and distinctness in the same way as for nodules.

There may be difficulty in classifying some glaebules as nodules or concretions because an almost complete range of fabrics—from numerous, thin, concentric laminae to undifferentiated fabrics with a single thin enclosing argillan—has been observed; the sharpness of the boundaries between the laminae is also quite variable. All glaebules with concentric fabric are classified as concretions, but an enclosing cutan of a plasma fraction different from that accumulated within the glaebule is not considered as part of the glaebule for purposes of classification; the concentric arrangement of these cutans is due to the shape of the already existing glaebule.

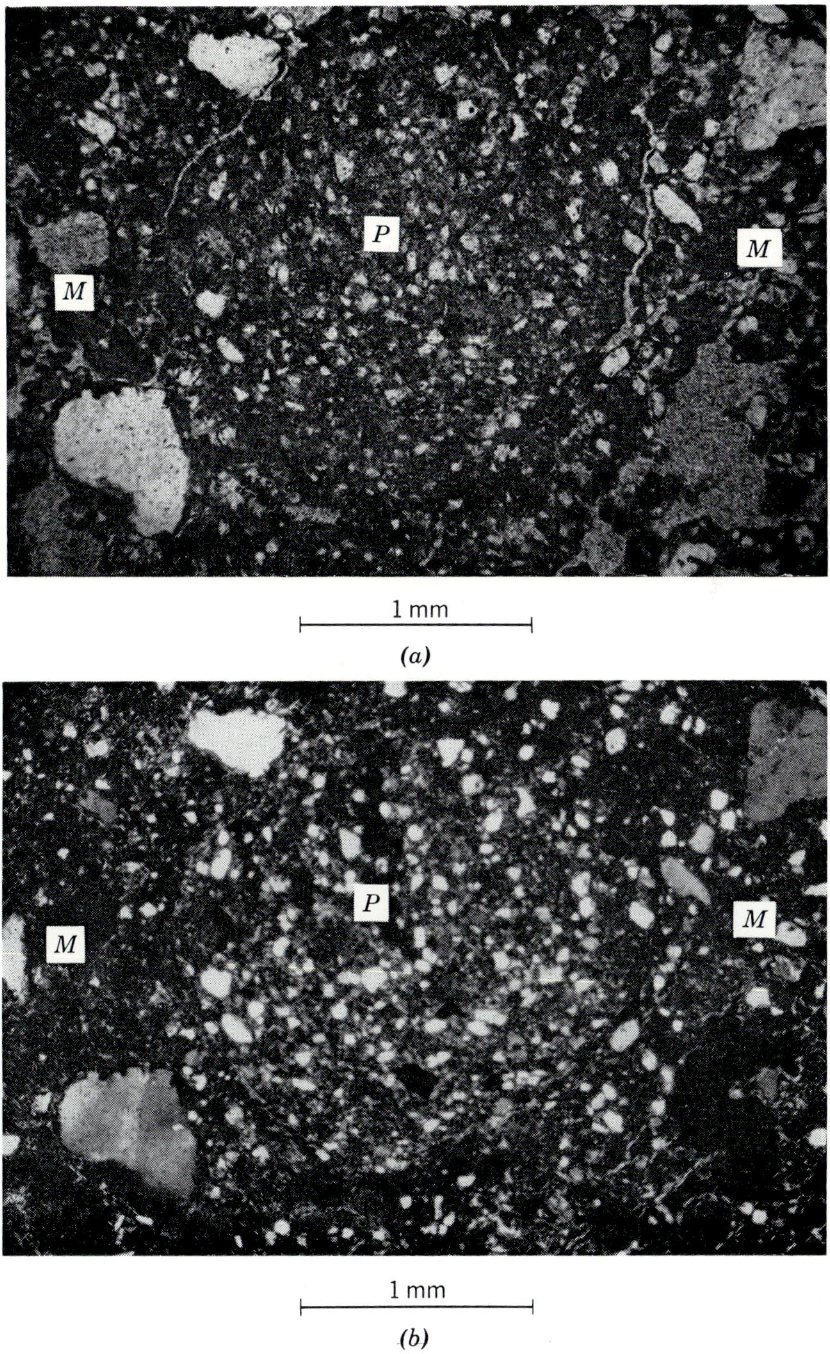

Fig. 75. Normal nodule with soil fabric (*P*—pedorelict). (a) Thin section in plain light. 30×. (*b*) Thin section under crossed polarizers. *M* is the s-matrix. 30×.

C. *Septaria* (Fig. 78)

Rice (1954; after Kemp, 1927 and Fay 1920), and Pettijohn (1957) defined septaria as nodules with a particular superimposed crack pattern, and so they can be defined as follows:

> *Septaria.* Glaebules with a series of radiating cracks crossed by a series of cracks concentric with the margins; the crack pattern is often highly irregular; they are usually spheroidal with sharp boundaries.

If the radial cracks are absent these features are known as *contraction spheroids* (after Pettijohn, 1957). If the outer margins of features with the typical crack pattern of septaria have concentric fabric, they are known as *concretionary septaria.* In some septaria both the radial and concentric series of cracks are filled with a crystalline deposit, but in many types from soil materials they are empty.

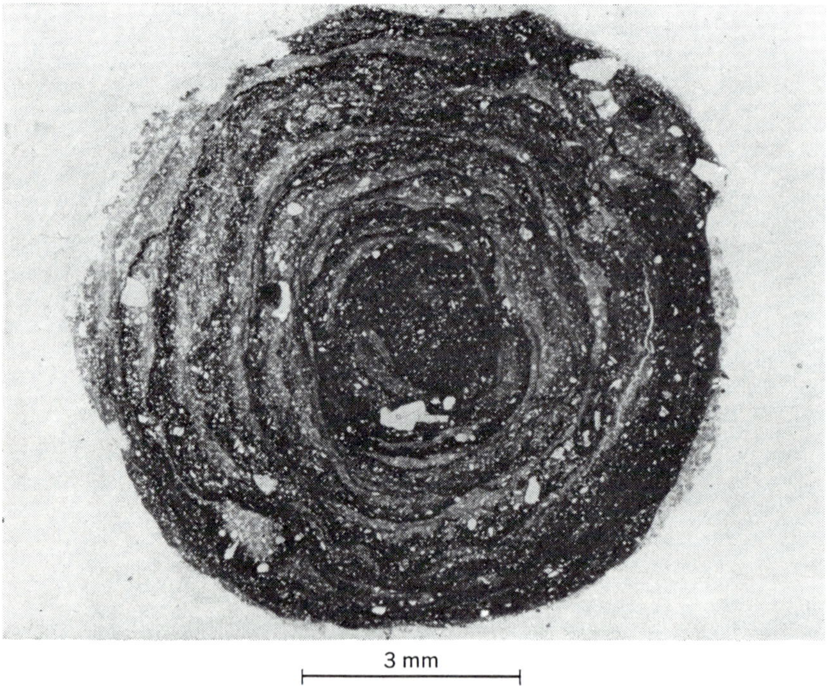

Fig. 76. Discrete, normal, sesquioxidic concretion (concentric fabric). Thin section in plain light (after Brewer and Sleeman, 1964). 9×.

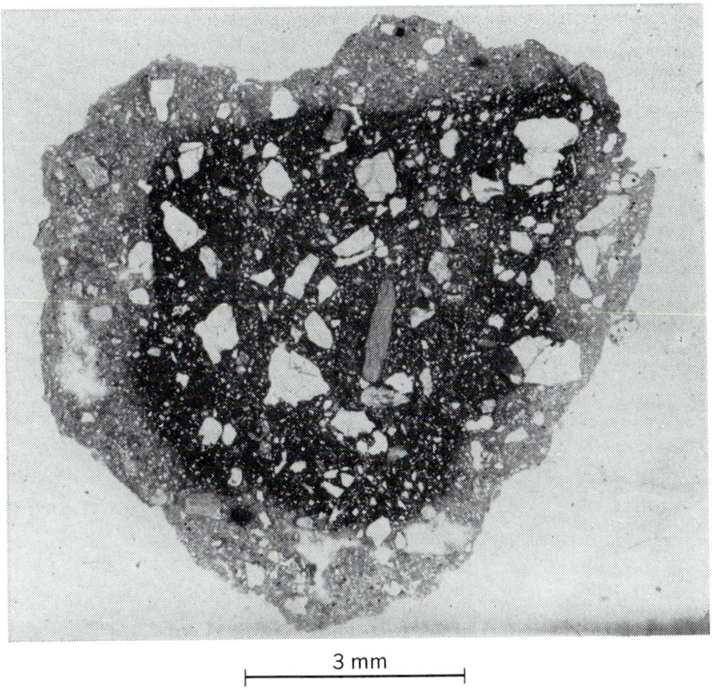

Fig. 77. Discrete, irregular, sesquioxidic concretion (concentric fabric). Thin section in plain light. Concentric fabric is expressed by a single outer band. 9×.

D. Pedodes (Figs. 79 and 80)

Geodes have been defined specifically as "hollow, globular bodies" with certain other significant features; the definition is very specific as regards structure, composition, and occurrence (Pettijohn, 1957). Some glaebules from soil materials conform in a general way to these specifications, but the outer chalcedonic layer is often absent, and the clay film between the walls and the enclosing soil material is not an obvious feature, possibly because the features usually occur in soil materials that have a high proportion of clay size particles. These features have been named *pedodes* as the pedological equivalent of geodes. They are defined as follows:

Pedodes. Glaebules with a hollow interior, often with a drusy lining of crystals; there may be associated veins or an outside layer of chalcedonic silica; they are usually spheroidal and discrete.

Fig. 78. Discrete, carbonate septarium (concentric and radial void patterns). Hand specimen (after Brewer and Sleeman, 1964). 1×.

Fig. 79. Discrete, carbonate pedode. Hand specimen (after Brewer and Sleeman, 1964). Broken to expose the central cavity which has a drusy lining. ⅞×.

Fig. 80. Discrete, carbonate pedode. Hand specimen (after Brewer and Sleeman, 1964). Broken to expose the central cavity which does not have a drusy lining. 1×.

Pedodes are classified as glaebules because, in spite of the central cavity, the concentrated plasma fraction has "invaded" the soil s-matrix, so the feature as a whole does not occur in a single void in the present soil material.

E. Glaebular Halos (Fig. 72)

A common feature associated with glaebules is a diffuse, irregularly shaped zone of material surrounding more strongly developed glaebules, for example, a diffuse zone of accumulation of manganese oxides surrounding a group of small manganiferous nodules with rather sharp boundaries. Because of the much weaker concentration in the zone surrounding the dense nodules and its diffuse boundaries, this feature is considered as a group of nodules with a halo rather than a compound nodule with very diffuse boundaries. Such halos can be defined as follows:

Glaebular halos. Weak accumulations of some fraction of the plasma surrounding much stronger glaebular features and having an undifferentiated fabric and very diffuse to diffuse external boundaries.

F. Papules (Figs. 81 and 82)

A number of glaebules that consist dominantly of clay minerals without skeleton grains have characteristic extinction patterns with crossed nicols under the petrological microscope. These features have been termed *papules** which may be defined as follows:

Papules. Glaebules composed dominantly of clay minerals with continuous and/or lamellar fabric; they have sharp external bound-

* Papule (plural form, *papules;* adjectival form, *papulic*) is derived from the Latin *papula,* a pustule or pimple (Gepp and Haigh, 1928).

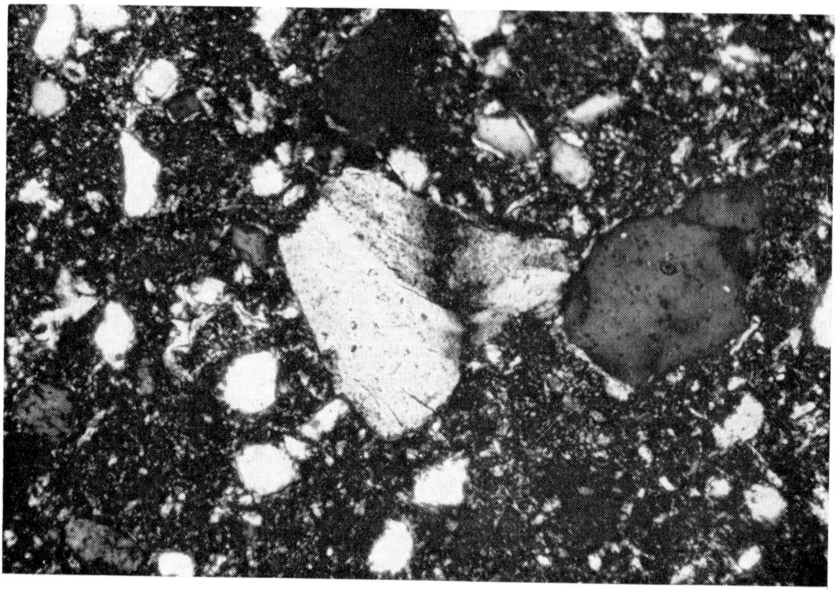

0.25 mm

Fig. 81. Papule (argillaceous glaebule with sharp boundaries). Thin section under crossed polarizers (after Brewer and Sleeman, 1964). Note the strong black extinction lines indicating continuous fabric. 135×.

Fig. 82. Two papules (argillaceous glaebules) with lamellar fabric. Thin section under crossed polarizers. 135×.

aries. Most commonly they are prolate to equant and somewhat rounded.

Included in papules are Pettijohn's (1957) clay galls: "markedly flattened fragments, somewhat rounded to subangular in outline." Some types contain a reasonably high proportion of silt size particles, and there seems to be an inverse correlation between degree of orientation within the laminae, as interpreted from extinction phenomena, and the amount of silt.

12.3 INTERPRETATION

There are two aspects of interpretation of glaebules: interpretation of the processes of formation of the individual types, and of their origin; the latter aspect refers to their classification as orthic, or *in situ* or transported inherited features (Section 7.5:B). There are several facets to interpretation of each of these aspects. To take the most complex case, suppose a particular glaebule can be diagnosed as having been inherited (origin) and formed by accretion (process); three problems of interpretation remain: the actual process and conditions of accretion, whether it was inherited from a rock (lithorelict) or a soil (pedorelict) or was formed during sedimentation (sedimentary relict), and whether it was formed *in situ* or has been transported. Such details of interpretation are not always possible from the morphological data available. Often, therefore, interpretations must be tentative at the present state of knowledge, and care must be taken in attributing the processes of formation to the soil material in which the glaebules occur.

Although these problems are complex some inferences can be drawn from the characteristics of the various kinds of glaebules, especially their internal fabric and composition, distinctness, and shape. Other data, such as their distribution and orientation patterns, may also be pertinent to their interpretation.

A. Internal Fabric and Composition

Internal fabric can give only limited evidence of origin. Taken in conjunction with the mineralogical composition and other characteristics, however, it may provide some evidence regarding both process of formation and origin.

(*a*) Undifferentiated Fabric. Many glaebules with undifferentiated fabric have an accumulation of some constituent as compared with the enclosing soil material. Some of these can be assumed to originate *in situ* in soil materials by accretion of soluble constituents which accumulate by diffusion or crystallization from solution in the numerous, very small voids in the s-matrix due to suitable local chemical conditions.

Drosdoff and Nikiforoff (1940) suggested that sesquioxidic accretions are formed by an initial drying that causes concentration of solutions in the small voids, thus initiating the formation of nuclei of deposition which include the soil s-matrix; these nuclei act as seeds and remove the accretionary constituent selectively from solution. Nuclei may be formed also by concentration and precipitation of soluble plasma fractions by organisms (Crawford, 1956; Aristovskaya, 1961). Humbert (1948) suggested that the process involved in the formation of sesquioxidic glaebules consisted of direct oxidation of iron released in the decomposition of ferromagnesian minerals to form a yellow to brown gel that orients itself around a nucleus, loses water by diffusion, and undergoes a gradual transformation to crystalline hematite. Yanitskii (1958) suggested that quartz grains surrounded by an envelope of dipole water molecules (with a negative charge) could act as nuclei causing adsorption of positively charged colloid particles of iron oxide hydrates (Fig. 83).

Undifferentiated fabric may indicate origin if compared with the fabric of the enclosing soil material. Nodules with high clay mineral content within *A* horizons of coarser material may be interpreted as remnants of an old soil horizon (pedorelict), especially if their internal fabric is similar to that of lower horizons in the profile; the normal soil fabric of these nodules indicates that the clay accumulation is not due to specific processes of concentration within the *A* horizon. Inherited nodules of parent rock or transported rock (lithorelicts) may be recognized as such by the similarity of their internal fabric with that of some specific rock fabric, and nodules of underlying buried soil materials (pedorelicts) may be recognized in the same way; for example, the number of skeleton grains per unit area

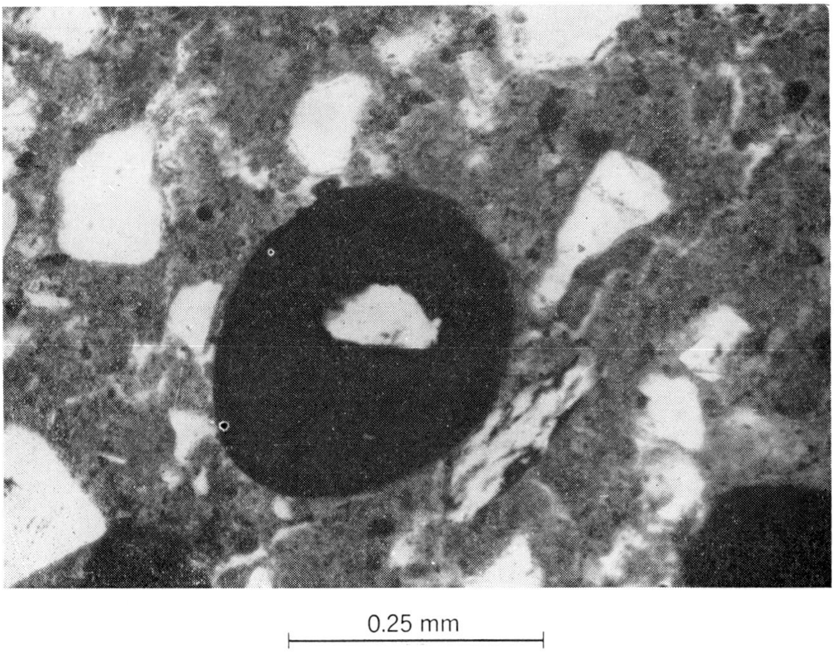

Fig. 83. Normal, sesquioxidic nodule with a central quartz grain. Thin section in plain light (after Brewer and Sleeman, 1964) 135×.

within the glaebule compared with the enclosing soil material may differentiate between formation *in situ* and transportation.

Glaebular halos with undifferentiated fabric indicate some change in conditions between formation of the halos and formation of the distinct glaebules which they surround; such halos are also diagnostic for formation *in situ.*

(*b*) ***Concentric Fabric.*** All the evidence available suggests that concretions are formed by accretion as proposed by Pettijohn (1957). There are several forms of concentric fabric, however, each of which probably requires some variation in interpretation. Thin, distinct laminae indicate formation in waves during alternating wet and dry periods with dessication following formation of each lamina. Alternatively, some process akin to the formation of liesegang rings may have been operative (Humbert, 1948; Stern, 1954; Carl and Amstutz, 1958). Thin laminae around a core with an undifferentiated fabric, and, sometimes, a lesser accumulation of the accreted material, indicate two phases of genesis. Broad laminae of different constituents or of different degrees of accumulation of the accreted constituent—for example, limonitic and hematitic iron oxides with rather diffuse boundaries between them—may be due to diffusion effects within a nodule or to the pattern of distribution of zones where conditions were suitable for deposition of the constituents.

Concentric fabric, of itself, provides no evidence of origin as orthic, *in situ,* or transported inherited features, but it is a useful fabric for revealing abrasion during transport which may wear a concretion to a shape that does not conform to the shape of the laminae; such a relationship is strong evidence for inheritance (Fig. 84).

(*c*) ***Lamellar and Continuous Fabrics.*** These fabrics are closely related; lamellar fabrics are due to distribution of the components in parallel sheets, while continuous fabrics are due to strong preferred orientation of anisotropic clay minerals. Both fabrics have been observed in illuviation cutans (Section 10.1:C) and are readily produced by sedimentation in the laboratory; they have been observed in puddles after rain and in surface wash materials after flood irrigation. There is considerable evidence, therefore, that they are produced by deposition from suspension. They may also be produced by weathering of mineral grains, especially micas (Carozzi, 1960, p. 131) or shale chips (lithorelicts). If glaebules with such fabrics have been formed by deposition, the deposit (whether argillans or sedimentary relicts) may have been broken up and redistributed since they now occur embedded in the soil material; such features may be

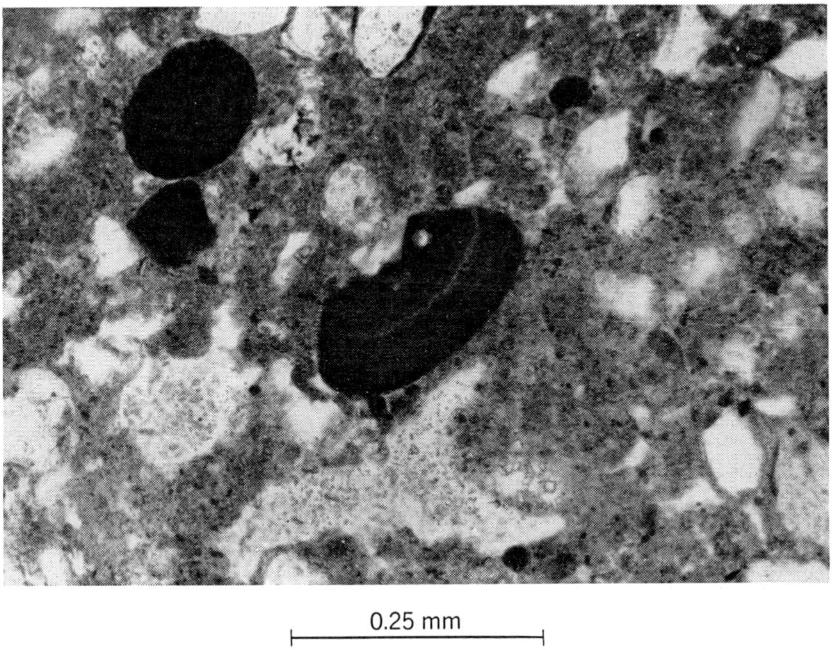

Fig. 84. Abraded, normal, sesquioxidic concretion with sharp boundaries. Thin section in plain light. 135×.

inherited or merely redistributed by disturbance of the soil material in which they were formed (pedoturbation—Hole, 1961); shape may help in distinguishing these two origins. If the fabric is due to weathering *in situ* thin sections throughout the profile will show a decreasing degree of weathering with increasing depth.

(d) Special Void Patterns. Pettijohn (1957; after Bassler, 1908) concluded that an initial cavity is "a necessary prerequisite for the formation of a geode," and he attributed the crack pattern of septaria (and contraction spheroids) to dehydration of a body of aluminous gel after case hardening; he also stated that "the mechanism involved in forming aluminous gel is most obscure. . . . The mechanism involved in case hardening, dehydration, and shrinkage pattern is irreversible chemical dessication." Pedodes and septaria from soil materials have sharp external boundaries and are case hardened; most are calcareous and contain other constituents of the enclosing soil material. If aluminous gel was the cause of the crack pattern in septaria it must have permeated the soil material (s-matrix).

The enclosed material, however, usually has a high proportion of clay size material, which is a normal part of the soil plasma, and drying of this, after case hardening, may be sufficient to produce the crack pattern. Similarly, the central cavity and irregular cracks around it in pedodes may be due to colloid contraction on drying, so it is unnecessary to assume large pressures to explain expansion; in geodes these pressures are attributed to osmosis between a fresh external solution and a connate salt solution in the initial cavity.

Since pedodes, septaria, and contraction spheroids are all case hardened, they may be inherited as well as formed *in situ.*

B. Distinctness

Accretionary glaebules with sharp external boundaries commonly occur in A_2 horizons that are subject to periodic desiccation, while those in lower horizons, which are less subject to strong drying, tend to have less sharp boundaries (Drosdoff and Nikiforoff, 1940). Sharp boundaries, therefore, may be due to "case hardening" ["irreversible chemical desiccation" (Pettijohn 1957)], while diffuse boundaries may be due simply to absence of this process. Alternatively, diffuse boundaries in glaebules formed *in situ* may occur because conditions within the glaebule, although suitable for deposition of the accreting constituent, were not greatly different from the surrounding soil material, so that deposition occurred over a relatively wide zone, increasing in intensity toward the focus where conditions were most suitable.

Glaebules with sharp external boundaries may have been transported or formed *in situ,* whereas those with diffuse boundaries have probably been formed *in situ* unless re-solution and deposition has occurred subsequent to transportation.

C. Shape

Shape is of limited use in interpretation of glaebules. Shapes bounded by relatively smooth curved faces generally have been correlated with formation by accretion (Pettijohn, 1957), thus, although roundness may indicate inheritance by indicating the effects of transport, care must be taken in making such interpretations; usually roundness due to the processes of formation and that due to abrasion can be distinguished, especially if the external shape is considered in relation to the internal fabric. It is reasonably certain that irregular glaebules have been formed *in situ.* Blumel (1962) found a correlation between shape of ferric hydroxide glaebules with occurrence in gley and pseudogley soil horizons. Rounded shapes occurred in groundwater gley and the *G* horizon of pseudogleys, and irregular shapes in the *G* horizon of slope gleys and of pseudogley—stagnogleys.

D. Associated Pedological Features and Soil Materials

The history of development of a soil material may also be revealed by the nature of the associated pedological features and by the structure and fabric of the soil material invaded by the accreting constituents in accretionary glaebules. Many compound sesquioxidic nodules include illuviation argillans with varying degrees of orientation on the walls of larger voids. If similar cutans occur in the enclosing soil material the question arises as to whether they were formed before or after the glaebule. Their relative thickness and frequency of occurrence within and without the glaebule may be a guide to diagnosis since it is logical that movement of clay minerals into the glaebule will be more restricted than movement through the enclosing soil material. This would be especially so in concretions where it can be assumed that the cutans preceded the glaebules unless evidence is found for movement of clay minerals through the outer concentric zones. Argillans conformable with the laminae are not uncommon in sesquioxidic concretions, indicating alternate depositions of clay minerals and iron oxides. The degree of staining of cutans by the accreting material in accretionary glaebules may also be diagnostic since, if the cutans precede the glaebule, they should be heavily stained and possibly less so if they succeed it. If such features as cutans occur within the glaebule but not in the enclosing soil material, the question arises as to whether the glaebule was developed *in situ* or is transported. At least the occurrence of cutans and glaebules together demonstrates the operation of two distinct processes during soil formation and possibly two different sets of circumstances during soil development.

Similar considerations can be put forward for the interpretation of other kinds of included pedological features and also those associated with glaebular surfaces (cutans).

E. Interpretation of Named Types

(*a*) ***Nodules.*** Nodules rich in soluble fractions of the plasma (e.g., calcite, iron hydroxides) are probably formed by accretion (plasma concentrations), whereas those with normal soil or rock fabrics are not. All types of accretionary nodules can be formed *in situ* (orthic or inherited), but only normal types are likely to be inherited by transport; those with soil or rock fabric are usually inherited (pedorelicts and lithorelicts), and degree of roundness may distinguish transported types from those formed *in situ.* Some normal nodules may be sedimentary features, for example, the *mud balls* of Pettijohn (1957); these have not been recorded in soil materials, but they have been observed in creek deposits.

(*b*) ***Concretions.*** Most concretions are accretionary. Normal and diffuse types can be formed *in situ* (orthic or inherited), but only the former are likely to be inherited by transport; inherited types can be lithorelicts or pedorelicts.

(*c*) ***Septaria and Pedodes.*** These are accretionary and usually formed *in situ;* septaria may be lithorelicts or pedorelicts; pedodes may be pedorelicts.

(*d*) ***Papules.*** Papules with continuous fabric occur as inherited lithorelicts, pedorelicts or sedimentary relicts, or as disrupted orthic pedological features (e.g., disrupted cutans), or disrupted *in situ* sedimentary relicts; lamellar fabric indicates origin as sedimentary features. Shape may indicate origin; discoidal shape suggests origin as a sedimentary feature, rounding suggests transport, and angularity suggests disruption of *in situ* features. The general mineralogy and fabric of papules indicate that they are not accretionary.

(*e*) ***Glaebular Halos.*** These halos are invariably accretionary, formed *in situ,* and they indicate that the associated, more distinct glaebules were also formed *in situ* by accretion.

13

Pedological Features IV: Crystallaria. Subcutanic Features. Fecal Pellets

13.1 CRYSTALLARIA

Kubiena (1938) described "crystal chambers," "crystal tubes," and "intercalary crystals," while Pettijohn (1957) described "spherulites" and "rosettes." All these features, except intercalary crystals, which are a special case, occur in relatively large voids and consist of crystalline, relatively pure minerals. Crystal chambers, crystal tubes, and crystal sheets are closely related to those cutans which are true coatings; they are crystallizations of relatively soluble fractions of the plasma on the walls of voids and represent extreme developments of cutans composed of particular kinds of constituents. Similarly the mode of occurrence of intercalary crystals is closely related to that of calcite crystals within calcareous nodules; they are relatively grossly developed crystals of pure minerals within the s-matrix of the soil material. Intercalary crystals, however, have a significant difference in distribution pattern of the crystals as compared with the distribution pattern of the crystals in glaebules. Glaebules (e.g., nodules) consist of groups of crystals concentrated closely together, whereas intercalary crystals may have a uniform or random distribution throughout the soil material.

Well-developed sheets, tubes, and groups of crystals are prominent morphological features in many soil materials, and, although they can be thought of as gross developments of cutans and crystals in voids in the s-matrix, there is considerable merit in recognizing them as a separate group of pedological features that can also include other crystallized forms, such as Pettijohn's (1957) spherulites and rosettes. The general term *crystallaria* is proposed for this group and is defined as follows:

> *Crystallaria.* Single crystals, or arrangements of crystals of relatively pure fractions of the plasma that do not enclose the s-matrix of the soil material but form cohesive masses; their morphology (especially shape and internal fabric) is consistent with their formation and present occurrence in original voids in the enclosing soil material.

A. Distinguishing Characteristics

By definition, crystallaria are cohesive masses of recognizable pure crystals or single crystals in voids; they can be subdivided on the following characteristics: (1) the internal fabric, (2) the kind of void in which they are formed, and (3) the mineralogical nature of the crystals. By using these characteristics crystallaria can be classified, named, and qualified by the addition of descriptive adjectives.

(*a*) *Internal Fabric.* Since crystallaria are composed entirely of relatively large crystals whose shape and arrangement can be observed, a significant feature of their fabric is the degree and kind of preferred orientation of the crystals. In addition, since they either occur in voids or have a sharp external boundary with the enclosing soil material, the referred orientation with regard to the walls of the void, or the contact of the crystals with the enclosing soil material, is an important aspect of their fabric. If the crystals are physically anisotropic, the preferred and referred orientation can be described and measured in reference to the direction of the long axes. If the crystals are optically anisotropic these characteristics can be measured also on the basis of the direction of the optical or crystallographic axes. As discussed earlier (Section 8.2), a strongly developed preferred orientation of individuals that is expressed by their referred orientation pattern may not be apparent in their orientation with regard to each other if the reference feature is a curved surface, as is very often the case with the walls of voids; individuals with a strongly developed preferred orientation with regard to the walls—that is, arranged tangential to or vertical to the curved walls—may show virtually no parallel orientation with regard to each other.

Crystallaria can be subdivided into five groups according to either the referred orientation of the crystals with regard to their contact with the enclosing soil material or the basic orientation of the individual crystals with regard to each other. These relationships can be expressed on a polar diagram: (1) by plotting the reference feature as a simple flat plane at the center of the diagram, irrespective of its actual shape, and the orientation of the individual grains with regard to that plane, or (2) by plotting the individual crystals in the normal way (Section 8.2).

The five groupings based on fabric are:

Parallel Arrangement. The crystals are arranged so that the major axis of each individual is approximately parallel to the most adjacent section of the wall of the void.

Normal Arrangement (Fig. 87). The crystals are arranged so that the major axis of each individual is approximately parallel to a radius of the void or perpendicular to the most adjacent section of the wall of the void.

Spherulitic Arrangement (Fig. 114). Pettijohn (1957) described spherulites from sedimentary rocks as "more or less spherical bodies" whose "constituents are arranged radially around one, or sometimes several, centers"; this is essentially a spherical crystallaria with spherulitic fabric. On the other hand, Rice (1954; after Kemp, 1927) defined spherulites as "rounded aggregates or rosettes, large or small, of acicular crystals that radiate from a center"; this is essentially a rounded crystallaria in which the crystals radiate from a single point. It is proposed to use *spherulite* in this latter restricted sense and *spherulitic* arrangement (fabric) in the broader sense applicable to all types of crystallaria, except perhaps intercalary crystals. In spherulitic arrangement, then, the crystals have their long axes radiating from one or more centers.

Crystallographic Arrangements. The arrangement of the crystals with regard to each other is characteristic for the mineral species, for example, twinning in gypsum.

Random Arrangement (Fig. 86). The crystals have no particular arrangement with regard to each other or to the walls of the void.

(*b*) *Kind of Void.* Although theoretically crystallaria can occur in voids of all the different kinds and shapes occurring in soil materials, they have been observed in only a limited variety. The actual mode of formation of the voids is relatively unimportant in the morphological classification of crystallaria as such, because the voids are only convenient spaces for crystallization and their genesis is not related to the processes of formation of the crystallaria. Pedological features are classified as crystallaria only when they are recognizable as concentrations of crystallized plasma in the larger voids, usually in channels, planes, vughs, vesicles, and chambers (Section 9.5).

(*c*) ***Mineralogical Nature of the Crystals.*** Crystallaria can be composed of any of the chemical compounds derived from the soil constituents that can crystallize from solution; they are usually composed of the more soluble fractions of the plasma, such as calcite, gypsum, halite, barite, etc. There is no agreement, however, as to whether certain clay minerals can be formed by crystallization from solution within soil materials. If they can there is no reason why they should not occur as crystallaria although they have not been reported as such. Nor is there any apparent reason why sesquioxides, manganese oxides, and such fractions of the plasma should not occur as crystallaria; relatively thick crystallizations of goethite in planes have been observed.

Identification of the constituents of crystallaria is relatively simple compared with most of the other pedological features; dissection and isolation from the soil material is restricted only by size of the crystallaria; identification of such materials optically, chemically, or by X-ray techniques presents few problems. In fact the constituents of many crystallaria are readily identifiable optically in thin section or even in hand specimen.

B. Classification of Crystallaria

Kubiena (1938) described and named several pedological features belonging to the group of crystallaria. Pettijohn (1957) discussed other features from sedimentary rocks whose counterparts have been observed in soil materials and which also belong to the group of crystallaria. All these features have been described on their internal fabric (the arrangement of the crystals) and the kind of void in which they occur, so the existing names can be used for groupings of crystallaria.

(*a*) ***Crystal Tubes*** (Fig. 85). Kubiena (1938) described and named crystal tubes, which consist of masses of crystals filling or partially filling relatively large tube-shaped or acicular voids in the soil material. He inferred that crystal tubes are formed by crystallization from the walls of the voids inward in that the examples he illustrated have central cavities with the crystals adhering to the walls. Such characteristics identify crystal tubes, which may be defined as follows:

Crystal Tubes. Crystallaria that occur in channels of simple or branching acicular shape; they usually show evidence of crystallization from the walls inward by preservation of a central void.

Crystal tubes are closely related to channel cutans composed of the more soluble fractions of the plasma (soluans). They are distinguished from such cutans primarily on degree of development, the crystal tubes being cohesive units.

Kubiena (1938) subdivided crystal tubes on the arrangement of the crystals into groups with *parallel* and *normal* arrangements (his cis-

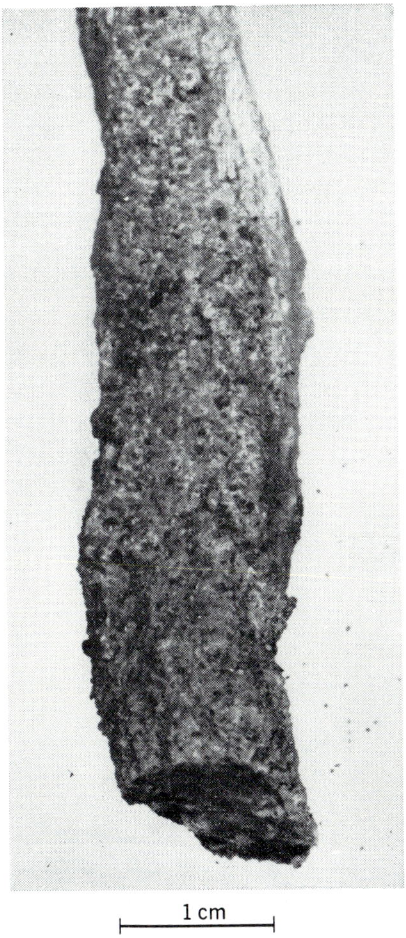

Fig. 85. Discrete, carbonate, crystal tube. Hand specimen. 2×.

tern-wall arrangement). Others have been observed with *random* arrangement of the crystals, and spherulitic arrangements may occur. They can be further subdivided on the mineralogical nature of the crystallized material.

(*b*) ***Crystal Chambers*** (Figs. 86 and 114). Kubiena (1938) also described and named crystal chambers, which differ from crystal tubes only in shape; they are approximately prolate to equant and can be defined as follows:

> *Crystal Chambers.* Usually prolate to equant crystallaria formed in vughs, vesicles, and chambers; they usually show evidence of crystallization from the walls inward by the preservation of a central void.

Crystal chambers are related to vugh cutans composed of the more soluble fractions of the plasma in the same way as crystal tubes are related to channel cutans.

Like crystal tubes, crystal chambers can be subdivided on the

Fig. 86. Gypseous, crystal chamber. Hand specimen. The individual crystals have a random arrangement. 2⅛×.

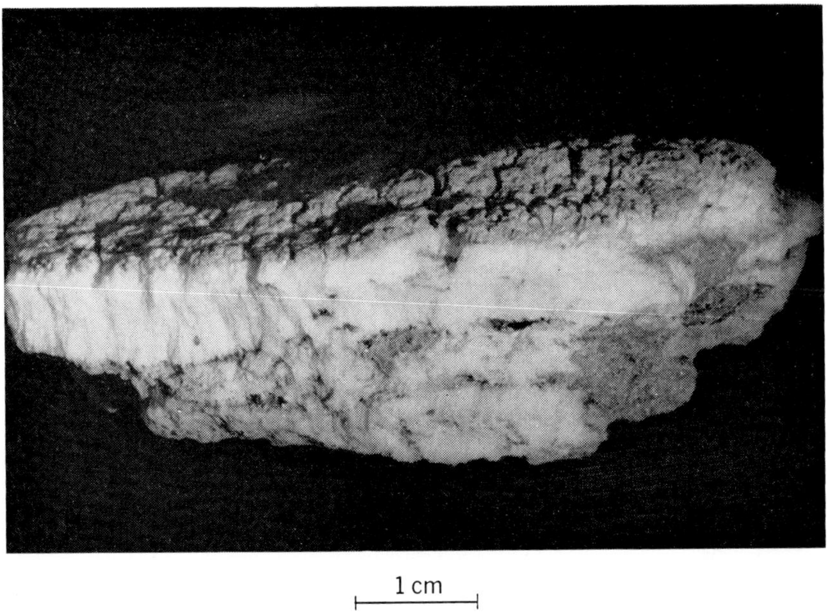

Fig. 87. Carbonate, crystal sheet. Hand specimen. Individual crystals have a normal arrangement with reference to the plane section of the sheet (the walls of the original planar void). 1½×.

arrangement of the crystals (parallel, normal, random, and spherulitic types) and on the mineralogical nature of the crystallized material. The specific type in which the crystals are arranged radially around a single center is termed a *spherulite;* a *rosette* is an incomplete spherulite (Pettijohn, 1957).

(*c*) ***Crystal Sheets*** (Fig. 87). A further variation of crystallaria, closely related to crystal tubes and crystal chambers, are cohesive sheets of crystals formed in planar voids in the soil material. These are very similar to crystal tubes, except for shape, and can be defined as follows:

Crystal Sheets. Planar shaped crystallaria formed in planes. There is no need for evidence of crystallization from the walls inward in crystal sheets because none of the possible crystal arrangements are likely to be confused with other types of crystallaria. Crystal sheets are related to plane cutans in the same way as crystal tubes are to channel cutans.

Fig. 88. Intercalary calcite crystals. Thin section under crossed polarizers. 80×.

As for crystal tubes, crystal sheets can be subdivided on the arrangement of the crystals (parallel, normal, random, or spherulitic types) and on the mineralogical nature of the crystallized material. The distinction between crystal sheets and plane cutans of the more soluble fractions of the plasma is quite narrow, and classification of doubtful cases as cutans or crystallaria is relatively unimportant.

(*d*) ***Intercalary Crystals*** (Figs. 88 and 89). Kubiena (1938) defined intercalary crystals as single crystals embedded in a dense soil matrix. They are classified as crystallaria mainly because they are relatively pure crystals, but they present a special case in that they do not obviously occur in voids. Such crystals have been reported from many Australian soils of fine particle size grading, such as the grey soils of heavy texture and the brown soils of heavy texture (Stephens, 1956). The crystals, however, often occur in groups rather than singly. To encompass such variations, intercalary crystals are defined as follows:

Intercalary Crystals. Crystallaria that consist of single large crystals or groups of a few large crystals set in the soil material and apparently not associated with voids of equivalent size or shape to that of the crystallaria as a whole; the crystals are euhedral to subhedral having at least some well developed crystal faces.

The qualification of intercalary crystals exhibiting crystal faces is inherent in Kubiena's description and is necessary to distinguish them from crystals of substances, such as calcite, which occur throughout many soil materials in a similar manner to intercalary crystals but as small, poorly developed crystals.

There is no evidence, so far, to necessitate subdivision of the intercalary crystals that occur in groups on arrangement of the crystals, except, perhaps, where particular crystallographic arrangements occur, but a distinction can be made between single crystals and groups of crystals. In agreement with Kubiena's (1938) observations, most Australian examples are crystals of gypsum (Fig. 89), but intercalary calcite (Fig. 88) has been observed.

Fig. 89. Intercalary gypsum crystals. Thin section under crossed polarizers. 135×.

C. Interpretation of Crystallaria

There is no apparent difficulty in interpretation of the formation of crystallaria in general terms. Crystal tubes, crystal chambers, and crystal sheets are undoubtedly due to crystallization of constituents from solution within voids in the soil material. There is no explanation of the different arrangements of the crystals. Spherulites and rosettes in soil materials usually have the crystals attached to the walls of the voids. Such an arrangement indicates radial growth of acicular crystals by secretion. As Pettijohn (1957) stated, such growths can be produced in the laboratory by normal crystallization. He also pointed out, however, that some workers attribute them to primary precipitation of a colloidal gel with subsequent crystallization. In both cases the internal fabric is due to crystallization of acicular crystals around a point. Intercalary crystals are probably due to the normal growth of larger crystals at the expense of smaller

Fig. 90. Compound crystal chamber. Thin section under crossed polarizers. The black central area is a void around which the calcite crystals are much larger than those away from the void. The thin white border to the void is an argillan with strong continuous orientation. 30×.

ones. The forces of crystallization would be sufficient to force the moist fine-grained soil material apart as the crystals grow. The occurrence of crystal faces and specific crystallographic arrangements indicate normal crystallization *in situ,* whereas it is most unlikely that voids with such shapes occurred in the soil material.

It is possible that features with most of the characteristics of intercalary crystals may be inherited. If they are transported, the crystals should show considerable rounding and loss of crystal faces, especially as gypsum and other salts are relatively soluble.

Compound crystallaria occur. Fig. 90 is a crystal chamber showing recrystallization of the calcite (large crystals around the void) and subsequent formation of illuviation argillans.

13.2 SUBCUTANIC FEATURES

A number of features observed in soil materials have definite relationships with natural surfaces, such as the surfaces of skeleton grains and the walls of voids, but do not occur immediately at the surfaces. Such phenomena constitute the group of subcutanic features defined as follows:

> *Subcutanic feature.* A pedological feature (recognized by a difference in texture, structure, or fabric as compared with the enclosing s-matrix) that has a consistent relationship with natural surfaces in the soil material, but does not occur immediately at the surfaces.

A. Classification of Subcutanic Features

Subcutanic features can be classified according to: (1) their relationships with the natural surfaces, (2) composition, and (3) internal fabric. Two broad groups have been recognized on their relationships with the natural surfaces, which are usually the walls of megascopic and microscopic voids; these are qualified according to the particular components of the soil material that are relatively concentrated or the particular structure or fabric by which they are recognized as pedological features.

(*a*) Neocutans. These occur subcutanically immediately adjoining the natural surfaces with which they are associated. Classification and nomenclature of neocutans follow the general principles used for cutans (Chapter 10) for the kind of surface with which they are associated, their composition, and internal fabric. Most of the varieties observed to date have been associated with the walls of megascopic

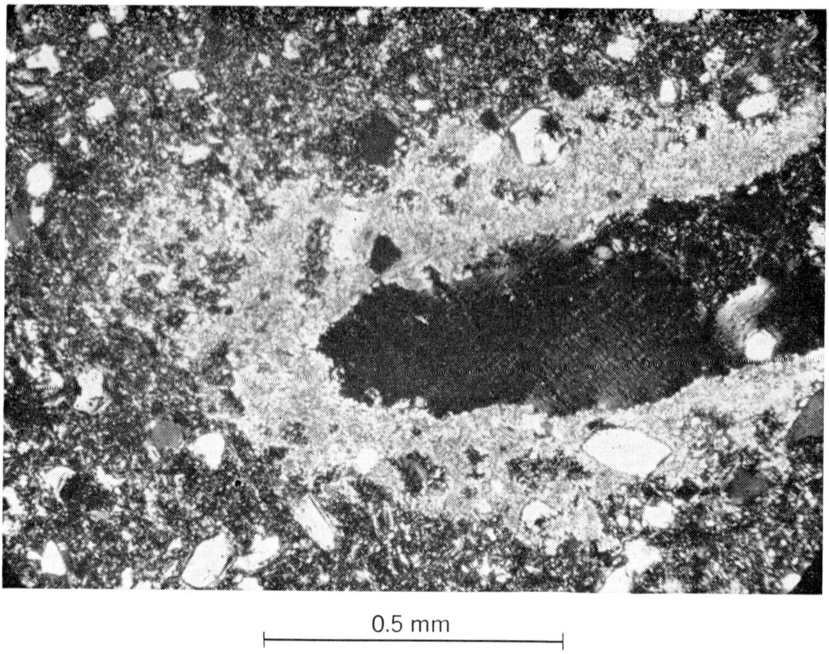

Fig. 91. Channel neocutan. Thin section under crossed polarizers. The section cuts across the channel giving a finger-shaped void surrounded by an almost white concentration of calcite which encloses skeleton grains (neocalcitan). 80×.

and microscopic voids, especially channels, planes and vughs, and peds. In the varieties that are recognized by a difference in composition from the s-matrix, the concentrated component can theoretically be any fraction of the plasma. Observed varieties consisted of accumulations of calcite, for example, *channel neocalcitans* (Fig. 91), sesquioxides and hydroxides (*neosesquans*), manganese oxides (*neomangans*), and organic matter (*neoorganans*—Fig. 96). Three varieties have been observed that are recognized essentially by a difference in fabric with the s-matrix: *neoskeletans,* which have a lower proportion of plasma to skeleton grains than the s-matrix, *neomatrans,* which have an increase in proportion of matrix, that is, a decrease in framework members (coarser skeleton grains) and/or denser packing than the s-matrix, and *neostrians* (Fig. 92), which have a typical striated orientation pattern of the plasma. The last are subcutanic varieties of plasma separations whose characteristics are used to define certain of the more common sepic plasmic fabrics, such as skelsepic and vosepic fabric (Section 14.1:B). Occasionally complex types occur, for example, with concentric fabric (Fig. 93). This classification can be qualified, as for cutans, according to the sharpness of boundary and adhesiveness of the neocutans. The sharpness of boundary of observed types is variable; the boundaries of neocalcitans are usually sharp, those of neosesquans and neoorganans are usually rather diffuse but may be sharp, those of neoskeletans are sharp to rather diffuse, and those of neomatrans and neostrians are usually rather diffuse. All neocutans are usually strongly adhesive.

Many cutans can be identified as plasma concentrations occurring as deposits of relatively pure fractions of the plasma on the surfaces of voids. The concentrated plasma fraction, however, may extend beyond the surface of the void into the s-matrix of the soil material, especially where the accumulating constituents are relatively soluble. This material then constitutes a subcutanic plasma concentration associated with a depositional cutan which may be an illuviation or

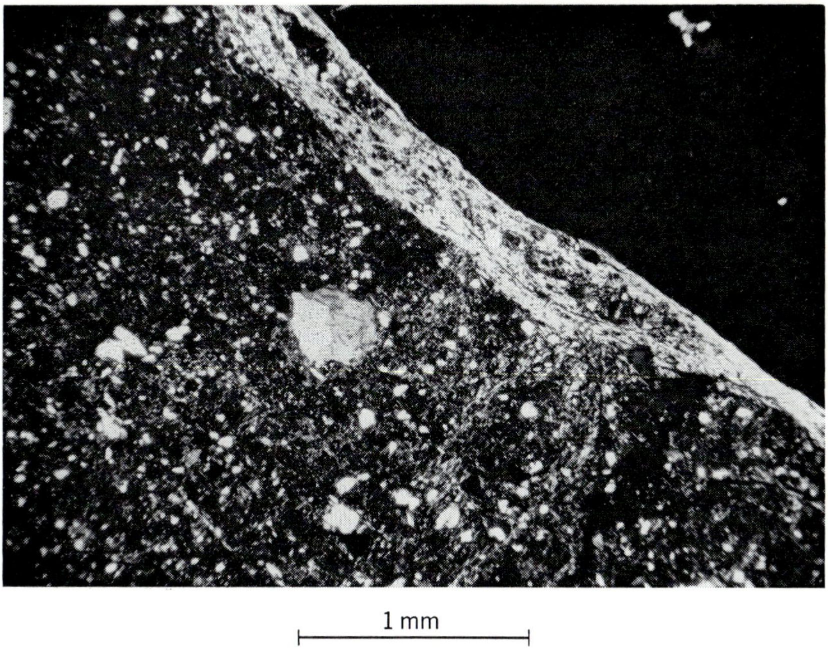

Fig. 92. Skew-plane neocutan. Thin section under crossed polarizers. The very pale, diagonal zone with a striated orientation pattern is a skew-plane neostrian within which the density of skeleton grains is the same as in the s-matrix. The very thin white line at the outer edge of the neostrian is a skew-plane palygorskan. 30×.

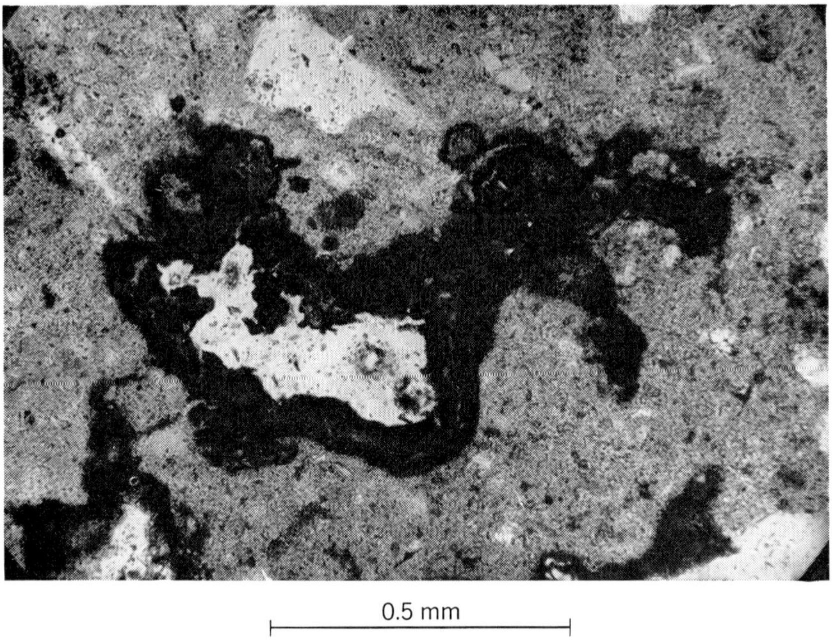

Fig. 93. Complex neocutan. Thin section in plain light. The white area left of the center is a void. The surrounding black area is a neosesquan, which has a concentric fabric, the laminae being approximately parallel to the walls of the void. 80×.

diffusion cutan. In the treatment of cutans, these subcutanic accumulations were not differentiated from the cutan because they are part of the same material and feature, and the feature as a whole was formed by the same general process. Description of such arrangements is made more specific if the concept of a neocutan is used and the whole feature is described, for example, as a mangan with associated neomangan (Fig. 94); cutans of one mineralogical composition may be associated with neocutans of a different composition (Fig. 92). Neostrians are usually associated with stress argillans in an analogous way, but they are dealt with in description and classification of plasmic fabric (Chapter 14).

(*b*) *Quasicutans.* These occur subcutanically to natural surfaces but not immediately adjoining them; they have some obvious relationship to natural surfaces (usually the walls of voids) in regard to shape and distribution but do not occur adjacent to or on the surfaces.

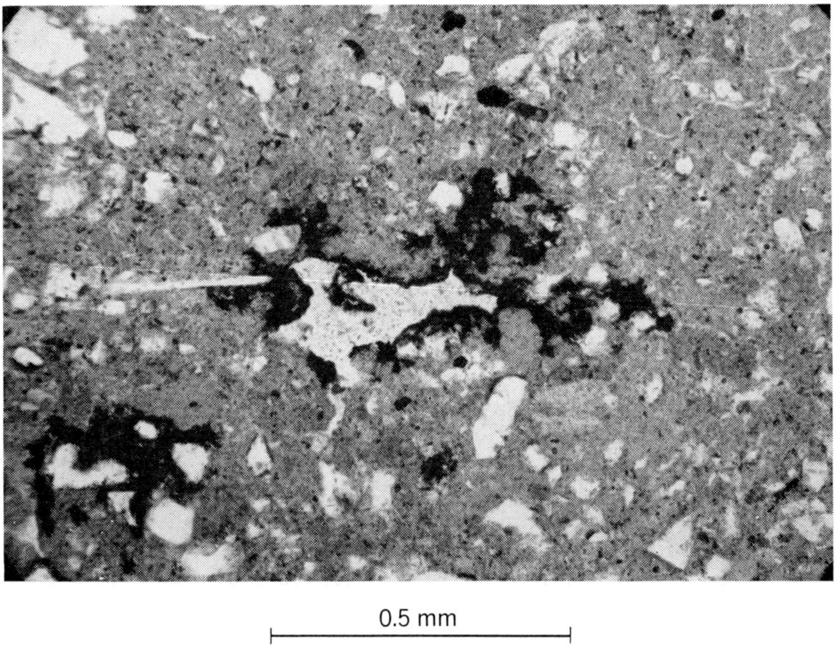

Fig. 94. Mangan–neomangan. Thin section in plain light. The central white area is a void. The black zone surrounding it is a mangan with an associated neomangan. 80×.

The most common expression of quasicutans are plasma concentrations in the form of bands of varying thickness whose internal and external surfaces have the same general shape as the void with which they are associated, for example, a cylindrical concentration around a channel and separated from the channel by a cylindrical zone of normal s-matrix that lacks the plasma concentration. Quasicutans can be classified according to the same principles used for cutans and neocutans. They are commonly composed of sesquioxides or hydroxides, for example, a *channel quasisesquan* (Fig. 95). Figure 93 could be regarded as a complex quasicutan–neocutan. Quasicutans are strongly adhesive, and their boundaries may be rather sharp to rather diffuse.

B. Interpretation of Subcutanic Features

The majority of neocutans and quasicutans are plasma concentrations, and the mechanism of formation of these types must be basically

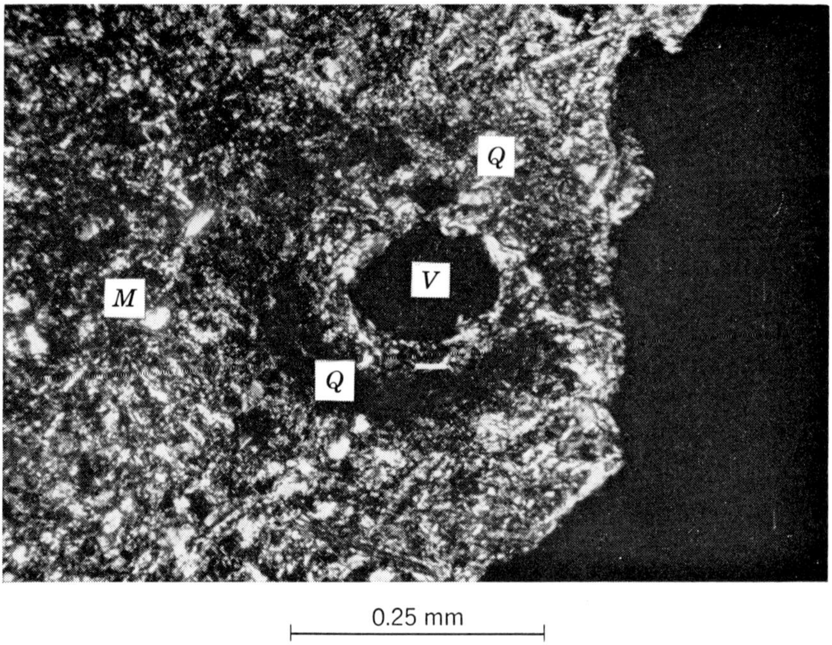

Fig. 95. Cross section of a channel quasicutan. Thin section under crossed polarizers. *V* is the section of the channel, *M* is the s-matrix, and *Q* is a quasi-sesquan; note the thin zone of s-matrix between *Q* and *V*. 135×.

one of crystallization of solubles from solution or deposition from colloiodal suspension. There must also have been movement of the concentrated plasma fraction toward the zone of accumulation. This could be achieved by either of two processes depending on the conditions within the soil material: by diffusion from within the body of the soil material toward the natural surfaces, which are usually the walls of voids which could act as drying surfaces, or by movement of suspensions or solutions along the void and penetration into the s-matrix of the soil material, which forms the walls of the voids (illuviation). The latter process would be facilitated by dessication of the soil material followed by sudden wetting and movement of solutions or suspensions from the void into the dry soil material surrounding it. The conditions necessary for deposition of each of the plasma fractions are different, and the reasons for these conditions occurring in a particular zone to cause the formation of quasicutans are obscure.

Neoskeletans are probably formed by eluviation of the plasma from what was originally normal s-matrix material. This appears to have been the process in domed columnar peds with bleached ped neoskeletans; Jongerius (1957) interpreted similar features in the soils of the Netherlands in the same way. The neomatrans that have been observed are associated with the walls of channels which, because of their shape, can be identified as either root channels or worm burrows; the denser packing is almost certainly due to pressures exerted during formation of the channels.

Interpretation of neostrians is discussed under interpretation of the formation of plasmic fabrics (Section 14.8).

13.3 FECAL PELLETS

The occurrence of faunal excreta, or fecal pellets, in soil materials has been discussed to some extent in the treatment of pedotubules (Chapter 11). At first sight it would seem reasonable to assume that recognizable fecal pellets must occur as pedotubules since the animal

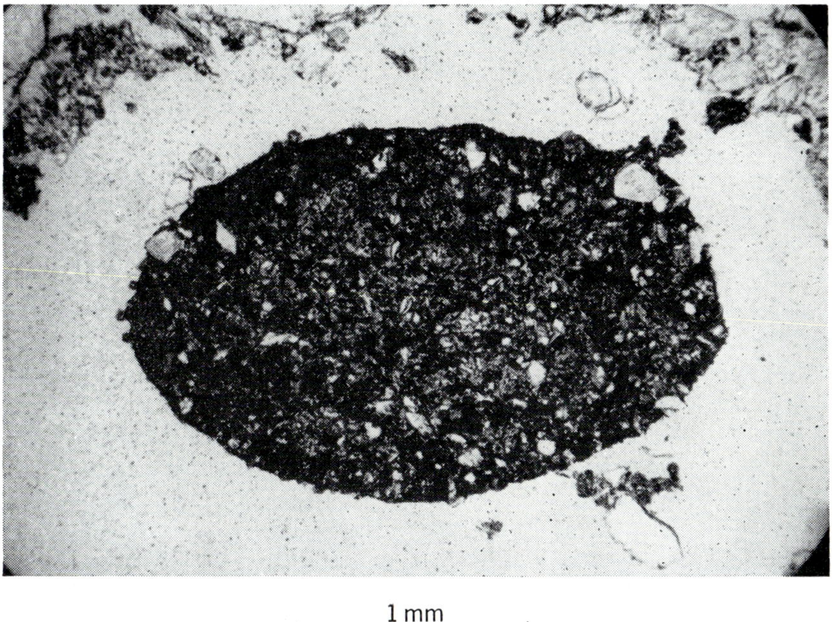

Fig. 96. Single fecal pellet. Thin section in plain light. The diffuse darker border constitutes a neoorganan. 30×.

responsible for their existence must have constructed a burrow or tunnel of some kind or used an already existing channel or chamber within which the fecal pellets are deposited. Disturbance of the soil material (pedoturbation—Hole, 1961), however, may destroy the external form of such pedotubules, especially if the tubulic material was not dense, for example, if it consisted of loosely packed fecal pellets. As a result of these processes fecal pellets may occur more or less embedded in the soil material, in which case there are sound grounds for classifying them as glaebules (Chapter 12). Bodies with the morphological characteristics of fecal pellets, however, sometimes occur in what appear to be the remnants of tunnels or burrows, and these are properly neither pedotubules nor glaebules. In addition, aggrotubules may consist of normal soil aggregates or of fecal pellets, and it is advantageous to distinguish these two types.

For these reasons it is necessary to recognize fecal pellets as such. Unfortunately little is known of the morphological characteristics of the fecal pellets of many of the fauna inhabiting soil materials. Some

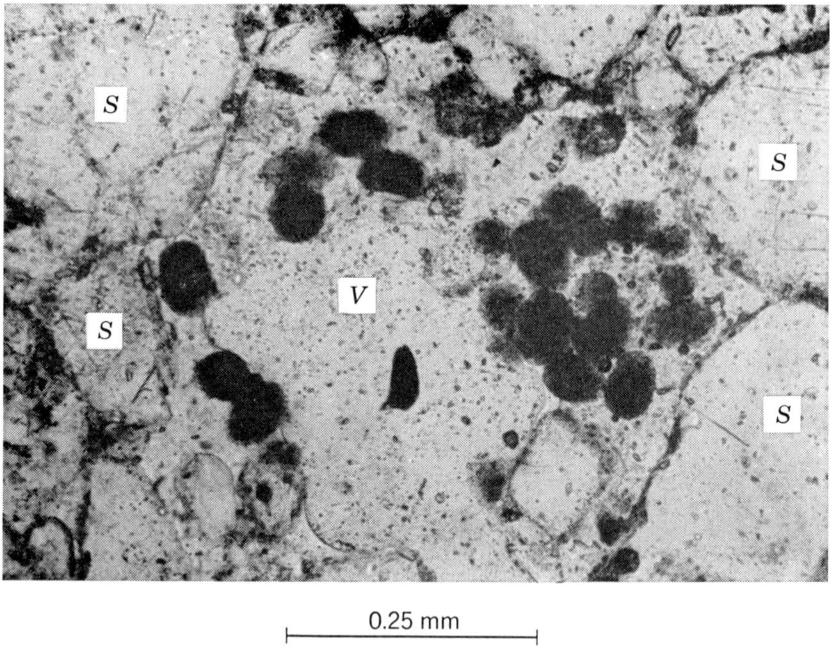

Fig. 97. Single fecal pellets. Thin section in plain light. *V* is a large void and *S* is skeleton grains. 135×.

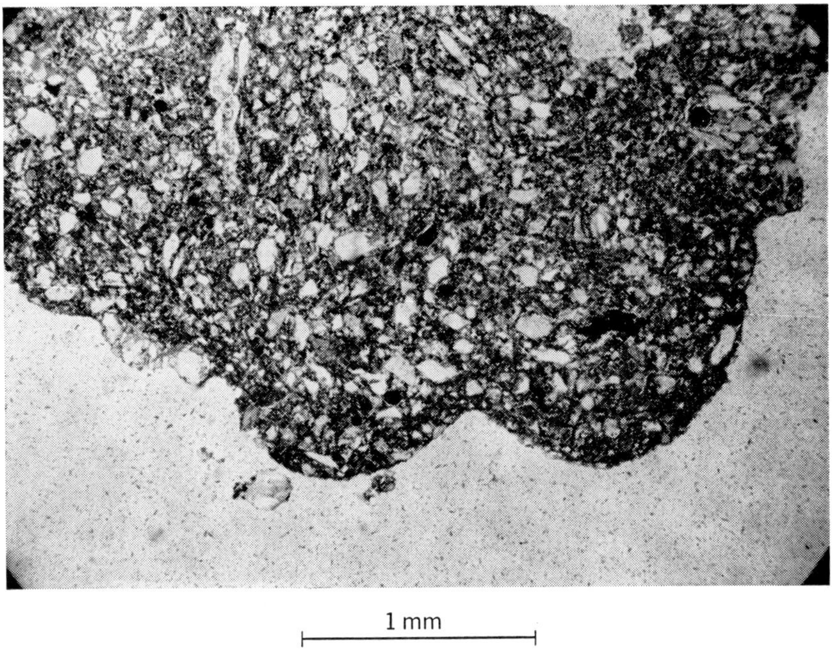

Fig. 98. Welded fecal pellets. Thin section in plain light. The mammillated outline suggests the original size and shape of individual pellets. 30×.

of the characteristics of earthworm excreta have been set out in the discussion of pedotubules (Section 11.2), but other types have not been described in detail. In general they are cylindrical, ellipsoidal, or spheroidal with relatively smooth walls, and they are usually denser and darker than the normal soil material although color is controlled to some extent by the color of the soil material which the animal ingested. Much more must be known of the various types of fecal pellets before their recognition can be certain and a classification can be set up.

For the present, one major subdivision on the basis of external shape seems appropriate. Fecal pellets occur commonly both as single pellets and as groups or aggregates of pellets which are "welded" together so that the individuals have lost their form over the areas where they adjoin each other. These two types are known as *single fecal pellets* and *welded fecal pellets* (Figs. 96–98). The convention may be adopted that they are single pellets unless specifically referred to as welded.

14

S-Matrix. Basic, Elementary, and Primary Structures

By definition, the s-matrix of a soil material is the material within peds, or composing unaggregated soil materials, in which the pedological features occur (Section 7.5:C). It includes the framework and matrix (material less than 30 μ diameter) of the sedimentary petrologists and consists essentially of skeleton grains, plasma, and intrapedal voids; it deals with plasma separations since they are simply variations in arrangement. Pedological features, too, have s-matrixes that have been used to classify pedological features into subclasses within the major broad classes, but the classes are based on broad fabric groups, some of which encompass only a single fabric feature, and none of which are sufficiently concise and detailed to satisfy the more specific needs for classification of s-matrixes or basic structure.

Voids, peds, and pedological features can generally be considered as individuals with characteristics of size, shape, and arrangement. On the other hand, s-matrixes merely form the background material within which pedological features occur; they are not entities with definite boundaries except insofar as certain ones may have definite limits with regard to others; for example, the s-matrix of a pedological feature has the limits of the pedological feature; of themselves they have no characteristics of size, shape, distribution, or orientation. S-matrixes, therefore, must be classified on their internal characteristics, that is, on the size, shape, and arrangement of the plasma (*plasmic structure*), skeleton grains, and voids, and the relationships of the plasma and skeleton grains. These characteristics are the elements of *basic structure,* which is in effect the structure of the s-matrix.

Elementary and primary structure both deal with the s-matrix and the pedological features associated with it, the former including only specific, characteristic pedological features and the latter including all intrapedal pedological features. Their description and classification, therefore, depends on classification of the s-matrix, or basic structure, and the relationships between the s-matrix and the appropriate pedological features.

14.1 PLASMIC STRUCTURE

The plasmic structure of soil materials and pedological features is the organization of the constituents of the plasma that have not been concentrated or crystallized to form pedological features and the associated very small voids which result from packing of plasma grains. Individual grains and voids are largely beyond the resolving power of normal optical microscopes, and special techniques, such as electron microscopy, are necessary for direct observation of them. Failing this the voids of the plasmic structure are assumed to be equivalent to packing voids and receive no further attention. A good deal of morphological description of plasmic fabric, however, is possible by optical methods in thin section on the petrological microscope, especially by interpretation of the extinction patterns of the various kinds and arrangements (distribution and orientation patterns) of plasma grains, if the nature of the constituents of the plasma is known (Sections 8.1 and 8.2).

Although the majority of the constituents of the plasma, which occur as grains less than 2 μ in diameter (clay minerals, oxides and hydroxides, soluble salts, and rock-forming minerals), are crystalline and anisotropic, they appear isotropic in thin section if the individual grains are unoriented with regard to their neighbors; this is because the grains are very small compared with the thickness of a thin section (25 μ). Some of the constituents are optically isotropic, and anisotropy can be masked by fractions of the humified organic matter and by oxides and hydroxides of such metal ions as iron, aluminium, and manganese, which are often amorphous (isotropic) to the petrological microscope and/or opaque or weakly translucent. Anisotropy of the plasma in thin section, therefore, is due either to translucent, anisotropic plasma grains that have some degree of preferred orientation with regard to their neighbors, or to larger crystals of anisotropic plasma; the former may form recognizable units which are apparently

Fig. 99. Argillasepic plasmic fabric. Thin section under crossed polarizers. Note the lack of plasma separations. 80×.

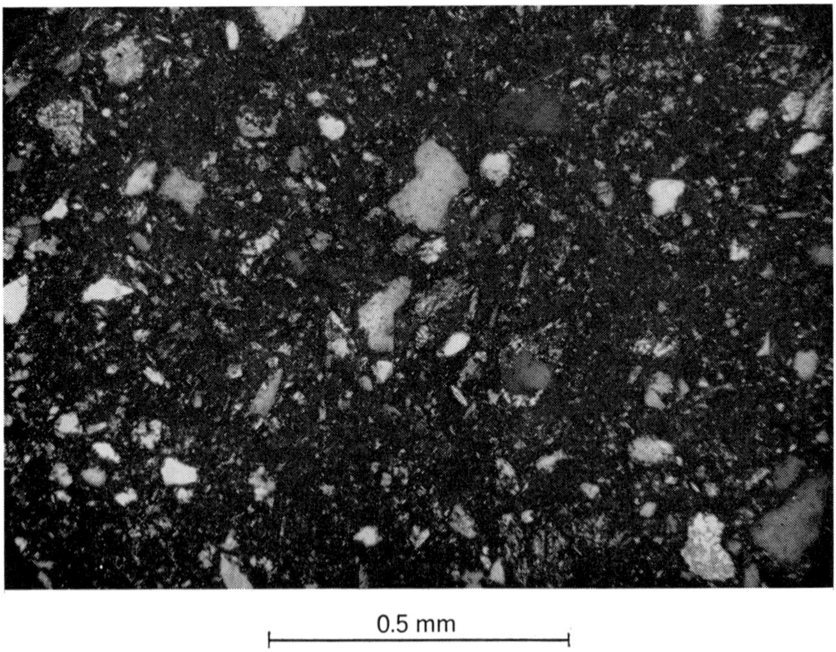

Fig. 100. Silasepic plasmic fabric. Thin section under crossed polarizers. Note the lack of plasma separations. 80×.

identical with the so-called *domains* (Aylmore and Quirk, 1959; Emerson, 1959).

Description of plasmic fabrics, therefore, is based on interpretation of optical properties under crossed nicols, especially extinction phenomena due to:

(1) visible crystals of plasma,

(2) the kind and degree of orientation of the plasma grains, including distinctions between opaque, isotropic, and anisotropic plasma,

(3) the kind and degree of preferred orientation of domains, if they occur, which cause flecked and striated basic orientation patterns (Section 8.2:A), and particularly

(4) the kind and degree of development of plasma separations, which can be described in terms of size, shape, basic and referred distribution pattern, degree of basic orientation, and referred orientation pattern especially with regard to the surfaces of skeleton grains and the walls of the larger voids.

We pointed out earlier that plasma that appears largely opaque or

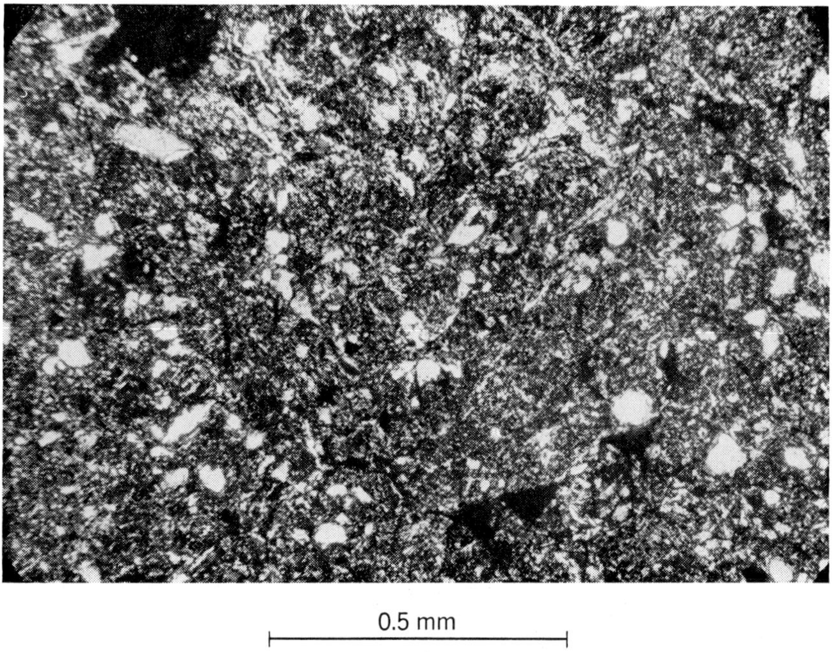

Fig. 101. Insepic plasmic fabric. Thin section under crossed polarizers. Plasma separations occur as isolated patches. 80×.

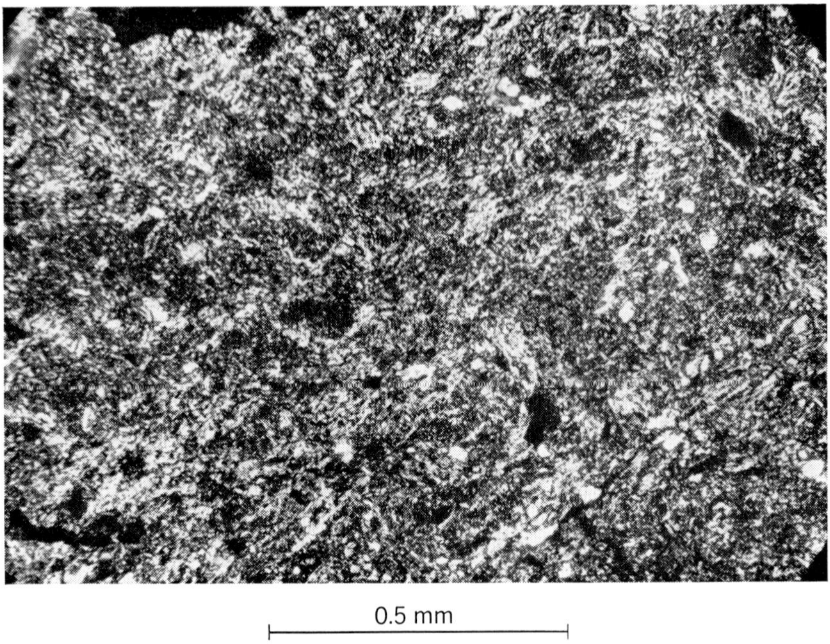

Fig. 102. Mosepic plasmic fabric. Thin section under crossed polarizers. Plasma separations occur as numerous patches. 80×.

isotropic at low magnifications may appear quite strongly anisotropic at high magnifications due to the higher concentrations of light obtained by using a converging condenser with the result that more light passes through masking constituents, such as humified organic matter and amorphous oxides and hydroxides of iron, than at low magnifications. In these instances, plasmic fabric is described and classified principally according to the characteristics observed at the highest magnifications, because the plasma consists essentially of anisotropic constituents whose characteristics were merely masked at lower magnifications. Similarly weakly developed plasma separations can be detected much more readily at high magnifications than at low magnifications; soil materials that appear to have an entirely flecked extinction pattern at low magnifications often exhibit some degree of development of plasma separations at high magnifications. It is important, therefore, to make observations at all possible magnifications to classify plasmic fabrics.

An important difficulty in classification of plasmic fabrics is the fact that their characteristics form a complete continuum. At any magnification there is a continuous gradation from plasma that appears opaque or isotropic to plasma that appears entirely anisotropic, and from fabrics that appear to have an entirely flecked orientation pattern to those that appear to consist entirely of a striated orientation pattern; in the latter the intermediate stages form a continuum both in degree of development of plasma separations with striated orientation and in the distribution patterns of the plasma separations. Similarly there is a gradual transition from plasmic fabrics consisting entirely of anisotropic domains (whether with striated or flecked extinction patterns), through plasmic fabrics in which the boundaries of the domains are not apparent but the plasma has an indistinct, undulose extinction pattern, to plasmic fabrics that are either opaque or isotropic even at the highest magnifications and with the highest possible light intensity. The best that can be done in the classifica-

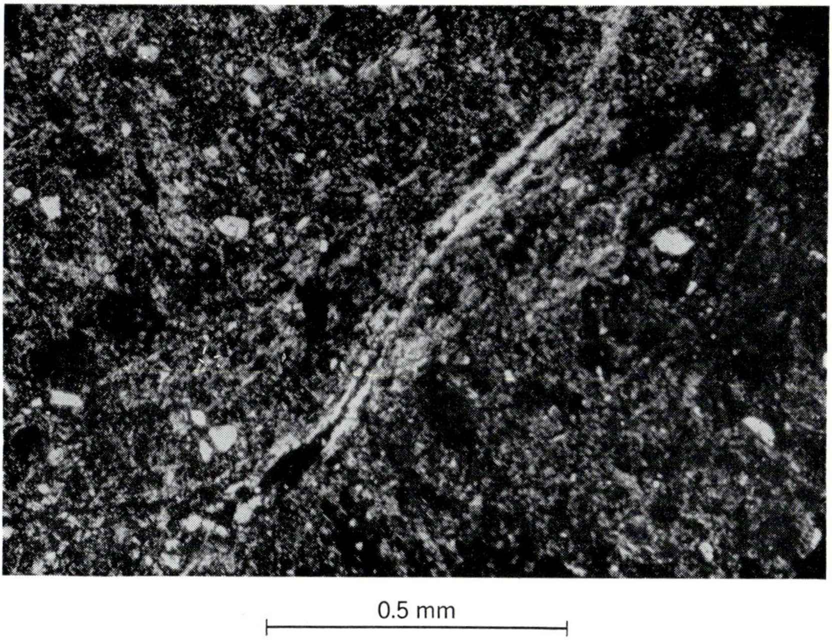

Fig. 103. Vosepic plasmic fabric. Thin section under crossed polarizers. More specifically vo-insepic fabric since the bulk of the s-matrix has an insepic plasmic fabric. The plasma separation associated with the void is specifically a plane neostrian. 80×.

TABLE 14-1. Morphological Classification of Typical Modal Plasmic Fabrics

Fabric Class	Extinction Pattern	Plasma Aggregates*	Plasma Separations
1. Asepic	Flecked		
(1) Argillasepic		Recognizable	Absent
(2) Silasepic		Present but difficult to distinguish from silt grains	Absent
2. Sepic	Flecked and striated		
(1) Insepic		Recognizable	As occasional isolated patches
(2) Mosepic			As numerous patches
(3) Vosepic			Associated with voids
(4) Skelsepic			Associated with skeleton grains
(5) Masepic			As elongated zones within the s-matrix
(6) Lattisepic			As two sets of very short discontinuous plasma separations
(7) Omnisepic	Dominantly striated		Strongly developed, complex striated pattern
3. (1) Undulic	Dominantly undulose	Absent	Absent
(2) Inundulic		Recognizable but rare	Rare isolated patches
4. Isotic	Isotropic (or opaque)	Absent	Absent
5. Crystic	Recognizable birefringent (or isotropic) crystals	Absent	Absent

* Plasma aggregates are recognized as small optically anisotropic units composed of clay minerals; they are apparently identical with *domains* (Aylmore and Quirk, 1959)

tion of plasmic fabrics, therefore, is to select typical modal fabrics that can be given appropriate names.

The following classification is proposed for the typical plasmic fabrics observed to date. They are classified first into four broad groups on their general anisotropy and extinction patterns in thin section under crossed nicols on the petrological microscope; each of these groups can be subdivided further according to the details of their extinction patterns. The classification is summarized in Table 14-1.

A. Asepic* Plasmic Fabrics

These fabrics have dominantly anisotropic plasma with anisotropic domains that are unoriented with regard to each other; that is, they

* The stem of the adjective "sepic" is derived from "separation" so that "asepic" means "without plasma separations," and "sepic" means with "plasma separations."

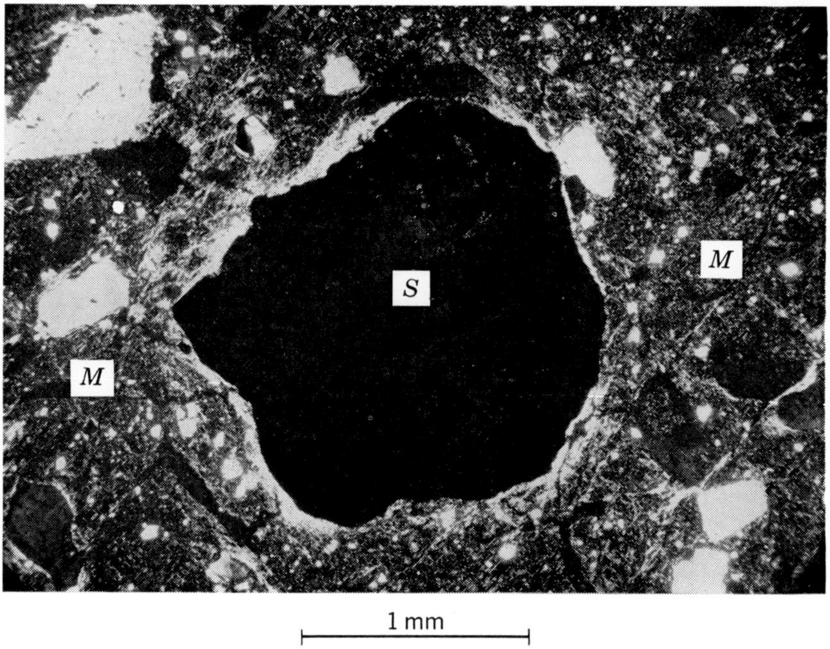

Fig. 104. Skelsepic plasmic fabric. Thin section under crossed polarizers. The large central, black area (*S*) is a skeleton grain at extinction. The white border to it is a plasma separation (an embedded grain neostrian). *M* is matrix. 30×.

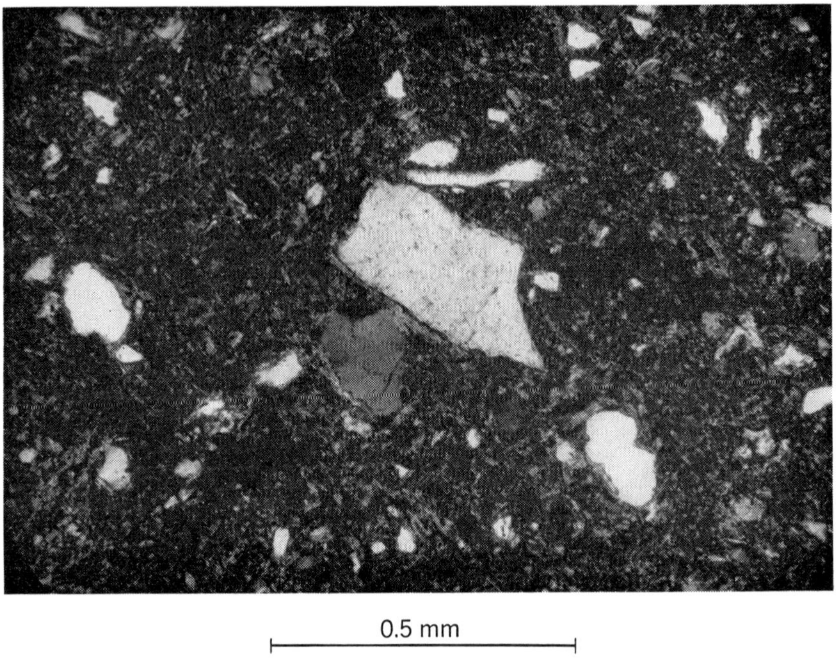

Fig. 105. Skel-insepic plasmic fabric. Thin section under crossed polarizers. The larger skeleton grains have associated plasma separations (embedded grain neostrians). The bulk of the s-matrix has an insepic plasmic fabric. 80×.

have a flecked extinction pattern; there are virtually no plasma separations. They are subdivided into two subgroups:

(*a*) ***Argillasepic* Fabric.*** (Fig. 99). The plasma of this fabric consists dominantly of anisotropic clay minerals and exhibits a flecked orientation pattern with recognizable domains.

(*b*) ***Silasepic† Fabric*** (Fig. 100). Soil materials with this fabric have a wide range of particle sizes with relatively high proportions of silt size grains so that domains are difficult to recognize; the matrix has a flecked extinction pattern.

* "Argill-" from "argillaceous."
† "Sil-" from "silt."

B. Sepic Plasmic Fabrics

These fabrics have recognizable anisotropic domains with various patterns of preferred orientation; that is, plasma separations with a striated extinction pattern are present. They can be subdivided on the characteristics of the plasma separations into the following:

(*a*) ***Insepic* Fabric*** (Fig. 101). The plasma separations with striated orientation occur as isolated patches, or islands, within the dominantly flecked plasma, which may have a very weak striated orientation at high magnifications; that is, the domains that exhibit a preferred orientation have a clustered distribution pattern.

(*b*) ***Mosepic† Fabric*** (Fig. 102). This is an extreme development of insepic fabric; the patches with striated orientation may

* "In-" from "insular," meaning "island."
† "Mo-" from "mosaic."

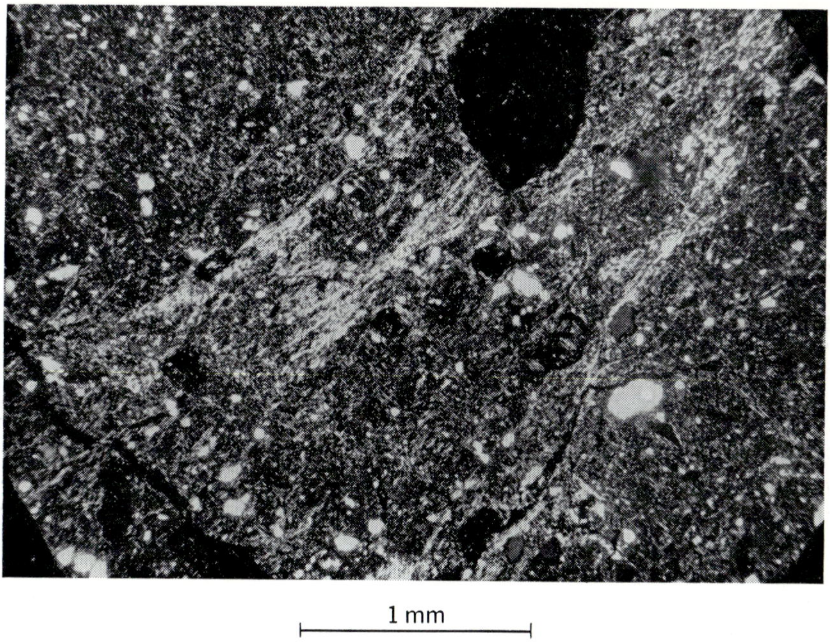

Fig. 106. Vo-masepic plasmic fabric. Thin section under crossed polarizers. Plasma separations with a striated orientation pattern occur as zones within the s-matrix and as neostrians associated with voids. 30×.

Fig. 107. Clino-bimasepic plasmic fabric. Thin section under crossed polarizers. Note the broad plasma separation with a striated orientation pattern within the s-matrix and the narrow, less well-developed plasma separations at approximately 60 degrees to it. 30×.

adjoin each other or be separated by small areas of plasma with flecked orientation but are unoriented with regard to each other.

*(c) **Vosepic* Fabric*** (Fig. 103). Part of the plasma has a flecked orientation pattern, but plasma separations with striated orientation occur subcutanically associated with the walls of voids; the striated orientation of the plasma separations is dominantly parallel to the walls of the voids, especially planes.

*(d) **Skelsepic† Fabric*** (Figs. 104 and 105). Part of the plasma has a flecked orientation pattern, but plasma separations with striated orientation occur subcutanically to the surfaces of skeleton grains; the

* "Vo-" from "void."
† "Skel-" from "skeleton grains."

striated orientation of the plasma separations is dominantly parallel to the surfaces of the skeleton grains.

(*e*) ***Masepic* Fabric*** (Figs. 106 and 107). Part of the plasma has a flecked orientation pattern, but plasma separations occur as zones within the s-matrix apparently not associated with the walls of voids or the surfaces of skeleton grains; the striated orientation is elongated parallel to the length of the zones. All the zones may be subparallel to each other (that is, there is a single direction of preferred orientation of the plasma separations and of the constituent domains) or the plasma separations may occur in two or three sets of subparallel zones, each set being inclined at a definite angle to the other sets. The latter are termed *bimasepic fabric* (Fig. 107) and *trimasepic fabric* where there are two and three sets of plasma separations, respectively;

* "Ma-" from "matrix."

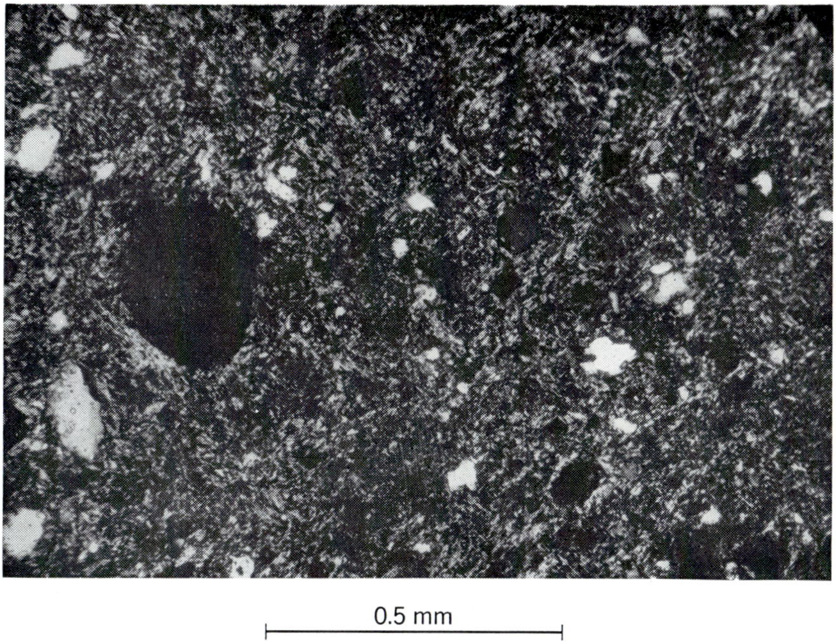

Fig. 108. Lattisepic plasmic fabric. Thin section under crossed polarizers. Note the lattice-like pattern of small prolate to acicular plasma separations. 80×.

Fig. 109. Omnisepic plasmic fabric. Thin section under crossed polarizers. The whole of the plasma of the s-matrix has a complex striated orientation pattern. 80×.

right and *clino-* are used to distinguish two sets of plasma separations at right angles to each other and two sets inclined at an acute angle to each other, respectively, that is, *right bimasepic fabric* and *clino-bimasepic fabric.*

(*f*) *Lattisepic* Fabric* (Fig. 108). This is essentially a variety of *bimasepic fabric;* the plasma generally has a flecked orientation pattern, but acicular and prolate domains occur in a lattice-like pattern; that is, there are two sets of very short, discontinuous plasma separations usually oriented approximately at right angles to each other.

(*g*) *Omnisepic† Fabric* (Figs. 109 and 110). All the plasma exhibits a complex striated orientation pattern; in thin section the striae have an intricate pattern as regards direction of preferred orientation.

* "Latti-" from "lattice."
† "Omni-" meaning "all."

Each of the named sepic plasmic fabrics is characterized by a single kind of plasma separation; this is an idealized concept. Soil materials usually exhibit combinations of different kinds of plasma separations. These more complex fabrics can be named rather specifically by using combinations of the prefixes used for the simple types, with the convention that the best developed fabric is written last, and qualification by addition of other appropriate prefixes. A commonly occurring plasmic fabric with well-developed plasma separations subcutanic to planar voids and less well-developed plasma separations associated with the surfaces of skeleton grains is adequately named as *skel-vosepic fabric.*

The named fabrics can be subdivided further according to details of the characteristics of the plasma separations, particularly the relative width of the zones of plasma separations in thin section as compared with the distance between them and the degree of development of the striated orientation pattern within them, that is, the degree of preferred orientation between domains. Much more experience is

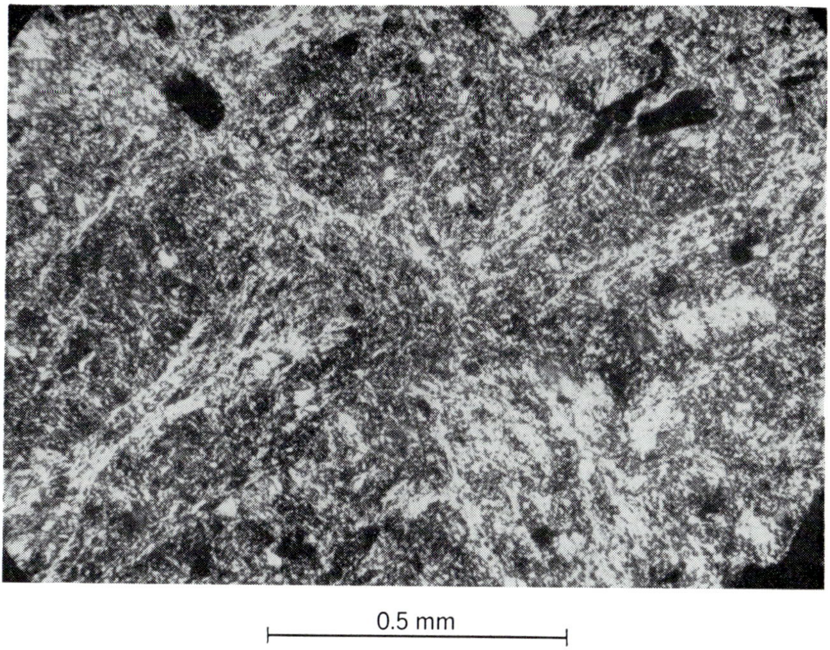

Fig. 110. Omnisepic plasmic fabric. Thin section under crossed polarizers. Essentially an extreme development of bimasepic plasmic fabric. 80×.

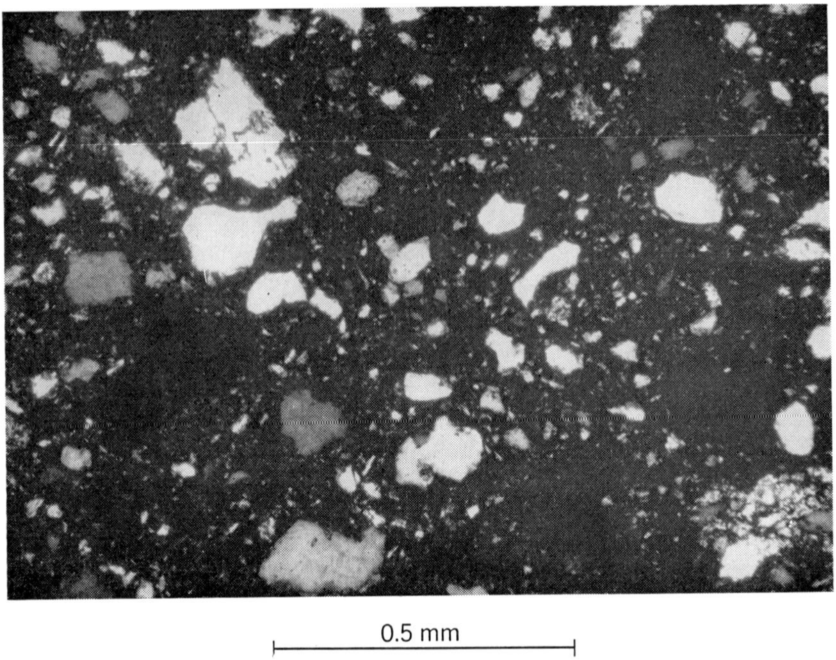

Fig. 111. Inundulic plasmic fabric. Thin section under crossed polarizers. Note the faint general anisotropy of the plasma of the s-matrix with some faintly distinguishable domains. 80×.

necessary before subgroups can be named according to these criteria. For the present, the named fabrics are qualified for the width of plasma separations (*broad:* wider than 200 μ; *normal:* between 200 and 25 μ; *narrow:* narrower than 25 μ) and degree of preferred orientation within plasma separations (weak, moderate, strong), for example, *broad, weak vosepic fabric* and *narrow, strong vosepic fabric.*

C. Undulic* Plasmic Fabrics

The plasma is indeterminate; that is, it is practically isotropic at low magnifications and weakly anisotropic with faint undulose extinction at high magnifications and high light intensities; domains are not recognizable, or are indistinct, even at high magnications. Observations on undulic fabrics are so sparse that little is known of the

* The stem of "undulic" is derived from "undulose," which describes the extinction pattern.

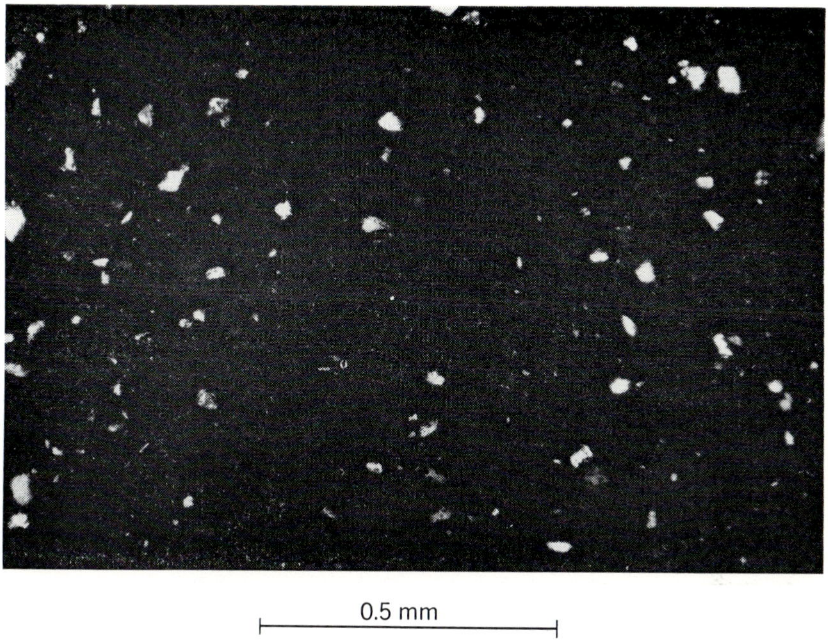

Fig. 112. Isotic plasmic fabric. Thin section under crossed polarizers. The plasma of the s-matrix is isotropic. 80×.

range of variation that may occur. One variation has been observed and named *inundulic fabric* (Fig. 111) in which the normal undulic fabric is modified by small, rather indistinct domains.

D. Isotic* Plasmic Fabrics (Fig. 112)

These fabrics have apparently isotropic plasma; that is, it is indeterminate at even the highest possible magnifications with high light intensity. It may be possible to subdivide isotic fabrics on the basis of the cause of the isotropic characteristics, that is, true isotropy of the minerals and opacity due to iron oxides, organic matter, etc.

E. Crystic† Plasmic Fabrics (Figs 113 and 114)

The plasma is usually anisotropic and consists of recognizable crystals, usually of the more soluble plasma fractions. Such fabrics

* The stem of "isotic" is derived from "isotropic."
† The stem of "crystic" is derived from "crystallized."

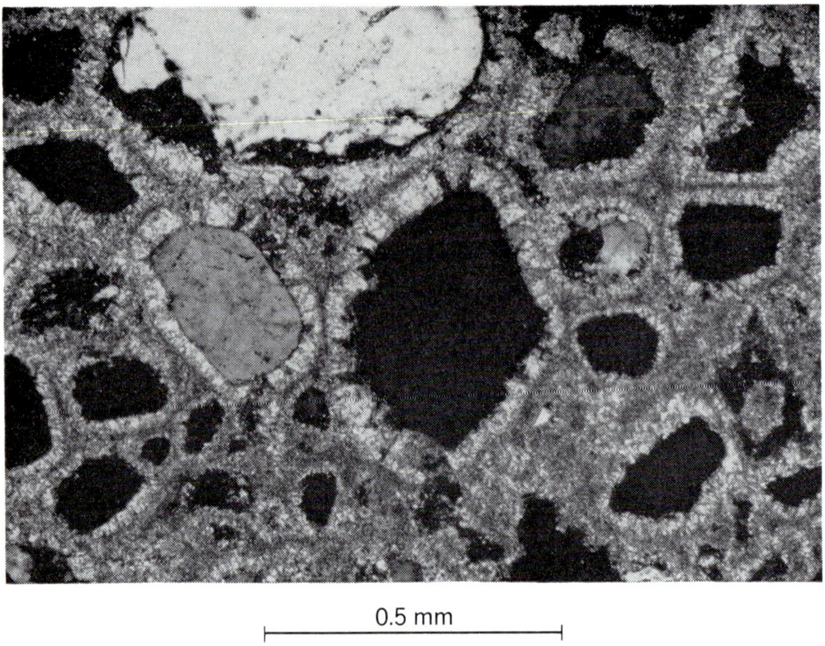

Fig. 113. Crystic plasmic fabric in a carbonate glaebule. Thin section under crossed polarizers. The carbonate crystals in the general s-matrix have a random arrangement while the larger crystals around the skeleton grains are oriented normal to the surface of the grains. 80×.

can be subdivided on the details of arrangement of the crystals; they occur most commonly in certain pedological features (Chapters 10 and 13).

F. Compound Plasmic Fabrics

Sepic fabrics have been defined as being dominantly anisotropic with recognizable domains and with various patterns of plasma separations; that is, they are basically asepic fabrics with superimposed plasma separations. Fabrics that are basically undulic or isotic usually do not exhibit plasma separations, but a number of such fabrics have been observed, particularly undulic fabrics with plasma separations associated with planar voids. Such a plasmic fabric can be classified quite adequately as *vosepic undulic,* and any other such

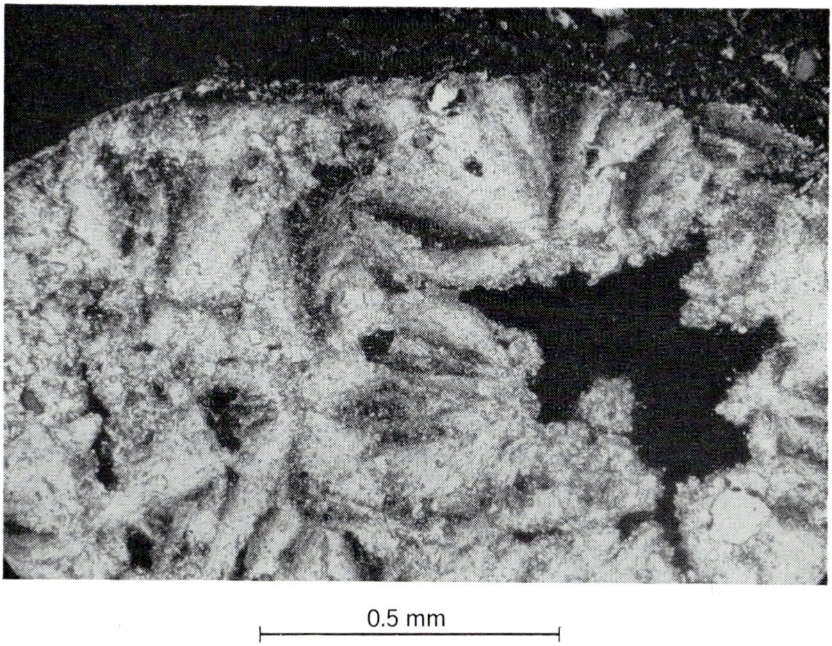

Fig. 114. Crystic plasmic fabric in a carbonate crystal chamber. Thin section under crossed polarizers. The dark zones in the matrix are extinction patterns which are typical of spherulitic fabric. 80×.

variants can be classified similarly by using the sepic plasmic names as adjectives qualifying undulic and isotic.

14.2 CHARACTERISTICS OF THE ASSOCIATED VOIDS

The voids that have relevance to the classification of s-matrixes by microscopic techniques are the intrapedal and transpedal voids that are visible at magnifications up to about 1200×, that is, the various varieties of vughs, vesicles, chambers, channels, and planes (Section 9.5).

It is not proposed, at this stage, to introduce new terms for the numerous possible combinations of the kinds of voids and plasmic fabrics; the plasmic fabric names are simply qualified by adjectives to denote the more important differences between the associated voids. The following adjectives are proposed:

Vughy. The associated voids are orthovughs.

Mammillated vughy. The associated voids are mammillated vughs.

Mammillated. The material is ramified by interconnected mammillated vughs.

Highly vughy. The soil material is ramified by interconnected vughs.

Vesicular. The associated voids are vesicles.

Chambered. The associated voids are chambers which are interconnected by channels.

Channelled. The associated voids belong to one of the systems of channels.

Jointed. The associated voids are sets or series of joint planes; if there are two sets the term *bijointed* may be used.

Fractured. The associated voids are skew planes.

Crazed. The associated voids are craze planes.

These adjectives do not describe the characteristics of the associated voids in all the detail used for their classification, but they are sufficiently specific to make broad, significant distinctions between soil materials with the same or similar plasmic fabrics, but different associated voids. More specific distinctions can be made by using the classification of voids set out earlier (Section 9.5).

14.3 CHARACTERISTICS OF THE SKELETON GRAINS

The skeleton grains in s-matrixes most commonly have a random distribution pattern without any specific arrangement with regard to the plasma and voids. It is sufficient, therefore, to describe the characteristics of the skeleton grains in the standard way. Of these characteristics the distribution pattern is the most important for the classification of s-matrixes; it can be described as *random, banded, clustered,* etc. (Section 8.3).

14.4 RELATED DISTRIBUTION OF PLASMA AND SKELETON GRAINS

The terms *porphyroskelic, agglomeroplasmic, intertextic,* and *granular* have been suggested for the principal broad related distribution patterns of plasma with regard to skeleton grains [Section 8.3:A(*c*)]. *Porphyroskelic* s-matrixes essentially have a relatively high proportion of ultramicroscopic voids; they may be vughy prophyroskelic, channelled porphyroskelic, jointed porphyroskelic, etc. *Agglomeroplasmic* fabrics may also have any of the different kinds of associated voids, but they also have within the plasma a relatively high proportion of small but recognizable voids which are essentially small interconnected vughs. *Intertextic* fabrics have ultramicroscopic voids within the plasma, but the arrangement of the plasma to form bridges between the skeleton grains results in a strongly interconnected set of voids which can be described best as an extreme development of interconnected vughs; this is a rather special arrangement, but intertextic fabrics may also contain other kinds of associated voids. *Granular* fabrics typically have simple packing voids between skeleton grains, but other kinds of associated voids can occur.

14.5 CLASSIFICATION AND NOMENCLATURE OF S-MATRIXES (BASIC STRUCTURE)

Classification of s-matrixes is, in fact, classification of basic structure, that is, classification of plasmic structure with qualification regarding the characteristics (size, shape, and arrangement) of associated voids and skeleton grains. The main objectives are to distinguish between s-matrixes whose characteristics suggest differences in genesis and/or possible differences in behavior (i.e., differences in physical and/or chemical characteristics) of soil materials. The most important characteristics in these regards are size, shape, and arrangement of plasma grains, domains, and associated voids. The characteristics of the skeleton grains are of lesser importance, but it is possible to make further subdivisions according to their size, shape, and arrangement, especially their distribution patterns. Since many of the characteristics of plasmic fabric form continua and there are no distinct breaks between the named fabrics, the proposed broad classes could be subdivided into innumerable subgroups on the basis of relatively minor differences, and these could be subdivided further according to the characteristics of the associated voids and skeleton grains. A decision must be made, therefore, as to how much detail is to be included in the classification.

The classification of s-matrixes should not be carried too far at the present state of knowledge. It is proposed that specific kinds of s-matrixes should not be given separate names at present, but their classification and nomenclature should be derived from the broad classification of plasmic fabric with qualification according to the characteristics of the voids and, less importantly, according to the characteristics of the skeleton grains; the terms for the broad groups of related distribution of plasma and skeleton grains are used where they are considered advantageous.

A number of conventions based on experience are useful to reduce the length of the descriptive names. First, in most soil materials the skeleton grains have a random distribution, and this is assumed unless specifically stated otherwise. Second, agglomeroplasmic and intertextic fabrics almost invariably have an asepic plasmic fabric, and each infers the occurrence of certain kinds of voids. When these terms are used, therefore, they are assumed to be asepic unless otherwise stated, and the kinds of voids inferred are not mentioned; for example, *channelled agglomeroplasmic s-matrix* (or basic fabric) denotes agglomeroplasmic fabric with random distribution of skeleton grains, asepic plasmic fabric and superimposed channels. Similarly, granular fabric is virtually a complete description without qualification since there is no plasma present, except that which occurs as pedological features, and the voids are dominantly simple packing voids; for example, *fractured granular s-matrix* denotes granular fabric with skew planes. Third, since agglomeroplasmic and intertextic are used without the plasmic fabric name, except where necessary, the term porphyroskelic usually need not be used; the convention can be adopted that the simple statement of the plasmic fabric refers to a porphyroskelic fabric; for example, *jointed vosepic s-matrix* denotes porphyroskelic fabric with random distribution of skeleton grains and ramified by subparallel planar voids (joint planes) which have plasma separations with a striated orientation pattern associated with their walls; *banded jointed vosepic s-matrix* has the additional qualification that the skeleton grains are concentrated in bands or layers.

Since this classification of s-matrixes is rather broad there is a need for expressing variations from the modal groups. This can be achieved most conveniently by using geographic names for specific s-matrixes which can then be defined in much more specific terms. Such named s-matrixes should be referred to a fixed horizon in a known soil profile, and through these to a type soil material and a type thin section, so that future collation of additional descriptive data can be readily achieved and the general classification of plasmic fabrics and s-matrixes can be broadened and extended as experience grows. It is proposed that such named variants should end in the suffix *ic* when it refers only to the plasmic fabric and in the suffix *an* when it refers to the whole s-matrix or basic structure; for example, the plasmic fabric of the soil material at a depth of 29 in. in profile P105 (thin section P105/13) has been named *Kempic fabric* (the profile occurs near Kempsey in New South Wales). This fabric is basically a skel-insepic fabric (porphyroskelic) in which the plasma separations around the skeleton grains occur only on the larger grains and the "islands" of material with striated orientation within the plasma often display a very weakly developed linear arrangement with regard to each other so that the fabric is an intergrade to masepic fabric. The *Kempan s-matrix* as a whole is a *vughy fractured Kempic s-matrix* in which the skeleton grains have an uneven size distribution and comprise a relatively low proportion of the s-matrix; the skew planes and vughs are both quite numerous, the latter varying from irregular, smoothed (some are mammillated) to regular orthovughs of the order of 0.4 mm in diameter.

14.6 ELEMENTARY AND PRIMARY STRUCTURE

Both elementary and primary fabrics deal with the s-matrix and the pedological features associated with it. Elementary fabric deals only with associations of specific pedological features with specific kinds of s-matrixes; primary fabric deals with all the pedological features (Section 7.6).

A. Elementary Fabric

Since basic fabrics have been named by qualification of the plasmic fabric name according to characteristics of the associated voids and skeleton grains (Section 14.5), the best approach to classification of elementary fabrics is by using additional qualifiers for specific pedological features. The following qualifiers are proposed:

(*a*) ***Cutanic.*** The significant pedological features associated with the basic fabric are cutans. The simple adjective "cutanic" is not very specific, but it can be made more so by including the specific name for the kind of cutan (to any required degree of specificity) in parentheses after the adjective, for example, cutanic (argillans) vughy vosepic fabric or, more specifically, cutanic (plane argillans) vughy vosepic fabric, which denotes a vosepic plasmic fabric with vughs as the dominant visible kind of void and with argillans on subsidiary planar voids. The nature of the cutans can be specified in complete detail by use of the code names (Section 10.3).

The adjective "cutanic" is taken to refer exclusively to illuviation and diffusion cutans which form definite coatings on the appropriate surfaces. This is legitimate because stress cutans have already been dealt with in the classification of the plasmic fabric by description of plasma separations with striated orientation, which are the most diagnostic characteristic of stress cutans, and because diffused or illuviated fractions of the plasma that occur subcutanically are encompassed by the term "subcutanic" which is defined subsequently.

(*b*) ***Pedotubulic.*** The significant pedological features associated with the basic fabric are pedotubules; this qualification can be made more specific by including the classification of the pedotubules, to any required degree of detail, in parentheses after the adjective; for example, pedotubulic (isotubules) jointed vosepic fabric denotes a jointed vosepic basic fabric with isotubules (Chapter 11).

(*c*) ***Glaebular.*** The significant pedological features associated with the basic fabric are glaebules; again the qualification can be made as specific as required by the classification of the glaebules (Chapter 12) placed in parentheses after the adjective, for example, glaebular (discrete sesquioxidic nodules) vughy masepic fabric.

(*d*) ***Crystallaric.*** The significant pedological features associated with the basic fabric are crystallaria; these, again, can be subdivided (Section 13.1).

(*e*) ***Subcutanic.*** The significant pedological features associated with the basic fabric occur subcutanically. Subcutanic features do not include plasma separations in this context since they are dealt with in the classification of plasmic fabric. Subcutanic fabrics are subdivided into neocutanic and quasicutanic fabrics and further according to the nature of the plasma fraction involved, for example, neocutanic (neocalcitans) vughy masepic fabric (Section 13.2).

(*f*) ***Fecal.*** Fecal pellets occur scattered within the s-matrix but not in recognizable pedotubules. Too little is known of the different kinds of fecal pellets that occur in soil materials to suggest significant subdivisions at this stage (Section 13.3).

Several different kinds of pedological features can occur together in any of the various s-matrixes, whereas elementary fabric deals only with the most significant group of pedological features. The difficulty, then, is to decide on the relative significance where more than one feature is present. It is proposed that the following general order should be adhered to:

(1) Cutans	(2) Crystallaria
(3) Subcutanic features	(4) Glaebules
(5) Fecal Pellets	(6) Pedotubules

The first four features in this list are records of the processes of leaching, fractionation of the plasma, and reorganization of the plasma fractions, that is, of the true soil-forming processes, while the last two are features imposed on the soil material by the activity of flora and fauna within the soil material. Cutans, as used in elementary fabric, refer to illuviation or diffusion cutans and are of first importance because they indicate translocation of colloidal and soluble constituents; argillans are especially important in this regard. Similarly crystallaria and subcutanic features indicate translocation in solution, especially of the soluble fractions of the plasma. Certain types of glaebules indicate reorganization of constituents by diffusion within the soil material, while others indicate inheritance or sedimentary processes. Fecal pellets within the s-matrix indicate a high degree of faunal activity, while pedotubules are often transpedal and merely indicate the activity of fauna and flora which use the soil material as a habitat. This order can be modified where the development of two features is such that the one lower on the list far outweighs that higher on the list in degree of development.

As experience grows, specific arrangements of characteristic pedological features may be found to occur consistently with certain basic fabrics; for example, one such relationship is the occurrence of fecal pellets and/or aggrotubules filled with fecal pellets in association with s-matrixes with asepic plasmic fabrics, random distribution of skeleton grains, and voids (macroscopic and microscopic) which are dominantly mammillated vughs. All these features are consistent with intense faunal activity.

The proposed adjectives qualifying the plasmic and/or basic fabric names subdivide elementary fabrics into broad groups. Details of variations within these groups can be described and named by using geographic names as was done with plasmic fabrics and s-matrixes. Such variants of elementary fabric should end in the suffix *us;* for example, *Kempus fabric* is an elementary fabric that has all the characteristics of the Kempan s-matrix and Kempic plasmic fabric (Section 14.5) and, in addition, has strongly oriented, yellow argillans on the walls of the vughs. In broad terms this elementary fabric is classified as "cutanic (vugh argillans) vughy fractured skel-insepic fabric."

B. Primary Fabrics

These fabrics can be very complex because of the occurrence of combinations of a variety of rather complex pedological features which differ particularly in the detail of their internal structure and fabric, in the sharpness of their boundaries, and in their distribution patterns.

It is unlikely, too, that specific arrays of pedological features occur consistently in association with specific plasmic and basic fabrics since inherited features can occur in soil materials with any plasmic or basic fabrics and in association with any of the orthic pedological features formed by soil-forming processes. Nor has experience shown any constant relationships in occurrence between any of the orthic pedological features; for example, strongly oriented argillans have been observed in soil materials with and without sesquioxidic nodules and with and without calcareous nodules, and manganese concentrations of various kinds occur independently of other pedological features, in association with sesquioxidic concentrations, and in association with calcite concentrations.

Because of this complexity and lack of understanding of the principles governing the association of various kinds of pedological features in the one soil material, primary fabrics cannot be organized satisfactorily into a classification system in which single or simple names indicate all the varieties that occur. It is not intended, therefore, to set up a system of classification and nomenclature for primary fabrics. They are dealt with by the simpler method of naming and classifying the basic or elementary fabric and adding a record of the pedological features which occur according to the classifications of those features as set out earlier (Chapters 10–13). Particular primary fabrics may be given geographical names as was done with plasmic and basic fabrics. Such names should end in the suffix *ac;* for example, *Kempac fabric* is a primary fabric with all the characteristics of *Kempus fabric* (the elementary fabric), *Kempan fabric* (the basic fabric), and *Kempic fabric* (the plasmic fabric), and with all the additional pedological features which occur in the s-matrix of the soil material.

14.7 APPLICATION TO SEDIMENTARY ROCKS

It has been argued that soils are in essence weathered rocks and that the vast majority of them are formed on bodies of unconsolidated sediments which are usually derived from hard rocks and have been transported varying distances (Sections 1.4 and 1.6). Fabrics of soil materials, therefore, can be expected to have many features in common with those of sedimentary rocks, particularly of unconsolidated sediments of similar mineralogy and grain size distribution, and so the system of description of soil fabric should have considerable application to the description of fabrics of sedimentary rocks. This is of special interest because the systematic morphological classification of fabrics of sedimentary rocks has not been developed to the level achieved for igneous rocks. In particular, the detailed description and classification of fabric for fine-grained (argillaceous) sediments leave much to be desired, and it is for just these rocks that the system for soil materials should prove most useful. All the levels of structure and fabric, from plasmic to primary, are applicable to sedimentary rocks by using the concept of sedimentary features (Shrock, 1948) instead of that of pedological features. Even higher levels of structure may be applicable to those sediments with units analogous to peds, for example, the units bounded by three sets of joint planes, which are not uncommon in consolidated sediments. There is a much smaller variety of such features in sediments, however, than in soil materials.

We have had the opportunity of studying only a very limited range of thin sections of sedimentary rocks. Even these, however, suggest that the concept of the s-matrix (which includes the framework and matrix of sediments; see Pettijohn, 1957) with included features is the most important aspect of the system of classification of soil fabric that is applicable to sediments, that is, the description and classification of plasmic, basic, and primary fabric. It is apparent, however, that the classification needs extending to include further classes specific to sedimentary rocks, just as it will require additions to include further fabrics of soil materials as experience grows.

The following discussion is restricted to the level of basic fabrics of sedimentary rocks, that is, essentially the fabric of the s-matrix. It is not intended to be exhaustive but rather to suggest a basis for classification of fabrics of sedimentary rocks and to point out the possible application of the system used for soil materials.

A. Plasmic Fabrics of Sedimentary Rocks

The classification of plasmic fabrics set out for soil materials (Section 14.1), as far as we know, is directly applicable to unconsolidated sediments. This is as would be expected since soil profiles are frequently formed on unconsolidated sediments so that the features of the sediments may be preserved in the soil materials of the profile. The soil-forming processes simply form pedological features and modify the plasmic fabric, their effects decreasing in intensity with depth in the profile.

Consolidated sediments, however, may exhibit significant variation from the plasmic fabrics of soil materials and of unconsolidated sediments because of the effects of the forces causing consolidation and, no doubt, due to diagenesis. These variations are best developed

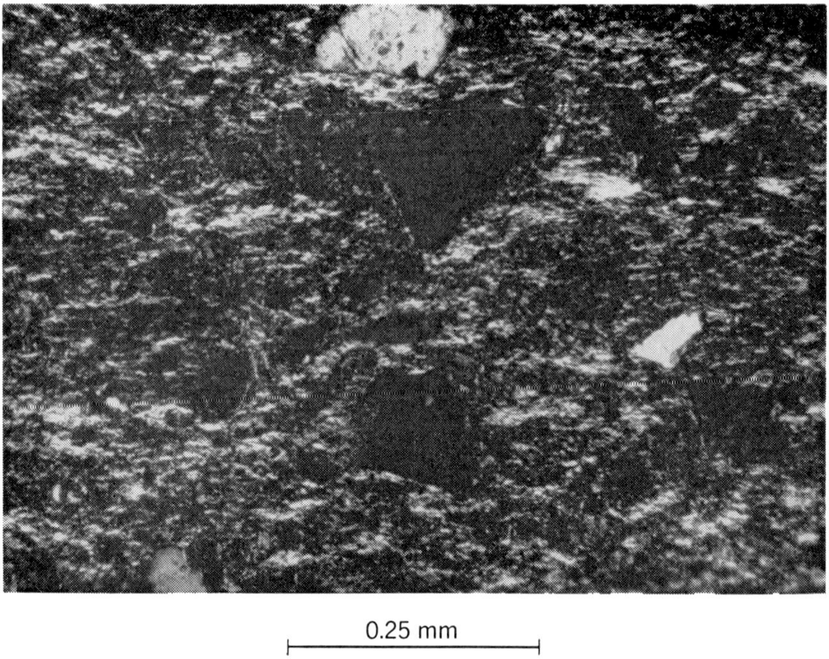

(a)

Fig. 115. Skel-lattisepic unistrial (moderate) plasmic fabric in a sedimentary rock. Thin section under crossed polarizers. (*a*) Section in a position of maximum illumination of the unistrial fabric. 135×. (*b*) Section at 45 degrees to (*a*)

in sediments with a fairly high proportion of plasma, especially argillaceous material. The most common variation is an overall, single direction of preferred orientation of the plasma grains. This is expressed in thin section under crossed nicols by a striated orientation pattern of varying degree of intensity developed in a single direction, usually parallel to the bedding planes of the sediment, but uniform throughout the whole of the material. The thin section transmits most light when this direction is placed at 45 degrees to the directions of vibration of the polarizer and analyzer; it transmits least light when this direction is parallel to the vibration direction of the polarizer or analyzer. Other sediments have a similar preferred orientation of the plasma grains developed in two directions, most commonly at approximately 45 degrees to each other, so that in thin section under crossed nicols one of these directions is at extinction when the other is showing maximum illumination due to the striated orientation pattern. These phenomena are overall plasmic fabrics

(b)

showing the skel-lattisepic fabric (basically lattisepic fabric with plasma separations also around the skeleton grains). 135×.

which are expressed to a similar degree throughout the whole section, so the striated extinction pattern cannot be properly classified as a plasma separation. For example, sediments with two directions of preferred orientation often also exhibit two sets of joint planes parallel to the directions of preferred orientation, but the plasmic fabric cannot be described as vosepic because the striated orientation occurs throughout the section and not particularly associated with the walls of the voids. Nor is it masepic fabric since the striated orientation pattern is not confined to zones within the s-matrix.

It is necessary, therefore, to define a new plasmic fabric for such material; the term *strial** is proposed:

> *Strial fabric.* A plasmic fabric in which the plasma as a whole exhibits preferred parallel orientation, which gives the specimen as a whole a unidirectional striated extinction pattern (*unistrial fab-*

* The stem of the term "strial" is derived from "striated" which describes the extinction pattern.

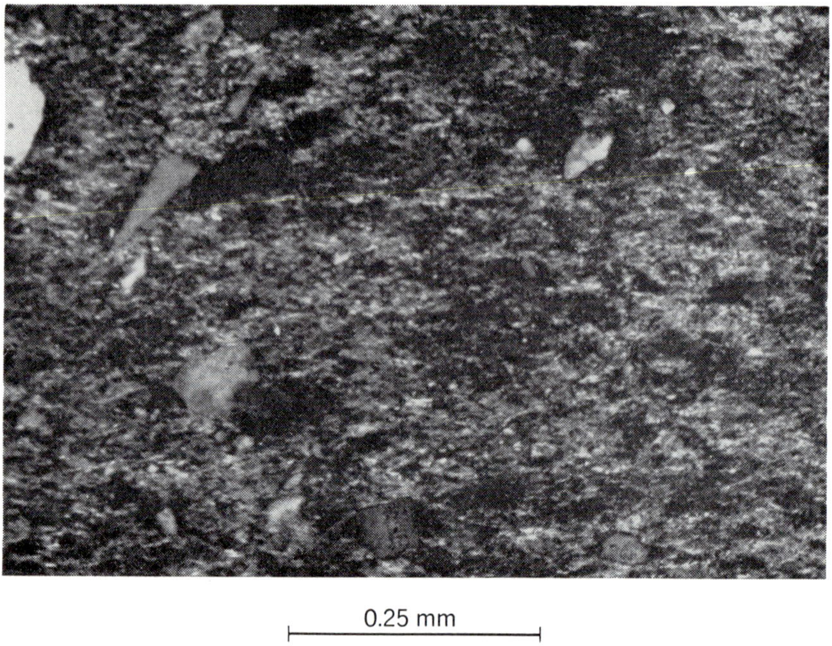

(a)

(b)

Fig. 116. Insepic (intergrading to lattisepic) unistrial plasmic fabric in a sedimentary rock. Thin section under crossed polarizers. (*a*) Section in a position of maximum illumination of the unistrial fabric. 135×. (*b*) Section at 45 degrees to (*a*) showing the insepic-lattisepic intergrade fabric. 135×.

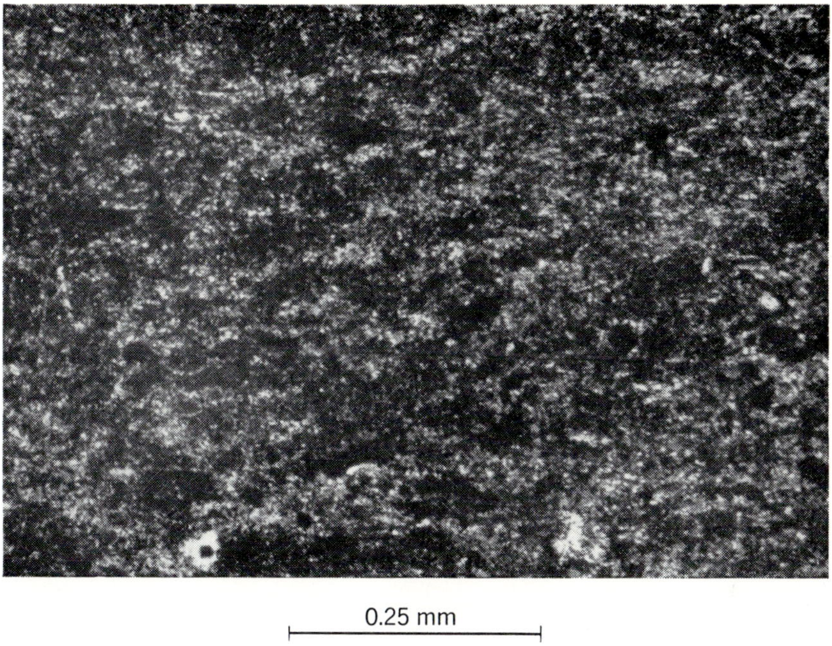

(a)

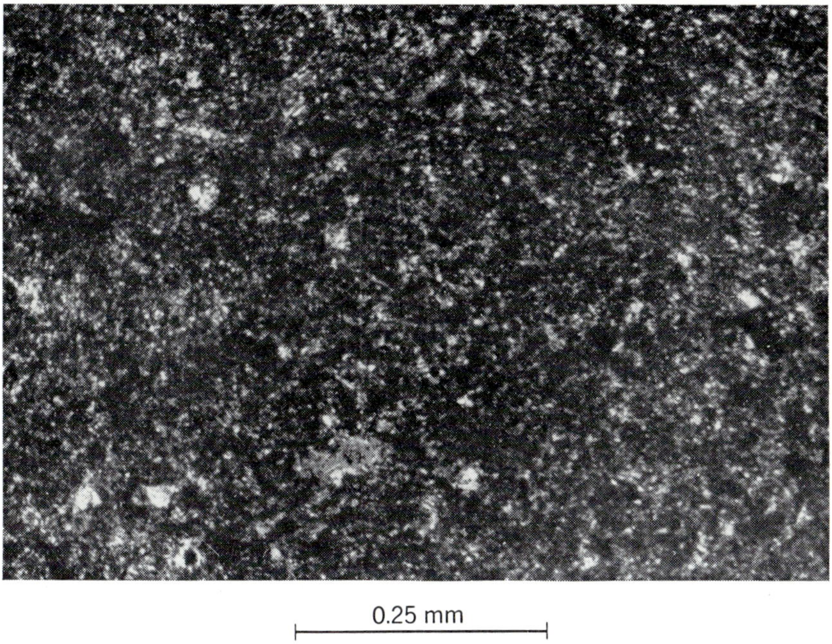

(b)

Fig. 117. Argillasepic unistrial (weak) plasmic fabric in a sedimentary rock. Thin section under crossed polarizers. (*a*) Section in a position of maximum illumination of the unistrial fabric. 135×. (*b*) Section at 45 degrees to (*a*) showing the argillasepic fabric (no plasma separations). 135×.

0.25 mm

(a)

Fig. 118. In-skelsepic unistrial (strong) plasmic fabric in a sedimentary rock. Thin section under crossed polarizers. (*a*) Section in a position of maximum

ric) or a striated extinction pattern in more than one direction (*bistrial fabric*—two directions).

Strial fabrics are further described in terms of degree of development of the striated pattern as very weak, weak, moderate, strong, or very strong. These terms are relative and based on the difference in illumination of the thin section under crossed nicols at the positions where the striated orientation pattern is at extinction and at maximum brightness. This difference varies from being just noticeable (*very weak*) to the other extreme where the section behaves almost as a single mica grain or, perhaps more descriptively, like a section parallel to the *C*-axes through a cluster of strongly oriented mica grains (*very strong*). *Weak* (Fig. 117), *moderate* (Fig. 115), and *strong* (Fig. 118) degrees of strial fabric describe the stages between these two extremes.

Sedimentary rocks, whether they have strial plasmic fabric or not,

(b)

illumination of the unistrial fabric. 135×. (*b*) Section at 45 degrees to (*a*) showing the in-skelsepic fabric. 135×.

can exhibit any of the plasmic fabrics described for soil materials. Strial fabric, if present, forms a general background on which may be superimposed the various possible other plasmic fabrics. In such materials the superimposed plasmic fabrics are seen best when the thin section is oriented under crossed nicols so that the strial fabric is at the position of minimum illumination. The standard technique for description and classification of plasmic fabrics of sediments, therefore, is first to classify according to the presence or absence of strial fabric and then to place the section in the position of minimum illumination of the strial fabric and classify according to any superimposed plasmic fabric. In naming the fabric the convention is adopted of putting the term "strial" last, if the fabric is such, and qualifying it with the term describing the superimposed plasmic fabric; for example, *skelsepic unistrial* fabric has plasma separations around the skeleton grains superimposed on a general unidirectional parallel orientation of the plasma grains. If strial fabric is absent

Fig. 119. Weak insepic (intergrading to weak lattisepic) plasmic fabric in a sedimentary rock. Thin section under crossed polarizers. An intergrade insepic-lattisepic plasmic fabric. Strial fabric is absent. 135×.

the plasmic fabric can be classified into the same groups used for soil materials. A number of fabrics of sedimentary rocks are shown in Figs. 115–120. Where strial fabric is present the micrographs show the thin sections under crossed nicols in the position of maximum illumination of the strial fabric and in the position of minimum illumination of the strial fabric to show the superimposed plasmic fabrics.

B. Voids

The voids within the s-matrix of sedimentary rocks can be described and classified in the same way as for soil materials. The classes set up for soil materials, if not entirely adequate, form a sound basis for classification of the kinds of voids observed in sediments. Certain classes, such as chambers, are unlikely to occur, and the sedimentary petrologist may prefer to distinguish other classes, such as bedding planes in the broad class of joint planes as defined for soil materials, where there is sufficient evidence for such an interpretive classification. Thus the classes of plasmic fabric can be qualified for the kinds of voids as was done for soil materials.

C. Skeleton Grains

Since the matrix of a sediment is commonly taken as including all grains up to 30 μ diameter, some of the skeleton grains are considered as part of the framework and some as part of the matrix. Even so, the size, shape, and arrangement of the skeleton grains can be described in the same terms as for soil materials. If necessary the descriptions can be made in two parts: the skeleton grains in the matrix, and the skeleton grains in the framework; the framework and matrix together comprise the s-matrix.

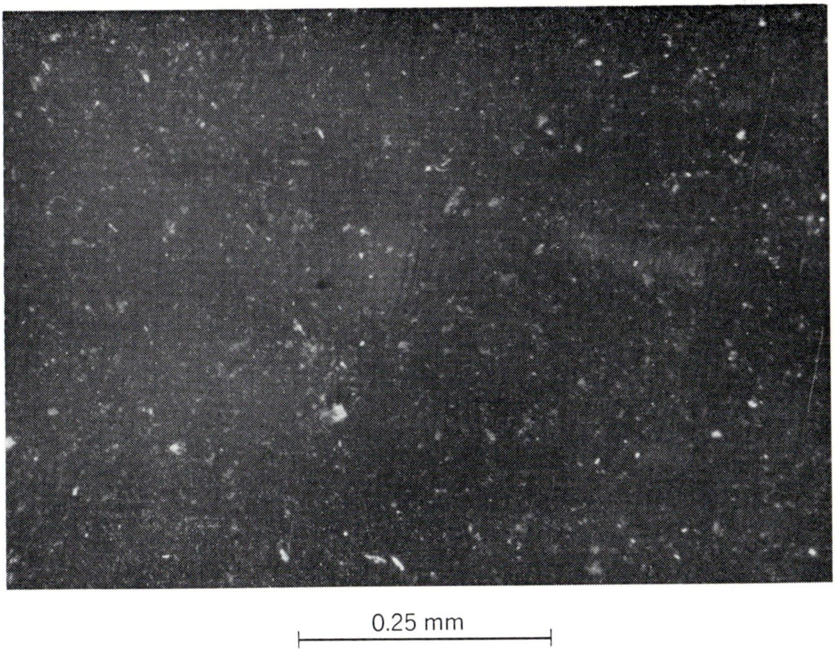

Fig. 120. Inundulic plasmic fabric in a sedimentary rock. Thin section under crossed polarizers. Strial fabric is absent. 135×.

14.8 INTERPRETATION

This discussion is restricted to soil materials; it does not include interpretation of s-matrixes of sedimentary rocks.

The genetic significance of the size, shape, distribution, and orientation patterns of the associated voids in s-matrixes has been considered (Section 9.6), as has interpretation of the size frequency distribution and shape characteristics of the skeleton grains (Chapter 2). The genesis of various kinds of pedological features has also been considered (Chapters 10–13). It remains to interpret the genesis of the arrangements of plasma grains, that is, plasmic fabrics.

Although any manifestation of optical anisotropy can be confidently attributed to preferred orientation of the plasma grains (Section 8.2), any certainty concerning the processes involved in producing such preferred orientation must await accumulation of basic data from experimentation and fundamental studies of the nature of the various constituents of the plasma and their behavior under different conditions, and on accumulation of a bulk of supporting observational data. A number of hypotheses, however, can be proposed for the formation of the broad named groups of plasmic fabrics on the basis of the little that is known.

Meade (1960) pointed out that preferred orientation of clay mineral grains in sediments can occur by three mechanisms: (1) as the result of deposition from suspension, (2) by mechanical reorientation due to pressure from an overburden, and (3) by recrystallization of unstable minerals to form more stable clay minerals and micas in the later stages of compaction. Similar processes are known to operate in soil materials, but the preferred orientation observed within the s-matrix is not due to deposition from suspension, which would produce pedological features, nor is it likely to be due to recrystallization except under special circumstances. It is probably due to the effect of pressure—not pressure due to a large overburden but rather pressures and tensions produced by wetting and drying. These are the bases on which the following hypotheses of the formation of the various plasmic fabrics are put forward.

A. *Asepic Fabrics*

The unoriented domains in asepic fabrics may be formed by sedimentation processes, by orientation of the plasma grains within wet soil materials, by direct inheritance, or by *in situ* weathering of primary mineral grains.

(a) Sedimentation Processes. The most probable cause of asepic fabrics are sedimentation processes. During transport of an alluvial sedimentary body the clay size material may be carried in a completely dispersed state or as floccules of different sizes, depending on the characteristics of the clay size minerals and of the water carrying them. Under either condition some degree of flocculation will occur either during transport or immediately after dropping of the load of sediment, and some degree of preferred orientation of the individual grains will probably take place within the floccules as they dry from a saturated state (due to drying tensions), whereas there is no reason to expect the floccules to become oriented with regard to each other. The size of the floccules and the degree of preferred orientation within them depend principally on the degree of flocculation and the kind of contaminants present. The greater the degree of flocculation the weaker is the degree of preferred orientation as judged by extinction phenomena in thin section (Brewer and Haldane, 1957) and the larger are the floccules. The presence of silt size material decreases the degree of orientation (Brewer and Haldane, *loc. cit.*), and the nature of the exchangeable cations and contaminants, such as organic matter and iron oxides, may affect the final degree of preferred orientation and size of the floccules.

Flocculation during transport and sedimentation of alluvial materials, therefore, may be expected to produce asepic plasmic fabrics, and it is relevant to note that many of the grey and brown soils of heavy texture (Stephens, 1956), which are formed on alluvial sediments in the Riverine Plain of western New South Wales, have dominantly asepic fabrics.

(b) Soil wetting. A possible cause of asepic fabrics is soil wetting. Unoriented domains may possibly be formed by wetting and drying within the soil mass, provided the soil material becomes wet enough to allow reorientation of the individual plasma grains and provided pressure and tension effects are not great enough to produce preferred orientation between domains. Disturbance of the soil material by soil fauna, heaving, soil creep, or mass movement may prevent preferred orientation between domains.

(c) Inheritance. Unoriented domains may be inherited. First, in colluvial or mass movement material, they may be inherited from a previously existing soil material, either because the older soil material contained unoriented domains or because the disturbance of the material caused disruption of previously existing orientation patterns. Second, they may be inherited from the parent rock (e.g., a shale) by physical breakdown with some subsequent disturbance; many sedimentary rocks contain anisotropic domains, which may persist after physical breakdown during which any orientation between domains may be destroyed.

(*d*) ***Weathering.*** The normal weathering of mineral grains often produces equant or prolate domains that are unoriented with regard to each other, but various kinds of preferred orientation between the domains may occur; disturbance again may destroy previous orientation patterns.

B. Sepic Fabrics

Sepic fabrics are allied to asepic fabrics insofar as a relatively high proportion of the plasma occurs as unoriented domains in most of them. The difference lies in that sepic fabrics have some proportion of acicular to prolate domains which have a preferred orientation with

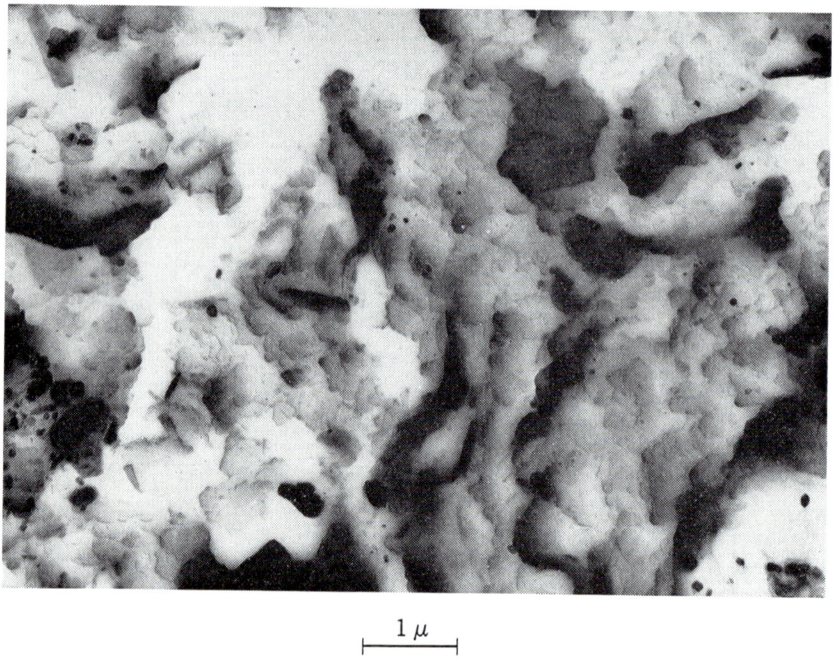

Fig. 121. Omnisepic plasmic fabric. Electron micrograph of the soil material shown in Fig. 109. Note the preferred parallel orientation and overlapping of the individual plasma (clay mineral) grains. *Courtesy Dr. G. H. Taylor, Division of Coal Research, C.S.I.R.O., Australia.* 12,000×.

regard to their neighbors. Figure 121 illustrates an electron micrograph of the omnisepic fabric shown in Fig. 109. It shows the high degree of preferred orientation of the individual clay mineral grains in the strongly developed plasma separations. It also demonstrates the efficacy of electron microscopy in the study of the details of plasmic fabric, which can only be inferred from optical properties in thin section. Such preferred orientation occurring as plasma separations may result from one or another of several processes.

(*a*) ***Pressure and Tension.*** There is strong evidence for attributing the striated orientation patterns of some of the plasma separations in sepic fabrics to the effects of pressures and tensions produced by wetting and drying (e.g., Rode, Yarilova, and Rashevskaya, 1960). In particular, there is little doubt that *vosepic* fabric is due to these effects. The planar voids in vosepic fabrics are the sites of maximum wetting and drying where pressure due to expansion and tension due to shrinkage are at a maximum; this is exemplified in slickensided planar voids which are indisputably due to pressures of swelling and are always accompanied by vosepic fabric or variants of it. *Masepic fabric* probably has a similar origin; the origin of planes has been discussed (Section 9.6), and if the forces responsible for their formation were just insufficient to cause fracturing of the soil material to form planar voids, some degree of orientation of the anisotropic plasma grains and/or domains would occur to form the various masepic fabrics, perhaps due to plastic flow. One or more directions of preferred orientation may be developed according to the conditions of shrinking and swelling. Mineralogy of the plasma may have some effect in the development of such sepic fabrics. Physically anisotropic plasma grains are more prone to reorientation by pressure than physically isotropic grains, and larger grains, such as illite, kaolinite, and chlorite, whose behavior is affected by gravitational forces, develop parallel orientation more readily than do smaller grains, such as montmorillonite, whose behavior is likely to be controlled entirely by forces related to the surface interaction of the grain and water (Meade, 1960).

Skelsepic fabric is also probably the result of swelling pressures since the skeleton grains present a solid surface against which pressures can be exerted. Lafeber (1962) suggested that the patterns of plasma separations associated with the skeleton grains in a number of skelsepic fabrics can be explained by rotation and/or translocation of physically anisotropic skeleton grains under pressure. He proposed that such differential movement of skeleton grains produces the striated orientation pattern in the plasma adjoining them and that the direction and kind of movement can be assessed from the pattern of the plasma separations. In the soil materials he studied the plasma separations had "tails" that extended from the surface of the skeleton grains into the plasma of the s-matrix. He attributed parallel tails on opposite sides of the one skeleton grain to translocation of the grain along a line, and somewhat curved tails pointing in opposite directions on opposite sides of the one grain to rotation, the direction of rotation being indicated by the aspect of the tails. Such a mechanism operating on a rather restricted scale could cause *lattisepic* fabric. We should note, however, that plasma separations associated with the surfaces of skeleton grains in other soil materials do not have such tails.

These hypotheses are not so satisfying for the formation of insepic and mosepic fabrics, in which only isolated patches within the mass of plasma have striated orientation, because the distribution of the patches has no apparent consistent relationships with locations that are propitious for larger pressures or tensions than the rest of the plasma. The extreme development of *omnisepic fabric,* in which all the plasma has a striated orientation pattern, may be possible under extreme conditions.

(*b*) *Inheritance.* A more readily acceptable origin for *mosepic* and *insepic fabrics* is by inheritance. Many sedimentary rocks have sepic fabrics sometimes superimposed on strial fabric (Figs. 115–120). If such rocks weather physically without drastic disturbance to form a soil material, a number of variants of mosepic fabric may result depending on the original pattern of orientation in the rock, the degree of disturbance, and the degree of alteration of the pattern of orientation due to wetting and drying within the soil material. If the parent material of the soil is an unconsolidated sediment that contains fragments of such fine-grained sediments, weathering, physical softening, and disintegration of the fragments may form insepic fabric. This process was observed in some terrace soils from Kempsey, New South Wales. Insepic fabrics, however, probably have other unexplained origins since not all of them can be explained definitely by inheritance.

C. Undulic Fabrics

Drying of highly flocculated colloids composed of anisotropic clay mineral grains produces a fabric that is essentially undulic (Brewer and Haldane, 1957), provided the proportion of skeleton grains to clay size material is relatively low. Undulic plasmic fabrics, therefore, could result from deposition of alluvial materials under appropriate conditions. Such materials, however, usually have an asepic fabric, while undulic fabric is common in such soils as residual krasnozems in which there is no suggestion of the parent material being alluvial in origin. In general there is a high correlation between undulic fabric and relatively high proportions of iron oxides, so the extinction phenomena observed may be due to the effects of high amounts of iron oxides or hydroxides. The mechanism is not understood, but iron oxides are known to flocculate and stabilize clay minerals. Certainly the low degree of anisotropy of such fabrics at low magnifications can be attributed to the presence of high proportions of amorphous or very finely crystalline, unoriented iron oxides, or hydroxides. *Inundulic fabric* is essentially an intermediate between undulic and insepic fabrics.

D. Isotic Fabrics

Plasmic fabrics that are apparently isotropic at all magnifications and levels of light intensity may be due either to the clay minerals themselves being isotropic or, if anisotropic, very fine-grained and unoriented with regard to their neighbors, or to the presence of opaque materials such as organic matter. No information is available as to the proportions, size or degree of orientation of fine-grained crystalline materials that will cause apparent isotropy in thin section.

E. Crystic Fabrics

Crystic plasmic fabrics are simply due to the crystallization of relatively soluble fractions of the plasma and can be explained by the general laws of nucleation and crystal growth.

15

Peds. Pedality. Secondary and Tertiary Structures

The basic unit of description of pedality and secondary structure is the primary ped, while tertiary structure deals with secondary peds; both secondary and tertiary structure also take account of the interpedal voids and pedological features. The nature of peds has been discussed (Section 7.5:A), and it was pointed out that they can be identified objectively by the characteristics of their faces, which are natural surfaces, as distinct from the method of inferring the occurrence of peds when application of a force to a soil mass gives a population of units of similar size and shape. Recognition by the characteristics of their surfaces allows the possibility of peds with a wide range of sizes and shapes occurring in the one soil material.

The objectives of studies of pedality and secondary and tertiary structure are primarily to study the genesis of peds, to characterize soil materials, and to correlate pedality and the structure levels with other properties. The first step, therefore, is to classify peds as units, and their most significant morphological characteristics are size, shape, and arrangement. Size and shape of peds can be measured according to the concepts of true nominal diameter, sphericity or maximum projection sphericity, and roundness (Section 2.1). These concepts have proved useful in studies of sediments in which size and shape of grains are used to interpret the mode and conditions of transport and deposition in the formation of sediments. Peds, however, are formed *in situ,* and the objectives of studies of peds are served better by some modifications of the approach to measurement of size and shape.

15.1 SIZE OF PEDS

Size of peds can be measured by the standard methods of sieving provided the individuals are sufficiently coherent to survive the necessary treatments, and true volume measurements of individuals can be made by displacement of water by wax-coated peds, by displacement of kerosene (McIntyre and Stirk, 1954), or by the various techniques of particle size analysis in section (Section 2.3). The last technique has been modified by using a Leitz TP200 projector to give an enlarged image on which areas can be measured directly by mechanical devices, such as a planimeter (Kubiena, Beckmann, and Geyger, 1961; Geyger, 1962). Data from these analyses can be presented and treated mathematically in the same way as size analyses of mineral grains (Section 2.5). Perhaps the most genetically significant characteristic of size of peds, however, is the length of the principal axes considered in conjunction with a classification of shape; these two characteristics together give as accurate a record of the principal characteristics of individual peds as is possible by using relatively few and simple measurements. An approximation of volume can be calculated from these data.

Measurement of size according to the principal axes gives three values for each ped, whereas size classes are assigned most readily to single values. If peds of a single shape class are considered, as is desirable, their sizes can be designated by measurement of a single axis. The most appropriate axis is the intermediate since it is part of both the ratios used to classify shapes according to the lengths of their axes. It is also convenient because sieving sorts according to the length of the intermediate axis so that the results obtained by sieving populations of peds of similar shape are comparable to those obtained by actual measurement of intermediate axes. Sieving, however, does not distinguish between peds of very different shape where the intermediate axes are of comparable length.

The size distribution of populations of peds of similar shape can be analyzed on these bases, and the pedality of pedal soil materials can be expressed in terms of shapes of peds and size distribution within the shape classes.

15.2 SHAPE OF PEDS

There is no great regularity in the shape of peds as in some other natural units, such as mineral crystals in which the various faces of any particular mineral species are set at fixed angles with regard to adjoining faces so that reference faces can be chosen and all other faces referred to them. Individual mineral crystals, therefore, have a basic regularity of shape. On the other hand, the shapes of peds depend on the mechanism of formation of the natural surfaces which enclose them, and the problem is to classify these shapes without having a fundamental understanding of the laws governing the arrangement of natural surfaces.

Certain fairly regular shapes of peds do frequently recur, and a useful shape classification can be devised by making due allowance for irregularities of the surfaces and for deviations from the ideal relationships of the faces with regard to each other. Natural surfaces in soil materials, and, therefore, the faces of peds, are rarely flat planes. They may be either *relatively* smooth, undulate surfaces that can be referred approximately to a plane, or compound relatively smooth curves that can be referred to a single curve. In setting up shape classes for peds such surfaces are considered as planes and simple curves, respectively. Nor can it be expected that the angles between adjacent faces of peds will be constant, as they are in crystal forms; for example, in a prismatic-shaped ped the angles between the typical parallelogram faces will not be consistently identical.

There are three essentially separate aspects of the classification of peds according to shape: *basic form* in terms of number and shape of faces and relationships of faces to each other, general shape in terms of the ratios of the lengths of the principal axes (*sphericity*), and *conformation,* or regularity, of outline in section (analogous to roundness). Each of these aspects is independent of the other two so that shape classes set up on one of them can be subdivided according to expressions of the other two. Most of the classifications in use at present (e.g., *U.S.D.A.*, 1951; Butler, 1955; Jongerius, 1957) are based primarily on basic form as referred to solid geometric figures, but they are more or less unsystematic in that, at any particular level of classification, one or more of the aspects of shape (basic form, ratios of axes, and conformation) may be used as well as other characteristics, such as degree of accommodation of the faces with adjoining peds and porosity of the peds. Accommodation is part of arrangement, and porosity is a part of the internal structure of the peds. In addition certain terms have been misused, and certain important differences in shape have not been differentiated.

It is proposed, therefore, to classify the shapes of peds at the highest level by reference to appropriate solid geometric figures which have been chosen purely because observation has suggested that most peds can be referred to one or another of them; other forms may oc-

TABLE 15-1. Classification of Shape for Peds

Characteristics	Basic Form — Classes*						
Flatness of faces	Polyhedral (all plane faces)			Spheral† (all curved faces) Polyspheral		Polysphedral (mixed curved & plane faces)	
Shape of principal sections‡	Polyhedric	Prismatic	Regular polyhedral	Polyspheric	Prismatic spheric	Polysphedric	Prismatic sphedric
Number of faces	Tetrahedric Pentahedric etc.	Triprismatic Tetraprismatic Hexaprismatic etc.	Tetrahedral Cubic Octahedral Dodecahedral Icosahedral	Tetraspheric Pentaspheric Hexaspheric etc.	Triprismatic spheric Tetraprismatic spheric	Tetrasphedric Pentasphedric Hexasphedric etc.	Tripris-matic sphedric etc.
Shape of faces	Trapezohedric	Parallelepipedal Trapezoprismatic Domed			Parallelepipedal spheric Trapezoprismatic		Parallelo- Trapezo-
Angle between bases and parallelogram faces		Right Clino-			Right Clino-		Right Clino-

* *Regular polyhedral* and *spherical* forms are equant by definition; all others may be qualified for ratios of axes into planar, equant, prolate, acicular, oblate, or triaxial

† *Spheroidal* and *spherical* forms have single curved faces

‡ Re-entrant angles between adjoining faces will usually be seen in the principal sections

cur, and other geometric shapes can be added as found necessary. The solid geometric figures proposed refer only to the *basic form* of peds, and these highest classes can be subdivided according to the ratios of the lengths of the principal axes, the number of faces, and the size of the angles between the faces, except that certain specific basic forms are defined on one or more of these characteristics. The proposed shape classification of peds is set out in Table 15-1. The terms do not coincide precisely with those used by previous workers, and certain terms are used for the first time.

A. Basic Form

Peds can be classified at the highest level into three groups according to flatness of the faces: *polyhedral, spheral,* and *polysphedral;* polyhedral is an accepted geometric term, while spheral and polysphedral are terms coined to describe major basic forms observed in peds but for which geometric terms do not exist.

(*a*) **Polyhedral.** This term is used according to its geometric and crystallographic definition, independent of the ratios of the principal axes, for the large group of peds bounded by plane faces. In its accepted meaning it embraces all forms bounded by plane faces, including Butler's (1955) "polyhedral" ("those polyhedrons having more than six sides"), blocky or cubic, prismatic, parallelepipedal, and platy forms, as well as the *U.S.D.A. Soil Survey Manual's* (1951) plate-like, prism-like, and block-like forms bounded by plane faces. Polyhedral forms can be subdivided according to the shapes of the principal sections* into *polyhedric, prismatic,* and *regular polyhedral.*

Polyhedric is proposed for polyhedral forms with unequal and dissimilar faces; the three principal sections are all irregular polygons. Polyhedric forms are subdivided simply on the number of faces. The choice of names for some of the subgroups presents a problem because tetrahedron, octahedron, dodecahedron, and icosahedron are specific names for regular polyhedrons. This is overcome by using the ending *ic* instead of *al* for the adjectival form of the irregular forms, thus: *tetrahedric, pentahedric, hexahedric,* etc., for the forms having four, five, and six faces, respectively. *Trapezohedric* is proposed for the special polyhedric form with six faces, four of which

* As for sphericity classes the principal axes are the *a* (longest), *c* (shortest), and *b* (intermediate) axes which are mutually at right angles. The principal sections are the three sections that contain the *a* and *b*, *b* and *c*, and *a* and *c* axes, respectively.

are trapeziums and the other two parallelograms so that two principal sections are trapeziums and the third is a parallelogram.

Prismatic is proposed for all shapes that can be referred to any form of prism; the principal sections are polygons, two of which are parallelograms. A prism is a "solid whose bases or ends are similar, equal, and parallel polygons, the faces being parallelograms" (Rice, 1954; after Fay, 1920), and rectangles and squares are specific cases of parallelograms (see *Shorter Oxford Dictionary*, 1956). This definition puts no limitations on the ratios of the lengths of the principal axes or on the inclination of the bases to the parallelogram faces. Previous uses of prismatic for classification of peds restricted it to prismatic forms with one axis considerably longer than the other two and with the long axis vertical in the soil material. Subdivisions of the basic prismatic form are made according to the number of parallelogram faces into *triprismatic, tetraprismatic, pentaprismatic,* etc. The principal section perpendicular to the axis of the zone of parallelogram faces in these forms has three, four, and five sides, respectively. There are two special cases of tetraprismatic form, *parallelepipedal* and *trapezoprismatic,* in which the principal sections perpendicular to the axis of the zone of parallelogram faces are a parallelogram and a trapezium, respectively. The next subdivision is made according to the inclination of the bases to the parallelogram faces: *right* is used as an adjective where they are at right angles and *clino-* as a prefix where they are inclined to each other; prismatic forms may be distinguished as *right triprismatic, clino-hexaprismatic,* etc. The adjective *right* may be deleted if the convention is used that any prismatic form not specifically stated as a clino-form is a right form. Further subdivisions are possible by actual measurement of the angle of inclination in the clino-forms.

"Columnar" has been used for a special form, related to the prismatic form, in which the axis of the zone of parallelogram faces is the longest physical axis and is vertical, with the additional specification that the upper face is domed (U.S.D.A., 1951). In normal English usage and in geology, columnar has the connotation of elongation along one axis but not of doming of one of the faces. In soil materials equant, and even planar, prismatic forms may have a domed upper face. The adjective *domed* is proposed, therefore, for general use for peds of any basic form whose upper face is convex upward even though such forms are strictly polysphedral [Section 15.2:A(*c*)].

Regular polyhedral forms are strictly limited in number and have been defined and named according to the number of faces. They have all their faces equal in all respects; they are the *tetrahedron* (four triangular faces), *cube* (six square faces), *octahedron* (eight triangular faces), *dodecahedron* (twelve five-sided faces), and *icosahedron* (twenty triangular faces). Such regular forms are not likely to occur in soil materials, but it is useful to point out that terms are available with specific meanings. In particular, tetrahedral and cubic should not be used simply for any four-sided and six-sided forms, respectively.

(*b*) *Spheral.* This term is proposed for forms that are bounded entirely by distinctly curved faces, for which there is no general geometric term although there are terms for specific ones. The form of such peds may consist essentially of a single continuous curved surface related to the sphere or ellipsoid or of a number of distinct curved surfaces. Spheral forms can be subdivided according to the shapes of the principal sections into *polyspheral, spheroidal,* and *spherical.*

Polyspheral is proposed for forms with a number of curved surfaces to distinguish them from forms consisting of a single curved surface. The principal sections are all plane figures with curved sides analogous to polygons. Such forms are described in the *U.S.D.A. Soil Survey Manual* (1951) as "polyhedrons having . . . curved surfaces," and are finally classified there as granular or crumb. Polyspheral forms are subdivided into *polyspheric* and *prismatic spheric,* the analogues of polyhedric and prismatic forms, respectively, in which the sides of the principal sections are curved. Polyspheric is subdivided further, according to the number of faces, into *dispheric, trispheric, tetraspheric,* etc. Prismatic spheric is subdivided according to the number of faces in the zone analogous to the zone of parallelogram faces in prismatic forms (e.g., tetraprismatic spheric), according to the shapes of the bases (parallelepipedal spheric, trapezoprismatic spheric, domed), and according to the inclination of the axis of the zone of faces to the bases (right, clino-). The kind of curvature of the faces can be described by recording the shapes of the principal sections.

Spheroidal is applied to forms that can be referred approximately to a spheroid, which is a solid figure generated by an ellipse rotating about its minor axis (oblate spheroid) or about its major axis (prolate spheroid); they consist of a single curved surface. Two principal sections are ellipses, and the third is a circle. Spheroidal forms cannot be subdivided further on the number or shape of faces, but they can be qualified according to the ratios of the principal axes.

Spherical is applied rather strictly to forms with three essentially circular principal sections. It is unlikely that truly spherical peds occur.

(*c*) ***Polysphedral.*** This term is applied to forms with mixed curved and plane faces; peds of this form are common in soil materials (U.S.D.A., 1951; Jongerius, 1957). Polysphedral forms are subdivided, according to the principles used for the polyhedral class, into *polysphedric* and *prismatic sphedric* which are analogous to polyhedric and prismatic forms, respectively. Polysphedric is subdivided according to number of faces (e.g., trisphedric, tetrasphedric), and more detail can be brought to the description by recording the numbers of curved and plane faces. Prismatic sphedric is subdivided according to the number of faces in the zone analogous to the parallelogram faces in prismatic forms (e.g., triprismatic sphedric, tetraprismatic sphedric) and can be further characterized as parallelepipedal, trapezoprismatic, and domed types; the terms right and clino- are also applicable as in true prismatic forms.

B. Sphericity

One of the principal criteria used for classification of peds is the ratios of the principal axes, or the relative length, breadth, and width of the peds. This factor of shape can be dealt with adequately by using the classes set up for sphericity in the system for general shape descriptions (Section 2.1:B; Table 2-1; and Fig. 1), and the shape terms may be used as adjectives to qualify the classes defined by basic form. All these terms are applicable to all the basic forms except regular polyhedral and spherical, which are equant by definition, and spheroidal in which b equals either a or c; for example, prismatic forms may be *planar prismatic* (e.g., planar tetraprismatic, planar hexaprismatic), *prolate prismatic, acicular prismatic,* etc.

C. Conformation

The classification of shape for peds to this stage has dealt with forms whose principal sections approximate, more or less closely, the shape of polygons of various kinds, circles, and ellipses. A further important group of variants is characterized by re-entrant angles between adjoining faces. Such forms are relatively common in the polyhedric, polyspheric, and polysphedric classes; Jongerius (1957) illustrated a number of these forms. They may also occur in the various shape classes referred to the prism. Peds displaying this phenomenon are considered to belong essentially to the group in which they would be classified if re-entrant angles were not present. Descriptions of this characteristic can be made specific by drawing the principal sections which, of course, will show the occurrence of re-entrant angles and the size of the angle. The frequency of occurrence of re-entrant angles can be expressed by the qualifying terms:

Weakly re-entrant. Less than one third of the interfacial angles are re-entrant angles.

Moderately re-entrant. Between one third and two thirds of the interfacial angles are re-entrant.

Strongly re-entrant. More than two thirds of the interfacial angles are re-entrant.

Alternatively the class names based on the lappungsquotient can be used as qualifiers (Table 9-2).

D. Specific Shape Names

The descriptive system and classification of shape set out in the preceding sections make it possible to describe the shapes of peds in considerable detail and to group them in a large number of classes. Although such flexibility is desirable it is convenient to use simple, specific names for shapes which occur very commonly or are of particular importance. This, in effect, is the technique used by the *U.S.D.A. Soil Survey Manual* (1951), Butler (1955), and Jongerius (1957), and to this extent their various terms are acceptable. The following terms and definitions are proposed as a preliminary list. Idealized examples of some of the forms are illustrated in Fig. 122; distinguishing characteristics are set out in Table 15-2.

Blocky. Equivalent to equant polyhedric, that is, $b/a > 2/3$, $c/b > 2/3$, and the ped has a number of plane faces; re-entrant angles between adjoining faces are absent. The term can be qualified by prefixes, according to the number of faces, into hexablocky (six faces), octablocky (eight faces), etc.

Subrounded Blocky. Equivalent to equant polysphedric. The specifications are as for blocky except that the ped has mixed curved and plane faces; re-entrant angles are absent or weakly developed. Qualification as regards number of faces can be used as for blocky.

Rounded Blocky. Equivalent to equant polyspheric. The specifications and qualifications are as for subrounded blocky except that all faces are curved; re-entrant angles are absent or weakly developed.

Multiangular Blocky. Equivalent to strongly re-entrant, equant polyhedric, or polysphedric. The faces are dominantly planes, but some may be curved.

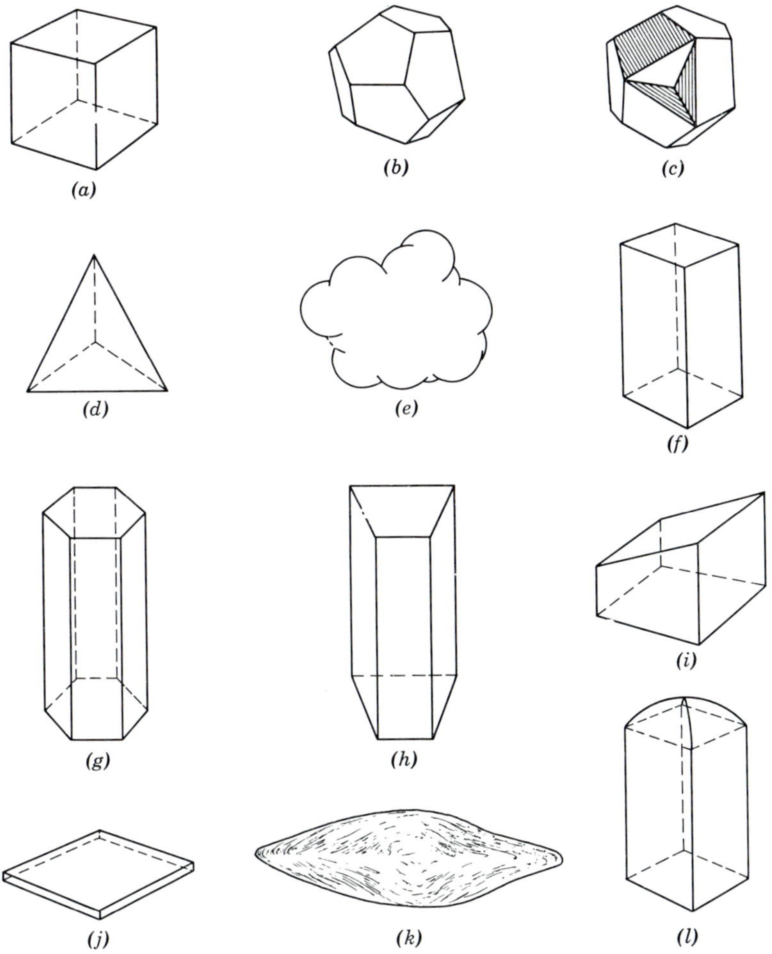

Fig. 122. Idealized ped shapes. (*a*) Hexablocky. (*b*) Dodecablocky. (*c*) Multiangular blocky (with re-entrant angles). (*d*) Pyramoidal (tetrablocky). (*e*) Mammillated. (*f*) Tetracolumnar. (*g*) Hexacolumnar. (*h*) Trapezocolumnar. (*i*) Cuneate. (*j*) Platy. (*k*) Lenticular. (*l*) Domed tetracolumnar.

Cuboid. Equivalent to hexablocky but the limits of the ratios of the principal axes are restricted to $b/a > 9/10$ and $c/b > 9/10$, and the angles between adjoining faces are approximately right angles.

Pyramoidal. Equivalent to tetrablocky; that is, the unit has four plane faces.

TABLE 15.2 Specific Shape Names for Peds*

Specific Name	Ratio of Axes	Shape of Faces	Re-entrant Angles	Qualifications
Blocky	Equant	Plane	Absent	As to number of faces (hexa-, octa-, etc.)
Subrounded blocky	Equant	Mixed curved and plane	Absent or weakly developed	Do
Rounded blocky	Equant	Curved	Do	Do
Multiangular blocky	Equant	Plane and/or curved	Strongly developed	Do
Cuboid	Equant	Plane	Absent	Six faces
Pyramoidal	Equant	Plane	Absent	Four faces (tetrablocky)
Mammillated	Equant or prolate	Curved	Strongly developed	As to number of faces
Columnar	Prolate or acicular	Plane	Absent or weakly developed	As to number of faces in the vertical zone
Parallelocolumnar	Do	Plane; bases are parallelograms	Do	Do
Trapezocolumnar	Do	Plane; bases are trapeziums	Do	Do
Cuneate	Usually equant or prolate	Plane; bases are trapeziums or triangles	Absent	The horizontal zone has three or four faces
Platy	Planar	Plane	Absent	Number of faces in the vertical zone
Lenticular	Usually prolate	Curved	Absent	Two faces only
Domed	A qualifying term for forms in which the *upper* face is convex upward			

* *Inclination:* all types may have any inclination

Mammillated. Equivalent to strongly re-entrant, equant, or prolate polyspheric. The faces are virtually all curved.

Columnar. Equivalent to prolate and acicular prismatic, that is, $b/a < 2/3$, $c/b > 2/3$. There are polygonal bases with a number of parallelogram plane faces arranged in a zone around an axis; re-entrant angles are absent or rare. Columnar can be qualified by prefixes according to the number of parallelogram faces in the zone, for example, tricolumnar, tetracolumnar, and hexacolumnar.

Parallelocolumnar. Equivalent to prolate and acicular parallelepipedal.

Trapezocolumnar. Essentially equivalent to prismatic in which the bases are trapeziums.

Cuneate. Wedge shaped; equivalent to trapezoprismatic or triprismatic.

Platy. Equivalent to planar prismatic, that is, $b/a > 2/3$, $c/b < 1/10$, and the long and intermediate axes are in a plane parallel to the bases.

Lenticular. "Shaped like a lentil or lens, double-convex" (*Oxford Shorter Dictionary,* 1956). Such forms consist of two curved faces which intersect in an ellipse or circle; one principal section is an ellipse or circle, and the other two are each composed of two curved lines. Lenticular, then, is equivalent to dispheric. Such forms have been reported from India (Krishna and Perumal, 1948) and also occur in Australian soils. They are usually prolate dispheric.

Domed. An adjective to qualify columnar, or other appropriate specific names, for forms which belong basically to the specific shape class, but in which the *upper* face is a curved surface, convex upward.

15.3 PEDALITY

The basic unit of description of pedality is the primary ped, and complete classification of pedality involves description of the size and shape of primary peds, their arrangement within the soil material, size, and shape of secondary peds, the arrangement of primary peds within secondary peds, size and shape of tertiary peds, and the arrangement of secondary peds within tertiary peds; interpedal pedological features and voids are ignored except insofar as the latter are inferred from descriptions of arrangement. Size and shape of peds have been dealt with earlier in this chapter. Size is not included in the following discussion because all the classes can be qualified according to size; size is not of itself a suitable characteristic for distinguishing classes.

A. Arrangement

There are three aspects of arrangement of peds at each level: *accommodation, packing,* and *inclination.*

(*a*) **Accommodation.** Accommodation of adjoining peds is a measure of the degree to which adjacent faces are molds of each other. Three degrees of accommodation have been defined:

> *Accommodated.* All the faces are accommodated by the faces of adjoining peds.
>
> *Partially Accommodated.* Some of the faces are accommodated by the faces of adjoining peds.
>
> *Unaccommodated.* Virtually none of the faces are accommodated by the faces of adjoining peds; adjoining peds touch only at points.

If it is considered worthwhile the partially accommodated class can be subdivided by stating the proportion of the faces accommodated.

Degree of accommodation generally has a close correlation with the shape and number of faces (basic form) of the peds. Peds bounded by plane faces generally have a high degree of accommodation although polyhedric forms with a relatively large number of faces may have lower degrees. Polysphedral forms often have a high degree of accommodation, the convex curved faces of one ped being accommodated by concave faces on the adjoining peds, but polysphedric forms with a relatively large number of faces may have lower degrees. Although spheral forms usually have low degrees of accommodation they can have quite high degrees as in some lenticular forms. In general, equant forms tend to have lower degrees and the more physically anisotropic forms higher degrees. The most common relationships of the shapes which have been given specific names are:

> Accommodated: blocky, multiangular blocky, cuboid, pyramoidal, all forms of columnar, cuneate, platy, lenticular.
>
> Partially accommodated: subrounded blocky.
>
> Unaccommodated: rounded blocky, mammillated.

(*b*) **Packing.** There are two broad classes of packing of peds: *regular* packing in which there is some orderly arrangement with regard to shape (the directions of the principal axes), and *random* packing in which there is no apparent orderly arrangement, usually because of the highly irregular shape of peds. Symmetrical forms with well-developed faces, such as blocky, cuboid, all forms of columnar, cuneate, platy, and lenticular, usually have a regular packing, while subrounded blocky, rounded blocky, mammillated, and multiangular blocky usually have random packing.

Of the regularly packed forms parallelocolumnar and trapezo-columnar usually have their axes aligned so that the parallelogram faces of peds which have their bases accommodated are in the one plane (Fig. 123), while lenticular peds usually have their longest axes parallel to, but slightly offset from, the longest axes of adjoining peds.

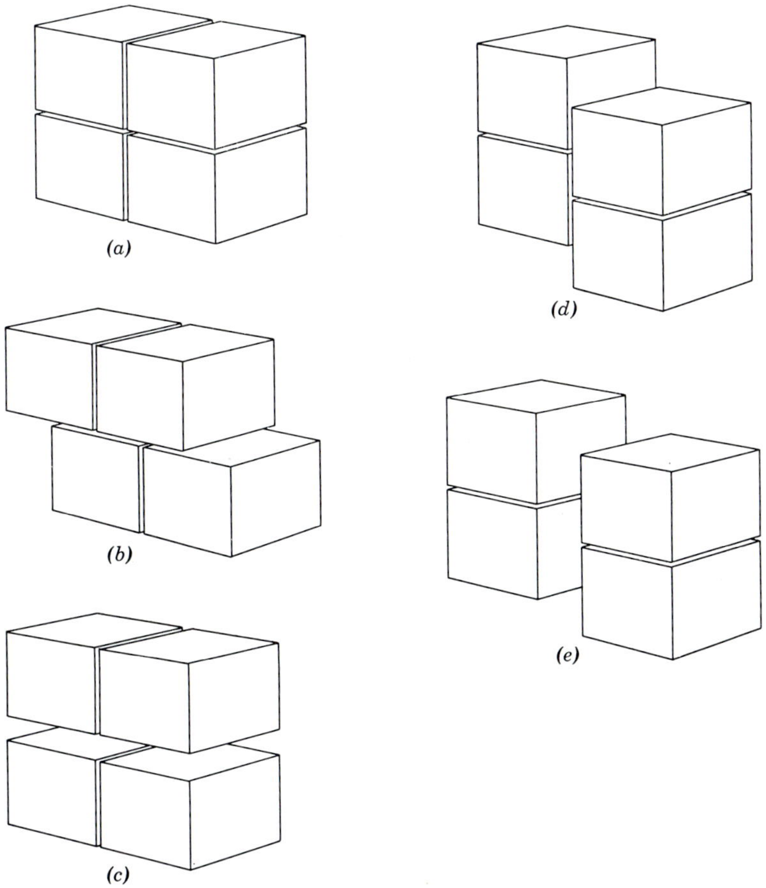

Fig. 123. Systems of packing of cubes. (*a*) Normal. (*b*) Offset. (*c*) Bi-offset. (*d*) Faulted. (*e*) Offset faulted.

Blocky, cuboid, columnar, and platy peds, however, can have a number of regular arrangements, five major variations of which are shown in Fig. 123. All the arrangements shown are with equally well-accommodated peds. The five methods of packing have been named:

Normal (Fig. 123*a*). The vertical axes of adjoining peds are in the same line, and the horizontal axes of horizontally adjoining peds are in the same plane.

Offset (Fig. 123*b*). The vertical axes of vertically adjoining peds are in the plane containing the vertical axis and one of the horizontal axes, but not in the same line.

Bi-offset (Fig. 123*c*). The vertical axes of vertically adjoining peds are neither in the plane containing the vertical axis and either of the horizontal axes, nor in the same line.

Faulted (Fig. 123*d*). The vertical axes of horizontally adjoining peds are in the plane containing the vertical axis and one of the horizontal axes, but the planes of the horizontal axes are displaced.

Offset Faulted (Fig. 123*e*). The vertical axes of horizontally adjoining peds are not in the plane containing the vertical axes and either of the horizontal axes, and the planes of the horizontal axes are displaced.

O'Neal (1949) described this aspect of packing simply as a percentage measure of the degree of overlap of vertically adjoining peds; this is a useful additional qualification of the systems of packing defined above.

(*c*) ***Inclination.*** Inclination of peds is their orientation in space with reference to the vertical or horizontal. It is usually defined by the orientation of the longest axis with reference to the vertical, but it is necessary to use the shortest axis or two axes for some forms. It is simply stated as the relationships of the appropriate principal axes with the vertical and/or horizontal, and so orientation classes can be established within the various shape classes. The following groupings are suggested:

Vertical. In all physically anisotropic forms, except planar (i.e., in prolate, acicular, triaxial, and oblate forms), the longest axis is vertical, in planar forms the plane containing the longest and intermediate axes is vertical, and in essentially equant prismatic forms the axis of the zone of parallelogram faces is vertical.

Horizontal. In all physically anisotropic forms, except planar, the longest axis is horizontal, in planar forms the plane containing the longest and intermediate axes is horizontal, and in essentially equant prismatic forms the axis of the zone of parallelogram faces is horizontal.

Inclined. In all forms the principal axes are inclined to horizontal and vertical.

The angle of inclination can be measured and quoted for inclined forms, the longest axis being used as the reference for measurement for prolate, acicular, triaxial and oblate forms, the plane of the longest and intermediate axes for planar forms, and any one of the principal axes for equant forms other than prismatic.

B. Classification and Nomenclature

Pedality is classified according to the size, shape, and arrangement of peds treated as units. The first subdivision can be into any of the classes in the comprehensive classification of peds according to shape or into the more restricted classes which have been given specific names (Section 15.2:D). Subdivisions within these groups can be made by detailed recording of all the appropriate characteristics, or a number of conventions can be adopted on the basis of the usual correlations of shape with packing, accommodation, and inclination. The two systems of classification and nomenclature are illustrated best by taking a particular example and dealing with it by both the detailed method and the method using all the possible conventions. Consider the tertiary ped shown in Fig. 28 and assume that the tertiary peds themselves are accommodated and have normal packing.

(*a*) **Detailed Method.** The shape of the tertiary ped is *prolate clino-tetraprismatic* with its long axis vertical, that is, *vertical prolate clino-tetraprismatic.* The two component secondary peds are also vertical prolate clino-tetraprismatic, and they have a *normal* packing. The component primary peds are dominantly *polyhedric* with re-entrant angles and *random* packing. All the peds at all levels are *accommodated.* Such complexity poses a difficulty in nomenclature. The tertiary ped can be described as vertical prolate clino-tetraprismatic composed of normally packed, accommodated, vertical, prolate clino-tetraprismatic secondary peds which, in turn, are composed of accommodated, randomly packed polyhedric primary peds with re-entrant angles. The pedality can then be described as normally packed, accommodated tertiary peds of the kind described. These are cumbersome descriptions that can be simplified by using a number of conventions.

(b) Conventional Method. Because of certain consistent correlations the following conventions can be used for "normal" cases, and qualifications can be made for the cases where one or more of the conventions does not hold:

(1) Use the specific name for shape.

(2) Assume that elongated forms (e.g., prolate, acicular), except lenticular and essentially equant prismatic forms, are vertical, and that planar and lenticular forms are horizontal, unless specifically stated otherwise.

(3) Leave out the shape adjective for primary or secondary peds if their shape is the same at all levels; otherwise include the shape names in brackets.

(4) Leave out the adjective for degree of accommodation if it conforms to the usual correlations with shape.

(5) Assume that packing is normal unless otherwise stated.

The nomenclature for the tertiary ped illustrated in Fig. 28 then becomes *clino-tetracolumnar,* the secondary peds are *clino-tetracolumnar,* and the primary peds are multiangular blocky; since the tertiary peds are accommodated and normally packed, the complete pedality is *clino-tetracolumnar* (*clino-tetracolumnar, multiangular blocky*).

(c) Specific Names. It should be possible to set up a more concise classification system based on all these morphological characteristics by defining specific names for the different levels of pedality which involve specific kinds of primary, secondary, and tertiary peds. This approach has been used by Jongerius (1957) to some extent, except that he too, as in the two methods just outlined, used a series of adjectives to qualify the shape classification of the secondary ped as a unit, for example, his "granular prismatic," "combined prismatic," "segmented prismatic," and "single prismatic." With more experience, as the kinds of arrangements of primary peds of various shapes which do occur are noted with more care, it should be possible to set up single names for specific combinations of all the distinguishing characteristics.

15.4 SECONDARY AND TERTIARY STRUCTURE

The four distinguishing characteristics of secondary and tertiary structures are:

(1) shape of peds,

(2) arrangement of peds,

(3) nature and arrangement of interpedal voids, and
(4) nature and arrangement of interpedal pedological features.

Description and classification of pedality effectively disposes of the most complex part of classification of secondary and tertiary structures by classification according to shape and arrangement of peds. Secondary and tertiary structures can then be classified by qualification of the classification of pedality according to the interpedal voids and pedological features.

A. *Interpedal Voids*

Voids between peds are restricted in kind and arrangement, by the shape and arrangement (accommodation and packing) of the peds, to planes (joint and craze planes) and compound packing voids (Section 9.5), and there are quite strong correlations between observed shapes of peds and the kind of interpedal voids. Joint planes usually occur between cuboid, blocky, pyramoidal, all types of columnar, cuneate and platy peds, craze planes between multiangular blocky and lenticular peds, and compound packing voids between subrounded blocky,* rounded blocky, and mammillated peds.

B. *Interpedal Pedological Features*

Pedological features between peds are virtually restricted to ped cutans and crystal sheets, both of which may be composed of any fraction of the plasma that normally occurs as such features (Chapters 10 and 13.1). Such pedological features do not always occur on *all* the faces of the peds, especially where the peds have regular forms with planar faces arranged approximately vertically and horizontally; the completeness and association of these features with specific ped faces should be recorded. Cutans may occur on the faces of any kind of ped irrespective of the shape of the interpedal voids, whereas crystal sheets occur only where the interpedal voids are planes.

C. *Classification and Nomenclature*

Secondary and tertiary structures are classified according to the principles used for pedality. In the tertiary ped shown in Fig. 28, whose pedality has been described (Section 15.3:B), the interpedal

* The interpedal voids between subrounded blocky peds are usually a combination of short planar and compound packing voids.

voids between primary peds are craze planes, and those between the secondary peds are joint planes. Suppose, also, that there are argillans on the horizontal faces of the secondary peds but not on the vertical faces; that is, the tertiary peds contain horizontal joint-plane argillans. The detailed description of the secondary structure is the internal structure of the secondary ped with description of the interpedal voids (craze planes) between the primary peds. The description of tertiary structure is the size, shape and arrangement of secondary peds and description of the interpedal voids (joint planes) and pedological features (horizontal joint-plane argillans) between the secondary peds. For the conventional method additional conventions can be used:

(1) Leave out the shape adjective for the interpedal voids if they conform to the usual correlation with shape of peds.
(2) Assume that there are no interpedal pedological features unless they are specifically mentioned.

The nomenclature for the secondary structure then is *multiangular blocky,* the tertiary structure is *cutanic* (*horizontal argillans*) *clino-tetracolumnar;* the quaternary structure (due to tertiary peds) is the same as the tertiary structure.

15.5 INTERPRETATION

Very little indeed is known about the processes of formation of peds of different size, shape, and arrangement. Most of the hypotheses put forward are little more than rough correlations of the occurrence of particular shapes with some other observable or measureable characteristic of the soil material.

Studies of columnar units in igneous rocks, especially in basalt flows, suggest that they are formed by thermal stresses produced during cooling and that their orientation in space is controlled by the orientation of surfaces of equal tensile stress (Spry, 1962). A similar explanation is probably valid for columnar peds in soil materials except that the tensile stress is produced by drying of the colloid size material (plasma).

Well-developed columnar peds, especially if they have domed upper faces, are widely attributed to solonization (e.g., Stephens, 1956; U.S.D.A., 1951) because of their correlation with the *B* horizons of soil profiles that have the chemical characteristics on which classification as solonetzic soils is partly based. These characteristics are, essentially, increasing pH and proportions of sodium ions on the exchange complex below the top of the *B* horizon. It is difficult, however, to accept these conditions as necessary for the formation of domed columnar peds.

On the one hand, profiles have been reported that have the chemical characteristics of solonetzic soils but not the domed columnar peds in the *B* horizons (e.g., Johnston, 1953). On the other hand, many profiles have been reported that have domed columnar peds in the *B* horizon but not the classical chemical characteristics of solodized and solonetzic soils (e.g., Riecken, 1943; Ellis and Caldwell, 1935; Joffe and Zimmerman, 1944; Eremin, 1952; White, 1961). The fact that a soil profile has been "solonized" and subsequently leached, therefore, is not a certain or absolutely necessary condition for development of domed columnar peds. Other conditions may cause the process, and these conditions may not. It is of interest to note that Wolkewitz (1958) found experimentally that calcium-saturated clays produced more regular crack patterns with wider spacings and wider cracks than did sodium-saturated clays; the domed columnar peds of solonetzic soils are the result of crack patterns which are, in fact, among the widest spaced and most regular patterns observed in soil materials. This is in contradiction to the generally accepted hypothesis that such peds are directly the result of sodium saturation.

Platy peds have often been ascribed to the physical action of ice, which may form horizontal wedges in the upper horizons of a soil profile thus forcing the soil apart along horizontal planes. McMillan and Mitchell (1953), however, ascribed platy peds in the *A* horizons of some podzolic and solonetzic soils from Saskatchewan to a "banded fabric" which consisted of a succession of subparallel horizontal bands, each of which showed a gradation in color and proportion of plasma from the bottom to the top; the upper part of each band was dark with a sharp boundary between it and the overlying band. Although there were no cracks between successive bands, there were surfaces of weakness between bands due to the contrast in fabric and amount of plasma between the top of one band and the bottom of the overlying band.

Kubiena (1938) proposed that this fabric resulted from drying of the soil from the top down, thus producing a zone of accumulation of colloidal material at the evaporating surface. "When the adjacent eluvial layer dries out, transportation is interrupted for a time, but will be resumed at a new surface within the soil system" (McMillan and Mitchell, 1953); thus, successive bands are formed, which tend to be self-perpetuating. In this instance, therefore, the immediate cause of platy peds is shown and a hypothesis for the mechanism of formation proposed. Other platy solonetzic and podzolic-solonetzic *A* horizons, however, showed "isoband fabric" in which there was no change in fabric or proportion of plasma throughout the thickness of the bands that were separated by subparallel, horizontal cracks. No explanation was proposed for this arrangement. It is of interest, also, to note that Peterson (1944) suggested, as a result of experimentation, that large proportions of kaolinite favored development of platy peds, whereas there was very little kaolinite in any of these Saskatchewan soils.

Shrinking and swelling, and possibly other processes, can cause rearrangement of the mobile constituents of the plasma to produce varying degrees of preferred orientation of the clay mineral grains with regard to each other and with regard to already existing surfaces of weakness. For example, the lower *B* horizons of some planosolic soils from Kempsey, New South Wales, consisted of very small, closely packed, lenticular peds with smooth, lustrous surfaces, and thin sections showed the material was composed of very small, prolate, anisotropic plasma aggregates which were quite strongly oriented with regard to each other. This plasmic fabric was the cause of the occurrence of closely packed, small, lenticular peds.

It seems, therefore, that the genetic processes controlling the formation of the various shapes, sizes, and arrangements of peds have not been elucidated for all occurrences of any of the shapes. One factor that has militated against detailed studies of processes of formation has been the current method of recognizing peds by the occurrence of a population of units of similar size and shape after application of a force to a mass of soil. The most revealing morphological information concerning the genesis of peds is to be obtained by a study of the kind and distribution of the surfaces of weakness between them, but this approach was not possible because the surfaces of weakness were merely assumed to be present, and no characteristics for recognizing them were quoted. By adopting Sleeman's (1963) definition of a ped (Section 7.5:A) the surfaces of weakness can be recognized and their distribution patterns studied. Such information is essential to successfully initiate and interpret experimentation in the reproduction of soil pedality.

An example of this approach is the study of crack patterns, or patterns of surfaces of weakness, in some soils from the Riverine Plain of New South Wales (Sleeman, 1963). In all these soils the shape and arrangement of the peds resulted directly from the formation of crack patterns, so they could be explained on the same basis as the genesis of those patterns. Sleeman reviewed the literature on the experimental production of crack patterns as a basis for interpretation; his findings were discussed earlier (Chapter 9: Voids). Briefly, the factors that control the nature and arrangement of cracks are: uniformity of the material, proportion of clay size material ($< 2\ \mu$ equivalent diameter), type of clay mineral, proportions of the various exchangeable cations, percentage of humic acid and iron hydroxides, and the uniformity and rate of drying (Winterkorn and Tschebotarioff, 1947; Zonneveld, 1956; Wolkewitz, 1958; Czeratzki and Frese, 1958; Winterkorn, 1958); also some patterns may be inherited from the parent material.

Interpretation of the formation of interpedal voids in secondary and tertiary structure is intimately connected with interpretation of the formation of peds; there can be no doubt that the same processes are responsible for the formation of both the peds and interpedal voids. The processes of formation of interpedal pedological features have been discussed under ped cutans and crystal sheets, but their occurrence as a part of secondary and tertiary structure can give some additional information for the soil profile; for example, ped cutans are formed subsequently to the peds. In some instances depositional ped cutans that occur on all the ped faces are interconnected with normal void cutans within the ped, indicating that the suspensions carrying the cutanic material permeated the whole soil material. In other instances cutans occur on the faces of peds within which there are no cutans, indicating that most of the water movement through the soil material occurs through the interpedal voids. Again the intrapedal soil material may have depositional void cutans while there are no ped cutans, indicating that illuviation was far advanced before the formation of peds. Some of the details of water movement may also be suggested by the occurrence of depositional cutans only on particular ped surfaces, for example, only on horizontal ped faces (Sleeman, 1963).

16

Scale and Sequence of Observations

In the preceding chapters no mention has been made of the scale at which particular observations should be made except in specific instances involving interpretation of phenomena that can be observed only in thin section with the petrological microscope, which has a limited resolving power and range of magnification. Nor has a sequence for observation of the units and levels of organization, or of particular characteristics, been suggested. The reasons for this are as follows:

(1) Many of the units of organization (including pedological features) occur in a wide range of sizes so that different scales of observation may be necessary for the same characteristics in different circumstances.

(2) The description and classification of the units and levels of organization have been set out generally as broad principles, and any scale of observation that is necessary and available should be used to satisfy these principles; for example, orientation patterns for megascopic individuals can be measured by direct methods, those for microscopic individuals can be inferred from extinction patterns in thin section under the petrological microscope if the individuals are anisotropic, and those for ultramicroscopic, opaque or isotropic microscopic individuals can be described only by special techniques, such as electron microscopy.

(3) The units and levels of organization have been chosen so that

each can be dealt with separately, without reference to other units and levels, thus a complete sequence of observation may not be necessary.

The natural and convenient sequence of making observations on soil materials and soil profiles is from the megascopic scale to the microscopic and ultramicroscopic scales, that is, commencing in the field and culminating in the most detailed possible observations by special techniques in the laboratory. Some characteristics, however, are observed best over a relatively limited scale. The following discussion, therefore, is intended simply to suggest a convenient procedure for observations leading to a complete description of structure and fabric of soil materials and soil profiles.

16.1 FIELD DESCRIPTION

Field description of structure and fabric follows the general principles laid down for their complete detailed description. The objective is to describe and classify, in as great detail as possible, the size, shape, and arrangement of the constituents, whether they be single grains or complex individuals. Because of the limitations of the techniques which can be used in the field, however, some modifications in approach are necessary, and descriptions and classifications cannot be carried as far as is possible by the more sensitive laboratory techniques. The definitions and classifications of the features observed in soil materials have been made with this difficulty in mind so that the individuals can be classified as far as possible on field data without danger of upsetting or cutting across the more complete and detailed classifications, which can be set up on the basis of more complete data.

It should be emphasized that classification in the field should not be carried beyond the positive data obtainable by field techniques. This means that not all features can be classified completely on field data alone, and doubtful classifications should be recorded as such rather than placing the particular feature in a certain class on too little data. This is particularly important because not all the possible kinds of features have been observed and classified; a continual search for new features is part of the field observer's activities. In addition, continual checking of uncertain classifications will help to establish reliable correlations between characteristics observed by more sensitive techniques and those observed in the field. In this way classifications made on field observations can be expected to become more certain and dependable as experience grows.

The first detailed observations in the field are made on a section on a pit face where the whole soil profile is exposed. As a preliminary step the profile is divided into soil horizons and/or soil materials; the divisions into horizons and soil materials may not coincide because certain characteristics, such as the more extensive voids, may extend beyond the limits of a soil horizon or a soil material defined on some other characteristics. Descriptions and classifications of the observable features are then made on the various soil materials, progressing from the grosser features to the finer.

A. Natural Surfaces

A great deal of the description and classification of structure and fabric in the field depends on recognition of natural surfaces (Section 7.5:A), particularly for recognition of voids, peds, and cutans. In practice the simplest procedure is to create a fracture surface in each of the soil materials to be examined and to compare the characteristics of this fracture surface, as observed in reflected light with a hand lens at magnifications up to 20×, with those of suspected natural surfaces. Of all the diagnostic characteristics for natural surfaces, luster and the characteristics of the skeleton grains at the surface are perhaps the most useful.

Several workers have distinguished two different kinds of plasma by their luster and surface smoothness in reflected light. Kubiena (1938) referred to these as "peptized" and "pectized" plasma, while Churchward (1959) named them "lac" and "floc," respectively. Kubiena used luster and degree of adhesion of the plasma to the surface of skeleton grains as the distinguishing characteristics, while Churchward used luster and roughness of surface on a microscale. Both agree that lac, or peptized plasma, has a high luster and a smooth appearance. Plasma with these characteristics, however, sometimes adheres to skeleton grains (as Kubiena's peptized plasma should) and sometimes does not. Lac and floc, therefore, are used here for plasma with the characteristics described by Churchward; they are defined as follows:

> *Lac.* Plasma which has a high luster and smooth surface in reflected light under a hand lens up to magnifications of approximately 20×.

Floc. Plasma which has a relatively low luster and a roughness of surface that gives it the appearance of clusters of silt size grains in reflected light under a hand lens up to magnifications of approximately 20×.

These terms are particularly suitable for describing the characteristics of natural surfaces in soil materials in the field; they can be termed *lac surfaces* and *floc surfaces* when they have the reflected light characteristics of lac and floc, respectively. A third kind of natural surface is a *skeletal surface,* which, in reflected light under a hand lens up to magnifications of 20×, has the appearance of a granular fabric, that is, packed skeleton grains with little or no plasma. Skeletal natural surfaces are often apparently bleached and are commonly paler than fracture surfaces in the same soil material.

B. Pedality

Perhaps the most obvious characteristic of soil materials exposed in section on a pit face is their pedality, or the size, shape, and arrangement of interpedal voids. Since the system of description of structure is based on the primary ped as a basic unit, and primary peds are usually megascopic units, it is convenient to base field observations on description and classification of pedality.

If the peds are large enough and separated by relatively large interpedal voids, their natural surfaces are recognizable as such *in situ* on the pit face, and their characteristics, including accommodation, packing, and inclination, can be described by judicious removal of individual peds. As each is removed, its surfaces can be examined to insure that it is a true ped completely enclosed by natural surfaces. In this way the pedality within secondary and tertiary peds, if they are present, can be described, and the peds of all levels classified. Smaller peds are more difficult to describe, especially where natural surfaces are not so obvious. The best approach is to gently break down the soil material by applying a manual force, examining the surfaces at each stage of breakdown to determine whether the units produced are in fact peds or fragments. Where the peds are so small that it is not possible to describe the details of their arrangement in the field, this must be done later in the laboratory by more sensitive techniques. It is possible that primary peds may be too small to identify in the field, and description of the details of pedality may have to await laboratory study.

C. Megascopic Voids

Field description is concerned only with larger voids that can be identified megascopically or under a hand lens. An absolute lower limit of size cannot be set because it depends on the shape of the void. Planar voids (planes) can be recognized and classified down to very small sizes because the traces of planes on the face of a pit, or on the surface of a hand specimen, are very obvious. There is a logical sequence for description of megascopic voids in the field.

The first phase is description of the interpedal voids, if peds are present, since these are the largest and most obvious voids; the voids between the peds of highest level are described first, that is, between tertiary peds, then those between secondary peds, and finally those between primary peds. If peds are absent, planes are described first because they are the most obvious voids of the s-matrix as seen on a pit face. Planes are often continuous over quite long distances, and joint planes and skew planes, particularly, can usually be classified as such on megascopic characteristics; craze planes may be developed on a much finer scale, and joint planes and skew planes occasionally occur on a scale too small for observation in the field. The grosser surface characteristics of planes can also be described in the field, such as their degree of curvature over relatively long distances, and often they can be classified as metaplanes or orthoplanes from field evidence. The smoothed or slickensided walls of well-developed metaplanes are usually obvious under a hand lens, but classification as orthoplanes on field evidence is less reliable.

The second phase is inspection of the pit face for megascopic, physically anisotropic voids other than planes, especially large channels, which may be transpedal or quite extensive in apedal soil materials. Description of the characteristics (especially branching patterns) of channels requires careful dissection of the soil material in the pit face. As for planes, the grosser characteristics of conformation of the walls and curvature of channels over relatively long distances can be described, but only the better developed degrees of smoothness of the walls can be recognized so that classification as metachannels or orthochannels is not always possible.

The third and final phase is careful inspection and description of the intrapedal voids and/or the more equidimensional voids in apedal soil materials, such as vughs, vesicles, and simple packing voids. These are described and classified from inspection of fracture surfaces on hand specimens using a hand lens. The completeness of the descriptions and classification depends on the size of the voids and the degree of development of the characteristics being described. A good deal of the detail of description will probably have to await laboratory examination.

D. S-matrix

It is, of course, impossible to recognize individual plasma grains in the field, or even the small aggregates of plasma grains (domains) that are recognizable in thin section under a petrological microscope. The concepts of floc and lac (Section 16.1:A), however, make it possible to distinguish very broadly the plasma of s-matrixes of soil materials that are significantly different, and some observations are possible on the relationships of plasma to skeleton grains in the s-matrix of the soil material even in the field.

Since natural surfaces are the most probable sites for deposition of illuviated material and *in situ* modification of fabric, both of which can cause considerable changes in characteristics observed in reflected light, descriptions of the s-matrix in the field are made on fracture surfaces of hand specimens. The first classification is simply on the characteristics of the plasma of the s-matrix. Soil materials whose plasma (excluding pedological features) is dominantly floc are referred to as *erdes,* and those whose plasma is dominantly lac are referred to as *lehms.* This nomenclature is derived from that used by Kubiena (1953), who referred to braunerde, braunlehm, roterde, and rotlehm. The next subdivision is on the broad relationships of plasma to skeleton grains. Granular and porphyroskelic fabrics (Section 14.4) are usually readily recognizable in the field with a hand lens, especially when the plasma in the latter is lac so that the soil material is relatively dense with the apparently porphyritic arrangement of the visible skeleton grains in the dense plasma. Intertextic and agglomeroplasmic fabrics, however, tend to grade into each other, and the latter grades into porphyroskelic fabric, especially when the plasma is floc.

There is no real classification of skeleton grains as such except in a mineralogical sense. Size distribution is assessed roughly by the method of estimating field grading, but additional data on the coarser fractions can be obtained by dispersing the material by trituration in the hand or a shallow dish, washing off the fine fractions, examining the coarser fractions under a hand lens, and comparing them with standard sieved samples. This also provides information on shape which may be valuable, especially as it is generally acknowledged that the largest grains most readily exhibit characteristics of rounding due to transport (Section 2.6:B). Distribution patterns, other than random patterns, may be observable on fracture surfaces with a hand lens if they are well developed, but orientation patterns are more difficult except where the individuals are quite large, and even then a large number of measurements of azimuth and dip should be made on the physically anisotropic individuals.

All these characteristics of the s-matrix can be described in much greater detail and with much greater certainty by more precise techniques in the laboratory, except where the skeleton grains are so large that their distribution and orientation patterns are most appropriately, indeed can only be, described in the field. Classifications of the s-matrix made in the field, therefore, should be regarded as tentative until more dependable correlations are established between the characteristics observed by laboratory techniques and those observed in the field.

E. Pedological Features

Having described pedality, the megascopic voids, and the s-matrix, attention can be given to the pedological features. Because of the wide variety of pedological features and their size and distribution, the opportunity should be taken to describe them at whatever stages they are recognized; in the field the scale of description varies from a section on a pit face for large features down to small, dissected hand specimens under a hand lens. Some features are so small that they are not recognizable by field techniques, but others are so large that their grosser characteristics are described most conveniently in the field. There should be little difficulty in classifying observed features into one of the six major groups to be discussed, except for integrades between certain groups, for example, crystallaria—cutan intergrades, pedotubule—cutan intergrades, fecal pellet—pedotubule intergrades, and subcutanic feature—cutan intergrades; the intergrade nature of the features should be recorded. The classification of pedological features can sometimes be carried to a much lower level within the major groups on field evidence alone.

(*a*) ***Cutans.*** Recognition of cutans in the field is an extension of the observations made to recognize natural surfaces as a basis for description of pedality and megascopic voids. Cutans are usually relatively thin and occur on natural surfaces, especially on the walls of voids. They are recognized in the field, therefore, by careful dissection of the soil material and inspection of the characteristics of the natural surfaces in reflected light under a hand lens. Additional information is obtained by breaking or cutting the soil material normal to the natural surfaces and inspecting the fracture or cut surface; this gives comparative data on the characteristics of the fracture surface and the natural surface, which is valuable for the recognition of cutans. By these techniques cutans can usually be subdivided in the field, according to the nature of the cutanic material, into argillans, sesquans, mangans, chalcedans, and soluans with some certainty if they are reasonably well developed. The distinction between illuviation, diffusion, and stress cutans, however, is often difficult in the field unless the illuviation types are quite thick and distinctive, especially as regards color. In particular the reflected light characteristics of illuviation and stress argillans are very similar, both producing lac surfaces. Similarly skeletans and neoskeletans both produce skeletal surfaces. On the other hand illuviation cutans may be so thick and distinctive that they can be recognized as definite coatings, and even successive layers of clay minerals may be recognizable in argillans; skeletans and neoskeletans may be distinguishable on sharpness of boundary and density of packing of skeleton grains.

Where cutans are recognizable in the field it is important to describe their distribution pattern throughout the whole profile, especially their relationships to peds and megascopic voids. Besides classification as ped cutans, plane cutans, etc., their vertical extent in the profile should be described since they may be associated with particular horizons and particular groups of peds or voids, or they may extend above and below particular horizons or soil materials.

(b) Pedotubules. Very considerable variation occurs in extent and in cross-sectional size of pedotubules, but they are commonly megascopic features whose grosser characteristics are described best in the field from a pit face. Careful dissection is the only method of describing their branching patterns, and dissection plus manipulation in the hand is the best method of assessing the degree of adhesion. Sharpness of boundaries also can usually be described in hand specimen with a hand lens.

The grosser characteristics of the internal fabric of pedotubules are usually sufficiently obvious under a hand lens to enable classification as granotubules, aggrotubules, isotubules, and striotubules in the field. It is often possible to describe even more detail of the internal fabric, especially the occurrence and distribution of particular kinds of included pedological features, and the s-matrix of the pedotubules can be described to a similar level of specification as was suggested for the s-matrix of soil materials (Section 16.1:D). Also, comparison of the grosser characteristics of the tubulic material with the surrounding soil material and with the soil materials from other horizons will usually place the pedotubule according to the nature of the tubulic material and into one of the groups of orthotubules, metatubules, or paratubules.

(*c*) ***Glaebules.*** Many glaebules are relatively small, and the grosser characteristics of individuals can be described best by dispersing the soil material by trituration, washing out the fine material, and thus concentrating the glaebules. Some types, however, may be destroyed or modified by such methods, for example, papules composed of dispersable clay minerals, types with diffuse boundaries, weak concentrations of certain fractions of the plasma, or strongly adhesive types. Others may be very large and can be examined best *in situ* in a pit face or by dissection. It is necessary, therefore, to search for glaebules at all stages of the examination, in the pit face, by dissection of hand specimens, and by concentration by dispersion and washing. Data from these three phases of inspection enable classification of the glaebules according to size, shape, sharpness of boundary, and adhesiveness.

The observable glaebules can often be classified at a lower level in the field. Composition of the concentrated plasma fraction can often be assessed in broad terms, so the glaebules can be classified as sesquioxidic, siliceous, carbonate, sulfate, and other types. Internal fabric can sometimes be assessed sufficiently well to place the glaebules as nodules, concretions, septaria, etc., and in large specimens even some of the included pedological features can be recognized and classified. This is done by examining fracture surfaces of the glaebules with a hand lens.

The distribution patterns of recognizable glaebules are described most reliably in the field from the pit face by careful dissection of the soil materials. In particular the referred distribution pattern may show relationships with features which are so large that the relationship can be observed only on the large scale involved in field description. On the other hand the details of internal fabric can be described only in the laboratory by more sensitive techniques, and assessments of composition and internal fabric made in the field should be checked by these techniques.

(*d*) ***Crystallaria.*** Usually crystallaria are megascopic features that are readily recognizable in the field. They vary considerably in size, however, and so should be looked for in the hand specimen under a hand lens as well as in the pit face. Occasionally they are too small to identify with certainty in the field. The most common types consist of carbonates and sulfates, which can be recognized in the field on morphology and reaction with dilute hydrochloric acid. Other mineralogical types, such as goethite, can be recognized tentatively if the kind of soil profile in which they occur is taken into consideration, but such tentative assessments should be checked and identifications of carbonates and sulfates should be made more specific later, by laboratory techniques. Thus crystallaria can be classified in the field as crystal sheets, crystal tubes, crystal chambers, and intercalary crystals and further into broad groups according to their composition. If the crystals are large enough even some of the details of their internal fabric, such as the orientation of the crystals with regard to the walls of the voids in which they are formed, can be described. As for glaebules and cutans the distribution pattern of recognizable crystal sheets should be described by dissection of the pit face.

(*e*) ***Subcutanic Features.*** Most subcutanic features, except certain expression of neoskeletans, are quite small, so the best method of observation in the field is by searching numerous fracture surfaces made normal to natural surfaces, especially the walls of voids, with a hand lens. It may not be possible to decide whether particular features are neocutans, cutans, or associations of neocutans with cutans, particularly where the concentrated plasma fraction is sesquioxides or manganese oxides; certainly neostrians cannot be identified positively in the field although their presence may be inferred with some certainty from the occurrence of smoothed and slickensided natural surfaces, especially when associated with planes. Doubtful identifications should be recorded as such and checked in the laboratory. Neoskeletans are usually identifiable in the field by the presence of a skeletal surface, which is usually bleached compared with the s-matrix of the soil material; they may be difficult to distinguish from skeletans or even from weakly developed granotubules. Quasicutans are not so difficult to recognize because there is little danger of confusion with cutans or pedotubules in the distribution of the concentrated plasma fraction with regard to the natural surface. Usually the concentrated plasma fraction can be identified in broad terms in the field.

When subcutanic features are relatively small their distribution is difficult to describe from a pit face, but when they are large this is the most convenient stage at which to describe the patterns.

(*f*) ***Fecal Pellets.*** Single and welded fecal pellets are usually quite small and can be recognized in the field only by careful dissection of the soil material under a hand lens. When they are relatively rare they are easily missed in field description, but when they are abundant their presence can be inferred from the occurrence of botryoidal masses of soil material and the mammillated conformation of the walls of the large voids as seen under a hand lens.

16.2 STEREOMICROSCOPIC OBSERVATIONS

The second stage of observations for description of structure and fabric is with a stereoscopic microscope, with magnifications up to about 200×, on undisturbed monoliths sampled from the pit face (see Appendix 1:A). In practice this stage is simply an extension of the observations made in the field, since the technique still depends on the characteristics which can be recognized in reflected light. Because of the magnifications that can be achieved, however, and the convenience of working on a laboratory bench with suitable equipment and microtools, much more detail and accuracy can be brought to the descriptions. The technique is limited mainly by the size of features too large to be sampled in a monolith and those too small to observe at the magnifications which can be achieved.

Of prime importance at this stage is recognition and description of the details of the characteristics of natural surfaces, especially those too small to recognize in the field. Lac, floc, and skeletal surfaces are usually readily recognized, small peds can be recognized on the characteristics of their surfaces, and the details of pedality can be described where the primary peds were too small to recognize in the field or their classification as true peds was doubtful because of the difficulty of identifying natural surfaces of very small extent in the field. Also, classification and description of the distribution and orientation patterns of voids can be carried a good deal further at this stage than was possible in the field; the sequence of description can be kept the same as for the field description, that is, interpedal voids, planes, transpedal voids, and intrapedal voids in that order. In particular more detail of the smoothness and conformation of the walls of voids can be described. Similarly classification of the soil material as erde or lehm and the s-matrix as granular, porphyroskelic, agglomeroplasmic, or intertextic on the relationships of plasma to skeleton grains (Section 14.4) can be made with greater certainty.

The distribution and orientation patterns of the smaller pedological features should be described in as great detail as possible at this stage. Many pedological features are not recognized readily enough in the field to describe these patterns with any great certainty. On the other hand a thin section is so small that, even if care is taken so that the orientation of the section and its position in the profile is known, such patterns can only be inferred without any great certainty. Stereomicroscopic studies of monoliths make it possible to dissect the soil material and so describe the characteristics of pedological features, such as cutans, pedotubules, crystal tubes, and crystal sheets, in three dimensions, which is not possible in thin section. In addition, size, shape, and degree of adhesion of the smaller pedological features are described best at this stage.

Typical internal fabrics of a number of pedological features may be recognized by stereomicroscopic examination if the features are large enough. The concentric fabric of many concretions is sufficiently well developed and on a large enough scale to recognize by this means, as are the void structures of pedodes and septaria, the internal fabrics of many pedotubules, and the crystal arrangements in some crystallaria; even the lamellar fabric of some papules is recognizable.

Although this stereomicroscopic examination is only an amplification of field observations its importance cannot be stressed too much. Besides the descriptive data obtained it forms the basis of subsequent observations on thin sections in that the specimens to be sectioned are chosen on the basis of the stereomicroscopic observations. Thus, if a natural surface cannot be recognized stereomicroscopically as having a cutan, but the presence of a cutan is suspected, the surface can be sampled and sectioned. Then, if a cutan is identified and classified from the thin section, its distribution can be described by returning to the stereomicroscopic examination of the monolith. The constancy of properties of the cutan with depth in the monolith can be checked by additional thin sections of selected areas of similar natural surfaces.

Stereomicroscopic techniques can also be used on polished blocks, with inclined illumination, to describe the two dimensional size, shape, conformation, and arrangement of natural surfaces and small features, and those characteristics of internal fabric which are apparent in reflected light, such as concentric and lamellar fabric. Serial observations can be made by successive grinding and polishing of the block.

16.3 PETROLOGICAL MICROSCOPIC OBSERVATIONS

Petrological microscopic observations on thin sections are the most appropriate means of describing the details of the internal fabric, more particularly the plasmic and basic fabrics, of the s-matrix and of individual pedological features. For this to be possible the constituents must be optically anisotropic and/or at least translucent. The greater part of the system of description and classification of these features, as set out in the earlier chapters, is based on interpretation of phenomena observed primarily under crossed nicols on a petrological microscope. In addition, the minute details of conformation and smoothness of natural surfaces, and of pedological features, can be described from thin sections, as well as details of the relationships between plasma and skeleton grains and the characteristics of very small voids.

Thin sections, however, suffer from a number of limitations, particularly for description of distribution patterns. They are quite small in extent and essentially two dimensional so that distribution and orientation patterns can be described only for very small individuals and only in two dimensions; description of three-dimensional patterns must depend on correlation of characteristics observed in thin sections with those observed stereomicroscopically in a monolith or in the field. This limitation can be overcome to some extent by careful selection of the specimens for thin sectioning; for example, referred orientation and distribution patterns can be described adequately in thin section by choosing sections with known orientations with regard to the specific reference features. Their limited extent also means that observation of the occurrence of relatively rare features may require examination of a large number of thin sections from the one soil material. In addition the important reflected light characteristics of natural surfaces are not observable in thin section, so natural surfaces are recognized only by the presence of small voids and by the details of their conformation and smoothness.

Microscopic observations can also be made on polished blocks with vertical illumination. In general the same characteristics can be observed by this technique as are observed under the stereomicroscope with inclined illumination, and again serial observations can be made.

16.4 ELECTRON MICROSCOPIC OBSERVATIONS

Electron microscopy is the only technique at present that enables observation of the size, shape, and arrangement of individual grains and features which are so small that they are beyond the resolving power of a light microscope. Description and classification of structure at the levels of plasmic and basic structure have been based, in the preceding chapters, on interpretation of the extinction characteristics of the plasma and s-matrix in thin section under crossed nicols. Although many such interpretations are soundly based on a knowledge of the effects of size, shape, and arrangement of grains that cannot be resolved by the light microscope, other interpretations are less reliable. Electron microscopy presents the opportunity of checking these interpretations. For example, Fig. 121 shows an electron micrograph of a slickensided surface obtained by the double replica method from the same soil material that had omnisepic fabric (Fig. 109). The high degree of preferred orientation of the individual plasma (clay mineral) grains is apparent in the electron micrograph, thus confirming the interpretation of the extinction phenomena. This level of observation has been virtually neglected.

PART V

THE ROLE OF FABRIC AND MINERAL ANALYSIS

17

The Role of Fabric and Mineral Analysis

The basis of pedographic studies is identification of the constituents and description of the size, shape, and arrangement of those constituents. It is logical to assume that the physical and chemical nature and behavior of soil materials depend on these characteristics and that the characteristics themselves result from modification of the characteristics of the parent rock, or parent material, by soil-forming processes. It is logical, therefore, to assume that considerable evidence of the kind and severity of soil-forming processes can be derived from comparison of the pedography of the parent material (or the petrography of the parent rock) and the pedography of the soil material. Integration of such studies on the soil materials which go to make up a soil profile provide an integrated history of development of the soil profile.

All this is not to say that pedography is an end in itself which will provide the complete solution to all problems concerning the nature, behavior and genesis of soil materials or soil profiles. Of themselves fabric and mineral analyses provide data concerning the identity and arrangement of the various constituents within soil materials and soil profiles. This information solves no problems unless the properties of the constituents, their behavior under various conditions, and the conditions and processes involved in producing the various kinds of arrangements of the constituents are known. It is possible, however, to identify all the constituents of soil materials and to study their properties and reactions. With this information and the data of fabric analysis concerning the distribution of these constituents in soil materials, it should be possible to predict the behavior of soil materials under the range of conditions for which the properties of the constituents, and their interactions, are known. Similarly fabric analysis of the parent rock, or parent material, and of the soil materials in the profile show the kind of reorganization that has been caused by the soil-forming processes. Knowing the properties of the constituents involved in this reorganization, the conditions and processes leading to the various changes in fabric from the parent rock to the soil material can be studied experimentally. As such data are obtained, fabric features can be interpreted directly in terms of soil-forming processes, that is, soil genesis.

The role of fabric and mineral analysis, therefore, is an intermediary one: identification of the constituents in soil materials provides the chemist and physicist with the materials which need to be studied if the nature and genesis of soil materials are to be understood, and description of the details of the arrangements of these constituents in soil materials, as compared with the arrangement in the parent material or parent rock, provides the experimentalist with a starting point for laboratory studies of the soil-forming processes. The experimentalist's problem is to explain the processes and conditions under which particular constituents have become arranged in particular ways. When the physicist, chemist, and experimentalist have made their contribution, pedography has a further contribution to make in attempting to predict the behavior of soil materials from their fabric and mineral analysis and to interpret processes of soil genesis directly from pedographic data.

Because of their intermediary role, fabric and mineral analyses hold implications for research workers in virtually every field of pedology, not only because of the positive data which they provide, but also because this data points directly to the avenues of research that are likely to be most rewarding in understanding the genesis and behavior of soil materials and soil profiles. The potential value of the techniques is illustrated by the following discussion under the headings: (1) description, (2) measurement of properties, and (3) classification. The discussion is an elaboration of a previous presentation (Brewer, 1960*a*).

17.1 DESCRIPTION

The basis for study of any naturally occurring material is description and classification according to description. For the results of measurements of properties or studies of genesis of the materials to have any breadth of application, it is necessary to be able to recognize materials which are similar to each other, those which are dissimilar, and to what extent and in what regard they are dissimilar. There is, therefore, an immediate need for consistency in concepts and descriptive methods. The alternative is that measurements and studies of genesis must be repeated on all samples of the materials whether they are identical, slightly dissimilar in a limited range of characteristics, or entirely dissimilar, and no correlations of measured properties or genesis with described characteristics are possible. Structure and fabric analyses provide descriptions of characteristics of soil materials which have been generally neglected by other descriptive systems; descriptions are incomplete, and can be inaccurate and misleading, when such characteristics are ignored.

Since the plasma is the mobile part of soil materials it can exhibit a wide variety of arrangements, and it is logical to assume that such differences in plasmic fabric as have been observed (Chapter 14) have considerable effects on the behavior of soil materials under different conditions. Many of the important characteristics of plasmic fabric are not observable in the field or in hand specimen; special techniques such as thin-section microscopy or electron microscopy are necessary to describe such features. This means that, without a complete structure and fabric analysis, it is probable that soil materials which differ significantly in plasmic fabric will be described and classified as the same kinds of materials; it cannot be expected that all measurements made on such materials will give consistent results, nor that correlations of measured properties with description and classification will be dependable.

The characteristics of basic structure that are additional to those of plasmic structure are also important. For example, description of a soil material as "porous" when it has a large number of visible voids, whether vughs, vesicles, or compound packing voids, and whether the vughs are interconnected or not, is insufficient. These three kinds of voids are likely to be correlated with differences in measured properties, such as permeability. Vesicles are interconnected only by very small voids, vughs may also be interconnected by very small voids or to different degrees by voids of similar size and compound packing voids are usually strongly interconnected. The three different kinds of voids are also indicative of different processes of formation (Chapter 9).

Similar considerations apply to pedological features, which vary widely in kind, composition, distribution within the soil material, and internal structure. Pedological features are most important as records of the kind and effectiveness of processes, and their nature and distribution patterns can profoundly affect the properties of soil materials under certain conditions. By far the greatest amount of data available concerns cutans or, more particularly, illuviation (depositional) argillans. The distinctions between varieties of illuviation cutans (location, composition, and internal fabric) and stress cutans are most important both for interpretation of the kinds and effectiveness of soil-forming processes and for predicting the behavior of soil materials by correlation of behavior with composition and structure. The amount of observational data available on other kinds of pedological features is not nearly so complete or useful; for example, glaebules, like cutans, vary widely in internal structure (Chapter 12), and yet they are commonly recorded as "nodules" or "concretions" irrespective of their internal structure, and the details of their distribution pattern, adhesion, and distinctness are rarely recorded. Unless the details of pedographic descriptions are available, interpretations of genesis and correlations of behavior of soil materials under different conditions must be incomplete and suspect.

The importance of consistency and suitability of concepts is illustrated by the situation regarding peds. Recognition of peds by the two different criteria (Sleeman, 1963; U.S.D.A., 1951; Section 7.5:A) results in two very different descriptions of pedality of a soil material, and the units recognized as peds are quite different in many soil materials. Those recognized by Sleeman's method are completely enclosed by recognizable surfaces of weakness, while those recognized by the U.S.D.A.'s method may have a number of faces that are natural surfaces and a number that are fracture surfaces. Since the latter method assumes a *regular* pattern of "surfaces of weakness" in pedal soil materials, it inhibits study of the true distribution and nature of natural surfaces. The implications of Sleeman's technique, which is a factual description of the distribution of recognizable natural surfaces, are illustrated in Figs. 13–18. Recognition and description of the characteristics and distribution of natural surfaces are most important aspects of structure analysis of soil materials.

17.2 MEASUREMENT OF PROPERTIES

Mineral analysis and particularly fabric analysis have focused attention on the highly organized and heterogeneous nature of soil materials, which are the significant features in interpretation of the data of measurement of properties. In the following discussion a number of examples have been taken as illustrations of the significance of the data of mineral and fabric analyses in each of the major disciplines concerned with measurement of properties of soils; the discussion is by no means exhaustive.

A. Soil Chemistry

It is common practice to reduce a soil material to a homogeneous mass by grinding and sieving prior to chemical analysis. The analytical results, therefore, are an average for the whole soil material (unless, indeed, some of it is discarded or lost preferentially during grinding and sieving) without reference to the form in which the constituents occur or to their distribution within the soil material.

The deficiencies in such methods are apparent when the fabrics of soil materials are considered. Many soil materials consist of an s-matrix of skeleton grains and clay minerals occurring as quite large domains, which are unoriented with regard to each other but within which the clay mineral grains are strongly oriented with regard to each other (argillasepic fabric; Fig. 99). Within this s-matrix there may be relatively large vughs and channels filled to varying degrees with chalcedony, strongly oriented clay minerals, manganese oxides, sesquioxides, calcite, etc., or combinations of these constitutents (Figs. 46–59, 85–87, and 90–94). Even though such constituents can be estimated quantitatively by chemical analysis, their occurrence as plasma concentrations (cutans and crystallaria), rather than as finely disseminated material in the s-matrix, is an important characteristic of the soil material. Disseminated chalcedony causes complete silicification of the soil material with resultant loss of plasticity and swelling properties, whereas chalcedans may have little effect on the bulk of the soil material. Disseminated calcite causes a generally high pH throughout the soil material, whereas calcitans cause high pH only in the vicinity of the voids. Both calcitans and calcareous nodules can occur in soil materials that are generally somewhat acid. Nor can the usual chemical analyses detect any differences in properties or constitution between clay minerals in the s-matrix and those which form argillans in the same soil material; significant differences have been demonstrated to occur (Buol, 1960; Buol and Hole, 1959, 1961; Grossman, 1959). Similarly Delecour and Manil (1962) used micromorphology (fabric analysis) to aid interpretation of analytical data for humus. Soil materials with similar proportions of a particular constituent, therefore, may have quite different chemical characteristics because of the distribution of that constituent, and so analytical results cannot be interpreted fully in terms of the nature of the soil material except in conjunction with fabric analysis. The possibilities of the application of the electron microprobe in this regard are illustrated in Fig. 124 which shows the trace for iron content across a ferri-argillan.

This factor of distribution has implications in soil-plant relationships. If a soil material consists simply of single grains of the constituents packed in a random manner, it behaves as a porous medium in which most of the constituents are accessible to plant roots. Soil formation, however, causes reorganization of the constituents to form specific patterns of voids with concentration and organization of the constituents into patterns which diverge widely from the pattern of normal random packing. Growth of plant roots, therefore, is conditioned by the size, shape, and arrangement of the voids and the relative accessibility of the constituents which now occur in specific parts of the soil material and in specific fabric patterns. If it is accepted that plant roots tend to follow existing void patterns, then the most accessible parts of the soil material are those immediately surrounding such voids. Cutans, therefore, should be of major significance to plant growth. Such coatings, especially argillans, may form

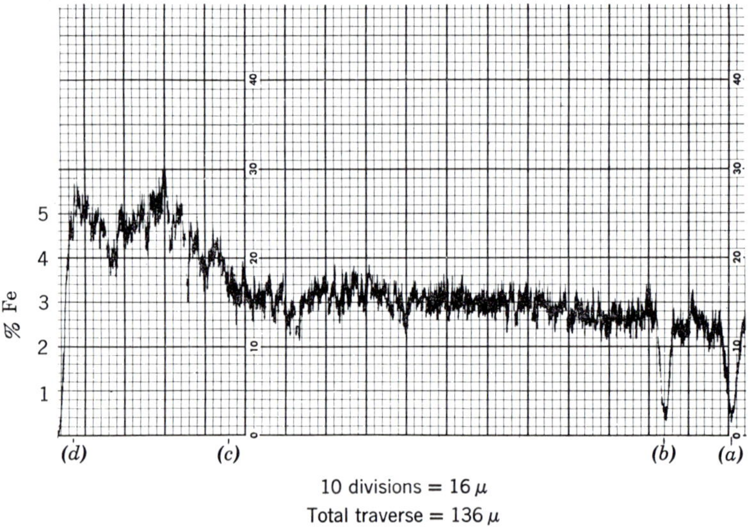

Fig. 124. Electron microprobe trace for iron in a skew-plane ferri-argillan. (*a*) A crack marking the boundary of the argillan with the s-matrix. (*b*) A secondary crack. (*c*) The boundary between the yellow argillan and its outer red rim. (*d*) The boundary of the argillan with the void. *Courtesy Dr. J. Lovering, School of Advanced Studies, Australian National University.*

a colloid membrane on practically all the voids so that the root environment is controlled by the characteristics of the cutans. Several such cases have been observed where all the present plant roots occurred in voids (channels and planes) which had quite thick argillans (Figs. 19 and 39).

Recently a series of pot trials were conducted to assess the effect of illuviation argillans on the growth of wheat and its uptake of potassium from the soil (Soileau, et al., 1964). Aggregates were prepared from New York Green Shale described as containing 85 percent illite. A proportion of these were coated with synthetic iron kaolinite to produce an analogue of strongly oriented illuviation cutans of thickness varying from 75 to 200 μ. The growth of wheat and its uptake of potassium were studied on these two materials in a replicated experiment. The data showed that the synthetic iron-kaolinite cutans, in fact, did protect the illitic aggregates from the roots; both growth rate and uptake of potassium were depressed in the pots containing the coated aggregates compared with those containing uncoated aggregates. It is interesting, too, to speculate on the reaction of plant roots which may try to penetrate along voids with cutans of manganese oxides, such as the mangans shown in Figs. 47 and 57.

It is pertinent to ask, therefore, whether the chemical properties of the clay minerals in either argillans or s-matrixes have any real relationship to the data obtained from the usual chemical analyses when such data are an average for a ground sample in which even the degree of orientation of the clay minerals may have been altered. The basic problem is to determine not only the absolute amounts of the various constituents in a soil material but also the effect of their mode of occurrence on the properties of the soil material.

B. Clay Mineralogy

Fabric analysis has also pointed to the need for identification of the mineral constituents which become concentrated in particular parts of soil materials. The problem is similar to that discussed for soil chemistry and for soil description. The coarser mineral grains usually can be identified in thin section under a petrological microscope, and their distribution can be studied by this technique. It is quite difficult to identify clay minerals in thin sections, however, except in the special circumstances where they occur as concentrations of relatively pure minerals; they are usually intimately intermixed. Normal methods of clay mineral analysis do not provide information concerning the distribution of mineral species in the soil material; they are usually either powder patterns obtained from a ground sample or patterns obtained from oriented films which are produced by preliminary dispersion and separation of the clay size fraction. Other techniques, therefore, must be used, and it is probable that the most useful results may be obtained by a combination of electron microscopy (Fig. 121) and analysis by the electron microprobe (Fig. 124). The former is a means of obtaining a fabric analysis at very high magnifications, while the latter may aid in identification of mineral species *in situ* by analysis of some of the elements in their crystal lattices.

Of particular interest is the identification of fractions of the plasma concentrated in the various pedological features; these are usually clay size mineral grains. That relatively pure concentrations do occur is demonstrated by the occurrence of layers of palygorskite and kaolinite (Beattie and Haldane, 1958) as compound plane cutans in a soil material from near Wagga Wagga in New South Wales (Fig. 49). This is an obvious example which can be noted in the field and in which the clay minerals were identified by peeling off the coatings and subjecting them to X-ray analysis; even optical methods in thin section are sufficient to identify the layers of palygorskite and note the occurrence of the underlying layers of a different clay mineral, which, however, could not be identified positively by this technique. Numerous other pedological features can be separated from soil materials, and so their constituents can be identified positively. Separation is not always so easy, however, and special techniques need to be developed to identify the constituents of such features to discover, for example, whether the mineralogy of certain argillans is the same as, or different from, the mineralogy of the plasma of the matrix.

Obvious questions which may be posed, therefore, and which may be answered by such techniques, are:

(1) Does fractionation in grain size and/or kind of material occur during the formation of pedological features?

(2) Are existing constituents simply concentrated, or are new mineral species produced?

(3) Do particular constituents tend to concentrate together under specific conditions?

C. Soil Physics

The organization within soil materials, as revealed by fabric analysis, is of considerable importance in interpretation of the data of physical analyses and as a guide to the kinds of analyses and research that are likely to be most rewarding in the study of the physical properties of soil materials. Winterkorn (1948) and Lafeber (1962) pointed out the importance of structure analysis in soil mechanics and engineering. The following have been selected from innumerable possible examples which illustrate these points.

(*a*) ***Voids.*** It is apparent that the size, shape, and arrangement of voids control the rate of movement of water through a soil material. The deficiencies inherent in the measurement of "pore size distribution" by tension and pressure methods due to the shape, arrangement, and degree of interconnection between the voids have been discussed to some extent (Section 9.2:A). Such measurements, too, do not take cognizance of the spatial orientation of voids, which may confer a high degree of physical anisotropy on the soil material. This may be of considerable importance in soil genesis and to plant growth since it may control, to a large degree, the rate of movement of water through the soil material and the soil profile. This in turn affects the degree of leaching of the profile and the availability of water to plant roots. To take a simple example consider the effects of two sets of parallel planar voids (joint planes), one set being horizontal and the other vertical. The different effects of these two sets of voids are obvious: the horizontal set is ineffective in conducting water through the soil material, the water thus being available for plant use and causing little or no leaching, while the vertical set conducts the bulk of the water out of the soil material quickly, thus causing considerable leaching, at least in the immediate vicinity of the voids, and allowing less water to be available for plant use between periods of rainfall. O'Neal (1949) considered some aspects of this problem. Similarly interconnected vughs are good conductors, while vughs connected only through much smaller voids are poor conductors.

A second aspect of the effect of voids on water movement and leaching is due to the physical nature of their walls. The degree and rate of wetting of a soil material after rain depend on the rate of vertical removal of water by "conducting" voids and the nature of the walls of the voids; that is, water moving down conducting voids penetrates the soil material forming the walls of those voids to a greater or lesser degree depending on the time of contact and the physicochemical nature of the walls.

These brief considerations suggest that the study of soil materials by flow methods may yield useful data and may be used to characterize soil materials, particularly if they are conducted on oriented samples in such a way as to show up any physical anisotropy. Even more important and fundamental is the need to study the physicochemical properties of natural surfaces, especially the walls of voids. Winterkorn (1948) pointed out that the physicochemical behavior of surfaces in soil materials cannot be predicted accurately by a summation of the behavior of the surfaces of the individual constituents because they are "functions of the constitution as well as the composition." He makes it quite clear that by "constitution" he means structure or, more specifically, fabric as defined in this work.

(*b*) ***Cutans.*** Cutans are essentially coatings on, or modifications in fabric at, natural surfaces in soil materials (Chapter 10). It is apparent, therefore, that they can effectively increase the physical anisotropy of soil materials, which is originally due to the size, shape, and arrangement of the voids within it. Differences in physicochemical properties between cutans and the normal soil material cause the properties of surfaces so affected to differ markedly from unaffected surfaces. The properties of such surfaces also depend on the nature and fabric of the cutans. It is important, therefore, to study the physicochemical properties of such surfaces, which vary from coatings of manganese oxides, iron oxides and hydroxides, calcite, and clay minerals of various types with different degrees of preferred orientation, to stress cutans with preferred orientation of the plasma grains at the surface associated with some degree of preferred orientation parallel to the surface for varying distances into the plasma of the s-matrix (neostrians).

Cutans, especially stress cutans and illuviation argillans, may be developed so extensively in some soil materials that certain of the properties, such as the movement of water from "conducting" voids into the soil mass, may be controlled by the properties of the cutans. Other physical properties, too, may be significantly affected; for example, breakdown of soil materials when wetted suddenly is very rapid where strongly oriented argillans are abundant (Brewer and Blackmore, 1956), free grain argillans, such as those in the *B* horizon of Tenningerie sand (Brewer, 1956; Fig. 44), significantly reduce permeability of coarse-grained soil materials, and cutans of other materials, especially of chalcedony and other forms of secondary silica, can act as very stable cementing materials, thus increasing hardness and stability of the soil material and decreasing its apparent plastic properties and permeability.

(*c*) ***Basic Fabric.*** The domain theory proposed that clay mineral grains in the matrix of soil materials occur and behave as aggregates, within which the individual grains have a high degree of preferred orientation, rather than as individuals (Aylmore and Quirk, 1959; Emerson, 1959). This suggests a correlation between the properties of soil materials and plasmic fabric or the arrangement of the individual grains within the plasma. Extinction phenomena in the plasma have been interpreted in terms of the degree of preferred orientation of the individual grains within recognizable aggregates and the degree of orientation of these aggregates with regard to each other (Section 8.2). It seems that the aggregates can be equated to domains exhibiting the attributes of size, shape, and kinds and degree of orientation with regard to each other. It is reasonable to assume that soil materials with domains that are different in size, shape and arrangement, and degree of preferred orientation of the individual grains within the domains, will exhibit different physical properties under various conditions—for example, the rate, amount, and mechanism of swelling on wetting. Other soil materials do not have recognizable domains (undulic fabric), and their physical behavior must surely differ widely from those with equant, unoriented domains (asepic fabrics), and both of these will differ from those with prolate or acicular, moderately to strongly oriented domains (the various sepic fabrics).

Other characteristics of the basic fabric can also be correlated with the physical behavior of soil materials. The effect of the arrangement of voids has been discussed, but other characteristics may be just as important. For example, two soil materials of approximately the same particle size distribution may differ widely in permeability simply because of differences in packing of the individual skeleton grains. Figure 125 is a fragipan from Queensland in which the size distribution, shape, and mode of packing of the grains all play an important role in producing a highly impermeable soil material, which is very hard and brittle when dry but quite soft when wet. It consists of a wide size range of angular grains down to silt size; there is very little clay size material, and no cementing material is visible in thin section. The physical properties of this soil material are explained adequately by the close, interlocked packing of the skeleton grains. Lafeber (1962) recognized the probable importance of the distribution of voids and skeleton grains in the s-matrix as a factor controlling the mechanical behavior of soil materials, and others (e.g., Knox, 1957; Grossman, Stephen, Fehrenbacker, and Beavers, 1959) have attempted to relate mechanical strength to the details of basic structure; Romashkevich (1962) related microstructure to erodibility.

Fig. 125. Close packing of skeleton grains in a fragipan. Thin section under crossed polarizers. Plasma is virtually absent. Void space is very low due to the system of packing and the size range of skeleton grains. 30×.

17.3 SOIL GENESIS

The genesis of soil materials and soil profiles consists fundamentally of weathering of igneous and sedimentary bodies and reorganization of the constituents by the soil-forming processes, which are essentially such physical and chemical processes as solution, leaching, diffusion, precipitation, shrinking and swelling, pedoturbation, etc. Fabric and mineral analyses are important techniques in the study of such processes, and they can be applied most effectively in the following three roles.

A. *Qualitative Study of Weathering*

Although the weathering products of minerals are often too fine grained or too intimately mixed to identify microscopically, microscopic techniques can identify progressive changes in mineral grains during weathering. In this way particular weathered grains can be isolated for further study by other techniques. A number of such studies have been quoted as examples of this kind of approach (Section 5.3). This phase of study would be aided greatly by electron microscopy and the electron microprobe. The weathering products of primary mineral species, at least in the early stages of weathering, are often formed within the original grains. Electron microscopy, then, would permit observation of the size, shape, and arrangement of the individual grains of secondary minerals, while the electron microprobe would enable a quantitative estimation of the change in chemical composition to be made at least for the most important elements. Figures 126–129 illustrate the kinds of observations possible in thin section on the petrological microscope. Figure 126 shows a felspar grain weathering to kaolinite in a yellow podzolic soil formed on granodiorite. Figure 127 shows a weakly weathered but highly exfoliated biotite grain from an alpine humus soil formed on granodiorite; Fig. 128 shows weathered grains of biotite, hornblende, and

Fig. 126. Weathered felspar (*F*) and hornblende (*H*) grains in the deeper horizons of a yellow podzolic soil profile formed on granodiorite. Thin section under crossed polarizers. 30×.

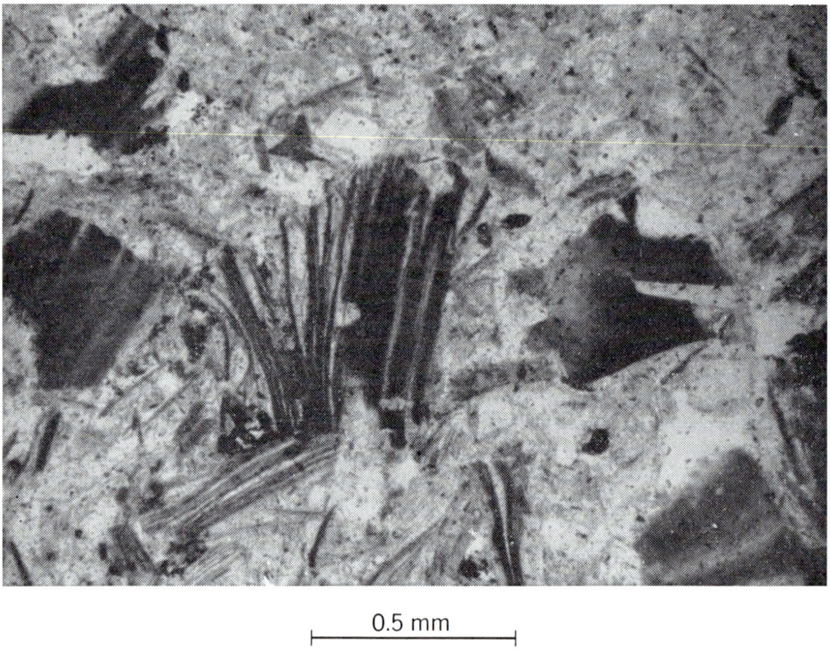

Fig. 127. Exfoliated, weakly weathered biotite in an alpine humus soil profile formed on granodiorite. Thin section in plain light. 55×.

felspar; and Fig. 129 shows a highly weathered, exfoliated biotite grain. Such grains can be isolated from the soil materials and studied by other techniques, or they could be studied *in situ* with the electron microscope and electron microprobe. A good deal of detailed microscopic work of this kind has been done on the laterites (e.g., Sivarajasingham, Alexander, Cady, and Cline, 1962; Alexander and Cady, 1962).

B. Calculations of Soil Formation

The principles and efficacy of quantitative calculations of soil formation have been discussed in some detail (Chapter 4). The calculations are usually based on estimations of the amounts of a resistant mineral in each horizon and in the parent material. From these, changes can be calculated in weight and thickness or amounts of any of the constituents for each horizon or for whole profiles. It is essential for these calculations to prove first that the parent material was uniform throughout the profile and that the parent material or parent rock is preserved.

The efficacy of this method, and the importance of the results, can be illustrated by considering the results obtained by Haseman and Marshall (1945) and Brewer (1955*b*). The two profiles studied had a similar absolute degree of particle size differentiation but markedly different morphology otherwise. The former workers studied a prairie soil and the latter a planosolic yellow podzolic; the prairie soil had amounts of clay size material varying from approximately 16 percent in the surface to a maximum of 41 percent at 22 to 32 in., while the yellow podzolic soil had 13.5 percent of particles less than 5 μ diameter in the A_2 horizon and 43 percent in the B_{21} horizon.

The calculations for the prairie soil showed that the profile had a net increase in volume and weight over the amount of parent material from which it was formed and that the *B* horizon in particular had become grossly swollen; this large increase in the *B* horizon could not

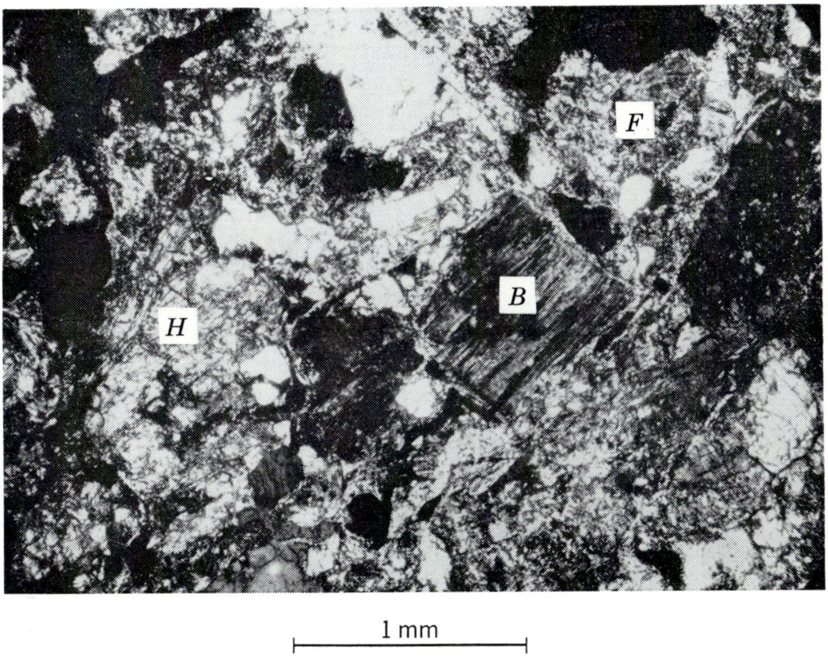

Fig. 128. Weathered grains of biotite (*B*), hornblende (*H*), and felspar (*F*) in a weakly weathered zone over granodiorite. Thin section under crossed polarizers. 30×.

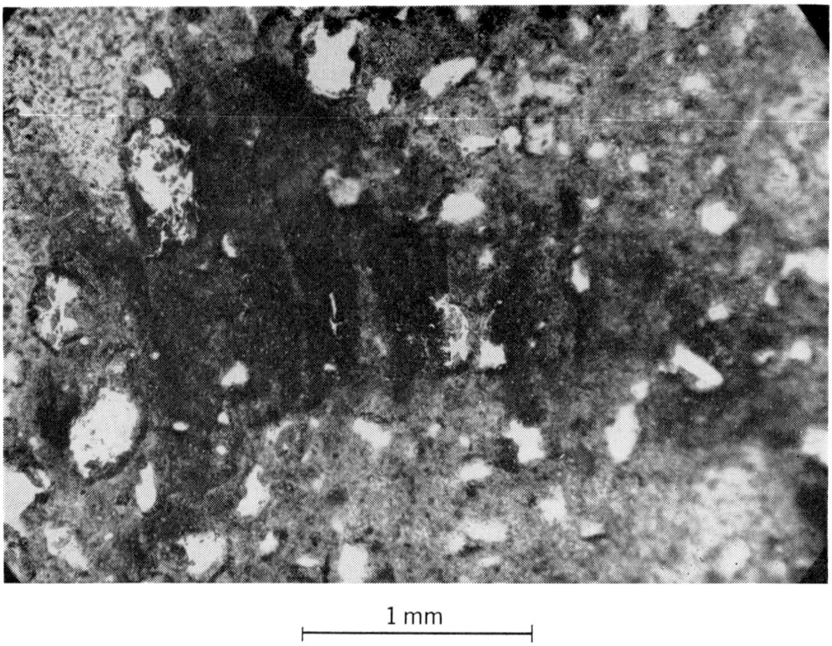

Fig. 129. Exfoliated, highly weathered biotite in the upper *B* horizons of a yellow podzolic soil profile formed on granodiorite. Thin section in plain light. 30×.

be accounted for entirely by deposition of illuviated clay size material in the cracks and interstices. The authors suggested that the increase in volume and weight was facilitated by the primary minerals weathering to beidellite-montmorillonite type clay minerals with little or no loss of constituents from the profile. Heavy mineral analyses indicated that mineral breakdown was vigorous in the 0 to 22 in. layers, fell off with increasing depth to 42 in. and was negligible below that.

In contrast to these results the calculations for the yellow podzolic soil showed that all horizons had decreased in weight and volume compared with the amount of parent material from which they were formed and that the amount of clay size material even in the B_{21} horizon could be explained by the amount of weathering of primary minerals without recourse to illuviation; lack of illuviation was supported by fabric studies of thin sections. The loss of weight and volume in this profile is explained in part by the considerable loss of weight in weathering of felspar, hornblende, and biotite (which were the principal weatherable primary minerals) to form kaolinite, which comprises a high proportion of the clay size fraction. On the basis of these calculations the theses were advanced that the *A* horizons were a younger deposit of weathered granite and that the particle size differentiation was partly due to vigorous weathering in the *A* horizons with loss of constituents in solution by lateral leaching.

C. Structure and Fabric Analysis

The kinds of soil-forming processes that have been operative can be inferred directly from structure and fabric analysis if sufficient background data are available. The differences between the structure of the soil material of a particular soil horizon and that of the parent material are a reflection of the effect of these processes, and it is a matter of interpreting these observed changes in terms of the processes which caused them. A number of hypotheses have been put forward concerning the processes involved in formation of the various kinds of plasmic fabrics, voids, peds, and pedological features (Part IV). The hypotheses are based on what appear to be logical inferences derived from first principles based on the known properties of the constituents involved and the broad processes for which there is good evidence, such as movement of solutions and suspensions through soil profiles, diffusion, shrinking and swelling with wetting and drying, etc. The bases for many of these hypotheses are unsure in that there is a large gap in knowledge of an observational kind, of the properties of the constituents in soil materials, and of the conditions and nature of soil-forming processes. The present role of structure and fabric analysis, therefore, is to propose possible hypotheses on logical grounds by interpretation of the phenomena observed, but, more particularly, it is to emphasize the kinds of phenomena which do occur so that the various conditions and processes that might lead to formation of the phenomena may be studied.

In recent years the number of such studies of particular kinds of profiles and particular soil-forming processes has increased rapidly. By far the greatest amount of interpretive work on fabric analysis has been concerned with recognition of translocation of constituents through the soil profile with subsequent deposition, that is, illuviation (e.g., Minashina, 1958; Grossman, Stephen, Fahrenbacher, and Beavers, 1959; Antipov-Karataev, and Tsyurupa, 1961; Gorbunov, 1961). A number of workers have attempted to correlate variations in fabrics which suggest illuviation with the processes of leaching ("lessivage") and podzolization. Fridland (1958) attempted to distinguish podozolization and "illimerization" (illuviation of clay minerals in suspension), Karpachevskii (1960) described the micromorphology (fabric) produced by leaching and podzolization under forest, and Manil (1962) defined "lessive" and attempted to place "sol lessives" in relation to podzolic soils, sols bruns lessives, and pseudogleys according to micromorphological expressions of the process. Similar studies of other soil groups based on interpretation of fabric have been published. Kubiena (1956) proposed two different processes for formation of red earths and laterites based on interpretation of fabric analyses, particularly the nature and distribution of the sesquioxides visible microscopically; Blumel, Janik, and Schiller (1959) interpreted the formation of oligotrophic braunerde, braunerde, sol lessive, pseudogley, and flood-plain soils from fabric data; Bourne and Whiteside (1962) made a similar study of a medial chernozem developed on loess; Alexander and Cady (1962) described the processes involved in genesis and hardening of laterite; and Gracanin (1926) used micromorphological data to interpret the genesis of a hillside peat. Gennart et al. (1961) interpreted the origin of the humus in the surfaces of some alluvial soils as coprogenic from its micromorphology.

Such hypotheses concerning the genesis of soil profiles and the processes of formation of particular fabric features provide a basis for the experimental work which should be done to reproduce the phenomena observed in soil materials. They also provide the means of checking the results of such experiments as to whether, indeed, the fabric features analogous to those occurring in soil materials have been produced by the experimentation. There are two possible phases to such experiments. The first is simply to attempt to reproduce fabric features according to the most probable hypotheses for their formation. The second is a more fundamental study of the details of the mechanisms of formation of the features. The production of illuviation cutans by movement of clay size material in suspension and deposition on drying (Brewer and Haldane, 1957) is an example of the first stage. More recently Hunter (1961) conducted preliminary experiments designed to elucidate the details of the mechanisms involved in deposition of clay size kaolinite from dilute suspensions on to the surfaces of sand size quartz grains; this is an example of the second phase of experimentation.

Quantitative fabric analyses provide evidence of the effectiveness of the various soil-forming processes in producing certain kinds of profiles, and they may provide evidence that certain features of the profile were formed by other processes. For example, Sleeman (in press) demonstrated the occurrence of a number of superimposed sedimentary layers in some soils from the Riverine plain of New South Wales by studying the quantitative distribution with depth of a number of pedological features, including papules, plasma separations, and cutans of various kinds. This work also demonstrated that, after a number of sedimentary layers had been laid down, a period of stability ensued during which soil formation occurred to a depth which included several of the layers. Subsequently, further sedimentary layers were deposited on this old soil profile, and soil formation again ensued. A similar study on the fragipan soils of Illinois indicated sedimentary layering even though mineralogy did not suggest lithological discontinuity (Grossman, Stephen, Fehrenbacher, and Beavers, 1959).

Another result of such quantitative fabric analyses has been a modification of the generally accepted hypothesis that the horizon with maximum clay size particles owes its particle size distribution to illuviation of clay size material. There is now considerable evidence that the greatest concentration of plasma that can be identified as illuviation cutans often occurs below the depth of the horizon of maximum clay size particles. In one instance the horizon with maximum clay size particles had only 5 percent illuviated material (at a depth of approximately 20 in.) while the lower horizons (at 72 in.) had approximately 30 percent illuviated material. In some differentiated profiles there is no evidence of illuviation (Brewer, 1955*b*), in others the proportion of illuviation cutans is insignificant compared with the total amount of clay size material, while in others again the horizon of maximum clay size material does coincide with maximum illuviation cutans.

17.4 SOIL CLASSIFICATION

The role of structure analysis in soil classification is obvious since such analyses are an integral part of description and they form a sound basis for the study of soil genesis. "Structure" in the old sense (U.S.D.A., 1951), that is, pedality, has been used as a characteristic for classification of soil profiles for some considerable time, as has a very broad description of voids. More recently (U.S.D.A., 1960) the occurrence of "clay coatings," or argillans, has been used to subdivide specific groups of soil profiles. The use of other features of fabric, such as plasmic fabric, in morphological classification systems is simply a logical extension of this trend. Present indications are that some groups of soil profiles embrace a number of quite distinct fabric features which indicate fundamental differences in composition and/or soil-forming processes. The use of such characteristics in classification, of course, involves a good deal more detailed observation than is made at present, but the fact that one characteristic is more difficult to observe than another is no reason why it should not be used for classification if it is considered to be of importance. This is parallel to the situation in petrology where a rock specimen often cannot be classified with certainty, or to a reasonably low level, without examination in thin section under the petrological microscope.

The use of structure and fabric analyses in soil classification is still in its infancy, and there are insufficient data as yet to evaluate the various proposed soil classification systems in terms of differences in structure and fabric between classes. It is becoming common practice, however, to include micromorphological descriptions as part of profile descriptions, and a number of studies of particular soil groups have been made (Johnston and Peterson, 1941; Laruelle, 1956, 1958; Minashina, 1958; Blumel, Janik, and Schiller, 1959; Marozova, 1962; Altemuller, 1962*a*; Sivarajasingham, Alexander, Cady, and Cline, 1962; Kubiena, 1962). In general these descriptions have been used simply as additional descriptive data of existing soil groups without thought of testing the homogeneity of the groups, except that Kubiena (1962) proposed fabric features of the kind and distribution of iron oxides as diagnostic criteria to distinguish braunlehm, pseudogley, rotlehm, roterde, and laterite.

Addition of all the characteristics of structure and mineral analyses to those used for classification will create a problem as to which are of greatest importance, that is, which should be used, how broad the groupings should be, and in what order the different characteristics should be used. Such difficulties should not stand in the way of attempting to use these characteristics and to evaluate their order of importance. A first attempt should be made at the level of soil materials rather than soil profiles. From present knowledge it would seem that the most rewarding results would be obtained by classifying soil materials at the highest level on their plasmic fabric. Subdivision could then be made in order as follows:

(1) Plasmic fabric.
(2) Intrapedal voids and skeleton grains (basic fabric).
(3) Pedality.
(4) Cutans.
(5) Other orthic pedological features.
(6) Inherited pedological features.

This order is suggested because (1) gives a description of the organization of the general mass of mobile constituents subject to reorganization, (1) and (2) describe the basic material (s-matrix) of the soil material, (3) describes the units into which this material is broken by soil-forming forces, (4) gives evidence of illuviation and general movement of the constituents in solution and suspension, (4) and (5) together describe the concentration and reorganization of the plasma *in situ*, and (6) describes accidental features.

The progression to classification of soil profiles depends on an integration of the classification of the soil materials that form the profiles. Those profiles would be classified together which had the same groups of soil materials arranged in the same order.

17.5 SUMMARY

It is apparent that much of the foregoing discussion of the uses and role of fabric and mineral analysis, as defined within the scope of pedography, is rather conjectural and inferential. This is because such studies as have been suggested are in their infancy, and the amount of data available are very limited indeed. In particular, the significance of fabric and structure analyses as applied to the properties and behavior of soil materials has been almost completely neglected, both regarding direct correlation of fabric and structure with properties of soil materials and interpretation of structure analysis as a means of explaining the properties and behavior of soil materials. The former correlations of fabric with soil properties should make it possible to predict the behavior of soil materials under various conditions from fabric analysis. The latter (interpretation of structure analyses) should lead to a better understanding of the mechanisms involved in the behavior of soil materials. For either of these aspects to have a useful application, it is necessary to accumulate a large amount of background data, much of which is not available at present. As an example of what might be achieved and of the approach envisaged, consider a hypothetical soil material consisting essentially of skeleton grains and montmorillonite. A good deal of work has been done on the swelling of montmorillonite under a considerable range of conditions. It would be expected, therefore, that the behavior of such a soil material during wetting and drying cycles could be predicted. Its behavior, however, will probably vary according to the arrangement of the individual montmorillonite particles, that is, according to the plasmic fabric, and according to the nature of the exchangeable cations. To predict the behavior of such a material during wetting and drying it is neessary to know the mineralogical nature of the material (mineral analysis), the chemical state of the active constituent, and the plasmic fabric. The role of fabric analysis in predicting and understanding the properties and behavior of soil materials, therefore, is not an independent one; it depends on basic knowledge of the nature and condition of the constituents.

Similarly interpretation of structure analysis in terms of the processes and conditions of formation of soil materials and soil profiles depends on identification of the constituents, their physical and chemical conditions in the soil material, and on a basic knowledge of the properties of those constituents under various sets of physical and chemical conditions. On the other hand mineral analysis alone can provide considerable data concerning the genesis of soil materials and soil profiles, but only as a factual measurement of the amount and kind of weathering that has taken place. Mineral analysis is the most dependable method of proving uniformity of parent material, and calculations of soil development based on estimation of a stable constituent are independent of other measurements to some degree. Such calculations, however, provide only limited understanding of the processes of soil formation as a whole. Although factual descriptions and measurements can be made their full significance cannot be assessed and understood without additional physical and chemical data, and, most particularly, without fundamental studies on the properties, reactions, and interactions of the constituents of the soil materials under a wide range of physical and chemical conditions.

It is apparent that fabric and mineral analyses are powerful tools for studying the genesis of soil profiles. It is reasonable, too, to assume that the physical and chemical characteristics of soil materials are affected by the nature and arrangement of their constituents, that is, their mineralogical composition and fabric. It is also reasonable to assume that soil-plant relationships are influenced by the material which the roots can contact, and this again depends on mineralogy and fabric. Finally, mineralogy and, particularly, fabric are part of description, and so are important in classification. Therefore, just as mineral and fabric analysis cannot be used to their fullest advantage without considerable knowledge in other fields of study of the characteristics and properties of soil materials and their constituents, the full significance of any of these other studies cannot be realized without the data from fabric and mineral analyses.

APPENDIX

Techniques for Fabric and Mineral Analysis of Soil Materials

Pedographic study of soil materials involves a number of preparations and treatments, most of which have been described. These techniques, however, are spread rather widely through the literature. For this reason, and because some modifications of the basic techniques are desirable for the study of soils, the more important techniques are set out here in some detail. They can be divided broadly into two groups: techniques for structure analysis and techniques for mineral analysis.

1. TECHNIQUES FOR STRUCTURE ANALYSIS

Structure analysis depends on examination and description of soil materials in the field, on hand specimens and in thin section, that is, on untreated soil materials and on mounted specimens. The following discussion deals only with laboratory specimens since description in the field requires no preparations or treatments other than the digging of a pit for observation.

A. Soil Monoliths

More specifically structure analysis depends on examination of undisturbed, oriented soil samples whose relationships to the soil profile must be known. The best method of achieving this is to sample the profile as a monolith which can be removed to the laboratory and studied under a stereomicroscope. Macroscopic characteristics can then be described, and undisturbed material can be selected whose orientation and relationship with the rest of the profile are known. Such monoliths provide an adequate volume of material from which to select representative samples of particular features for thin sectioning or other analyses; they permit study of transitional zones between horizons and eliminate most of the need for careful marking and orientation when sampling in the field.

The method described here is essentially a modification of that used by Kubiena (1953). Two sizes of boxes are used according to requirements. Both are 6 in. wide and 4 in. deep; one is 2 ft long and the other is 1 ft long (internal measurements). The boxes are made from 1-in. tongue-and-groove flooring boards and are open on one of the 2 ft $\times$ 6-in. sides; one of the 2 $\times$ 4-in. sides is hinged to allow ready access to the soil material in the laboratory (Fig. 130*b*).

The profile block is prepared in much the same way as described by Kubiena (1953). A vertical wall of the profile pit is smoothed down and marked out by placing the box against the wall, and then the soil column is cut to shape with wood chisels and knives. Initially the sides are pared back roughly to 2 in. beyond the depth of the box. Then the base of the column is treated in the same way, and finally the whole column is trimmed to fit the box snugly (Fig. 130*a*). A small try square is used to keep the cuts at right angles to the vertical face. Then the box is pushed firmly onto the soil column and held while a 4-in. auger hole is dug 3 to 4 in. behind the box. A crowbar is driven into the base of the auger hole, and the monolith in the box is gradually eased off the face of the pit by leverage on the bar. Alternatively the soil can be cut back into the auger hole to free the monolith completely; it can then be lifted out of the pit. This technique provides a monolith that fills the box and protrudes some inches beyond it. Prizing off the soil column by the crowbar usually gives natural faces and fractures over most of the exposed surface; a small amount of judicious picking gives a surface completely free of shearing. Any gaps between the monolith and the sides of the box are packed with cotton wool.

A wide range of soils under varying moisture conditions have been collected by this method. Although the task is easier and quicker if the soil is moist and stone-free, very dry soils with considerable amounts of soft stone or concretions have been collected successfully. The greatest difficulties are encountered with very dry, strongly pedal clay horizons and dry loose sands.

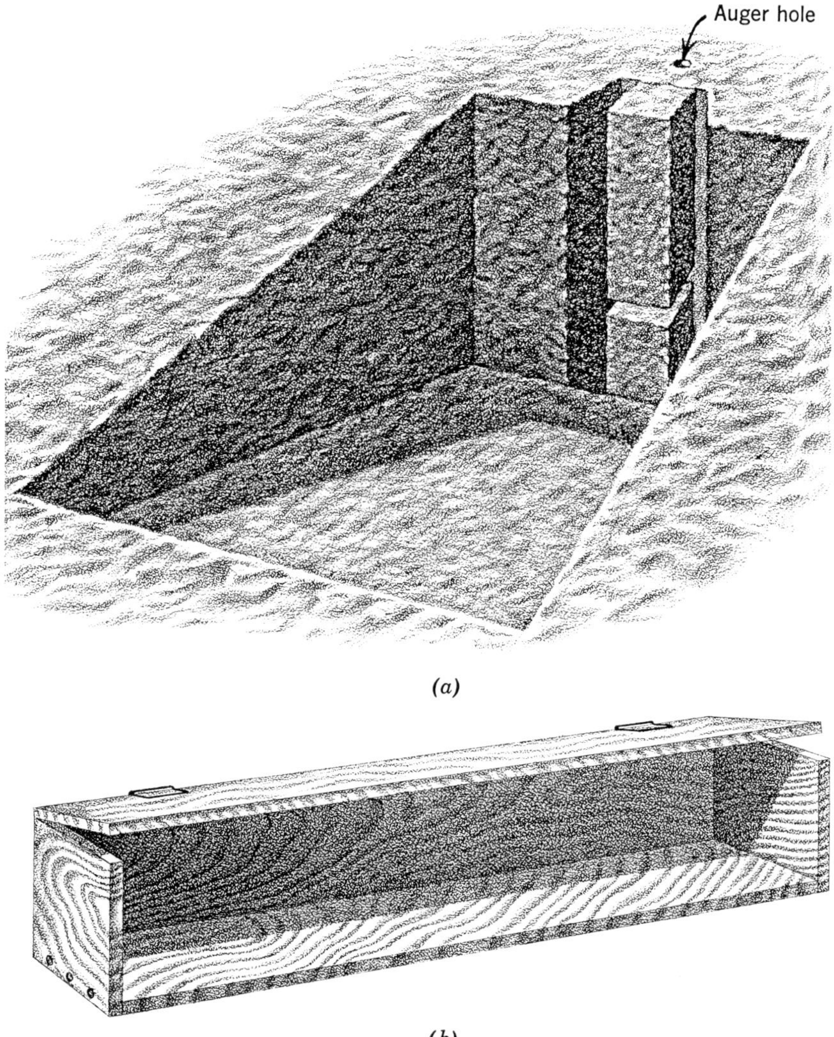

(*a*)

(*b*)

Fig. 130. Taking a soil monolith. (*a*) Pit showing the prepared monolith block. (*b*) A monolith box, internal measurements 24″ × 6″ × 4″.

In the laboratory the monolith boxes are marked in inches or millimeters down the side and across the top so that grid references can be given for features removed for more detailed study by thin sectioning, chemical analysis, or other appropriate techniques. These features are sampled with microtools (Kubiena, **1938**), the most useful of which are several very finely pointed forceps with variously shaped arms, which enable small pieces of soil material to be removed, and a series of fine probes. Variously shaped probes can be manufactured by sharpening stainless steel needles in different ways on a diamond or fine carborundum lap. Specimens removed in this way may include pedological features, particular mineral grains, and small blocks of soil material (about 1 $in.^3$) from representative parts of the monolith. These specimens are given a grid reference, marked for orientation, and described in detail, especially their relationships to the surrounding soil material and/or specific reference features.

B. Impregnation of Specimens

Most of the specimens obtained from soil materials are too friable to be sectioned or polished without stabilization by impregnation with some binding material. The ideal impregnating material should have as low a viscosity as possible under the conditions of impregnation, should set with a minimum of volume change, and should be nonpolar. The conditions of impregnation should not be so extreme as to risk any changes in optical properties or arrangement of the constituents of the soil material. When set, the impregnating material should be relatively hard and brittle at room temperature, colorless and isotropic in thin section, with a refractive index as close as possible to that of canada balsam, the classical mounting medium for thin sections of rocks. Within limits the refractive index of the impregnating material is the least important of these requirements since the objective usually is the study of fabric rather than mineral identification.

(*a*) *Impregnating Materials.* The following materials have been used successfully:

(*1*) *Natural Resins.* Canada balsam, dissolved in xylol, and kollolith (Kubiena, 1938) have been the most popular natural resins. Impregnation with these requires rather prolonged heating with the danger of boiling the resin and disturbing the specimen. This danger is reduced and impregnation is more effective if the procedure is carried out under vacuum. Cutting, grinding, and polishing can be troublesome when resins are used. Often water cannot be used as a lubricant and coolant because it may cause dispersion of the fine-grained parts of the soil material. On the other hand, many nonpolar-liquids, such as kerosene, cannot be used because they dissolve the resins. Consequently grinding often must be done dry, which is a time-consuming procedure requiring cooling of the grinding plate with an ice-salt mixture to prevent heating and flowing of the resin.

Against these disadvantages the resins generally have approximately the correct refractive index, and, more important, the specimens do not need to be thoroughly dried because the water will boil off during impregnation.

(*2*) *Plastics.* Use of the cold polymerizing plastics, such as the polystyrenes and epoxy resins, overcomes most of the difficulties encountered with the natural resins. Most of the methods now in use follow that described by Bourbeau and Berger (1947) for "Castolite," modified by using various plastics, various rates of polymerization as controlled by the proportion of catalyst added, and various diluents (e.g., styrene) to reduce viscosity (e.g., Buol and Fodness, 1961; Kawai, 1961; Pol'skii, 1962; Altemuller, 1962*b*; Wells, 1962; Reineck, 1962; Borchert, 1962). Where diluents are used, complete polymerization may require a long time, and care must be taken that the nature of the polymer formed is not adversely affected by the diluent. In general, plastics will not polymerize satisfactorily in the presence of moisture, and impregnation is always superior if done under vacuum. A vacuum equivalent to about 27 in. of mercury is adequate; this can be achieved with a water vacuum pump. High vacuums are more efficient but may be unsatisfactory because of boiling of the plastic or of the added catalyst and/or accelerator.

Impregnation with most plastics is improved by slowing the rate of setting by reducing the amount of hardener, or catalyst and accelerator, to lower concentrations than those recommended by the manufacturers. This offsets the increased rate of setting that usually results from impregnation under vacuum and reduces the risk of cracking. Initial setting should take at least 4 hr, and much longer periods are sometimes advantageous. The proportion of hardener, or catalyst and accelerator, necessary for this rate of setting must be determined for the particular plastic used. Setting too quickly, or wrong curing procedure, can produce strain and consequent birefringence effects under the polarization microscope.

Most plastics shrink slightly on setting, but they differ in their adhesion to various kinds of surfaces. Many do not adhere to glass, so ordinary glass tubes can be used as molds. These are relatively cheap and satisfactory even if they must be broken to remove the block of impregnated material. For plastics which adhere strongly to glass, molds can be made of any material provided they are coated with a release agent. Alternatively molds of rubber or polythene are satisfactory for most plastics, and these need not be destroyed to remove the impregnated block. The following plastics are used in this laboratory according to requirements:

*Plastrene 47** is a polyester type plastic similar to "Castolite" (Bourbeau and Berger, 1947) and is relatively cheap but somewhat viscous. It is suitable for normal, relatively permeable soil materials; the samples must be oven-dried or freeze-dried. The following mixture gives a suitable setting rate: 10 ml Plastrene 47, 2 drops of catalyst *G*, 1 drop of accelerator *H*.

Epirez No. 2659E† (*Epimount*) is an epoxy type plastic used instead of Plastrene 47 for specific purposes, such as impregnating samples for work with the electron probe for which the polyester types are too volatile. Like Plastrene 47, it is rather viscous, and the samples must be oven-dried or freeze-dried. The following mixture gives a suitable setting rate: 10 ml Epirez, 1 ml of hardener.

Durcupan‡ is an epoxy type plastic of low viscosity which is particularly useful for dense soil materials that are difficult to impregnate; it can also be used successfully on air-dried or slightly moist soil materials, especially if they are permeable. The manufacturers claim that it can be used for impregnating wet materials by replacement of water by one of the constituents that is miscible with water. The procedure is tedious, however, and the replacing constituent is somewhat polar and tends to make swelling soils decrepitate. Durcupan must be mixed according to the manufacturers specifications.

The impregnating plastics can be colored with a number of *pigments*. This is particularly useful for the study of voids. "Fluorescent green" (Werner, 1962) is very useful; the impregnating plastic is mixed with fluorescent green which causes the voids to shine under ultraviolet light.

(*3*) *Carbowax 6000*§ can be used for wet specimens (Mitchell 1956). It is hard at room temperature but melts at 55°C and is easily soluble in water in all proportions. The specimen is immersed in melted carbowax at 60°C and held at that temperature for three days in an oven. The sample is then removed from the oven and cooled. Specimens impregnated in this way are not very hard (about the hardness of talc). The greatest disadvantage, however, is that carbowax is anisotropic with low birefringence; under crossed nicols it has the appearance of clusters of small crystals.

(*4*) *Latex Sols* (Hunter and Brewer, in press) can stabilize wet soils

* Polymer Corporation Pty Ltd, Chemical Manufacturers, Homebush, N.S.W., Australia.

† Indelab Pty Ltd, Liverpool, N.S.W. Australia.

‡ Fluka A. G., Chemische Fabrik, Buchs. S.G., Switzerland.

§ Carbide and Carbon Chemical Company.

by allowing them to take up latex from a latex sol by capillarity and diffusion. The sample is placed on a grid (wire mesh) with its base in contact with the sol, left for several days, and then dried. Such stabilized specimens require subsequent impregnation with one of the plastics.

(*b*) ***Standard Impregnation Procedure.*** Satisfactory impregnation of most specimens from soil materials can be achieved by the following procedure. The apparatus shown in Fig. 131 is simple and effective.

(1) Place the sample in a mold and dry it in an oven at 105°C for 12 hr.*

(2) Place the sample and mold in a dessicator (silica gel) until lukewarm (approximately 1 hr).

* If Durcupan is to be used thoroughly air-dried samples are satisfactory.

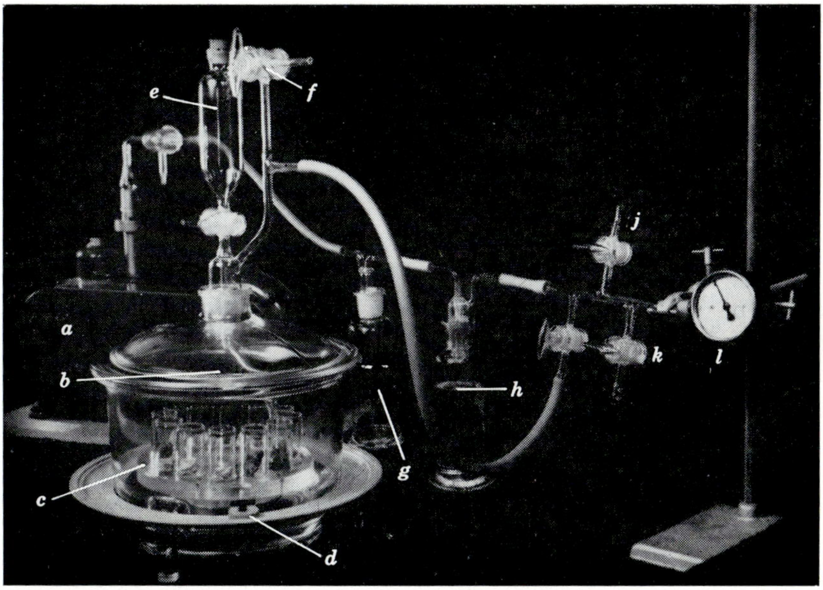

Fig. 131. Apparatus for vacuum impregnation of soil materials with cold polymerizing plastics. (*a*) Vacuum pump. (*b*) Vacuum dessicator. (*c*) Rotatable sample holder. (*d*) Permanent magnet. (*e*) Funnel for plastic. (*f*) Three-way tap. (*g*) Safety trap. (*h*) Vapor trap (water if a water pump is used, vacuum pump oil if a vacuum pump is used). (*j*) Air bleed. (*k*) Tap for second vacuum dessicator. (*l*) Vacuum guage.

(3) While the sample is cooling, mix the required amount of plastic and evacuate it in a vacuum dessicator until frothing ceases, to remove dissolved air.

(4) Place the mold on the rotatable specimen holder *D** in the vacuum dessicator.

(5) Evacuate the whole system by means of the vacuum pump *F* connected through the trap *E* to a vacuum of 25 to 27 in. of mercury.

(6) Allow the plastic to run into the mold by using the three-way tap *B*. By rotating *D*, a number of samples can be impregnated without releasing the vacuum.

(7) Keep the samples under vacuum for approximately half an hour.

(8) Stand the samples in a dessicator (silica gel) until initial setting takes place.

(9) Cure at 30°C for 12 hr, and then very slowly increase the temperature (over a period of 3 to 4 hr) to 50°C and hold it for 6 hr.

(10) Cure further at 70°C for 3 hr, at 90°C for 2 hr, and finally at 100°C for 1 hr.

(*c*) ***Impregnating Procedure for Moist Samples.*** Carbowax, latex sols and Durcupan can all be used to impregnate moist soils; carbowax has the disadvantages of being rather soft and anisotropic, latex sols can be used only for saturated soils, and the replacement of water by the miscible component of Durcupan is tedious and may cause decrepitation. The most effective procedure, therefore, is to freeze-dry the sample. This can be done on samples of any moisture content, its only disadvantage being that it is effective only to a depth of about 2 mm into the sample, so considerable care is necessary in choosing and preparing the sample. The following procedure has proved effective (after Hunter and Brewer, in press; Mackenzie and Dawson, 1961; Burges and Nicholas, 1961). The freezing apparatus consists simply of an open Dewar-type vessel (300 ml capacity) into which fits a metal cup; the drying apparatus is shown in Fig. 132.

(1) Place the metal cup in the Dewar vessel, surround it with liquid air, and allow Freon 12, from a cylinder, to condense against the side of the cup through a glass tube of 3 to 4 mm bore. When about 50 to 60 ml of Freon have been collected remove the cup from the Dewar vessel and allow the Freon to melt.

(2) Immerse the soil sample, supported on a brass gauze, in the

* *D* can be rotated while the whole system is under vacuum by attaching a small permanent magnet to its outer edge and using another magnet to rotate it and to hold it in any position.

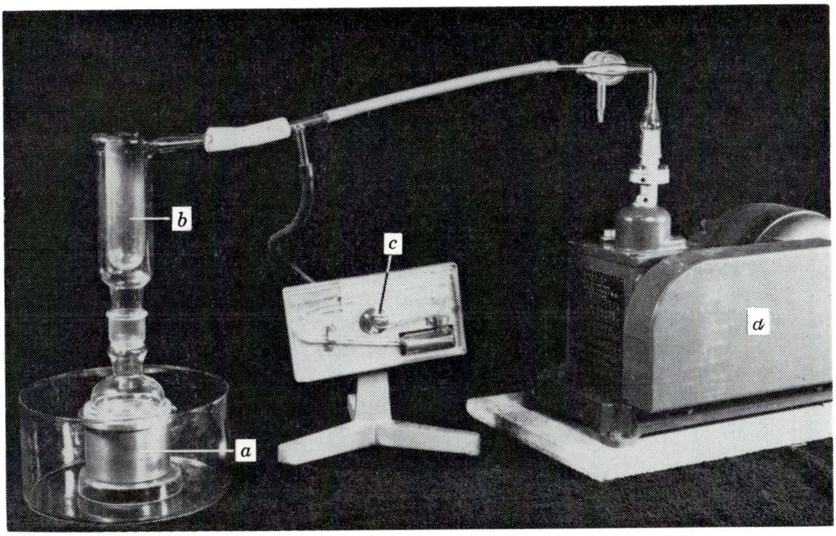

Fig. 132. Apparatus for freeze drying small samples. (*a*) Brass sample container. (*b*) Cold finger (liquid air). (*c*) Vacuum gauge. (*d*) Vacuum pump.

liquid for 3 to 4 min, and then transfer it to a previously chilled specimen tube of 1-in. diameter.

(3) Place the tube in the brass block *A*, the flange of which was previously lubricated to facilitate sealing to the remainder of the apparatus.

(4) Evacuate to 0.02 mm of mercury and immerse the brass block in a salt-ice bath; the block can be chilled quickly with a little liquid air when it is certain that a good seal has been attained.

(5) Keep under vacuum for approximately 8 hr.

(6) Remove the dried specimen and impregnate it according to the standard procedure [see 1:B(*b*), steps (4) to (10) inclusive].

Note: Steps (3) and (4) should be made as quickly as possible.

C. Preparation of Polished Blocks

Many soil materials are quite dense, and the impregnating material probably will not penetrate all the extremely fine voids between the smaller constituents, such as the clay mineral grains, which are also often quite reactive, especially with polar liquids; water may disperse such materials even when they are impregnated. Sections of such materials can be made successfully by using a nonpolar liquid, such as kerosene,* as the lubricant and coolant. Of course, water cannot be used with specimens impregnated with carbowax. The following procedure is effective using kerosene as a lubricant and coolant at all stages.

(1) Expose a section of the impregnated specimen by cutting a thin slice in the desired direction with a thin diamond saw; discard the cut-off slice. If the specimen was freeze-dried it may be cut through the center to show whether the drying was effective through the whole volume of the sample. If it was not the specimen is ground back (on a medium grain size diamond lap) from the center toward the outside until inspection shows that a properly freeze-dried zone has been reached. Alternatively the sample is ground carefully from the outside until soil material that has not been disturbed during preparation of the specimen is exposed.

(2) Condition the sample on a hotplate at 85 to 95°C for 2 to 3 hr; this prevents buckling during subsequent heating procedures in making thin sections.

(3) Grind the exposed surface flat on a medium grain size (D150) diamond lap, or a zinc lap with medium grain size carborundum, or a lap with carborundum paper.

(4) Inspect the ground surface for efficiency of impregnation; unsatisfactory impregnation is revealed by roughness. This can often be remedied by running a thin coating of activated plastic on the surface, placing the specimen under vacuum for 15 min, and curing as before. The reimpregnated surface is then ground carefully on a fine grain size diamond lap (D70) to re-expose the surface. If this is unsuccessful, or if impregnation has been particularly poor, a thin slice (about ⅛-in. thick) is cut from the block with the diamond saw, reimpregnated by the normal procedure for a new specimen, and ground down on the medium grain size diamond lap to re-expose the surface.

(5) Continue grinding on the fine grain size diamond lap.

(6) Polish for about 12 hr in a Syntron† using a slurry of fine‡ aluminium oxide in kerosene.

* Kerosene may have harmful effects if contact with it is prolonged, or if large quantities are inhaled.

† International Combustion Company (Aust.) Ltd, Rydalmere, N.S.W., Australia.

‡ Two sizes of aluminium oxide can be used (Linde A and Linde B) in two separate bowls, depending on the degree of polish required (Linde Company, Union Carbide Corporation, East Chicago, U.S.A.).

(7) Clean the surface with a stream of petroleum ether followed by compressed air, and allow it to dry.

D. Preparation of Thin Sections

Thin sections are prepared from the polished blocks. The procedure involves fixing the polished face of the block to a glass slip, cutting and grinding to the required thickness, and affixing a cover glass.

Thermoplastic cements, such as Lakeside 70, resins, such as kollolith, santolite, and canada balsam, and some of the cold polymerizing plastics have all been used successfully for fixing the blocks to the glass slips and fixing the cover glasses. For fixing the blocks to the glass slips canada balsam has the disadvantages of danger of overcooking and solubility in the nonpolar liquids used in grinding. Thermoplastic cements make a strong bond, do not overcook readily, and are not soluble in kerosene. Plastics have the advantage of cold setting over a reasonably long period, which allows time for manipulation to remove bubbles. Those with strong adhesive properties for glass that have been used so far (e.g., Epirez No. 2659E) have the disadvantage of having a rather high refractive index (R.I. $\doteqdot$ 1.58) and are so strongly adhesive that it is virtually impossible to remount the section if the slide is broken accidentally.

To affix the cover glasses the resins and thermoplastic cements require heating, which may cause loosening of the section from the glass slip with the danger of warping and/or disintegration. To overcome this some workers have used the resins dissolved in xylene; the cover glass is pressed on at room temperature and the whole section dried in an oven at low temperatures. This procedure is not very satisfactory because drying cures the resin around the edges of the cover glass but not in the center. A rather more successful method is to coat the section with a thin film of gum dissolved in amyl acetate, dry, and then affix the cover glass with canada balsam with the normal cooking procedure used in rock sectioning. The gum holds the section together during cooking and manipulation to remove bubbles; this procedure is unnecessary if the section has been fixed to the glass slip with a plastic. The cold polymerizing plastics have obvious advantages for affixing cover glasses; they require no heating and allow ample time for manipulation to remove bubbles. Their disadvantages are as quoted before, that is, high refractive index and difficulty of remounting damaged slides.

Different workers have aimed at preparing sections of various thicknesses according to their objectives. Sections comparable in thickness to the standard petrological slide (i.e., 25 to 30 μ) are ideal for most purposes, especially for studying the optical properties of the soil constituents and the structure and fabric of the specimens as reflected by optical properties. Sometimes it is useful to prepare sections rather thicker than this, of the order of 50 to 100 μ, for work on fabric analysis, especially in conjunction with the use of a stereomicroscope (e.g., Kubiena, 1938; McMillan and Mitchell, 1953).

The heating procedures necessary for mounting the sections tend to warp specimens impregnated with plastics. Steps are taken to prevent this in the following procedures, both of which have been used successfully in this laboratory:

(*a*) *Procedure 1.*

(1) Heat the impregnated block on a hotplate at 70 to 80°C and warm a glass slip.

(2) Coat the center of the glass slip thinly with activated plastic (Epirez No. 2659E) and press the block firmly on to the glass slip with the polished surface down. Manipulate the block to remove bubbles and set the mount aside to harden.

(3) Cut the block on a diamond saw so that a thin slice (about $\frac{1}{16}$-in. thick) is left affixed to the glass slip; affixing the block to the glass slip *before* cutting prevents warping.

(4) Affix the mount to a glass holder ($\frac{1}{4}$-in. plate glass 3 in. $\times$ 2 in.) by melting paraffin wax on the holder, lying the glass slip on it, and allowing it to cool, or by coating the holder with a thin film of collodion, lying the glass slip on it, and drying under vacuum.

(5) Grind the mount to a thickness of approximately 60 μ (quartz showing second-order interference colors) on the medium grain size diamond lap.*

(6) Continue grinding to approximately 40 μ thick (quartz showing high first-order interference colors) on the fine grain size diamond lap.

(7) Inspect the surface for any voids not filled with the impregnating material. If such voids are present run on a thin coating of activated plastic, place the mount under vacuum for 15 min, cure, and then repeat step (6) until the surface is just exposed.

(8) Grind to the required thickness (25 to 30 μ: quartz showing first-order grey, white, and very pale yellow interference colors) on a glass plate with fine aloxite (4 μ).

* A number of automatic grinding machines are now available (e.g., Freeman and Gardner, 1962; Altemuller, 1962), but the saving in time must be balanced against the considerable cost of this equipment.

(9) Clean the mount thoroughly with a stream of petroleum ether, followed by compressed air.

(10) Polish in the Syntron for 12 hr.*

(11) Repeat step (9) and dry the mount thoroughly.

(12) Place the mount on a hotplate thermostatically controlled at 90°C. Place a drop of canada balsam in the center of the mount and cook it. Meanwhile warm a cover glass of suitable size on the side of the hotplate.

13) When the balsam is cooked (just brittle when cold) place the cover glass on the mount allowing one edge to touch first so that the glass settles slowly, extruding the bubbles. Then press the cover glass on firmly with a slight backward and forward motion to extrude bubbles and excess balsam. If examination of the sections at high magnifications are required, cover glasses of thickness No. 1 should be used.

(14) Remove the mount, cool, and clean it by scraping off the excess balsam and washing the mount in xylene followed by soap and warm water.

Note: The cover glass can be mounted with Epirex No. 2659E as in step (2).

(*b*) *Procedure 2.*

(1) Place the impregnated block, with the polished face upward, on a hotplate at 110°C, place small chips of Lakeside 70 on the surface and allow it to melt and spread evenly over the surface; covering the sample with an inverted pyrex evaporating dish, or "brushing" the Lakeside with the flame of a bunsen burner, speeds up melting and spreading.

(2) Remove the block from the hotplate and allow it to cool.

(3) Warm a glass slip on the hotplate and place the Lakeside covered surface of the block on it in a central position. As the Lakeside melts press the block down gently but firmly with a very slight rotary movement to release bubbles.

(4) Remove the mount from the hotplate and allow it to cool.

(5) Follow procedure 1 through steps (3) to (14) inclusive, except that if the cover glass is mounted with canada balsam the mount must first be coated with a solution of gum in amyl acetate, applied with a fine camel hair brush, and allowed to dry.

* Although polishing improves the appearance of thin sections quite markedly, it is not absolutely necessary unless special techniques, such as phase contrast microscopy, are to be used.

2. TECHNIQUES FOR MINERAL ANALYSIS

Mineral analysis of soil materials depends on separation and concentration of the mineral species. The principal techniques are grain size, specific gravity, and magnetic separations, but various other specialized techniques are used for specific purposes (Milner, 1962). Quantitative estimation of the mineral species depends on counting methods applied to mounts of these separates. The procedure for mineral analysis, therefore, involves breakdown of the soil material into its ultimate particles, separation of the grains into various classes, cleaning the grains, and mounting and identifying them.

A. *Grain Size Separates*

Many soil materials consist of compound particles or aggregates and contain constituents that militate against their breakdown into ultimate particles. The most common of these constituents are organic matter, carbonates and sulfates of calcium and magnesium, and oxides and hydroxides of iron and aluminium. Standard procedures for dispersion of soil materials have been devised (e.g., Piper, 1942; Hutton, 1950, 1955; Jackson, 1958). Anderson (1963*a*) considered the effects of a number of procedures on dispersion and concluded that the various treatments gave significantly different results but that the best or most valid pretreatments cannot be selected at the present state of knowledge.

The pretreatments used vary according to the objectives, and the different procedures have different effects on the constituents of the soil materials. The usual aim in pedographic work is to break the soil material down to its ultimate particles without breaking down inherited features and without affecting the properties of the minerals present. Inherited features, however, are often too akin in properties to the soil material being treated to be able to preserve them while breaking down the rest of the soil material; lithorelicts and some pedorelicts, such as iron nodules, are exceptions to this, but preservation of these means that similar features formed *in situ* will be preserved also. It may be advantageous sometimes to preserve the carbonates in soil materials whether they occur as discrete crystals or as features, such as nodules, enclosing both plasma and skeleton grains. In other instances it may be necessary to release all the enclosed grains from concentrations of all kinds, carbonate, iron oxide, manganese oxide, etc.

A single procedure for the breakdown of soil materials, therefore, cannot meet all needs. The following methods applied in various combinations meet most cases. The quantities of reagents used are based on a sample of 100 g of soil material except where otherwise stated.

(*a*) ***Sampling.*** Wherever possible sampling should be done on an exposed profile, whether in a pit, cutting, quarry, or similar exposure, rather than from auger borings. Samples from auger borings are dangerous for detailed work because of contamination and the difficulty of recognizing the horizons. The exposed profile is examined carefully and distinct or typical horizons marked; this is a matter of experience, and no hard and fast rules for the recognition of such horizons can be laid down. Four samples of each horizon are taken close together, centered about the middle of the horizon; the depth over which the samples are taken is governed by the thickness of the horizon but normally should not exceed about 6 in. These samples can be taken in volume weight tins (Appendix, Section 3) or by the channel sampling method. If the latter method is used, volume weight samples should be taken also if the nature of the soil material allows. Alternatively a single large sample extending over an area of at least 2 ft^2 is taken from each horizon, thoroughly mixed, and subsampled to an appropriate size. Transition zones between the marked horizons are sampled in the same way so that there are no more than about 3 in. of soil material unsampled between any two consecutive samples. Preferably the whole profile should be sampled even though all samples may not be used in the laboratory; subsequent work may indicate the necessity of examining some of the material which would not have been sampled.

In addition to this detailed sampling, corresponding samples should be collected, centered around the middle of the chief horizons, in profiles approximately 6 ft away from that which was sampled, so that lateral variation can be checked. This can be done in an L-shaped pit, as used by Haseman and Marshall (1945), or on the face of other exposures.

The size of sample used for grain size analysis depends on the size of the coarsest grains. Wentworth (1926) concluded that the sample should be large enough to include a sufficient number of grains of the coarsest fragments so that "the probability of a serious accidental deviation from the normal number of such fragments in a sample collected by a reliable random method is small." His findings in regard to size of sample in relation to size of coarsest particles are summarized in Table 1 (after Krumbein and Pettijohn, 1938). The field

TABLE 1. Practical Sample Weights (after Wentworth, 1926, as modified by Krumbein and Pettijohn, 1938)

Diameter of Coarsest Size (mm)	Suggested Weight of Sample	Approximate Volume of Sample
Cobbles 128–64 mm	32 kg	—
Pebbles 64–4 mm	16–2 kg	—
Granules 4–2 mm	1 kg	1 liter
Sand 2–$\frac{1}{16}$ mm	500–125 g	500 cc
Silt $\frac{1}{16}$–$\frac{1}{256}$ mm	125 g	250 cc
Clay Under $\frac{1}{256}$ mm	125 g	250 cc

sample should be approximately ten times the size of the sample used for grain size analysis.

(*b*) ***Removal of Carbonates and Gypsum.*** There are two common methods for removal of carbonates and gypsum: dilute hydrochloric acid (Piper, 1942; Hutton, 1950) and 1N sodium acetate (Jackson, 1958). These two methods have different effects both on the carbonates and on the other constituents of the soil material. Hydrochloric acid dissolves all the carbonates whether they occur as fine discrete crystals or as concentrations, such as nodules; it may also dissolve some minerals, such as apatite, and affect the properties of others, such as biotite. If the concentration and total amount of the acid are controlled so that the action is not too vigorous and only a slight excess is added above what is necessary to dissolve carbonates and gypsum, then its adverse effects on most minerals are slight. The sodium acetate treatment has little effect on primary minerals, but it is ineffective in dissolving carbonates and gypsum that occur as concentrations or coarse crystals. According to requirements one of the following two methods is used.

Method 1. Hydrochloric acid treatment (after Piper, 1942).

(1) Place the 100 g sample in a 3-liter pyrex beaker.

(2) Add 1 liter of water and then 100 ml of 2N hydrochloric acid and an extra 5 ml of 2N acid for each percent of calcium carbonate above 2 percent. Stir occasionally over a period of 30 min.

(3) Fill the beaker with water and wash by repeated sedimentation and decantation until the supernatant solution is nearly free from acid. Filtering through a buchner funnel, or flocculation with sodium

chloride and repeated washing, facilitates removal of all the acid where the clay size material is to be retained.

Method 2. Sodium acetate treatment (after Jackson, 1958).

(1) Place the 100 g sample in a 3-liter pyrex beaker.

(2) Add 600 ml of 1N sodium acetate buffer (pH 5), bring the soil into suspension by stirring, cover the beaker, and digest for 30 min on a water or sand bath.

(3) Allow the suspension to settle and decant the supernatant liquid.

(4) Wash by sedimentation and decantation, once with sodium acetate solution to remove excess calcium salts, and once with water.

Note: If necessary the procedure can be repeated. As for Method 1, filtering, or flocculation with sodium chloride and repeated washing by decantation, facilitates removal of excess reagent (acetate) where the clay size material is to be retained.

(*c*) ***Removal of Organic Matter.*** The standard method of removal of organic matter prior to granulometric analysis is by treatment with hydrogen peroxide (e.g., Piper, 1942; Truog et al., 1937). The method is effective only in slightly acid solution (McLean, 1931; Scharrer, 1933; Jackson, 1958) and in the absence of excess calcium so that a prerequisite is removal of carbonates and slight acidification of the solution. This is achieved best by Method 1 for removal of carbonates and gypsum. The result is that the carbonates *are* removed, and the acidification may affect some of the primary minerals. The method may also cause formation of calcium oxalate (Martin, 1954), and hydrogen peroxide may cause excessive exfoliation of some of the micas, especially if they are partially weathered (Drosdoff and Miles, 1938). Exfoliation may not be very serious if the granulometric analysis is done by sieving methods since exfoliation of platy minerals does not change their effective size significantly as determined by sieving. It vastly increases the number of individual grains, however, and affects the equivalent size fraction in which the mica appears when elutriation or sedimentation methods are used. It is sometimes preferable, therefore, to use sodium hypobromite or sodium hypochlorite (Anderson, 1963*a*,*b*). Both reagents are just as effective as the peroxide treatment in removing organic matter, and they do not require prior removal of carbonates and gypsum.

Method 1. Hydrogen peroxide treatment (after Hutton, 1950).

(1) To the material remaining after treatment by Method 1 for the removal of carbonates (originally 100 g of soil), add 300 ml of

6 percent hydrogen peroxide and allow the suspension to stand overnight.

(2) Add 300 ml of water and very carefully bring the contents of the beaker to the boil. Boil for 10 min, or until effervescence ceases, and allow the suspension to cool.

(3) Wash several times by sedimentation and decantation, flocculating with sodium chloride if it is necessary to retain the clay size fraction.

Method 2. Sodium hypobromite treatment (after Anderson, 1963*a*).

(1) To the material remaining after treatment by Methods 1 or 2 for the removal of carbonates, or the untreated original sample (100 g of soil), add 400 ml of freshly prepared sodium hypobromite solution, and allow the reaction to continue at room temperature for 2 hr with occasional stirring.

(2) Add a second increment of approximately 400 ml of sodium hypobromite solution, stir vigorously, and allow the suspension to stand overnight (or 12 hr) at room temperature.

(3) Wash by sedimentation and decantation, or by filtering, with flocculation with sodium chloride if the clay size material is to be retained.

Preparation of Sodium Hypobromite Solution. Dissolve 20 ml of bromine (2.5 ml per 100 ml of sodium hydroxide solution) in 800 ml of approximately 0.9N sodium hydroxide solution and adjust the pH to about 9.5 with additional sodium hydroxide.

Notes. (i) If there is much manganese dioxide present it should be reduced before treatment for removal of organic matter by adding an appropriate amount of sodium bisulfite (2 to 8 g per 100 g of soil), and boiling for 30 min to 1 hr. (ii) Anderson (1963*a*) added sodium carbonate to the sodium hypobromite solution to make a 2 percent solution of pH about 10.5; this usually caused dispersion. After treatment with sodium hypobromite solution washing was done with 2 percent sodium carbonate solution (step 3 above).

Method 3. Sodium hypochlorite treatment (after Anderson, 1963*b*).

(1) To the material remaining after treatment by Methods 1 or 2 for the removal of carbonates, or the untreated original sample (100 g of soil), add 200 ml of sodium hypochlorite solution freshly adjusted to pH 9.5.

(2) Place the beaker in a boiling water bath for 30 min (Anderson suggested 15 min but the large bulk used here requires longer heating).

(3) Wash three times with water by sedimentation and decantation, flocculating with sodium chloride if the clay size fraction is to be retained, or filter.

(4) Repeat steps (1) to (3) twice.

Sodium Hypochlorite Solution. The solution used by Anderson was purchased, already made up, as free of calcium and phosphorus and containing 5 to 6 percent available chlorine.

Note: If manganese dioxide is present it should be oxidized before treatment as in Method 2.

(d) Removal of sesquioxides. Some of the most difficult soil materials to disperse are those which contain high proportions of sesquioxides and hydroxides. In many procedures these are largely removed by treatment with hydrochloric acid (Piper, 1942; Hutton, 1950). This is to be avoided where mineral analyses are to be made because of its effects on carbonates, gypsum, and some of the primary minerals. Most of the alternative methods depend on reduction and solution by the action of the hydrogen ion in acid solution (e.g., Hendrick and Newlands, 1928; Tamm, 1934; Chakraborty, 1943; Jeffries, 1946), and these have been summarized by Jeffries (1946), whose magnesium ribbon-oxalic acid treatment is the most suitable of these methods. Williams (1950) made a critical study of the last method and found it very efficient, the only disadvantage being some slight attack on biotite. A more recent method using dithionite and a chelating agent (Aguilera and Jackson, 1953) at pH 7.0 to 7.5 is most effective for solution of sesquioxides and the least harmful to the minerals; bentonite cleaned in this way showed no signs of any effect due to the treatment (M. L. Jackson, priv. com.). This method is simple to use on large samples, so it can be used as a pretreatment to dispersion. Both the dithionite method and Jeffries' method are generally ineffective in dissolving sesquioxidic concentrations such as nodules and concretions. This is an advantage in some respects because such pedological features are preserved, but it necessitates a further, more drastic treatment if release of primary minerals enclosed in such features is desired. A further method of cleaning mineral grains with detergents and considerable trituration has been used (Matelski, 1953) but is relatively ineffective.

Method 1. Dithionite treatment (modified from Aguilera and Jackson, 1953; and Jackson, 1958).

(1) After washing following removal of organic matter add 300 ml of water to the material left in the 3-liter beaker.

(2) Add 24 g of solid sodium citrate and 2.8 g of solid sodium bicarbonate.

(3) Heat on a sand bath to 75 to 80°C with stirring to insure complete solution.

(4) Add 7 g of solid sodium hyposulfite and stir constantly for 5 min and occasionally for 10 min.

(5) Wash by repeated sedimentation and decantation, or filtering. *Method 2.* Oxalic acid-magnesium ribbon treatment (after Jeffries, 1946; as modified by Williams, 1950) for small samples of grain separates.

(1) If carbonates have not been removed treat with the appropriate amount of hydrochloric acid [see Appendix 2:A(*b*)].

(2) Place 2 g of carbonate-free sample in a 150 ml pyrex beaker.

(3) Add 40 ml of potassium oxalate solution (103.7 g per liter) and heat to 80°C on a sand bath.

(4) Add 10 ml of oxalic acid solution (95 g per liter), stir the suspension, and heat to 90°C.

(5) Add about 0.2 g of magnesium ribbon (about 8 in.), stir, and keep at 90 to 95°C for 3 to 5 min. This is done best by bending a stirring rod into a loop at one end and wrapping the magnesium ribbon around the loop. Stirring then introduces the ribbon and it is simply removed by removing the stirrer.

(6) Add an additional 5 ml of oxalic acid solution and continue heating for 3 min or until the material is white or grey.

(7) Wash by repeated sedimentation and decantation, or filtering, the first washings being with N/10 hydrochloric acid.

(*e*) ***Dispersion.*** Most methods of dispersion depend on saturation of the colloid complex with sodium ions and some degree of mechanical agitation; lithium saturation is generally more effective and is necessary for dispersing some soils (K. Norrish, priv. comm.). If carbonates, gypsum, organic matter, and sesquioxides are all removed by pretreatments, dispersion can usually be achieved by addition of dilute sodium hydroxide solution and agitation (Piper, 1942). If organic matter and sesquioxides are removed by the sodium hypobromite (or hypochlorite) and dithionite methods so that carbonates and gypsum are retained, it is advisable to use some method of precipitating soluble calcium ions to achieve dispersion even though Troell (1939) claimed that no further treatment was necessary after pretreatment with sodium hypobromite. Several reagents have been used to precipitate calcium ions. Wintermeyer and Kinter (1954) studied the efficiency of many of them; sodium hexametaphosphate ("Calgon") is quite effective (Hutton, 1950) in conjunction with dilute sodium hydroxide solution.

It is preferable to use the least severe method of mechanical agitation possible for mineral analysis; removal of sesquioxides and organic matter often eliminates the necessity for any agitation apart from stirring and gentle rubbing with a rubber policeman. If this is not sufficient end-over-end shaking (Piper, 1942) is theoretically preferable to agitation by a mechanical beater of the propeller type. In practice, however, there is little real damage to most grains by using the propeller type beater, and this is the quickest and most effective method.

Method.

(1) To the material remaining in the 3-liter beaker (originally 100 g of soil) after the pretreatments, add approximately 1 liter of water, 40 ml of 1N sodium hydroxide solution, and 20 ml of 10 percent calgon solution.

(2) Agitate for 10 min on a mechanical beater of the propeller type.

(3) Wash through a 70-mesh sieve into a sedimentation cylinder.

(4) Dry the sieve on an aluminium tray and shake it over the tray until no more material passes through.

(5) Add the material from the tray to the sedimentation cylinder and retain the material on the sieve for further sieving.

(6) Shake the sedimentation cylinder on an end-over-end shaker for 12 to 16 hr (overnight).

Notes: (i) If gentle dispersion is essential agitation on a mechanical beater can be eliminated. Where the mechanical beater is used end-over-end shaking is often unnecessary. (ii) Calgon is not necessary if carbonates have been removed.

(*f*) ***Granulometric Analysis.*** In general, grains larger than approximately 50 μ diameter are separated into size classes by sieving and those smaller than 50 μ equivalent diameter by elutriation or sedimentation and decantation. Although this change in procedure gives an artificial break in the grain size distribution curve it is unavoidable because of the limitations of the two methods. In practice it is convenient to divide the sample at about 44 μ because this is approximately the smallest size for which sieves are dependable and it is about the smallest standard size in the Tyler, A.S.T.M., and B.S. series.

Sieving is done best on a mechanical device, such as the Ro-tap shaker, which accommodates a nest of sieves and achieves efficient results with approximately 10 min sieving. The efficiency depends on the size of the initial sample, the diameter of the sieves, and the grain size distribution of the sample. The most efficient load and sieving time should be ascertained for the equipment used. Standard Tyler, A.S.T.M., or B.S. sieves are suitable; they are all based approximately on a geometric scale. Most granulometric analyses can be made with one or other of the sets of sieves tabulated in Table 2.

The efficiency of elutriators, within the size range for which they

TABLE 2. Standard Sieves (after Milner, 1962)

	Aperture, in mm			
Sieve Number	I.M.M.	Tyler	U.S. Standard (A.S.T.M.)	B.S.
325	—	—	0.044	—
300	—	—	—	0.053
270	—	—	0.053	—
240	—	—	—	0.066
230	—	—	0.062	—
200	0.063	0.074	0.074	0.076
170	—	0.088	0.088	0.089
150	0.084	0.104	—	0.104
140	—	—	0.105	—
120	0.107	—	0.125	0.124
115	—	0.124	—	—
100	0.127	0.147	0.149	0.152
90	0.139	—	—	—
85	—	—	—	0.178
80	0.157	0.175	0.177	—
72	—	—	—	0.211
70	0.180	—	0.210	—
65	—	0.208	—	—
60	0.211	0.246	0.250	0.251
52	—	—	—	0.295
50	0.254	—	0.297	—
48	—	0.295	—	—
45	—	—	0.35	—
44	—	—	—	0.353
42	—	0.351	—	—
40	0.317	—	0.42	—
36	—	—	—	0.422
35	—	0.417	0.50	—
32	—	0.495	—	—
30	0.421	—	0.59	0.500
28	—	0.589	—	—
25	—	—	0.71	0.599
24	—	0.701	—	—
22	—	—	—	0.699
20	0.635	0.833	0.84	—
18	—	—	1.00	0.853
16	0.792	0.991	1.19	1.003
14	—	1.168	1.41	1.204
12	1.056	1.397	1.68	1.405

TABLE 2. Standard Sieves (after Milner, 1962) (*Continued*)

	Aperture, in mm			
Sieve Number	I.M.M.	Tyler	U.S. Standard (A.S.T.M.)	B.S.
10	1.270	1.651	2.00	1.676
9	—	1.981	—	—
8	1.574	2.362	2.38	2.057
7	—	2.794	2.83	2.411
6	—	3.327	3.36	2.812
5	2.540	3.962	4.00	3.353
4	—	4.699	4.76	—
$3\frac{1}{2}$	—	5.613	—	—
3	—	6.680	—	—
$2\frac{1}{2}$	—	7.925	—	—

are designed, depends to some extent on the size of the sample, for which there is a maximum for the design of the elutriator. For sedimentation methods this restriction is overcome by increasing the volume of the sedimentation vessel. Piper (1942) set out sedimentation times for some of the grain size classes for a range of temperatures and a standard height of sedimentation; a more complete list is given in Table 3. Hutton (1955) described a sand washing machine which depends on a constant water temperature, mechanical stirring, and automatic siphoning; such machines are time-saving. Elutriators, too, are time-saving, but are rather restricted as to size of sample, and they are subject to errors due to fluctuations in temperature and water pressure unless rather expensive installations are made.

Certain standard procedures simplify the complete granulometric analysis of the rather large samples used for soil materials. The dispersion procedure [Appendix 2:A(*e*)] reduces the bulk of sample to be separated by sedimentation methods by removing the fraction held on a 70-mesh sieve (larger than 200 μ diameter). Then, for sedimentation methods, it is most convenient to make the separations consecutively starting with the smallest size first; this also removes the smaller sizes before separation of the coarser fractions by sieving.

Method 1. Sedimentation.

(1) Half fill the sedimentation cylinder containing the dispersed soil material with water, stir the suspension without swirling, and

TABLE 3. Sedimentation Time for 10-cm Fall

	Particle Size, in μ												
	2		5			10		20		30		50	
Temp., C°	hr,	min	hr,	min,	sec	min,	sec	min,	sec	min,	sec	min,	sec
5	12	05	1	56	00	29	00	7	15	3	13	1	10
6	11	45	1	52	50	28	10	7	00	3	08	1	08
7	11	20	1	49	00	27	20	6	50	3	02	1	05
8	11	00	1	45	50	26	30	6	35	2	56	1	04
9	10	45	1	43	00	25	40	6	25	2	52	1	02
10	10	25	1	40	00	25	00	6	15	2	47	1	00
11	10	10	1	37	00	24	20	6	05	2	42		58
12	9	50	1	34	20	23	40	5	55	2	37		57
13	9	35	1	31	10	23	00	5	45	2	33		55
14	9	20	1	29	40	22	20	5	35	2	29		54
15	9	05	1	27	10	21	50	5	25	2	25		52
16	8	50	1	24	50	21	10	5	20	2	21		51
17	8	35	1	22	40	20	40	5	10	2	18		50
18	8	25	1	20	50	20	10	5	05	2	15		48
19	8	10	1	18	40	19	40	4	55	2	11		47
20	8	00*	1	16	50	19	10	4	48*	2	08		46
21	7	50	1	15	00	18	40	4	40	2	05		45
22	7	40	1	13	00	18	20	4	35	2	02		44
23	7	25	1	11	30	17	50	4	30	1	59		43
24	7	15	1	09	50	17	30	4	20	1	56		42
25	7	05	1	08	20	17	00	4	15	1	54		41
26	6	55	1	06	40	16	40	4	10	1	51		40
27	6	45	1	05	20	16	20	4	05	1	49		39
28	6	40	1	04	00	16	00	4	00	1	47		38
29	6	30	1	02	40	15	40	3	55	1	44		38
30	6	25	1	01	20	15	20	3	50	1	42		37

* International Standard Definition of 2 μ and 20 μ by sedimentation velocity

allow it to settle. Mark the cylinder at a height of 30 cm above the surface of the sediment.

(2) Fill the cylinder to the mark with water, stir, and allow the suspension to sediment for the time appropriate for a 30-cm fall, the temperature of the water, and the grain size to be poured off (Table 3).

(3) Decant or siphon off the suspension into a 5-liter beaker, taking care not to disturb the sediment in the bottom of the cylinder.

(4) Repeat steps (2) and (3) until the liquid above the sediment is clear.

(5) Allow the material in the 5-liter beaker to sediment completely (after flocculation with sodium chloride if necessary), decant the supernatant liquid, dry the sediment at 105°C, and weigh and bottle it.

(6) Repeat steps (2) to (5) inclusive, adjusting the time of sedimentation to separate the next coarser fraction, until all fractions have been collected.

Notes. (i) If there is a high proportion of clay size material present originally, several decantations or syphonings should be made at approximately twice the calculated time to reduce the density of the suspension which may upset sedimentation times. (ii) If decantation is used, the height to which the cylinder is filled with water should be checked constantly because the amount of sediment in the bottom of the cylinder is being successively reduced. (iii) If siphoning is used, the cylinder should be marked at approximately 32 cm above the surface of the sediment, and the bottom end of the siphon tube fixed at 30 cm below this mark. (iv) Efficient mixing without swirling can be obtained by an upward and downward movement of a stirrer which consists of a perforated disc, slightly smaller in diameter than the inside diameter of the sedimentation cylinder, attached to the end of a long stirring rod of glass or stainless steel.

Method 2. Elutriation.

(1) Introduce the dispersed soil material in the sedimentation cylinder into the elutriator, taking care that the load is not greater than that specified for the capacity of the elutriator used.

(2) Adjust the flow rate for separation of the coarser limit of the smallest size class required and collect the material carried over in a 5-liter, or larger, beaker.

(3) When no more particles are being carried over increase the flow rate to separate the next coarser size class; collect the material carried over in another large beaker.

(4) Repeat step (3) until all the separations that can be handled by the elutriator are made.

(5) Run the material remaining in the elutriator vessel into a separate beaker.

(6) Dry each fraction at 105°C, and weigh and bottle them.

Note: The procedure described is for a single bowl elutriator where the flow rate is controlled by capillary tubes of various diameters and lengths; temperature of the water must be kept constant. Some elutriators consist of several bowls and make several separations at fixed intervals simultaneously.

Method 3. Sieving.

(1) Transfer all the material remaining after separation by the sedimentation methods to an evaporating dish and dry at 105°C.

(2) Transfer the material to the top (coarsest) sieve of a nest selected to give the required size ranges. Place the lid on the nest of sieves, clamp them in the mechanical shaker, and shake for 10 min.

(3) Inspect each sieve and, if the load on any one is considered too great, repeat the shaking for a further 10 min either with or without addition of appropriate intermediate sieves to reduce the load on one or more particular sieves.

(4) Upturn each sieve on a sheet of glazed paper and gently tap and brush the sieve to remove adhering grains.

(5) Weigh and bottle each fraction.

B. Specific Gravity Separations

The greatest variety of mineral species occur in the fractions of higher specific gravity, so it is advantageous to separate the minerals according to their specific gravity. Some studies may require pure separates of particular mineral species. Specific gravity separations may achieve this or be a step in the procedure. The separations are made by means of liquids of various specific gravities. For normal studies many workers use bromoform (s.g. 2.84 ± 0.02) or *S*-tetrabromethane (s.g. 2.95); the latter is sometimes mixed with nitrobenzene to give other specific gravities; for example, some workers make separations at 2.70 and 2.95 as a routine. For size fractions separated from water suspensions, specific gravity separations are facilitated if the fractions are dried from acetone to prevent caking, especially in the finer grades. After the final water decantation the sediment is washed three times by decantation in relatively large volumes of acetone, the volume depending on the bulk of the sediment.

Separations can be made in an evaporating dish, in glass funnels, in normal centrifuge tubes, and in special centrifuge tubes. The following methods are effective and relatively simple for their specific purposes.

(a) Separation in an Evaporating Dish. For rough work or for coarse grains the separations can be made in evaporating dishes according to the following procedure:

(1) Place the sample to be separated in an evaporating dish.

(2) Gently pour the specific gravity liquid on to the grains, thus forming a suspension.

(3) Cover the dish and allow the sample to sediment for 10 min.

(4) Stir the floating grains, cover the dish, and allow the sample to sediment for a further 30 min or until there are two distinct layers of minerals separated by clear liquid.

(5) Remove the bulk of the floating grains by gently drawing a teaspoon across the surface of the liquid and place them in a separate dish.

(6) Swirl the dish gently by a rotary action in the hand so that most of the remaining floating grains adhere to the sides of the dish above the level of the liquid.

(7) Pour the liquid off into a filter funnel fitted with a fast filter paper, leaving the heavy fraction behind. At the same time brush the adhering and floating grains toward the lip of the dish and into the filter funnel, using a small camel hair brush.

(8) Add the grains removed by the teaspoon to the filter funnel.

(9) Wash both fractions with a solvent appropriate for the liquid used (alcohol for bromoform).

(10) Dry, weigh, and bottle the fractions.

Note: The method has the advantage of a large surface area of specific gravity liquid, and so relatively large samples can be used; the maximum load is about 0.2 g/cm^2 of surface area of liquid so that an evaporating dish with an internal diameter of 8 cm will usually give a fairly clean separation if the load does not exceed 8 g and the grain size is not less than 50 μ.

(b) Separation by Centrifugation. Haseman and Marshall (1945) described a method of specific gravity separation by centrifuging in a special centrifuge tube, but the method is similar irrespective of the tube used. They recommended a maximum load of ⅔ g of material per cm^2 of surface area of liquid with three centrifugings for grains larger than 25 μ diameter, four centrifugings for grains between 11 and 25 μ diameter, and a repetition of four centrifugings on the light fraction for grains smaller than 11 μ diameter; each centrifuging was at 1000 to 1500 rpm for 10 min. The effectiveness of the separations can be increased by reducing the load per unit surface area of liquid. In this laboratory the load is limited to 1 g per 10 cm^2 surface area of liquid, which is a load of 2 g in normal centrifuge tubes of internal diameter of 5 cm. The following procedure is used:

(1) Place the sample in the bottom of the tubes and gently pour in

the liquid to a depth of about 7 cm; the load is not to exceed 1 g per 10 cm^2 surface area of liquid.

(2) Balance the tubes in their cups by adding drops of the heavy liquid as necessary.

(3) Stopper the tubes, shake them vigorously, and centrifuge immediately for 10 min at 1000 to 1500 rpm.

(4) For grains larger than 50 μ diameter reshake the tubes and recentrifuge. In practice it is necessary to reshake only the light fraction, and this can be done by stirring the light fraction floating at the top of the liquid without removing the tubes from the centrifuge.

(5) For grains between 50 and 20 μ diameter, repeat step (4) once.

(6) For grains between 20 and 10 μ diameter, repeat step (4) twice.

(7) For grains between 5 and 10 μ diameter, repeat step (4) twice, recover the light fraction, and recentrifuge it four times.

(8) Recover the two fractions and weigh and bottle them.

Notes. (i) Drying the grains by washing in benzene before centrifuging often facilitates the separation; benzene is highly inflammable and poisonous. (ii) The procedure for recovering the minerals from the centrifuge tubes varies according to the type of tube and the specific gravity liquid used. The following methods have proven satisfactory:

Freezing Technique for Normal Tubes. After centrifuging, the tubes are stood with their bottoms in a freezing mixture until the heavy minerals are embedded in a solid cake of frozen liquid. The light minerals are then poured off into an evaporating dish and the sides of the tubes freed from adhering grains by washing quickly with a jet of solvent suitable for the specific gravity liquid used. Then the frozen liquid in the tube is allowed to melt, and the heavy minerals are washed into another evaporating dish. Some of the specific gravity liquids commonly used are well suited to this method; bromoform will freeze in a salt-ice mixture, but tetrabromethane is rather more difficult to freeze.

Special Centrifuge Tube (after Haseman and Marshall, 1945; Fig. 133). To recover the heavy minerals the stopper in the bottom of the tube is removed, and the heavy minerals are ejected into an evaporating dish simply by pushing the upper stopper tightly into the tube. Then the light fraction is washed out of the tube into another evaporating dish.

Polythene Centrifuge Tubes. Soft polythene centrifuge tubes that fit snugly into the normal glass or metal tubes have been used. Centrifuging is done in the normal way. Then the polythene tube is withdrawn, and a laboratory clip is fitted between the heavy and light

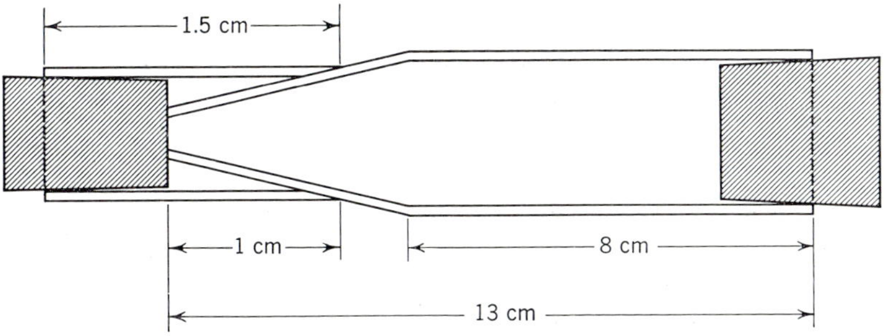

Fig. 133. Special centrifuge tube (after Haseman and Marshall, 1945).

minerals. The tube is held over an evaporating dish, and the clip is tightened allowing the light minerals to run into the dish. Then the adhering grains are washed into the dish with the appropriate solvent for the liquid used. Finally the clip is released, and the heavy minerals are washed into a separate evaporating dish.

Recovery of the Separated Fractions. The separate fractions obtained in these various ways are washed several times by adding the solvent, allowing the grains to settle, and then pouring off the solvent; all washings should be retained for recovery of the specific gravity liquid. Finally the fractions are dried in an oven at 105°C, weighed, and bottled. This method is preferable for quantitative work to washing through a filter paper where losses may occur in recovering the grains from the filter paper. Most of the solvents are of very low specific gravity, and even the light minerals settle quickly in them.

Recovery of Specific Gravity Liquids. Most of the liquids of high specific gravity used in these separations are quite expensive, and relatively large quantities are necessary for a single separation. It is necessary, therefore, to recover as much of the liquid as possible after the separations have been made. In effect this means separating the pure liquid from the solvents used for washing the separated minerals. Bromoform is recovered by shaking the bromoform-alcohol mixture in a large volume of water in a separating funnel. The alcohol dissolves in the water preferentially, and the bromoform settles as the lower layer and is run off. Water is removed from the recovered bromoform by shaking with anhydrous calcium chloride and filtering. Acetone is a suitable solvent for *S*-tetrabromethane which is recovered by distilling off the acetone in a hot water bath; the recovered liquid is brought up to its former density by heating to about 90°C and cooling several times.

C. Magnetic Separations

Although there are several methods of separating mineral grains according to their magnetic and electrical properties, the magnetic and electromagnetic methods are by far the most useful. Practically pure concentrates of particular minerals can be obtained by these methods with suitable equipment. By comparison electrostatic and dielectric methods are cumbersome and of restricted use; these methods are described by Milner (1962). The details of the procedures used for magnetic separations depend on the instruments available and the mineral species present, except for the one general rule that the minerals should be separated in order of decreasing magnetic strength.

The most versatile instrument is the isodynamic magnetic separator. It consists essentially of an automatic feed fitted to a vibrator table which lies between two elongated poles of a powerful electromagnet. A wide range of mineral species can be separated according to their magnetic strength by adjusting the slope (in the long axis of the instrument), the tilt (in the short axis at right angles to the slope) of the instrument, and the strength of the magnetic field (by adjusting the current in the coils). Strongly magnetic minerals, such as magnetite, can be separated by using a vertical feed clamped to the outside of the instrument. If such an instrument is not available a number of separations can be made using a permanent magnet, preferably of the horseshoe-type with adjustable pole pieces, and a simple bench-type electromagnet with adjustable pole pieces and current.

(a) Separation by the Isodynamic Magnetic Separator.

(1) Check with a permanent magnet whether there are any very strongly magnetic mineral species present.

(2) If there are, set the instrument in a vertical position with the supplementary vibrator table and hopper fitted for vertical feed.

(3) By adjusting the strength of the magnetic field separate the strongly magnetic minerals into the required number of fractions, depending on the variety of minerals and the amounts of each species, removing the most magnetic species first.

(4) Check the completeness of each separation by repeating it until an insignificant amount of the required separate is being recovered.

(5) Set the instrument up in its normal position with the vibrator table sloping between the poles.

(6) Adjust the slope of the vibrator table to give a smooth even flow of grains for the grain size being separated.

(7) By adjusting the strength of the magnetic field and the tilt of the vibrator table, make the required number of separations according to the variety of minerals and the amount of each species, removing the most magnetic first.

(8) Check the completeness of each separation as before.

(9) Weigh and bottle the separates.

Note: The setting of the slope of the vibrator table depends on the grain size of the fraction and the rate of feed required; the slower the rate of feed the more complete is the separation, and so the fewer repetitions required. The settings of the side tilt and field strength (current in the coils) depend on the species present and the separation desired. Rosenblum (1958) has published instrumental settings for the Frantz isodynamic magnetic separator for about fifty mineral species.

(b) Separation with Simple Equipment.

(1) Spread the grains uniformly on a flat glossy surface.

(2) Separate the strongly magnetic minerals by passing a bar or horseshoe magnet systematically just above the grains, constantly turning the grains over if there is more than a single layer; a small strip of thin tissue paper held over the poles facilitates release of the attracted grains.

(3) Repeat step (2) until all the strongly magnetic grains have been removed.

(4) If the magnet is a horseshoe-type with adjustable poles repeat steps (2) and (3) with the poles progressively closer together to make additional separates of the strongly magnetic minerals.

(5) Respread the grains on the glossy surface.

(6) Adjust the current and distance between the poles of the simple bench-type electromagnet to remove the next most magnetic mineral grains.

(7) Pass the grains systematically under the poles, releasing the attracted grains into a basin at the end of each pass by switching off the current.

(8) Repeat step (7) until all the grains of that particular magnetic strength have been collected.

(9) Repeat steps (7) and (8) to collect as many separates as considered desirable.

(10) Weigh and bottle the separates.

D. Subsampling for Grain Mounts

Usually it is impossible to count all the grains in any of the separates. The accuracy of mineral estimations, therefore, depends on how closely the subsample approximates the whole separate in mineral composition.

The simplest procedure for subsampling is by the standard method of quartering. The sample is placed in a conical pile on a flat glossy surface and cut into four quarters with a sharp straight edge, such as a cord or thin spatula. Opposite quarters are rejected and the other two combined and split again. This is repeated until the desired size of sample is obtained. Pettijohn (1931) performed this operation by overlapping four rectangular sheets of paper to form a square composed of one-fourth of each sheet; the sample is poured on the center of the square and spread into a circular heap, and then the papers are pulled apart. Opposite quarters are recombined and the procedure repeated until a sufficiently small sample is obtained.

A number of mechanical subsamplers have been designed specifically for samples of mineral grains. These are discussed by Milner (1962) in regard to their construction and efficiency and the difficulties involved in their use for small samples, such as many of the mineral separates are.

E. Mounting Grains

Of the many methods and materials available for mounting mineral grains for microscopic examination, the following are the most convenient and effective for the various needs of mineral analysis of soil materials.

(*a*) ***Preliminary Exploratory Examination.*** Spread the subsample on a glass slide, add a drop of eugenol (R.I., 1.54), and cover the sample with a cover glass. By this simple method the composition of the material can be examined and the necessity for more elaborate mountings judged. Since the grains are quite free they can be turned, if necessary, usually by pushing on the cover glass with a pencil or probe. The grains are easily recovered by washing with xylol or alcohol. It may not be necessary to make other mounts, and grain counts can be made directly on this simple mount.

(*b*) ***Mounts for R.I. Liquids.*** The refractive index of the mounting medium can be changed by washing the loose grains with xylol and remounting in another liquid or by fixing the minerals to the glass slide with a very thin film which does not cover their upper surfaces so that the grains can be washed, without moving them, and remounted in another liquid.

Method 1. Loose grains.

(1) Place the grains on a recessed glass slide and add a drop of the first refractive index liquid.

(2) Carefully place a drop of xylol at the edge of the drop of liquid. Surface tension effects between the two liquids drive the refractive index liquid away from the grains and up the edges of the recess.

(3) Remove the displaced refractive index liquid by absorption on a filter paper.

(4) Repeat steps (2) and (3) until all the refractive index liquid is removed.

(5) Allow the grains to dry and add the next refractive index liquid.

Method 2. Grains fixed with gelatin (after Fairbairn, 1943).

(1) Place one drop of solution *A* on a clean dry slide and dry at 80°C.

(2) Place one drop of solution *B* directly on the film and spread the grains evenly over the drop with a needle. Dry as before.

(3) Place one drop of the first refractive index liquid on the grains.

(4) Remove the refractive index liquid with a gentle stream of a suitable solvent (xylol, acetone, alcohol) directed against the slide just above the mount.

(5) Allow the grains to dry and add the next refractive index liquid.

Solution A. 0.1 percent gelatin solution
Solution B. Mix 10 ml of distilled water, 5 ml of acetone and 2 ml of 2 percent formalin.

Notes. (i) Large grains may require a stronger gelatin solution. (ii) The solutions should be freshly made.

Method 3. Grains fixed with gum tragacanth.

(1) Spread a drop of the gum tragacanth solution on a clean dry slide, using a glass rod held on its side. Warm gently to dry.

(2) Breathe on the dry gum coating to moisten it and sprinkle on the grains.

(3) Add and remove the refractive index liquids as in Method 2.

Gum tragacanth solution. Make a 1 percent solution of gum tragacanth by boiling the gum in distilled water. Add 2 to 3 drops of ammonium hydroxide per 150 ml as a preservative, and 2 drops of "teepol" per 100 ml to aid spreading on the slide.

Note. The strength of the gum solution should be increased for large grains, and decreased for very small grains.

Method 4. Grains fixed with plastics (after Langford, 1962).

(1) Spread a very thin film of activated Epriez No. 2659E on a clean dry slide.

(2) Allow the plastic to partially polymerize, that is, to become tacky.

(3) Sprinkle on the grains.

(4) Add and remove the refractive index liquids as in Method 2.

Note: Considerable care is required in using this method to insure that the grains are not covered by the plastic.

(*c*) *Permanent Mounts.* The following methods are examples using particular natural resins or plastics, but many other similar materials are suitable.

Method 1. Canada balsam.

(1) Fix the grains to a glass slide with gelatin or gum tragacanth (Appendix 2:E(*b*), Methods 2 and 3).

(2) Add a drop of eugenol.

(3) Remove the eugenol by absorption with a coarse filter paper.

(4) Add a drop of canada balsam and cook it on a hotplate.

(5) Press on a warmed cover glass.

Notes. (i) Loose grains can be mounted in the same way, but fixing the grains to the slide simplifies the procedure. (ii) Wetting with eugenol [steps (2) and (3)] makes it easier to eliminate bubbles.

Method 2. Epirez No. 2659E.

(1) Proceed according to Method 1 (1).

(2) Heat the mount on a hotplate at 80°C.

(3) Add a drop of activated Epirez No. 2659E.

(4) Press on the cover glass and set aside to harden.

Note: It is often advantageous to section coarse grains. This is done by setting the grains in a plastic according to Method 2 and then proceeding according to the method for making thin sections of impregnated soil materials. Such mounts are sections of loosely packed grains, and so, if they are ground to a thickness of 25 μ, many of the sections of the grains will be thinner than that; also, many of the grains will appear smaller than their original size.

F. Flow Sheet for Mineral Analysis

(1) Subsample the bulk sample to obtain 100 g of material which is representative of the bulk sample.

(2) Transfer the subsample to a 3-liter beaker.

(3) Remove organic matter.

(4) Remove sesquioxides.

(5) Disperse.

(6) Wet sieve through a 70-mesh sieve into a 1250 ml sedimentation cylinder. Dry the sieve on an aluminium tray. Shake the sieve and transfer any grains passing it to the sedimentation cylinder.

(7) Separate the required size fractions from the material in the sedimentation cylinder by sedimentation and decantation, or by elutriation, separating the finer fractions first.

(8) Transfer the sediment remaining in the cylinder to an evaporating dish, add the material held on the 70-mesh sieve, and dry in an oven at 105°C.

(9) Transfer the dried coarse material to the top sieve of a nest selected for the ranges required and sieve mechanically for 10 min.

(10) Subsample the fractions to be separated further.

(11) Make the appropriate specific gravity separations.

(12) Make the appropriate magnetic separations, separating the most strongly magnetic minerals first.

(13) Subsample the separates to be examined microscopically.

(14) Mount the subsamples in an appropriate medium.

(15) Estimate mineral ratios.

3. BULK DENSITY MEASUREMENTS

Quantitative calculations of soil formation require measurement of the bulk density of soil materials. There are three common methods for such measurements.

A. Bulk Density Samples

The soil materials are sampled with a Coile sampler which takes a sample container of known or measureable internal volume. The sampler consists of a cylindrical cutting device recessed behind the cutting edge to take the container. The sampler is pushed into the soil material exposed on the face of a pit by means of a jack hammer or hydraulic jack. The sampler is then dug out with a spade, the

Antipov-Karataev, I. N., and I. G. Tsyurupa (1961). Forms and conditions of migration of substances in the soil profile. *Pochvovedenie No. 8,* 1–12.

Aomine, S., and K. Wada (1962). Differential weathering of volcanic ash and pumice, resulting in formation of hydrated halloysite. *Am. Mineralogist* **47,** 1024–1048.

Aristovskaya, T. V. (1961). Accumulation of iron during decomposition of organo-mineral complexes of humic substances by microorganisms. *Dokl. Akad. Nauk* **136,** 954–957.

Aylmore, L. A. G., and J. P. Quirk (1959). Swelling of clay-water systems. *Nature* **183,** 1752–1753.

Bailey, E. H., and R. E. Stevens (1960). Selective staining of *K* felspar and plagioclase on rock slabs and thin sections. *Am. Mineralogist* **45,** 1020–1025.

Baker, G., and G. W. Leeper (1958). Phytoliths in Victorian soils. *Aust. J. Sci.* **21,** 84.

Barley, K. P. (1959). Earthworms and soil fertility, IV: The influence of earthworms on the physical properties of a red-brown earth. *Aust. J. Agr. Res.* **10,** 371.

Barshad, I. (1955). *Chemistry of the Soil.* (F. E. Baer, ed.), I: Soil development (Reinhold Publ. Corp., New York).

Barshad, I. (1959). Factors affecting clay formation. *Proc. Sixth Natl. Conf. Clays and Clay Minerals, Berkeley,* **1957,** 110–132.

Bassler, R. S. (1908). The formation of geodes, with remarks on the silicification of fossils. *U.S. Nat. Museum Proc.* **35,** 133–154.

Baver, L. D. (1948). *Soil Physics.* (Chapman-Hall, Ltd., London).

Beattie, J. A., and A. D. Haldane (1958). The occurrence of palygorskite and barytes in certain parna soils of the Murrumbidgee region, New South Wales. *Aust. J. Sci.* **20,** 274.

Beckmann, W. (1962). On the micromorphometric investigation of cavities and aggregates in soils. *Z. PflErnahr. Dung., Bodenkunde* **99,** 129–139.

Billings, M. P. (1942). *Structural Geology.* (Prentice-Hall, New York).

Bloomfield, C. (1953*a*). A study of podzolisation, I: The mobilisation of iron and aluminium by Scots pine needles. *J. Soil Sci.* **4,** 5.

Bloomfield, C. (1953*b*). A study of podzolisation, II: The mobilisation of iron and aluminium by the leaves and bark of Agathis Australis (Kauri). *J. Soil Sci.* **4,** 17.

Bloomfield, C. (1953*c*). Sesquioxide immobilisation and clay movement in podzolised soils. *Nature* **172,** 958.

Bloomfield, C. (1956*a*). The experimental production of podzolisation. *Rept. Sixth Intl. Cong. Soil Sci., Paris, Com. V,* 21.

Bloomfield, C. (1956*b*). The deflocculation of kaolinite by aqueous leaf extracts: The role of certain constituents of the extracts. *Rept. Sixth Intl. Cong. Soil Sci., Paris, Com. I,* 27.

Blumel, F. (1962). Forms of iron-hydroxide precipitates in gleys and pseudogleys. *Z. PflErnahr. Dung., Bodenkunde* **98,** 258–264.

Blumel, F., V. Janik, and H. Schiller (1959). The micromorphology and colloid condition of different types of soil. *Testchr. sechsigjahr. Best. landw.-chem. Bundes-Vers Anst. Linz,* 89–120.

Bolter, E. (1961). The decomposition products of olivine feldspar basalts. *Beitr. Mineral. u. Petrograph* **8,** 111–140.

Borchert, H. (1962). The preparation of thin sections of soils with the fixation medium Plexigum M7466. *Z. PflErnahr. Dung., Bodenkunde* **99,** 159–164.

Bourbeau, G. A., and K. C. Berger (1947). Thin sections of soils and friable materials prepared by impregnation with the plastic "Castolite." *Soil Sci. Soc. Am. Proc.* **12,** 409.

Bourne, W. C., and E. P. Whiteside (1962). A study of the morphology and pedogenesis of a medial chernozem developed in loess. *Soil Sci. Soc. Am. Proc.* **26,** 484–490.

Bouyoucos, G. J. (1927). The hydrometer as a new method for the mechanical analysis of soils. *Soil Sci.* **23,** 343–353.

Bouyoucos, G. J. (1934). A comparison between pipette and hydrometer methods for making mechanical analyses of soil. *Soil Sci.* **38,** 335.

Bouyoucos, G. J. (1937). A sensitive hydrometer for determining small amounts of clay or colloids in soils. *Soil Sci.* **44,** 245–246.

Bowen, N. L. (1922). The reaction principle in petrogenesis. *J. Geol.* **30,** 177–198.

Brammall, A. (1928). Dartmoor detritals: a study in provenance. *Proc. Geol. Assoc.* **37,** 27.

Breese, B. F. (1960). Quartz overgrowths as evidence of silica deposition in soils. *Aust. J. Sci.* **23,** 18.

Brewer, R. (1948). Mineralogical examination of soils developed on the Prospect Hill intrusion, New South Wales. *J. Proc. Roy. Soc. N.S.W.* **82,** 272–285.

Brewer, R. (1954). Soil parent material. *Aust. J. Sci.* **16,** 1–4.

Brewer, R. (1955*a*). The occurrence of diatom skeletons and sponge spicules in the soils of New South Wales. *Aust. J. Sci.* **17,** 177–179.

Brewer, R. (1955*b*). Mineralogical examination of a yellow podzolic soil formed on granodiorite. *C.S.I.R.O. (Aust.) Soil Publ. No. 5.,* 28 pp.

Brewer, R. (1956). A petrographic study of two soils in relation to their origin and classification. *J. Soil Sci.* **7,** 268–279.

Brewer, R. (1960*a*). The petrographic approach to the study of soils. *Trans. Seventh Intl. Cong. Soil Sci., Madison* **1,** 1–13.

Brewer, R. (1960*b*). Cutans: their definition, recognition, and classification. *J. Soil Sci.* **11,** 280–292.

Brewer, R., and A. V. Blackmore (1956). The effects of entrapped air and optically oriented clay on aggregate breakdown and soil consistence. *Aust. J. Appl. Sci.* **7,** 59–68.

Brewer, R., and B. E. Butler (1953). Pedology of part of the upper Hunter Valley, N.S.W. *C.S.I.R.O. (Aust.) Soil Publ. No. 2,* 104 pp.

Brewer, R., and A. D. Haldane (1957). Preliminary experiments in the development of clay orientation in soils. *Soil Sci.* **84,** 301.

Brewer, R., and J. R. Sleeman (1960). Soil structure and fabric: their definition and description. *J. Soil Sci.* **11,** 172–185.

Brewer, R., and J. R. Sleeman (1963). Pedotubules: their definition, classification, and interpretation. *J. Soil Sci.* **14,** 156–166.

Brewer, R., and J. R. Sleeman (1964). Glaebules: their definition, classification and interpretation. *J. Soil Sci.* **15,** 66–78.

Briggs, L. I., D. S. McCulloch, and F. Moser (1962). The hydraulic shape of sand particles. *J. Sed. Petrol.* **32,** 645–656.

Bryan, W. A. (1952). Soil nodules and their significance. *Sir Douglas Mawson Anniversary Volume,* 43–53, (University of Adelaide, Aust.)

Bryson, H. R. (1939). The identification of soil insects by their burrow characteristics. *Trans. Kans. Acad. Sci.* **42,** 245–254.

Buol, S. W. (1960). Clay skin genesis in Wisconsin soils. *Diss. Abstrs.* **21**, 389–390. Univ. Wisconsin. Ph.D. Thesis, Univ. Wisconsin.

Buol, S. W., and D. M. Fodness (1961). A new method of impregnating fragile material for thin sectioning *Soil Sci. Soc. Am. Proc.* **25**, 253.

Buol, S. W., and F. D. Hole (1959). Some characteristics of clay skins on peds in the *B* horizon of a grey-brown podzolic soil. *Soil Sci. Am. Proc.* **23**, 239–241.

Buol, S. W., and F. D. Hole (1961). Clay skin genesis in Wisconsin soils. *Soil Sci. Soc. Am. Proc.* **25**, 377–379.

Burges, A., and D. P. Nicholas (1961). Use of soil sections in studying the amount of fungal hyphae in soil. *Soil Sci.* **92**, 25.

Butler, B. E. (1950). A theory of prior streams as a causal factor of soil occurrence in the Riverine Plain of south-eastern Australia. *Aust. J. Agr. Res.* **1**, 23.

Butler, B. E. (1955). A system for the description of soil structure and consistence in the field. *J. Aust. Inst. Agr. Sci.* **21**, 239–249.

Butler, B. E. (1956). Parna: an aeolian clay. *Aust. J. Sci.* **18**, 145.

Butler, B. E. (1958). Depositional systems of the Riverine Plain in relation to soils. *C.S.I.R.O. (Aust.) Soil Publ. No. 10,* 35 pp.

Butler, B. E. (1959). Periodic phenomena in landscapes as a basis for soil studies. *C.S.I.R.O. (Aust.) Soil Publ. No. 14,* 20 pp.

Butler, B. E., and J. T. Hutton (1956). Parna in the Riverine Plain of south-eastern Australia and the soils thereon. *Aust. J. Agr. Res.* **7**, 536–553.

Butler, J. R. (1953). The geochemistry and mineralogy of rock weathering, I: The Lizard area, Cornwall. *Geochim. et Cosmochim. Acta* **4**, 157–178.

Butterfield, J. A. (1938). Outgrowths on zircon. *Geol. Mag.* **73**, 511–516.

Cady, J. G. (1940). Soil analyses significant in forest soils investigations and methods of determination, 3: Some mineralogical characteristics of podzol and brown forest soil profiles. *Soil Sci. Soc. Am. Proc.* **5**, 353.

Cady, J. G. (1950). Rock weathering and soil formation in the North Carolina piedmont region. *Soil Sci. Soc. Am. Proc.* **15**, 337–342.

Caillere, S., S. Henin, and J. Mering (1947). Experimental transformation of montmorillonite into a phyllite with a constant (14Å) spacing. *Compt. Rend.* **224**, 842–843.

Carl, J. D., and G. C. Amstutz (1958). Three-dimensional Liesegang rings by diffusion in a colloidal matrix, and their significance for the interpretation of geological phenomena. *Bull. Geol. Soc. Am.* **66**, 1467.

Carroll, D. (1937). Some aspects of soil mineralogy. *J. Roy. Soc. W.A.* **23**, 7–12.

Carroll, D. (1938). Recording the results of heavy mineral analyses. *J. Sed. Petrol.* **8**, 3–9.

Carroll, D. (1953). Weatherability of zircon. *J. Sed. Petrol.* **23**, 100–116.

Carroll, D. (1957). A statistical study of heavy minerals in sands of the South River, Augusta County, Virginia. *J. Sed. Petrol.* **27**, 387–404.

Carroll, D., R. Brewer, and J. E. Harley (1950). Pebbles from the upper Hunter Valley, N.S.W. *J. Proc. Roy. Soc. N.S.W.* **83**, 251–262.

Carozzi, A. V. (1960). *Microscopic Sedimentary Petrography.* (John Wiley and Sons, New York).

Chakraborty, J. N. (1943). Preliminary treatment of red soil separates, as obtained by mechanical analysis for mineralogical examination. *Indian J. Agric. Sci.* **13**, 609.

Chayes, F. (1956). *Petrographic Modal Analysis.* (John Wiley and Sons, New York).

Childs, E. C. (1957). The anisotropic conductivity of soil. *J. Soil Sci.* **8,** 42–47.

Childs, E. C., and N. Collis-George (1950). The permeability of porous materials. *Proc. Roy. Soc.* **A 201,** 392–405.

Churchward, H. M. (1959). Pedological significance of groundsurface development in an aeolian landscape: a study of soils near Swan Hill. M. Sc. Thesis, University of Sydney.

Churchward, H. M. (1961). Soil studies at Swan Hill, Victoria, Australia, I: Soil layering. *J. Soil Sci.* **12,** 73–86.

Claisse, G. (1953). Study of weathering of charnockite of the Man region. *Ann. Agron.* **4** (n.s.), 45–52.

Clark, J. S., C. A. Gobin, and P. N. Sprout (1961). Yellow mottles in some poorly drained soils of the lower Fraser valley, British Columbia. *Can. J. Soil Sci.* **41,** 218–227.

Clarke, F. W. (1924). The data of geochemistry. *U.S. Geol. Surv. Bull.* **770,** 841 pp.

Correns, C. W. (1962). Formation and transformation of clay minerals during the decomposition of basalts. *Colloq. Intern. Centre Natl. Rech. Sci.* (*Paris*) *No. 105,* 109–120.

Cox, G. H., C. L. Dake, and G. A. Mulenberg (1921). *Field Methods in Petroleum Geology.* (McGraw-Hill Book Co., New York).

Crawford, D. V. (1956). Microbiological aspects of podzolization. *Repts. Sixth Intl. Cong. Soil Sci. Paris, Com III,* 197–202.

Crocker, R. L. (1952). Soil genesis and the pedogenic factors. *Quart. Rev. Biol.* **27,** 139–168.

Crocker, R. L. (1959). The plant factor in soil formation. *Aust. J. Sci.* **21,** 180–192.

Crompton, E. (1960). The significance of the weathering/leaching ratio in the differentiation of major soil groups, with particular reference to some very strongly leached brown earths of the hills of Britain. *Trans. Seventh Intl. Cong. Soil Sci., Madison, Com. V,* 406–412.

Czeratzki, W., and H. Frese (1958). Water and its conduction in soils. *Highway Res. Board Spec. Rept.* **40,** 200.

Dalrymple, J. B. (1962). Some micromorphological implications of time as a soil forming factor, illustrated from sites in south-eastern England. *Z. PflErnahr. Dung., Bodenkunde* **98,** 232–239.

Day, P. R. (1950). Physical basis of particle size analysis by the hydrometer method. *Soil Sci.* **70,** 363–374.

Day, P. R. (1953). Experimental confirmation of hydrometer theory. *Soil Sci.* **75,** 181–186.

Delecour, F., and G. Manil (1962). A micromorphological contribution to the knowledge of the acid brown soils on the Belgian Ardennes. *Z. PflErnahr. Dung., Bodenkunde* **98,** 219–224.

Denison, I. A., W. H. Fry, and P. L. Gile (1929). Alteration of muscovite and biotite in soil. *U.S.D.A. Tech. Bull.* **128,** 32 pp.

De Vore, G. W. (1959). The surface chemistry of feldspars as an influence on their decomposition products. *Proc. Sixth Natl. Conf. Clays and Clay Minerals, Berkeley,* **1957,** 26–41.

Dijk, D. C. van (1958). Principles of soil distribution in the Griffith-Yenda area, N.S.W. *C.S.I.R.O. (Aust.) Soil Publ. No. 11,* 24 pp.

Dijk, D. C. van (1959). Soil features in relation to erosional history in the vicinity of Canberra. *C.S.I.R.O. (Aust.) Soil Publ. No. 13,* 41 pp.

Doeglas, D. J. (1946). Interpretation of the results of mechanical analyses. *J. Sed. Petrol.* **16,** 19–40.

Drosdoff, M., and E. F. Miles (1938). Action of hydrogen peroxide on weathered mica. *Soil Sci.* **46,** 391.

Drosdoff, M., and C. C. Nikiforoff (1940). Iron-manganese concretions in Dayton soils. *Soil Sci.* **49,** 333.

Droste, J. B. (1959). Clay minerals in playas of the Mojave Desert, California. *Science* **130,** 100.

Druif, J. H. (1937). Aantasting von mineralen in den boden von Deli. *Bull. 38 von Set Deli Praefstation to Medon,* Sumatra.

Dryden, L. (1931). Accuracy in percentage representation of heavy mineral frequencies. *Proc. Nat. Acad. Sci.* **17,** 233–238.

Dryden, L. (1935). A statistical method for the comparison of heavy mineral suites. *Am. J. Sci.* **29,** 393–408.

Dryden, L., and C. Dryden (1946). Comparative rates of weathering of some common heavy minerals. *J. Sed. Petrol.* **16,** 91–96.

Ellis, B. S. (1952). Genesis of a tropical red soil. *J. Soil Sci.* **3,** 52.

Ellis, J. H., and O. G. Caldwell (1935). Magnesium clay solonetz. *Trans. Third Intl. Cong. Soil Sci., Oxford, Com. V,* 348.

Emerson, W. W. (1959). The structure of soil crumbs. *J. Soil Sci.* **10,** 235–244.

Engelhardt, W. von (1937). Grain size distribution, mineral composition and chemical relations of four soil horizons of southern Mechlenberg, Germany. *Chem. Erde* **11,** 17–37.

Eremin, G. G. (1952). Solonetz and solonetz-like soils of the Krazno-Perekop section of the Crimean region and their improvement. *Vestnik Moskov Univ. 7, No. 12; Ser. Fiz. Mat. i. Estestven Nauk No. 8,* 99–113.

Evans, P., R. J. Hayman, and M. A. Majeed (1933). The graphical representation of heavy mineral analyses. *World Petrol. Cong. Proc.* **1,** 251–256.

Fairbairn, H. W. (1943). Gelatinized slides for R.I. immersion mounting. *Am. Mineralogist* **28,** 396.

Fairbanks, E. E. (1925). A modification of Lemberg's staining methods. *Am. Mineralogist* **10,** 126–127.

Fay, A. H. (1920). Glossary of the mining and mineral industry. *U.S. Bur. Mines, Bull.* **95,** 754 pp.

Feigl, F. (1954). *Spot Tests.* Vol. 1. (Elsevier Publ. Co., London).

Fieldes, M., and L. D. Swindale (1954). Chemical weathering of silicates in soil formation. *N.Z. J. Sci. Tech.* **36B,** 140–154.

Fieldes, M., I. F. Walker, and P. P. Williams (1956). Clay mineralogy of New Zealand soils. Infra-red absorption spectra on soil clays. *N.Z. J. Sci. Tech.* **38,** 31.

Folk, R. L. (1954). The distinction between grain size and mineral composition in sedimentary rock nomenclature. *J. Geol.* **62,** 344–359.

Folk, R. L. (1955). Student operator error in determination of roundness, sphericity and grain size. *J. Sed. Petrol.* **25,** 297–301.

Folk, R. L., and W. Ward (1957). Brazos River bar: a study in the significance of grain size parameters. *J. Sed. Petrol.* **27,** 3–26.

Ford, W. E. (1947). *Dana's Textbook of Mineralogy.* (John Wiley and Sons, New York).

Freeman, R. S., and M. W. Gardner (1962). Production of thin rock sections using metal plates on a surface grinding machine. *N.Z. J. Geol. and Geophysics* **5,** 662–665.

Frei, E., and M. G. Cline (1949). Profile studies of normal soils of New York, II: Micromorphological studies of the grey-brown podzolic–brown podzolic sequence. *Soil Sci.* **68,** 333–345.

Fridland, V. M. (1958). Podzolization and illimerization. *Pochvovedenie No. 1,* 24–32.

Friedman, G. M. (1958). Determination of sieve-size distribution from thin-section data for sedimentary petrological studies. *J. Geol.* **66,** 394.

Friedman, G. M. (1962). On sorting coefficients, and the lognormality of the grain size distribution of sandstones. *J. Geol.* **70,** 737–753.

Fry, W. H. (1933). Petrographic methods for soil laboratories. *U.S.D.A. Tech. Bull.* **344,** 96 pp.

Fuechtbauer, H. (1959). Zur Nomenklatur der Sedimentgesteine. *Erdsel und Kohle* **12,** 605–613.

Gabriel, S., and E. P. Cox (1929). A staining method for the quantitive determination of certain rock minerals. *Am. Mineralogist* **14,** 290.

Gennart, M., M. Kurbanovic, A. Nanson, et. al. (1961). A micromorphological contribution to the bio-dynamic study of equatorial forest soil. *Pedologie* **11,** 110–122.

Gepp, G. G., and A. F. Haigh (1928). *Latin-English Dictionary.* (Longmans, Green and Co., London).

Geyger, E. (1962). On the method of micro-morphometric soil investigation. *Z. PflErnahr. Dung., Bodenkunde* **99,** 118–129.

Goldich, S. S. (1938). A study in rock weathering. *J. Geol.* **46,** 17–58.

Gorbunov, N. I. (1961). Migration of colloidal and clay particles in soils. *Pochvovedenie No. 7,* 13–28.

Gotschalk, Y. J. (1957). Remarks on insufficiently studied problems of soil hydrology. *Pochvovedenie No. 1,* 76–81.

Grabau, A. W., and H. W. Shimer (1909–1910). *North American Index Fossils–Invertebrates.* (A. G. Seiler and Co., New York).

Gracanin, Z. (1962). The genesis, morphology and micromorphology of hill-side peats on limestone in Croatia. *Z. PflErnahr. Dung., Bodenkunde* **98,** 264–272.

Graham, E. R. (1950). The plagioclase felspars as an index to soil weathering. *Soil Sci. Soc. Am. Proc.* **14,** 300–302.

Graham, E. R. (1955). Rapid determination of quartz, potash minerals and plagioclase felspars in soils. *Chem. Anal.* **44,** 37–38.

Grim, R. E. (1953). *Clay Mineralogy.* (McGraw-Hill Book Co., London).

Gritsaenko, G. S., S. J. Berkhin, and E. S. Rudnitskaya (1956). Dependence of the composition of clayey minerals on nature of the environment. *Kora Vyvetirivaniya (Moscow: Akad. Nauk SSSR) Sbornik* **2,** 101–106.

Grogan, R. M. (1945). Shape variation of some Lake Superior beach pebbles. *J. Sed. Petrol.* **15,** 3–10.

Grossman, R. B. (1959). Characterisation of ped surfaces in contrast to ped interiors from B horizons of some major great soil groups in Illinois. Ph.D. Thesis, University of Illinois.

Grossman, R. B., I. Stephen, J. B. Fehrenbacher, and A. H. Beavers (1959). Fragipan soils of Illinois, III: Micromorphological studies of Hosmer silt loam. *Soil Sci. Soc. Am. Proc.* **23,** 73–75.

Gruner, J. W., and G. A. Thiel (1937). The occurrence of fine-grained authigenic felspar in shales and silts. *Am. Mineralogist* **22,** 842–846.

Hall, A. D. (1945). *The Soil. An Introduction to the Scientific Study of the Growth of Crops.* (John Murray, London).

Hallsworth, E. G. (1952). An interpretation of the soil formations found on basalt in the Richmond-Tweed region of New South Wales. *Aust. J. Agr. Res.* **2,** 411.

Harrison, J. L., and H. H. Murray (1959). Clay mineral stability and formation during weathering. *Proc. Sixth Nat. Conf. Clays and Clay Minerals, Berkeley,* **1957,** 144–153.

Haseman, J. F., and C. E. Marshall (1945). The use of heavy minerals in studies of the origin and development of soils. *Mo. Agr. Exp. Sta. Res. Bull.* **387,** 75 pp.

Hawkins, R. H., and E. R. Graham (1950). Mineral contents of the silt separates of some Missouri soils as these indicate the fertility level and degree of weathering. *Soil Sci. Soc. Am. Proc.* **15,** 308.

Hendrick, J., and G. Newlands (1928). The mineral composition of the soil as a factor in soil classification. *Proc. First Intl. Cong. Soil Sci., Washington,* **4,** 104–107.

Herdan, G. (1960). *Small Particle Statistics.* (Butterworths, London).

Hilgard, E. W. (1914). *Soils.* (Macmillan Book Co., New York).

Hill, G. F. (1942). Termites (Isoptera) from the Australian Region. *Monograph, Counc. Sci. Ind. Res. (Aust.)*, Melbourne.

Hoehne, K. (1954). The genesis of quartz layers and similar formations in coal beds of the Ruhr Carboniferous. *Chem. Erde* **17,** 63–72.

Hole, F. D. (1961). A classification of pedoturbations and some other processes and factors of soil formation in relation to isotropism and anisotropism. *Soil Sci.* **19,** 375–377.

Holmes, A. (1928). *The Nomenclature of Petrology.* (Thomas Murby and Co., London).

Hseung, Y., and M. L. Jackson (1952). Mineral composition of the clay fraction, III: Of some main soil groups of China. *Soil Sci. Soc. Am. Proc.* **16,** 294–297.

Hugie, V. K., and H. B. Passey (1963). Cicadas and their effect upon soil genesis in certain soils in southern Idaho, northern Utah, and northeastern Nevada. *Soil Sci. Soc. Am. Proc.* **27,** 78–82.

Humbert, R. (1948). The genesis of laterite. *Soil Sci.* **65,** 281.

Humbert, R. P., and C. E. Marshall (1943). Mineralogical and chemical studies of soil formation from acid and basic igneous rocks in Missouri. *Mo. Agr. Exp. Sta. Res. Bull.* **359,** 60 pp.

Hunter, R. J. (1961). Flow of colloidal suspensions through porous media. Ph.D. Thesis, University of Sydney.

Hutton, J. T. (1950). Detailed method of particle size analysis of soils using mechanical dispersion and a plummet balance. *C.S.I.R.O. (Aust.) Div. Soils, Tech. Mem. 7/50.*

Hutton, J. T. (1955). A method of particle size analysis of soils. *C.S.I.R.O. (Aust.) Div. Soils, Div. Rept. 11/55.*

Inman, D. L. (1952). Measures for describing the size distribution of sediments. *J. Sed. Petrol.* **22,** 125–145.

Jackson, M. L. (1958). *Soil Chemical Analysis.* (Constable and Co. Ltd., London).

Jackson, M. L. (1959). Frequency distribution of clay minerals in major great soil groups as related to the factors of soil formation. *Proc. Sixth Natl. Conf. Clays and Clay Minerals, Berkeley,* **1957,** 133–143.

Jackson, M. L., and G. D. Sherman (1953). Chemical weathering of minerals in soils. *Advan. Agron.* **5,** 219–318.

Jacot, A. P. (1936). Soil structure and soil biology. *Ecology* **17,** 359.

Jacot, A. P. (1940). The fauna of soil. *Quart. Rev. Biol.* **15,** 28.

Jeffries, C. D. (1946). A rapid method for the removal of free iron oxides in soil prior to petrographic analysis. *Soil Sci. Soc. Am. Proc.* **11,** 211.

Jeffries, G. D., B. N. Rolfe, and G. W. Kunze (1953). Mica weathering sequence in the Highfield and Chester soil profiles. *Soil Sci. Soc. Am. Proc.* **17,** 337.

Jeffries, C. D., and J. W. White (1937). Some mineralogical and chemical characteristics of a Hagerstown profile. *Soil Sci. Soc. Am. Proc.* **2,** 133.

Jenny, H. (1941). *Factors of Soil Formation.* (McGraw-Hill Book Co., New York).

Jessup, R. W. (1960). Identification and significance of buried soils of Quaternary age in the south-eastern portion of the Australian arid zone. *J. Soil Sci.* **11,** 197–205.

Joffe, J. S. (1936). *Pedology.* (Rutgers Univ. Press, New Brunswick, N.J.).

Joffe, J. S., and M. Zimmerman (1944). Sodium, calcium, and magnesium ratios in the exchange complex. *Soil Sci. Soc. Am. Proc.* **9,** 51.

Johannsen, A. (1939). *A Descriptive Petrography of the Igneous Rocks.* Vol. 1 (Chicago Univ. Press, Chicago).

Johnson, J. R., and J. B. Peterson (1941). Microscopic study of soils from five great soil groups. *Soil Sci. Soc. Am. Proc.* **6,** 360–367.

Johnson, W. M., J. E. McClelland, S. A. McCaleb, et al. (1960). Classification and description of soil pores. *Soil Sci.* **89,** 319–321.

Johnston, E. J. (1953). Pedology of the Deniboota irrigation district, N.S.W. *C.S.I.R.O. (Aust.) Soil Publ. No. 1,* 92 pp.

Jongerius, A. (1957). Morphologic investigations of soil structure. *Bodemkundige Studies No. 2.,* Mededelingen van de Stickting voor Bodemkartering. Wageningen.

Karpachevskii, L. O. (1960). Micromorphological investigation of the processes of leaching and podzolisation of soils under forest. *Pochvovedenie No. 5,* 43–52.

Katchinski, N. A. (1956). Soil structure and differential porosity. *Rept. Sixth Intl. Cong. Soil Sci., Paris, Com.* **I,** 35–52.

Kawai, K. (1961). A method of preparing thin sections of soil with polyester resin. *J. Sci. Soil Tokyo* **33,** 190–194.

Keller, W. D. (1955). *The Principles of Chemical Weathering.* (Lucas Bros, Columbia, Missouri).

Keller, W. D. (1958). Argillation and direct bauxitization in terms of concentrations of hydrogen and metal cations at surface of hydrolysing aluminium silicates. *Bull. Am. Assoc. Petrol. Geologists* **42,** 233–245.

Kemp, J. F. (1927). *A Handbook of Rocks.* (D. van Nostrand Co., Amsterdam).

Kennedy, G. C. (1947). Charts for correlation of optical properties with chemical composition of some common rock-forming minerals. *Am. Mineralogist* **32,** 561.

Kevan, O. K. McE. (1955). *Soil Zoology.* (Butterworths, London).

Khan, D. H. (1959). Studies on translocation of chemical constituents in some red-brown soils, terra rossas, and rendzinas, using zirconium as a weathering index. *J. Soil Sci.* **88,** 196–200.

Knox, E. G. (1957). Fragipan horizons in New York soils, III: The basis of rigidity. *Soil Sci. Soc. Am. Proc.* **21,** 326.

Kovda, V. A., B. A. Zimovets, and A. G. Amchislavskaya (1958). The hydromorphic accumulation of compounds of silica and sesquioxides in the soils of the Amur region. *Pochvovedenie No. 5,* 1–11.

Kreveld, A. van (1942). The size distribution of fat globules in milk. *Rec. Trav. Chim. Pays-Bas* **61,** 29–40.

Krishna, P. G., and S. Perumal (1948). Structure in black cotton soils of the Nizamsayar project area, Hyderabad State, India. *Soil Sci.* **66,** 29.

Krumbein, W. C. (1934). Size frequency distributions of sediments. *J. Sed. Petrol.* **4,** 65–77.

Krumbein, W. C. (1941). Measurement and geological significance of shape and roundness of sedimentary particles. *J. Sed. Petrol.* **11,** 64–72.

Krumbein, W. C. (1942). Settling-velocity and flume-behavior of nonspherical particles. *Trans. Am. Geophys. Union,* 621–633.

Krumbein, W. C., and F. J. Pettijohn (1938). *Manual of Sedimentary Petrography.* (Appleton-Century Co., New York).

Krumbein, W. C., and L. L. Sloss (1955). *Stratigraphy and Sedimentation.* (W. H. Freeman and Co., California).

Krynine, P. D. (1950). Microscopic morphology of quartz types. *Annals Second Pan. Am. Cong. Min. Eng. Geol.* **3** (2nd Comm.), 35–49.

Kubiena, W. L. (1931). Micropedological studies. *Wiss. Arch. Pflanzenbau* **5,** 613–648.

Kubiena, W. L. (1937). Method of preparing thin slices of soils in an undisturbed state. *Zeiss-Nachr.* **2,** 81.

Kubiena, W. L. (1938). *Micropedology.* (Collegiate Press Inc., Ames, Iowa).

Kubiena, W. L. (1953). *The Soils of Europe.* (Thomas Murby and Co., London).

Kubiena, W. L. (1956*a*). Zur Mikormorphologie, Systematik und Entwicklung der rezenten und fossilen Lossboden. *Eiszeitalter und Gegenwart, Ohringen* **7,** 102–112.

Kubiena, W. L. (1956*b*). Red earth formation and lateritization: their differentiation by micromorphological characteristics. *Repts. Sixth Intl. Cong. Soil Sci., Paris, Com. V,* 247–249.

Kubiena, W. L. (1962). Taxonomic importance of the type and formation of iron-hydroxide minerals in tropical soils. *Z. PflErnahr. Dung., Bodenkunde* **98,** 205–213.

Kubiena, W. L., W. Beckmann, and E. Geyger (1961). Method of photogrammetric analysis of soil structure. *Z. PflErnahr. Dung., Bodenkunde* **92,** 116–126.

Kuenen, Ph. H. (1958). Some experiments on fluviatile rounding. *Proc. Kon. Ned. Akad. Wetensch. Amsterdam, Ser. B,* **61,** 47–53.

Kuenen, Ph. H. (1959). Experimental abrasion, III: Fluviatile action on sand. *Am. J. Sci.* **257**, 172–190.

Kuenen, Ph. H. (1960). Experimental abrasion, IV: Aeolian action. *J. Geol.* **68**, 427–449.

Lafeber, D. (1962). Aspects of stress-induced differential movements of fabric elements in mineral soils. *Proc. First Conf. Aust. Road Res. Board* **1**, 1059–1067.

Langford, F. F. (1962). Epoxy resins for oil immersion and heavy mineral studies. *Am. Mineralogist* **47**, 1478–1480.

Lapham, M. H. (1932). Genesis and morphology of desert soils. *Rept. Twelfth Ann. Meeting Am. Soil Surv. Assoc. Bull.* **13**, 34–52.

Laruelle, J. (1956). Some aspects of the micro-structure of soils in the north-east of the Belgian Congo. *Pedologie* **6**, 38–54.

Laruelle, J. (1958). Micromorphology of the soils of Belgium. *Pedologie* **8**, 79–102.

Leeper, G. W. (1952). *Introduction to Soil Science.* (Melbourne Univ. Press).

Low, A. J. (1954). Study of soil structure in the field and the laboratory. *J. Soil Sci.* **5**, 57–74.

McAleese, D. M., and W. A. Mitchell (1958). Studies on the basaltic soils of Northern Ireland, IV: Mineralogical study of the clay separates. *J. Soil Sci.* **9**, 76–80.

McCaleb, S. B. (1954). Profile studies of normal soils of New York, IV: Mineralogical properties of the grey-brown podzolic-brown podzolic soil sequence. *Soil Sci.* **77**, 319–333.

McCammon, R. B. (1962). Efficiencies of percentile measures for describing mean size and sorting of sedimentary particles. *J. Geol.* **70**, 453–465.

MacEwen, D. M. C. (1954). Cordenite, a trioctahedral montmorillonoid derived from biotite. *Clay Minerals Bull.* **2**, 120–125.

McIntyre, D. S., and G. B. Stirk (1954). A method for determination of apparent density of soil aggregates. *Aust. J. Agr. Res.* **5**, 291–296.

McLean, W. (1931). Effect of hydrogen peroxide on soil organic matter. *J. Agr. Sci.* **21**, 251–261.

McMillan, N. J., and J. Mitchell (1953). Microscopic study of platy and concretionary structures in certain Saskatchewan soils. *Sci. Agr.* **33**, 178–183.

Macaulay Institute (1948). The soil clays of north-east Scotland. *Macaulay Institute Rpt.* 1947–1948, 10 pp.

Mackenzie, A. F., and J. E. Dawson (1961). The preparation and study of thin sections of wet organic soil materials. *J. Soil Sci.* **12**, 142–144.

Manil, G. (1962). Discussion of the term "lessive" on the basis of micromorphological observations. *Z. PflErnahr. Dung., Bodenkunde* **98**, 214–218.

Marbutt, C. F. (1927). A scheme for soil classification. *Proc. First Intl. Cong. Soil Sci., Washington, Com.* **V**, 6.

Marel, H. W. van der (1949). Mineralogical composition of a heath podzol. *Soil Sci.* **67**, 193–207.

Marshall, C. E. (1935). The importance of the lattice structure of the clays for the study of soils. *J. Soc. Chem. Ind.,* **54**, 393.

Marshall, C. E., and C. D. Jeffries (1946). Mineralogical methods in soil research, I: The correlation of soil types and parent materials with supplementary information on weathering processes. *Soil Sci. Soc. Am. Proc.* **10**, 397.

Marshall, T. J. (1959). *Relations between Water and Soil.* Comm. Bur. Soils, Tech. Com. No. 50, 91 pp. (Comm. Agric. Bur., Harpenden).

Martin, R. T. (1954). Calcium oxalate formation in soil from hydrogen peroxide treatment. *Soil Sci.* **77,** 143–145.

Matelski, R. P. (1953). Removal of coatings from soil particles for petrographic analysis. *Soil Sci. Soc. Am. Proc.* **17,** 103–106.

Matelski, R. P., and L. M. Turk (1947). Heavy minerals in some podzolic soil profiles in Michigan. *Soil Sci.* **64,** 469.

Meade, R. H. (1960). Compaction and development of preferred orientation in clayey sediments. Ph.D. Thesis, Stanford Univ.; *Dissert. Absts.* **21,** 3421 (1961).

Merrill, G. P. (1921). *Rocks, Rock Weathering, and Soils.* (Macmillan Book Co., New York).

Merwe, H. van der, and H. Heysteck (1952). Clay minerals of South African soil groups, I: Laterites and related soils. *Soil Sci.* **74,** 383–401.

Milner, H. B. (1962). *Sedimentary Petrography.* (George Allen and Unwin Ltd., London).

Minashina, N. G. (1958). Optically oriented clays in soils. *Soviet Soil Sci. No. 4,* 424–430.

Mitchell, B. D. (1962). Effect of soil-forming factors on clay genesis. *Colloq. Intern. Centre Natl. Rech. Sci. (Paris) No. 105,* 139–47.

Mitchell, J. K. (1956). The fabric of natural clays and its relationship to engineering properties. *Proc. Highway Res. Board* **35,** 693.

Mohr, E. J. (1944). *Soils of Equatorial Regions.* (Edwards Bros., Ann Arbor, Michigan).

Moreland, M. M., K. Lawton, and G. Uehara (1956). Alteration of biotite to vermiculite by plant growth. *Soil Sci.* **82,** 477–481.

Morozova, T. D. (1962). Use of the micromorphological method for studying fossil soils. *Izv. Akad. Nauk Ser. Gergr. No. 1,* 109–114.

Muir, A. (1951). Notes on the soils of Syria. *J. Soil Sci.* **2,** 163–182.

Neuwirth, E. (1954). The weathering of felspar and biotite in the Karalin district (Styria, Austria). *Ber. -u Luttenmann. Monatsch. Montan Hochschule Leoben* **99,** 166–171.

Nicholls, A. (1936). The mineralogy of the sand fractions of some Victorian soils. *Proc. Roy. Soc. Vic.* **49** (n.s.), 17–35.

Nikiforoff, C. C., and M. Drosdoff (1943). Genesis of a clay-pan soil. *Soil Sci.* **55,** 459.

Nikiforoff, C. C., R. P. Humbert, and J. G. Cady (1948). The hardpan in certain soils of the Coastal Plain. *Soil Sci.* **65,** 135–153.

Oertel, A. C. (1961). Pedogenesis of some red brown soils based on trace element profiles. *J. Soil Sci.* **12,** 242–258.

O'Neal, A. M. (1949). Soil characteristics significant in evaluating permeability. *Soil Sci.* **67,** 403–409.

Paletskaya, L. N., A. P. Lavrov, and Sh. I. Kogan (1958). The question of the formation of porosity of takyr crusts. *Pochvovedenie No. 3,* 34–41.

Peterson, J. B. (1937). The micromorphology of some loessial soils of Iowa. *Soil Sci. Soc. Am. Proc.* **2,** 9–13.

Peterson, J. B. (1944). The effect of montmorillonitic and kaolinitic clays on the formation of platy structures. *Soil Sci. Soc. Am. Proc.* **9,** 37.

Pettijohn, F. J. (1931). Petrography of the beach sands of southern Lake Michigan. *J. Geol.* **39,** 432–455.

Pettijohn, F. J. (1941). Persistence of heavy minerals and geologic age. *J. Geol.* **49,** 610–625.

Pettijohn, F. J. (1957). *Sedimentary Rocks.* (Harper Bros., New York) (Second Ed.).

Pettijohn, F. J. and A. C. Lundahl (1943). Shape and roundness of Lake Erie beach sands. *J. Sed. Petrol.* **13,** 69–78.

Piper, C. S. (1942). *Soil and Plant Analysis.* (Hassell Press, Adelaide).

Plas, L. van der (1962). Preliminary note on the granulometric analysis of sedimentary rocks. *Sedimentology* **1,** 145–157.

Poldervaart, A. (1956). Zircons in rocks, 2: Igneous rocks. *Am. J. Sci.* **254,** 521–554.

Pol'skii, M. N. (1962). Some new methods of preparing thin sections of soils. *Pochvovedenie No. 10,* 104–111.

Polynov, B. P. (1944). Total analyses of soils and their interpretation. *Pedology (U.S.S.R.)* **10,** 482–490.

Povoas, J. C., and J. B. Macedo (1959). The influence of drainage on the type of clay mineral. *Proc. Third Interafr. Soils Conf., Dalaba,* **1,** 155–156.

Powers, M. (1953). A new roundness scale for sedimentary particles. *J. Sed. Petrol.* **23,** 117–119.

Prescott, J. A. (1949). A climatic index for the leaching factor in soil formation. *J. Soil Sci.* **1,** 9.

Quirk, J. P., and C. R. Panabokke (1962). Pore volume-size distribution and swelling of natural soil aggregates. *J. Soil Sci.* **13,** 71–81.

Raeside, J. D. (1959). Stability of index minerals in soils with particular reference to quartz, zircon, and garnet. *J. Sed. Petrol.* **29,** 493–502.

Rankama, K., and T. G. Sahama (1950). *Geochemistry.* (Univ. Chicago Press, Chicago).

Ratcliffe, F. N., and T. Greaves (1940). The subterranean foraging galleries of Coptotermes lacteus. *J. Counc. Sci. Ind. Res. (Aust.)* **13,** 150.

Rayleigh, Lord (1942). The ultimate shape of pebbles, natural and artificial. *Proc. Roy. Soc.* **A 181,** 107.

Rayleigh, Lord (1944). Pebbles of regular shape, and their reproduction in experiment. *Nature* **154,** 169.

Read, H. H. (1948). *Rutley's Elements of Mineralogy.* (Thomas Murby and Co., London).

Reineck, H. E. (1962). Relief casts of undisturbed samples of sand. *Z. Pfl-Ernahr. Dung., Bodenkunde* **99,** 151–153.

Rice, C. M. (1954). *Dictionary of Geological Terms.* (Edwards Bros., Ann Arbor, Michigan).

Richard, J. A., and R. F. Chandler, Jr. (1943). Some physical and chemical properties of mature podzol profiles. *Soil Sci. Soc. Am. Proc.* **8,** 379.

Rieken, F. F. (1943). Some considerations in the magnesium cycle of weathering in solonetz soils. *Soil Sci. Soc. Am. Proc.* **8,** 391.

Robinson, G. W. (1922). A new method for the mechanical analysis of soils and other dispersions. *J. Agr. Sci.* **12,** 306–321.

Robinson, G. W. (1949). *Soils. Their Origin, Constitution and Classification.* (Thomas Murby and Co., London).

Rode, A. A., Ye. A. Yarilova, and I. M. Rashevskaya (1960). Certain genetic characteristics of dark soils. *Soviet Soil Sci. No. 8,* 799–809.

Rolfe, B. N., and C. D. Jeffries (1952). A new criterion for weathering in soils. *Soil Sci.* **116,** 599.

Romashkevich, A. I. (1962). Microstructure and microaggregation of soils in relation to erosion and formation of deposits. *Pochvovedenie No. 10,* 56–61.

Rosenblum, S. (1958). Magnetic susceptibilities of minerals in the Frantz isodynamic magnetic separator. *Am. Mineralogist* **43,** 170–173.

Rosiwal, A. (1898). Ueber geometrische Gesteinsanalysen. *Ein enfacher Weg zuziffernmassigen Festellung des Quantitatsverhaltnisses der Mineralbestandtheile gemengter Gesteine: Verhandl K. K. Geol. Reichsanstalt, Vienna,* p. 143.

Rubey, W. W. (1933). The size distribution of heavy minerals within a water-laid sandstone. *J. Sed. Petrol.* **3,** 3–29.

Ruhe, R. V. (1956). Geomorphic surfaces and the nature of soils. *Soil Sci.* **82,** 441–455.

Russell, R. D., and R. E. Taylor (1937). Roundness and shape of Mississippi River sands. *J. Geol.* **45,** 225–267.

Scharrer, K. (1933). Reaction Kinetics of the hydrogen-peroxide-decomposition properties of soils. *Z. PflErnahr. Dung., Bodenkunde* **31A,** 27–36.

Schneiderhoehn, P. (1953). Untersuchungen zur Siebanalyse von Sanden und zur Darstellung ihrer Ergebnisse. *Neues Jb. Mineralog., Abb.* **85,** 141–202.

Scott, W. B. (1943). *An Introduction to Geology.* (Macmillan Book Co., New York).

Semenov, E. I. (1959). Zirconium content in titanium minerals. *Trudy Inst. Mineral., Geokhim i Kristallokhim. Redkikh Elementov, Voprosy Mineral., Geokhim. i. Genezisa Mestarozhdenii Redkikh Elementov, No. 2,* 87–92.

Sharp, W. E., and P. F. Fan (1963). A sorting index. *J. Geol.* **71,** 76–84.

Shapiro, L., and W. W. Brannock (1956). Rapid analysis of silicate rocks. *U.S. Geol. Surv. Bull.* **1036-C,** 56 pp.

Shorter Oxford Dictionary (1956). (Oxford Clarendon Press.)

Shrock, R. F. (1948). *Sequence in Layered Rocks.* (McGraw-Hill Book Co., London).

Shukevich, M. M. (1954). Problems of the determination of clay minerals by dye-staining methods. *Doklady Akad. Nauk S.S.R.* **94,** 327–328.

Sigmond, A. A. J. de (1938). *The Principles of Soil Science.* (Thomas Murby and Co., London).

Simon, L. E. (1941). *Engineers' Manual of Statistical Methods.* (John Wiley and Sons, New York) (out of print).

Sivarajasingham, S., L. T. Alexander, J. G. Cady, and M. G. Cline (1962). Laterite. *Advan. in Agron.* **14,** 1–60.

Sleeman, J. R. (1963). Cracks, peds, and their surfaces in some soils of the Riverine Plain, N.S.W. *Aust. J. Soil Res.* **1,** 91–102.

Smith, R. (1949). The zonal concept and soil classification. *J. Aust. Inst. Agr. Sci.* **15,** 108.

Smithson, F. (1941). Alteration of the detrital minerals in the Mesozoic rocks of Yorkshire. *Geol. Mag.* **78,** 97–112.

Smithson, F. (1956*a*). Plant opal in soil. *Nature* **178,** 107.

Smithson, F. (1956*b*). Silica particles in some British soils. *J. Soil Sci.* **7,** 122–129.

Smithson, F. (1958). Grass opal in British soils. *J. Soil Sci.* **9,** 148.
Smithson, F. (1959). Opal sponge spicules in soils. *J. Soil Sci.* **10,** 105.
Sneed, E. D., and R. L. Folk (1958). Pebbles in the lower Colorado River, Texas: a study in particle morphogenesis. *J. Geol.* **66,** 114–150.
Soileau, J. M., W. A. Jackson, and R. J. McCracken (1964). Cutans (clay films) and potassium availability to plants. *J. Soil Sci.* **15,** 116–123.
Sorby, H. C. (1880). On the structure and origin of non-calcareous stratified rocks. *Quart. J. Geol. Soc. London* **36,** 43–54.
Spencer, D. W. (1963). The interpretation of grain size distribution curves of clastic sediments. *J. Sed. Petrol.* **33,** 180.
Springer, M. E. (1958). Desert pavement and vesicular layer of some soils of the desert of the Lahontan Basin Nevada. *Soil Sci. Soc. Am. Proc.* **22,** 63–66.
Spry, A. (1962). The origin of columnar jointing, particularly in basalt flows. *J. Geol. Soc. Aust.* **8,** 191–216.
Stephen, I. (1952*a*). A study of rock weathering with reference to the soils of the Malvern Hills, I: Weathering of biotitite and granite. *J. Soil Sci.* **3,** 20.
Stephen, I. (1952*b*). A study of rock weathering with reference to the soils of the Malvern Hills, II: Weathering of appinite and "Ivy-Scar rock." *J. Soil Sci.* **3,** 219.
Stephen, I. (1959). Pedology. Pedogenic weathering. *Sci. Progress* **47,** 306–313.
Stephen, I. (1960). Pedology. Clay orientation in soils. *Sci. Progress* **48,** 322–331.
Stephens, C. G. (1947). Functional synthesis in pedo-genesis. *Trans. Roy. Soc. S. Australia* **71,** 168–181.
Stephens, C. G. (1956). *A Manual of Australian Soils.* (C.S.I.R.O. [Aust.], Melbourne).
Stern, K. H. (1954). The Liesegang phenomenon. *Chem. Rev.* **54,** 79–99.
Stirk, G. B. (1954). Some aspects of soil shrinkage and the effect of cracking upon water entry into soil. *Aust. J. Agr. Res.* **5,** 279.
Stow, M. H. (1932). Authigenic tourmaline in the Oriskany sandstone. *Am. Mineralogist* **17,** 150.
Swanson, C. L. W., and J. B. Peterson (1942). The use of the micrometric and other methods for the evaluation of soil structure. *Soil Sci.* **53,** 173–183.
Swindale, L. D., and M. L. Jackson (1956). Genetic processes in some residual podzolised soils of New Zealand. *Rept. Sixth Int. Cong. Soil Sci., Paris, Com.* **V,** 233–234.
Tamm, O. (1934). The oxalate method of soil analysis. *Medd. Skogsforsoksanst.* **27,** 1–20.
Tanner, C. B., and M. L. Jackson (1947). Nomographs of sedimentation times for soil particles under gravity or centrifugal acceleration. *Soil Sci. Soc. Am. Proc.* **12,** 60–65.
Tester, A. C., and G. I. Atwater (1934). The occurrence of authigenic felspars in sediments. *J. Sed. Petrol.* **4,** 23–31.
Thiel, G. A. (1940). The relative resistance to abrasion of mineral grains of sand size. *J. Sed. Petrol.* **10,** 103–124.
Thorp, J., L. E. Strong, and E. E. Gamble (1957). Experiments in soil genesis—the role of leaching. *Soil Sci. Soc. Am. Proc.* **21,** 99–102.
Trask, P. D. (1932). *Origin and Environment of Source Sediments of Petroleum.* (Gulf Publ. Co., Houston, Texas).

Tricart, J. (1958). Methods amelioree pour l'etude des sables. *Rev. Geomorf. Dynam.* **9,** 43–54.

Troell, E. (1931). The use of sodium hypobromite for the oxidation of organic matter in the mechanical analysis of soils. *J. Agr. Sci.* **5,** 476–484.

Truog, E., J. R. Taylor, R. W. Pearson, M. E. Weeks, and R. W. Simonson (1937). Procedure for special type of mechanical and mineralogical soil analysis. *Soil Sci. Soc. Am. Proc.* **1,** 101.

Twenhofel, W. H. (1950). *Principles of Sedimentation.* (McGraw-Hill Book Co., New York).

U.S.D.A. (1951). *Soil Survey Manual.* U.S. Dept. Agr. Handbook No. 18.

U.S.D.A. (1960). *Soil Classification. A Comprehensive System.* Seventh Approximation. U.S. Dept. Agr., Soil Conserv. Service, Soil Survey Staff.

U.S. Engineers (1943). A study of methods used in measurement and analysis of sediment loads in streams. *Rept No. 7:* A study of new methods of size analysis of suspended sediment samples. (St. Paul, U.S. Engineers District Sub-Office, Hydraulic Lab., Univ. of Iowa, Iowa City).

U.S. Engineers (1953). A study of methods used in measurement and analysis of sediment loads in streams. *Rept No. 10:* Accuracy of sediment size analyses made by the bottom withdrawal tube method. (St. Paul, U.S. Engineers District Sub-Office, Hydraulic Lab., Univ. of Iowa, Iowa City).

Vermaat, J. G., and C. F. Bentley (1955). The age and channelling of Ceylon laterite. *Soil Sci.* **79,** 239.

Wadell, H. (1932). Volume, shape and roundness of rock particles. *J. Geol.* **40,** 443–451.

Wadell, H. (1933). Sphericity and roundness of rock particles. *J. Geol.* **41,** 310–331.

Wadell, H. (1935). Volume, shape, and roundness of quartz particles. *J. Geol.* **43,** 250–279.

Walger, E. (1962). Microscopic determination of rock-forming carbonates in thin sections. *Neues Jahrb. Mineral., Monatsh,* 182–187.

Walker, G. F. (1949). The decomposition of biotite in the soil. *Mineral. Mag.* **28,** 693–703.

Walker, G. F. (1950). Trioctahedral minerals in the soil clays of northeast Scotland. *Mineral. Mag.* **29,** 72–84.

Walker, P. H. (1959). A study of cyclic soils, their relation to landscape and their development on the south coast of New South Wales. M.Sc. Thesis, University of Sydney.

Walker, P. H. (1962). Soil layers on hill-slopes: A study at Nowra, N.S.W., Australia. *J. Soil Sci.* **13,** 167–177.

Warne, S. St.G. (1962). A quick field or laboratory staining scheme for the differentiation of the major carbonate minerals. *J. Sed. Petrol.* **32,** 29–38.

Washington, H. S. (1930). *The Chemical Analysis of Rocks.* (Fourth ed.) (John Wiley and Sons, New York) (out of print).

Wells, C. B. (1962). Resin impregnation of soil samples. *Nature* **193,** 804.

Wentworth, C. K. (1922). The shapes of pebbles. *U.S. Geol. Surv. Bull.* **730C,** 91–114.

Wentworth, C. K. (1926). Methods of mechanical analysis of sediments. *Univ. Iowa, Studies in Nat. Hist.* Vol. 11, No. 11.

Werner, J. (1962). On the preparation of fluorescing soil sections. *Z. PflErnahr. Dung., Bodenkunde* **99,** 144–150.

Weyl, R. (1952). Studies of heavy minerals in soil profiles. *Z. PflErnahr. Dung., Bodenkunde* **57,** 135–141.

Wheeler, W. M. (1960). *Ants.* (Columbia University Press, New York).

White, E. M. (1961). Calcium-solodi or planosol genesis from solodized-solonetz. *Soil Sci.* **91,** 175–177.

White, E. M., and R. G. Bonestell (1960). Some gilgaied soils in South Dakota. *Soil Sci. Soc. Am. Proc.* **24,** 305–309.

White, J. L., G. W. Bailey, and J. U. Anderson (1960). The influence of parent material and topography on soil genesis in the Midwest. *Purdue Agr. Exp. Sta. Res. Bull.* **693,** 20 pp.

Whiteside, E. P. (1953). Some relationships between the classification of rocks by geologists and the classification of soils by soil scientists. *Soil Sci. Soc. Am. Proc.* **17,** 138–143.

Wild, A. (1961*a*). A pedological study of phosphorus in 12 soils derived from granite. *Aust. J. Agr. Res.* **12,** 286–299.

Wild, A. (1961*b*). Loss of zirconium from 12 soils derived from granite. *Aust. J. Agr. Res.* **12,** 300–305.

Williams, C. H. (1950). An examination of Jeffries' magnesium ribbon-potassium oxalate method for the removal of free iron oxides in soils. *Aust. Jour. Agr. Res.* **1,** 156.

Winchell, A. N. (1951). *Elements of Optical Mineralogy.* (John Wiley and Sons, New York).

Winterkorn, H. F. (1948). Physicochemical properties of soils. *Proc. Second Intl. Cong. Soil Mechs. and Foundation Engineering, Rotterdam,* **1,** 23–29.

Winterkorn, H. F. (1958). Water and its conduction in soils. *Highway Res. Board Spec. Rept.* **40,** 324.

Winterkorn, H. F., and G. P. Tschebotarioff (1947). Sensitivity of clay to remoulding and its possible causes. *Proc. Highway Res. Board* **27,** 435.

Wintermeyer, A. M., and E. B. Kinter (1954). A study of dispersing agents for particle-size analysis of soils. *Public Roads* **28,** 55–62.

Wolkewitz, H. von (1958). Cracking of drying clays and its significance. *Z. PflErnahr. Dung., Bodenkunde* **82,** 17.

Yamasaki, T., and T. Yoshizawa (1961). Concretion of ferrous carbonate (siderite) in paddy soils in Japan, I: Occurrence and constituents of ferrous carbonate and the mechanism of its formation. *Bull. Hokuriku Agric. Exp. Sta. No. 2,* 1–14.

Yanitskii, A. L. (1958). Formation of oolitic iron ores of the Lisakovsk (Aralo-Turgai) type. *Trudy Ob' edinennoi Kustanaiskoi Nauch. Sessii Posvyaschchennoi Problem. Turgaisk. Regional—Ekonom. Kompleksa* **2,** 92–116.

Zakharov, S. A. (1927). Achievements of Russian science in the morphology of soils. *Acad. Sci. U.S.S.R. (Leningrad) Russ. Ped. Inst. No. 2,* 47.

Zingg, Th. (1935). Beitrag zur Schotteranalyse. *Schweiz. Mineral. Petrog. Mitt.* **15,** 39–140.

Zonneveld, I. S. (1956). The physical ripening of fresh tidal sediments. *Repts. Sixth Int. Cong. Soil Sci., Paris, Com I* 281.

Index

The page numbers in bold face type refer to the definition of the entry.

Systemy uczące się

Mirosław Krzyśko
Waldemar Wołyński
Tomasz Górecki
Michał Skorzybut

Wydawnictwa Naukowo-Techniczne
Warszawa

Systemy uczące się

rozpoznawanie wzorców
analiza skupień
i redukcja wymiarowości

Recenzenci: *dr hab. Jan Mielniczuk*
prof. dr hab. Roman Zmyślony

Redaktor *Lilianna Szymańska*
Projekt okładki i stron tytułowych *Paweł G. Rubaszewski*
Redaktor techniczny *Ewa Kosińska*
Korekta *Halina Stykowska, Gabriela Szpunar*
Przygotowanie do druku *IMTEX*

Książka dotowana przez Uniwersytet im. Adama Mickiewicza w Poznaniu

Wydawnictwa Naukowo-Techniczne Sp. z o.o.
00-048 Warszawa, ul. Mazowiecka 2/4
tel. 0-22 826 72 71, e-mail: wnt@wnt.pl
www.wnt.pl

ISBN 978-83-204-3459-0

Spis treści

Przedmowa

Niniejsza książka jest adresowana do wszystkich Czytelników zainteresowanych najnowszymi metodami eksploracji danych. Materiał zawarty w książce jest dobrze skorelowany z programem nauczania czwartego i piątego roku studiów matematycznych, informatycznych i technicznych, gdzie analiza danych jest bądź przedmiotem obowiązkowym, bądź fakultatywnym. Stąd, książka ta może również służyć wielu studentom jako podręcznik. Uwzględnia ona większość nowych, najbardziej obiecujących metod powstałych w ostatnim dziesięcioleciu.

Składa się z dwóch zasadniczych części: *uczenia się pod nadzorem* (supervised learning) i *bez nadzoru* (unsupervised learning). Problem uczenia się pod nadzorem, inaczej nazywany uczeniem się z nauczycielem lub uczeniem się na przykładach, sprowadza się do określenia przydziału obiektów opisanych za pomocą wartości wielu cech (atrybutów) do jednej z wcześniej ustalonych klas (grup) decyzyjnych. Klasom są przypisane kody zwane etykietami. Pierwszym elementem systemu uczącego się pod nadzorem jest *próba ucząca* składająca się z n niezależnych par zmiennych $(\boldsymbol{X}_1, Y_1), \ldots, (\boldsymbol{X}_n, Y_n)$, gdzie $\boldsymbol{X}_i$ jest wektorem zaobserwowanych cech (atrybutów) na obiekcie pochodzącym z klasy o etykiecie Y_i, $i = 1, \ldots, n$.

Drugim elementem systemu uczącego się pod nadzorem jest funkcja określona na przestrzeni wartości cech o wartościach w zbiorze etykiet, zwana *klasyfikatorem*, skonstruowana na bazie próby uczącej. Trzecim elementem systemu uczącego się jest *ocena skuteczności działania klasyfikatora* (ocena wartości błędu klasyfikowania lub trafności klasyfikowania). Ocenę wartości błędu klasyfikowania przeprowadza się na zbiorze przykładów testowych, w odniesieniu do których jest znana ich rzeczywista przynależność do klas i które nie są częścią zbioru uczącego. Przykłady testowe są klasyfikowane za pomocą klasyfikatora skonstruowanego ze zbioru uczącego.

Umożliwia to porównanie proponowanej przez klasyfikator decyzji z rzeczywistą i stwierdzenie, czy decyzja klasyfikatora jest poprawna, czy błędna. W sytuacji, gdy nie dysponujemy niezależną próbą testową, a wielkość próby uczącej nie pozwala na jej podział na część uczącą i testową, można posłużyć się różnorodnymi ocenami błędu klasyfikacji uzyskanymi z próby uczącej.

Systemy uczące się pod nadzorem mogą być stosowane między innymi w:

- wspomaganiu diagnozowania chorób na podstawie symptomów,
- automatycznym rozpoznawaniu mowy ludzkiej,
- rozpoznawaniu pisma na podstawie przykładów (np. automatyczne rozpoznawanie ręcznie pisanych kodów pocztowych).

Omówieniu systemów uczących się pod nadzorem, w tym omówieniu konstrukcji z próby uczącej różnorodnych klasyfikatorów, poczynając od klasycznych, a kończąc na najnowszych, takich jak np. klasyfikatory oparte na wektorach nośnych, poświęconych jest dziesięć pierwszych rozdziałów książki. Omówione są również metody wzmacniania i łączenia (agregacji) klasyfikatorów.

W uczeniu bez nadzoru dysponujemy zbiorem przykładów opisanych jedynie wektorami wartości cech (atrybutów) bez podania ich przynależności do klas. Obserwując te przykłady, system ma odkryć regularności w ich zbiorze, np. przeprowadzić grupowanie przykładów w klasy, czy też wykryć przykłady nietypowe, odstające od pozostałych.

W czterech ostatnich rozdziałach książki opisano następujące metody uczenia się bez nadzoru: analizę składowych głównych, analizę skupień, skalowanie wielowymiarowe i analizę korespondencji (odpowiedniości).

Proponowany Czytelnikowi podręcznik wymaga elementarnej wiedzy z teorii prawdopodobieństwa, statystyki matematycznej oraz algebry liniowej (głównie algebry macierzy).

Tematyka poruszana w podręczniku ma określoną tradycję w środowisku poznańskim. Z tematyki tej wcześniej ukazały się książki Mirosława Krzyśki *Analiza dyskryminacyjna* (1990) i *Wielowymiarowa analiza statystyczna* (2000) oraz książka Krzysztofa Krawca i Jerzego Stefanowskiego *Uczenie maszynowe i sieci neuronowe* (2004). W polskojęzycznej literaturze z tej dziedziny godne polecenia są książki: Jacka Koronackiego i Jana Ćwika *Statystyczne systemy uczące się* (2005), Eugeniusza Gatnara *Symboliczne metody klasyfikacji danych* (1998) i *Nieparametryczna metoda dyskryminacji i regresji* (2000), Pawła Cichosza *Systemy uczące się* (2000) i tłumaczenie na język polski książki Daniela T. Larose'a *Odkrywanie wiedzy z danych. Wprowadzenie do eksploracji danych* (2006).

Istnieje bardzo bogata literatura anglojęzyczna. Szczególnie ważne są monografie: Fukunagi (1990), McLachlana (1992), Vapnika (1995), Hastiego, Tibshiraniego i Friedmana (2001), Dudy, Harta i Storka (2001), Webba (2003), Alpaydina (2004), Maimona i Rokacha (2005) oraz Bishopa (2006).

Zdecydowana większość opisanych metod została już zaimplementowana w powszechnie dostępnym oprogramowaniu statystycznym. Szczególnie interesujący jest darmowy pakiet statystyczny R[1] oraz system do obliczeń naukowo-technicznych Matlab[2] wraz z wieloma dostępnymi dla niego specjalistycznymi narzędziami (toolboxes). Przykłady zamieszczone w książce zostały wykonane przy użyciu tego oprogramowania.

Traktując tę książkę jako dodatkową pomoc, obok wcześniej wymienionych podręczników, polecamy ją życzliwej uwadze studentów i pracowników naukowych prowadzących zajęcia z analizy danych.

Serdecznie dziękujemy wszystkim tym, których życzliwe uwagi przyczyniły się do udoskonalenia tej książki, a w szczególności Panom Recenzentom Janowi Mielniczukowi i Romanowi Zmyślonemu.

Dziękujemy także Pani Redaktor Liliannie Szymańskiej za okazaną nam przychylność oraz Władzom Wydziału Matematyki i Informatyki UAM za wsparcie finansowe wydania tej książki.

Poznań, styczeń 2008 r. *Autorzy*

---

[1] http://www.r-project.org/.
[2] http://www.mathworks.com/.

Spis oznaczeń

$\boldsymbol{Z} = (\boldsymbol{X},\ Y)$	obiekt
K	liczba klas (grup)
p	liczba cech
$\boldsymbol{X}$	wektor obserwowanych cech obiektu $\boldsymbol{Z}$
Y	etykieta obiektu $\boldsymbol{Z}$
$\mathcal{X}$	przestrzeń próby
$\mathcal{Y}$	zbiór etykiet
π_i	prawdopodobieństwo a priori przynależności obiektu do i-tej klasy
$d,\ d(\boldsymbol{x})$	klasyfikator (reguła klasyfikacyjna)
$f_i(\boldsymbol{x})$	gęstość wektora X w i-tej klasie
$r(\boldsymbol{x})$	funkcja regresji Y względem $\boldsymbol{X}$
$d_B,\ d_B(\boldsymbol{x})$	klasyfikator bayesowski
$\mathcal{L}_n$	n-elementowa próba ucząca
$\boldsymbol{Z}_i = (\boldsymbol{X}_i,\ Y_i)$	i-ty element próby uczącej
$\mathcal{T}_m$	m-elementowa próba testowa
$\boldsymbol{Z}_i^t = (\boldsymbol{X}_i^t,\ Y_i^t)$	i-ty element próby testowej
$\mathcal{L}_n^{(-j)}$	próba ucząca bez j-tej obserwacji
$\tilde{\mathcal{L}}_n^{(j)}$	j-ta część próby uczącej w metodzie CV
$\tilde{\mathcal{L}}_n^{(-j)}$	j-ta próba ucząca w metodzie CV
$\mathcal{L}_n^{\star b}$	b-ta próba bootstrapowa
$\hat{d},\ \hat{d}(\boldsymbol{x}),\ \hat{d}(\boldsymbol{x};\mathcal{L}_n)$	klasyfikator oparty na próbie uczącej $\mathcal{L}_n$
$\hat{d}_F(\boldsymbol{x})$	klasyfikator Fishera
$\hat{d}_{NB}(\boldsymbol{x})$	naiwny klasyfikator bayesowski
$\hat{d}_T(\boldsymbol{x})$	klasyfikator bazujący na drzewie klasyfikacyjnym T
$\hat{d}_{OPC}(\boldsymbol{x})$	klasyfikator metody OPC

$\hat{d}_{ECOC}(\boldsymbol{x})$	klasyfikator metody ECOC
$\hat{d}_{Bag1}(\boldsymbol{x})$	klasyfikator uzyskany metodą bagging
$\hat{d}_{Bag2}(\boldsymbol{x})$	klasyfikator uzyskany ważoną metodą bagging
$\hat{d}_{AdaB}(\boldsymbol{x})$	klasyfikator uzyskany metodą AdaBoost
$\hat{d}_{rAdaB}(\boldsymbol{x})$	klasyfikator uzyskany metodą AdaBoost (wersja rzeczywista)
$\hat{d}_{ARCx4}(\boldsymbol{x})$	klasyfikator uzyskany metodą ARCx4
$\hat{d}_{V}(\boldsymbol{x})$	klasyfikator metody głosowania
$\hat{d}_{HT}(\boldsymbol{x})$	klasyfikator Hastiego i Tibshiraniego
$\hat{d}_{WHT}(\boldsymbol{x})$	ważony klasyfikator Hastiego i Tibshiraniego
$\hat{d}_{JS}(\boldsymbol{x})$	klasyfikator Jelonka i Stefanowskiego
$\hat{\pi}_i$	estymator prawdopodobieństwa a priori przynależności obiektu do i-tej klasy oparty na próbie uczącej $\mathcal{L}_n$
$\hat{f}_i(\boldsymbol{x})$	estymator gęstości wektora X w i-tej klasie oparty na próbie uczącej $\mathcal{L}_n$
$\hat{r}(\boldsymbol{x})$	estymator funkcji regresji oparty na próbie uczącej $\mathcal{L}_n$
$e(d)$	rzeczywisty poziom błędu klasyfikatora d
$e^\star$	błąd bayesowski
$e(\hat{d})$	aktualny poziom błędu klasyfikatora $\hat{d}$
$\hat{e}$, $\hat{e}(\hat{d})$	błąd klasyfikacji
$\hat{e}_{\mathcal{T}}$, $\hat{e}_{\mathcal{T}}(\hat{d})$	błąd klasyfikacji uzyskany za pomocą próby testowej $\mathcal{T}$
$\hat{e}_R$, $\hat{e}_R(\hat{d})$	błąd klasyfikacji uzyskany metodą ponownego podstawiania
$\hat{e}_{CV}$, $\hat{e}_{CV}(\hat{d})$	błąd klasyfikacji uzyskany metodą sprawdzania krzyżowego
$\hat{e}_{vCV}$, $\hat{e}_{vCV}(\hat{d})$	błąd klasyfikacji uzyskany v-krokową metodą sprawdzania krzyżowego
$\hat{e}_B$, $\hat{e}_B(\hat{d})$	błąd klasyfikacji uzyskany metodą prób bootstrapowych
$H+$, $H-$	hiperpłaszczyzny kanoniczne w metodzie SVM
$\mathcal{SV}$	zbiór wszystkich wektorów nośnych w metodzie SVM
C	strata w metodzie SVM
$K = K(\cdot, \cdot)$	funkcja jądrowa
VC	wymiar Vapnika–Chervonenkisa
L	funkcja Lagrange'a
L_D	forma dualna funkcji Lagrange'a
T	drzewo klasyfikacyjne
T_{max}	maksymalne drzewo klasyfikacyjne

$\lvert T\rvert$	liczba węzłów drzewa klasyfikacyjnego T
$\tilde{T}$	liczba liści drzewa klasyfikacyjnego T
$R_\alpha(T)$	miara błędu i złożoności drzewa klasyfikacyjnego T
$T' \prec T$	poddrzewo drzewa klasyfikacyjnego T
$I(T)$	miara niejednorodności elementów drzewa klasyfikacyjnego T
t	węzeł drzewa klasyfikacyjnego
t_L	lewy węzeł potomek węzła t
t_R	prawy węzeł potomek węzła t
$i(t)$	miara niejednorodności elementów węzła t
ind	funkcja przyporządkowująca liściom drzewa klasyfikacyjnego etykiety klas
$s(t)$	funkcja podziału węzła t
$\boldsymbol{w}$	wektor wag synaptycznych
$\boldsymbol{W}$	macierz wag synaptycznych
s	sygnał wyjściowy neuronu
$d(\boldsymbol{x}_i)$	wzorzec obiektu $\boldsymbol{x}_i$ w sieci neuronowej
$Q(\boldsymbol{w})$	funkcja błędu sieci neuronowej
η	współczynnik uczenia
γ	współczynnik zapominania
α	momentum (pęd)
$\boldsymbol{g}(\boldsymbol{w})$	gradient funkcji błędu Q
$\boldsymbol{H}(\boldsymbol{w})$	hesjan funkcji błędu Q
$\boldsymbol{G}(\boldsymbol{w})$	przybliżona wartość hesjanu $\boldsymbol{H}(\boldsymbol{w})$
$\boldsymbol{V}(t)$	odwrotność macierzy $\boldsymbol{G}(\boldsymbol{w})$
T_k	schemat wyżarzania
$E(\boldsymbol{W})$	stan energetyczny sieci
$\boldsymbol{X}^+$	odwrotność Moore'a–Penrose'a macierzy $\boldsymbol{X}$
$\boldsymbol{v}_i$	wektor Voronoia
V_i	obszar Voronoia
$G(j,\ \boldsymbol{x})$	funkcja sąsiedztwa dla j-tego neuronu
$\varphi(\boldsymbol{x},\ c)$	radialna funkcja bazowa
s_j	współczynnik wrażliwości
$\rho(R,\ S)$	miara niepodobieństwa pomiędzy skupieniami
C_K	funkcja przyporządkowująca obiektom numer skupienia
$W(C_K)$	macierz zmienności wewnątrz skupień
$B(C_K)$	macierz zmienności pomiędzy skupieniami
$CH(K)$	indeks Calińskiego–Harabasza
$Gap(K)$	statystyka odstępu

CZĘŚĆ I

Statystyczne systemy uczące się pod nadzorem

Rozpoznawanie wzorców

1

Probabilistyczne metody klasyfikacyjne

Uczenie się pod nadzorem lub uczenie się z przykładów jest procesem budowy, na bazie dostępnych danych wejściowych $\boldsymbol{X}_i$ oraz wyjściowych Y_i, $i = 1, 2, \ldots, n$, reguły klasyfikacyjnej zwanej inaczej klasyfikatorem, służącej do predykcji etykiety Y grupy, do której należy obserwacja $\boldsymbol{X}$. W rozdziale tym pokażemy różnorodne konstrukcje klasyfikatorów oparte na przesłankach probabilistycznych.

Czytelników zainteresowanych poszerzeniem wiedzy z zakresu tego rozdziału odsyłamy do książek: Bishopa (2006), Koronackiego i Ćwika (2005) oraz Webba (2003).

1.1. Wektory losowe i ich rozkłady prawdopodobieństwa

Niech $\boldsymbol{X} = (X_1, X_2, \ldots, X_p)'$ będzie wektorem losowym o p składowych, gdzie znak prim oznacza transpozycję. Umawiamy się, że wszystkie wektory są wektorami kolumnowymi. Wektor wierszowy jest zatem transpozycją wektora kolumnowego i odwrotnie.

Wartością oczekiwaną wektora losowego $\boldsymbol{X}$ jest wektor postaci

$$\mathrm{E}(\boldsymbol{X}) = (\mathrm{E}(X_1), \mathrm{E}(X_2), \ldots, \mathrm{E}(X_p))'.$$

Wariancją wektora losowego $\boldsymbol{X}$ jest macierz stopnia p postaci

$$\mathrm{Var}(\boldsymbol{X}) = (\mathrm{Cov}(X_i, X_j)),$$

gdzie

$$\mathrm{Cov}(X_i, X_j) = \mathrm{E}\{[X_i - \mathrm{E}(X_i)][X_j - \mathrm{E}(X_j)]\}$$

jest kowariancją między składowymi X_i oraz X_j wektora losowego $\boldsymbol{X}$, $i,\ j = 1,\ 2,\ \ldots,\ p$.

Ponieważ elementami macierzy kwadratowej $\mathrm{Var}(\boldsymbol{X})$ są kowariancje między składowymi wektora $\boldsymbol{X}$, więc jest ona nazywana *macierzą kowariancji*. Macierz kowariancji $\mathrm{Var}(\boldsymbol{X})$ możemy równoważnie zapisać wzorem

$$\mathrm{Var}(\boldsymbol{X}) = \mathrm{E}\left\{[\boldsymbol{X} - \mathrm{E}(\boldsymbol{X})][\boldsymbol{X} - \mathrm{E}(\boldsymbol{X})]'\right\}. \tag{1.1}$$

Zauważmy, że $\mathrm{Cov}(X_i,\ X_i) = \mathrm{Var}(X_i)$, $i = 1,\ 2,\ \ldots,\ p$, gdzie $\mathrm{Var}(X)$ oznacza wariancję zmiennej losowej X.

Stąd na głównej przekątnej macierzy kowariancji znajdują się wariancje poszczególnych składowych wektora losowego $\boldsymbol{X}$. Wiemy również, że $\mathrm{Cov}(X_i,\ X_j) = \mathrm{Cov}(X_j,\ X_i)$, dla $i,\ j = 1,\ 2,\ \ldots,\ p$, $j \neq i$. Zatem macierz kowariancji jest macierzą symetryczną. Można również udowodnić (patrz np. Krzyśko (2000), s. 15), że macierz kowariancji jest nieujemnie określona, tj. dla każdego wektora $\boldsymbol{a} \in \mathbb{R}^p$, $\boldsymbol{a}'\mathrm{Var}(\boldsymbol{X})\boldsymbol{a} \geq 0$.

Zauważmy jeszcze, że jeżeli wszystkie składowe wektora losowego $\boldsymbol{X}$ są zmiennymi losowymi niezależnymi, to $\mathrm{Cov}(X_i,\ X_j) = 0$, dla $i,\ j = 1,\ 2,\ \ldots,\ p$, $j \neq i$. W tym przypadku macierz kowariancji jest macierzą diagonalną o elementach diagonalnych równych wariancjom poszczególnych składowych wektora losowego $\boldsymbol{X}$.

Łatwo można pokazać (patrz np. Krzyśko (2000), s. 17), że jeżeli $\mathrm{Var}(\boldsymbol{X}) = \boldsymbol{\Sigma}$ i $\boldsymbol{Y} = \boldsymbol{C}\boldsymbol{X}$, gdzie $\boldsymbol{C}$ jest macierzą stałą rozmiaru $r \times p$, to $\mathrm{Var}(\boldsymbol{Y}) = \boldsymbol{C}\boldsymbol{\Sigma}\boldsymbol{C}'$ jest macierzą stopnia r.

Mówimy, że wektor losowy $\boldsymbol{X} = (X_1,\ X_2,\ \ldots,\ X_p)'$ ma p-wymiarowy rozkład normalny, jeżeli każda funkcja liniowa $a_1X_1 + a_2X_2 + \cdots + a_pX_p$ ma jednowymiarowy rozkład normalny.

Dowodzi się, że jeżeli wektor losowy $\boldsymbol{X}$ ma p-wymiarowy rozkład normalny, to $\mathrm{E}(\boldsymbol{X}) = \boldsymbol{\mu}$ oraz $\mathrm{Var}(\boldsymbol{X}) = \boldsymbol{\Sigma}$ istnieją i rozkład prawdopodobieństwa wektora $\boldsymbol{X}$ jest jednoznacznie wyznaczony przez $\boldsymbol{\mu}$ i $\boldsymbol{\Sigma}$. Fakt, że wektor $\boldsymbol{X} = (X_1,\ X_2,\ \ldots,\ X_p)'$ ma p-wymiarowy rozkład normalny z $\mathrm{E}(\boldsymbol{X}) = \boldsymbol{\mu}$ i $\mathrm{Var}(\boldsymbol{X}) = \boldsymbol{\Sigma}$ zapisywać będziemy następująco: $\boldsymbol{X} \sim N_p(\boldsymbol{\mu},\ \boldsymbol{\Sigma})$.

Jeżeli $\boldsymbol{X} \sim N_p(\boldsymbol{\mu}, \boldsymbol{\Sigma})$ i macierz kowariancji $\boldsymbol{\Sigma}$ jest dodatnio określona, tj. dla każdego niezerowego wektora $\boldsymbol{a} \in \mathbb{R}^p$, $\boldsymbol{a}'\boldsymbol{\Sigma}\boldsymbol{a} > 0$, to gęstość wektora losowego $\boldsymbol{X}$ jest równa

$$f(\boldsymbol{x}) = (2\pi)^{-\frac{p}{2}}|\boldsymbol{\Sigma}|^{-\frac{1}{2}} \exp\left[-\frac{1}{2}(\boldsymbol{x} - \boldsymbol{\mu})'\boldsymbol{\Sigma}^{-1}(\boldsymbol{x} - \boldsymbol{\mu})\right], \tag{1.2}$$

gdzie $|\boldsymbol{\Sigma}|$ oznacza wyznacznik macierzy $\boldsymbol{\Sigma}$, natomiast $\boldsymbol{\Sigma}^{-1}$ jest macierzą odwrotną macierzy $\boldsymbol{\Sigma}$, tj. taką, że $\boldsymbol{\Sigma}\boldsymbol{\Sigma}^{-1} = \boldsymbol{\Sigma}^{-1}\boldsymbol{\Sigma} = \mathbf{I}_p$, gdzie $\mathbf{I}_p$ jest macierzą jednostkową stopnia p.

Można udowodnić (patrz np. Krzyśko (2000), s. 19), że jeżeli $\boldsymbol{X} \sim N_p(\boldsymbol{\mu},\ \boldsymbol{\Sigma})$, to dla każdego wektora $\boldsymbol{a} \in \mathbb{R}^k$ i każdej stałej macierzy $\boldsymbol{A}$ rozmiaru $k \times p$, wektor losowy $\boldsymbol{Y} = \boldsymbol{AX} + \boldsymbol{a} \sim N_k(\boldsymbol{A\mu} + \boldsymbol{a},\ \boldsymbol{A\Sigma A}')$.

Jeżeli $\boldsymbol{X} \sim N_p(\boldsymbol{\mu}, \boldsymbol{\Sigma})$ oraz $\boldsymbol{X}$, $\boldsymbol{\mu}$ i $\boldsymbol{\Sigma}$ są podzielone następująco:

$$\boldsymbol{X} = \begin{bmatrix} \boldsymbol{X}_1 \\ \boldsymbol{X}_2 \end{bmatrix}, \qquad \boldsymbol{\mu} = \begin{bmatrix} \boldsymbol{\mu}_1 \\ \boldsymbol{\mu}_2 \end{bmatrix}, \qquad \boldsymbol{\Sigma} = \begin{bmatrix} \boldsymbol{\Sigma}_{11} & \boldsymbol{\Sigma}_{12} \\ \boldsymbol{\Sigma}_{21} & \boldsymbol{\Sigma}_{22} \end{bmatrix},$$

gdzie $\boldsymbol{X}_1$ i $\boldsymbol{\mu}_1$ są wektorami o k składowych, a $\boldsymbol{\Sigma}_{11}$ jest macierzą stopnia k, to wartość oczekiwana rozkładu warunkowego $\boldsymbol{X}_2|\boldsymbol{X}_1 = \boldsymbol{x}_1$ jest równa

$$\mathrm{E}(\boldsymbol{X}_2|\boldsymbol{X}_1 = \boldsymbol{x}_1) = \boldsymbol{b} + \boldsymbol{B}'\boldsymbol{x}_1, \tag{1.3}$$

gdzie

$$\boldsymbol{b} = \boldsymbol{\mu}_2 - \boldsymbol{\Sigma}_{21}\boldsymbol{\Sigma}_{11}^{-1}\boldsymbol{\mu}_1, \qquad \boldsymbol{B} = \boldsymbol{\Sigma}_{11}^{-1}\boldsymbol{\Sigma}_{12}. \tag{1.4}$$

Funkcja dana wzorem (1.3) nazywa się *liniową funkcją regresji* wektora $\boldsymbol{X}_2$ względem wektora $\boldsymbol{X}_1$. Macierz $\boldsymbol{B}$ jest macierzą współczynników regresji. Funkcja (1.3) jest wykorzystywana do prognozowania wartości wektora $\boldsymbol{X}_2$ na podstawie wartości $\boldsymbol{x}_1$ wektora $\boldsymbol{X}_1$.

Jeżeli $\boldsymbol{X}_1,\ \boldsymbol{X}_2,\ \ldots,\ \boldsymbol{X}_n$ są niezależnymi wektorami losowymi o identycznym rozkładzie $N_p(\boldsymbol{0},\ \boldsymbol{\Sigma})$, $\boldsymbol{\Sigma} > 0$, to rozkład prawdopodobieństwa symetrycznej macierzy losowej

$$\boldsymbol{A} = \sum_{i=1}^{n} \boldsymbol{X}_i\boldsymbol{X}_i'$$

nazywamy *rozkładem Wisharta* z n stopniami swobody i parametrem skali $\boldsymbol{\Sigma}$ i piszemy $\boldsymbol{A} \sim W_p(n,\ \boldsymbol{\Sigma})$, gdzie p jest stopniem macierzy $\boldsymbol{A}$.

Niech $\boldsymbol{X} \sim N_p(\boldsymbol{\mu},\ \boldsymbol{\Sigma})$, $\boldsymbol{A} \sim W_p(n,\ \boldsymbol{\Sigma})$, gdzie $\boldsymbol{\Sigma} > 0$ i $n > p$ oraz niech $\boldsymbol{X}$ i $\boldsymbol{A}$ będą niezależne. Rozkład prawdopodobieństwa zmiennej losowej

$$T^2 = n\boldsymbol{X}'\boldsymbol{A}^{-1}\boldsymbol{X}$$

nazywamy *rozkładem* T^2 *Hotellinga* z n stopniami swobody i piszemy $T^2 \sim T^2_{p,n}$.

Dowodzi się, że

$$\frac{n-p+1}{np}T^2_{p,n} \sim F_{p,n-p+1}(\delta^2),$$

gdzie $\delta^2 = \boldsymbol{\mu}'\boldsymbol{\Sigma}^{-1}\boldsymbol{\mu}$ jest *parametrem niecentralności*. Jeżeli $\boldsymbol{\mu} = \boldsymbol{0}$, to $\delta^2 = 0$ i wówczas

$$\frac{n-p+1}{np}T^2_{p,n} \sim F_{p,n-p+1}$$

(centralny rozkład F Snedecora z p oraz $n-p+1$ stopniami swobody).

Jeżeli $\boldsymbol{A}$ i $\boldsymbol{B}$ są niezależnymi macierzami losowymi takimi, że $\boldsymbol{A} \sim W_p(\nu_A, \boldsymbol{\Sigma})$, $\boldsymbol{B} \sim W_p(\nu_B, \boldsymbol{\Sigma})$, $\nu_A \geq p$ oraz $\lambda_1, \lambda_2, \ldots, \lambda_p$ są wartościami własnymi macierzy $\boldsymbol{A}^{-1}\boldsymbol{B}$, to rozkład prawdopodobieństwa statystyki:

1. $$\Lambda = \frac{|\boldsymbol{A}|}{|\boldsymbol{A}+\boldsymbol{B}|} = \prod_{i=1}^{p} \frac{1}{1+\lambda_i}$$

   nazywamy *rozkładem Λ Wilksa* z p, ν_B i ν_A stopniami swobody i piszemy $\Lambda \sim W_{p,\nu_B,\nu_A}$.

2. $$T^2 = \operatorname{tr}(\boldsymbol{A}^{-1}\boldsymbol{B}) = \sum_{i=1}^{p} \lambda_i$$

   nazywamy *rozkładem T^2 Lawleya–Hotellinga* z p, ν_B i ν_A stopniami swobody i piszemy $T^2 \sim T^2_{p,\nu_B,\nu_A}$.

3. $$V = \operatorname{tr}[\boldsymbol{B}(\boldsymbol{A}+\boldsymbol{B})^{-1}] = \sum_{i=1}^{p} \frac{\lambda_i}{1+\lambda_i}$$

   nazywamy *rozkładem V Pillaia* z p, ν_B i ν_A stopniami swobody i piszemy $V \sim V_{p,\nu_B,\nu_A}$.

Jeżeli $\nu_A \geq p$, to z prawdopodobieństwem jeden $\boldsymbol{A} > 0$ (macierz $\boldsymbol{A}$ jest dodatnio określona).

1.2. Pojęcia wstępne

Załóżmy, że dysponujemy K niezależnymi, prostymi próbami losowymi o liczebnościach, odpowiednio, $n_1, n_2, \ldots, n_K$, pobranymi z K różnych populacji (klas, grup):

$\boldsymbol{X}_{11}, \boldsymbol{X}_{12}, \ldots, \boldsymbol{X}_{1n_1}$ – z populacji 1

$\boldsymbol{X}_{21}, \boldsymbol{X}_{22}, \ldots, \boldsymbol{X}_{2n_2}$ – z populacji 2

...

$\boldsymbol{X}_{K1}, \boldsymbol{X}_{K2}, \ldots, \boldsymbol{X}_{Kn_K}$ – z populacji K, gdzie $\boldsymbol{X}_{ij} = (X_{ij1},\ X_{ij2},\ \dots,\ X_{ijp})'$ jest j-tą obserwacją z i-tej populacji zawierającą p obserwowanych cech, $i = 1,\ 2,\ \dots,\ K$, $j = 1,\ 2,\ \dots,\ n_i$.

Powyższe dane można wygodniej zapisać w innej postaci, a mianowicie w postaci jednego ciągu n uporządkowanych par losowych

$$(\boldsymbol{X}_1,\ Y_1),\ \dots,\ (\boldsymbol{X}_n,\ Y_n),$$

gdzie $\boldsymbol{X}_i = (X_{i1},\ X_{i2},\ \dots,\ X_{ip})' \in \mathcal{X} \subset \mathbb{R}^p$ jest i-tą obserwacją, natomiast Y_i jest etykietą populacji, do której ta obserwacja należy, przyjmującą wartości w pewnym skończonym zbiorze $\mathcal{Y}$, $i = 1,\ 2,\ \dots,\ n$, $n = n_1 + n_2 + \dots + n_K$.

Składowe wektora $\boldsymbol{X}_i = (X_{i1},\ X_{i2},\ \dots,\ X_{ip})'$ będziemy nazywać *cechami* lub *zmiennymi* (nazewnictwo przyjęte w statystyce matematycznej). W eksploracji danych jest używana zamiennie nazwa *atrybuty*.

Próbę $\mathcal{L}_n = \{(\boldsymbol{X}_1,\ Y_1),\ \dots,\ (\boldsymbol{X}_n,\ Y_n)\}$ będziemy nazywać *próbą uczącą*.

Interesuje nas problem predykcji etykiety Y na podstawie wektora cech $\boldsymbol{X}$.

Problem ten jest nazywany klasyfikacją, dyskryminacją, uczeniem się pod nadzorem lub rozpoznawaniem wzorców.

Reguła klasyfikacyjna, zwana krótko *klasyfikatorem*, jest funkcją $d : \mathcal{X} \to \mathcal{Y}$. Gdy obserwujemy nowy wektor $\boldsymbol{X}$, to prognozą etykiety Y jest $d(\boldsymbol{X})$.

Przykład 1.1.
Na rysunku 1.1 jest pokazanych 20 punktów. Wektor cech $\boldsymbol{X} = (X_1,\ X_2)'$ jest dwuwymiarowy, a etykieta $Y \in \mathcal{Y} = \{1,\ 0\}$. Wartości cechy $\boldsymbol{X}$ dla $Y = 0$ są reprezentowane przez trójkąty, a dla $Y = 1$ przez kwadraty. Linia przerywana reprezentuje liniową regułę klasyfikacyjną postaci

$$d(\boldsymbol{x}) = \begin{cases} 1, \text{ jeżeli } a + b_1 x_1 + b_2 x_2 > 0, \\ 0, \text{ poza tym.} \end{cases}$$

Każdy punkt leżący poniżej tej linii jest klasyfikowany do grupy o etykiecie 0 oraz każdy punkt leżący powyżej tej linii jest klasyfikowany do grupy o etykiecie 1.

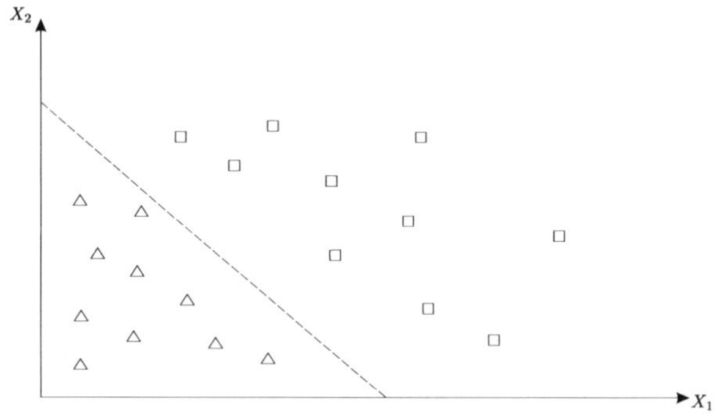

Rys. 1.1. Dwie cechy i liniowa granica decyzyjna. □ oznacza, że $Y = 1$, △ oznacza, że $Y = 0$. Te dwie grupy są dokładnie oddzielone przez liniową granicę decyzyjną

1.3. Rzeczywisty i aktualny poziom błędu, klasyfikator bayesowski

Klasyczny problem klasyfikacji polega na predykcji nieznanej etykiety $Y \in \mathcal{Y}$ na podstawie wektora cech $\boldsymbol{X} \in \mathcal{X} \subseteq \mathbf{R}^p$ klasyfikowanego obiektu. Naszym celem jest znalezienie takiego klasyfikatora $d : \mathcal{X} \to \mathcal{Y}$, który daje dokładną predykcję. Miarą jakości klasyfikatora jest jego rzeczywisty poziom błędu.

Definicja 1.1.
Rzeczywisty poziom błędu[1] klasyfikatora d jest równy

$$e(d) = P(d(\boldsymbol{X}) \neq Y). \tag{1.5}$$

Weźmy najpierw pod uwagę przypadek dwóch klas, tj. gdy $\mathcal{Y} = \{1, 0\}$. Niekiedy klasom będziemy nadawać również etykiety $+1$ i -1.

W celu stworzenia modelu klasyfikacji przyjmiemy pewne założenia probabilistyczne, a mianowicie załóżmy, że $(\boldsymbol{X}, Y)$ jest parą losową o wartościach w $\mathbf{R}^p \times \{1, 0\}$ oraz że jej rozkład prawdopodobieństwa opisuje para (μ, r), gdzie μ jest miarą probabilistyczną wektora $\boldsymbol{X}$ oraz r jest regresją Y względem $\boldsymbol{X}$. Bardziej precyzyjnie, dla borelowskiego zbioru $A \subseteq \mathbf{R}^p$,

$$\mu(A) = \mathrm{P}(\boldsymbol{X} \in A)$$

[1] Po ang. the true error rate.

oraz dla każdego $\boldsymbol{x} \in \mathbf{R}^p$,

$$r(\boldsymbol{x}) = \mathrm{E}(Y|\boldsymbol{X} = \boldsymbol{x}) = 1 \cdot \mathrm{P}(Y = 1|\boldsymbol{X} = \boldsymbol{x}) + 0 \cdot \mathrm{P}(Y = 0|\boldsymbol{X} = \boldsymbol{x}) = \\ = \mathrm{P}(Y = 1|\boldsymbol{X} = \boldsymbol{x}). \tag{1.6}$$

Zatem $r(\boldsymbol{x})$ jest prawdopodobieństwem warunkowym, że $Y = 1$, gdy $\boldsymbol{X} = \boldsymbol{x}$.

Rozkład prawdopodobieństwa pary $(\boldsymbol{X}, Y)$ wyznacza para (μ, r). Funkcja $r(\boldsymbol{x})$ nazywa się *funkcją regresji* lub *prawdopodobieństwem a posteriori*.

Z twierdzenia Bayesa mamy

$$\begin{aligned} r(\boldsymbol{x}) &= \mathrm{P}(Y = 1|\boldsymbol{X} = \boldsymbol{x}) = \\ &= \frac{f(\boldsymbol{x}|Y = 1)\mathrm{P}(Y = 1)}{f(\boldsymbol{x}|Y = 1)\mathrm{P}(Y = 1) + f(\boldsymbol{x}|Y = 0)\mathrm{P}(Y = 0)} = \\ &= \frac{\pi_1 f_1(\boldsymbol{x})}{\pi_1 f_1(\boldsymbol{x}) + \pi_0 f_0(\boldsymbol{x})}, \end{aligned} \tag{1.7}$$

gdzie

$$f_1(\boldsymbol{x}) = f(\boldsymbol{x}|Y = 1), \qquad f_0(\boldsymbol{x}) = f(\boldsymbol{x}|Y = 0),$$
$$\pi_1 = \mathrm{P}(Y = 1), \qquad \pi_0 = \mathrm{P}(Y = 0), \qquad \pi_1 + \pi_0 = 1$$

są prawdopodobieństwami a priori dwóch klas o etykietach jeden i zero.

Definicja 1.2.
Klasyfikator postaci

$$d_B(\boldsymbol{x}) = \begin{cases} 1, \text{ jeżeli } r(\boldsymbol{x}) > \frac{1}{2}, \\ 0, \text{ poza tym} \end{cases} \tag{1.8}$$

będziemy nazywać *klasyfikatorem bayesowskim.*

Klasyfikator bayesowski zapisać można w innych równoważnych postaciach:

$$d_B(\boldsymbol{x}) = \begin{cases} 1, \text{ jeżeli } \mathrm{P}(Y = 1|\boldsymbol{X} = \boldsymbol{x}) > \mathrm{P}(Y = 0|\boldsymbol{X} = \boldsymbol{x}), \\ 0, \text{ poza tym} \end{cases} \tag{1.9}$$

lub

$$d_B(\boldsymbol{x}) = \begin{cases} 1, \text{ jeżeli } \pi_1 f_1(\boldsymbol{x}) > \pi_0 f_0(\boldsymbol{x}), \\ 0, \text{ poza tym.} \end{cases} \tag{1.10}$$

Definicja 1.3.
Zbiór postaci

$$\{\boldsymbol{x} : \mathrm{P}(Y = 1|\boldsymbol{X} = \boldsymbol{x}) = \mathrm{P}(Y = 0|\boldsymbol{X} = \boldsymbol{x})\}$$

będziemy nazywać *powierzchnią rozdzielającą* grupy 1 i 0.

TWIERDZENIE 1.1.

Klasyfikator bayesowski jest optymalny, tj. jeżeli d jest jakimkolwiek innym klasyfikatorem, to $e(d_B) \leq e(d)$, gdzie $e(d)$ jest rzeczywistym poziomem błędu klasyfikatora d danym wzorem (1.5).

Niestety, klasyfikator bayesowski zależy od rozkładu prawdopodobieństwa pary $(\boldsymbol{X}, Y)$. Jeżeli rozkład ten jest znany, to możemy posługiwać się klasyfikatorem bayesowskim d_B. Najczęściej rozkład ten nie jest znany i stąd również nie jest znany klasyfikator bayesowski d_B.

Pojawia się zatem problem skonstruowania klasyfikatora $\hat{d}(\boldsymbol{x}) = \hat{d}(\boldsymbol{x}; \mathcal{L}_n)$ opartego na próbie uczącej $\mathcal{L}_n = \{(\boldsymbol{X}_1, Y_1), \ldots, (\boldsymbol{X}_n, Y_n)\}$ zaobserwowanej w przeszłości.

Proces konstruowania klasyfikatora $\hat{d}$ jest nazywany *uczeniem się*, *uczeniem pod nadzorem* lub *uczeniem się z nauczycielem*.

W naszym modelu klasyfikacyjnym zakładamy, że próba ucząca $\mathcal{L}_n = \{(\boldsymbol{X}_1, Y_1), \ldots, (\boldsymbol{X}_n, Y_n)\}$ jest ciągiem niezależnych par losowych o identycznym rozkładzie prawdopodobieństwa takim, jak rozkład pary $(\boldsymbol{X}, Y)$.

Jakość klasyfikatora $\hat{d}$ mierzona jest za pomocą warunkowego prawdopodobieństwa błędu

$$e(\hat{d}) = \mathrm{P}(\hat{d}(\boldsymbol{X}) \neq Y|\mathcal{L}_n),$$

gdzie para losowa $(\boldsymbol{X}, Y)$ jest niezależna od próby uczącej $\mathcal{L}_n$. Wielkość $e(\hat{d})$ nazywamy *aktualnym poziomem*[2] *błędu* klasyfikatora $\hat{d}$.

[2]Po ang. the actual error rate.

Różnorodnym metodom szacowania tego prawdopodobieństwa jest poświęcony podrozdz. 1.10.

Chcemy znaleźć taki klasyfikator $\hat{d}$, dla którego $e(\hat{d})$ jest bliskie $e(d_B)$. Jednakże $e(\hat{d})$ jest zmienną losową, ponieważ zależy od losowej próby uczącej $\mathcal{L}_n$.

Klasyfikator $\hat{d}$ będziemy nazywać *zgodnym*, jeżeli $e(\hat{d})$ zbiega wg prawdopodobieństwa do $e(d_B)$, tzn. jeśli dla każdego $\varepsilon > 0$

$$\lim_{n\to\infty} \mathrm{P}(|e(\hat{d}) - e(d_B)| \geq \varepsilon) = 0$$

oraz *mocno zgodnym*, jeżeli

$$\mathrm{P}(\lim_{n\to\infty} e(\hat{d}) = e(d_B)) = 1.$$

Znane są trzy główne podejścia do konstrukcji klasyfikatora z próby uczącej.

1. Wybieramy pewną klasę $\mathcal{D}$ klasyfikatorów $\hat{d} : \mathcal{X} \to \mathcal{Y}$ wyznaczonych przy użyciu próby uczącej $\mathcal{L}_n$ oraz ustalamy miarę mierzącą ryzyko związane z użyciem klasyfikatora $\hat{d}$. Oznaczmy ją przez $R(\hat{d})$.
   Wybieramy ten z klasyfikatorów z rodziny $\mathcal{D}$, który minimalizuje ryzyko $R(\hat{d})$, tzn.
   $$\hat{d}_0 = \arg\min_{\hat{d}\in\mathcal{D}} R(\hat{d}).$$
2. Znajdujemy ocenę $\hat{r}$ funkcji regresji r i definiujemy
   $$\hat{d}(\boldsymbol{x}) = \begin{cases} 1, \text{ jeżeli } \hat{r}(\boldsymbol{x}) > \dfrac{1}{2}, \\ 0, \text{ poza tym.} \end{cases}$$
3. Estymujemy f_1 z tych $\boldsymbol{X}_i$, dla których $Y_i = 1$, estymujemy f_0 z tych $\boldsymbol{X}_i$, dla których $Y_i = 0$. Niech $\hat{\pi}_1 = \dfrac{1}{n}\sum_{i=1}^{n} Y_i$. Definiujemy
   $$\hat{r}(\boldsymbol{x}) = \hat{\mathrm{P}}(Y = 1|\boldsymbol{X} = \boldsymbol{x}) = \frac{\hat{\pi}_1 \hat{f}_1(\boldsymbol{x})}{\hat{\pi}_1 \hat{f}_1(\boldsymbol{x}) + (1 - \hat{\pi}_1)\hat{f}_0(\boldsymbol{x})}$$
   i
   $$\hat{d}(\boldsymbol{x}) = \begin{cases} 1, \text{ jeżeli } \hat{r}(\boldsymbol{x}) > \dfrac{1}{2}, \\ 0, \text{ poza tym.} \end{cases}$$

Dotychczasowe rozważania uogólnimy na przypadek, gdy Y może przyjmować więcej niż dwie wartości.

Definicja 1.4.
Załóżmy, że $Y \in \mathcal{Y} = \{1, \ldots, K\}$. Klasyfikator postaci

$$d_B(\boldsymbol{x}) = \arg\max_k \mathrm{P}(Y = k | \boldsymbol{X} = \boldsymbol{x}) = \arg\max_k \pi_k f_k(\boldsymbol{x}) \tag{1.11}$$

nazywamy *klasyfikatorem bayesowskim*, gdzie

$$\mathrm{P}(Y = k | \boldsymbol{X} = \boldsymbol{x}) = \frac{\pi_k f_k(\boldsymbol{x})}{\sum_{i=1}^{K} \pi_i f_i(\boldsymbol{x})}, \tag{1.12}$$

$$\pi_i = \mathrm{P}(Y = i), \qquad f_i(\boldsymbol{x}) = f(\boldsymbol{x} | Y = i)$$

oraz $\arg\max_k$ oznacza tę wartość k, która maksymalizuje dane wyrażenie.

TWIERDZENIE 1.2.

Klasyfikator bayesowski d_B dany wzorem (1.12) jest optymalny, tj. jeżeli d jest jakimkolwiek innym klasyfikatorem, to $e(d_B) \leq e(d)$, gdzie $e(d)$ jest rzeczywistym poziomem błędu klasyfikatora d danym wzorem (1.5). Ponadto

$$e(d_B) = 1 - \int_{\mathbb{R}^p} \max_{1 \leq j \leq K} [f_j(\boldsymbol{x}) \pi_j] d\boldsymbol{x}. \tag{1.13}$$

Dowód. Niech $\Omega_i = \{\boldsymbol{x} \in \mathcal{X} : d(\boldsymbol{x}) = i\}$, $i = 1, 2, \ldots, K$.
Mamy

$$\begin{aligned}
\mathrm{P}(d(\boldsymbol{X}) = Y) &= \sum_{j=1}^{K} \mathrm{P}(d(\boldsymbol{X}) = j | Y = j) \pi_j = \\
&= \sum_{j=1}^{K} \int_{\Omega_j} f_j(\boldsymbol{x}) \pi_j d\boldsymbol{x} = \\
&= \int_{\mathbb{R}^p} \left[\sum_{j=1}^{K} I(d(\boldsymbol{x}) = j) f_j(\boldsymbol{x}) \pi_j \right] d\boldsymbol{x},
\end{aligned}$$

gdzie $I(A)$ jest indykatorem zdarzenia A.

Dla ustalonego $\boldsymbol{x}$,

$$\sum_{j=1}^{K} I(d(\boldsymbol{x}) = j) f_j(\boldsymbol{x})\pi_j \leq \max_j [f_j(\boldsymbol{x})\pi_j].$$

Równość jest osiągana wówczas, gdy $d(\boldsymbol{x})$ równa się tej wartości j, dla której $f_j(\boldsymbol{x})\pi_j$ jest maksymalne. Stąd klasyfikator dany wzorem (1.11) ma tę własność, że dla każdego innego klasyfikatora d

$$\mathrm{P}(d(\boldsymbol{X}) = Y) \leq \mathrm{P}(d_B(\boldsymbol{X}) = Y) = \int_{\mathbb{R}^p} \max_{1 \leq j \leq K} [f_j(\boldsymbol{x})\pi_j] d\boldsymbol{x}.$$

To pokazuje optymalność klasyfikatora bayesowskiego d_B oraz prawdziwość wzoru (1.13). ∎

Niech

$$p_{ij} = P(d(\boldsymbol{X}) = j | Y = i), \qquad i \neq j \tag{1.14}$$

oznacza prawdopodobieństwo błędnego zaklasyfikowania obserwacji $\boldsymbol{X}$ do grupy j-tej wówczas, gdy w rzeczywistości należy ona do grupy i-tej. Mamy wówczas:

$$p_{ij} = P(\boldsymbol{X} \in \Omega_j | Y = i) = \int_{\Omega_j} f_i(\boldsymbol{x}) d\boldsymbol{x}, \qquad i \neq j. \tag{1.15}$$

Niech ponadto

$$p_{i\cdot} = \sum_{j=1,\, j\neq i}^{K} P(d(\boldsymbol{X}) = j | Y = i) \tag{1.16}$$

oznacza prawdopodobieństwo błędnej klasyfikacji obserwacji z grupy i-tej. Rzeczywisty poziom błędu (1.5) klasyfikatora bayesowskiego d_B nazywamy *błędem bayesowskim*[3] i oznaczamy przez e^*. Jest on równy:

$$\begin{aligned} e^* = e(d_B) &= P(d(\boldsymbol{X}) \neq Y) = \\ &= \sum_{i=1}^{K} \pi_i p_{i\cdot} = \sum_{i=1}^{K} \pi_i \sum_{j=1,\, j\neq i}^{K} P(d(\boldsymbol{X}) = j | Y = i) = \\ &= \sum_{i=1}^{K} \sum_{j=1,\, j\neq i}^{K} \int_{\Omega_j} \pi_i f_i(\boldsymbol{x}) d\boldsymbol{x}. \end{aligned} \tag{1.17}$$

[3]Po ang. Bayes error rate.

1.4. Klasyfikatory gaussowskie

Wydaje się, że najprostszym podejściem do zagadnienia klasyfikacji jest wykorzystanie strategii estymacji funkcji gęstości i przyjęcie modelu parametrycznego dla gęstości, tj. przyjęcie założenia, że znana jest postać gęstości z wyjątkiem tkwiących w niej parametrów.

Załóżmy, że $\mathcal{Y} = \{1, 0\}$ i $f_1(\boldsymbol{x}) = f(\boldsymbol{x}|Y = 1)$ oraz $f_0(\boldsymbol{x}) = f(\boldsymbol{x}|Y = 0)$ są gęstościami p-wymiarowego rozkładu normalnego (gaussowskiego):

$$f_k(\boldsymbol{x}) = (2\pi)^{-\frac{p}{2}}|\boldsymbol{\Sigma}_k|^{-\frac{1}{2}} \exp\left\{-\frac{1}{2}(\boldsymbol{x} - \boldsymbol{\mu}_k)'\boldsymbol{\Sigma}_k^{-1}(\boldsymbol{x} - \boldsymbol{\mu}_k)\right\}, \qquad k = 1,\ 0.$$

Wówczas $\boldsymbol{X}|Y = 1 \sim N_p(\boldsymbol{\mu}_1,\ \boldsymbol{\Sigma}_1)$ i $\boldsymbol{X}|Y = 0 \sim N_p(\boldsymbol{\mu}_0,\ \boldsymbol{\Sigma}_0)$.

TWIERDZENIE 1.3.

Jeżeli $\boldsymbol{X}|Y = 1 \sim N_p(\boldsymbol{\mu}_1,\ \boldsymbol{\Sigma}_1)$ *i* $\boldsymbol{X}|Y = 0 \sim N_p(\boldsymbol{\mu}_0,\ \boldsymbol{\Sigma}_0)$, *to klasyfikator bayesowski ma postać*

$$d_B(\boldsymbol{x}) = \begin{cases} 1, \text{ jeżeli } r_1^2 < r_0^2 + 2\ln\left(\dfrac{\pi_1}{\pi_0}\right) + \ln\left(\dfrac{|\boldsymbol{\Sigma}_0|}{|\boldsymbol{\Sigma}_1|}\right), \\ 0, \text{ poza tym}, \end{cases} \tag{1.18}$$

gdzie

$$r_i^2 = (\boldsymbol{x} - \boldsymbol{\mu}_i)'\boldsymbol{\Sigma}_i^{-1}(\boldsymbol{x} - \boldsymbol{\mu}_i), \qquad i = 1,\ 0 \tag{1.19}$$

jest kwadratem odległości Mahalanobisa.

Klasyfikator bayesowski (1.18) możemy zapisać w następującej równoważnej postaci:

$$d_B(\boldsymbol{x}) = \arg\max_k \delta_k(\boldsymbol{x}),$$

gdzie

$$\delta_k(\boldsymbol{x}) = -\frac{1}{2}\ln|\boldsymbol{\Sigma}_k| - \frac{1}{2}(\boldsymbol{x} - \boldsymbol{\mu}_k)'\boldsymbol{\Sigma}_k^{-1}(\boldsymbol{x} - \boldsymbol{\mu}_k) + \ln\pi_k. \tag{1.20}$$

Funkcja (1.20) jest funkcją kwadratową argumentu $\boldsymbol{x}$ i jest nazywana *kwadratową funkcją klasyfikującą* grupy k. Procedura klasyfikacji oparta na tej funkcji nosi nazwę *kwadratowej analizy dyskryminacyjnej* (QDA)[4].

[4] Po ang. Quadratic Discriminant Analysis (QDA).

Ponieważ parametry π_1, π_0, $\boldsymbol{\mu}_1$, $\boldsymbol{\mu}_0$, $\boldsymbol{\Sigma}_1$, $\boldsymbol{\Sigma}_0$ nie są znane, więc w praktyce zastępujemy je ich estymatorami z próby uczącej $\mathcal{L}_n$. Estymatory te mają następującą postać:

$$\hat{\pi}_1 = \frac{1}{n}\sum_{i=1}^{n} Y_i, \qquad \hat{\pi}_0 = \frac{1}{n}\sum_{i=1}^{n}(1 - Y_i),$$

$$\hat{\boldsymbol{\mu}}_1 = \frac{1}{n_1}\sum_{Y_i=1} \boldsymbol{X}_i, \qquad \hat{\boldsymbol{\mu}}_0 = \frac{1}{n_0}\sum_{Y_i=0} \boldsymbol{X}_i,$$

$$\hat{\boldsymbol{\Sigma}}_1 = \boldsymbol{S}_1 = \frac{1}{n_1 - 1}\sum_{Y_i=1}(\boldsymbol{X}_i - \hat{\boldsymbol{\mu}}_1)(\boldsymbol{X}_i - \hat{\boldsymbol{\mu}}_1)',$$

$$\hat{\boldsymbol{\Sigma}}_0 = \boldsymbol{S}_0 = \frac{1}{n_0 - 1}\sum_{Y_i=0}(\boldsymbol{X}_i - \hat{\boldsymbol{\mu}}_0)(\boldsymbol{X}_i - \hat{\boldsymbol{\mu}}_0)',$$

gdzie $n_1 = \sum_{i=1}^{n} Y_i$, $n_0 = \sum_{i=1}^{n}(1 - Y_i)$.

Procedura ta znacznie się uprości, jeżeli dodatkowo założymy równość macierzy kowariancji w dwóch grupach, tj. jeżeli założymy, że $\boldsymbol{\Sigma}_0 = \boldsymbol{\Sigma}_1 = \boldsymbol{\Sigma}$. W tym przypadku klasyfikator bayesowski ma postać

$$d_B(\boldsymbol{x}) = \arg\max_k \delta_k(\boldsymbol{x}), \tag{1.21}$$

gdzie teraz

$$\delta_k(\boldsymbol{x}) = \boldsymbol{x}'\boldsymbol{\Sigma}^{-1}\boldsymbol{\mu}_k - \frac{1}{2}\boldsymbol{\mu}_k'\boldsymbol{\Sigma}^{-1}\boldsymbol{\mu}_k + \ln \pi_k, \qquad k = 1,\ 0. \tag{1.22}$$

Funkcja (1.22) jest funkcją liniową $\boldsymbol{x}$ i nosi nazwę *liniowej funkcji klasyfikującej* grupy k. Powierzchnia rozdzielająca $\{\boldsymbol{x} : \delta_1(\boldsymbol{x}) = \delta_0(\boldsymbol{x})\} = \{\boldsymbol{x} : \delta_{10}(\boldsymbol{x}) = \delta_1(\boldsymbol{x}) - \delta_0(\boldsymbol{x}) = 0\}$ jest w tym przypadku hiperpłaszczyzną.

Funkcja $\delta_{10}(\boldsymbol{x}) = \delta_1(\boldsymbol{x}) - \delta_0(\boldsymbol{x})$ nazywa się *liniową funkcją dyskryminacyjną*. Ma ona postać

$$\delta_{10}(\boldsymbol{x}) = \left(\boldsymbol{x} - \frac{1}{2}(\boldsymbol{\mu}_1 + \boldsymbol{\mu}_0)\right)' \boldsymbol{\Sigma}^{-1}(\boldsymbol{\mu}_1 - \boldsymbol{\mu}_0) + \ln\left(\frac{\pi_1}{\pi_0}\right). \tag{1.23}$$

Obserwację $\boldsymbol{x}$ klasyfikujemy do grupy 1 wówczas, gdy $\delta_{10}(\boldsymbol{x}) > 0$. W przeciwnym razie klasyfikujemy do grupy 0. Procedura klasyfikacji bazująca na liniowej funkcji klasyfikującej lub liniowej funkcji dyskryminacyjnej nosi nazwę *liniowej analizy dyskryminacyjnej* (LDA)[5].

[5] Po ang. Linear Discriminant Analysis (LDA).

W przypadku nieznanych parametrów π_1, π_0, $\boldsymbol{\mu}_1$, $\boldsymbol{\mu}_0$, $\boldsymbol{\Sigma}$ estymujemy je z próby uczącej tak jak poprzednio, z tą różnicą, że estymator parametru $\boldsymbol{\Sigma}$ jest równy

$$\hat{\boldsymbol{\Sigma}} = \boldsymbol{S} = \frac{1}{n_1 + n_0 - 2}[(n_1 - 1)\boldsymbol{S}_1 + (n_0 - 1)\boldsymbol{S}_0].$$

Przykład 1.2.
Wyznaczmy rzeczywisty poziom błędu klasyfikatora bayesowskiego w problemie dyskryminacji dwóch populacji o rozkładach normalnych $N_p(\boldsymbol{\mu}_0,\ \boldsymbol{\Sigma})$, $N_p(\boldsymbol{\mu}_1,\ \boldsymbol{\Sigma})$ odpowiednio.

Wiemy, że w tym przypadku powierzchnia rozdzielająca jest hiperpłaszczyzną o równaniu (1.23):

$$\delta_{10} = \left[\boldsymbol{x} - \frac{1}{2}(\boldsymbol{\mu}_1 + \boldsymbol{\mu}_0)\right]' \boldsymbol{\Sigma}^{-1}(\boldsymbol{\mu}_1 - \boldsymbol{\mu}_0) + \ln\left(\frac{\pi_1}{\pi_0}\right),$$

natomiast klasyfikator bayesowski d_B ma postać:

$$d_B(\boldsymbol{x}) = \begin{cases} 1\ , \text{ jeżeli } \delta_{10} > 0, \\ 0\ , \text{ poza tym.} \end{cases}$$

Rozważmy zmienną losową

$$W = \left[\boldsymbol{X} - \frac{1}{2}(\boldsymbol{\mu}_1 + \boldsymbol{\mu}_0)\right]' \boldsymbol{\Sigma}^{-1}(\boldsymbol{\mu}_1 - \boldsymbol{\mu}_0).$$

Oczywiście ma ona 1-wymiarowy rozkład normalny. Mamy ponadto:

$$\begin{aligned} E\,[W|\boldsymbol{X} \sim N_p(\boldsymbol{\mu}_1,\ \boldsymbol{\Sigma})] &= \frac{1}{2}(\boldsymbol{\mu}_1 - \boldsymbol{\mu}_0)'\boldsymbol{\Sigma}^{-1}(\boldsymbol{\mu}_1 - \boldsymbol{\mu}_0) = \frac{1}{2}\Delta^2, \\ E\,[W|\boldsymbol{X} \sim N_p(\boldsymbol{\mu}_0,\ \boldsymbol{\Sigma})] &= \frac{1}{2}(\boldsymbol{\mu}_0 - \boldsymbol{\mu}_1)'\boldsymbol{\Sigma}^{-1}(\boldsymbol{\mu}_1 - \boldsymbol{\mu}_0) = -\frac{1}{2}\Delta^2, \\ Var\,[W|\boldsymbol{X} \sim N_p(\boldsymbol{\mu}_i,\ \boldsymbol{\Sigma})] &= (\boldsymbol{\mu}_1 - \boldsymbol{\mu}_0)'\boldsymbol{\Sigma}^{-1}(\boldsymbol{\mu}_1 - \boldsymbol{\mu}_0) = \Delta^2, \qquad i = 0,\ 1, \end{aligned}$$

gdzie Δ jest odległością Mahalanobisa między populacjami.

Mamy:

$$p_{10} = P\left(W < \ln\frac{\pi_0}{\pi_1} \mid \boldsymbol{X} \sim N_p(\boldsymbol{\mu}_1, \boldsymbol{\Sigma})\right) =$$

$$= P\left(\frac{W - \frac{1}{2}\Delta^2}{\Delta} < \frac{\ln\frac{\pi_0}{\pi_1} - \frac{1}{2}\Delta^2}{\Delta} \mid \boldsymbol{X} \sim N_p(\boldsymbol{\mu}_0, \boldsymbol{\Sigma})\right) =$$

$$= \Phi\left(\frac{\ln\frac{\pi_0}{\pi_1} - \frac{1}{2}\Delta^2}{\Delta}\right),$$

$$p_{01} = P\left(W > \ln\frac{\pi_0}{\pi_1} \mid \boldsymbol{X} \sim N_p(\boldsymbol{\mu}_0, \boldsymbol{\Sigma})\right) =$$

$$= 1 - \Phi\left(\frac{\ln\frac{\pi_0}{\pi_1} + \frac{1}{2}\Delta^2}{\Delta}\right) = \Phi\left(-\frac{\ln\frac{\pi_0}{\pi_1} + \frac{1}{2}\Delta^2}{\Delta}\right).$$

Zatem:

$$e^* = \pi_0\Phi\left(-\frac{\ln\frac{\pi_0}{\pi_1} + \frac{1}{2}\Delta^2}{\Delta}\right) + \pi_1\Phi\left(\frac{\ln\frac{\pi_0}{\pi_1} - \frac{1}{2}\Delta^2}{\Delta}\right).$$

Jeżeli $\pi_0 = \pi_1 = 1/2$, to

$$e^* = \Phi\left(-\frac{1}{2}\Delta\right).$$

Zauważmy, że im bardziej odległe będą od siebie rozpatrywane populacje, tym mniejsza będzie wartość błędu bayesowskiego.

Niestety, już w przypadku populacji o rozkładach normalnych z różnymi macierzami kowariancji metody wyznaczania wielkości p_{ij} nie są obecnie znane, a oszacowanie ich jest bardzo trudne (Krzyśko, 2000).

Podamy teraz numeryczny sposób wyznaczania błędu bayesowskiego dla przypadku dwóch populacji.
Mamy:

$$e^* = \pi p_{10} + (1 - \pi)p_{01},$$

gdzie:

$$p_{10} = \int_{\Omega_0} f_1(\boldsymbol{x})d\boldsymbol{x}, \qquad p_{01} = \int_{\Omega_1} f_0(\boldsymbol{x})d\boldsymbol{x}.$$

Ponieważ zwykle całkowanie odbywa się po bardzo skomplikowanych obszarach, wygodniej jest całkować względem jednowymiarowej funkcji:

$$h(\boldsymbol{x}) = -\ln\frac{f_1(\boldsymbol{x})}{f_0(\boldsymbol{x})}.$$

Mamy:

$$p_{10} = \int_{\ln\frac{\pi}{1-\pi}}^{\infty} g_1(h)dh, \tag{1.24}$$

$$p_{01} = \int_{-\infty}^{\ln\frac{\pi}{1-\pi}} g_0(h)dh, \tag{1.25}$$

gdzie $g_i(h)$ jest gęstością zmiennej losowej $h(\boldsymbol{X})|Y = i$, $i = 0,\ 1$.

Odtąd będziemy zakładać, że $\pi = 1/2$. Zdefiniujmy funkcje charakterystyczne zmiennej losowej $h(\boldsymbol{X})|Y = i$, $i = 0,\ 1$:

$$\phi_1(t) = \int_{\mathbb{R}^p} \exp\{ith(\boldsymbol{x})\}f_1(\boldsymbol{x})d\boldsymbol{x} = \int_{-\infty}^{\infty} \exp\{ith\}g_1(h)dh,$$

$$\phi_0(t) = \int_{\mathbb{R}^p} \exp\{ith(\boldsymbol{x})\}f_0(\boldsymbol{x})d\boldsymbol{x} = \int_{-\infty}^{\infty} \exp\{ith\}g_0(h)dh.$$

Widzimy, że funkcje charakterystyczne mogą być obliczone albo przez całkowanie wielokrotne przy użyciu gęstości $f_i(\boldsymbol{x})$, albo przez całkowanie jednokrotne za pomocą funkcji $g_i(h)$, $i = 0,\ 1$.
Zależności (1.24) i (1.25) przyjmują postać:

$$p_{10} = \frac{1}{2} + \frac{1}{\pi}\int_0^{\infty} \frac{|\phi_1(t)|}{t}\sin(\phi_1(t))dt, \tag{1.26}$$

$$p_{01} = \frac{1}{2} - \frac{1}{\pi}\int_0^{\infty} \frac{|\phi_0(t)|}{t}\sin(\phi_0(t))dt. \tag{1.27}$$

Znając postać funkcji $\phi_1(t)$ oraz $\phi_0(t)$, wielkości (1.26) i (1.27) obliczamy, stosując metody numeryczne całkowania funkcji.

Poniższy przykład został zaczerpnięty z pracy Góreckiego (2005c).

Przykład 1.3.
Wyznaczmy błąd bayesowski w problemie dyskryminacji dwóch populacji o rozkładach normalnych $N_8(\boldsymbol{\mu}_1,\ \boldsymbol{\Sigma}_1)$, $N_8(\boldsymbol{\mu}_0,\ \boldsymbol{\Sigma}_0)$, gdzie:

$$\begin{aligned}
\boldsymbol{\mu}_1 &= \mathbf{0}_8, \qquad \boldsymbol{\mu}_0 = (3.86,\ 3.10,\ 0.84,\ 0.84,\ 1.64,\ 1.08,\ 0.26,\ 0.01)',\\
\boldsymbol{\Sigma}_1 &= \boldsymbol{I}_8, \qquad \boldsymbol{\Sigma}_0 = \mathrm{diag}(8.41,\ 12.06,\ 0.12,\ 0.22,\ 1.49,\ 1.77,\ 0.35,\ 2.73).
\end{aligned}$$

Mamy:

$$\begin{aligned}
h(\boldsymbol{x}) &= -\ln\frac{f_1(\boldsymbol{x})}{f_0(\boldsymbol{x})} = \\
&= \frac{1}{2}\left[(\boldsymbol{x}-\boldsymbol{\mu}_1)'\boldsymbol{\Sigma}_1^{-1}(\boldsymbol{x}-\boldsymbol{\mu}_1) - (\boldsymbol{x}-\boldsymbol{\mu}_0)'\boldsymbol{\Sigma}_0^{-1}(\boldsymbol{x}-\boldsymbol{\mu}_0) + \ln\frac{|\boldsymbol{\Sigma}_1|}{|\boldsymbol{\Sigma}_0|}\right] = \\
&= \frac{1}{2}\left[\boldsymbol{x}'\boldsymbol{x} - (\boldsymbol{x}-\boldsymbol{\mu}_0)'\boldsymbol{\Sigma}_0^{-1}(\boldsymbol{x}-\boldsymbol{\mu}_0) + \ln\frac{1}{|\boldsymbol{\Sigma}_0|}\right].
\end{aligned}$$

Stąd

$$\begin{aligned}
\phi_1(t) = {}& \left(\frac{1}{2\pi}\right)^4 \int_{\mathbb{R}^8} \exp\left\{-\frac{1}{2}\left[(1-it)\boldsymbol{x}'\boldsymbol{x} + it(\boldsymbol{x}-\boldsymbol{\mu}_0)'\boldsymbol{\Sigma}_0^{-1}(\boldsymbol{x}-\boldsymbol{\mu}_0)\ +\right.\right.\\
&+\ \left.\left. it\ln|\boldsymbol{\Sigma}_0|\right]\right\} d\boldsymbol{x}.
\end{aligned}$$

Oznaczając

$$it\ln|\boldsymbol{\Sigma}_0| = it\prod_{j=1}^{8}\lambda_j = c,$$

otrzymujemy:

$$\phi_1(t) = \tag{1.28}$$

$$= \left(\frac{1}{2\pi}\right)^4 \int_{\mathbb{R}^8} \exp\left\{-\frac{1}{2}\left[(1-it)\sum_{j=1}^{8} x_j^2 + it\sum_{j=1}^{8}(x_j-\mu_{0j})^2\lambda_j^{-1} + c\right]\right\} d\boldsymbol{x}.$$

Zauważmy, że dla $j = 1, 2, \ldots, 8$ mamy:

$$\int_{-\infty}^{\infty} \exp\left\{-\frac{1}{2}\left[(1-it)x_j^2 + it(x_j^2 - 2x_j\mu_{0j} + \mu_{0j}^2)\lambda_j^{-1}\right]\right\} dx_j =$$

$$= \int_{-\infty}^{\infty} \exp\left\{-\frac{1}{2}\left[\frac{\left(x_j - \dfrac{\mu_{0j} it\lambda_j^{-1}}{1-it+it\lambda_j^{-1}}\right)^2}{\sigma_j^2} + \right.\right.$$

$$\left.\left. + \frac{\mu_{0j}^2 t^2 \lambda_j^{-2}}{1-it+it\lambda_j^{-1}} + it\mu_{0j}^2\lambda_j^{-1}\right]\right\} dx_j =$$

$$= \left(\frac{1}{2\pi}\right)^{-\frac{1}{2}} \sigma_j \exp\left\{-\frac{1}{2}\left[\frac{\mu_{0j}^2 t^2\lambda_j^{-2}}{1-it+it\lambda_j^{-1}} + it\mu_{0j}^2\lambda_j^{-1}\right]\right\},$$

gdzie $\sigma_j^2 = (1 - it + it\lambda_j^{-1})^{-1}$.

Uwzględniając ten fakt w zależności (1.28), otrzymujemy:

$$\phi_1(t) = \exp\left\{-\frac{1}{2} it \ln \prod_{j=1}^{8} \lambda_j\right\} \times$$

$$\times \prod_{j=1}^{8} \left\{\sigma_j \exp\left[-\frac{1}{2}\left(\frac{\mu_{0j}^2 t^2 \lambda_j^{-2}}{1-it+it\lambda_j^{-1}} + it\mu_{0j}^2\lambda_j^{-1}\right)\right]\right\}.$$

Rozumując analogicznie można wykazać, że:

$$\phi_0(t) = \left(\prod_{j=1}^{8} \lambda_j\right)^{-\frac{1}{2}} \exp\left\{-\frac{1}{2} it \ln \prod_{j=1}^{8} \lambda_j\right\} \times$$

$$\times \prod_{j=1}^{8} \left\{\delta_j \exp\left[-\frac{1}{2}\left(-\frac{\mu_{0j}^2\lambda_j^{-2}(1+it)^2}{(1+it)\lambda_j^{-1} - it} + (1+it)\mu_{0j}^2\lambda_j^{-1}\right)\right]\right\},$$

gdzie $\delta_j^2 = [(1+it)\lambda_j^{-1} - it]^{-1}$.

Stosując metody całkowania numerycznego, ze wzorów (1.26) i (1.27) otrzymujemy:

$$p_{10} = 0.0146, \qquad p_{01} = 0.0214.$$

Stąd ostatecznie błąd bayesowski wynosi

$$e^* = 0.018.$$

Uzyskane wyniki uogólnimy na przypadek większej liczby grup, tj. gdy $Y \in \{1, \ldots, K\}$.

TWIERDZENIE 1.4.

Załóżmy, że $Y \in \{1, \ldots, K\}$. Jeżeli $f_k(\boldsymbol{x}) = f(\boldsymbol{x}|Y = k)$ jest gęstością p-wymiarowego rozkładu normalnego (gaussowskiego) $N_p(\boldsymbol{\mu}_k, \boldsymbol{\Sigma}_k)$, to klasyfikator bayesowski ma postać

$$d_B(\boldsymbol{x}) = \arg\max_k \delta_k(\boldsymbol{x}), \tag{1.29}$$

gdzie

$$\delta_k(\boldsymbol{x}) = -\frac{1}{2}\ln|\boldsymbol{\Sigma}_k| - \frac{1}{2}(\boldsymbol{x} - \boldsymbol{\mu}_k)'\boldsymbol{\Sigma}_k^{-1}(\boldsymbol{x} - \boldsymbol{\mu}_k) + \ln\pi_k. \tag{1.30}$$

Jeżeli ponadto wszystkie macierze kowariancji są sobie równe i równe macierzy $\boldsymbol{\Sigma}$, to

$$\delta_k(\boldsymbol{x}) = \boldsymbol{x}'\boldsymbol{\Sigma}^{-1}\boldsymbol{\mu}_k - \frac{1}{2}\boldsymbol{\mu}_k'\boldsymbol{\Sigma}^{-1}\boldsymbol{\mu}_k + \ln\pi_k. \tag{1.31}$$

Funkcja (1.30) jest funkcją kwadratową i nosi nazwę *kwadratowej funkcji klasyfikującej k-tej grupy*. Powierzchnia rozdzielająca grupy i oraz j ma postać $\{\boldsymbol{x} : \delta_i(\boldsymbol{x}) = \delta_j(\boldsymbol{x})\} = \{\boldsymbol{x} : \delta_{ij}(\boldsymbol{x}) = \delta_i(\boldsymbol{x}) - \delta_j(\boldsymbol{x}) = 0\}$, gdzie

$$\begin{aligned}\delta_{ij}(\boldsymbol{x}) = {} & \frac{1}{2}\boldsymbol{x}'(\boldsymbol{\Sigma}_j^{-1} - \boldsymbol{\Sigma}_i^{-1})\boldsymbol{x} + (\boldsymbol{\mu}_i'\boldsymbol{\Sigma}_i^{-1} - \boldsymbol{\mu}_j'\boldsymbol{\Sigma}_j^{-1})\boldsymbol{x} + \\ & + \frac{1}{2}\boldsymbol{\mu}_j'\boldsymbol{\Sigma}_j^{-1}\boldsymbol{\mu}_j - \frac{1}{2}\boldsymbol{\mu}_i'\boldsymbol{\Sigma}_i^{-1}\boldsymbol{\mu}_i + \frac{1}{2}\ln\left(\frac{|\boldsymbol{\Sigma}_j|}{|\boldsymbol{\Sigma}_i|}\right) + \ln\left(\frac{\pi_i}{\pi_j}\right), \\ & i, j = 1, 2, \ldots, K, \quad j \neq i. \end{aligned} \tag{1.32}$$

Funkcja $\delta_{ij}(\boldsymbol{x})$ dana wzorem (1.32) nosi nazwę *kwadratowej funkcji dyskryminacyjnej* grup i oraz j. Równanie $\delta_{ij}(\boldsymbol{x}) = 0$ przedstawia hiperpowierzchnię algebraiczną stopnia drugiego. W przypadku dwuwymiarowym równanie to przedstawia elipsę, parabolę lub hiperbolę.

Procedura klasyfikacji bazująca na kwadratowej funkcji klasyfikującej postaci (1.30) nosi nazwę *kwadratowej analizy dyskryminacyjnej*.

Funkcja (1.31) jest funkcją liniową i nosi nazwę *liniowej funkcji klasyfikującej związanej z k-tą grupą*. Powierzchnia rozdzielająca grupy i oraz j ma w tym przypadku postać:

$$\{\boldsymbol{x} : \delta_i(\boldsymbol{x}) = \delta_j(\boldsymbol{x})\} = \{\boldsymbol{x} : \delta_{ij}(\boldsymbol{x}) = \delta_i(\boldsymbol{x}) - \delta_j(\boldsymbol{x}) = 0\},$$

gdzie

$$\delta_{ij}(\boldsymbol{x}) = \boldsymbol{x}'\boldsymbol{\Sigma}^{-1}(\boldsymbol{\mu}_i - \boldsymbol{\mu}_j) - \frac{1}{2}(\boldsymbol{\mu}_i + \boldsymbol{\mu}_j)'\boldsymbol{\Sigma}^{-1}(\boldsymbol{\mu}_i - \boldsymbol{\mu}_j) + \ln\left(\frac{\pi_i}{\pi_j}\right),$$
$$i,\ j = 1,\ 2,\ \ldots,\ K, \quad j \neq i. \tag{1.33}$$

Funkcja $\delta_{ij}(\boldsymbol{x})$ dana wzorem (1.33) nosi nazwę *liniowej funkcji dyskryminacyjnej* grup i oraz j. Równanie $\delta_{ij}(\boldsymbol{x}) = 0$ jest równaniem hiperpłaszczyzny. Procedura klasyfikacji bazująca na liniowej funkcji klasyfikującej postaci (1.31) nosi nazwę *liniowej analizy dyskryminacyjnej*.

Parametry π_k, $\boldsymbol{\mu}_k$, $\boldsymbol{\Sigma}_k$, $\boldsymbol{\Sigma}$, $k = 1, 2, \ldots, K$, występujące w kwadratowej funkcji klasyfikującej (1.23) i w liniowej funkcji klasyfikującej (1.30) nie są zazwyczaj znane i w praktyce należy zastąpić je ich estymatorami z próby uczącej. Jeżeli próba ucząca zawiera n_i obserwacji z i-tej grupy, $n_1 + n_2 + \cdots + n_K = n$ oraz $\boldsymbol{X}_{ij}$ jest j-tą obserwacją z i-tej grupy, to estymatory nieznanych parametrów są równe:

$$\hat{\pi}_k = \frac{n_k}{n}, \tag{1.34}$$

$$\hat{\boldsymbol{\mu}}_k = \bar{\boldsymbol{X}}_k = \frac{1}{n_k}\sum_{j=1}^{n_k}\boldsymbol{X}_{kj}, \tag{1.35}$$

$$\hat{\boldsymbol{\Sigma}}_k = \boldsymbol{S}_k = \frac{1}{n_k - 1}\sum_{j=1}^{n_k}(\boldsymbol{X}_{kj} - \bar{\boldsymbol{X}}_k)(\boldsymbol{X}_{kj} - \bar{\boldsymbol{X}}_k)', \tag{1.36}$$

$$\hat{\boldsymbol{\Sigma}} = \boldsymbol{S} = \frac{1}{n - K}\sum_{k=1}^{K}\sum_{j=1}^{n_k}(\boldsymbol{X}_{kj} - \bar{\boldsymbol{X}}_k)(\boldsymbol{X}_{kj} - \bar{\boldsymbol{X}}_k)'. \tag{1.37}$$

Estymator $\boldsymbol{S}_k$ jest nieobciążonym estymatorem macierzy kowariancji $\boldsymbol{\Sigma}_k$ dodatnio określonej. Zatem estymator $\boldsymbol{S}_k$ powinien być również macierzą dodatnio określoną. Można udowodnić (patrz np. Krzyśko (2000), s. 34), że gdy $n_k > p$, to z prawdopodobieństwem jeden macierz $\boldsymbol{S}_k$ jest dodatnio określona. Próba ucząca musi być zatem dostatecznie liczna, by w każdej z K grup warunek ten mógłby być spełniony. Małoliczne próby skutkują złą jakością estymatorów. Mogą one charakteryzować się nadmierną zmiennością. Chcąc tego uniknąć, możemy posłużyć się *estymatorami ściągającymi*[6] zaproponowanymi przez Friedmana (1989). Estymator ściągający macierzy kowariancji $\boldsymbol{\Sigma}_k$ ma postać:

$$\hat{\boldsymbol{\Sigma}}_k(\lambda,\ \gamma) = (1-\gamma)\hat{\boldsymbol{\Sigma}}_k(\lambda) + \gamma p^{-1}\mathrm{tr}(\hat{\boldsymbol{\Sigma}}_k(\lambda))\boldsymbol{I}_p, \tag{1.38}$$

gdzie $\boldsymbol{I}_p$ jest macierzą jednostkową stopnia p, $\mathrm{tr}(\boldsymbol{A})$ oznacza ślad macierzy $\boldsymbol{A}$ (sumę elementów diagonalnych macierzy $\boldsymbol{A}$) oraz

$$\hat{\boldsymbol{\Sigma}}_k(\lambda) = \frac{(1-\lambda)(n_k-1)\boldsymbol{S}_k + \lambda(n-K)\boldsymbol{S}}{(1-\lambda)(n_k-1) + \lambda(n-K)}, \tag{1.39}$$
$$0 \le \lambda \le 1, \quad 0 \le \gamma \le 1.$$

Wartości parametrów λ i γ zmieniają się w kwadracie jednostkowym postaci:

$$\{(\lambda,\ \gamma): \quad 0 \le \lambda \le 1, \quad 0 \le \gamma \le 1\}.$$

Wierzchołek $(\lambda = 0, \gamma = 0)$ odpowiada kwadratowej funkcji dyskryminacyjnej, wierzchołek $(\lambda = 1,\ \gamma = 0)$ odpowiada liniowej funkcji dyskryminacyjnej, wierzchołek $(\lambda = 1,\ \gamma = 1)$ odpowiada regule klasyfikacyjnej opartej na minimalnej odległości euklidesowej między grupami, natomiast wierzchołek $(\lambda = 0,\ \gamma = 1)$ odpowiada regule klasyfikacyjnej opartej na ważonej minimalnej odległości euklidesowej, gdzie wagi grup są odwrotnie proporcjonalne do $\mathrm{tr}(\hat{\boldsymbol{\Sigma}}_k)/p$. Jeżeli $\gamma = 0$, a λ zmienia się, to uzyskuje się reguły pośrednie między kwadratowymi i liniowymi. Jeżeli $\lambda = 1$, a γ zmienia się, to uzyskujemy pewną analogię *regresji grzbietowej*[7] dla liniowej analizy dyskryminacyjnej.

W praktyce optymalne wartości parametrów regularyzujących λ i γ nie są znane i należy je wyestymować z próby uczącej. Optymalną parą $(\lambda,\ \gamma)$ jest ta para, dla której ocena aktualnego poziomu błędu, uzyskana metodą sprawdzania krzyżowego, jest minimalna. Ocenom aktualnego poziomu błędu jest poświęcony podrozdz. 1.10. Analiza dyskryminacyjna wykorzystująca estymatory (1.38) została nazwana przez Friedmana *regularyzowaną*[8] *analizą dyskryminacyjną* (RDA).

[6]Po ang. shrinkage.
[7]Po ang. ridge regression.
[8]Po ang. Regularized Discriminant Analysis (RDA).

Siatka punktów (λ, γ) nałożona na jednostkowy kwadrat zawiera zazwyczaj od 25 do 50 różnych kombinacji parametrów regularyzujących λ i γ. Ten problem dwuparametrowej optymalizacji będzie wymagał nadmiernych obliczeń, jeżeli zostanie zaimplementowany w sposób naiwny. Jeżeli przez N_p oznaczymy liczbę punktów na siatce, to w tej procedurze musimy wykonać $N_p n$ obliczeń wskaźników dyskryminacyjnych.

Jednakże Friedman podał wzory przystosowane do obliczeń regularyzowanego estymatora macierzy kowariancji i jego odwrotności w przypadku, gdy z próby uczącej sukcesywnie usuwamy pojedyncze i różne obserwacje (tak jak to się dzieje w metodzie sprawdzania krzyżowego).

Przewaga metody RDA nad metodą LDA i QDA rośnie wraz ze wzrostem wymiaru przestrzeni cech. Ponadto regularyzacja okazuje się potrzebna, jeśli liczebności próbek uczących w grupach są małe w porównaniu z wymiarem przestrzeni cech.

Pewną niedogodnością jest fakt, że optymalny poziom błędu jest często osiągany dla kilku różnych punktów (λ, γ) na siatce. W takim przypadku wybieramy taki punkt, aby dla największej dopuszczalnej wartości λ wybrać największą wartość γ. Dodatkowo okazuje się, że w przeciwieństwie do LDA oraz QDA metoda RDA nie jest niezmiennicza ze względu na skalowanie danych. Ten brak niezmienniczości jest spowodowany wprowadzeniem parametru γ. Jednak, jak wykazują badania, usunięcie tego parametru z modelu prowadzi do spadku efektywności metody RDA.

Zainteresowanych Czytelników odsyłamy do oryginalnej pracy Friedmana (1989).

Przykład 1.4.

Dane zaczerpnięte ze strony 501 monografii Johnsona i Wicherna (1982) zawierają wyniki dwóch testów wykonanych na grupie $n = 85$ kandydatów do szkoły handlowej, którzy zostali zaklasyfikowani do trzech grup: przyjęci, nieprzyjęci i przypadki wątpliwe o liczebnościach odpowiednio $n_1 = 31$, $n_2 = 28$, $n_3 = 26$.

Wektory wartości średnich i macierze kowariancji z prób dla tych trzech grup są równe:

$$\bar{\boldsymbol{X}}_1 = (3.4,\ 561.2)', \quad \bar{\boldsymbol{X}}_2 = (2.5,\ 447.1)', \quad \bar{\boldsymbol{X}}_3 = (3.0,\ 446.2)',$$

$$\boldsymbol{S}_1 = \begin{bmatrix} 0.04 & -0.28 \\ -0.28 & 461.25 \end{bmatrix}, \qquad \boldsymbol{S}_2 = \begin{bmatrix} 0.03 & -1.19 \\ -1.19 & 3891.25 \end{bmatrix},$$

$$\boldsymbol{S}_3 = \begin{bmatrix} 0.03 & -5.40 \\ -5.40 & 2246.90 \end{bmatrix}.$$

Przy założeniu, że macierz kowariancji w trzech grupach jest identyczna, jej wspólna ocena z trzech prób jest równa

$$S = \begin{bmatrix} 0.04 & -2.02 \\ -2.02 & 3655.90 \end{bmatrix}.$$

Na rysunku 1.2 są pokazane trzy obszary klasyfikacji wydzielone za pomocą liniowych i kwadratowych funkcji dyskryminacyjnych.

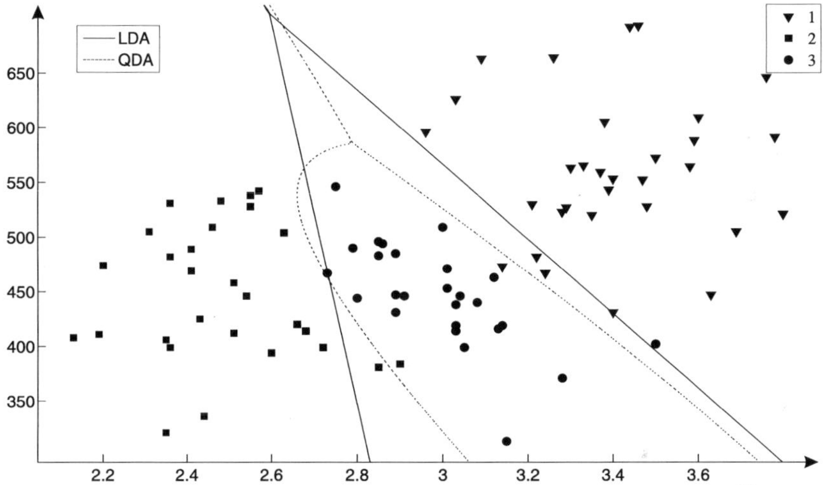

Rys. 1.2. Obszary klasyfikacji z przykładu 1.4 wydzielone za pomocą liniowych i kwadratowych funkcji dyskryminacyjnych

1.5. Naiwny klasyfikator bayesowski

Reguła bayesowska ma postać $d_B(\boldsymbol{x}) = \arg\max_k \pi_k f_k(\boldsymbol{x})$. Jeżeli mamy możliwość estymacji π_k i f_k, to mamy możność estymacji klasyfikatora bayesowskiego. Estymacja π_k jest łatwa, ale co z estymacją funkcji gęstości f_k? Wcześniej założyliśmy, że f_k jest gęstością rozkładu normalnego. Inna możliwość to skonstruowanie nieparametrycznego estymatora funkcji f_k, np. estymatora jądrowego. O takim estymatorze będziemy mówić w podrozdz 2.4. Problem estymacji funkcji f_k znacznie się upraszcza, jeżeli założymy, że wszystkie składowe obserwowanego wektora losowego $\boldsymbol{X} = (X_1, X_2, \ldots, X_p)'$ są niezależne. Wówczas $f_k(x_1, x_2, \ldots, x_p) = \prod_{j=1}^{p} f_{kj}(x_j)$. Problem został zredukowany do estymacji p jednowymiarowych funkcji gęstości w każdej z K grup. Uzyskany w ten sposób klasyfikator nazywa się *naiwnym klasyfikatorem bayesowskim*. Założenie, że składowe wektora losowego $\boldsymbol{X}$ są niezależne jest zazwyczaj nieprawdziwe, ale uzyskany przy tym założeniu klasyfikator może dawać dobre rezultaty. Poszczególne kroki budowy naiwnego klasyfikatora bayesowskiego są podane w postaci algorytmu.

Naiwny estymator bayesowski

1. Dla każdej grupy k konstruujemy estymator $\hat{f}_{kj}$ gęstości f_{kj} zmiennej losowej X_j, wykorzystując do tego celu te elementy próby uczącej, dla których $Y_i = k$.
2. Niech
$$\hat{f}_k(\boldsymbol{x}) = \hat{f}_k(x_1, x_2, \ldots, x_p) = \prod_{j=1}^{p} \hat{f}_{kj}(x_j).$$
3. Niech
$$\hat{\pi}_k = \frac{1}{n}\sum_{i=1}^{n} I(Y_i = k),$$
gdzie $I(Y_i = k) = 1$, jeżeli $Y_i = k$ oraz $I(Y_i = k) = 0$, jeżeli $Y_i \neq k$.
4. Niech
$$\hat{d}_{NB}(\boldsymbol{x}) = \arg\max_k \hat{\pi}_k \hat{f}_k(\boldsymbol{x}).$$

Naiwny klasyfikator bayesowski jest szczególnie popularny wówczas, gdy $\boldsymbol{X}$ jest wektorem dyskretnym o dużej liczbie składowych. W tym przypadku estymatory $\hat{f}_{kj}(x_j)$ mają szczególnie prostą postać.

1.6. Liniowa funkcja dyskryminacyjna Fishera

Pierwsza wersja liniowej analizy dyskryminacyjnej w przypadku dwóch grup pochodzi od Fishera (1936). Idea Fishera polegała na zredukowaniu wektora cech do jednego wymiaru przez zrzutowanie danych na prostą. Algebraicznie oznacza to zastąpienie wektora $\boldsymbol{X} = (X_1, X_2, \ldots, X_p)'$ kombinacją liniową $U = \boldsymbol{w}'\boldsymbol{X} = \sum_{j=1}^{p} w_j X_j$. Naszym celem jest taki dobór wektora $\boldsymbol{w}$, by jak najlepiej były od siebie oddzielone grupy.

Zatem chcemy zbudować klasyfikator na bazie jednej zmiennej U zamiast wektora $\boldsymbol{X}$. Zdefiniujmy teraz pojęcie oddzielania grup. Niech $\boldsymbol{\mu}_j$ oznacza wektor wartości oczekiwanych wektora $\boldsymbol{X}$ dla $Y = j$ i niech $\boldsymbol{\Sigma}$ będzie macierzą kowariancji wektora $\boldsymbol{X}$ dla $Y = j$. Wówczas $\mathrm{E}(U|Y = j) =$

$\mathrm{E}(\boldsymbol{w}'\boldsymbol{X}|Y = j) = \boldsymbol{w}'\boldsymbol{\mu}_j$ oraz $\mathrm{Var}(U) = \mathrm{Var}(\boldsymbol{w}'\boldsymbol{X}) = \boldsymbol{w}'\boldsymbol{\Sigma}\boldsymbol{w}$ (zakładamy, że w obydwu grupach wektor $\boldsymbol{X}$ ma identyczną macierz kowariancji $\boldsymbol{\Sigma}$).

Jako miarę oddzielania dwóch grup 1 oraz 0 przyjmujemy

$$J(\boldsymbol{w}) = \frac{[\mathrm{E}(U|Y = 1) - \mathrm{E}(U|Y = 0)]^2}{\boldsymbol{w}'\boldsymbol{\Sigma}\boldsymbol{w}} = \frac{(\boldsymbol{w}'\boldsymbol{\mu}_1 - \boldsymbol{w}'\boldsymbol{\mu}_0)^2}{\boldsymbol{w}'\boldsymbol{\Sigma}\boldsymbol{w}}.$$

W istocie $J(\boldsymbol{w})$ jest kwadratem odległości Mahalanobisa między dwiema grupami. Im $J(\boldsymbol{w})$ ma większą wartość, tym lepiej grupy są oddzielone. Miarę J estymujemy następująco. Niech $n_j = \sum_{i=1}^{n} I(Y_i = j)$ będzie liczbą obserwacji z grupy j oraz niech $\bar{\boldsymbol{X}}_j$ będzie wektorem średnich z próby z grupy j oraz niech $\boldsymbol{S}_j$ będzie macierzą kowariancji z próby z grupy j. Definiujemy

$$\hat{J}(\boldsymbol{w}) = \frac{[\boldsymbol{w}'(\bar{\boldsymbol{X}}_1 - \bar{\boldsymbol{X}}_0)]^2}{\boldsymbol{w}'\boldsymbol{S}_W\boldsymbol{w}}, \tag{1.40}$$

gdzie

$$\boldsymbol{S}_W = \frac{1}{n_1 + n_0 - 2}[(n_1 - 1)\boldsymbol{S}_1 + (n_0 - 1)\boldsymbol{S}_0].$$

TWIERDZENIE 1.5.

Wektor $\boldsymbol{w}$ maksymalizujący miarę $\hat{J}(\boldsymbol{w})$ jest równy

$$\boldsymbol{w} = \boldsymbol{S}_W^{-1}(\bar{\boldsymbol{X}}_1 - \bar{\boldsymbol{X}}_0)$$

lub dowolnej wielokrotności $\boldsymbol{S}_W^{-1}(\bar{\boldsymbol{X}}_1 - \bar{\boldsymbol{X}}_0)$.

Dowód. Różniczkujemy miarę (1.40) względem wektora $\boldsymbol{w}$ (jeżeli $\boldsymbol{A}$ jest macierzą symetryczną, to $\partial(\boldsymbol{x}'\boldsymbol{A}\boldsymbol{x})/\partial\boldsymbol{x} = 2\boldsymbol{A}\boldsymbol{x}$) i uzyskaną pochodną przyrównujemy do wektora zerowego:

$$\frac{\partial\hat{J}(\boldsymbol{w})}{\partial\boldsymbol{w}} =$$
$$= \frac{\boldsymbol{w}'\boldsymbol{S}_W\boldsymbol{w}\; 2\boldsymbol{w}'(\bar{\boldsymbol{X}}_1 - \bar{\boldsymbol{X}}_0)(\bar{\boldsymbol{X}}_1 - \bar{\boldsymbol{X}}_0) - \left[\boldsymbol{w}'(\bar{\boldsymbol{X}}_1 - \bar{\boldsymbol{X}}_0)\right]^2 2\boldsymbol{S}_W\boldsymbol{w}}{(\boldsymbol{w}'\boldsymbol{S}_W\boldsymbol{w})^2} = \boldsymbol{0}.$$

Powyższa równość jest równoważna równości

$$\boldsymbol{w}'\boldsymbol{S}_W\boldsymbol{w}(\bar{\boldsymbol{X}}_1 - \bar{\boldsymbol{X}}_0) - \boldsymbol{w}'(\bar{\boldsymbol{X}}_1 - \bar{\boldsymbol{X}}_0)\boldsymbol{S}_W\boldsymbol{w} = \boldsymbol{0}.$$

Mnożąc ostatnią równość lewostronnie przez $\boldsymbol{S}_W^{-1}$, otrzymujemy:

$$\boldsymbol{w}'\boldsymbol{S}_W\boldsymbol{w}\boldsymbol{S}_W^{-1}(\bar{\boldsymbol{X}}_1 - \bar{\boldsymbol{X}}_0) - \boldsymbol{w}'(\bar{\boldsymbol{X}}_1 - \bar{\boldsymbol{X}}_0)\boldsymbol{w} = \boldsymbol{0}.$$

Stąd

$$\boldsymbol{w} = \frac{\boldsymbol{w}'\boldsymbol{S}_W\boldsymbol{w}}{\boldsymbol{w}'(\bar{\boldsymbol{X}}_1 - \bar{\boldsymbol{X}}_0)}\boldsymbol{S}_W^{-1}(\bar{\boldsymbol{X}}_1 - \bar{\boldsymbol{X}}_0) = c\boldsymbol{S}_W^{-1}(\bar{\boldsymbol{X}}_1 - \bar{\boldsymbol{X}}_0).$$ ■

Widzimy, że wektor $\boldsymbol{w}$ maksymalizujący miarę (1.40) nie jest wyznaczony jednoznacznie, ale jego kierunek jest jednoznaczny: ilorazy liczb w_1, w_2, $\ldots$, w_p są jednoznaczne.

Wyrażenie

$$u = \boldsymbol{w}'\boldsymbol{x} = (\bar{\boldsymbol{X}}_1 - \bar{\boldsymbol{X}}_0)'\boldsymbol{S}_W^{-1}\boldsymbol{x} \tag{1.41}$$

nazywamy *liniową funkcją dyskryminacyjną Fishera.*

Wypada zauważyć, że kryterium Fishera nie prowadzi do pełnej reguły klasyfikacyjnej, ale jedynie do redukcji przestrzeni cech do jednego wymiaru. Jeżeli chcemy uzyskać regułę klasyfikacyjną, to musimy wybrać taki punkt graniczny w_0, że jeżeli $\boldsymbol{w}'\boldsymbol{x} \geq w_0$, to obserwację $\boldsymbol{x}$ klasyfikujemy do grupy pierwszej. Najczęściej za w_0 wybiera się środek odcinka łączącego punkty $\boldsymbol{w}'\bar{\boldsymbol{X}}_1 = (\bar{\boldsymbol{X}}_1 - \bar{\boldsymbol{X}}_0)'\boldsymbol{S}_W^{-1}\bar{\boldsymbol{X}}_1$ oraz $\boldsymbol{w}'\bar{\boldsymbol{X}}_0 = (\bar{\boldsymbol{X}}_1 - \bar{\boldsymbol{X}}_0)'\boldsymbol{S}_W^{-1}\bar{\boldsymbol{X}}_0$. Środek ten jest równy

$$m = \frac{1}{2}(\boldsymbol{w}'\bar{\boldsymbol{X}}_1 + \boldsymbol{w}'\bar{\boldsymbol{X}}_0) = \frac{1}{2}(\bar{\boldsymbol{X}}_1 - \bar{\boldsymbol{X}}_0)'\boldsymbol{S}_W^{-1}(\bar{\boldsymbol{X}}_1 + \bar{\boldsymbol{X}}_0).$$

Wówczas reguła klasyfikacyjna Fishera będzie miała postać

$$\hat{d}_F(\boldsymbol{x}) = \begin{cases} 1, \text{ jeżeli } \boldsymbol{w}'\boldsymbol{x} > m, \\ 0, \text{ jeżeli } \boldsymbol{w}'\boldsymbol{x} \leq m. \end{cases} \tag{1.42}$$

Regułę tę uzyskaliśmy bez zakładania, że wektor cech $\boldsymbol{X}$ ma wielowymiarowy rozkład normalny (założyliśmy jedynie, że macierze kowariancji w dwóch grupach są identyczne).

Funkcja $\boldsymbol{w}'\boldsymbol{x} - m$ jest równa estymatorowi liniowej funkcji dyskryminacyjnej (1.23), gdy $\hat{\pi}_1 = \hat{\pi}_0 = \frac{1}{2}$. Posługując się funkcją $\boldsymbol{w}'\boldsymbol{x} - m$ oraz estymatorem funkcji (1.23), uzyskujemy identyczne wyniki klasyfikacji. Zauważmy, że funkcja (1.23) została skonstruowana przy założeniu normalności wektora $\boldsymbol{X}$ oraz równości macierzy kowariancji w grupach.

1.7. Zmienne dyskryminacyjne

Zajmiemy się teraz uogólnieniem liniowej funkcji dyskryminacyjnej Fishera na przypadek wielu grup. Niech $\boldsymbol{X}_{i1}$, $\boldsymbol{X}_{i2}$, ..., $\boldsymbol{X}_{in_i}$ będzie próbą z grupy i, gdzie $i = 1, 2, \ldots, K$ i niech $n_1 + n_2 + \cdots + n_k = n$. Z próby tej obliczamy wektor średnich $\bar{\boldsymbol{X}}_i = \frac{1}{n_i}\sum_{j=1}^{n_i} \boldsymbol{X}_{ij}$ oraz macierz

$$\boldsymbol{S}_i = \frac{1}{n_i - 1}\sum_{j=1}^{n_i}(\boldsymbol{X}_{ij} - \bar{\boldsymbol{X}}_i)(\boldsymbol{X}_{ij} - \bar{\boldsymbol{X}}_i)',\ i = 1,\ 2,\ \ldots,\ K.$$

Z całej próby uczącej liczącej n elementów obliczamy

$$\bar{\boldsymbol{X}} = \frac{1}{n}\sum_{i=1}^{K} n_i \bar{\boldsymbol{X}}_i, \tag{1.43}$$

$$\boldsymbol{B} = \sum_{i=1}^{K} n_i(\bar{\boldsymbol{X}}_i - \bar{\boldsymbol{X}})(\bar{\boldsymbol{X}}_i - \bar{\boldsymbol{X}})', \tag{1.44}$$

$$\boldsymbol{W} = \sum_{i=1}^{K}(n_i - 1)\boldsymbol{S}_i. \tag{1.45}$$

Macierz $\boldsymbol{B}$ nosi nazwę *macierzy zmienności międzygrupowej*, natomiast $\boldsymbol{W}$ jest *macierzą zmienności wewnątrzgrupowej*.

Niech

$$\boldsymbol{S}_B = \frac{1}{K-1}\boldsymbol{B}, \qquad \boldsymbol{S}_W = \frac{1}{n-K}\boldsymbol{W}. \tag{1.46}$$

Poszukujemy zbioru wektorów $\boldsymbol{a}_i$, które maksymalizują miarę zróżnicowania K grup postaci

$$J(\boldsymbol{a}) = \frac{\boldsymbol{a}'\boldsymbol{S}_B\boldsymbol{a}}{\boldsymbol{a}'\boldsymbol{S}_W\boldsymbol{a}}, \tag{1.47}$$

przy dodatkowym założeniu ograniczającym

$$\boldsymbol{a}_i'\boldsymbol{S}_W\boldsymbol{a}_j = \delta_{ij} \tag{1.48}$$

oznaczającym, że zmienne $u_i = \boldsymbol{a}_i'\boldsymbol{x}$ w przekształconej przestrzeni cech są nieskorelowane i mają jednostkowe wariancje.

Niech $\boldsymbol{a}_i$ będzie wektorem własnym macierzy $\boldsymbol{S}_W^{-1}\boldsymbol{S}_B$ odpowiadającym wartości własnej $\lambda_i = \lambda_i(\boldsymbol{S}_W^{-1}\boldsymbol{S}_B)$. Liczba niezerowych wartości własnych jest równa $s = \min(p,\ K-1)$, gdzie $K - 1$ jest rzędem macierzy $\boldsymbol{S}_B$.

Wówczas

$$\boldsymbol{S}_W^{-1}\boldsymbol{S}_B\boldsymbol{a}_i = \lambda_i\boldsymbol{a}_i$$

lub równoważnie

$$(\boldsymbol{S}_B - \lambda_i\boldsymbol{S}_W)\boldsymbol{a}_i = \boldsymbol{0}, \qquad i = 1,\ 2,\ \ldots,\ s. \tag{1.49}$$

Wartości własne spełniające równanie (1.49) są niekiedy nazywane wartościami własnymi macierzy $\boldsymbol{S}_B$ w metryce $\boldsymbol{S}_W$. Zauważmy, że jeżeli lewostronnie pomnożymy (1.49) przez $\boldsymbol{a}_i$ i z równości tej obliczymy λ_i, to

$$\lambda_i(\boldsymbol{S}_W^{-1}\boldsymbol{S}_B) = \frac{\boldsymbol{a}_i'\boldsymbol{S}_B\boldsymbol{a}_i}{\boldsymbol{a}_i'\boldsymbol{S}_W\boldsymbol{a}_i}, \qquad i = 1,\ 2,\ \ldots,\ s.$$

Niech $\lambda_1 \geq \lambda_2 \geq \cdots \geq \lambda_s > 0$. Wektorem, który maksymalizuje miarę (1.47) jest wektor $\boldsymbol{a}_1$ odpowiadający maksymalnej wartości własnej λ_1.

W celu pokazania tego faktu zróżniczkujmy miarę (1.47) względem wektora $\boldsymbol{a}$ (jeżeli $\boldsymbol{A}$ jest macierzą symetryczną, to $\partial(\boldsymbol{x}'\boldsymbol{A}\boldsymbol{x})/\partial\boldsymbol{x} = 2\boldsymbol{A}\boldsymbol{x}$) i uzyskaną pochodną przyrównajmy do wektora zerowego:

$$\frac{\partial\hat{J}(\boldsymbol{a})}{\partial\boldsymbol{a}} = \frac{\boldsymbol{a}'\boldsymbol{S}_W\boldsymbol{a}2\boldsymbol{S}_B\boldsymbol{a} - \boldsymbol{a}'\boldsymbol{S}_B\boldsymbol{a}2\boldsymbol{S}_W\boldsymbol{a}}{(\boldsymbol{a}'\boldsymbol{S}_W\boldsymbol{a})^2} = \boldsymbol{0}.$$

Mnożąc to wyrażenie przez $\boldsymbol{a}'\boldsymbol{S}_W\boldsymbol{a}$, otrzymujemy:

$$\frac{\boldsymbol{a}'\boldsymbol{S}_W\boldsymbol{a}}{\boldsymbol{a}'\boldsymbol{S}_W\boldsymbol{a}}\boldsymbol{S}_B\boldsymbol{a} - \frac{\boldsymbol{a}'\boldsymbol{S}_B\boldsymbol{a}}{\boldsymbol{a}'\boldsymbol{S}_W\boldsymbol{a}}\boldsymbol{S}_W\boldsymbol{a} = \boldsymbol{0}.$$

Z definicji λ_1 oraz wektora $\boldsymbol{a}_1$ wyrażenie to możemy zapisać w postaci

$$\boldsymbol{S}_B\boldsymbol{a}_1 - \lambda_1\boldsymbol{S}_W\boldsymbol{a}_1 = \boldsymbol{0}$$

lub $(\boldsymbol{S}_B - \lambda_1\boldsymbol{S}_W)\boldsymbol{a}_1 = \boldsymbol{0}$. Mnożąc to wyrażenie lewostronnie przez $\boldsymbol{S}_W^{-1}$, otrzymujemy

$$(\boldsymbol{S}_W^{-1}\boldsymbol{S}_B - \lambda_1\boldsymbol{I})\boldsymbol{a}_1 = \boldsymbol{0}$$

co oznacza, że wektor $\boldsymbol{a}_1$ odpowiadający maksymalnej wartości własnej λ_1 macierzy $\boldsymbol{S}_W^{-1}\boldsymbol{S}_B$ maksymalizuje miarę $J(\boldsymbol{a})$ daną wzorem (1.47).

Wektor $\boldsymbol{a}_1$ normujemy tak, by $\boldsymbol{a}_1'\boldsymbol{S}_W\boldsymbol{a}_1 = 1$. Funkcję liniową $u_1 = \boldsymbol{a}_1'\boldsymbol{x}$ będziemy nazywać *pierwszą zmienną dyskryminacyjną*[9]. Ma ona jednostkową wariancję.

[9]Stosowany jest również termin *zmienna kanoniczna*.

Następnie obliczamy wektor $\boldsymbol{a}_2$ odpowiadający wartości własnej λ_2 i normalizujemy go tak, by $\boldsymbol{a}_1\boldsymbol{S}_W\boldsymbol{a}_2 = 0$ oraz $\boldsymbol{a}_2\boldsymbol{S}_W\boldsymbol{a}_2 = 1$. Funkcję liniową $u_2 = \boldsymbol{a}_2'\boldsymbol{x}$ nazywamy *drugą zmienną dyskryminacyjną*. Ma ona jednostkową wariancję i jest nieskorelowana z pierwszą zmienną dyskryminacyjną u_1.

Kontynuując ten proces, uzyskamy s zmiennych dyskryminacyjnych $u_1, u_2, \ldots, u_s$ odpowiadających s niezerowym wartościom własnym $\lambda_1 \geq \lambda_2 \geq \cdots \geq \lambda_s > 0$ macierzy $\boldsymbol{S}_W^{-1}\boldsymbol{S}_B$. Zmienne te są wzajemnie nieskorelowane i mają jednostkowe wariancje.

Uwaga 1.1. Zauważmy, że równanie (1.49) można zapisać w postaci

$$(\boldsymbol{B} - \lambda_i^\star \boldsymbol{W})\boldsymbol{a}_i = \boldsymbol{0}, \tag{1.50}$$

gdzie $\lambda_i^\star = \dfrac{K-1}{n-K}\lambda_i$, $i = 1, 2, \ldots, s$.

Widzimy, że wektory własne nie ulegają zmianie, lecz tylko zmieniają się wartości własne. Zatem zmienne dyskryminacyjne możemy wyznaczyć z pary macierzy $(\boldsymbol{S}_B, \boldsymbol{S}_W)$ lub pary $(\boldsymbol{B}, \boldsymbol{W})$.

Uwaga 1.2. Jeżeli macierz $\boldsymbol{W}$ nie jest źle uwarunkowana ze względu na jej odwrotność, to wektory własne spełniające równanie (1.49) możemy wyznaczyć przez rozwiązanie równoważnego układu równań

$$\boldsymbol{W}^{-1}\boldsymbol{B}\boldsymbol{a}_i = \lambda_i\boldsymbol{a}_i, \qquad i = 1, 2, \ldots, s. \tag{1.51}$$

Jednakże macierz $\boldsymbol{W}^{-1}\boldsymbol{B}$ nie jest macierzą symetryczną. Wiele algorytmów obliczania wartości własnych i wektorów własnych akceptuje tylko macierze symetryczne. Z problemem tym można sobie poradzić, biorąc pod uwagę rozkład Cholesky'ego macierzy $\boldsymbol{W}$ postaci $\boldsymbol{W} = \boldsymbol{U}'\boldsymbol{U}$, gdzie $\boldsymbol{U}$ jest macierzą dolnotrójkątną. Wówczas wartości własne macierzy symetrycznej $(\boldsymbol{U}^{-1})'\boldsymbol{B}\boldsymbol{U}^{-1}$ i macierzy $\boldsymbol{W}^{-1}\boldsymbol{B}$ są identyczne. Jeśli $\boldsymbol{b}$ jest wektorem własnym macierzy $(\boldsymbol{U}^{-1})'\boldsymbol{B}\boldsymbol{U}^{-1}$, to $\boldsymbol{a} = \boldsymbol{U}^{-1}\boldsymbol{b}$ jest wektorem własnym macierzy $\boldsymbol{W}^{-1}\boldsymbol{B}$.

Uwaga 1.3. Najmniejsza spośród liczebności $n_1, n_2, \ldots, n_K$ grup powinna być większa od liczby cech p. Nie jest to wymóg modelowy, ale warunek ten prowadzi do większej stabilności zmiennych dyskryminacyjnych.

Uwaga 1.4. Zmienne dyskryminacyjne są nieskorelowane, jednakże nie są one ortogonalne. W praktyce jednakże zwyczajną procedurą jest rysowanie zmiennych dyskryminacyjnych w prostokątnym układzie współrzędnych. Otrzymane zniekształcenie nie jest na ogół poważne. Chcąc by elementy próby zajmowały centralne położenie na rysunku, należy posłużyć się układem zmiennych $z_j = \boldsymbol{a}_j'(\boldsymbol{x} - \bar{\boldsymbol{x}})$, $j = 1, 2, \ldots, s$, gdzie $\bar{\boldsymbol{x}}$ jest średnią z próby uczącej. Nie prowadzi to do straty ogólności.

Uwaga 1.5. Bezwzględne wartości współczynników danej zmiennej dyskryminacyjnej są wykorzystywane do uporządkowania zmiennych pierwotnych $x_1, x_2, \ldots, x_p$ w porządku ich wkładu w rozdzielczość grup przez daną zmienną dyskryminacyjną.

W celu wyrównania różnic skal między zmiennymi pierwotnymi, współczynniki zmiennych dyskryminacyjnych oraz zmienne pierwotne $x_1, x_2, \ldots, x_p$ powinny być standaryzowane. Jeżeli $\boldsymbol{x}_{ij} = (x_{ij1}, x_{ij2}, \ldots, x_{ijp})'$ jest j-tą obserwacją w i-tej grupie, to zmienną dyskryminacyjną $u_k(\boldsymbol{x}_{ij})$ możemy wyrazić w terminach zmiennych standaryzowanych następująco:

$$u_k(\boldsymbol{x}_{ij}) = a_{k1}^{\star}\frac{x_{ij1} - \bar{x}_{i1}}{s_1} + a_{k2}^{\star}\frac{x_{ij2} - \bar{x}_{i2}}{s_2} + \cdots + a_{kp}^{\star}\frac{x_{ijp} - \bar{x}_{ip}}{s_p},$$
$$k = 1, 2, \ldots, s,$$

gdzie $\bar{\boldsymbol{x}}_i = (\bar{x}_{i1}, \bar{x}_{i2}, \ldots, \bar{x}_{ip})'$ jest wektorem średnich w i-tej grupie, $i = 1, 2, \ldots, K$, $j = 1, 2, \ldots, n_i$ oraz s_r jest pierwiastkiem kwadratowym z r-tego elementu diagonalnego macierzy $\boldsymbol{S}_W = \frac{1}{n-K}\boldsymbol{W}$, $r = 1, 2, \ldots, p$, natomiast $a_{kr}^{\star} = s_r a_{kr}$, $k = 1, 2, \ldots, s$ lub w formie wektorowej $\boldsymbol{a}_k^{\star} = (\mathrm{diag}\boldsymbol{S}_W)^{\frac{1}{2}}\boldsymbol{a}_k$.

Wartości bezwzględne współczynnków standaryzowanych $a_{k1}^{\star}$, $a_{k2}^{\star}$, $\ldots$, $a_{kp}^{\star}$ poprawnie odzwierciedlają wkład zmiennych pierwotnych x_1, $x_2, \ldots, x_p$ w rozdzielczość K grup przez k-tą zmienną dyskryminacyjną, $k = 1, 2, \ldots, s = \min(K-1, p)$. Wektor współczynników zmiennej dyskryminacyjnej u_k jest k-tym wektorem własnym macierzy $\boldsymbol{W}^{-1}\boldsymbol{B}$ i jako taki bierze pod uwagę skorelowanie zmiennych pierwotnych. Stąd wartości bezwzględne tych współczynników pokazują siłę rozdzielczą poszczególnych zmiennych pierwotnych w obecności pozostałych zmiennych (patrz np. Rencher (1998), s. 211).

W wielu pakietach statystycznych oblicza się współczynniki korelacji między daną zmienną dyskryminacyjną a każdą zmienną pierwotną z osobna i współczynniki te zaleca się interpretować jako miarę siły dyskryminacyjnej poszczególnych zmiennych pierwotnych. Jednakże współczynniki te nie pokazują wielowymiarowego wkładu zmiennych w zróżnicowanie grup, ale podają tylko jednowymiarową informację, jak każda ze zmiennych z osobna rozdziela grupy, ignorując obecność pozostałych zmiennych.

Uwaga 1.6. Wygodną miarą względnej siły dyskryminacyjnej i-tej zmiennej dyskryminacyjnej $u_i = \boldsymbol{a}_i'\boldsymbol{x}$ jest wielkość

$$\frac{\lambda_i}{\sum_{j=1}^{s} \lambda_j} \cdot 100\%, \qquad i = 1,\ 2,\ \ldots,\ s.$$

Uwaga 1.7. Zmienna dyskryminacyjna $u_i = \boldsymbol{a}_i'\boldsymbol{x}$ nie jest użyteczna w procesie klasyfikacji, jeżeli wartość własna $\lambda_i(\boldsymbol{W}^{-1}\boldsymbol{B})$ nie różni się istotnie od zera.

Z prawdopodobieństwem jeden rząd$(\boldsymbol{B}) = K - 1$ i stąd z prawdopodobieństwem jeden liczba niezerowych wartości własnych macierzy $\boldsymbol{W}^{-1}\boldsymbol{B}$ jest równa $s = \min(K-1,\ p)$. Odpowiednikiem macierzy $\boldsymbol{W}^{-1}\boldsymbol{B}$ w populacji jest macierz $\boldsymbol{\Sigma}^{-1}\boldsymbol{\Omega}$, gdzie $\boldsymbol{\Omega} = \sum_{i=1}^{K} n_i(\boldsymbol{\mu}_i - \bar{\boldsymbol{\mu}})(\boldsymbol{\mu}_i - \bar{\boldsymbol{\mu}})'$, $\bar{\boldsymbol{\mu}} = \frac{1}{K}\sum_{i=1}^{K} \boldsymbol{\mu}_i$, $\boldsymbol{\mu}_1,\ \boldsymbol{\mu}_2,\ \ldots,\ \boldsymbol{\mu}_K$ są wektorami średnich p-cech w K populacjach (grupach) oraz $\boldsymbol{\Sigma}$ jest macierzą kowariancji identyczną w każdej z K grup. Powiedzmy, że rząd$(\boldsymbol{\Omega}) = r \le K - 1$, przy czym równość zachodzi wtedy, gdy wektory $\boldsymbol{\mu}_1,\ \boldsymbol{\mu}_2,\ \ldots,\ \boldsymbol{\mu}_K$ są liniowo niezależne. Zatem liczba niezerowych wartości własnych macierzy $\boldsymbol{\Sigma}^{-1}\boldsymbol{\Omega}$ wynosi $d = \min(r,\ p)$ i może być mniejsza od $s = \min(K-1,\ p)$.

Stąd nieznaną liczbę d niezerowych wartości własnych macierzy $\boldsymbol{\Sigma}^{-1}\boldsymbol{\Omega}$ musimy ustalić za pomocą odpowiedniego testu istotności. Niech

$$\gamma_1 \ge \gamma_2 \ge \cdots \ge \gamma_d > \gamma_{d+1} = \cdots = \gamma_p = 0$$

będą wartościami własnymi macierzy $\boldsymbol{\Sigma}^{-1}\boldsymbol{\Omega}$, gdzie d nie jest nam znaną liczbą.

Chcemy sekwencyjnie testować układ hipotez

$$H_d:\ \gamma_{d+1} = \gamma_{d+2} = \cdots = \gamma_p = 0, \qquad d = 0,\ 1,\ \ldots,\ p-1.$$

Najpierw testujemy hipotezę H_0, że $d = 0$. Jeżeli hipotezę tę odrzucimy, to testujemy H_1, że $d = 1$. Proces ten kontynuujemy do momentu, w którym po raz pierwszy nie odrzucimy hipotezy H_d. Wówczas stwierdzamy, że $p - d$ pozostałe wartości własne są równe zeru.

Do testowania hipotez H_d $(d = 0, 1, \ldots, p-1)$ możemy użyć trzech następujących statystyk lub ich odpowiednich funkcji, znanych z wielowymiarowej analizy wariancji:

$$\Lambda_d = \prod_{i=d+1}^{s} \frac{1}{1+\lambda_i} \quad \text{(statystyka } \Lambda \text{ Wilksa)},$$

$$T_d^2 = \sum_{i=d+1}^{s} \lambda_i \quad \text{(statystyka śladowa Bartletta–Lawleya–Hotellinga)},$$

$$V_d = \sum_{i=d+1}^{s} \frac{\lambda_i}{1+\lambda_i} \quad \text{(statystyka Bartletta–Nandy–Pillaia)},$$

gdzie $d = 0, 1, \ldots, s-1$ oraz $\lambda_1 > \lambda_2 > \cdots > \lambda_s > 0$ są niezerowymi wartościami własnymi macierzy $\boldsymbol{W}^{-1}\boldsymbol{B}$.

Fujikoshi (1977) wykazał, że funkcje testowe

$$\chi_B^2 = -\left[\nu_W - \frac{1}{2}(p - \nu_B + 1)\right] \ln \Lambda_d,$$

$$\chi_L^2 = -\left[\nu_W - d\frac{1}{2}(p - \nu_B + 1) + \sum_{i=1}^{d} \theta_i^{-1}\right] \ln \Lambda_d,$$

$$\chi_{BLH}^2 = -[\nu_W - (p - \nu_B + 1) + \sum_{i=1}^{d} \theta_i^{-1}] \ln T_d^2,$$

$$\chi_{BNP}^2 = -[\nu_W - 2d + \sum_{i=1}^{d} \theta_i^{-1}] \ln V_d,$$

mają asymptotycznie rozkład $\chi^2_{(\nu_b - d)(p-d)}$, gdzie $\nu_W = n - K$, $\nu_B = K - 1$ oraz θ_i są rozwiązaniami równania $|\boldsymbol{B} - \theta(\boldsymbol{B} + \boldsymbol{W})| = 0$.

Statystyki χ_B^2 i χ_L^2 zostały po raz pierwszy zaproponowane przez Bartletta (1947) i Lawleya (1959). Fujikoshi (1977) zweryfikował czynnik korygujący Lawleya i znalazł aproksymację za pomocą rozkładu chi-kwadrat statystyki śladowej Bartletta–Lawleya–Hotellinga i statystyki Bartletta–Nandy–Pillaia.

Jeżeli którakolwiek z funkcji testowych χ_B^2, χ_L^2, χ_{BLH}^2 lub χ_{BNP}^2 jest większa od wartości krytycznej $\chi^2_{(\nu_b-d)(p-d)}(1-\alpha)$, gdzie $\chi^2_{(\nu_b-d)(p-d)}(1-\alpha)$ jest kwantylem rzędu $1-\alpha$ rozkładu chi-kwadrat z $(\nu_b - d)(p-d)$ stopniami swobody dla poziomu istotności α, to hipotezę H_d odrzucamy.

Jako wymiar przestrzeni dyskryminacyjnej przyjmuje się wartość d, dla której po raz pierwszy nie odrzucimy hipotezy H_d.

Na każdym etapie testowania sekwencyjnego testujemy daną hipotezę na poziomie istotności α. Jednakże nie mamy gwarancji, że poziom ten zostanie zachowany dla całej procedury. Wady tej nie ma *zamknięta procedura testowa* zaproponowana przez Calińskiego i Lejeunea (1998). Procedura ta bazuje na statystyce śladowej Bratletta–Lawleya–Hotellinga:

$$T_d^2 = \sum_{i=d+1}^{s} \lambda_i,$$

gdzie λ_i są wartościami własnymi macierzy $\boldsymbol{W}^{-1}\boldsymbol{B}$ oraz $s = \min(K-1,\ p)$. Jeżeli hipoteza H_0 jest prawdziwa, to statystyka T_0^2 ma rozkład Lawleya–Hotellinga $T^2_{p,\ \nu_B,\ \nu_W}$, gdzie $\nu_B = K-1$, $\nu_W = n-K$.

Dokładny rozkład prawdopodobieństwa statystyki T_d^2 dla $d > 0$ nie jest znany. Caliński i Lejeune (1998) pokazali, że jeżeli hipoteza H_d jest prawdziwa, to rozkład Lawleya–Hotellinga indeksowany trójką $(p-d,$ $\nu_B - d,\ \nu_W)$ jest optymalnym ograniczeniem górnym nieznanego rozkładu statystyki T_d^2, tj.

$$\mathrm{P}(T_d^2 \geq T^2_{p-d,\ \nu_B-d,\ \nu_W}(\alpha)|H_d) \leq \alpha,$$

gdzie $T^2_{p-d,\ \nu_B-d,\ \nu_W}(\alpha)$ jest górną wartością krytyczną rozkładu $T^2_{p-d,\ \nu_B-d,\ \nu_W}$ Lawleya–Hotellinga. Rodzina hipotez H_d jest testowana za pomocą zamkniętej procedury testowej, podanej przez Marcusa i innych (1976), na poziomie istotności α, która polega na testowaniu tych hipotez sekwencyjnie za pomocą statystyki T_d^2, dopóki nie będzie spełniona nierówność $T_d^2 \leq T^2_{p-d,\ \nu_B-d,\ \nu_W}(\alpha)$.

Jeżeli T_d^2 jest pierwszą wartością, która spełnia powyższą nierówność, to stwierdzamy, że liczba niezerowych wartości własnych macierzy $\boldsymbol{\Sigma}^{-1}\boldsymbol{\Omega}$ wynosi d.

Prawdopodobieństwo niepopełnienia błędu I rodzaju, tj. nie przeestymowania prawdziwej liczby niezerowych wartości własnych przez zbyt późne zatrzymanie procedury sekwencyjnej jest równe co najmniej $1-\alpha$.

Górne wartości krytyczne statystyki $\dfrac{\nu_W}{\nu_B}T^2$, gdzie $\nu_W \geq p$ i $\nu_B \geq p$ są zamieszczone, między innymi, w książce Krzyśki (2000), Tablica 6. Jeżeli $\nu_B < p$, to korzystamy z faktu, że

$$T^2_{p,\ \nu_B,\ \nu_W} = T^2_{\nu_B,\ p,\ \nu_W+\nu_B-p}.$$

Znana jest także bardzo dokładna aproksymacja rozkładu $T^2_{p,\ \nu_B,\ \nu_W}$ za pomocą rozkładu F-Snedecora pochodząca od McKeona (1974):

$$F = \frac{1}{c}T^2_{p,\ \nu_B,\ \nu_W} \dot{\sim} F_{a,\,b},$$

$$\text{gdzie} \quad a = p\nu_B, \quad b = 4 + \frac{a+2}{B-1}, \quad c = \frac{a(b-2)}{b(\nu_W - p - 1)},$$

$$B = \frac{(\nu_W + \nu_B - p - 1)(\nu_W - 1)}{(\nu_W - p - 3)(\nu_W - p)}.$$

Uwaga 1.8. Przekształcenie pierwotnej przestrzeni cech $x_1, x_2, \ldots, x_p$ w nową przestrzeń zmiennych dyskryminacyjnych $u_1, u_2, \ldots, u_s$ nie prowadzi jeszcze do reguły klasyfikacyjnej.

W tej nowej przestrzeni zmiennych nieskorelowanych o jednostkowych wariancjach można zastosować dowolną regułę klasyfikacyjną. Reguły te przyjmują prostą postać. Jeśli np. użyjemy liniowej funkcji klasyfikującej postaci (1.31), to jej ocena w przestrzeni zmiennych dyskryminacyjnych $u_i(\boldsymbol{x}) = \boldsymbol{a}_i'\boldsymbol{x}$, $i = 1, 2, \ldots, s$, ma szczególnie prostą postać (patrz np. Srivastava (2002), s. 261):

$$\hat{\delta}_k(\boldsymbol{x}) = \sum_{i=1}^{s} [\boldsymbol{a}_i'(\boldsymbol{x} - \bar{\boldsymbol{x}}_k)]^2, \qquad k = 1, 2, \ldots, K.$$

Przykład 1.5.

Dane zostały zaczerpnięte z pracy Lubischewa (1962) i dotyczą pomiarów sześciu cech na okazach męskich trzech gatunków chrząszczy skaczących: 21 okazów gatunku Chaetocnema concinna, 31 okazów gatunku Chaetocnema heikertingeri oraz 22 okazy gatunku Chaetocnema heptapotamica.

Macierze zmienności międzygrupowej $\boldsymbol{B}$ oraz wewnątrzgrupowej $\boldsymbol{W}$ są postaci:

$$\boldsymbol{B} = \begin{bmatrix} 51704.0 & -3690.5 & -2045.2 & -9059.7 & 3581.9 & -19049.0 \\ -3690.5 & 1367.3 & 349.0 & 2899.9 & -127.7 & 3472.5 \\ -2045.2 & 349.0 & 118.3 & 772.8 & -118.2 & 1142.1 \\ -9059.7 & 2899.9 & 772.8 & 6186.7 & -366.5 & 7650.3 \\ 3581.9 & -127.7 & -118.2 & -366.5 & 263.0 & -1074.7 \\ -19049.0 & 3472.5 & 1142.1 & 7650.3 & -1074.7 & 11062.0 \end{bmatrix},$$

$$\boldsymbol{W} = \begin{bmatrix} 11448.0 & 4170.3 & 1479.6 & 1615.3 & 11.4 & 1489.3 \\ 4170.3 & 3883.5 & 798.0 & 698.5 & -34.3 & 851.7 \\ 1479.6 & 798.0 & 434.6 & 460.1 & -16.5 & 341.0 \\ 1615.3 & 698.5 & 460.1 & 1634.7 & -39.7 & 830.0 \\ 11.4 & -34.3 & -16.5 & -39.7 & 72.0 & 4.1 \\ 1489.3 & 851.7 & 341.0 & 830.0 & 4.1 & 3875.9 \end{bmatrix}.$$

Niezerowe wartości własne macierzy $\boldsymbol{W}^{-1}\boldsymbol{B}$ są równe: $\lambda_1 = 17.779$, $\lambda_2 = 3.885$. Odpowiadające tym wartościom wektory własne są równe:

$$\boldsymbol{a}_1 = (0.093, -0.059, -0.137, -0.065, 0.288, -0.047)',$$

$$\boldsymbol{a}_2 = (0.014, 0.042, -0.290, 0.202, 0.523, 0.011)'.$$

W budowie pierwszej zmiennej dyskryminacyjnej $u_1 = \boldsymbol{a}_1'\boldsymbol{x}$ największy udział ma zmienna x_5, ponieważ moduł stojącego przy niej współczynnika jest maksymalny. Podobnie w budowie drugiej zmiennej dyskryminacyjnej największy udział ma zmienna x_5. Na rysunku 1.3 zamieszczono 74 punkty reprezentujące osobniki należące do trzech gatunków chrząszczy skaczących.

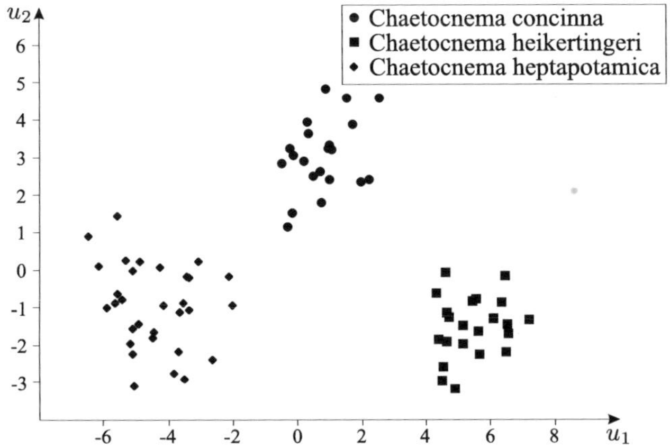

Rys. 1.3. Punkty reprezentujące okazy chrząszczy skaczących należących do trzech grup w układzie dwóch pierwszych zmiennych dyskryminacyjnych

1.8. Klasyfikatory liniowe Andersona–Bahadura

W podrozdziale 1.4 została podana konstrukcja klasyfikatorów bayesowskich w modelu normalnym.

Założenie $\boldsymbol{X}|Y = i \sim N_p(\boldsymbol{\mu}_i,\ \boldsymbol{\Sigma}_i)$, $(i = 1,\ 0)$ doprowadziło do wniosku, że optymalna bayesowska funkcja klasyfikująca, w przypadku gdy $\boldsymbol{\Sigma}_1 \neq \boldsymbol{\Sigma}_0$, jest funkcją kwadratową argumentu $\boldsymbol{x}$, czyli powierzchnia rozdzielająca jest hiperpowierzchnią stopnia drugiego.

W dalszym ciągu, utrzymując założenie normalności rozkładów populacji, będziemy poszukiwać takiego klasyfikatora, dla którego powierzchnia rozdzielająca będzie hiperpłaszczyzną.

Ogólna postać takiego liniowego klasyfikatora jest następująca:

$$d(\boldsymbol{x}) = \begin{cases} 1, \text{ jeżeli } \boldsymbol{w}'\boldsymbol{x} > m, \\ 0, \text{ jeżeli } \boldsymbol{w}'\boldsymbol{x} \leq m. \end{cases} \tag{1.52}$$

Niech $U = \boldsymbol{w}'\boldsymbol{X}$, $\boldsymbol{w} \neq \boldsymbol{0}$. Wtedy

$$U \sim N(\boldsymbol{w}'\boldsymbol{\mu}_i,\ \boldsymbol{w}'\boldsymbol{\Sigma}_i\boldsymbol{w}), \qquad i = 1,\ 0.$$

Prawdopodobieństwa błędnej klasyfikacji mają wtedy postać:

$$\begin{aligned}
\mathrm{P}(d(\boldsymbol{X})=0|Y=1) &= \mathrm{P}(\boldsymbol{w}'\boldsymbol{X}\le m|Y=1)= \\
&= \Phi\left(\frac{m-\boldsymbol{w}'\boldsymbol{\mu}_1}{\sqrt{\boldsymbol{w}'\boldsymbol{\Sigma}_1\boldsymbol{w}}}\right)=1-\Phi\left(\frac{\boldsymbol{w}'\boldsymbol{\mu}_1-m}{\sqrt{\boldsymbol{w}'\boldsymbol{\Sigma}_1\boldsymbol{w}}}\right)= \\
&= 1-\Phi(z_1)
\end{aligned}$$

oraz

$$\begin{aligned}
\mathrm{P}(d(\boldsymbol{X})=1|Y=0) &= \mathrm{P}(\boldsymbol{w}'\boldsymbol{X}> m|Y=0)= \\
&= 1-\Phi\left(\frac{m-\boldsymbol{w}'\boldsymbol{\mu}_0}{\sqrt{\boldsymbol{w}'\boldsymbol{\Sigma}_0\boldsymbol{w}}}\right)= \\
&= 1-\Phi(z_0).
\end{aligned}$$

Zainteresowani jesteśmy minimalizacją tych prawdopodobieństw lub równoważnie maksymalizacją

$$z_1=\frac{\boldsymbol{w}'\boldsymbol{\mu}_1-m}{\sqrt{\boldsymbol{w}'\boldsymbol{\Sigma}_1\boldsymbol{w}}}\quad\text{oraz}\quad z_0=\frac{m-\boldsymbol{w}'\boldsymbol{\mu}_0}{\sqrt{\boldsymbol{w}'\boldsymbol{\Sigma}_0\boldsymbol{w}}}. \tag{1.53}$$

Ustalmy z_0 i poszukajmy maksymalnej wartości z_1. Eliminując stałą m z (1.53), otrzymujemy:

$$z_1=(\boldsymbol{w}'\boldsymbol{\gamma}-z_0\sqrt{\boldsymbol{w}'\boldsymbol{\Sigma}_0\boldsymbol{w}})/\sqrt{\boldsymbol{w}'\boldsymbol{\Sigma}_1\boldsymbol{w}},$$

gdzie $\boldsymbol{\gamma}=\boldsymbol{\mu}_1-\boldsymbol{\mu}_0$.

Po zróżniczkowaniu względem $\boldsymbol{w}$ mamy:

$$\begin{aligned}
\frac{\partial z_1}{\partial \boldsymbol{w}} &= (\boldsymbol{\gamma}-z_0/\sqrt{\boldsymbol{w}'\boldsymbol{\Sigma}_0\boldsymbol{w}}\cdot\boldsymbol{\Sigma}_0\boldsymbol{w})/\sqrt{\boldsymbol{w}'\boldsymbol{\Sigma}_1\boldsymbol{w}}- \\
&- (\boldsymbol{w}'\boldsymbol{\gamma}-z_0\sqrt{\boldsymbol{w}'\boldsymbol{\Sigma}_0\boldsymbol{w}})(\boldsymbol{w}'\boldsymbol{\Sigma}_1\boldsymbol{w})^{-3/2}\boldsymbol{\Sigma}_1\boldsymbol{w}.
\end{aligned}$$

Przyrównując pochodne do zera oraz podstawiając

$$\begin{aligned}
t_1 &= \frac{\boldsymbol{w}'\boldsymbol{\gamma}-z_0\sqrt{\boldsymbol{w}'\boldsymbol{\Sigma}_0\boldsymbol{w}}}{\boldsymbol{w}'\boldsymbol{\Sigma}_1\boldsymbol{w}}, \\
t_0 &= \frac{z_0}{\sqrt{\boldsymbol{w}'\boldsymbol{\Sigma}_0\boldsymbol{w}}},
\end{aligned}$$

otrzymujemy warunek

$$(t_1\boldsymbol{\Sigma}_1+t_0\boldsymbol{\Sigma}_0)\boldsymbol{w}=\boldsymbol{\gamma}. \tag{1.54}$$

Korzystając z (1.53), mamy ponadto

$$m=\boldsymbol{w}'\boldsymbol{\mu}_1-t_1\boldsymbol{w}'\boldsymbol{\Sigma}_1\boldsymbol{w}=\boldsymbol{w}'\boldsymbol{\mu}_0+t_0\boldsymbol{w}'\boldsymbol{\Sigma}_0\boldsymbol{w}. \tag{1.55}$$

Stąd procedura doboru wektora $\boldsymbol{w}$ i stałej m jest następująca:

1. Obieramy t_1 i t_0 i rozwiązujemy układ równań (1.54).
2. Stałą m wyznaczamy z jednego z warunków (1.55).

Dalsze rozważania dotyczące doboru t_1 i t_0 poprzedźmy następującą definicją.

Dla klasyfikatorów d_1 i d_2 powiemy, że d_1 *jest lepszy niż* d_2, gdy oba prawdopodobieństwa błędnych klasyfikacji dla d_1 są nie większe, niż odpowiadające im prawdopodobieństwa błędnych klasyfikacji dla d_2 oraz przynajmniej jedno z nich jest mniejsze.

Klasyfikator d, dla którego nie istnieje klasyfikator lepszy, będziemy nazywać *dopuszczalnym.*

Anderson i Bahadur (1962) podali następujące twierdzenie charakteryzujące liniowe klasyfikatory dopuszczalne.

TWIERDZENIE 1.6.

Dla dowolnych t_1 i t_0 takich, że macierz $t_1\boldsymbol{\Sigma}_1+t_0\boldsymbol{\Sigma}_0$ jest dodatnio określona, klasyfikator liniowy, dla którego

$$\begin{aligned} \boldsymbol{w} &= (t_1\boldsymbol{\Sigma}_1 + t_0\boldsymbol{\Sigma}_0)^{-1}\boldsymbol{\gamma}, \\ m &= \boldsymbol{w}'\boldsymbol{\mu}_1 - t_1\boldsymbol{w}'\boldsymbol{\Sigma}_1\boldsymbol{w} = \boldsymbol{w}'\boldsymbol{\mu}_0 + t_0\boldsymbol{w}'\boldsymbol{\Sigma}_0\boldsymbol{w} \end{aligned}$$

jest dopuszczalny.

Jak widać z tw. 1.6 istnieje wiele „dobrych” możliwości doboru stałych t_1 i t_0. W szczególności, gdy $t_1 = t_0 = \frac{1}{2}$, $\boldsymbol{\Sigma}_1 = \boldsymbol{\Sigma}_0 = \boldsymbol{\Sigma}$, po zastąpieniu nieznanych parametrów ich estymatorami otrzymujemy:

$$\boldsymbol{w} = \boldsymbol{S}_{\boldsymbol{w}}^{-1}(\bar{\boldsymbol{X}}_1 - \bar{\boldsymbol{X}}_0)$$

oraz

$$m = \frac{1}{2}(\bar{\boldsymbol{X}}_1 - \bar{\boldsymbol{X}}_0)'\boldsymbol{S}_{\boldsymbol{w}}^{-1}(\bar{\boldsymbol{X}}_1 + \bar{\boldsymbol{X}}_0).$$

Jest to reguła klasyfikacyjna Fishera opisana w tw. 1.5.

Poniżej są podane trzy podstawowe sposoby doboru stałych t_1 i t_0.

Zwróćmy najpierw uwagę, że bez straty ogólności możemy przyjąć:

$$t_1 = t, \qquad t_0 = 1 - t, \qquad 0 \le t \le 1.$$

SPOSÓB 1:
Niech

$$\mathrm{P}(d(\boldsymbol{X}) = 1 | Y = 0) = \alpha, \qquad 0 < \alpha \le \frac{1}{2} \tag{1.56}$$

lub, co jest równoważne: $z_0 = \Phi^{-1}(1 - \alpha)$.

Przy tym warunku t wyznaczamy tak, aby zminimalizować $\mathrm{P}(d(\boldsymbol{X}) = 0|Y = 1)$ lub równoważnie zmaksymalizować z_1.

Z warunku (1.56) wynika, że

$$z_0 = \Phi^{-1}(1 - \alpha) \ge 0.$$

Zatem poszukiwana wartość t musi spełniać warunek

$$\Phi^{-1}(1 - \alpha) = (1 - t)\sqrt{\boldsymbol{w}'\boldsymbol{\Sigma}_0\boldsymbol{w}},$$

przy czym $\boldsymbol{w} = [t\boldsymbol{\Sigma}_1 + (1 - t)\boldsymbol{\Sigma}_0]^{-1}\boldsymbol{\gamma}$.

Wyznaczenie stałej t nie jest skomplikowane, gdyż funkcja

$$f(t) = (1 - t)\sqrt{\boldsymbol{w}'\boldsymbol{\Sigma}_0\boldsymbol{w}}$$

jest malejąca oraz dla $t = 0$, $f(t) = \sqrt{\boldsymbol{w}'\boldsymbol{\Sigma}_0\boldsymbol{w}} = \sqrt{\boldsymbol{\gamma}'\boldsymbol{\Sigma}_0^{-1}\boldsymbol{\gamma}}$, gdzie $\boldsymbol{\Sigma}_0\boldsymbol{w} = \boldsymbol{\gamma}$, a dla $t = 1$, $f(t) = 0$.

SPOSÓB 2: (Rozwiązanie minimaksowe)

Niech

$$\mathrm{P}(d(\boldsymbol{X}) = 0|Y = 1) = \mathrm{P}(d(\boldsymbol{X}) = 1|Y = 0)$$

lub równoważnie $z_1 = z_0$. Wtedy

$$\begin{aligned} 0 = z_1^2 - z_0^2 &= t^2\boldsymbol{w}'\boldsymbol{\Sigma}_1\boldsymbol{w} - (1 - t)^2\boldsymbol{w}'\boldsymbol{\Sigma}_0\boldsymbol{w} = \\ &= \boldsymbol{w}'[t^2\boldsymbol{\Sigma}_1 - (1 - t)^2\boldsymbol{\Sigma}_0]\boldsymbol{w}. \end{aligned} \tag{1.57}$$

Ponieważ $t^2\boldsymbol{w}'\boldsymbol{\Sigma}_1\boldsymbol{w}$ rośnie wraz ze wzrostem t, a $(1-t)^2\boldsymbol{w}'\boldsymbol{\Sigma}_0\boldsymbol{w}$ maleje, zatem istnieje dokładnie jedno rozwiązanie równania (1.57).

Do wyznaczenia stałej t możemy również wykorzystać równania (1.53). Po wyeliminowaniu stałej m otrzymujemy:

$$z_1 = z_0 = \frac{\boldsymbol{w}'\boldsymbol{\gamma}}{\sqrt{\boldsymbol{w}'\boldsymbol{\Sigma}_1\boldsymbol{w}} + \sqrt{\boldsymbol{w}'\boldsymbol{\Sigma}_0\boldsymbol{w}}}. \tag{1.58}$$

Optymalne t wyznaczamy, maksymalizując (1.58) przy warunku $\boldsymbol{w} = [t\boldsymbol{\Sigma}_1 + (1-t)\boldsymbol{\Sigma}_0]^{-1}\boldsymbol{\gamma}$.

SPOSÓB 3: (Rozwiązanie bayesowskie)

Niech π będzie prawdopodobieństwem a priori tego, że obserwacja $\boldsymbol{x}$ pochodzi z populacji 1, $1-\pi$ prawdopodobieństwem a priori tego, że pochodzi ona z populacji 0. Wtedy prawdopodobieństwo błędnej klasyfikacji wynosi:

$$\begin{aligned}\pi \mathrm{P}(d(\boldsymbol{X}) = 0|Y = 1) + (1-\pi)\mathrm{P}(d(\boldsymbol{X}) = 1|Y = 0) &= \\ = \pi(1 - \Phi(z_1)) + (1-\pi)(1 - \Phi(z_0)) &= \\ = 1 - [\pi\Phi(z_1) + (1-\pi)\Phi(z_0)]&.\end{aligned}$$

Aby zminimalizować powyższe prawdopodobieństwo, analogicznie jak w poprzednim rozwiązaniu, podstawiamy

$$z_1 = t\sqrt{\boldsymbol{w}'\boldsymbol{\Sigma}_1\boldsymbol{w}}, \qquad z_0 = (1-t)\sqrt{\boldsymbol{w}'\boldsymbol{\Sigma}_0\boldsymbol{w}},$$

gdzie $\boldsymbol{w} = [t\boldsymbol{\Sigma}_1 + (1-t)\boldsymbol{\Sigma}_0]^{-1}\boldsymbol{\gamma}$.

Po zróżniczkowaniu i przyrównaniu pochodnej do zera otrzymujemy warunek:

$$\pi\phi(z_1)\frac{dz_1}{dt} + (1-\pi)\phi(z_0)\frac{dz_0}{dt} = 0, \tag{1.59}$$

gdzie $\phi(z) = \frac{1}{\sqrt{2\pi}}\exp\left(-\frac{z^2}{2}\right)$.

Rozwiązanie równania (1.59) nie jest łatwe, gdyż lewa strona tego równania nie jest funkcją monotoniczną. Co więcej równanie (1.59) może mieć wiele różnych rozwiązań (dających minima lokalne).

1.9. Klasyfikatory liniowe maksymalizujące odległości probabilistyczne

W przypadku klasyfikatora liniowego Fishera hiperpłaszczyzna rozdzielająca populacje była wyznaczona przez maksymalizację miary

$$J(\boldsymbol{w}) = \frac{(\boldsymbol{w}'\boldsymbol{\mu}_1 - \boldsymbol{w}'\boldsymbol{\mu}_0)^2}{\boldsymbol{w}'\boldsymbol{\Sigma}\boldsymbol{w}} \tag{1.60}$$

oddzielającej te populacje, gdzie $\boldsymbol{\mu}_1$ i $\boldsymbol{\mu}_0$ były wartościami oczekiwanymi obserwacji $\boldsymbol{X}$ w populacjach 1 i 0 odpowiednio, natomiast $\boldsymbol{\Sigma}$ wspólną dla obu populacji macierzą kowariancji.

W dalszym ciągu zamiast miary (1.60) będziemy używać tzw. odległości probabilistycznych między populacjami.

Niech zatem f_1 i f_0 oznaczają funkcje gęstości populacji 1 i 0 odpowiednio.

Do najczęściej wykorzystywanych odległości probabilistycznych należą:

- *odległość Chernoffa* postaci:

$$\rho_1(f_1,\ f_0) = -\ln \int_{\mathbb{R}^p} \big(f_1(\boldsymbol{x})\big)^{1-s}\big(f_0(\boldsymbol{x})\big)^{s} d\boldsymbol{x}, \qquad s \in [0,\ 1].$$

  Szczególnie znany i wykorzystywany jest przypadek, gdy $s = \frac{1}{2}$. Odległość ρ_1 jest wtedy zwana *odległością Bhattacharyyi*,

- *odległość Morisity* postaci:

$$\rho_2(f_1,\ f_0) = -\ln \frac{2\Delta(f_1,\ f_0)}{\Delta(f_1) + \Delta(f_0)},$$

  gdzie

$$\Delta(f_1,\ f_0) = \int_{\mathbb{R}^p} f_1(\boldsymbol{x}) f_0(\boldsymbol{x}) d\boldsymbol{x}$$

  oraz

$$\Delta(f_i) = \Delta(f_i,\ f_i), \qquad i = 1,\ 0.$$

  Zwróćmy uwagę, że odległość ta jest ściśle związana z klasyczną odległością

$$\Big(\int_{\mathbb{R}^p} \big(f_1(\boldsymbol{x}) - f_0(\boldsymbol{x})\big)^2 d\boldsymbol{x}\Big)^{\frac{1}{2}},$$

- *odległość Kullbacka–Leiblera* postaci:

$$\rho_3(f_1,\ f_0) = \int_{\mathbb{R}^p} [f_1(\boldsymbol{x}) - f_0(\boldsymbol{x})] \ln \frac{f_1(\boldsymbol{x})}{f_0(\boldsymbol{x})} d\boldsymbol{x}.$$

Niech $\boldsymbol{\mu}_i$ oraz $\boldsymbol{\Sigma}_i$ oznaczają wektor wartości oczekiwanych oraz macierz kowariancji wektora $\boldsymbol{X}$ dla $Y = i$, $i = 1,\ 0$. Zredukujmy zagadnienie do jednego wymiaru przez zrzutowanie danych na prostą.

Niech zatem $U = \boldsymbol{w}'\boldsymbol{X}$, $\boldsymbol{w} \neq \boldsymbol{0}$. Wtedy

$$\begin{aligned} \mathrm{E}(U|Y = i) &= \boldsymbol{w}'\boldsymbol{\mu}_i, \\ \mathrm{Var}(U|Y = i) &= \boldsymbol{w}'\boldsymbol{\Sigma}_i\boldsymbol{w}, \qquad i = 1,\ 0 \end{aligned}$$

oraz prawdziwe jest następujące twierdzenie.

TWIERDZENIE 1.7.

Jeżeli $U|Y = i \sim N(\boldsymbol{w}'\boldsymbol{\mu}_i, \boldsymbol{w}'\boldsymbol{\Sigma}_i\boldsymbol{w})$, $i = 1,\ 0$, *to*

$$\begin{aligned} \rho_1(f_1,\ f_0) = {} & \frac{1}{2}s(1-s)\frac{(\boldsymbol{w}'\boldsymbol{\gamma})^2}{(1-s)\boldsymbol{w}'\boldsymbol{\Sigma}_1\boldsymbol{w} + s\boldsymbol{w}'\boldsymbol{\Sigma}_0\boldsymbol{w}} + \\ & + \frac{1}{2}\ln[(1-s)\boldsymbol{w}'\boldsymbol{\Sigma}_1\boldsymbol{w} + s\boldsymbol{w}'\boldsymbol{\Sigma}_0\boldsymbol{w}] - \\ & - \frac{1}{2}(1-s)\ln(\boldsymbol{w}'\boldsymbol{\Sigma}_1\boldsymbol{w}) - \frac{1}{2}s\ln(\boldsymbol{w}'\boldsymbol{\Sigma}_0\boldsymbol{w}), \\ \rho_2(f_1,\ f_0) = {} & \frac{1}{2}\frac{(\boldsymbol{w}'\boldsymbol{\gamma})^2}{\boldsymbol{w}'(\boldsymbol{\Sigma}_1 + \boldsymbol{\Sigma}_0)\boldsymbol{w}} + \frac{1}{2}\ln[\boldsymbol{w}'(\boldsymbol{\Sigma}_1 + \boldsymbol{\Sigma}_0)\boldsymbol{w}] + \\ & + \ln[(\boldsymbol{w}'\boldsymbol{\Sigma}_1\boldsymbol{w})^{\frac{1}{2}} + (\boldsymbol{w}'\boldsymbol{\Sigma}_0\boldsymbol{w})^{\frac{1}{2}}] - \\ & - \frac{1}{2}\ln[(\boldsymbol{w}'\boldsymbol{\Sigma}_1\boldsymbol{w})(\boldsymbol{w}'\boldsymbol{\Sigma}_0\boldsymbol{w})] - \ln(2\sqrt{2}), \\ \rho_3(f_1,\ f_0) = {} & \frac{1}{2}\Big[(\boldsymbol{w}'\boldsymbol{\gamma})^2\Big(\frac{1}{\boldsymbol{w}'\boldsymbol{\Sigma}_1\boldsymbol{w}} + \frac{1}{\boldsymbol{w}'\boldsymbol{\Sigma}_0\boldsymbol{w}}\Big) + \\ & + \boldsymbol{w}'(\boldsymbol{\Sigma}_1 - \boldsymbol{\Sigma}_0)\boldsymbol{w}\Big(\frac{1}{\boldsymbol{w}'\boldsymbol{\Sigma}_1\boldsymbol{w}} - \frac{1}{\boldsymbol{w}'\boldsymbol{\Sigma}_0\boldsymbol{w}}\Big)\Big]. \end{aligned}$$

Zwróćmy uwagę, że wszystkie trzy odległości z tw. 1.7 są niezmiennicze ze względu na pomnożenie wektora $\boldsymbol{w}$ przez dowolny skalar.

Korzystając ze związku

$$\frac{\partial \rho_i(\boldsymbol{w})}{\partial(\boldsymbol{w})} = 0, \qquad i = 1,\ 2,\ 3,$$

otrzymujemy optymalną postać wektora $\boldsymbol{w}$.

TWIERDZENIE 1.8.

Wektory

$$\boldsymbol{w}_i = (\boldsymbol{\Sigma}_1 + \theta_i \boldsymbol{\Sigma}_0)^{-1}\boldsymbol{\gamma}, \tag{1.61}$$

gdzie stała

$$\theta_i = \frac{a_2(\boldsymbol{w}_i)}{a_1(\boldsymbol{w}_i)} \tag{1.62}$$

jest taka, że macierz $\boldsymbol{\Sigma}_1 + \theta_i \boldsymbol{\Sigma}_0$ *jest nieosobliwa i*

$$a_1(\boldsymbol{w}_1) = (1-s)\Big(A_1 + \frac{1}{\boldsymbol{w}_1'\boldsymbol{\Sigma}_1\boldsymbol{w}_1}\Big),$$

$$a_2(\boldsymbol{w}_1) = s\Big(A_1 + \frac{1}{\boldsymbol{w}_1'\boldsymbol{\Sigma}_0\boldsymbol{w}_1}\Big),$$

$$A_1 = \frac{s(1-s)(\boldsymbol{w}_1'\boldsymbol{\gamma})^2}{[(1-s)\boldsymbol{w}_1'\boldsymbol{\Sigma}_1\boldsymbol{w}_1 + s\boldsymbol{w}_1'\boldsymbol{\Sigma}_0\boldsymbol{w}_1]^2} - \\ - \frac{1}{(1-s)\boldsymbol{w}_1'\boldsymbol{\Sigma}_1\boldsymbol{w}_1 + s\boldsymbol{w}_1'\boldsymbol{\Sigma}_0\boldsymbol{w}_1}, \qquad s \in [0,\ 1],$$

$$a_1(\boldsymbol{w}_2) = A_2 - B_2(\boldsymbol{w}_2'\Sigma_1\boldsymbol{w}_2)^{-\frac{1}{2}} + (\boldsymbol{w}_2'\Sigma_1\boldsymbol{w}_2)^{-1},$$

$$a_2(\boldsymbol{w}_2) = A_2 - B_2(\boldsymbol{w}_2'\boldsymbol{\Sigma}_0\boldsymbol{w}_2)^{-\frac{1}{2}} + (\boldsymbol{w}_2'\boldsymbol{\Sigma}_0\boldsymbol{w}_2)^{-1},$$

$$A_2 = \frac{(\boldsymbol{w}_2'\boldsymbol{\gamma})^2}{[\boldsymbol{w}_2'(\boldsymbol{\Sigma}_1 + \boldsymbol{\Sigma}_0)\boldsymbol{w}_2]^2} - [\boldsymbol{w}_2'(\boldsymbol{\Sigma}_1 + \boldsymbol{\Sigma}_0)\boldsymbol{w}_2]^{-1},$$

$$B_2 = \Big((\boldsymbol{w}_2'\boldsymbol{\Sigma}_1\boldsymbol{w}_2)^{\frac{1}{2}} + (\boldsymbol{w}_2'\boldsymbol{\Sigma}_0\boldsymbol{w}_2)^{\frac{1}{2}}\Big)^{-1},$$

$$a_1(\boldsymbol{w}_3) = A_3 - (\boldsymbol{w}_3'\boldsymbol{\Sigma}_0\boldsymbol{w}_3)^{-1}, \qquad a_2(\boldsymbol{w}_3) = B_3 - (\boldsymbol{w}_3'\boldsymbol{\Sigma}_1\boldsymbol{w}_3)^{-1},$$

$$A_3 = \frac{(\boldsymbol{w}_3'\boldsymbol{\gamma})^2 + \boldsymbol{w}_3'\boldsymbol{\Sigma}_0\boldsymbol{w}_3}{(\boldsymbol{w}_3'\boldsymbol{\Sigma}_1\boldsymbol{w}_3)^2}, \qquad B_3 = \frac{(\boldsymbol{w}_3\boldsymbol{\gamma})^2 + \boldsymbol{w}_3'\boldsymbol{\Sigma}_1\boldsymbol{w}_3}{(\boldsymbol{w}_3'\boldsymbol{\Sigma}_0\boldsymbol{w}_3)^2},$$

maksymalizują odległości ρ_i, $i = 1,\ 2,\ 3$.

Pozostaje jeszcze problem wyznaczania wektorów $\boldsymbol{w}_i$ opisanych równaniem (1.61). Zagadnienie nie jest, niestety, proste, gdyż występujące w tym równaniu θ_i również zależy od wektora $\boldsymbol{w}_i$.

Rozwiązanie równania (1.61) sprowadzimy do zagadnienia wyznaczania punktu stałego pewnego odwzorowania ψ.

Ponieważ $\boldsymbol{\Sigma}_1$ i $\boldsymbol{\Sigma}_0$ są macierzami dodatnio określonymi, więc istnieje taka nieosobliwa macierz $\boldsymbol{P}$, że

$$\boldsymbol{\Sigma}_1 = \boldsymbol{P}'\boldsymbol{P}, \quad \boldsymbol{\Sigma}_0 = \boldsymbol{P}'\boldsymbol{\Lambda}\boldsymbol{P},$$

gdzie $\boldsymbol{\Lambda} = \operatorname{diag}(\lambda_1, \ldots, \lambda_p)$ oraz λ_i $(i = 1, \ldots, p)$ są wartościami własnymi macierzy $\boldsymbol{\Sigma}_0\boldsymbol{\Sigma}_1^{-1}$.

Wtedy

$$\boldsymbol{w}_i'\boldsymbol{\Sigma}_1\boldsymbol{w}_i = \boldsymbol{\beta}_i'\boldsymbol{\beta}_i,$$

gdzie

$$\boldsymbol{P}\boldsymbol{w}_i = \boldsymbol{\beta}_i, \qquad \boldsymbol{w}_i'\boldsymbol{\Sigma}_0\boldsymbol{w}_i = \boldsymbol{\beta}_i'\boldsymbol{\Lambda}\boldsymbol{\beta}_i, \qquad \boldsymbol{w}_i'\boldsymbol{\gamma} = \boldsymbol{\beta}_i'\boldsymbol{\eta}$$

oraz

$$\boldsymbol{\gamma} = \boldsymbol{P}'\boldsymbol{\eta}.$$

Zatem

$$\begin{aligned}\boldsymbol{\beta}_i &= \boldsymbol{P}\boldsymbol{w}_i = \boldsymbol{P}(\boldsymbol{\Sigma}_1 + \theta_i\boldsymbol{\Sigma}_0)^{-1}\boldsymbol{\gamma} = \\ &= \boldsymbol{P}(\boldsymbol{P}'\boldsymbol{P} + \theta_i\boldsymbol{P}'\boldsymbol{\Lambda}\boldsymbol{P})^{-1}\boldsymbol{\gamma} = \\ &= (\mathbf{I} + \theta_i\boldsymbol{\Lambda})^{-1}\boldsymbol{\eta}.\end{aligned}$$

Stąd równanie (1.62) przyjmuje postać:

$$\theta_i = \psi_i(\theta_i), \tag{1.63}$$

gdzie

$$\begin{aligned}\psi_1(\theta_1) &= \frac{s}{1-s}\frac{\boldsymbol{\beta}_1'\boldsymbol{\beta}_1}{\boldsymbol{\beta}_1'\boldsymbol{\Lambda}\boldsymbol{\beta}_1} + \Big(\frac{s}{1-s} - \theta_1\Big)A_1(\boldsymbol{\beta}_1'\boldsymbol{\beta}_1),\\ A_1 &= \frac{s(1-s)(\boldsymbol{\beta}_1'\boldsymbol{\eta})^2}{[(1-s)\boldsymbol{\beta}_1'\boldsymbol{\beta}_1 + s\boldsymbol{\beta}_1'\boldsymbol{\Lambda}\boldsymbol{\beta}_1]^2} - \frac{1}{(1-s)\boldsymbol{\beta}_1'\boldsymbol{\beta}_1 + s\boldsymbol{\beta}_1'\boldsymbol{\Lambda}\boldsymbol{\beta}_1},\\ \psi_2(\theta_2) &= \Big[\frac{B_2}{A_2}(\boldsymbol{\beta}_2'\boldsymbol{\beta}_2)^{-\frac{1}{2}} - (A_2\boldsymbol{\beta}_2'\boldsymbol{\beta}_2)^{-1}\Big]\theta_2 + 1 - \frac{B_2}{A_2}(\boldsymbol{\beta}_2'\boldsymbol{\Lambda}\boldsymbol{\beta}_2)^{-\frac{1}{2}} + \\ &\quad + [A_2(\boldsymbol{\beta}_2'\boldsymbol{\Lambda}\boldsymbol{\beta}_2)]^{-1},\end{aligned}$$

$$A_2 = \frac{(\boldsymbol{\beta}_2'\boldsymbol{\eta})^2}{[\boldsymbol{\beta}_2'(\mathbf{I}+\boldsymbol{\Lambda})\boldsymbol{\beta}_2]^2} - (\boldsymbol{\beta}_2'(\mathbf{I}+\boldsymbol{\Lambda})\boldsymbol{\beta}_2)^{-1},$$

$$B_2 = \left((\boldsymbol{\beta}_2'\boldsymbol{\beta}_2)^{\frac{1}{2}} + (\boldsymbol{\beta}_2'\boldsymbol{\Lambda}\boldsymbol{\beta}_2)^{\frac{1}{2}}\right)^{-1},$$

$$\psi_3(\theta_3) = A_3(\boldsymbol{\beta}_3'\boldsymbol{\Lambda}\boldsymbol{\beta}_3)\theta_3 - B_3(\boldsymbol{\beta}_3'\boldsymbol{\Lambda}\boldsymbol{\beta}_3) + (\boldsymbol{\beta}_3'\boldsymbol{\Lambda}\boldsymbol{\beta}_3)(\boldsymbol{\beta}_3'\boldsymbol{\beta}_3)^{-1},$$

$$A_3 = \frac{(\boldsymbol{\beta}_3'\boldsymbol{\eta})^2 + \boldsymbol{\beta}_3'\boldsymbol{\Lambda}\boldsymbol{\beta}_3}{(\boldsymbol{\beta}_3'\boldsymbol{\beta}_3)^2}, \qquad B_3 = \frac{(\boldsymbol{\beta}_3'\boldsymbol{\eta})^2 + \boldsymbol{\beta}_3'\boldsymbol{\beta}_3}{(\boldsymbol{\beta}_3'\boldsymbol{\Lambda}\boldsymbol{\beta}_3)^2}.$$

Do rozwiązania równania (1.63) możemy zastosować następującą procedurę iteracyjną. Znajdujemy przedział $[a, b]$ taki, że dla każdego punktu tego przedziału jest spełniony warunek $|\psi'(\theta)| < 1$. Obieramy dowolny punkt θ_0 z przedziału $[a, b]$. Można pokazać, że granicą ciągu (θ_n), gdzie $\theta_n = \psi(\theta_{n-1})$ jest pierwiastek równania (1.63) (Vilenkin, 1979).

Pozostał jeszcze problem doboru stałej m (wartości progowej). Z twierdzenia 1.8 wynika, że stała θ_i $(i = 1,\ 2,\ 3)$ jest wyznaczona tak, aby macierz $\boldsymbol{\Sigma}_1 + \theta_i\boldsymbol{\Sigma}_0$, była dodatnio określona. Zatem, jeśli przyjmiemy

$$m = \boldsymbol{w}'\boldsymbol{\mu}_1 - \boldsymbol{w}'\boldsymbol{\Sigma}_1\boldsymbol{w} = \boldsymbol{w}'\boldsymbol{\mu}_0 + \theta_i\boldsymbol{w}'\boldsymbol{\Sigma}_0\boldsymbol{w},$$

to z tw. 1.6 skonstruowane w ten sposób klasyfikatory liniowe będą dopuszczalne.

W szczególnym przypadku, gdy $\boldsymbol{\Sigma}_1 = \boldsymbol{\Sigma}_0 = \boldsymbol{\Sigma}$, dla odległości Bhattacharyya, Morisity oraz Kulbacka–Leiblera mamy

$$\theta_i = 1, \qquad i = 1,\ 2,\ 3.$$

Zatem

$$\boldsymbol{w}_i = \frac{1}{2}\boldsymbol{\Sigma}^{-1}\boldsymbol{\gamma},$$

$$m_i = \frac{1}{4}(\boldsymbol{\mu}_1 - \boldsymbol{\mu}_0)'\boldsymbol{\Sigma}^{-1}(\boldsymbol{\mu}_1 + \boldsymbol{\mu}_0), \qquad i = 1,\ 2,\ 3.$$

Po zastąpieniu nieznanych parametrów przez ich estymatory otrzymujemy liniową funkcję klasyfikacyjną Fishera.

Rozważane w podrozdziałach 1.8 i 1.9 liniowe reguły klasyfikacyjne zostały wyznaczone dla przypadku jedynie dwóch populacji. Do ich uogólnienia na przypadek wielu populacji zastosować można techniki omówione w rozdz. 8.

1.10. Estymacja aktualnego poziomu błędu

Niech $\hat{d}(\boldsymbol{x}) = \hat{d}(\boldsymbol{x};\ \mathcal{L}_n)$ oznacza klasyfikator skonstruowany za pomocą próby uczącej $\mathcal{L}_n$. W podrozdziale 1.4 przyjęliśmy za miarę jakości tego klasyfikatora aktualny poziom błędu określony wzorem

$$e(\hat{d}) = \mathrm{P}(\hat{d}(\boldsymbol{X}) \neq Y | \mathcal{L}_n).$$

Niech $\hat{e} \equiv \hat{e}(\hat{d})$ oznacza ocenę aktualnego poziomu błędu klasyfikatora $\hat{d}$. Ocenę tę będziemy nazywać *błędem klasyfikacji*. W sytuacjach, kiedy na populacje nie narzuca się żadnej konkretnej rodziny rozkładów, jedyną drogą oceny prawdopodobieństwa $e(\hat{d})$ jest użycie metod estymacji nieparametrycznej.

W najlepszej sytuacji jesteśmy wtedy, gdy dysponujemy m-elementową *próbą testową* $\mathcal{T}_m$ niezależną od próby uczącej $\mathcal{L}_n$.

Niech zatem $\mathcal{T}_m = \{(\boldsymbol{X}_1^t,\ Y_1^t), \ldots, (\boldsymbol{X}_m^t,\ Y_m^t)\}$. Wtedy za estymator aktualnego poziomu błędu klasyfikatora $\hat{d}$ przyjmujemy:

$$\hat{e}_{\mathcal{T}} = \frac{1}{m} \sum_{j=1}^{m} I\left(\hat{d}(\boldsymbol{X}_j^t; \mathcal{L}_n) \neq Y_j^t\right).$$

W przypadku gdy nie dysponujemy niezależną próbą testową, do estymacji używamy jedynie próby uczącej.

Metoda ponownego podstawienia

Naturalną oceną aktualnego poziomu błędu jest wtedy wartość *estymatora ponownego podstawienia* (resubstytucji)[10]

$$\hat{e}_R = \frac{1}{n} \sum_{j=1}^{n} I\left(\hat{d}(\boldsymbol{X}_j; \mathcal{L}_n) \neq Y_j\right). \tag{1.64}$$

Wartość estymatora (1.64) uzyskuje się przez klasyfikację regułą $\hat{d}$ tych samych obserwacji, które służyły do jej konstrukcji. Oznacza to, iż próba ucząca jest zarazem próbą testową. Estymator ten jest więc obciążonym estymatorem wielkości $e(\hat{d})$ i zaniża jej rzeczywistą wartość. Uwidacznia się to szczególnie w przypadku złożonych klasyfikatorów opartych na relatywnie małych próbach uczących.

Redukcję obciążenia można uzyskać, stosując poniższe metody estymacji.

[10]Po ang. resubstitution estimator lub apparent error rate.

Metoda sprawdzania krzyżowego

Jednym ze sposobów redukcji obciążenia estymatora $\hat{e}_R$ przy zastosowaniu próby uczącej jako jednocześnie próby testowej jest tzw. metoda *podziału próby*[11] na dwa podzbiory: próbę uczącą i próbę testową. Wówczas klasyfikator konstruuje się za pomocą pierwszego z nich, drugi natomiast służy do konstrukcji estymatora danego wzorem (1.64).
Wykorzystanie tylko części informacji w celu uzyskania reguły klasyfikacyjnej prowadzi jednak często do zawyżenia wartości estymatora błędu. Rozwiązaniem tego problemu jest metoda *sprawdzania krzyżowego*[12].

Oznaczmy przez $\mathcal{L}_n^{(-j)}$ próbę uczącą $\mathcal{L}_n$ z której usunięto obserwację $\boldsymbol{Z}_j = (\boldsymbol{X}_j, Y_j)$. Klasyfikator konstruuje się, wykorzystując próbę $\mathcal{L}_n^{(-j)}$, a następnie testuje się go na pojedynczej obserwacji $\boldsymbol{Z}_j$[13]. Czynność tę powtarza się n razy, dla każdej obserwacji $\boldsymbol{Z}_j$ z osobna.
Odpowiedni estymator ma postać:

$$\hat{e}_{CV} = \frac{1}{n}\sum_{j=1}^{n} I\left(\hat{d}(\boldsymbol{X}_j; \mathcal{L}_n^{(-j)}) \neq Y_j\right). \tag{1.65}$$

Procedura ta w każdym z n etapów jest w rzeczywistości metodą podziału próby dla przypadku jednoelementowego zbioru testowego. Każda obserwacja próby jest użyta do konstrukcji klasyfikatora $\hat{d}$. Każda z nich jest też (dokładnie jeden raz) elementem testującym.

Estymator (1.65), choć granicznie nieobciążony, ma większą wariancję (McLachlan, 1992). Ponadto wymaga on konstrukcji n klasyfikatorów, co dla dużych n oznacza znaczący wzrost obliczeń. Rozwiązaniem pośrednim jest *metoda rotacyjna*, zwana też *v-krokową metodą sprawdzania krzyżowego*[14]. Polega ona na losowym podziale próby na v podzbiorów, przy czym $v-1$ z nich tworzy próbę uczącą, natomiast pozostały — próbę testową. Procedurę tę powtarza się v razy, dla każdego podzbioru rozpatrywanego kolejno jako zbiór testowy.
Odpowiedni estymator ma postać:

$$\hat{e}_{vCV} = \frac{1}{n}\sum_{i=1}^{v}\sum_{j=1}^{n} I(\boldsymbol{Z}_j \in \tilde{\mathcal{L}}_n^{(i)}) I\left(\hat{d}(\boldsymbol{X}_j; \tilde{\mathcal{L}}_n^{(-i)}) \neq Y_j\right), \tag{1.66}$$

[11] Po ang. holdout method.
[12] Po ang. cross-validation (CV).
[13] Stąd inna nazwa – ang. leave-one-out (LOO).
[14] Po ang. v-fold cross-validation method.

gdzie $\tilde{\mathcal{L}}_n^{(1)}, \tilde{\mathcal{L}}_n^{(2)}, \ldots, \tilde{\mathcal{L}}_n^{(v)}$ jest losowym v-podziałem próby $\mathcal{L}_n$ na równoliczne podzbiory, a $\tilde{\mathcal{L}}_n^{(-i)} = \mathcal{L}_n \backslash \tilde{\mathcal{L}}_n^{(i)}$, $i = 1, 2, \ldots, v$.
Metoda ta daje mniejsze obciążenie błędu niż metoda podziału próby i wymaga mniejszej liczby obliczeń w porównaniu ze sprawdzaniem krzyżowym (jeśli tylko $v < n$). W zagadnieniu estymacji aktualnego poziomu błędu zalecane jest obranie wartości $v = 10$ (Webb, 2003).

Odnotujmy jeszcze, że metoda sprawdzania krzyżowego jest powszechnie wykorzystywana w zagadnieniu wyboru modelu (m.in. w optymalizacji parametrów klasyfikatorów jądrowych czy w procesie przycinania drzewa decyzyjego). Z rodziny klasyfikatorów opisanej parametrycznie wybieramy wtedy klasyfikator, dla którego błąd klasyfikacji ma wartość najmniejszą.

Metoda *jackknife*

Niech S_n będzie statystyką z n-elementowej próby $\mathcal{L}_n$. Załóżmy, że dla dużych n wartość oczekiwana statystyki S_n ma postać:

$$E(S_n) = \theta + \frac{a_1(\theta)}{n} + \frac{a_2(\theta)}{n^2} + O(n^{-3}), \tag{1.67}$$

gdzie a_1, a_2 nie zależą od n.
Niech $S_n^{(-j)}$ będzie statystyką z próby $\mathcal{L}_n^{(-j)}$ oraz niech

$$S_n^{(\cdot)} = \frac{1}{n} \sum_{j=1}^{n} S_n^{(-j)}.$$

Wówczas

$$\begin{aligned} E(S_n^{(\cdot)}) &= \frac{1}{n} \sum_{j=1}^{n} \left(\theta + \frac{a_1(\theta)}{n-1} + O(n^{-2}) \right) = \\ &= \theta + \frac{a_1(\theta)}{n-1} + O(n^{-2}). \end{aligned} \tag{1.68}$$

Ze wzorów (1.67) i (1.68) wynika, że kombinacja liniowa

$$S_J = nS_n - (n-1)S_n^{(\cdot)} \tag{1.69}$$

ma obciążenie rzędu n^{-2} (Webb, 2003). S_J nazywamy *estymatorem typu jackknife* statystyki S_n.

Biorąc jako S_n estymator $\hat{e}_R$, mamy:

$$\hat{e}_J = n\hat{e}_R - (n-1)\hat{e}_R^{(\cdot)} = \hat{e}_R + (n-1)(\hat{e}_R - \hat{e}_R^{(\cdot)}), \tag{1.70}$$

gdzie

$$\hat{e}_R^{(\cdot)} = \frac{1}{n} \sum_{j=1}^{n} \hat{e}_R^{(-j)}, \tag{1.71}$$

natomiast $\hat{e}_R^{(-j)}$ jest estymatorem postaci (1.64) z próby $\mathcal{L}_n^{(-j)}$, tj.

$$\hat{e}_R^{(-j)} = \frac{1}{n-1} \sum_{k=1,\, k \neq j}^{n} I\left(\hat{d}(\boldsymbol{X}_k; \mathcal{L}_n^{(-j)}) \neq Y_k\right). \tag{1.72}$$

Niestety, estymator (1.70) aktualnego poziomu błędu ma nadal obciążenie rzędu n^{-1} (McLachlan, 1992). Redukcję obciążenia do rzędu drugiego można uzyskać przez zastosowanie estymatora postaci:

$$\hat{e}_{J'} = \hat{e}_R + (n-1)(\tilde{e}_R - \hat{e}_R^{(\cdot)}), \tag{1.73}$$

gdzie

$$\tilde{e}_R = \frac{1}{n^2} \sum_{j=1}^{n} \sum_{k=1}^{n} I\left(\hat{d}(\boldsymbol{X}_k; \mathcal{L}_n^{(-j)}) \neq Y_k\right). \tag{1.74}$$

Metoda prób bootstrapowych

Definicja 1.5.
Próbą bootstrapową nazywamy próbę n-elementową pobraną z n-elementowej próby uczącej w procesie n-krotnego losowania pojedynczych obserwacji ze zwracaniem.

Niech $\mathcal{L}_n^{*1}$, $\mathcal{L}_n^{*2}$, ..., $\mathcal{L}_n^{*B}$ będzie ciągiem kolejno pobranych B prób bootstrapowych.
Znane są następujące oceny aktualnego poziomu błędu (Wehberg, Schumacher, 2004):

$$\hat{e}_{B_1} = \hat{e}_R + \text{obciążenie}, \tag{1.75}$$

gdzie

$$\text{obciążenie} = \frac{1}{Bn} \left[\sum_{b=1}^{B} \sum_{j=1}^{n} I\left(\hat{d}(\boldsymbol{X}_j; \mathcal{L}_n^{*b}) \neq Y_j\right) - \right.$$
$$\left. - \sum_{b=1}^{B} \sum_{j=1, \boldsymbol{Z}_j \in \mathcal{L}_n^{*b}}^{n} I\left(\hat{d}(\boldsymbol{X}_j; \mathcal{L}_n^{*b}) \neq Y_j\right) \right] \tag{1.76}$$

oraz

$$\hat{e}_{B2} = \frac{1}{B}\sum_{b=1}^{B}\frac{\sum\limits_{j=1}^{n} I\left(\boldsymbol{Z}_j \notin \mathcal{L}_n^{*b}\right) I\left(\hat{d}(\boldsymbol{X}_j; \mathcal{L}_n^{*b}) \neq Y_j\right)}{\sum\limits_{j=1}^{n} I\left(\boldsymbol{Z}_j \notin \mathcal{L}_n^{*b}\right)}. \tag{1.77}$$

Widać, że powyższa ocena aktualnego poziomu błędu jest uzyskana metodą sprawdzania krzyżowego zastosowaną do prób bootstrapowych. Ponadto estymator ten jest zgodnym estymatorem aktualnego poziomu błędu $e(\hat{d})$, jeżeli tylko $B \to \infty$.

W celu dalszej redukcji obciążenia estymatorów (1.75) i (1.77) Efron (1983) zaproponował estymator postaci:

$$\hat{e}_{.632} = 0.368\,\hat{e}_R + 0.632\,\hat{e}_{B_2}. \tag{1.78}$$

Waga 0.368 jest przybliżoną wartością wielkości $e^{-1} = \lim_{n\to\infty}(1 - 1/n)^n$ i jest graniczną wartością prawdopodobieństwa nie wylosowania obserwacji z próby uczącej do próby bootstrapowej.

Kolejne udoskonalenie oceny aktualnego poziomu błędu podali Efron i Tibschirani (1997):

$$\hat{e}_{.632+} = (1-\omega)\hat{e}_R + \omega\hat{e}_{B_2}, \tag{1.79}$$

gdzie

$$\omega = \frac{0.632}{1 - 0.368R}, \tag{1.80}$$

$$R = \begin{cases} 1\,, & \text{jeżeli} \quad \gamma \leq \hat{e}_{B_2}, \\ \dfrac{\hat{e}_{B_2} - \hat{e}_R}{\gamma - \hat{e}_R}\,, & \text{jeżeli} \quad \gamma, \hat{e}_{B_2} > \hat{e}_R, \\ 0\,, & \text{poza tym}, \end{cases} \tag{1.81}$$

$$\gamma = \frac{1}{n^2}\sum_{j=1}^{n}\sum_{k=1}^{n} I\left(\hat{d}(\boldsymbol{X}_j; \mathcal{L}_n) \neq Y_k\right). \tag{1.82}$$

Wielkości R oraz γ nazywamy *względnym współczynnikiem obciążenia*[15] i *nieinformującym współczynnikiem błędu*[16] odpowiednio.
Zauważmy, że wraz ze wzrostem wartości współczynnika R, ocena aktualnego poziomu błędu $\hat{e}_{.632+}$ przebiega wartości od $\hat{e}_{.632}$ (dla $R = 0$) do $\hat{e}_{B_2}$ (dla $R = 1$).

[15] Po ang. related overfitting rate.
[16] Po ang. no-informative error rate.

Przykład 1.6.
Rozważmy zadanie dyskryminacji dwóch klas z przykł. 1.3. Z podrozdziału 1.4. wiemy, że kwadratowa funkcja klasyfikująca dana równaniem (1.20) jest w tym przypadku klasyfikatorem bayesowskim. Jego rzeczywisty poziom błędu – wyznaczony numerycznie w przykł. 1.3 – wynosi $e^* = 0,018$.
Wygenerowano kolejno $n_0 = n_1 = n/2 = 50,\ 100,\ 250,\ 500$ obserwacji dla każdej z 2 klas[17]. Następnie skonstruowano kwadratowe funkcje klasyfikujące $\hat{d}(\boldsymbol{x};\ \mathcal{L}_n)$, $n = 100,\ 200,\ 500,\ 1000$, dla parametrów $\hat{\boldsymbol{\mu}}_i$, $\hat{\boldsymbol{\Sigma}}_i$ $(i = 0,\ 1)$ wyestymowanych na podstawie wygenerowanych kolejno prób.
Oceny aktualnego poziomu błędu klasyfikatorów $\hat{d}_n = \hat{d}(\boldsymbol{x}; \mathcal{L}_n)$ zamieszczono w tab. 1.1[18].

Tab. 1.1. Wartości błędów klasyfikacji dla różnych liczebności próby uczącej

	$\hat{d}_{100}$	$\hat{d}_{200}$	$\hat{d}_{500}$	$\hat{d}_{1000}$
$\hat{e}_R$	0.0065	0.0110	0.0148	0.0164
$\hat{e}_{CV}$	0.0115	0.0150	0.0242	0.0200
$\hat{e}_{10CV}$	0.0138	0.0157	0.0239	0.0197
$\hat{e}_{J'}$	0.0143	0.0151	0.0247	0.0198
$\hat{e}_{B_1}$	0.0125	0.0224	0.0362	0.0332
$\hat{e}_{B_2}$	0.0194	0.0325	0.0323	0.0288
$\hat{e}_{.632}$	0.0147	0.0246	0.0280	0.0250
$\hat{e}_{.632+}$	0.0149	0.0248	0.0282	0.0251
$\hat{e}_{\mathcal{T}}$	0.0262	0.0215	0.0197	0.0187

Ostatni wiersz tab. 1.1 zawiera estymator $\hat{e}_{\mathcal{T}}$ aktualnego poziomu błędu klasyfikatora uzyskany z niezależnie wygenerowanej próby testowej o liczbie elementów $m = 100\,000$. Nie będzie więc dużym nadużyciem (ze względu na liczebność próby testowej), jeżeli ten błąd klasyfikacji przyjmiemy za aktualny poziom błędu klasyfikatorów $\hat{d}_n$.

Wobec powyższej uwagi zauważmy, że dla $n = 100$ wszystkie estymatory zaniżają (niedoestymowują) aktualny poziom błędu. To znaczne obciążenie wszystkich ocen jest efektem zbyt małej próby (pamiętajmy, że obserwacje pochodzą z przestrzeni 8-wymiarowej). Dla $n = 200$ niedoszacowanie aktualnego poziomu błędu obserwujemy dla estymatora ponownego podstawienia, sprawdzania

[17] Dla każdej rozpatrywanej liczebności próby n, próbę generowano 10-krotnie, a wszystkie otrzymane wyniki uśredniono.

[18] W metodzie krokowej sprawdzania krzyżowego obrano liczbę kroków $v = 10$. Rozpatrzono $B = 50$ prób bootstrapowych.

krzyżowego oraz dla estymatora typu jackknife, natomiast w przypadku większych prób – dla wszystkich ocen z wyjątkiem estymatora resubstytucji.

Ponadto, dla każdej z rozpatrywanych wartości n, wartości ocen uzyskane metodą sprawdzania krzyżowego są mniejsze od uzyskanych metodami prób bootstrapowych.

Na koniec zauważmy, że wraz ze wzrostem liczby elementów próby uczącej n ocena $\hat{e}_{\mathcal{T}}$ zbiega do błędu bayesowskiego e^*, co jest skutkiem uzyskiwania coraz to lepszych estymatorów parametrów $\boldsymbol{\mu}_i$ oraz $\boldsymbol{\Sigma}_i$, $i = 1,\ 2$.

Obok opisanych powyżej metod estymacji aktualnego poziomu błędu znane jest również podejście oparte na nierównościach probabilistycznych. Wymaga ono jednak znajomości pojęcia wymiaru Vapnika–Chervonenkisa, stąd zostanie omówione w podrozdz. 4.4.

2

Metody regresyjne

Prostym podejściem do klasyfikacji jest estymacja funkcji regresji $r(\boldsymbol{x}) = \mathrm{E}(Y|\boldsymbol{X} = \boldsymbol{x})$ z pominięciem estymacji gęstości f_k. Zacznijmy od przypadku dwóch grup, tj. gdy $\mathcal{Y} = \{1, 0\}$. Wówczas $r(\boldsymbol{x}) = \mathrm{P}(Y = 1|\boldsymbol{X} = \boldsymbol{x})$ i po uzyskaniu estymatora $\hat{r}$ możemy posłużyć się klasyfikatorem postaci

$$\hat{d}(\boldsymbol{x}) = \begin{cases} 1, \text{ jeżeli } \hat{r}(\boldsymbol{x}) > \dfrac{1}{2}, \\ 0, \text{ poza tym.} \end{cases} \tag{2.1}$$

2.1. Regresja liniowa

Najprostszym modelem regresji jest *model regresji liniowej*

$$Y = r(\boldsymbol{x}) + \varepsilon = \beta_0 + \sum_{j=1}^{p} \beta_j x_j + \varepsilon, \tag{2.2}$$

w którym $\mathrm{E}(\varepsilon) = 0$.

Model ten nie może być poprawny, ponieważ nie wymusza on $Y = 1$ lub $Y = 0$. Niemniej w wielu przypadkach prowadzi on do przyzwoitego klasyfikatora.

Przypomnijmy sobie, że estymator najmniejszych kwadratów wektora $\boldsymbol{\beta} = (\beta_0, \beta_1, \ldots, \beta_p)'$ minimalizuje resztową sumę kwadratów

$$RSK(\boldsymbol{\beta}) = \sum_{i=1}^{n} (Y_i - \beta_0 - \sum_{j=1}^{p} x_{ij}\beta_j)^2.$$

Niech $\boldsymbol{Y} = (Y_1, Y_2, \ldots, Y_n)'$ i niech $\boldsymbol{X}$ oznacza macierz $n \times (p+1)$ postaci

$$\boldsymbol{X} = \begin{bmatrix} 1 \; x_{11} \; \ldots \; x_{1p} \\ 1 \; x_{21} \; \ldots \; x_{2p} \\ \ldots\ldots\ldots\ldots \\ 1 \; x_{n1} \; \ldots \; x_{np} \end{bmatrix}. \tag{2.3}$$

Wówczas

$$RSK(\boldsymbol{\beta}) = (\boldsymbol{Y} - \boldsymbol{X}\boldsymbol{\beta})'(\boldsymbol{Y} - \boldsymbol{X}\boldsymbol{\beta})$$

i model regresji liniowej możemy zapisać w postaci

$$\boldsymbol{Y} = \boldsymbol{X}\boldsymbol{\beta} + \boldsymbol{\varepsilon}, \tag{2.4}$$

gdzie $\boldsymbol{\varepsilon} = (\varepsilon_1, \ldots, \varepsilon_n)'$.

Estymator najmniejszych kwadratów minimalizujący $RSK(\boldsymbol{\beta})$ jest równy

$$\hat{\boldsymbol{\beta}} = (\boldsymbol{X}'\boldsymbol{X})^{-1}\boldsymbol{X}'\boldsymbol{Y}. \tag{2.5}$$

Predyktor wektora $\boldsymbol{Y}$ jest równy $\hat{\boldsymbol{Y}} = \boldsymbol{X}\hat{\boldsymbol{\beta}}$. Estymator $\hat{r}(\boldsymbol{x})$ funkcji regresji ma postać

$$\hat{r}(\boldsymbol{x}) = \hat{\beta}_0 + \sum_{j=1}^{p} \hat{\beta}_j x_j. \tag{2.6}$$

Klasyfikator $\hat{d}(\boldsymbol{x})$ dany wzorem (2.1) jest więc całkowicie określony.

Rozważania te uogólnimy teraz na przypadek, gdy liczba grup $K \geq 2$ i $\mathcal{Y} = \{1, 2, \ldots, K\}$. W tym przypadku etykietę Y o wartościach 1 lub 0 musimy zastąpić etykietą wektorową $\boldsymbol{Y} = (Y_1, Y_2, \ldots, Y_K)'$ o wartości $(0, \ldots, 0, 1, 0, \ldots, 0)'$, gdzie jedynka występuje na k-tym miejscu, jeśli obserwacja $\boldsymbol{x}_{ij} = (x_{ij1}, x_{ij2}, \ldots, x_{ijp})'$ pochodzi z k-tej grupy.

Model regresji liniowej (2.4) należy wówczas zastąpić wielowymiarowym modelem liniowym regresji postaci

$$\boldsymbol{Y} = \boldsymbol{X}\boldsymbol{B} + \boldsymbol{E}, \tag{2.7}$$

gdzie $\boldsymbol{Y}$ jest macierzą rozmiaru $n \times K$ postaci

$$\boldsymbol{Y} = \begin{bmatrix} Y_{11} \; Y_{12} \; \ldots \; Y_{1K} \\ Y_{12} \; Y_{22} \; \ldots \; Y_{2K} \\ \ldots\ldots\ldots\ldots\ldots \\ Y_{n1} \; Y_{n2} \; \ldots \; Y_{nK} \end{bmatrix}, \tag{2.8}$$

macierz $\boldsymbol{X}$ rozmiaru $n \times (p+1)$ ma postać (2.3), $\boldsymbol{B}$ jest macierzą rozmiaru $(p+1) \times K$ nieznanych współczynników regresji, natomiast $\boldsymbol{E}$ jest macierzą błędów rozmiaru $n \times K$ taką, że jej wartością oczekiwaną jest macierz zerowa.

Estymator $\hat{\boldsymbol{B}}$ macierzy $\boldsymbol{B}$ uzyskany metodą najmniejszych kwadratów jest równy

$$\hat{\boldsymbol{B}} = (\boldsymbol{X}'\boldsymbol{X})^{-1}\boldsymbol{X}'\boldsymbol{Y}, \tag{2.9}$$

a prognoza macierzy $\boldsymbol{Y}$ ma postać

$$\hat{\boldsymbol{Y}} = \boldsymbol{X}\hat{\boldsymbol{B}} = \boldsymbol{X}(\boldsymbol{X}'\boldsymbol{X})^{-1}\boldsymbol{X}'\boldsymbol{Y}. \tag{2.10}$$

Zauważmy, że i-ty wiersz macierzy $\boldsymbol{X}$ ma postać $[1,\ \boldsymbol{x}_i']$, gdzie $\boldsymbol{x}_i' = (x_{i1},\ x_{i2},\ \ldots,\ x_{ip})$, $i = 1,\ 2,\ \ldots,\ n$. Stąd nowej obserwacji $\boldsymbol{x} = (x_1,\ x_2,\ \ldots,\ x_p)'$ odpowiada prognoza

$$\hat{\boldsymbol{Y}}(\boldsymbol{x}) = (\hat{\boldsymbol{Y}}_1(\boldsymbol{x}),\ \hat{\boldsymbol{Y}}_2(\boldsymbol{x}),\ \ldots,\ \hat{\boldsymbol{Y}}_K(\boldsymbol{x}))' = (1,\ \boldsymbol{x}')\hat{\boldsymbol{B}}. \tag{2.11}$$

Jeżeli $Y \in \{1,\ 2,\ \ldots,\ K\}$, to $\mathrm{E}(Y_k|\boldsymbol{X} = \boldsymbol{x}) = \mathrm{P}(Y = k|\boldsymbol{X} = \boldsymbol{x})$. Zatem k-ta składowa wektora (2.11) jest prognozą (oceną) prawdopodobieństwa a posteriori $\mathrm{P}(Y = k|\boldsymbol{X} = \boldsymbol{x})$ przynależności obserwacji $\boldsymbol{x}$ do k-tej grupy, $k = 1,\ 2,\ \ldots,\ K$.

Chociaż $\hat{\boldsymbol{Y}}_k(\boldsymbol{x})$ jest prognozą (oceną) prawdopodobieństwa $\mathrm{P}(Y = k|\boldsymbol{X} = \boldsymbol{x})$, to wartości $\hat{\boldsymbol{Y}}_k(\boldsymbol{x})$ mogą wykraczać poza przedział $[0,\ 1]$. Można natomiast pokazać, że dla każdego wektora $\boldsymbol{x}$, $\sum_{k=1}^{K} \hat{\boldsymbol{Y}}_k(\boldsymbol{x}) = 1$.

Uzyskaliśmy zatem klasyfikator postaci

$$\hat{d}(\boldsymbol{x}) = \arg\max_k \hat{\boldsymbol{Y}}_k(\boldsymbol{x}), \tag{2.12}$$

gdzie $\hat{\boldsymbol{Y}}_k(\boldsymbol{x})$ jest k-tą składową wektora $\hat{\boldsymbol{Y}}(\boldsymbol{x})$ danego wzorem (2.11).

2.2. Regresja logistyczna

Alternatywą dla modelu regresji liniowej jest *model regresji logistycznej*.

Niech $(\boldsymbol{X}_i,\ Y_i)$, gdzie $Y_i \in \{1,\ 0\}$, $i = 1,\ 2,\ \ldots,\ n$ będzie próbą uczącą. Jeżeli $\boldsymbol{X}_i$ jest wektorem p-wymiarowym, to model regresji logistycznej ma postać

$$p_i = p_i(\boldsymbol{x}_i^\star; \boldsymbol{\beta}) = \mathrm{P}(Y_i = 1 | \boldsymbol{X}_i = \boldsymbol{x}_i) =$$

$$= \frac{\exp\left(\beta_0 + \sum_{j=1}^{p} \beta_j x_{ij}\right)}{1 + \exp\left(\beta_0 + \sum_{j=1}^{p} \beta_j x_{ij}\right)} = \frac{\exp\left(\boldsymbol{\beta}' \boldsymbol{x}_i^\star\right)}{1 + \exp\left(\boldsymbol{\beta}' \boldsymbol{x}_i^\star\right)}, \tag{2.13}$$

gdzie $\boldsymbol{\beta} = (\beta_0, \beta_1, \ldots, \beta_p)'$, $\boldsymbol{x}_i^\star = (1, \boldsymbol{x}_i')'$, $i = 1, 2, \ldots, n$.

Model (2.13) możemy równoważnie zapisać w postaci

$$\mathrm{logit}(p_i) = \beta_0 + \sum_{j=1}^{p} \beta_j x_{ij} = \boldsymbol{\beta}' \boldsymbol{x}_i^\star, \tag{2.14}$$

gdzie

$$\mathrm{logit}(p) = \ln\left(\frac{p}{1-p}\right). \tag{2.15}$$

Nazwa „regresja logistyczna" pochodzi stąd, że funkcja $e^x/(1+e^x)$ jest nazywana *funkcją logistyczną*. Jej wykres jest pokazany na rys. 2.1. Metoda ta pochodzi od Coxa (1966) oraz Daya i Kerridge'a (1967), jednakże w pełni została rozwinięta przez Andersona (1982).

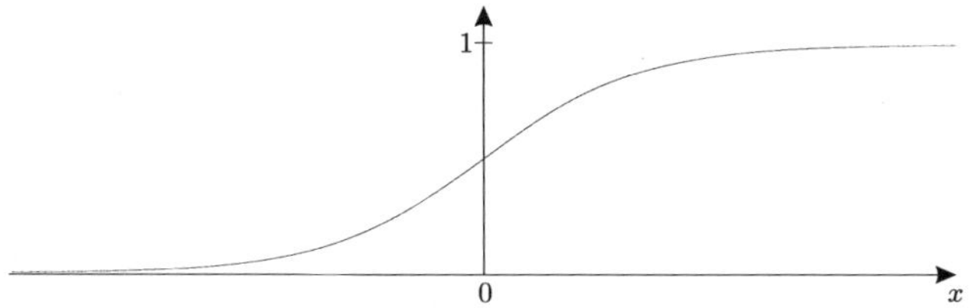

Rys. 2.1. Wykres funkcji logistycznej $f(x) = \mathrm{e}^x/(1 + \mathrm{e}^x)$

Zauważmy, że $1-p = \mathrm{P}(Y = 0 | \boldsymbol{X} = \boldsymbol{x})$ oraz że zgodnie ze wzorem (1.7)

$$\mathrm{logit}(p) = \ln\left(\frac{p}{1-p}\right) = \ln\left(\frac{\pi f_1(\boldsymbol{x})}{(1-\pi) f_0(\boldsymbol{x})}\right) =$$

$$= \ln\left(\frac{f_1(\boldsymbol{x})}{f_0(\boldsymbol{x})}\right) + \ln\left(\frac{\pi}{1-\pi}\right),$$

gdzie $\pi = \mathrm{P}(Y = 1)$, $1 - \pi = \mathrm{P}(Y = 0)$.

Optymalna reguła bayesowska zależy tylko od ilorazu gęstości $f_1(\boldsymbol{x})/f_0(\boldsymbol{x})$. Stąd iloraz ten możemy modelować bez specyfikowania indywidualnych gęstości $f_i(\boldsymbol{x})$, $i = 1, 0$, za pomocą modelu logistycznego. Istnieje stosunkowo szeroka klasa rozkładów prawdopodobieństwa spełniających warunek (2.14), w tym klasa wielowymiarowych rozkładów normalnych z równymi macierzami kowariancji.

Omówimy teraz estymację parametrów $\beta_0, \beta_1, \ldots, \beta_p$ z próby uczącej $(\boldsymbol{X}_i, Y_i)$, $i = 1, 2, \ldots, n$. Ponieważ Y_i są zmiennymi losowymi o rozkładzie binarnym, więc rozkład warunkowy

$$Y_i|\boldsymbol{X}_i = \boldsymbol{x}_i$$

jest rozkładem Bernoulliego z parametrem p_i, $i = 1, \ldots, n$.

Stąd (warunkowa) funkcja wiarogodności jest równa

$$L(\boldsymbol{\beta}) = \prod_{i=1}^{n} p_i(\boldsymbol{x}_i^\star; \boldsymbol{\beta})^{Y_i}(1 - p_i(\boldsymbol{x}_i^\star; \boldsymbol{\beta}))^{1-Y_i} \tag{2.16}$$

oraz

$$\begin{aligned} \ln L(\boldsymbol{\beta}) = l(\boldsymbol{\beta}) &= \sum_{i=1}^{n} [Y_i \ln p_i(\boldsymbol{x}_i^\star; \boldsymbol{\beta}) + (1 - Y_i)\ln(1 - p_i(\boldsymbol{x}_i^\star; \boldsymbol{\beta}))] = \\ &= \sum_{i=1}^{n} \left[Y_i \boldsymbol{\beta}' \boldsymbol{x}_i^\star - \ln\left(1 + \exp(\boldsymbol{\beta}' \boldsymbol{x}_i^\star)\right)\right], \end{aligned} \tag{2.17}$$

gdzie $\boldsymbol{\beta} = (\beta_0, \beta_1, \ldots, \beta_p)'$, $\boldsymbol{x}_i^\star = (1, \boldsymbol{x}_i')'$, $i = 1, 2, \ldots, n$.

W celu zmaksymalizowania funkcji (2.17) przyrównajmy jej pochodne do zera. Mamy

$$\frac{\partial l(\boldsymbol{\beta})}{\partial \boldsymbol{\beta}} = \sum_{i=1}^{n} \boldsymbol{x}_i^\star (Y_i - p_i(\boldsymbol{x}_i^\star; \boldsymbol{\beta})) = \mathbf{0}. \tag{2.18}$$

Otrzymaliśmy $p+1$ równań nieliniowych względem $\boldsymbol{\beta}$. Ponieważ pierwsza składowa wektora $\boldsymbol{x}_i^\star$ jest równa 1, więc równanie pierwsze ma postać

$$\sum_{i=1}^{n} Y_i = \sum_{i=1}^{n} p_i(\boldsymbol{x}_i^\star; \boldsymbol{\beta}).$$

Aby rozwiązać układ (2.18) wykorzystamy *algorytm Newtona–Raphsona*, który wymaga znajomości drugiej pochodnej (hesjanu):

$$\boldsymbol{H}(\boldsymbol{\beta}) = \frac{\partial^2 l(\boldsymbol{\beta})}{\partial \boldsymbol{\beta} \partial \boldsymbol{\beta}'} = -\sum_{i=1}^{n} \boldsymbol{x}_i^\star (\boldsymbol{x}_i^\star)' p_i(\boldsymbol{x}_i^\star; \boldsymbol{\beta})(1 - p_i(\boldsymbol{x}_i^\star;\ \boldsymbol{\beta})). \tag{2.19}$$

Niech $\boldsymbol{\beta}^{\text{stare}}$ oznacza początkową (startową) wartość parametru $\boldsymbol{\beta}$. Rozwijając funkcję (2.17) w szereg Taylora w punkcie $\boldsymbol{\beta}^{\text{stare}}$, otrzymujemy:

$$l(\boldsymbol{\beta}) \simeq l(\boldsymbol{\beta}^{\text{stare}}) + \boldsymbol{U}(\boldsymbol{\beta}^{\text{stare}})'(\boldsymbol{\beta} - \boldsymbol{\beta}^{\text{stare}}) + \\ + \frac{1}{2}(\boldsymbol{\beta} - \boldsymbol{\beta}^{\text{stare}})'\boldsymbol{H}(\boldsymbol{\beta}^{\text{stare}})(\boldsymbol{\beta} - \boldsymbol{\beta}^{\text{stare}}),$$

gdzie

$$\boldsymbol{U}(\boldsymbol{\beta}) = \frac{\partial l(\boldsymbol{\beta})}{\partial \boldsymbol{\beta}}.$$

Różniczkując ostatnią równość i przyrównując pochodne do zera, otrzymujemy układ równań:

$$\boldsymbol{U}(\boldsymbol{\beta}) \simeq \boldsymbol{U}(\boldsymbol{\beta}^{\text{stare}}) + \boldsymbol{H}(\boldsymbol{\beta}^{\text{stare}})(\boldsymbol{\beta} - \boldsymbol{\beta}^{\text{stare}}) = \mathbf{0}.$$

Rozwiązując ten układ, otrzymujemy

$$\boldsymbol{\beta}^{\text{nowe}} = \boldsymbol{\beta}^{\text{stare}} - \boldsymbol{H}(\boldsymbol{\beta}^{\text{stare}})^{-1}\boldsymbol{U}(\boldsymbol{\beta}^{\text{stare}}). \tag{2.20}$$

Powtarzając ten proces, uzyskamy ciąg estymatorów $(\hat{\boldsymbol{\beta}}^k)$ parametru $\boldsymbol{\beta}$ zbieżny do estymatora największej wiarogodności (rozwiązania układu (2.18)), gdy $k \to \infty$.

Niech $\boldsymbol{Y} = (Y_1, Y_2, \ldots, Y_n)'$, niech $\boldsymbol{X}$ będzie macierzą rozmiaru $n \times (p+1)$ i postaci (2.3), niech

$$\boldsymbol{p} = (p_1(\boldsymbol{x}_1^\star; \boldsymbol{\beta}^{\text{stare}}),\ p_2(\boldsymbol{x}_2^\star;\ \boldsymbol{\beta}^{\text{stare}}),\ \ldots,\ p_n(\boldsymbol{x}_n^\star;\ \boldsymbol{\beta}^{\text{stare}}))'$$

oraz niech $\boldsymbol{W}$ będzie macierzą diagonalną stopnia n z i-tym elementem diagonalnym równym $p_i(\boldsymbol{x}_i^\star; \boldsymbol{\beta}^{\text{stare}})(1 - p_i(\boldsymbol{x}_i^\star; \boldsymbol{\beta}^{\text{stare}}))$. Wówczas

$$\boldsymbol{U}(\boldsymbol{\beta}^{\text{stare}}) = \boldsymbol{X}'(\boldsymbol{Y} - \boldsymbol{p}) \quad \text{oraz} \quad \boldsymbol{H}(\boldsymbol{\beta}^{\text{stare}}) = -\boldsymbol{X}'\boldsymbol{W}\boldsymbol{X}.$$

Krok procedury Newtona–Raphsona jest równy

$$\begin{aligned} \boldsymbol{\beta}^{\text{nowe}} &= \boldsymbol{\beta}^{\text{stare}} + (\boldsymbol{X}'\boldsymbol{W}\boldsymbol{X})^{-1}\boldsymbol{X}'(\boldsymbol{Y} - \boldsymbol{p}) = \\ &= (\boldsymbol{X}'\boldsymbol{W}\boldsymbol{X})^{-1}\boldsymbol{X}'\boldsymbol{W}(\boldsymbol{X}\boldsymbol{\beta}^{\text{stare}} + \boldsymbol{W}^{-1}(\boldsymbol{Y} - \boldsymbol{p})) = \\ &= (\boldsymbol{X}'\boldsymbol{W}\boldsymbol{X})^{-1}\boldsymbol{X}'\boldsymbol{W}\boldsymbol{Z}, \end{aligned} \tag{2.21}$$

gdzie $\boldsymbol{Z} = \boldsymbol{X}\boldsymbol{\beta}^{\text{stare}} + \boldsymbol{W}^{-1}(\boldsymbol{Y} - \boldsymbol{p})$.

Algorytm ten jest nazywany *iteracyjnie ważoną metodą najmniejszych kwadratów*[1], ponieważ w każdym kroku jest rozwiązywany ważony problem najmniejszych kwadratów:

$$\boldsymbol{\beta}^{\text{nowe}} \leftarrow \arg\min_{\boldsymbol{\beta}} (\boldsymbol{Z} - \boldsymbol{X}\boldsymbol{\beta})'\boldsymbol{W}(\boldsymbol{Z} - \boldsymbol{X}\boldsymbol{\beta}). \tag{2.22}$$

Wydaje się, że dobrą wartością początkową w tym algorytmie iteracyjnym jest wartość $\boldsymbol{\beta} = \boldsymbol{0}$, chociaż nie daje to gwarancji zbieżności. Zazwyczaj algorytm jest zbieżny, ponieważ funkcja $l(\boldsymbol{p})$ jest wklęsła, ale może pojawić się przestrzelenie.

Algorytm iteracyjnej ważonej metody najmniejszych kwadratów

Wybieramy wartość początkową $\hat{\boldsymbol{\beta}}_0$ i obliczamy p_i^0 z równania (2.18) dla $i = 1, 2, \ldots, n$. Przyjmujemy $s = 0$ i iterujemy następne kroki do momentu uzyskania zbieżności.

1. Przyjmujemy
$$Z_i = \text{logit}(p_i^s) + \frac{Y_i - p_i^s}{p_i^s(1 - p_i^s)}, \qquad i = 1, 2, \ldots, n.$$
2. Niech $\boldsymbol{W}$ będzie macierzą diagonalną z elementem diagonalnym (i, i) równym $p_i^s(1 - p_i^s)$, $i = 1, 2, \ldots, n$.
3. Przyjmujemy
$$\hat{\boldsymbol{\beta}}^s = (\boldsymbol{X}'\boldsymbol{W}\boldsymbol{X})^{-1}\boldsymbol{X}'\boldsymbol{W}\boldsymbol{Z},$$
gdzie macierz $\boldsymbol{X}$ jest dana wzorem (2.3) i $\boldsymbol{Z} = (Z_1, Z_2, \ldots, Z_n)'$, tak jak w modelu ważonej regresji liniowej $\boldsymbol{Z}$ względem $\boldsymbol{X}$.
4. Podstawiamy $s = s + 1$ i wracamy do kroku pierwszego.

Ostatecznie klasyfikator ma postać

$$\hat{d}(\boldsymbol{x}) = \begin{cases} 1, \text{ jeżeli } \hat{\beta}_0 + \hat{\beta}_1 x_1 + \cdots + \hat{\beta}_p x_p > 0, \\ 0, \text{ poza tym.} \end{cases} \tag{2.23}$$

Zdefiniujmy teraz model logistyczny w przypadku liczby grup większej od dwóch ($K > 2$).

[1] Po ang. Iteratively Reweighted Least Squares (IRLS).

Niech

$$p_{ik} = \mathrm{P}(Y_i = k|\boldsymbol{X}_i = \boldsymbol{x}_i) = \frac{\exp\left(\beta_{0k} + \sum_{j=1}^{p} \beta_{jk} x_{ij}\right)}{1 + \exp\left(\beta_{0k} + \sum_{j=1}^{p} \beta_{jk} x_{ij}\right)}, \tag{2.24}$$

$i = 1, 2, \ldots, n$, $k = 1, 2, \ldots, K$.

Wybierając umownie grupę K jako bazową, model logistyczny przybiera postać

$$\operatorname{logit}(p_{ik}) = \ln\left(\frac{p_{ik}}{p_{iK}}\right) = \beta_{0k} + \sum_{j=1}^{p} \beta_{jk} x_{ij}, \tag{2.25}$$

$i = 1, 2, \ldots, n$, $k = 1, 2, \ldots, K$. Zauważmy, że $\sum_{k=1}^{K} p_{ik} = 1$. Stąd $p_{iK} = 1 - \sum_{k=1}^{K-1} p_{ik}$. Tak jak poprzednio, estymatory parametrów $\beta_{0k}, \beta_{1k}, \ldots, \beta_{pk}$, $k = 1, 2, \ldots, K-1$, uzyskuje się numerycznie. W tym ogólnym przypadku, model regresji logistycznej wygodniej jest rozpatrywać jako rodzaj uogólnionego modelu liniowego (McCullagh i Nelder, 1989).

2.3. Związek między regresją logistyczną i liniową analizą dyskryminacyjną

Liniowa analiza dyskryminacyjna i regresja logistyczna są prawie tym samym. Z jednej strony, jeżeli założymy, że w każdej grupie obserwowany wektor losowy $\boldsymbol{X}$ ma p-wymiarowy rozkład normalny z tą samą macierzą kowariancji $\boldsymbol{\Sigma}$, to

$$\begin{aligned}\ln\left(\frac{\mathrm{P}(Y = 1|\boldsymbol{X} = \boldsymbol{x})}{\mathrm{P}(Y = 0|\boldsymbol{X} = \boldsymbol{x})}\right) &= \ln\left(\frac{\pi}{1-\pi}\right) - \frac{1}{2}(\boldsymbol{\mu}_1 + \boldsymbol{\mu}_0)'\boldsymbol{\Sigma}^{-1}(\boldsymbol{\mu}_1 - \boldsymbol{\mu}_0) + \\ &+ \boldsymbol{x}'\boldsymbol{\Sigma}^{-1}(\boldsymbol{\mu}_1 - \boldsymbol{\mu}_0) \equiv \\ &\equiv \alpha_0 + \alpha_1 x_1 + \cdots + \alpha_p x_p.\end{aligned}$$

Z drugiej strony model logistyczny ma postać

$$\ln\left(\frac{\mathrm{P}(Y = 1|\boldsymbol{X} = \boldsymbol{x})}{\mathrm{P}(Y = 0|\boldsymbol{X} = \boldsymbol{x})}\right) = \beta_0 + \beta_1 x_1 + \cdots + \beta_p x_p.$$

W istocie są to te same modele, ponieważ prowadzą do klasyfikatora liniowego względem wektora $\boldsymbol{x} = (x_1, x_2, \ldots, x_p)'$. Różnica polega tylko na sposobie estymacji parametrów. Łączny rozkład pojedynczej obserwacji $(\boldsymbol{X}, Y)$ ma gęstość

$$f(\boldsymbol{x}, y) = f(\boldsymbol{x}|y)f(y) = f(y|\boldsymbol{x})f(\boldsymbol{x}).$$

W liniowej analizie dyskryminacyjnej estymujemy cały rozkład łączny poprzez maksymalizację funkcji wiarogodności

$$\prod_{i=1}^{n} f(\boldsymbol{x}_i, y_i) = \underbrace{\prod_{i=1}^{n} f(\boldsymbol{x}_i|y_i)}_{\text{rozkład normalny}} \quad \underbrace{\prod_{i=1}^{n} f(y_i)}_{\text{rozkład Bernoulliego}}.$$

W regresji logistycznej maksymalizujemy warunkową funkcję wiarogodności $\prod_{i=1}^{n} f(y_i|\boldsymbol{x}_i)$, ale pomijamy drugi czynnik $\prod_{i=1}^{n} f(\boldsymbol{x}_i)$:

$$\prod_{i=1}^{n} f(\boldsymbol{x}_i, y_i) = \underbrace{\prod_{i=1}^{n} f(y_i|\boldsymbol{x}_i)}_{\text{rozkład logistyczny}} \quad \underbrace{\prod_{i=1}^{n} f(\boldsymbol{x}_i)}_{\text{pomijamy}}.$$

Ponieważ reguła klasyfikacyjna wymaga znajomości tylko $f(y|\boldsymbol{x})$, więc faktycznie nigdy nie musimy estymować całego łącznego rozkładu prawdopodobieństwa. Regresja logistyczna pomija estymację gęstości $f(\boldsymbol{x})$ rozkładu brzegowego.

Model regresji logistycznej jest zatem bardziej ogólny niż model liniowej analizy dyskryminacyjnej, ponieważ wymaga mniej założeń.

Przykład 2.1.
W celu zobrazowania różnic między liniową, kwadratową oraz logistyczną funkcją dyskryminacyjną wygenerowano dane typu Highleymana. W tego typu danych obie klasy pochodzą z rozkładu normalnego dwuwymiarowego, przy czym dla pierwszej klasy mamy $\boldsymbol{\mu}_1 = (1, 1)'$ oraz $\boldsymbol{\Sigma}_1 = \begin{bmatrix} 1 & 0 \\ 0 & 0.25 \end{bmatrix}$. Natomiast dla klasy drugiej mamy $\boldsymbol{\mu}_2 = (2, 0)'$ oraz $\boldsymbol{\Sigma}_2 = \begin{bmatrix} 0.01 & 0 \\ 0 & 4 \end{bmatrix}$. Do zbioru uczącego wygenerowano po 50 obserwacji z każdej klasy, podobnie do zbioru testowego. Obszary klasyfikacyjne zostały przedstawione na rys. 2.2.

Błąd klasyfikacji na zbiorze testowym wyniósł 7.00% dla metody QDA, 15.00% dla metody LDA oraz 17.00% dla metody logistycznej.

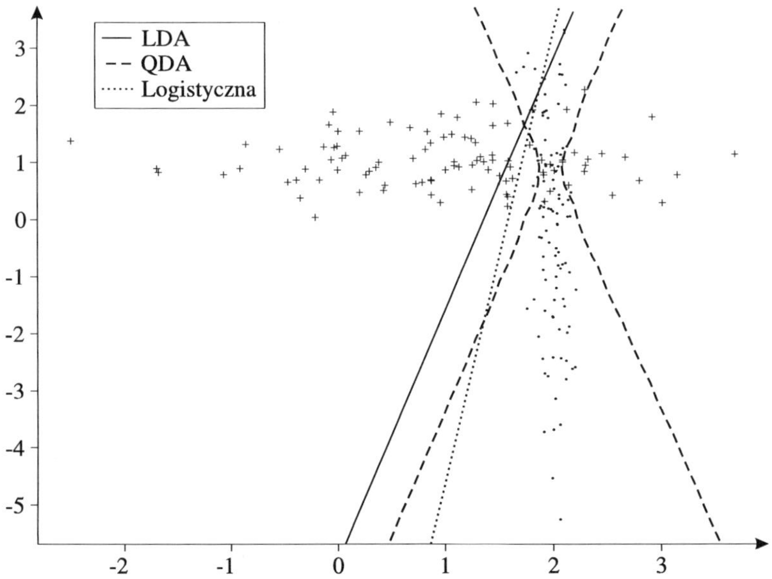

Rys. 2.2. Obszary klasyfikacji dla metody QDA, LDA oraz logistycznej

2.4. Estymatory jądrowe gęstości

Optymalny klasyfikator bayesowski dany wzorem (1.11) wykorzystuje prawdopodobieństwa a priori oraz funkcje gęstości. Jeżeli wielkości te nie są znane, możemy je zastąpić ich estymatorami z próby uczącej. W podrozdziale tym omówimy *estymatory jądrowe* funkcji gęstości.

Niech $X_1, X_2, \ldots, X_n$ będzie próbą z rozkładu o gęstości f.

Definicja 2.1.
Jądrem będziemy nazywać każdą gładką funkcję K taką, że $K(x) \geq 0$, $\int K(x)dx = 1$, $\int xK(x)dx = 0$ i $\sigma_K^2 = \int x^2 K(x)dx > 0$.

Przykładami jąder są:

1. Jądro jednostajne:

$$K(x) = \frac{1}{2} I(x). \tag{2.26}$$

2. Jądro gaussowskie (normalne):

$$K(x) = (2\pi)^{-\frac{1}{2}} \exp\left(-\frac{x^2}{2}\right). \tag{2.27}$$

3. Jądro Epanechnikowa:

$$K(x) = \frac{3}{4}(1 - x^2)I(x). \tag{2.28}$$

4. Jądro stopnia trzeciego:

$$K(x) = \frac{70}{81}(1 - |x|^3)^3 I(x), \tag{2.29}$$

gdzie

$$I(x) = \begin{cases} 1, \text{ jeżeli } |x| \leq 1, \\ 0, \text{ jeżeli } |x| > 1. \end{cases}$$

Wykresy tych jąder są pokazane na rys. 2.3.

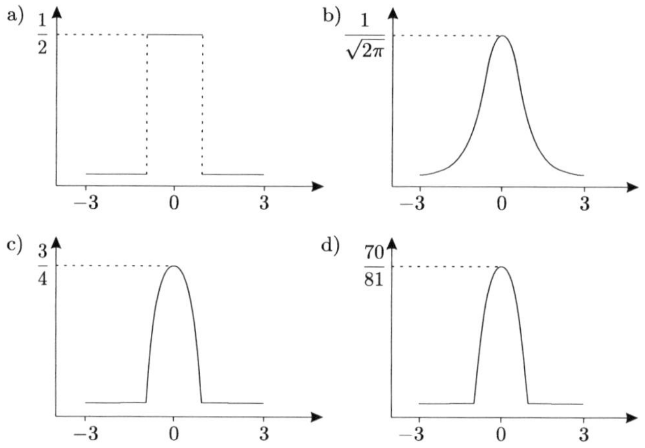

Rys. 2.3. Przykłady jąder: a) jądro jednostajne, b) jądro gaussowskie, c) jądro Epanechnikowa, d) jądro stopnia trzeciego

Definicja 2.2.
Dla danego jądra K i dodatniej liczby h, zwanej współczynnikiem gładkości, *jądrowy estymator Rosenblatta–Parzena* gęstości f jest równy

$$\hat{f}_n(x) = \frac{1}{n}\sum_{i=1}^{n} \frac{1}{h} K\left(\frac{x - X_i}{h}\right). \tag{2.30}$$

W celu skonstruowania jądrowego estymatora gęstości, musimy wybrać jądro K i współczynnik gładkości h. Można pokazać teoretycznie i empirycznie, że wybór jądra K nie jest sprawą kluczową. Jednakże wybór współczynnika gładkości h jest bardzo istotny.

Zauważmy, że estymator $\hat{f}_n(x)$ rozpatrywany w punkcie x jest zmienną losową. Estymator jest zmienną losową, ponieważ zależy od próby X_1, X_2, ..., X_n.

Jako funkcję straty wybierzmy scałkowany błąd średniokwadratowy postaci

$$L(f,\ \hat{f}_n) = \int (f(x) - \hat{f}_n(x))^2 dx. \tag{2.31}$$

Ryzyko estymatora $\hat{f}_n$ jest równe wartości oczekiwanej funkcji L, tj.

$$R(f,\ \hat{f}_n) = \mathrm{E}(L(f,\ \hat{f}_n)). \tag{2.32}$$

Ryzyko R można zapisać w postaci

$$R(f,\ \hat{f}_n) = \int b^2(x)dx + \int v(x)dx, \tag{2.33}$$

gdzie

$$b(x) = \mathrm{E}(\hat{f}_n(x)) - f(x) \tag{2.34}$$

jest obciążeniem estymatora $\hat{f}_n(x)$ w ustalonym punkcie x oraz

$$v(x) = \mathrm{Var}(\hat{f}_n(x)) = \mathrm{E}\left[(\hat{f}_n(x) - \mathrm{E}(\hat{f}_n(x))^2\right] \tag{2.35}$$

jest jego wariancją.

O zależności ryzyka estymatora $\hat{f}_n$ od współczynnika gładkości h mówi następujące twierdzenie.

TWIERDZENIE 2.1.

Jeżeli $f''(x)$ jest funkcją absolutnie ciągłą, $\int (f'''(x))^2 dx < \infty$ i $K(x)$ spełnia def. 2.1, to

$$R(f,\ \hat{f}_n) \approx \frac{1}{4}\sigma_K^4 h^4 \int (f''(x))^2 dx + \frac{1}{nh}\int K^2(x)dx, \tag{2.36}$$

gdzie $\sigma_K^2 = \int x^2 K(x)dx$.

Optymalna wartość współczynnika gładkości jest równa

$$h^\star = \frac{c_1^{-\frac{2}{5}} c_2^{\frac{1}{5}} c_3^{-\frac{1}{5}}}{n^{\frac{1}{5}}}, \tag{2.37}$$

gdzie

$$c_1 = \int x^2 K(x)dx, \qquad c_2 = \int K^2(x)dx \qquad \text{i} \qquad c_3 = \int (f''(x))^2 dx.$$

Dla tak wybranego h,

$$R(f,\ \hat{f}_n) \approx \frac{c_4}{n^{4/5}} \tag{2.38}$$

przy pewnej stałej $c_4 > 0$.

Dowód. Niech

$$K_h(x,\ X) = h^{-1} K((x - X)/h)$$

i niech

$$\hat{f}(x) = n^{-1} \sum_{i=1}^{n} K_h(x,\ X_i).$$

Wówczas

$$\mathrm{E}[\hat{f}(x)] = \mathrm{E}[K_h(x,\ X)] \qquad \text{oraz} \qquad \mathrm{Var}[\hat{f}(x)] = n^{-1}\mathrm{Var}[K_h(x,\ X)].$$

Mamy

$$\begin{aligned}
\mathrm{E}[K_h(x,\ X)] &= \int \frac{1}{h} K\left(\frac{x-t}{h}\right) f(t)dt = \\
&= \int K(u) f(x - hu) du = \\
&= \int K(u)[f(x) - huf'(x) + \frac{h^2u^2}{2} f''(x) + \cdots]du = \\
&= f(x) + \frac{1}{2} h^2 f''(x) \int u^2 K(u) du + \cdots ,
\end{aligned}$$

ponieważ

$$\int K(x)dx = 1 \qquad \text{i} \qquad \int xK(x)dx = 0.$$

Wykorzystaliśmy tu rozwinięcie funkcji $f(x - hu)$ w szereg potęgowy. Stąd obciążenie estymatora $\hat{f}_n(x)$ jest równe

$$\mathrm{E}[K_h(x,\ X)] - f(x) \approx \frac{1}{2}\sigma_k^2 h^2 f''(x).$$

Stosując podobne rozumowanie, otrzymujemy

$$\mathrm{Var}[\hat{f}_n(x)] \approx \frac{f(x) \int K^2(x)dx}{nh}.$$

Tezę twierdzenia uzyskujemy po scałkowaniu kwadratu obciążenia estymatora, scałkowaniu wariancji estymatora i dodaniu scałkowanych wyrażeń. ∎

Z powyższego twierdzenia widzimy, że szybkość zbieżności estymatora jądrowego jest rzędu $n^{-4/5}$. Można pokazać, że przy słabych założeniach nie istnieje estymator nieparametryczny, którego szybkość zbieżności byłaby większa od $n^{-4/5}$.

Praktyczny sposób wyboru optymalnego parametru gładkości h polega na estymacji funkcji ryzyka i jej minimalizacji względem h.

Zauważmy, że funkcja straty, którą teraz zapiszemy jako funkcję h, ma postać

$$L(h) = \int (\hat{f}_n(x) - f(x))^2 dx = \\ = \int \hat{f}_n^2(x)dx - 2\int \hat{f}_n^2(x)f(x)dx + \int f^2(x)dx.$$

Ponieważ ostatni wyraz nie zależy od współczynnika gładkości h, więc minimalizacja ryzyka jest równoważna minimalizacji wartości oczekiwanej wyrażenia

$$J(h) = \int \hat{f}_n^2(x)dx - 2\int \hat{f}_n^2(x)f(x)dx.$$

Wyrażenie $\mathrm{E}(J(h))$ będziemy nazywać *ryzykiem*, chociaż różni się ono od prawdziwego ryzyka o wyraz stały $\int f^2(x)dx$.

Definicja 2.3.
Estymator kroswalidacyjny (estymator sprawdzania krzyżowego)[2] funkcji ryzyka jest równy

$$\hat{J}(h) = \int \hat{f}_n^2(x)dx - \frac{2}{n}\sum_{i=1}^{n} \hat{f}_{(-i)}(X_i), \tag{2.39}$$

gdzie $\hat{f}_{(-i)}$ jest estymatorem jądrowym gęstości uzyskanym z próby, w której pominięto i-tą obserwację.

[2] Po ang. the cross-validation estimator.

Chcąc obliczyć $\hat{J}(h)$ powinniśmy w zasadzie n-krotnie obliczać estymator gęstości i to dla wszystkich zadanych wartości h. Na szczęście możemy skorzystać ze skróconego wzoru.

TWIERDZENIE 2.2.

Dla każdego $h > 0$,

$$\mathrm{E}[\hat{J}(h)] = \mathrm{E}[J(h)]$$

oraz

$$\hat{J}(h) \approx \frac{1}{hn^2}\sum_i \sum_j K^\star\left(\frac{X_i - X_j}{h}\right) + \frac{2}{nh}K(0), \tag{2.40}$$

gdzie $K^\star(x) = K^{(2)}(x) - 2K(x)$ *oraz* $K^{(2)}(z) = \int K(z-y)K(y)dy$.

W szczególności, jeżeli K *jest gęstością rozkładu* $N(0,\ 1)$, *to* $K^{(2)}(x)$ *jest gęstością rozkładu* $N(0,\ 2)$.

Optymalnym współczynnikiem gładkości jest współczynnik h_n, który minimalizuje $\hat{J}(h)$. Uzasadnienie takiego postępowania jest zawarte w godnym uwagi twierdzeniu Stone'a (1977).

TWIERDZENIE 2.3. (Stone'a, 1977)

Załóżmy, że f *jest funkcją ograniczoną. Niech* $\hat{f}_h$ *oznacza estymator jądrowy ze współczynnikiem gładkości* h *i niech* h_n *oznacza współczynnik gładkości wybrany metodą sprawdzania krzyżowego.*

Wówczas

$$\frac{\int (f(x) - \hat{f}_{h_n}(x))^2 dx}{\inf_h \int (f(x) - \hat{f}_h(x))^2 dx} \longrightarrow 1\ (p.w.) \tag{2.41}$$

Sheather i Jones (1991) zaproponowali estymator $\hat{h}$ z próby wpółczynnika gładkości h o lepszych własnościach od estymatora kroswalidacyjnego.

Niech

$$\phi(x) = (2\pi)^{-\frac{1}{2}} \exp\left(-\frac{1}{2}x^2\right),$$

$R(K) = \int K^2(x)dx$ oraz $\sigma_K^2 = \int x^2 K(x)dx$.

Estymator $\hat{h}$ Sheathera i Jonesa jest rozwiązaniem równania

$$\left[R(K)/\{\sigma_K^4 \hat{S}_D(\hat{\alpha}_2(h))\}\right]^{1/5} n^{-1/5} - h = 0,$$

gdzie

$$\hat{\alpha}_2(h) = 1.357\{\hat{S}_D(a)/\hat{T}_D(b)\}^{1/7} h^{5/7},$$

$$\hat{S}_D(\alpha) = \{n(n-1)\}^{-1}\alpha^{-5}\sum_{i=1}^{n}\sum_{j=1}^{n}\phi^{(4)}(\alpha^{-1}(X_i - X_j)),$$

$$\hat{T}_D(b) = -\{n(n-1)\}^{-1}b^{-7}\sum_{i=1}^{n}\sum_{j=1}^{n}\phi^{(6)}(b^{-1}(X_i - X_j)),$$

$$a = 0.920\hat{\lambda}n^{-1/7}, \quad b = 0.912\hat{\lambda}n^{-1/9},$$

$\hat{\lambda}$ jest rozstępem międzykwartylowym z próby, tj. różnicą między trzecim i pierwszym kwartylem z próby.

Równanie, z którego uzyskuje się $\hat{h}$, rozwiązuje się metodą Newtona–Raphsona.

Estymator jądrowy można łatwo uogólnić na p-wymiarów. Estymator jądrowy p-wymiarowy gęstości $f(\boldsymbol{x})$ ma postać

$$\hat{f}_n(\boldsymbol{x}) = \frac{1}{n}\sum_{i=1}^{n}\prod_{j=1}^{p}\frac{1}{h_j}K\left(\frac{x_j - X_{ij}}{h_j}\right), \tag{2.42}$$

gdzie $\boldsymbol{x} = (x_1, x_2, \ldots, x_p)'$, X_{ij} jest j-tą składową wektora $\boldsymbol{X}_i = (X_{i1}, \ldots, X_{ip})'$, $i = 1, 2, \ldots, n$, $K(z)$ jest jądrem jednowymiarowym oraz $h_1, \ldots, h_p$ są współczynnikami gładkości związanymi z poszczególnymi zmiennymi.

Dla prostoty możemy przyjąć, że $h_j = hs_j$, gdzie s_j jest odchyleniem standardowym z próby j-tej zmiennej. W takim przypadku optymalizujemy tylko jeden współczynnik gładkości. Wykonując obliczenia podobne do przypadku jednowymiarowego można pokazać, że ryzyko tego estymatora ma postać

$$R(f, \hat{f}_n) \approx \frac{1}{4}\sigma_K^4\left[\sum_{j=1}^{p} h_j^4 \int f_{jj}^2(x)dx + \sum_{j\neq k} h_j^2 h_k^2 \int f_{jj}f_{kk}dx\right] + \\ + \frac{\left(\int K^2(x)dx\right)^p}{Nh_1\cdots h_p},$$

gdzie f_{jj} jest pochodną cząstkową rzędu drugiego funkcji f.

Optymalny współczynnik gładkości ma postać $h_i \approx c_1 n^{-1/(4+p)}$, co prowadzi do ryzyka rzędu $n^{-4/(4+p)}$.

Widzimy, że ryzyko rośnie bardzo szybko wraz z wymiarem p przestrzeni cech. Aby zobaczyć jak poważna jest ta zależność przyjrzyjmy się tabeli zaczerpniętej z monografii Silvermana (1986), która pokazuje niezbędną wielkość próby, by zapewnić to, że względny błąd średnio-kwadratowy w zerze będzie mniejszy od 0.1, gdy h jest gęstością p-wymiarowego rozkładu normalnego i posługujemy się optymalnym współczynnikiem gładkości.

Wymiar	Liczebność próby
1	4
2	19
3	67
4	223
5	768
6	2 790
7	10 700
8	43 700
9	187 000
10	842 000

Z tabeli tej wynika niepocieszający fakt. Widzimy, że posiadanie 842 000 obserwacji w przestrzeni dziesięciowymiarowej odpowiada faktowi posiadania czterech obserwacji jednowymiarowych. Zjawisko to jest nazywane „przekleństwem wielowymiarowości".

Do estymacji wielowymiarowych gęstości możemy również użyć jąder wielowymiarowych. Wówczas estymator jądrowy p-wymiarowej gęstości będzie postaci

$$\hat{f}_n(\boldsymbol{x}) = \frac{1}{nh^p} \sum_{i=1}^{n} K\left(\frac{\boldsymbol{x} - \boldsymbol{X}_i}{h}\right). \tag{2.43}$$

Jako jądro K możemy wybrać wielowymiarowe jądro Epanechnikova postaci

$$K(\boldsymbol{z}) = \begin{cases} \dfrac{(p+2)\Gamma\left(1+\dfrac{p}{2}\right)}{2\pi^{p/2}}(1-\boldsymbol{z}'\boldsymbol{z}), & \text{jeżeli } \boldsymbol{z}'\boldsymbol{z} < 1, \\ 0, & \text{poza tym.} \end{cases} \tag{2.44}$$

Jest to jednomodalna funkcja gęstości o symetrii kołowej w przestrzeni p-wymiarowej.

Inny możliwy wybór jądra, to $K(\boldsymbol{z})$ postaci

$$K(\boldsymbol{z}) = (2\pi)^{-\frac{p}{2}} \exp(\Big(-\frac{1}{2}\boldsymbol{z}'\boldsymbol{z}\Big), \qquad \boldsymbol{z} \in \mathbb{R}^p. \tag{2.45}$$

Jest to gęstość p-wymiarowego standaryzowanego rozkładu normalnego.

Klasyfikacja z wykorzystaniem nieparametrycznych estymatorów gęstości

Niech $(\boldsymbol{X}_i, Y_i)$, $i = 1, 2, \ldots, n$, będzie próbą uczącą, gdzie $\boldsymbol{X}_i \in \mathbb{R}^p$ oraz $Y_i \in \{1, 2, \ldots, K\}$. Niech

$$\hat{f}_k^{(h)}(\boldsymbol{x}) = \frac{1}{n_k h^p} \sum_{i=1}^{n} K\left(\frac{\boldsymbol{x} - \boldsymbol{X}_i}{h}\right) I(Y_i = k) \tag{2.46}$$

będzie nieparametrycznym estymatorem gęstości $f_k(\boldsymbol{x})$ w grupie k i niech $\hat{\pi}_k = n_k/n$ będzie estymatorem prawdopodobieństwa a priori π_k, gdzie $n_k = \sum_{i=1}^{n} I(Y_i = k)$, $k = 1, 2, \ldots, K$.

Obserwację $\boldsymbol{x}$ klasyfikujemy do grupy k wówczas, gdy

$$k = d_h(\boldsymbol{x}) = \arg \max_{k=1, \ldots, K} \hat{\pi}_k \hat{f}_k^{(h)}(\boldsymbol{x}). \tag{2.47}$$

Optymalną wartość parametru gładkości h dobieramy metodą *sprawdzania krzyżowego*[3], tj. poszukujemy takiej wartości h, która minimalizuje

$$\hat{e}_{CV}(h) = n^{-1} \sum_{j=1}^{n} I\left(\hat{d}_h(\boldsymbol{X}_j;\ \mathcal{L}_n^{(-j)}) \neq Y_j\right), \tag{2.48}$$

gdzie $\hat{d}_h(\boldsymbol{X}_j;\ \mathcal{L}_n^{(-j)})$ oznacza regułę klasyfikacyjną z próby, z której usunięto j-tą obserwację (porównaj ze wzorem (1.65)).

2.5. Regresja nieparametryczna

Danych jest n par obserwacji $(\boldsymbol{X}_1, Y_1), \ldots, (\boldsymbol{X}_n, Y_n)$ związanych zależnością

$$Y_i = r(\boldsymbol{X}_i) + \varepsilon_i, \tag{2.49}$$

[3]Po ang. Cross-Validation (CV).

gdzie $\boldsymbol{X}_i \in \mathbb{R}^p$, $Y_i \in \mathbb{R}$, $\boldsymbol{X}_i$ oraz ε_i są niezależne oraz $\varepsilon_1, \ldots, \varepsilon_n$ są niezależnymi błędami losowymi takimi, że $\mathrm{E}(\varepsilon_i|\boldsymbol{X}_i = \boldsymbol{x}) = 0$ oraz $\mathrm{Var}(\varepsilon_i|\boldsymbol{X}_i = \boldsymbol{x}) = \sigma^2(\boldsymbol{x})$, $i = 1, \ldots, n$; $r(\boldsymbol{x})$ jest funkcją regresji postaci

$$r(\boldsymbol{x}) = \mathrm{E}(Y|\boldsymbol{X} = \boldsymbol{x}).$$

Naszym celem jest skonstruowanie nieparametrycznego estymatora funkcji regresji $r(\boldsymbol{x})$. Mamy

$$r(\boldsymbol{x}) = \mathrm{E}(Y|\boldsymbol{X} = \boldsymbol{x}) = \int_{-\infty}^{\infty} y f_Y(y|\boldsymbol{x})dy =$$

$$= \frac{\int_{-\infty}^{\infty} y f(\boldsymbol{x}, y)dy}{f_{\boldsymbol{X}}(\boldsymbol{x})} = \frac{\int_{-\infty}^{\infty} y f(\boldsymbol{x}, y)dy}{\int_{-\infty}^{\infty} f(\boldsymbol{x}, y)dy}.$$

Zastąpimy teraz łączną gęstość $f(\boldsymbol{x}, y)$ pary $(\boldsymbol{X}, Y)$ estymatorem jądrowym postaci

$$\hat{f}_n(\boldsymbol{x}, y) = \frac{1}{nh^{p+1}} \sum_{i=1}^{n} K\left(\frac{\boldsymbol{x} - \boldsymbol{X}_i}{h}\right) K\left(\frac{y - Y_i}{h}\right),$$

gdzie $K(\cdot)$ jest odpowiednim jądrem, a h parametrem gładkości.

Korzystając z tego, że $\int K(x)dx = 1$ oraz $\int xK(x)dx = 0$ mamy

$$\int_{-\infty}^{\infty} \hat{f}_n(\boldsymbol{x}, y)dy = \frac{1}{nh^p} \sum_{i=1}^{n} K\left(\frac{\boldsymbol{x} - \boldsymbol{X}_i}{h}\right)$$

oraz

$$\int_{-\infty}^{\infty} y\hat{f}_n(\boldsymbol{x}, y)dy = \frac{1}{nh^p} \sum_{i=1}^{n} K\left(\frac{\boldsymbol{x} - \boldsymbol{X}_i}{h}\right) Y_i.$$

Stąd

$$\hat{r}(\boldsymbol{x}) = \frac{\sum_{i=1}^{n} K\left(\frac{\boldsymbol{x} - \boldsymbol{X}_i}{h}\right) Y_i}{\sum_{i=1}^{n} K\left(\frac{\boldsymbol{x} - \boldsymbol{X}_i}{h}\right)}.$$

Definicja 2.4.
Dla danego jądra K i dodatniego współczynnika gładkości h, *jądrowy estymator Nadarayi–Watsona* funkcji regresji $r(\boldsymbol{x})$ ma postać

$$\hat{r}(\boldsymbol{x}) = \sum_{i=1}^{n} w_i(\boldsymbol{x}) Y_i, \tag{2.50}$$

gdzie wagi $w_i(\boldsymbol{x})$ mają postać

$$w_i(\boldsymbol{x}) = \frac{K\left(\dfrac{\boldsymbol{x}-\boldsymbol{X}_i}{h}\right)}{\sum_{i=1}^{n} K\left(\dfrac{\boldsymbol{x}-\boldsymbol{X}_i}{h}\right)}, \qquad i = 1,\ 2,\ \ldots,\ n. \tag{2.51}$$

Szybkość zbieżności estymatora $\hat{r}(\boldsymbol{x})$ do funkcji regresji $r(\boldsymbol{x})$ jest rzędu $O(n^{-4/(4+p)})$, gdzie p jest wymiarem wektora $\boldsymbol{X}$.

To implikuje, że duża wartość p spowalnia szybkość zbieżności i jedynym wyjściem jest znaczne zwiększenie liczebności próby n. Jest to dobrze znane przekleństwo wymiarowości.

Klasyfikacja z wykorzystaniem regresji nieparametrycznej

Niech $(\boldsymbol{X}_i,\ Y_i)$, $i = 1,\ 2,\ \ldots,\ n$, będzie próbą uczącą, gdzie $\boldsymbol{X}_i \in \mathbb{R}^p$ oraz $Y_i \in \{1,\ 2,\ \ldots,\ K\}$.

Niech $\hat{r}_h^{(k)}(\boldsymbol{x})$ będzie estymatorem nieparametrycznym funkcji regresji w grupie k:

$$\hat{r}_h^{(k)}(\boldsymbol{x}) = \frac{\sum_{i=1}^{n} K\left(\dfrac{\boldsymbol{x}-\boldsymbol{X}_i}{h}\right) I(Y_i = k)}{\sum_{i=1}^{n} K\left(\dfrac{\boldsymbol{x}-\boldsymbol{X}_i}{h}\right)}, \qquad k = 1,\ 2,\ \ldots,\ K. \tag{2.52}$$

Obserwację $\boldsymbol{x}$ klasyfikujemy do grupy k wówczas, gdy

$$k = d_h(\boldsymbol{x}) = \arg \max_{k=1,\ \ldots,\ K} \{\hat{r}_h^{(k)}(\boldsymbol{x})\}. \tag{2.53}$$

Optymalną wartość parametru gładkości h dobieramy *metodą sprawdzania krzyżowego*, tj. poszukujemy takiej wartości h, która minimalizuje wyrażenie $\hat{e}_{CV}(h)$ dane wzorem (2.48).

3

Krzywe ROC

Niech etykieta $Y \in \mathcal{Y} = \{1,\ 0\}$ i niech X będzie obserwowaną zmienną absolutnie ciągłą.

Reguła klasyfikacyjna ma postać

$$d(x) = \begin{cases} 1, \text{ jeżeli } x > c, \\ 0, \text{ jeżeli } x \leq c, \end{cases} \tag{3.1}$$

gdzie c jest wartością progową.

Dla wybranego progu c $(-\infty \leq c \leq +\infty)$ niech

$$\mathrm{P}(d(X) = 1|Y = 1) = \mathrm{P}(X > c|Y = 1) = \bar{G}(c) = 1 - G(c), \tag{3.2}$$

gdzie $\bar{G}$ jest funkcją przeżycia, a G dystrybuantą zmiennej X w grupie o etykiecie 1 oraz niech

$$\mathrm{P}(d(X) = 1|Y = 0) = \mathrm{P}(X > c|Y = 0) = \bar{F}(c) = 1 - F(c), \tag{3.3}$$

gdzie $\bar{F}$ jest funkcją przeżycia, a F dystrybuantą zmiennej X w grupie o etykiecie 0.

Definicja 3.1.
Krzywa ROC[1] ma postać

$$\mathrm{ROC}(\cdot) = \{(1 - F(c),\ 1 - G(c)) :\ -\infty \leq c \leq \infty\}.$$

[1] Po ang. the Receiver Operating Characteristic curves (ROC curves). Krzywe te zastosowano po raz pierwszy w celu określenia jak dobrze sprzęt radiowy w czasie II wojny światowej odróżniał zakłócenia przypadkowe od sygnałów prawidłowo wskazujących samoloty nieprzyjaciela. To tłumaczy użycie terminu „charakterystyka odbiornika” w nazwie tych krzywych.

W terminologii statystyki matematycznej krzywa ROC jest wykresem zależności mocy od rozmiaru testu statystycznego z obszarem odrzucenia $\{x : x > c\}$, przy zmieniającym się progu c.

Ponieważ $F(+\infty) = G(+\infty) = 1$ oraz $F(-\infty) = G(-\infty) = 0$, więc krzywa ROC$(\cdot)$ łączy wierzchołki $(0, 0)$ i $(1, 1)$ jednostkowego kwadratu (rys. 3.1).

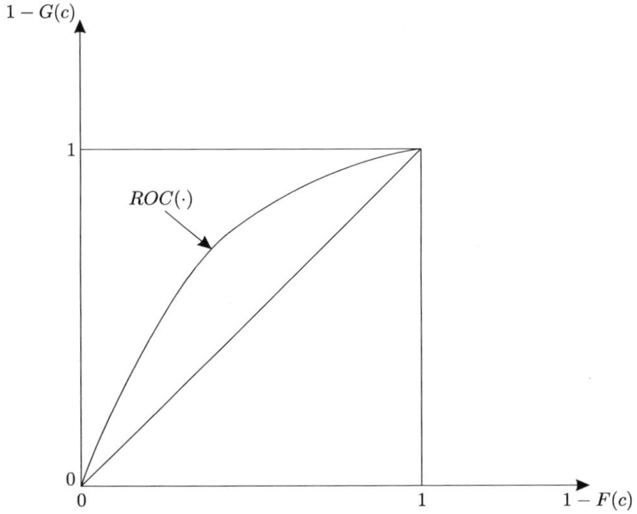

Rys. 3.1. Przykład krzywej ROC

Krzywe ROC znajdują liczne zastosowania w diagnostyce technicznej i medycznej. Na przykład w medycynie grupa o etykiecie $Y = 0$ może oznaczać grupę ludzi zdrowych (grupa kontrolna), a grupa o etykiecie $Y = 1$ grupę ludzi chorujących na określoną chorobę. Zmienną X możemy interpretować jako wartość (wynik) pewnego testu diagnostycznego, a c jako wartość progową w tym teście.

W diagnostyce medycznej prawdopodobieństwo

$$\mathrm{P}(d(X) = 1|Y = 1) = \mathrm{P}(X > c|Y = 1)$$

nosi nazwę *czułości*[2] reguły diagnostycznej i jest oznaczane przez SE(c). Jego oceną z próby uczącej jest frakcja osób chorych, u których prawidłowo rozpoznano chorobę[3]. Frakcja ta jest oznaczana przez TPF(c). Wyrażenie

[2]Po ang. sensitivity.
[3]Po ang. True Positive Fraction (TPF).

$$\mathrm{P}(d(X) = 1|Y = 0) = 1 - \mathrm{P}(X > c|Y = 0)$$

nazywa się *swoistością*[4] reguły diagnostycznej i jest oznaczane przez $\mathrm{SP}(c)$.

Oceną prawdopodobieństwa $\mathrm{P}(d(X) = 1|Y = 0) = 1 - \mathrm{P}(X > c|Y = 0)$ z próby uczącej jest frakcja osób zdrowych, u których błędnie stwierdzono występowanie choroby[5]. Frakcja ta jest oznaczana przez $\mathrm{FPF}(c)$.

Wróćmy teraz do ogólnych własności krzywych ROC. Jeżeli cecha X jest nieinformująca, tj. rozkłady prawdopodobieństwa cechy X w dwóch grupach o etykietach 0 i 1 są identyczne ($F(c) = G(c)$), to krzywa ROC ma postać $\mathrm{ROC}(t) = t$, dla $t \in [0,\ 1]$. Jest to przekątna kwadratu jednostkowego łącząca punkty $(0,\ 0)$ i $(1,\ 1)$.

Ale również dla zmiennej idealnie oddzielającej grupy oraz wybranego progu c mamy $\mathrm{P}(X > c|Y = 1) = 1$ oraz $\mathrm{P}(X > c|Y = 0) = 0$ lub $1 - G(c) = 1$ oraz $1 - F(c) = 0$, tj. krzywa ROC pokrywa się z bokami kwadratu łączącymi punkty $(0,\ 0)$ i $(0,\ 1)$ oraz $(0,\ 1)$ i $(1,\ 1)$. Dla większości cech krzywa ROC leży między krzywymi odpowiadającymi cechom nieinformującym i idealnie rozdzielającym grupy. Cechy silnie dyskryminujące grupy mają krzywe ROC leżące blisko wierzchołka $(0,\ 1)$ kwadratu.

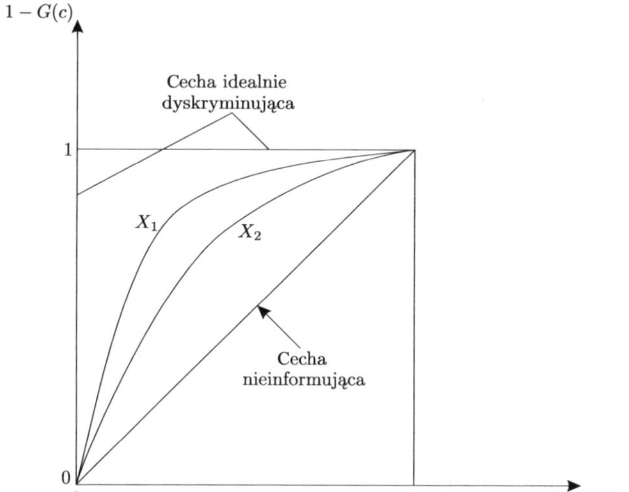

Rys. 3.2. Krzywe ROC dla dwóch cech, gdzie cecha X_1 jest cechą jednoznacznie silniej dyskryminującą od cechy X_2. Dla porównania pokazano również krzywe ROC dla cech: nieinformującej oraz idealnie dyskryminującej

[4]Po ang. specificity.

[5]Po ang. False Positive Fraction (FPF).

Na rysunku 3.2 cecha X_1 jest cechą jednoznacznie silniej dyskryminującą od cechy X_2. Pokazano również dla porównania krzywą ROC dla cechy nieinformującej i cechy idealnie dyskryminującej.

Przytoczymy pewne własności charakteryzujące krzywe ROC.

WŁASNOŚĆ 1.

Krzywe ROC są niezmiennicze względem ściśle rosnących przekształceń cechy X.

Dowód. Niech $(1 - F(c),\ 1 - G(c))$ będzie punktem na krzywej ROC dla cechy X. Niech h będzie przekształceniem ściśle rosnącym i niech $W = h(X)$ oraz $d = h(c)$. Wówczas $\mathrm{P}(W > d|Y = 0) = \mathrm{P}(X > c|Y = 0)$ i $\mathrm{P}(W > d|Y = 1) = \mathrm{P}(X > c|Y = 1)$.

Zatem ten sam punkt leży na krzywej ROC dla cechy W. Podobnie pokazuje się, że każdy punkt na krzywej ROC dla W jest również punktem na krzywej ROC dla X. ■

WŁASNOŚĆ 2.

Krzywa ROC ma następującą reprezentację:

$$\mathrm{ROC}(t) = \bar{G}(\bar{F}^{-1}(t)),\ t \in [0,\ 1] \tag{3.4}$$

lub

$$\mathrm{ROC}(t) = 1 - G(F^{-1}(1 - t)),\ t \in [0,\ 1]. \tag{3.5}$$

Dowód. Niech $t = \bar{F}(c)$. Wówczas $c(t) = \bar{F}^{-1}(t)$. Stąd dla $t \in [0,\ 1]$ otrzymujemy:

$$\mathrm{ROC}(t) = \bar{G}(c(t)) = \bar{G}(\bar{F}^{-1}(t)).$$

Podobnie oznaczając $t = 1 - F(c)$, otrzymujemy $F(c) = 1 - t$ i $c(t) = F^{-1}(1 - t)$. Stąd dla $t \in [0,\ 1]$, $\mathrm{ROC}(t) = 1 - G(c(t)) = 1 - G(F^{-1}(1 - t))$. ■

Im rozkłady zmiennych X_1 i X_0 będą bardziej zróżnicowane, tym klasyfikator będzie lepszy. Miarą zróżnicowania rozkładów w dwóch grupach jest pole pod krzywą ROC.

WŁASNOŚĆ 3.

Niech $\mathrm{AUC}^6 = \int_0^1 \mathrm{ROC}(t)dt$ *będzie polem pod krzywą ROC. Wówczas*

$$\mathrm{AUC} = \mathrm{P}(X_1 > X_0), \tag{3.6}$$

gdzie X_1 *i* X_0 *są zmiennymi losowymi niezależnymi oraz*

$$X_1 = (X|Y = 1) \sim G, \qquad X_0 = (X|Y = 0) \sim F.$$

Dowód. Mamy

$$\begin{aligned}\mathrm{AUC} = \int_0^1 \mathrm{ROC}(t)dt &= \int_0^1 \bar{G}(\bar{F}^{-1}(t))dt = \\ &= \int_{-\infty}^{\infty} \bar{G}(y)d\bar{F}(y) = \\ &= \int_{-\infty}^{\infty} \mathrm{P}(X_1 > y)f_0(y)dy,\end{aligned}$$

gdzie dokonano zamiany zmiennych $y = \bar{F}^{-1}(t)$ oraz f_0 oznacza gęstość zmiennej losowej X_0.

Stąd z niezależności X_1 i X_0 mamy

$$\mathrm{AUC} = \int_{-\infty}^{\infty} \mathrm{P}(X_1 > y,\ X_0 = y)dy = \mathrm{P}(X_1 > X_0).$$

■

W literaturze można spotkać tzw. *indeks Giniego*. Ma on ścisły związek z polem pod krzywą ROC. Weźmy pod uwagę pole zakreskowanego obszaru na rys. 3.3.

Obszar ten od góry jest ograniczony krzywą ROC, a od dołu jej symetrycznym odbiciem względem punktu $\left(\frac{1}{2}, \frac{1}{2}\right)$. Pole tego obszaru nosi nazwę indeksu Giniego. Między indeksem Giniego (WG) a polem pod krzywą ROC (AUC) zachodzi zależność:

$$\mathrm{WG} + 1 = 2\mathrm{AUC}.$$

[6] Po ang. The area under the ROC curve (AUC).

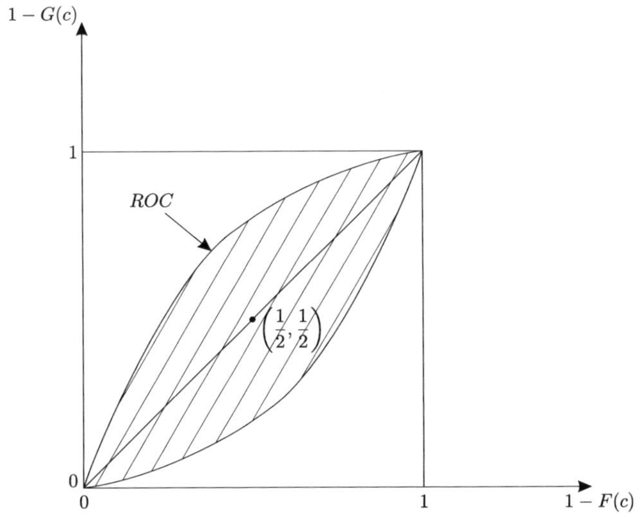

Rys. 3.3. Indeks Giniego równy polu zakreskowanego obszaru

Zajmiemy się teraz estymacją krzywej ROC z próby uczącej.

Niech $X_{i1}, X_{i2}, \ldots, X_{iN_i}$ będzie próbą uczącą z grupy o etykiecie i, gdzie $i = 0, 1$. Zakładamy, że próby te są niezależne. Estymatorem z próby krzywej ROC jest

$$\widehat{\mathrm{ROC}}(t) = 1 - \hat{G}_{N_1}(\hat{F}_{N_0}^{-1}(1-t)), \qquad t \in [0, 1], \tag{3.7}$$

gdzie $\hat{G}_{N_1}$ i $\hat{F}_{N_0}$ są dystrybuantami z próby postaci:

$$\hat{G}_{N_1}(t) = \frac{1}{N_1}\sum_{j=1}^{N_1} \mathrm{I}_{(-\infty,\,t]}(X_{1j}), \ \hat{F}_{N_0}(t) =$$

$$= \frac{1}{N_0}\sum_{j=1}^{N_0} \mathrm{I}_{(-\infty,\,t]}(X_{0j}),$$

gdzie

$$\mathrm{I}_{(-\infty,\,t]}(X_{ij}) = \begin{cases} 1, \text{ gdy } X_{ij} \le t, \\ 0, \text{ gdy } X_{ij} > t, \end{cases}$$

dla $j = 1, 2, \ldots, N_i$, $i = 0, 1$.

Estymator $\widehat{\mathrm{ROC}}(t)$ jest niemalejącą funkcją schodkową. Z własności 3 wiemy, że jeżeli X_1 i X_0 są niezależnymi zmiennymi losowymi takimi, że $X_1 = (X|Y=1) \sim G$ i $X_0 = (X|Y=0) \sim F$, to $\mathrm{AUC} = \mathrm{P}(X_1 > X_0)$.

Stąd

$$\widehat{\text{AUC}} = \frac{1}{N_0 N_1} \sum_{i=1}^{N_1} \sum_{j=1}^{N_0} \text{I}(X_{1i} > X_{0j}). \tag{3.8}$$

Wyrażenie to jest znaną *statystyką U Manna–Whitneya.*

WŁASNOŚĆ 4.

Jeżeli $X_1 \sim N(\mu_1,\ \sigma_1^2)$ *i* $X_0 \sim N(\mu_0,\ \sigma_0^2)$*, to*

$$\text{ROC}(t) = \Phi(a + b\Phi^{-1}(t)), \tag{3.9}$$

gdzie

$$a = \frac{\mu_1 - \mu_0}{\sigma_1}, \qquad b = \frac{\sigma_0}{\sigma_1} \tag{3.10}$$

oraz Φ *oznacza dystrybuantę rozkładu* $N(0,\ 1)$.

Dowód. $\text{ROC}(t) = \bar{G}(\bar{F}^{-1}(t)),\ t \in [0,\ 1]$.
Niech

$$t = \bar{F}(c) = 1 - F(c) = 1 - \text{P}(X_0 \le c) =$$
$$= 1 - \text{P}\left(\frac{X_0 - \mu_0}{\sigma_0} \le \frac{c - \mu_0}{\sigma_0}\right) = 1 - \Phi\left(\frac{c - \mu_0}{\sigma_0}\right) = \Phi\left(\frac{\mu_0 - c}{\sigma_0}\right).$$

Wówczas

$$\frac{\mu_0 - c}{\sigma_0} = \Phi^{-1}(t)$$

lub

$$c(t) = \mu_0 - \sigma_0\Phi^{-1}(t).$$

Stąd

$$\text{ROC}(t) = \bar{G}(c(t)) = 1 - G(c(t)) =$$
$$= \Phi\left(\frac{\mu_1 - c(t)}{\sigma_1}\right) = \Phi\left(\frac{\mu_1 - \mu_0 + \sigma_o\Phi^{-1}(t)}{\sigma_1}\right) =$$
$$= \Phi(a + b\Phi^{-1}(t)),$$

gdzie a i b są dane wzorem (3.10). ∎

WŁASNOŚĆ 5.

Pole pod krzywą ROC postaci (3.9) jest równe

$$\mathrm{AUC} = \Phi\left(\frac{a}{\sqrt{1+b^2}}\right), \tag{3.11}$$

gdzie a i b są dane wzorem (3.10).

Dowód. Wiemy, że $\mathrm{AUC} = \mathrm{P}(X_1 > X_0) = \mathrm{P}(X_1 - X_0 > 0)$.

Niech $W = X_1 - X_0$. Wówczas $W \sim N(\mu_1 - \mu_0,\ \sigma_1^2 + \sigma_0^2)$.

Stąd

$$\mathrm{P}(W > 0) = 1 - \Phi\left(\frac{\mu_0 - \mu_1}{\sqrt{\sigma_1^2 + \sigma_0^2}}\right) = \Phi\left(\frac{\mu_1 - \mu_0}{\sqrt{\sigma_1^2 + \sigma_0^2}}\right) =$$

$$= \Phi\left(\frac{\mu_1 - \mu_0}{\sigma_1} \Big/ \sqrt{1 + \frac{\sigma_0^2}{\sigma_1^2}}\right) = \Phi\left(\frac{a}{\sqrt{1+b^2}}\right).$$

■

Widzimy, że AUC jest rosnącą funkcją a i malejącą funkcją b.

Wiemy, że krzywa ROC jest niezmiennicza ze względu na ściśle rosnące przekształcenie zmiennej X. Stąd, jeżeli X_1 i X_0 mają rozkłady normalne i zmienne te przekształcimy w nowe zmienne $W_1 = h(X_1)$ i $W_0 = h(X_0)$, gdzie h jest funkcją ściśle rosnącą, to krzywa ROC dla W_1 i W_0 ma znowu postać (3.9) taką, jak dla X_1 i X_0.

WŁASNOŚĆ 6.

Jeżeli krzywa ROC dla X_1 i X_0 ma postać (3.9) i zdefiniujemy

$$h(x) = -\Phi^{-1}(\bar{F}(x)),$$

gdzie $\bar{F}$ jest funkcją ciągłą przeżycia dla zmiennej X_0, to

$$h(X_0) \sim N(0,\ 1), \qquad h(X_1) \sim N\left(\frac{a}{b},\ \frac{1}{b^2}\right).$$

Dowód. Mamy

$$\mathrm{P}(h(X_0) < x) = \mathrm{P}[\bar{F}(X_0) > \Phi(-x)] = \mathrm{P}[F(X_0) < \Phi(x)] = \Phi(x),$$

ponieważ $F(X_0) \equiv 1 - \bar{F}(X_0)$ ma rozkład jednostajny.

Zatem

$$\begin{aligned}P[h(X_1) > x] &= P[X_1 > \bar{F}(\Phi(-x))] = \bar{F}\bar{F}^{-1}(\Phi(-x)) = \\ &= \mathrm{ROC}(\Phi(-x)) = \Phi(a - bx) = 1 - \Phi\left(\frac{a - a/b}{1/b}\right).\end{aligned}$$

Stąd $h(X_0) \sim N(0,\ 1)$ i $h(X_1) \sim N\left(\frac{a}{b},\ \frac{1}{b^2}\right)$. ▒

Z własności 6 wynika bardzo istotna informacja: zmienna X może nie mieć rozkładu normalnego, ale wnioskowanie oparte na krzywej ROC zbudowanej dla rozkładu normalnego może dawać dobre wyniki, ponieważ dla krzywych ROC nie mają znaczenia poszczególne rozkłady, lecz związki między tymi rozkładami.

Uogólnimy teraz własności 4 i 5 na przypadek wielowymiarowy. Zamiast obserwowanej zmiennej losowej X weźmiemy pod uwagę wektor losowy $\boldsymbol{X} = (X_1,\ X_2,\ \ldots,\ X_p)'$ i założymy, że wektor ten w grupie o etykiecie 1 ma p-wymiarowy rozkład normalny $N_p(\boldsymbol{\mu}_1,\ \boldsymbol{\Sigma}_1)$ oraz w grupie o etykiecie 0 ma p-wymiarowy rozkład normalny $N_p(\boldsymbol{\mu}_0,\ \boldsymbol{\Sigma}_0)$, gdzie $\boldsymbol{\Sigma}_1$ i $\boldsymbol{\Sigma}_0$ są macierzami kowariancji określonymi dodatnio ($\boldsymbol{\Sigma}_1 > 0,\ \boldsymbol{\Sigma}_0 > 0$).

Weźmiemy pod uwagę regułę klasyfikacyjną opartą na kombinacji liniowej

$$U(\boldsymbol{a}) = \boldsymbol{a}'\boldsymbol{X} = a_1X_1 + a_2X_2 + \cdots + a_pX_p$$

postaci

$$d(\boldsymbol{x}) = \begin{cases} 1, \text{ jeżeli } u(\boldsymbol{a}) = \boldsymbol{a}'\boldsymbol{x} > c, \\ 0, \text{ jeżeli } u(\boldsymbol{a}) = \boldsymbol{a}'\boldsymbol{x} \leq c, \end{cases} \tag{3.12}$$

gdzie $-\infty \leq c \leq +\infty$ jest wartością progową.

Niech U_1 i U_0 będą niezależnymi zmiennymi losowymi reprezentującymi kombinację liniową $U(\boldsymbol{a}) = \boldsymbol{a}'\boldsymbol{X}$ w grupie o etykietach 1 i 0. Wtedy

$$U_1(\boldsymbol{a}) \sim N(\boldsymbol{a}'\boldsymbol{\mu}_1,\ \boldsymbol{a}'\boldsymbol{\Sigma}_1\boldsymbol{a}) \quad \text{i} \quad U_0(\boldsymbol{a}) \sim N(\boldsymbol{a}'\boldsymbol{\mu}_0,\ \boldsymbol{a}'\boldsymbol{\Sigma}_0\boldsymbol{a}).$$

Wówczas zgodnie z własnościami 4 i 5, krzywa ROC dla zmiennych U_1 i U_0 ma postać (3.9), a pole pod nią jest równe (3.11), gdzie

$$a = \frac{\boldsymbol{a}'\boldsymbol{\mu}_1 - \boldsymbol{a}'\boldsymbol{\mu}_0}{\boldsymbol{a}'\boldsymbol{\Sigma}_1\boldsymbol{a}}, \qquad b = \frac{\boldsymbol{a}'\boldsymbol{\Sigma}_0\boldsymbol{a}}{\boldsymbol{a}'\boldsymbol{\Sigma}_1\boldsymbol{a}}. \tag{3.13}$$

WŁASNOŚĆ 7.

Wektor $\boldsymbol{a}$, *dla którego pole* $\mathrm{AUC}(\boldsymbol{a})$ *pod krzywą ROC jest maksymalne ma postać*

$$\boldsymbol{a} = (\boldsymbol{\Sigma}_1 + \boldsymbol{\Sigma}_0)^{-1}\boldsymbol{\delta}, \tag{3.14}$$

gdzie

$$\boldsymbol{\delta} = \boldsymbol{\mu}_1 - \boldsymbol{\mu}_0. \tag{3.15}$$

Dowód. Mamy

$$\mathrm{AUC}(\boldsymbol{a}) = \Phi\left(\frac{\boldsymbol{a}'\boldsymbol{\delta}}{(\boldsymbol{a}'\boldsymbol{\Sigma}_1\boldsymbol{a} + \boldsymbol{a}\boldsymbol{\Sigma}_0\boldsymbol{a})^{\frac{1}{2}}}\right).$$

Różniczkując dystrybuantę Φ względem wektora $\boldsymbol{a}$ i przyrównując pochodną do wektora zerowego, otrzymujemy:

$$\frac{\partial \mathrm{AUC}(\boldsymbol{a})}{\partial \boldsymbol{a}} = \frac{1}{\sqrt{2\pi}} \exp\left(-\frac{1}{2}\frac{(\boldsymbol{a}'\boldsymbol{\delta})^2}{\boldsymbol{a}'\boldsymbol{\Sigma}_1\boldsymbol{a} + \boldsymbol{a}\boldsymbol{\Sigma}_0\boldsymbol{a}}\right) \times$$

$$\times \frac{\boldsymbol{\delta}(\boldsymbol{a}'\boldsymbol{\Sigma}_0\boldsymbol{a} + \boldsymbol{a}'\boldsymbol{\Sigma}_1\boldsymbol{a})^{\frac{1}{2}} - \frac{1}{2}\boldsymbol{a}'\boldsymbol{\delta}(\boldsymbol{a}'\boldsymbol{\Sigma}_0\boldsymbol{a} + \boldsymbol{a}'\boldsymbol{\Sigma}_1\boldsymbol{a})^{-\frac{1}{2}}(2\boldsymbol{\Sigma}_0\boldsymbol{a} + 2\boldsymbol{\Sigma}_1\boldsymbol{a})}{\boldsymbol{a}'\boldsymbol{\Sigma}_0\boldsymbol{a} + \boldsymbol{a}'\boldsymbol{\Sigma}_1\boldsymbol{a}} = \mathbf{0}$$

lub

$$\boldsymbol{\delta}(\boldsymbol{a}'\boldsymbol{\Sigma}_0\boldsymbol{a} + \boldsymbol{a}'\boldsymbol{\Sigma}_1\boldsymbol{a}) - \boldsymbol{a}'\boldsymbol{\delta}(\boldsymbol{\Sigma}_0 + \boldsymbol{\Sigma}_1)\boldsymbol{a} = 0,$$

lub

$$\boldsymbol{a}'\boldsymbol{\delta}(\boldsymbol{\Sigma}_0 + \boldsymbol{\Sigma}_1)\boldsymbol{a} = (\boldsymbol{a}'\boldsymbol{\Sigma}_0\boldsymbol{a} + \boldsymbol{a}'\boldsymbol{\Sigma}_1\boldsymbol{a})\boldsymbol{\delta},$$

lub

$$(\boldsymbol{\Sigma}_0 + \boldsymbol{\Sigma}_1)t\boldsymbol{a} = \boldsymbol{\delta},$$

gdzie

$$t = \frac{\boldsymbol{a}'\boldsymbol{\delta}}{\boldsymbol{a}'\boldsymbol{\Sigma}_0\boldsymbol{a} + \boldsymbol{a}'\boldsymbol{\Sigma}_1\boldsymbol{a}},$$

lub

$$(\boldsymbol{\Sigma}_0 + \boldsymbol{\Sigma}_1)\boldsymbol{a}^\star = \boldsymbol{\delta},$$

gdzie $\boldsymbol{a}^\star = t\boldsymbol{a}$.

Ponieważ $\boldsymbol{\Sigma}_0 > 0$ i $\boldsymbol{\Sigma}_1 > 0$, więc $\boldsymbol{\Sigma}_1 + \boldsymbol{\Sigma}_0 > 0$ i stąd

$$\boldsymbol{a}^\star = (\boldsymbol{\Sigma}_0 + \boldsymbol{\Sigma}_1)^{-1}\boldsymbol{\delta}.$$

Ale pole AUC($\boldsymbol{a}$) jest niezmiennicze względem mnożenia wektora $\boldsymbol{a}$ przez skalar t. Stąd otrzymujemy wzór (3.14). W celu pokazania, że wektor $\boldsymbol{a}$ postaci (3.14) istotnie maksymalizuje pole AUC($\boldsymbol{a}$) wystarczy pokazać, że

$$\Phi((\boldsymbol{\delta}'(\boldsymbol{\Sigma}_0 + \boldsymbol{\Sigma}_1)^{-1}\boldsymbol{\delta})^{\frac{1}{2}}) - \Phi\left(\frac{\boldsymbol{a}'\boldsymbol{\delta}}{(\boldsymbol{a}'\boldsymbol{\Sigma}_0\boldsymbol{a} + \boldsymbol{a}\boldsymbol{\Sigma}_1\boldsymbol{a})^{\frac{1}{2}}}\right) \geq 0,$$

dla każdego wektora $\boldsymbol{a} \in \mathbb{R}^p$, gdzie $\Phi\left((\boldsymbol{\delta}'(\boldsymbol{\Sigma}_0 + \boldsymbol{\Sigma}_1)^{-1}\boldsymbol{\delta})^{\frac{1}{2}}\right)$ jest maksymalnym polem pod krzywą ROC uzyskanym dla $\boldsymbol{a}$ postaci (3.14).

Ponieważ Φ jest funkcją ściśle rosnącą, więc wystarczy pokazać, że

$$(\boldsymbol{\delta}'(\boldsymbol{\Sigma}_0 + \boldsymbol{\Sigma}_1)^{-1}\boldsymbol{\delta})^{\frac{1}{2}} - \frac{\boldsymbol{a}'\boldsymbol{\delta}}{(\boldsymbol{a}'\boldsymbol{\Sigma}_0\boldsymbol{a} + \boldsymbol{a}\boldsymbol{\Sigma}_1\boldsymbol{a})^{\frac{1}{2}}} \geq 0,$$

dla każdego $\boldsymbol{a} \in \mathbb{R}^p$.

Tę ostatnią nierówność możemy równoważnie napisać w postaci:

$$\boldsymbol{a}'\boldsymbol{\delta} \leq (\boldsymbol{a}'\boldsymbol{\Sigma}_0\boldsymbol{a} + \boldsymbol{a}'\boldsymbol{\Sigma}_1\boldsymbol{a})^{\frac{1}{2}}(\boldsymbol{\delta}'(\boldsymbol{\Sigma}_0 + \boldsymbol{\Sigma}_1)^{-1}\boldsymbol{\delta})^{\frac{1}{2}},$$

dla każdego $\boldsymbol{a} \in \mathbb{R}^p$ lub

$$(\boldsymbol{a}'\boldsymbol{\delta})^2 \leq (\boldsymbol{a}'\boldsymbol{\Sigma}_0\boldsymbol{a} + \boldsymbol{a}'\boldsymbol{\Sigma}_1\boldsymbol{a})(\boldsymbol{\delta}'(\boldsymbol{\Sigma}_0 + \boldsymbol{\Sigma}_1)^{-1}\boldsymbol{\delta}),$$

dla każdego $\boldsymbol{a} \in \mathbb{R}^p$.

Ponieważ macierz $\boldsymbol{\Sigma}_0 + \boldsymbol{\Sigma}_1$ jest macierzą symetryczną, dodatnio określoną, to istnieje nieosobliwa macierz $\boldsymbol{C}$ taka, że

$$\boldsymbol{\Sigma}_0 + \boldsymbol{\Sigma}_1 = \boldsymbol{C}'\boldsymbol{C}.$$

Stąd

$$(\boldsymbol{a}'\boldsymbol{\delta})^2 \leq (\boldsymbol{a}'\boldsymbol{C}'\boldsymbol{C}\boldsymbol{a})(\boldsymbol{\delta}'(\boldsymbol{C}'\boldsymbol{C})^{-1}\boldsymbol{\delta}),$$

dla każdego $\boldsymbol{a} \in \mathbb{R}^p$.

Oznaczając $\boldsymbol{u} = \boldsymbol{C}\boldsymbol{a}$, $\boldsymbol{v} = (\boldsymbol{C}')^{-1}\boldsymbol{\delta}$, otrzymujemy

$$(\boldsymbol{u}'\boldsymbol{v})^2 \leq (\boldsymbol{u}'\boldsymbol{u})(\boldsymbol{v}'\boldsymbol{v}),$$

dla każdego $\boldsymbol{u}$, $\boldsymbol{v} \in \Re^p$.

Ostatnia nierówność jest dobrze znaną nierównością Cauchy'ego–Schwartza. To kończy dowód własności 7. ∎

Zajmiemy się teraz dopuszczalnością naszej reguły klasyfikacyjnej. Prawdopodobieństwa błędnych zaklasyfikowań oparte na regule (3.12) są równe:

$$\mathrm{P}(d(\boldsymbol{X}) = 1|Y = 0) = \mathrm{P}(U > c|Y = 0) = 1 - F(c) = \\ = 1 - \Phi\left(\frac{c - \boldsymbol{a}'\boldsymbol{\mu}_0}{(\boldsymbol{a}'\boldsymbol{\Sigma}_0\boldsymbol{a})^{\frac{1}{2}}}\right)$$

oraz

$$\mathrm{P}(d(\boldsymbol{X}) = 0|Y = 1) = \mathrm{P}(U \leq c|Y = 1) = G(c) = \Phi\left(\frac{c - \boldsymbol{a}'\boldsymbol{\mu}_1}{(\boldsymbol{a}'\boldsymbol{\Sigma}_1\boldsymbol{a})^{\frac{1}{2}}}\right).$$

Przypomnijmy (patrz rozważania poprzedzające tw. 1.6), że reguła klasyfikacyjna z progiem c_1 jest lepsza od reguły z progiem c_2, jeżeli prawdopodobieństwa błędnych zaklasyfikowań spełniają nierówności:

$$1 - F(c_1) \leq 1 - F(c_2) \quad \text{i} \quad G(c_1) \leq G(c_2)$$

oraz co najmniej jedna z nich jest ostra.

Drugą nierówność możemy równoważnie zapisać w postaci

$$1 - G(c_1) \geq 1 - G(c_2).$$

Zauważmy, że punkty $(1 - F(c),\ 1 - G(c))$ przy zmieniającym się c definiują krzywą ROC. Przypomnijmy, że reguła klasyfikacyjna z progiem c jest dopuszczalna, jeżeli nie istnieje reguła od niej lepsza.

WŁASNOŚĆ 8.

Reguła klasyfikacyjna (3.12), gdzie wektor $\boldsymbol{a}$ *jest dany wzorem (3.14) oraz*

$$c = \boldsymbol{a}'\boldsymbol{\mu}_0 + \boldsymbol{a}'\boldsymbol{\Sigma}_0\boldsymbol{a} = \boldsymbol{a}'\boldsymbol{\mu}_1 - \boldsymbol{a}'\boldsymbol{\Sigma}_1\boldsymbol{a} \tag{3.16}$$

jest regułą dopuszczalną w klasie reguł liniowych.

Dowód. Wystarczy skorzystać z wyniku Andersona i Bahadura (1962) przytoczonego w tw. 1.6. W naszym przypadku $t_0 = t_1 = 1$ i ponieważ $\boldsymbol{\Sigma}_0 > 0$ i $\boldsymbol{\Sigma}_1 > 0$, to $\boldsymbol{\Sigma}_1 + \boldsymbol{\Sigma}_0 > 0$. Zatem nasza reguła klasyfikacyjna jest dopuszczalna w klasie reguł liniowych. ∎

Wektor $\boldsymbol{a}$ jest zadany równością

$$\boldsymbol{a} = (\boldsymbol{\Sigma}_0 + \boldsymbol{\Sigma}_1)^{-1}\boldsymbol{\delta}$$

lub

$$(\boldsymbol{\Sigma}_0 + \boldsymbol{\Sigma}_1)\boldsymbol{a} = \boldsymbol{\delta},$$

lub

$$\boldsymbol{a}'(\boldsymbol{\Sigma}_0 + \boldsymbol{\Sigma}_1)\boldsymbol{a} = \boldsymbol{a}'\boldsymbol{\delta}.$$

Jeżeli $\boldsymbol{\Sigma}_0 + \boldsymbol{\Sigma}_1 > 0$, to reguła klasyfikacyjna jest dopuszczalna. Macierz $\boldsymbol{\Sigma}_0 + \boldsymbol{\Sigma}_1$ jest dodatnio określona wówczas, gdy dla każdego wektora $\boldsymbol{a} \neq \boldsymbol{0}$

$$\boldsymbol{a}'(\boldsymbol{\Sigma}_0 + \boldsymbol{\Sigma}_1)\boldsymbol{a} > 0.$$

Stąd $\boldsymbol{a}'\boldsymbol{\delta} \neq 0$, dla każdego $\boldsymbol{a} \neq \boldsymbol{0}$ lub $\boldsymbol{a}'\boldsymbol{\mu}_0 \neq \boldsymbol{a}\boldsymbol{\mu}_1$ lub $\boldsymbol{\mu}_0 \neq \boldsymbol{\mu}_1$. Oznacza to, że w każdej regule dopuszczalnej $\boldsymbol{\mu}_0 \neq \boldsymbol{\mu}_1$.

Prawdopodobieństwa błędnych zaklasyfikowań w liniowej regule dopuszczalnej są równe:

$$\mathrm{P}(U > c|Y = 0) = 1 - \Phi(\sqrt{\boldsymbol{a}'\boldsymbol{\Sigma}_0\boldsymbol{a}})$$

oraz

$$\mathrm{P}(U \leq c|Y = 1) = \Phi\left(-\sqrt{\boldsymbol{a}'\boldsymbol{\Sigma}_1\boldsymbol{a}}\right) = 1 - \Phi\left(\sqrt{\boldsymbol{a}'\boldsymbol{\Sigma}_1\boldsymbol{a}}\right).$$

Zauważmy ponadto, że jeżeli $\boldsymbol{\Sigma}_0 = \boldsymbol{\Sigma}_1 = \boldsymbol{\Sigma}$, to

$$\boldsymbol{a} = \frac{1}{2}\boldsymbol{\Sigma}^{-1}\boldsymbol{\delta},$$

$$c = \boldsymbol{a}'\boldsymbol{\mu}_0 + \boldsymbol{a}'\boldsymbol{\Sigma}\boldsymbol{a} = \boldsymbol{a}'\boldsymbol{\mu}_1 - \boldsymbol{a}'\boldsymbol{\Sigma}\boldsymbol{a}$$

oraz

$$\begin{aligned}\boldsymbol{a}'\boldsymbol{x} - c = \boldsymbol{a}'(\boldsymbol{x} - \boldsymbol{\mu}_0 - \boldsymbol{\Sigma}\boldsymbol{a}) &= \frac{1}{2}\boldsymbol{\delta}'\boldsymbol{\Sigma}^{-1}\left(\boldsymbol{x} - \boldsymbol{\mu}_0 - \frac{1}{2}\boldsymbol{\Sigma}\boldsymbol{\Sigma}^{-1}\boldsymbol{\delta}\right) = \\ &= \frac{1}{2}\boldsymbol{\delta}'\boldsymbol{\Sigma}^{-1}\left(\boldsymbol{x} - \boldsymbol{\mu}_0 - \frac{1}{2}(\boldsymbol{\mu}_1 - \boldsymbol{\mu}_0)\right) = \\ &= \frac{1}{2}(\boldsymbol{\mu}_1 - \boldsymbol{\mu}_0)'\boldsymbol{\Sigma}^{-1}\left[\boldsymbol{x} - \frac{1}{2}(\boldsymbol{\mu}_0 + \boldsymbol{\mu}_1)\right].\end{aligned}$$

Przyrównując to wyrażenie do zera, otrzymamy równanie hiperpłaszczyzny rozdzielającej grupy o etykietach zero i jeden. Identyczny wynik uzyskuje się w przypadku jednakowych prawdopodobieństw a priori, przyrównując liniową funkcję dyskryminacyjną (1.23) do zera.

Chcąc w praktyce posłużyć się regułą klasyfikacyjną (3.12), gdzie wektor $\boldsymbol{a}$ jest dany wzorem (3.14), natomiast skalar c jest dany wzorem (3.16), musimy nieznane parametry $\boldsymbol{\mu}_0$, $\boldsymbol{\mu}_1$, $\boldsymbol{\Sigma}_0$ i $\boldsymbol{\Sigma}_1$ zastąpić ich ocenami z próby uczącej.

Zajmiemy się teraz porównaniem jakości klasyfikatorów za pomocą krzywych ROC.

Pola pod krzywymi ROC konstruowanymi dla różnych klasyfikatorów wykorzystuje się do porównania między sobą tych klasyfikatorów. Im większe pole, tym klasyfikator lepszy. Ustalmy klasyfikator oraz $t \in (0, 1)$. Dla każdej obserwacji z próby testowej obliczamy prawdopodobieństwo a posteriori przynależności tego elementu do grupy 1. Jeżeli prawdopodobieństwo to jest większe od t, to element klasyfikujemy do grupy 1, w przeciwnym razie do grupy 0. Dla ustalonego $t \in (0, 1)$ otrzymujemy macierz błędów postaci

Tab. 3.1. Macierz błędów

Wskazanie	Prawdziwa grupa	
	0	1
0	a	c
1	b	d
Razem	$a+b$	$c+d$

Z macierzy błędów obliczamy oceny prawdopodobieństw

$$\alpha = \hat{\mathrm{P}}(d(\boldsymbol{X}) = 1 | Y = 0) = \frac{b}{a+b} \quad \text{i} \quad \beta = \hat{\mathrm{P}}(d(\boldsymbol{X}) = 1 | Y = 1) = \frac{d}{c+d}.$$

Dla danego $t \in (0,\ 1)$ otrzymujemy punkt $(\alpha(t),\ \beta(t))$ na krzywej ROC. Następnie zmieniamy wartość t (wartości t ustalamy z zadanym krokiem, np. 0.01) i wszystkie czynności powtarzamy. Otrzymamy w ten sposób empiryczną krzywą ROC. Pole pod krzywą ROC obliczamy metodą trapezów.

Niech $\Delta\alpha = \alpha_i - \alpha_{i-1}$ oraz $\Delta\beta = \beta_i - \beta_{i-1}$. Wówczas

$$\mathrm{AUC} = \sum_i \left(\beta_{i-1} \Delta\alpha + \frac{1}{2} \Delta\beta \cdot \Delta\alpha \right). \tag{3.17}$$

Metoda trapezów systematycznie niedoestymowuje pole pod krzywą ROC. Pole to możemy również uzyskać, obliczając wartość statystyki U Manna–Whitneya danej wzorem (3.8).

Do porównywania metod klasyfikacji przez porównanie pól pod krzywymi ROC służy następująca procedura testowa opisana przez Bradleya (1997). Rozważmy następujący układ hipotez:

$$H_0 : AUC_1 = AUC_2$$
$$H_1 : AUC_1 \neq AUC_2.$$

Do jego sprawdzenia służy następująca statystyka testowa:

$$z = \frac{\widehat{AUC_1} - \widehat{AUC_2}}{\sqrt{SE^2(\widehat{AUC_1}) + SE^2(\widehat{AUC_2})}}, \tag{3.18}$$

gdzie

$$SE(\widehat{AUC}) = \sqrt{\frac{\theta(1-\theta) + (n_1 - 1)(Q_1 - \theta^2) + (n_2 - 1)(Q_2 - \theta^2)}{n_1 n_2}},$$

$$Q_1 = \frac{\theta}{2-\theta},$$
$$Q_2 = \frac{2\theta_2}{1+\theta}$$

oraz θ jest prawdziwą wartością pola pod krzywą ROC (w praktyce jest używany estymator), n_1, n_2 są liczebnościami odpowiednio pierwszej i drugiej klasy. Statystyka (3.18) ma rozkład $N(0,\ 1)$. Hipotezę zerową odrzucamy wówczas, gdy $|z| > z_\alpha$.

Przykład 3.1.
Wygenerowano po 50 obserwacji do zbioru uczącego oraz po 100 do zbioru testowego z każdej z dwóch klas, z których obie mają rozkłady normalne z następującymi parametrami:

$$\boldsymbol{\mu}_1 = \boldsymbol{\mu}_0 = \mathbf{0}, \qquad \boldsymbol{\Sigma}_1 = \mathbf{I}_8, \qquad \boldsymbol{\Sigma}_0 = 4\mathbf{I}_8.$$

Zbiór ten został zaczerpnięty z książki Fukunagi (1990) i jest oznaczany $I-4I$. Na rysunku 3.4 są pokazane krzywe ROC dla czterech metod łączonych opisanych w rozdz. 10.

W tabeli 3.2 znajdują się poziomy krytyczne (p-wartości) omawianego testu.

Hipotezę o równości pól pod krzywą ROC odrzucamy wówczas, gdy $\alpha > p$. Zauważmy, że na poziomie istotności $\alpha = 5\%$ okazuje się, że pola pod krzywymi ROC dla metody sekwencyjnej, sumacyjnej oraz regresji stosowej różnią się istotnie od pola dla metody iloczynowej.

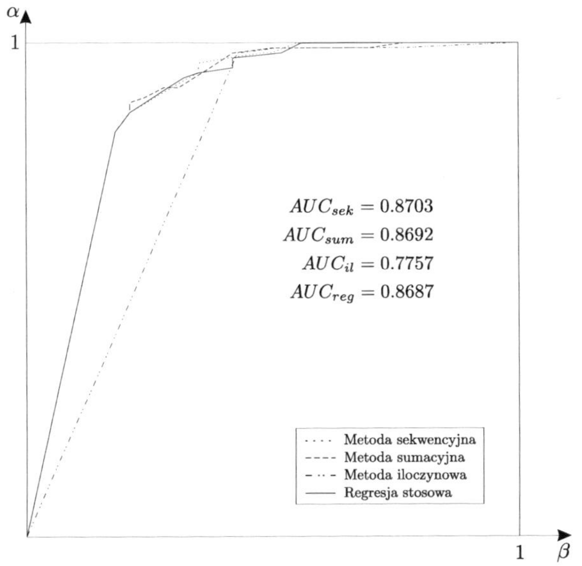

Rys. 3.4. Krzywa ROC dla zbioru $I - 4I$

Tab. 3.2. p-wartości dla testu równości AUC

	$I - 4I$
$\widehat{AUC}_{sek} = \widehat{AUC}_{sum}$	0.9748
$\widehat{AUC}_{sek} = \widehat{AUC}_{il}$	**0.0235**
$\widehat{AUC}_{sek} = \widehat{AUC}_{reg}$	0.9650
$\widehat{AUC}_{sum} = \widehat{AUC}_{il}$	**0.0255**
$\widehat{AUC}_{sum} = \widehat{AUC}_{reg}$	0.9902
$\widehat{AUC}_{il} = \widehat{AUC}_{reg}$	**0.0263**

Na zakończenie zauważmy jeszcze, że krzywe ROC są konstruowane dla klasyfikatorów binarnych. W przypadku liczby grup większej od dwóch Hand i Till (2001) zaproponowali porównywanie jakości klasyfikatorów wielogrupowych za pomocą nowego wskaźnika M.

Załóżmy, że $c > 2$ grup ma etykiety $0, 1, 2, \ldots, c-1$. Dany klasyfikator daje nam oceny $\hat{p}(i|\boldsymbol{x})$ prawdopodobieństw a posteriori przynależności każdej obserwacji testowej $\boldsymbol{x}$ do każdej z c grup, dla $i = 0, 1, \ldots, c-1$. Dla każdej pary (i, j) grup możemy obliczyć ocenę $\hat{A}(i|j)$ prawdopodobieństwa, że losowo wybrany reprezentant grupy j będzie miał mniejsze prawdopodobieństwo przynależności do grupy i, niż losowo wybrany reprezentant grupy i. W przypadku dwóch grup $\hat{A}(0|1) = \hat{A}(1|0)$.

W przypadku $c > 2$, $\hat{A}(i|j) \neq \hat{A}(j|i)$. Niech

$$\hat{A}(i,\ j) = \frac{1}{2}(\hat{A}(i|j) + \hat{A}(j|i))$$

będzie miarą rozdzielczości grup i oraz j.

Miarą dobroci klasyfikatora wielogrupowego jest wówczas miara postaci

$$M = \frac{2}{c(c-1)} \sum_{i<j} \hat{A}(i,\ j). \tag{3.19}$$

4

Metoda wektorów nośnych

Rozwój metody wektorów nośnych[1] (SVM) przebiegał niejako w kierunku odwrotnym w stosunku do rozwoju sieci neuronowych. Te ostatnie przeszły drogę od zastosowań i wyczerpujących obliczeniowo eksperymentów, aż po teorię. Metoda wektorów nośnych wypłynęła natomiast z rozważań teoretycznych, przez implementację, aż po eksperymenty. Interesującym jest fakt, iż silne podwaliny teoretyczne nie od początku szły w parze z uznaniem dla SVM. Doniosłość pierwszych prac autorstwa Vapnika i Chervonenkisa (1971, 1974) była niedostrzeżona aż do lat dziewięćdziesiątych XX wieku. Obecnie wiadomo, że metoda Vapnika wektorów nośnych realizująca zasadę SRM[2], daje w wielu zagadnieniach najlepsze rezultaty (Kecman, 2005).

4.1. Model liniowy

Rozpatrzmy najpierw zagadnienie dyskryminacji dwóch populacji liniowo separowalnych $\mathcal{G}_i$ $(i = 1,\ 2)$, to jest takich, których obserwacje można rozgraniczyć za pomocą klasyfikatora liniowego.

Niech $\mathcal{L}_n = \{(\boldsymbol{x}_1,\ y_1),\ \ldots,\ (\boldsymbol{x}_n,\ y_n)\}$ będzie n-elementową próbą uczącą, gdzie $\boldsymbol{x}_j$ jest wektorem obserwowanych cech, a y_j etykietą klasy $(j = 1,\ 2,\ \ldots,\ n)$.

Przyjmijmy, że

$$y_j = \begin{cases} +1, & \text{jeżeli} \quad \boldsymbol{x}_j \in \mathcal{G}_1 \\ -1, & \text{jeżeli} \quad \boldsymbol{x}_j \in \mathcal{G}_2 \end{cases}, \qquad j = 1,\ 2,\ \ldots,\ n.$$

[1]Po ang. Support Vector Machine (SVM).
[2]Po ang. Structural Risk Minimization Principle (SRM).

Niech

$$g(\boldsymbol{x}) = \boldsymbol{w}'\boldsymbol{x} + w_0.$$

Ponadto oznaczmy przez $\hat{d}(\boldsymbol{x})$ klasyfikator liniowy postaci:

$$\hat{d}(\boldsymbol{x}) = \begin{cases} +1, & \text{jeżeli} \quad g(\boldsymbol{x}) > 0, \\ -1, & \text{jeżeli} \quad g(\boldsymbol{x}) \leq 0. \end{cases}$$

Stąd każdy element próby zostanie poprawnie zaklasyfikowany, jeżeli będzie spełniony warunek:

$$y_j(\boldsymbol{w}'\boldsymbol{x_j} + w_0) > 0, \qquad j = 1,\ 2,\ \ldots,\ n. \tag{4.1}$$

Ideą metody wektorów nośnych jest obranie tej liniowej reguły klasyfikacyjnej spośród spełniających warunek (4.1), dla której odpowiednia hiperpłaszczyzna $g(\boldsymbol{x}) = 0$ jest maksymalnie odległa od najbliższej jej obserwacji pochodzącej z próby uczącej $\mathcal{L}_n$. Innymi słowy – poszukujemy hiperpłaszczyzny $g(\boldsymbol{x}) = 0$ maksymalizującej odległość ρ (rys. 4.1)[3].

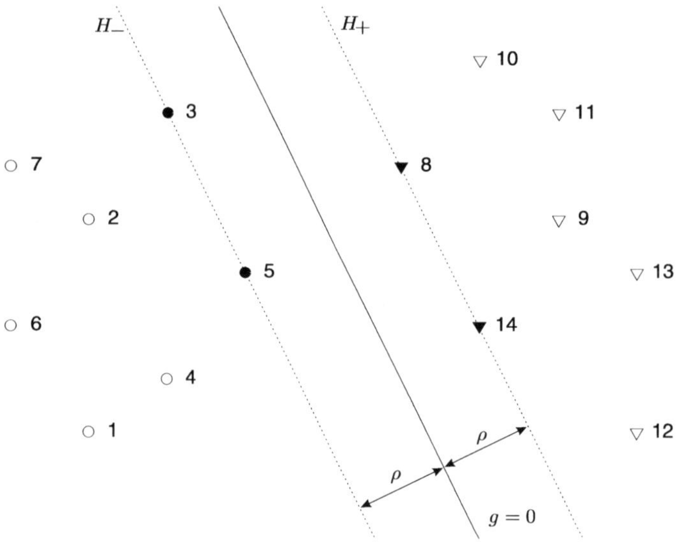

Rys. 4.1. Optymalna hiperpłaszczyzna oddzielająca dwie populacje liniowo separowalne

[3]Stąd w literaturze anglojęzycznej obok nazwy *Support Vector Machine* równolegle używa się innego określenia opisywanej tu metody: *Maximal Margin Classifier*.

Wprowadzając w (4.1) stałą $b > 0$ taką, że:

$$y_j(\boldsymbol{w}'\boldsymbol{x_j} + w_0) \geq b, \qquad j = 1,\ 2,\ \ldots,\ n,$$

otrzymujemy

$$\rho = \frac{b}{||\boldsymbol{w}||}.$$

Bez straty ogólności możemy przyjąć $b = 1$ (przy jednoczesnym przeskalowaniu wektora wag $\boldsymbol{w}$ i stałej w_0). Hiperpłaszczyzny o równaniach:

$$H+\colon \boldsymbol{w}'\boldsymbol{x} + w_0 = +1$$

$$H-\colon \boldsymbol{w}'\boldsymbol{x} + w_0 = -1$$

nazywamy *hiperpłaszczyznami kanonicznymi* (rys. 4.1).
Wobec powyższego, zagadnienie maksymalizacji odległości ρ jest równoważne minimalizacji wielkości $||\boldsymbol{w}||$ przy warunku pobocznym:

$$y_j(\boldsymbol{w}'\boldsymbol{x_j} + w_0) \geq 1, \qquad j = 1,\ 2,\ \ldots,\ n. \tag{4.2}$$

Stosując metodę mnożników Lagrange'a, mamy:

$$L = L(\boldsymbol{w},\ w_0; \boldsymbol{\alpha}) = \frac{1}{2}\boldsymbol{w}'\boldsymbol{w} - \sum_{j=1}^{n} \alpha_j [y_j(\boldsymbol{w}'\boldsymbol{x_j} + w_0) - 1], \tag{4.3}$$

gdzie $\boldsymbol{\alpha} = (\alpha_1,\ \alpha_2,\ \ldots,\ \alpha_n)' \geq 0$.
Różniczkując funkcję Lagrange'a L kolejno względem $\boldsymbol{w}$ oraz w_0 i przyrównując pochodne cząstkowe do zera, otrzymujemy odpowiednio równości:

$$\boldsymbol{w} = \sum_{j=1}^{n} \alpha_j y_j \boldsymbol{x}_j, \tag{4.4}$$

$$\sum_{j=1}^{n} \alpha_j y_j = 0, \qquad \boldsymbol{\alpha} \geq 0. \tag{4.5}$$

Podstawiając równości (4.4) i (4.5) do (4.3), otrzymujemy tzw. formę dualną funkcji L:

$$L_D = L_D(\boldsymbol{\alpha}) = \sum_{j=1}^{n} \alpha_j - \frac{1}{2} \sum_{j=1}^{n} \sum_{k=1}^{n} \alpha_j \alpha_k y_j y_k \boldsymbol{x}_j' \boldsymbol{x}_k. \tag{4.6}$$

Zauważmy, że formułując problem w postaci formy dualnej L_D zwiększamy wymiar przestrzeni parametrów z $p+1$ (wektor wag $\boldsymbol{w}$ oraz stała w_0) do n (mnożniki Lagrange'a α_j), tj. liczebności próby.

Zaletą formy dualnej jest to, że wyraża ona kryterium optymalizacji w terminach iloczynów skalarnych obserwacji $\boldsymbol{x}_j$. Cecha ta okaże się istotną przy uogólnieniu metody na model nieliniowy.

Stosując notację macierzową, formę dualną (4.6) możemy zapisać w postaci:

$$L_D = \boldsymbol{\alpha}'\mathbf{1} - \frac{1}{2}\boldsymbol{\alpha}'\boldsymbol{D}\boldsymbol{\alpha}, \tag{4.7}$$

gdzie **1** jest n-wymiarowym wektorem jednostkowym, zaś $\boldsymbol{D}$ macierzą symetryczną stopnia n o elementach:

$$d_{jk} = y_j y_k \boldsymbol{x}_j' \boldsymbol{x}_k, \qquad j,\ k = 1,\ 2,\ \ldots,\ n.$$

Szukamy więc punktu $\boldsymbol{\alpha} \in \mathbb{R}^n$, dla którego funkcja L_D osiąga maksimum przy warunku pobocznym:

$$\boldsymbol{\alpha}'\boldsymbol{y} = 0, \qquad \boldsymbol{\alpha} \geq 0, \tag{4.8}$$

gdzie $\boldsymbol{y} = (y_1,\ y_2,\ \ldots,\ y_n)'$.

Zgodnie z twierdzeniem Karusha–Kühna–Tuckera (Karush, 1939; Kühn i Tucker, 1951) rozwiązanie powyższego problemu związane jest dodatkową zależnością postaci[4]:

$$\alpha_j[y_j(\boldsymbol{w}'\boldsymbol{x_j} + w_0) - 1] = 0, \qquad j = 1,\ 2,\ \ldots,\ n. \tag{4.9}$$

Wynika z niej w szczególności następująca implikacja:

$$\alpha_j > 0 \quad \Longrightarrow \quad y_j(\boldsymbol{w}'\boldsymbol{x_j} + w_0) = 1.$$

Zatem wszystkie obserwacje $\boldsymbol{x}_j$, którym odpowiadają niezerowe mnożniki Lagrange'a leżą na hiperpłaszczyznach kanonicznych. Obserwacje te nazywamy *wektorami nośnymi*[5].
Na rysunku 4.1 wektorami nośnymi są obserwacje oznaczone numerami 3, 5, 8 i 14.

Po wyznaczeniu wielkości α_j wektor wag $\boldsymbol{w}$ jest dany zależnością (patrz (4.4)):

$$\boldsymbol{w} = \sum_{\boldsymbol{x}_j \in \mathcal{SV}} \alpha_j y_j \boldsymbol{x}_j, \tag{4.10}$$

[4]Po ang. *KKT* complementarity condition.
[5]Po ang. support vectors.

gdzie $\mathcal{SV}$ oznacza zbiór wszystkich wektorów nośnych. Ma on normę (w przestrzeni L_2) równą

$$||\boldsymbol{w}|| = (\boldsymbol{w}'\boldsymbol{w})^{1/2} = \left(\sum_{\boldsymbol{x}_j \in \mathcal{SV}} \sum_{\boldsymbol{x}_k \in \mathcal{SV}} \alpha_j \alpha_k y_j y_k \boldsymbol{x}_j \boldsymbol{x}_k \right)^{1/2}.$$

Dla dowolnego $\boldsymbol{x}_k \in \mathcal{SV}$ z (4.9) mamy:

$$y_k(\boldsymbol{w}'\boldsymbol{x}_{\boldsymbol{k}} + w_0) = 1$$
$$w_0 = y_k - \boldsymbol{w}'\boldsymbol{x}_k$$

(ponieważ $\alpha_k > 0$, $y_k \in \{+1, \; -1\}$).
Wtedy jako optymalną wartość współczynnika w_0 obieramy:

$$w_0 = \frac{1}{2}\left[(y_{k_1} - \boldsymbol{w}'\boldsymbol{x}_{k_1}) + (y_{k_2} - \boldsymbol{w}'\boldsymbol{x}_{k_2})\right] = -\frac{1}{2}\boldsymbol{w}'(\boldsymbol{x}_{k_1} + \boldsymbol{x}_{k_2}), \tag{4.11}$$

gdzie $\boldsymbol{x}_{k_1}$, $\boldsymbol{x}_{k_2} \in \mathcal{SV}$ są takie, że $y_{k_1} y_{k_2} = -1$.
Z punktu widzenia obliczeń numerycznych zaleca się wyznaczanie stałej w_0, uśredniając jej wartość po wszystkich wektorach nośnych:

$$w_0 = \frac{1}{|\mathcal{SV}|}\left\{ \sum_{\boldsymbol{x}_j \in \mathcal{SV}} y_j - \sum_{\boldsymbol{x}_j \in \mathcal{SV}} \sum_{\boldsymbol{x}_k \in \mathcal{SV}} \alpha_j y_j \boldsymbol{x}'_j \boldsymbol{x}_k \right\}. \tag{4.12}$$

Z równań (4.10), (4.12) wynika, że o ostatecznej postaci funkcji dyskryminacyjnej g decydują wyłącznie wektory nośne. Im większa wartość α_j wektora nośnego $\boldsymbol{x}_j$, tym większy jest jego wpływ na kształt granic decyzyjnych. Usunięcie z próby którejkolwiek (a stąd wszystkich) z pozostałych obserwacji nie wpłynie na postać hiperpłaszczyzny $g(\boldsymbol{x}) = 0$.

W ogromnej większości przypadków populacje nie są separowalne w sposób liniowy, tj. nie istnieje reguła klasyfikacyjna liniowa, która klasyfikowałaby bezbłędnie wszystkie elementy próby. W takich sytuacjach przedstawione postępowanie nie doprowadzi nas do poszukiwanego rozwiązania – funkcja L_D określona wzorem (4.6) przyjmuje wartości dowolnie duże.

Metodę wektorów nośnych stosuje się więc przy warunku słabszym od (4.2), wprowadzając zmienną $\boldsymbol{\xi} = (\xi_1, \; \xi_2, \; \ldots, \; \xi_n)' \geq 0$ taką, że:

$$\boldsymbol{\xi}: \begin{array}{ll} \boldsymbol{w}'\boldsymbol{x}_{\boldsymbol{j}} + w_0 \geq +1 - \xi_j & \text{dla} \quad \boldsymbol{x}_j \in \mathcal{G}_1 \\ \boldsymbol{w}'\boldsymbol{x}_{\boldsymbol{j}} + w_0 \leq -1 + \xi_j & \text{dla} \quad \boldsymbol{x}_j \in \mathcal{G}_2. \end{array}$$

Tym samym, dla pewnych elementów próby, dopuszcza się możliwość błędnej klasyfikacji. Warunek ten można łącznie zapisać jako:

$$\boldsymbol{\xi} \geq 0, \qquad y_j(\boldsymbol{w}'\boldsymbol{x_j} + w_0) \geq 1 - \xi_j. \tag{4.13}$$

Stąd obserwacja $\boldsymbol{x}_j$ zostanie błędnie zaklasyfikowana, jeśli $\xi_j > 1$.

Poszukujemy liniowej funkcji dyskryminacyjnej $g(\boldsymbol{x}) = \boldsymbol{w}'\boldsymbol{x} + w_0$ minimalizującej wyrażenie:

$$\frac{1}{2}\boldsymbol{w}'\boldsymbol{w} + C\sum_{j=1}^{n}\xi_j, \tag{4.14}$$

gdzie C jest nieznanym parametrem, przy warunku pobocznym (4.13). Drugi składnik powyższej sumy należy rozumieć jako dodatkowy koszt związany z błędną klasyfikacją. Im większa wartość parametru C, tym większa strata związana z każdą obserwacją próby, dla której $\xi_j > 0$, a stąd mniejsza odległość hiperpłaszczyzn kanonicznych od hiperpłaszczyzny $g(\boldsymbol{x}) = 0$.

Funkcja Lagrange'a L przyjmuje postać:

$$\begin{aligned} L &= L(\boldsymbol{w},\ w_0,\ \boldsymbol{\xi};\ \boldsymbol{\alpha},\ \boldsymbol{\beta}) = \\ &= \frac{1}{2}\boldsymbol{w}'\boldsymbol{w} + C\sum_{j=1}^{n}\xi_j - \sum_{j=1}^{n}\alpha_j[y_j(\boldsymbol{w}'\boldsymbol{x_j} + w_0) - 1 + \xi_j] - \sum_{j=1}^{n}\beta_j\xi_j, \end{aligned} \tag{4.15}$$

gdzie $\boldsymbol{\alpha} \geq 0$, $\boldsymbol{\beta} \geq 0$ są mnożnikami Lagrange'a.
Różniczkując funkcję Lagrange'a L względem $\boldsymbol{w}$, w_0, ξ_j $(j = 1,\ 2,\ \ldots,\ n)$ i przyrównując pochodne cząstkowe do zera, otrzymujemy kolejno:

$$\boldsymbol{w} = \sum_{j=1}^{n}\alpha_j y_j \boldsymbol{x}_j, \tag{4.16}$$

$$\sum_{j=1}^{n}\alpha_j y_j = 0, \tag{4.17}$$

$$\alpha_j + \beta_j = C. \tag{4.18}$$

Podstawiając (4.16), (4.17), (4.18) do (4.15), otrzymujemy postać dualną L_D daną ponownie równaniem (4.6).

Analogicznie – zagadnienie sprowadza się do znalezienia punktu $\boldsymbol{\alpha} \in \mathbb{R}^n$, dla którego funkcja L_D osiąga maksimum, przy czym warunki poboczne są teraz postaci:

$$\sum_{j=1}^{n}\alpha_j y_j = 0, \qquad 0 \leq \alpha_j \leq C, \qquad j = 1,\ 2,\ \ldots,\ n \tag{4.19}$$

(ponieważ z (4.18) mamy $\alpha_j = C - \beta_j$ oraz $\beta_j \geq 0$). Stąd jedyną zmianą jest pojawienie się ograniczenia górnego C dla mnożników Lagrange'a α_j.

Zgodnie z twierdzeniem Karusha–Kühna–Tuckera rozwiązanie spełnia warunki:

$$\alpha_j[y_j(\boldsymbol{w}'\boldsymbol{x_j} + w_0) - 1 + \xi_j] = 0 \tag{4.20}$$

$$\beta_j\xi_j = (C - \alpha_j)\xi_j = 0. \tag{4.21}$$

Obserwacje, dla których $\alpha_j > 0$ będziemy nazywać wektorami nośnymi. Z zależności (4.18)–(4.21) wynikają następujące trzy wnioski:

- jeśli $\alpha_j = 0$, to $\xi_j = 0$ i $\boldsymbol{x}_j$ leży w odległości większej niż $1/||\boldsymbol{w}||$ od hiperpłaszczyzny $g(\boldsymbol{x}) = 0$ (poprawna klasyfikacja);
- jeśli $0 < \alpha_j < C$, to $\xi_j = 0$ i $\boldsymbol{x}_j$ leży na hiperpłaszczyźnie kanonicznej (poprawna klasyfikacja);
- $\xi_j \neq 0$ tylko jeśli $\alpha_j = C$; wtedy obserwacja $\boldsymbol{x}_j$ jest albo błędnie ($\xi_j > 1$), albo poprawnie zaklasyfikowana, lecz leży w odległości mniejszej niż $1/||\boldsymbol{w}||$ od hiperpłaszczyzny $g(\boldsymbol{x}) = 0$ ($0 < \xi_j \leq 1$).

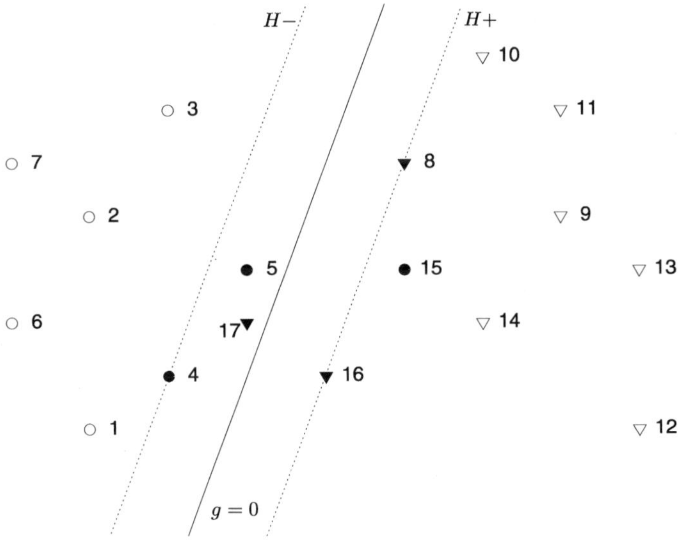

Rys. 4.2. Optymalna hiperpłaszczyzna dla dwóch populacji liniowo nieseparowalnych ($C = 10$)

Dla danych z rys. 4.2 wektorami nośnymi są obserwacje oznaczone numerami: 4, 8, 16 oraz 5, 15 i 17. Dla pierwszych trzech z nich mamy $0 < \alpha < C$, dla pozostałych trzech $\alpha = C$. Wektorom o etykietach 15 i 17 odpowiadają wartości $\xi > 1$ (błędna klasyfikacja).

Rozwiązaniem zagadnienia jest klasyfikator liniowy o wektorze wag $\boldsymbol{w}$ danym wzorem (4.10) oraz stałej w_0 równej:

$$w_0 = \frac{1}{|\mathcal{SV}^*|}\left\{\sum_{\boldsymbol{x}_k \in \mathcal{SV}^*} y_k - \sum_{\boldsymbol{x}_j \in \mathcal{SV}} \sum_{\boldsymbol{x}_k \in \mathcal{SV}^*} \alpha_j y_j \boldsymbol{x}_j' \boldsymbol{x}_k\right\}, \tag{4.22}$$

gdzie $\mathcal{SV}^*$ oznacza zbiór wektorów nośnych, dla których $0 < \alpha_j < C$. Stąd – analogicznie jak w przypadku grup liniowo separowalnych – postać funkcji dyskryminacyjnej determinują wyłącznie wektory nośne.

Na koniec odnotujmy, że w modelu liniowym metoda wektorów nośnych może być narzędziem w zagadnieniu wyboru cech[6]. Rangowania zmiennych dokonuje się na podstawie wielkości modułu współrzędnych wektora $\boldsymbol{w}$.

4.2. Model nieliniowy

W poprzednim podrozdziale przedstawiliśmy metodę wektorów nośnych jako technikę wyznaczania optymalnej hiperpłaszczyzny dla grup liniowo separowalnych, a następnie uogólniliśmy jej ideę na przypadek grup liniowo nieseparowalnych.
Najczęściej jednak metodę tę stosuje się nie w samej przestrzeni próby $\mathcal{X}$, a w przestrzeni $\boldsymbol{\varphi}(\mathcal{X})$, dla pewnego nieliniowego przekształcenia $\boldsymbol{\varphi}$. Podejście takie jest podstawą nieliniowych metod dyskryminacyjnych: obserwacje próby przetransformuj nieliniowo z wyjściowej przestrzeni cech do przestrzeni wyższego wymiaru (nierzadko nieskończonego), a następnie zastosuj w niej model liniowy.

Niech

$$g(\boldsymbol{x}) = \boldsymbol{w}'\boldsymbol{\varphi}(\boldsymbol{x}) + w_0,$$

gdzie $\boldsymbol{\varphi} = (\varphi_1, \varphi_2, \ldots, \varphi_q)' \colon \mathbb{R}^p \to \mathbb{R}^q$ jest pewnym przekształceniem nieliniowym oraz niech $\hat{d}_{\boldsymbol{\varphi}}(\boldsymbol{x})$ oznacza regułę klasyfikacyjną taką, że:

$$\hat{d}_{\boldsymbol{\varphi}}(\boldsymbol{x}) = \begin{cases} +1, & \text{jeżeli} \quad g(\boldsymbol{x}) > 0, \\ -1, & \text{jeżeli} \quad g(\boldsymbol{x}) \le 0. \end{cases}$$

[6]Po ang. feature selection.

Z rozumowania przeprowadzonego w podrozdz. 4.1 wynika, że funkcja L_D dana równaniem (4.6) zależy od obserwacji próby tylko przez ich iloczyny skalarne. Stąd:

$$L_D = L_D(\boldsymbol{\alpha}) = \sum_{j=1}^{n} \alpha_j - \frac{1}{2}\sum_{j=1}^{n}\sum_{k=1}^{n} \alpha_j \alpha_k y_j y_k \boldsymbol{\varphi}'(\boldsymbol{x}_j)\boldsymbol{\varphi}(\boldsymbol{x}_k). \tag{4.23}$$

Postępując analogicznie – poszukujemy punktu $\boldsymbol{\alpha} \in \mathbb{R}^n$, dla którego funkcja L_D osiąga maksimum przy warunku pobocznym (4.19). Prowadzi to do pojęcia wektorów nośnych, tj. wektorów $\boldsymbol{x}_j$ o mnożnikach Lagrange'a $\alpha_j > 0$, oraz do rozwiązania $g(\boldsymbol{x}) = \boldsymbol{w}'\boldsymbol{\varphi}(\boldsymbol{x}) + w_0$, gdzie:

$$\boldsymbol{w} = \sum_{\boldsymbol{x}_j \in \mathcal{SV}} \alpha_j y_j \boldsymbol{\varphi}(\boldsymbol{x}_j) \tag{4.24}$$

$$w_0 = \frac{1}{|\mathcal{SV}^*|}\left\{\sum_{\boldsymbol{x}_k \in \mathcal{SV}^*} y_k - \sum_{\boldsymbol{x}_j \in \mathcal{SV}} \sum_{\boldsymbol{x}_k \in \mathcal{SV}^*} \alpha_j y_j \boldsymbol{\varphi}'(\boldsymbol{x}_j)\boldsymbol{\varphi}(\boldsymbol{x}_k)\right\}. \tag{4.25}$$

Występujące w równaniu (4.25) iloczyny skalarne $\boldsymbol{\varphi}'(\boldsymbol{x}_j)\boldsymbol{\varphi}(\boldsymbol{x}_k)$ w $\mathbb{R}^q$ (ogólnie – w dowolnej przestrzeni euklidesowej $\mathcal{E}$) chcemy wyrazić jako funkcje jądra K:

$$K(\boldsymbol{x}_j,\ \boldsymbol{x}_k) = \boldsymbol{\varphi}'(\boldsymbol{x}_j)\boldsymbol{\varphi}(\boldsymbol{x}_k). \tag{4.26}$$

Funkcje K, dla których istnieje co najmniej jedna para $\{\mathcal{E},\ \boldsymbol{\varphi}\}$ taka, że zachodzi zależność (4.26), muszą spełniać tzw. *warunek Mercera* (patrz np. Vapnik, 1995):

TWIERDZENIE 4.1. (Mercera, 1909)

Na to, by symetryczna funkcja $K(\boldsymbol{u},\ \boldsymbol{v}) \in L_2$, $\boldsymbol{u},\ \boldsymbol{v} \in \mathbb{R}^p$, miała rozwinięcie:

$$K(\boldsymbol{u},\ \boldsymbol{v}) = \sum_{k=1}^{\infty} a_k \varphi_k(\boldsymbol{u})\varphi_k(\boldsymbol{v}), \qquad a_k > 0$$

potrzeba i wystarcza, aby dla dowolnej funkcji $0 \not\equiv f \in L_2$ był spełniony warunek:

$$\iint K(\boldsymbol{u},\ \boldsymbol{v})f(\boldsymbol{u})f(\boldsymbol{v})d\boldsymbol{u}d\boldsymbol{v} > 0. \tag{4.27}$$

Przykład 4.1.
Rozważmy jądro postaci $K(\boldsymbol{u},\ \boldsymbol{v}) = (1 + \boldsymbol{u}'\boldsymbol{v})^2$, gdzie $\boldsymbol{u},\ \boldsymbol{v} \in \mathbb{R}^2$. Mamy:

$$\begin{aligned} K(\boldsymbol{u},\ \boldsymbol{v}) &= (1 + u_1v_1 + u_2v_2)^2 = \\ &= 1 + 2u_1v_1 + 2u_2v_2 + 2u_1u_2v_1v_2 + u_1^2v_1^2 + u_2^2v_2^2 = \\ &= \varphi'(\boldsymbol{u})\varphi(\boldsymbol{v}), \end{aligned}$$

gdzie $\varphi(\boldsymbol{u}) = (1,\ \sqrt{2}u_1,\ \sqrt{2}u_2,\ \sqrt{2}u_1u_2,\ u_1^2,\ u_2^2)'$.
Zatem dla jądra K będącego wielomianem stopnia drugiego, φ jest transformacją wyjściowej obserwacji do przestrzeni 5-wymiarowej. Ogólnie – dla jądra wielomianowego stopnia q $K(\boldsymbol{u},\ \boldsymbol{v}) = (1 + \boldsymbol{u}'\boldsymbol{v})^q$, docelowa przestrzeń ma wymiar $\binom{p+q}{q} - 1$.

Można pokazać, że funkcje zawarte w tab. 4.1 spełniają warunek Mercera (4.27)[7].

Tab. 4.1. Dopuszczalne typy jąder dla metody wektorów nośnych

Nazwa jądra	Zależność funkcyjna
wielomianowe	$K(\boldsymbol{u},\ \boldsymbol{v}) = (1 + \boldsymbol{u}'\boldsymbol{v})^q$
normalne (Gaussa)	$K(\boldsymbol{u},\ \boldsymbol{v}) = \exp\left(-\frac{\lVert\boldsymbol{u} - \boldsymbol{v}\rVert^2}{2\sigma^2}\right)$
sigmoidalne	$K(\boldsymbol{u},\ \boldsymbol{v}) = \mathrm{tgh}(\kappa\,\boldsymbol{u}'\boldsymbol{v} - \delta)$

Zastosowanie jądra sprawia, że nie zachodzi konieczność znajomości jawnej postaci przekształcenia φ, jak również obliczania wartości $\varphi(\boldsymbol{x})$ w procesie konstrukcji funkcji dyskryminacyjnej g (co jest szczególnie istotne dla przekształceń φ o wartościach w przestrzeni nieskończenie wymiarowej). Ponadto wektor wag $\boldsymbol{w}$ nie musi być wyznaczany z zależności (4.24) – funkcja dyskryminacyjna przyjmuje bowiem postać:

$$g(\boldsymbol{x}) = \sum_{\boldsymbol{x}_j \in \mathcal{SV}} \alpha_j y_j K(\boldsymbol{x}_j,\ \boldsymbol{x}) + w_0. \tag{4.28}$$

4.3. Zagadnienie K klas

Naturalnym sposobem dyskryminacji K klas ($K > 2$) jest rozszerzenie przedstawionego modelu binarnego przez konstrukcję K funkcji dyskrymi-

[7] Jądro sigmoidalne tylko dla pewnych wartości parametrów κ i δ (Burges, 1998).

nacyjnych g_i[8]:

$$g_i(\boldsymbol{x}) = (\boldsymbol{w}^i)'\varphi(\boldsymbol{x}) + w_0^i, \qquad i = 1,\ 2,\ \ldots,\ K.$$

Poszukujemy funkcji g_i minimalizujących wyrażenie:

$$\sum_{i=1}^{K}(\boldsymbol{w}^i)'\boldsymbol{w}^i + C\sum_{i=1}^{K}\sum_{j=1}^{n}\xi_j^i \tag{4.29}$$

przy warunku pobocznym:

$$(\boldsymbol{w}^m)'\varphi(\boldsymbol{x}_j) + w_0^m - [(\boldsymbol{w}^i)'\varphi(\boldsymbol{x}_j) + w_0^i] \geq 1 - \xi_j^i, \qquad \xi_j^i \geq 0 \tag{4.30}$$

$(m = 1,\ 2,\ \ldots,\ K,\ i \neq m,\ j = 1,\ 2,\ \ldots,\ n)$.
Zatem w uogólnionej funkcji straty (4.29) rozpatruje się równocześnie wszystkie funkcje dyskryminacyjne g_i. Ponadto osłabiony zostaje warunek poboczny: nie wymaga się, by funkcje g_i zerowały się na granicach decyzyjnych, a jedynie by wartość funkcji dyskryminacyjnej odpowiadającej rzeczywistej klasie była większa od wartości pozostałych funkcji o wielkość $1 - \xi_j^i$.

Reguła klasyfikacyjna dla przypadku K klas ma postać:

$$\hat{d}_\varphi(\boldsymbol{x}) = \arg\max_{1\leq i\leq K} g_i(\boldsymbol{x}).$$

4.4. *VC* wymiar i zasada SRM

Miarę złożoności klasyfikatora utożsamia się często z liczbą jego parametrów. Vapnik i Chervonenkis rozwinęli nową teorię, wprowadzając bardziej uniwersalną miarę złożoności danej klasy funkcji. Zgodnie z nią powstała metoda wektorów nośnych.

Niech $\mathcal{A}$ oznacza pewną rodzinę zbiorów. Ponadto niech dany będzie skończony zbiór $S = \{x_1,\ x_2,\ \ldots,\ x_n\}$ n elementów pewnej przestrzeni $\mathcal{X}$. Wprowadźmy wielkość $N_{\mathcal{A}}(S)$ następująco:

$$N_{\mathcal{A}}(S) := \#\{S \cap A\colon A \in \mathcal{A}\}. \tag{4.31}$$

[8]Alternatywne metody polegające na łączeniu klasyfikatorów binarnych zostaną przedstawione w rozdz. 8.

Definicja 4.1.
Współczynnikiem rozbicia[9] nazywamy liczbę:

$$s(\mathcal{A},\ n) := \max_{S \in \mathcal{S}_n} N_{\mathcal{A}}(S), \tag{4.32}$$

gdzie $\mathcal{S}_n$ jest rodziną wszystkich zbiorów n-elementowych zawartych w $\mathcal{X}$.

Definicja 4.2.
Mówimy, że rodzina zbiorów $\mathcal{A}$ ma *VC wymiar (wymiar Vapnika–Chervonenkisa)*[10] $VC(\mathcal{A}) = h < \infty$, jeżeli:

1. $s(\mathcal{A},\ n) = 2^n$ dla każdego $n \leq h$,
2. $s(\mathcal{A},\ h+1) < 2^{h+1}$.

W przeciwnym przypadku $VC(\mathcal{A}) = \infty$.

Przykład 4.2.
Niech $\mathcal{A} = \{(-\infty,\ a] \colon a \in \mathbb{R}\}$. Wówczas dla dowolnego punktu $x \in \mathbb{R}$ mamy:

$$N_{\mathcal{A}}(\{x\}) = \#\{\emptyset,\ \{x\}\} = 2.$$

Zatem $s(\mathcal{A},\ 1) = 2$. Ponadto dla każdego zbioru 2-elementowego $\{x,\ y\} \subset \mathbb{R}$: $s(\mathcal{A},\ 2) < 2^2$. Stąd $VC(\mathcal{A}) = 1$.

Niech teraz $\mathcal{F}$ będzie pewną klasą funkcji (klasyfikatorów). Wówczas rodzinę $\mathcal{A}$ definiujemy jako:

$$\mathcal{A} := \{\{x \colon d(x) = 1\} \colon d \in \mathcal{F}\}, \tag{4.33}$$

zaś

$$s(\mathcal{F},\ n) := s(\mathcal{A},\ n). \tag{4.34}$$

Zwróćmy uwagę, że rodzina $\mathcal{F}$ funkcji klasyfikujących ma VC wymiar równy h, gdy istnieje zbiór h punktów $\{x_j\}_{j=1}^{h}$, który może być zaklasyfikowany do dwóch grup za pomocą funkcji z $\mathcal{F}$ na każdy z 2^h możliwych sposobów, oraz nie istnieje zbiór $h+1$ punktów mający powyższą własność. W przeciwnym przypadku $VC(\mathcal{F}) = \infty$.

---
[9] Po ang. shatter coefficient.
[10] Po ang. Vapnik–Chervonenkis dimension.

Przykład 4.3.
Rozpatrzmy rodzinę $\mathcal{F}$ klasyfikatorów liniowych (hiperpłaszczyzn zorientowanych) w przestrzeni $\mathbb{R}^2$. Istnieją trzy punkty na płaszczyźnie (niewspółliniowe) mogące być dowolnie przydzielone do dwóch grup za pomocą funkcji liniowych (rys. 4.3). Zauważmy jednocześnie, że własności tej nie ma żadna czwórka punktów. Stąd $VC(\mathcal{F}) = 3$.

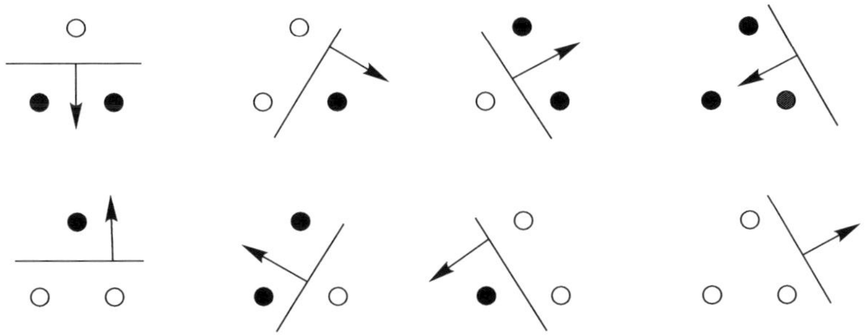

Rys. 4.3. Trzy punkty klasyfikowane za pomocą zorientowanych hiperpłaszczyzn w $\mathbb{R}^2$ na 8 różnych sposobów liniowo nieseparowalnych ($C = 10$)

Ogólniej – rozważmy zbiór hiperpłaszczyzn (zorientowanych) w przestrzeni $\mathbb{R}^p$. Prawdziwe jest następujące twierdzenie (Burges, 1998):

TWIERDZENIE 4.2.

Niech będzie dany zbiór m punktów w przestrzeni $\mathbb{R}^p$. Obierzmy jeden z nich jako zero tej przestrzeni. Wówczas rodzina hiperpłaszczyzn w $\mathbb{R}^p$ ma VC wymiar $h \geq m$ wtedy i tylko wtedy, gdy pozostały zbiór $m - 1$ punktów jest liniowo niezależny.

Ponieważ w $\mathbb{R}^p$ można znaleźć maksymalnie p punktów liniowo niezależnych, więc na mocy powyższego twierdzenia rodzina liniowych funkcji dyskryminacyjnych ma wymiar *VC* równy $p + 1$.

Wyżej przeprowadzone rozumowanie sugerować by mogło, iż wymiar *VC* jest związany monotoniczną zależnością (rosnącą) z liczbą nieznanych parametrów klasyfikatora. Z poniższego przykładu wynika, że własność ta nie jest jednak prawdziwa.

Przykład 4.4.
Rozpatrzmy rodzinę funkcji:

$$d_\lambda(x) = \operatorname{sgn}(\sin(\lambda x)), \qquad x \in \mathbb{R},$$

gdzie $\lambda \in \mathbb{R}$ jest parametrem, natomiast

$$\mathrm{sgn}(u) = \begin{cases} -1, & \text{gdy} \quad u < 0 \\ +1, & \text{gdy} \quad u \geq 0. \end{cases}$$

Weźmy ciąg punktów $x_j = 10^{-j}$, $j = 1, 2, \ldots, m$. Może on być przydzielony dowolnie do jednej z dwóch grup przez następujący wybór parametru:

$$\lambda = \pi \left(1 + \sum_{j=1}^{m} \frac{(1 - y_j)10^j}{2} \right),$$

gdzie

$$y_j = \begin{cases} +1, & \text{gdy} \quad x_j \in \mathcal{G}_1 \\ -1, & \text{gdy} \quad x_j \in \mathcal{G}_2. \end{cases}$$

Przykład ten – pochodzący od Vapnika – wskazuje, że istnieją rodziny funkcji o jednym parametrze i wymiarze *VC* nieskończonym.

Powrócimy teraz do zagadnienia estymacji aktualnego poziomu błędu (patrz podrozdz. 1.10).

Niech $\mathcal{F}$ będzie ustaloną rodziną klasyfikatorów. Ponadto niech

$$\hat{d}_0 = \arg\min_{\hat{d} \in \mathcal{F}} \hat{e}_R(\hat{d}). \tag{4.35}$$

Jak wiemy, $\hat{e}_R(\hat{d}_0)$ będzie zazwyczaj niedoestymowaną oceną aktualnego poziomu błędu $e(\hat{d}_0)$. Celem naszym jest oszacowanie wielkości tego niedoestymowania.

Załóżmy wpierw, że rodzina $\mathcal{F} = \{\hat{d}_1, \hat{d}_2, \ldots, \hat{d}_m\}$ składa się ze skończonej liczby m klasyfikatorów. Oczywiście dla każdego $i = 1, 2, \ldots, m$ wielkość $\hat{e}_R(\hat{d}_i)$ (przy liczebności próby $n \to \infty$) zbiega do $e(\hat{d}_i)$ z prawdopodobieństwem 1.

Zachodzi twierdzenie mocniejsze – o zbieżności jednostajnej:

TWIERDZENIE 4.3.

Niech $\mathcal{F}$ będzie zbiorem m-elementowym ($m < \infty$). Wówczas:

$$\mathrm{P}(\max_{\hat{d} \in \mathcal{F}} |\hat{e}_R(\hat{d}) - e(\hat{d})| > \varepsilon) \leq 2me^{-2n\varepsilon^2}. \tag{4.36}$$

Dowód. Przypomnijmy najpierw następujący:

LEMAT 4.1. (nierówność Hoeffdinga).
Jeżeli $X_1, X_2, \ldots, X_n$ *jest próbą z rozkładu Bernoulliego* $b(p)$, *to dla dowolnego* $\varepsilon > 0$ *mamy:*

$$\mathrm{P}(|\widehat{p} - p| > \varepsilon) \leq 2e^{-2n\varepsilon^2}, \tag{4.37}$$

gdzie

$$\widehat{p} = \frac{1}{n}\sum_{i=1}^{n} X_i.$$

Korzystając z powyższego lematu, mamy:

$$\begin{aligned}\mathrm{P}(\max_{\hat{d}\in\mathcal{F}} |\hat{e}_R(\hat{d}) - e(\hat{d})| > \varepsilon) &= \mathrm{P}(\bigcup_{\hat{d}\in\mathcal{F}} |\hat{e}_R(\hat{d}) - e(\hat{d})| > \varepsilon) \leq \\ &\leq \sum_{\hat{d}\in\mathcal{F}} \mathrm{P}(|\hat{e}_R(\hat{d}) - e(\hat{d})| > \varepsilon) \leq \\ &\leq \sum_{\hat{d}\in\mathcal{F}} 2e^{-2n\varepsilon^2} = 2me^{-2n\varepsilon^2}.\end{aligned}$$

■

WNIOSEK 4.1.

Niech:

$$\varepsilon = \sqrt{\frac{1}{2n}\ln\left(\frac{2m}{\alpha}\right)}. \tag{4.38}$$

Wówczas $\hat{e}_R(\hat{d}_0) \pm \varepsilon$ *jest przedziałem ufości dla* $e(\hat{d}_0)$ *na poziomie ufności* $1 - \alpha$.

Dowód. Na mocy tw. 4.3 mamy:

$$\begin{aligned}\mathrm{P}(|\hat{e}_R(\hat{d}_0) - e(\hat{d}_0)| > \varepsilon) &\leq \mathrm{P}(\max_{\hat{d}\in\mathcal{F}} |\hat{e}_R(\hat{d}) - e(\hat{d})| > \varepsilon) \leq \\ &\leq 2me^{-2n\varepsilon^2} = \alpha.\end{aligned}$$

■

Zauważmy, że wraz ze wzrostem mocy rodziny $\mathcal{F}$ wzrasta długość przedziału ufności. Im więcej klasyfikatorów zawiera rodzina $\mathcal{F}$, tym bowiem bardziej możliwe staje się przeestymowanie aktualnego poziomu błędu $e(\hat{d}_0)$.

W praktyce najczęściej rozpatrywana rodzina $\mathcal{F}$ zawiera nieskończenie wiele funkcji klasyfikujących, np. $\mathcal{F}$ jest rodziną wszystkich klasyfikatorów liniowych. W tym przypadku chcielibyśmy oszacować z góry $P(\sup_{\hat{d}\in\mathcal{F}} |\hat{e}_R(\hat{d}) - e(\hat{d})| > \varepsilon)$ przez pewną, „niezbyt dużą" wielkość.

Prawdziwe jest następujące (Wasserman, 2004):

TWIERDZENIE 4.4. (Vapnika–Chervonenkisa, 1971)

Dla dowolnego $\varepsilon > 0$:

$$P(\sup_{\hat{d}\in\mathcal{F}} |\hat{e}_R(\hat{d}) - e(\hat{d})| > \varepsilon) \leq 8s(\mathcal{F},\ n)e^{-n\varepsilon^2/32}. \tag{4.39}$$

Stąd $\hat{e}_R(\hat{d}) \pm \varepsilon_n$ jest przedziałem ufości dla $e(\hat{d})$ na poziomie ufności $1-\alpha$, gdzie

$$\varepsilon_n^2 = (32/n)\ln\left(8s(\mathcal{F},\ n)/\alpha\right). \tag{4.40}$$

Powyższe twierdzenie jest użyteczne tylko w przypadku, gdy współczynnik rozbicia s nie rośnie zbyt szybko wraz ze wzrostem n. Poniższe twierdzenie mówi, że jeżeli rodzina $\mathcal{F}$ ma *VC* wymiar skończony, to wielkości $s(\mathcal{F},\ n)$ rosną w sposób wielomianowy:

TWIERDZENIE 4.5.

Jeżeli $\mathcal{F}$ ma VC wymiar $h < \infty$, to

$$s(\mathcal{F},\ n) \leq n^h + 1. \tag{4.41}$$

WNIOSEK 4.2.

Ponieważ rodzina klasyfikatorów liniowych w przesterzeni $\mathbb{R}^p$ ma VC wymiar równy $p+1$ (patrz tw. 4.2), więc na mocy tw. 4.4 i 4.5 przedziałem ufości dla aktualnego poziomu błędu $e(\hat{d}_0)$ na poziomie ufności $1-\alpha$ jest $\hat{e}_R(\hat{d}_0) \pm \varepsilon_n$, gdzie:

$$\varepsilon_n^2 = \frac{32}{n}\ln\frac{8(n^{p+1}+1)}{\alpha}. \tag{4.42}$$

Odnotujmy w tym miejscu fakt, że oszacowania otrzymane z nierówności Vapnika–Chervonenkisa (4.39) są słabe i z reguły prowadzą do relatywnie szerokich przedziałów ufności.

Znane jest następujące oszacowanie aktualnego poziomu błędu (Vapnik, 1995):

TWIERDZENIE 4.6.

Z prawdopodobieństwem $1 - \varepsilon$ prawdziwa jest nierówność:

$$e(\hat{d}) \leq \hat{e}_R(\hat{d}) + c(h), \tag{4.43}$$

gdzie

$$c(h) = \sqrt{\frac{h\left(\ln \frac{2n}{h} + 1\right) - \ln \frac{\varepsilon}{4}}{n}}. \tag{4.44}$$

Zaletą oszacowania (4.43) jest to, że wielkość $c(h)$ nie zależy od rozkładu i może być łatwo wyznaczona, jeśli tylko jest znany wymiar *VC*. Im mniejszy wymiar h, tym mniej złożona jest funkcja g, co prowadzi do wzrostu wielkości $\hat{e}_R(\hat{d})$. Jednocześnie, im mniejsze h, tym mniejsza wartość $c(h)$ (rys. 4.4).

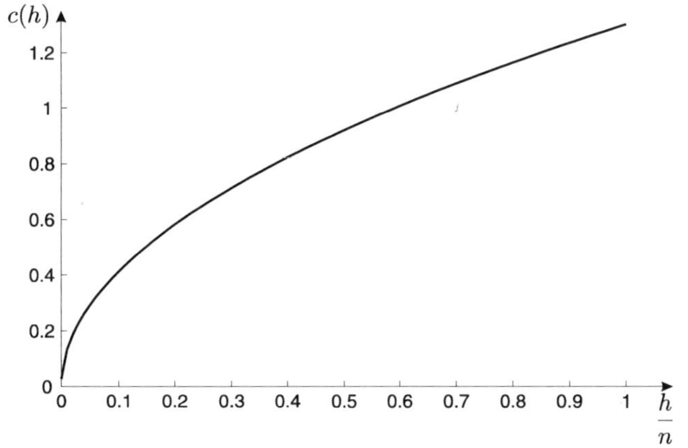

Rys. 4.4. Wielkość $c(h)$ jest monotoniczną funkcją argumentu h ($n = 200$, $\varepsilon = 0,05$)

Niech $\mathcal{F}_h$ oznacza rodzinę klasyfikatorów o wymiarze *VC* równym h. Zachodzą inkluzje:

$$\mathcal{F}_1 \subset \mathcal{F}_2 \subset \ldots \subset \mathcal{F}_\infty.$$

Zasada SRM polega na wskazaniu tej z rodzin $\mathcal{F}_h$ (a w konsekwencji na zastosowaniu funkcji klasyfikującej $\hat{d} \in \mathcal{F}_h$), dla której prawa strona nierówności (4.43) osiąga minimum.

W jaki sposób metoda wektorów nośnych realizuje powyższą zasadę? Niech $||\boldsymbol{w}|| < A$. Ponieważ

$$d(d; \boldsymbol{\varphi}(\boldsymbol{x})) = \frac{|\boldsymbol{w}'\boldsymbol{\varphi}(\boldsymbol{x}) + w_0|}{||\boldsymbol{w}||}, \tag{4.45}$$

więc z uwagi na warunek (4.2) w przestrzeni $\mathcal{E}$ mamy:

$$d(d; \boldsymbol{\varphi}(\boldsymbol{x})) \geq \frac{1}{A}. \tag{4.46}$$

Zatem dopuszczamy tylko te z hiperpłaszczyzn g w przestrzeni $\mathcal{E}$, które leżą w odległości nie mniejszej niż $1/A$ od każdej obserwacji próby (w przestrzeni $\mathcal{E}$). Ograniczenie to redukuje zbiór możliwych rozwiązań, zmniejszając wymiar *VC*, a tym samym złożoność funkcji dyskryminacyjnej (rys. 4.5).

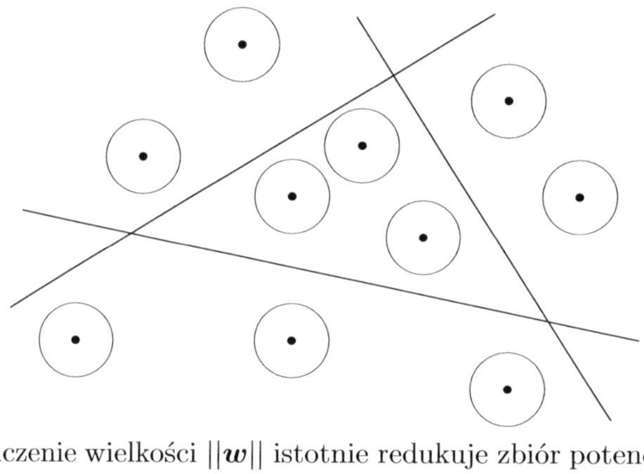

Rys. 4.5. Ograniczenie wielkości $||\boldsymbol{w}||$ istotnie redukuje zbiór potencjalnych rozwiązań

Vapnik pokazał, iż dla hiperpłaszczyzn w przestrzeni q-wymiarowej wymiar *VC* spełnia nierówność:

$$h \leq \min(R^2A^2,\ q) + 1, \tag{4.47}$$

gdzie R jest promieniem kuli $K(R)$ takiej, że $\boldsymbol{\varphi}(\boldsymbol{x}_j) \in K(R)$, dla każdego $j = 1, 2, \ldots, n$.

Jak wyznaczyć minimalną wielkość R? Zagadnienie to jest równoważne minimalizacji wielkości R^2 przy warunku pobocznym:

$$||\boldsymbol{\varphi}(\boldsymbol{x}_j) - \boldsymbol{S}||^2 \leq R^2, \qquad j = 1,\ 2,\ \ldots,\ n, \tag{4.48}$$

gdzie $\boldsymbol{S} \in \mathcal{E}$ jest środkiem kuli $K(R)$.
Funkcja Lagrange'a ma postać:

$$L = L(R,\ \boldsymbol{S}; \boldsymbol{\alpha}) = R^2 - \sum_{j=1}^{n} \alpha_j (R^2 - ||\boldsymbol{\varphi}(\boldsymbol{x}_j) - \boldsymbol{S}||^2). \tag{4.49}$$

Różniczkując kolejno względem R i $\boldsymbol{S}$, otrzymujemy formę dualną postaci:

$$L_D = \sum_{j=1}^{n} \alpha_j K(\boldsymbol{x}_j,\ \boldsymbol{x}_j) - \sum_{j=1}^{n}\sum_{k=1}^{n} \alpha_j \alpha_k K(\boldsymbol{x}_j,\ \boldsymbol{x}_k) \tag{4.50}$$

$(K(\boldsymbol{x}_j,\ \boldsymbol{x}_k) = \boldsymbol{\varphi}'(\boldsymbol{x}_j)\boldsymbol{\varphi}(\boldsymbol{x}_k))$ z warunkiem pobocznym:

$$\sum_{j=1}^{n} \alpha_j = 1, \qquad \alpha_j \geq 0 \tag{4.51}$$

oraz rozwiązaniem:

$$\boldsymbol{S} = \sum_{j=1}^{n} \alpha_j \boldsymbol{\varphi}(\boldsymbol{x}_j). \tag{4.52}$$

Wówczas promień kuli wyznaczamy z zależności:

$$\begin{aligned} R^2 &= \max_{1 \leq j \leq n} ||\boldsymbol{\varphi}(\boldsymbol{x}_j) - \boldsymbol{S}||^2 = \\ &= \max_{1 \leq j \leq n} \left[\boldsymbol{\varphi}'(\boldsymbol{x}_j)\boldsymbol{\varphi}(\boldsymbol{x}_j) - 2\boldsymbol{\varphi}'(\boldsymbol{x}_j)\boldsymbol{\varphi}(\boldsymbol{s}) + \boldsymbol{\varphi}'(\boldsymbol{s})\boldsymbol{\varphi}(\boldsymbol{s})\right] = \\ &= \max_{1 \leq j \leq n} \left(K(\boldsymbol{x}_j, \boldsymbol{x}_j) - 2K(\boldsymbol{x}_j,\ \boldsymbol{s}) + K(\boldsymbol{s},\ \boldsymbol{s})\right), \end{aligned} \tag{4.53}$$

gdzie $\boldsymbol{\varphi}(\boldsymbol{s}) = \boldsymbol{S}$.

W przypadku grup liniowo separowalnych warunek (4.2) sprawia, że estymator błędu $\hat{e}_R$ jest równy 0. Następnie przez minimalizację wielkości $||\boldsymbol{w}||$ (równoważnie – maksymalizację odległości pomiędzy hiperpłaszczyznami kanonicznymi) minimalizuje się wymiar *VC*, a stąd wielkość $c(h)$ i w konsekwencji rzeczywisty poziom błędu L. Jednakże, nawet w przypadku klas liniowo separowalnych, można uzyskać mniejsze ograniczenie górne błędu w (4.43), dopuszczając błędną klasyfikację pewnych elementów próby uczącej. Swoistą równowagę między złożonością klasyfikatora i błędem $\hat{e}_R$ uzyskuje się odpowiednio wybierając parametr C. Może on być estymowany np. metodą sprawdzania krzyżowego (patrz podrozdz. 1.10).

Interesującym jest fakt, że za pomocą metody sprawdzania krzyżowego można ograniczyć z góry poziom błędu – a mianowicie, wielkością $|\mathcal{SV}|/n$. Istotnie, obserwacje próby niebędące wektorami nośnymi nie wpływają na ostateczne rozwiązanie zadania optymalizacyjnego. Stąd obserwacje te – klasyfikowane poprawnie przez klasyfikator skonstruowany z całej n-elementowej próby – będą poprawnie klasyfikowane w procesie sprawdzania krzyżowego. Oszacowanie to uzasadnia dobre własności reguł klasyfikacyjnych o relatywnie małym odsetku wektorów nośnych, i to – co znamienne – nawet w przestrzeniach o bardzo wysokim wymiarze.

4.5. Optymalizacja

Rozwiązanie metody wektorów nośnych sprowadza się do minimalizacji formy kwadratowej $||\boldsymbol{w}||^2$ przy warunku pobocznym (4.2). Jest to zatem kwadratowe zadanie optymalizacyjne z ograniczeniem liniowym.
Ponieważ norma w przestrzeni l_2 jest funkcją wypukłą oraz warunki poboczne (4.2) są liniowe[11], można zastosować numeryczne metody programowania wypukłego.
Analogicznie, wypukłe kwadratowe zadanie optymalizacyjne otrzymujemy w przypadku grup liniowo nieseparowalnych. Większa wartość parametru C dopuszcza większy obszar możliwych rozwiązań ($0 \leq \alpha_j \leq C$). To z kolei przyczynia się do wzrostu liczby obliczeń.

Minimalizacja funkcji straty (4.14)[12] jest równoważna maksymalizacji funkcji:

$$L_D = \boldsymbol{\alpha}'\mathbf{1} - \frac{1}{2}\boldsymbol{\alpha}'\boldsymbol{D}\boldsymbol{\alpha}, \tag{4.54}$$

[11] Każde z ograniczeń liniowych definiuje zbiór wypukły. Warunek (4.2) jako przekrój n zbiorów wypukłych jest zbiorem wypukłym.

[12] Cortes i Vapnik (Burges, 1998) zdefiniowali funkcję straty postaci $\frac{1}{2}\boldsymbol{w}'\boldsymbol{w}+$ $+C\left(\sum_{j=1}^{n}\xi_j\right)^k$. Dla naturalnego k otrzymujemy wypukłe zadanie optymalizacyjne. W szczególności, gdy $k \in \{1,\ 2\}$ jest to (wypukłe) kwadratowe zadanie optymalizacyjne. Dla ułatwienia przyjmuje się zwykle $k = 1$ (przy przejściu do formy dualnej L_D znikają wówczas zmienne ξ_j oraz ich mnożniki Lagrange'a β_j).

gdzie $\boldsymbol{D}$ jest macierzą o elementach:

$$d_{jk} = y_j y_k K(\boldsymbol{x}_j,\ \boldsymbol{x}_k), \qquad j,\ k = 1,\ 2,\ \ldots,\ n,$$

przy warunkach pobocznych:

$$\boldsymbol{\alpha}'\boldsymbol{z} = 0, \qquad 0 \le \boldsymbol{\alpha} \le C. \tag{4.55}$$

Zadanie to ma dokładnie jedno rozwiązanie, jeśli tylko L_D jest ściśle wypukła (równoważnie – gdy hesjan jest dodatnio określony[13]).

Czas trwania obliczeń w głównej mierze zależy od liczby obserwacji n ciągu uczącego. Jest ona równa liczbie niewiadomych mnożników Lagrange'a α_j. Jednocześnie, standardowa implementacja kwadratowego zadania optymalizacyjnego wymaga przechowywania w pamięci komputera elementów macierzy $\boldsymbol{D}$. Niezbędnym okazuje się wówczas tzw. *algorytm rozkładu* (*dekompozycji*), w którym zadanie optymalizacji funkcji L_D zastępuje się ciągiem podzadań o mniejszym rozmiarze.

Niech

$$\boldsymbol{\alpha} = (\boldsymbol{\alpha}'_W,\ \boldsymbol{\alpha}'_N)',$$

gdzie

$$\boldsymbol{\alpha}_W = (\alpha_{j_1},\ \alpha_{j_2},\ \ldots,\ \alpha_{j_s})'$$

jest wektorem zmiennych, natomiast

$$\boldsymbol{\alpha}_N = (\alpha_{j_{s+1}},\ \alpha_{j_{s+2}},\ \ldots,\ \alpha_{j_n})'$$

wektorem ustalonych mnożników Lagrange'a α_j.
Zadanie maksymalizacji funkcji (4.54) przy warunkach pobocznych (4.55) zastępujemy zadaniem maksymalizacji funkcji:

$$L^*_{\boldsymbol{D}}(\boldsymbol{\alpha}_W) = \boldsymbol{\alpha}'_W\mathbf{1} + \boldsymbol{\alpha}'_N\mathbf{1} - \frac{1}{2}[\boldsymbol{\alpha}'_W\boldsymbol{D}_{WW}\boldsymbol{\alpha}_W + \boldsymbol{\alpha}'_W\boldsymbol{D}_{WN}\boldsymbol{\alpha}_N + \\ + \boldsymbol{\alpha}'_N\boldsymbol{D}_{NW}\boldsymbol{\alpha}_W + \boldsymbol{\alpha}'_N\boldsymbol{D}_{NN}\boldsymbol{\alpha}_N], \tag{4.56}$$

gdzie

$$\boldsymbol{D} = \begin{pmatrix} \boldsymbol{D}_{WW} & \boldsymbol{D}_{WN} \\ \boldsymbol{D}_{NW} & \boldsymbol{D}_{NN} \end{pmatrix},$$

[13] Jednoznaczność rozwiązania można otrzymać także w pewnych przypadkach dodatniej półokreśloności hesjanu (Burges, 1998).

przy warunkach pobocznych:

$$\begin{aligned} &\boldsymbol{\alpha}'_W \boldsymbol{z}_W + \boldsymbol{\alpha}'_N \boldsymbol{z}_N = 0 \\ &0 \leq \boldsymbol{\alpha}_W \leq C. \end{aligned} \tag{4.57}$$

Ponieważ $\boldsymbol{D}$ jest macierzą symetryczną, zaś składnik $\boldsymbol{\alpha}'_N \mathbf{1} - \frac{1}{2}\boldsymbol{\alpha}'_N \boldsymbol{D}_{NN} \boldsymbol{\alpha}_N$ jest stałą, przyjmujemy:

$$L^*_{\boldsymbol{D}}(\boldsymbol{\alpha}_W) = \boldsymbol{\alpha}'_W \mathbf{1} - \frac{1}{2}\left[\boldsymbol{\alpha}'_W \boldsymbol{D}_{WW} \boldsymbol{\alpha}_W + 2\boldsymbol{\alpha}'_W \boldsymbol{D}_{WN} \boldsymbol{\alpha}_N\right]. \tag{4.58}$$

Algorytm dekompozycji (rozkładu)

1. Wybierz losowo s ($s \ll n$) elementów ze zbioru $\{\alpha_j\}_{j=1}^n$. Niech $W = \{\alpha_{j_1}, \alpha_{j_2}, \ldots, \alpha_{j_s}\}$, $N = \{\alpha_j\}_{j=1}^n \setminus W$.
2. Znajdź maksimum funkcji (4.58) przy warunkach pobocznych (4.57).
3. Jeżeli istnieje element $\alpha_k \in N$ niespełniający warunku Karusha–Kühna–Tuckera, tzn. dla którego:

$$\begin{aligned} &y_k g(\boldsymbol{x}_k) < 1, \text{ jeśli } \alpha_k = 0 \\ &y_k g(\boldsymbol{x}_k) > 1, \text{ jeśli } \alpha_k = C \\ &y_k g(\boldsymbol{x}_k) \neq 1, \text{ jeśli } 0 < \alpha_k < C \end{aligned}$$

(gdzie rozwiązanie $g(\boldsymbol{x}_k)$ jest dane równaniem (4.28)), weź $W = (W \cup \{\alpha_k\}) \setminus \{\alpha_l\}$, $N = \{\alpha_j\}_{j=1}^n \setminus W$ i krok 2.

Można pokazać (Osuna, Freund, Girosi, 1997), że wartość funkcji $L^*_{\boldsymbol{D}}$ rośnie z każdą iteracją, tj. powyższy algorytm zbiega do optymalnego rozwiązania w skończonej liczbie kroków.

Metoda wektorów nośnych ma tylko dwa parametry. Są to: wielkość C oraz funkcja jądra K.

Z podrozdziału 4.4 wiemy, że przez regulację wielkości C możemy osiągnąć optymalną zależność między złożonością klasyfikatora a oceną aktualnego poziomu błędu uzyskaną metodą resubstytucji. Błędne jest bowiem postępowanie polegające na stosowaniu wyszukanych przekształceń $\boldsymbol{\varphi}$ w celu uzyskania liniowej separowalności grup w docelowej przestrzeni euklidesowej $\mathcal{E}$. Prowadzi to zwykle do przeestymowania granic decyzyjnych i w konsekwencji do dużej wartości oceny z próby testowej aktualnego poziomu błędu.

Sformułowany model nieliniowy nie zawiera jednak jakiegokolwiek kryterium wyboru jądra K. Nawet w przypadku ustalenia typu jądra zachodzi potrzeba estymacji jego parametrów. Potencjalnym kryterium może być nierówność (4.43) wraz z zależnością (4.47). Wymaga ona jednak wyznaczenia promienia R wyrażonego w terminach $\varphi(\boldsymbol{x}_j)$. Stąd estymacji parametrów funkcji jądrowej dokonuje się najczęściej empirycznie – na drodze minimalizacji oceny z próby testowej aktualnego poziomu błędu bądź w przypadku jej braku – na drodze minimalizacji oceny z próby uczącej aktualnego poziomu błędu uzyskanej np. metodą sprawdzania krzyżowego.

Przykład 4.5.
Metodę wektorów nośnych zilustrujemy, wykorzystując rzeczywiste dane doświadczalne Wisconsin Breast Cancer[14]. Analizowana próba obejmowała 699 kobiet z wykrytym guzem piersi – u 458 z nich stwierdzono guza łagodnego (klasa 1), a u pozostałych 241 – guza złośliwego (klasa 2). Dokonano pomiarów 9 różnych cech ilościowych o wartościach ze zbioru $\{1, 2, \ldots, 10\}$. Niezaobserwowane wartości cechy szóstej u 19 kobiet zostały zastąpione wartością średnią. Oryginalne dane przekształcono w składowe główne. Aby możliwa była graficzna prezentacja wyników na płaszczyźnie, wykorzystano dwie pierwsze składowe główne, które zawierają łącznie blisko 80% informacji o zmienności 9 cech pierwotnych (patrz rozdz. 11).
Zastosowano metodę wektorów nośnych z użyciem jądra normalnego.
W tabeli 4.2 podano wyniki otrzymane dla różnych wartości parametru σ. Optymalny, aktualny poziom błędu, estymowany 10-krokową metodą sprawdzania krzyżowego otrzymano dla wartości $\sigma = 7$ (błąd predykcji poniżej 3%). Wektory nośne stanowią w tym przypadku 7% wszystkich obserwacji.

Tab. 4.2. Wyniki otrzymane dla różnych wartości parametru σ $(C = 10, s = 50)$.

σ	# iteracji	$\lvert\mathcal{SV}\rvert$	$\hat{e}_{10CV}$
1	46	140	0.174
2	35	79	0.084
3	11	64	0.080
4	8	60	0.033
5	9	51	0.036
6	9	50	0.030
7	**6**	**49**	**0.029**
8	6	53	0.030
9	6	53	0.032
10	6	54	0.032
15	6	59	0.083
20	5	61	0.086
30	5	66	0.069

[14]Dane pochodzą ze składnicy danych Uniwersytetu Kalifornijskiego w Irvine. http://www.ics.uci.edu/~mlearn/MLRepository.html.

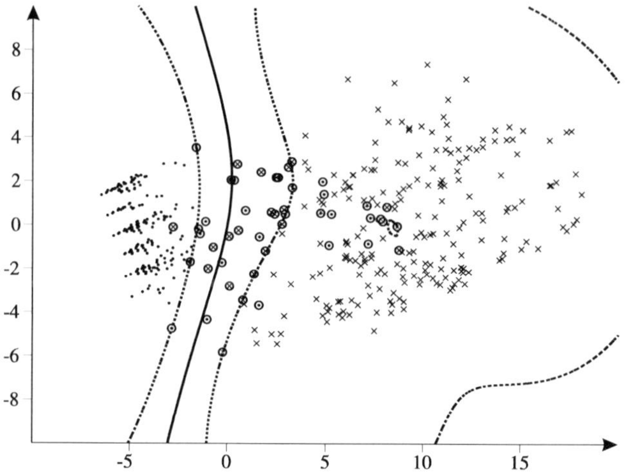

Rys. 4.6. Optymalne granice decyzyjne

Na rysunkach 4.6 i 4.7 przedstawiono odpowiednio optymalną postać granic decyzyjnych (wektory nośne oznaczono symbolem $\odot$) oraz wartości mnożników Lagrange'a α_j ($j = 1, 2, \ldots, 699$) dla poszczególnych elementów próby.

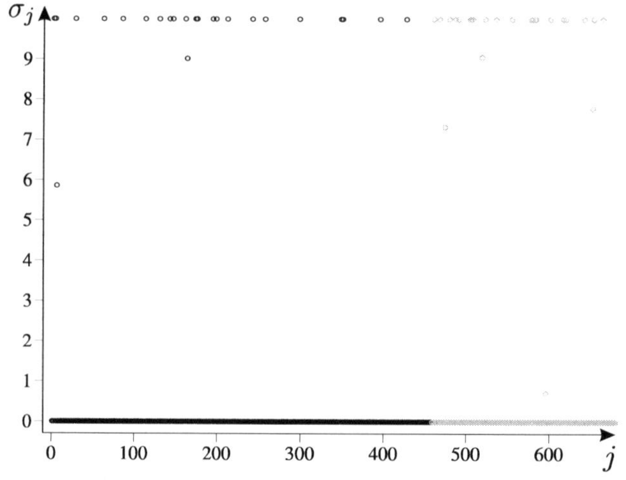

Rys. 4.7. Wartości mnożników Lagrange'a α_j dla poszczególnych obserwacji próby uczącej (optymalna wartość σ)

Granice decyzyjne oraz wartości mnożników Lagrange'a dla $\sigma = 1$ i $\sigma = 30$ przedstawiono na rys. 4.8 oraz 4.9.

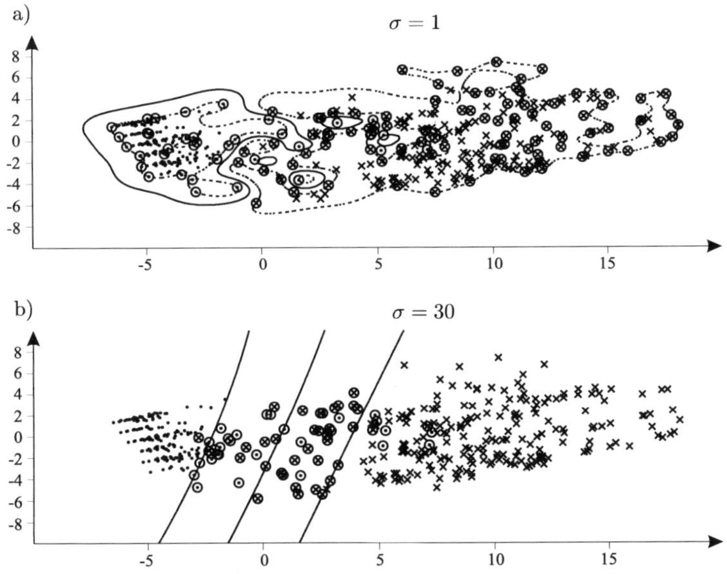

Rys. 4.8. Przeestymowane (a) i niedoestymowane (b) granice decyzyjne otrzymane w wyniku błędnego wyboru parametru σ

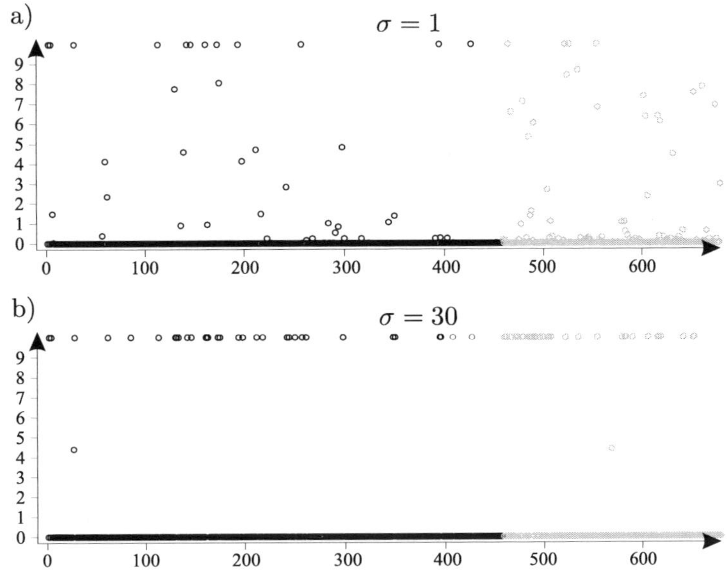

Rys. 4.9. Wartości mnożników Lagrange'a α_j dla poszczególnych obserwacji próby uczącej w przypadku przeestymowania (a) i niedoestymowania (b) granic decyzyjnych

Wybór zbyt małej wartości parametru σ prowadzi do przeestymowania granic decyzyjnych (rys. 4.8a), a w konsekwencji do wzrostu błędu. Wzrasta wtedy również liczba wektorów nośnych (20% obserwacji dla $\sigma = 1$). Znaczna część z nich leży na hiperpłaszczyznach kanonicznych. Oznacza to, że odpowiadające im mnożniki Lagrange'a α_j przyjmują wartości z przedziału (0; 10) (rys. 4.9a). Ale również wybór zbyt dużej wartości parametru σ prowadzi do niedoestymowania granic decyzyjnych (rys. 4.8b), wzrostu błędu, jak też liczby wektorów nośnych (9.5% obserwacji dla $\sigma = 30$). Większość z nich leży w obszarze pomiędzy hiperpłaszczyznami kanonicznymi, tzn. odpowiednie mnożniki Lagrange'a α_j osiągają górną wartość $C = 10$ (rys. 4.9b). Oczywiście, im bardziej regularne będą granice decyzyjne, tym mniejsza liczba iteracji algorytmu rozkładu. Standardowo, metody numeryczne rozwiązywania kwadratowego zadania optymalizacyjnego są inicjowane z wartościami startowymi $\alpha_j = 0$. Należy mieć na uwadze fakt, że część elementów próby pozostanie wektorami nośnymi nawet po zmianie wartości parametru (np. jądra). W tabeli 4.3 podano przykładowe wyniki eksperymentu przeprowadzonego przy zastosowaniu algorytmu rozkładu z wartościami startowymi α_j uzyskanymi na wyjściu algorytmu dla wartości $\sigma = 30$. Zauważmy, że liczba wektorów nośnych zwiększyła się nieznacznie. Istotnie zmniejszyła się natomiast liczba iteracji algorytmu ($\sigma = 1$ i $\sigma = 2$).

Tab. 4.3. Odpowiedni wybór wartości startowych α_j zwiększa szybkość algorytmu rozkładu

σ	# iteracji	$\|\mathcal{SV}\|$
1	26	146
2	13	83
3	9	63
7	6	49

W tabeli 4.4 podano wyniki eksperymentu uzyskane dla zmiennej wartości parametru C. Obranie zbyt małej straty związanej z błędną klasyfikacją ($C = 0.1$) prowadzi do niedoestymowania granic decyzyjnych (rys. 4.10a) oraz wysokim odsetkiem wektorów nośnych (15%).

Tab. 4.4. Wyniki otrzymane dla różnych wartości parametru C ($\sigma = 7$)

C	# iteracji	$\|\mathcal{SV}\|$
0.1	7	105
1	6	67
10	7	49
100	9	46
1000	21	49
10 000	18	55
100 000	22	55

Narzucenie zbyt dużej straty ($C = 100\,000$) daje w efekcie przeestymowane granice decyzyjne (rys. 4.10b). Skutkuje to wzrostem liczby iteracji algorytmu, a stąd czasu trwania obliczeń. Nieznacznie zwiększyła się natomiast liczba wektorów nośnych.

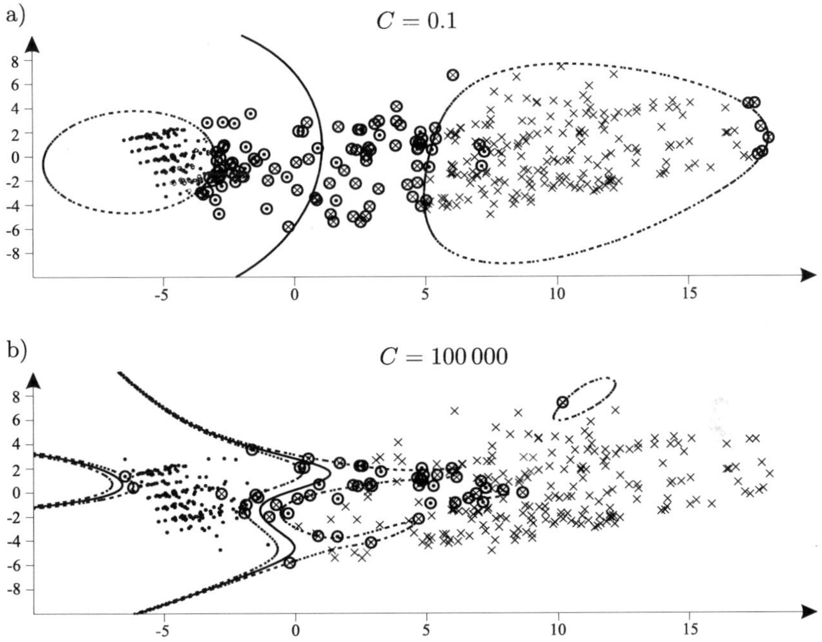

Rys. 4.10. Granice decyzyjne przy zastosowaniu różnych wartości parametru C ($\sigma = 7$)

Nierzadko natura przeprowadzanego eksperymentu wymaga, by (w miarę możności) nie dopuszczać do błędnej klasyfikacji obserwacji pochodzących z pewnej klasy. Takim przykładem może być właśnie rozpatrywany zbiór danych. Chcąc ograniczyć liczbę błędnych diagnoz dla obserwacji z klasy 1 (obiekty z nowotworem łagodnym), możemy zwiększyć wartość straty C dla tej klasy. Na rysunku 4.11 przedstawiono granice decyzyjne ze stratą C równą 1 dla klasy 1 oraz równą 100 dla klasy 2. Porównując rysunki 4.6 i 4.11, widzimy, że – zgodnie z intencją – nastąpiło wyraźne przesunięcie granic decyzyjnych w kierunku obserwacji z klasy 2.

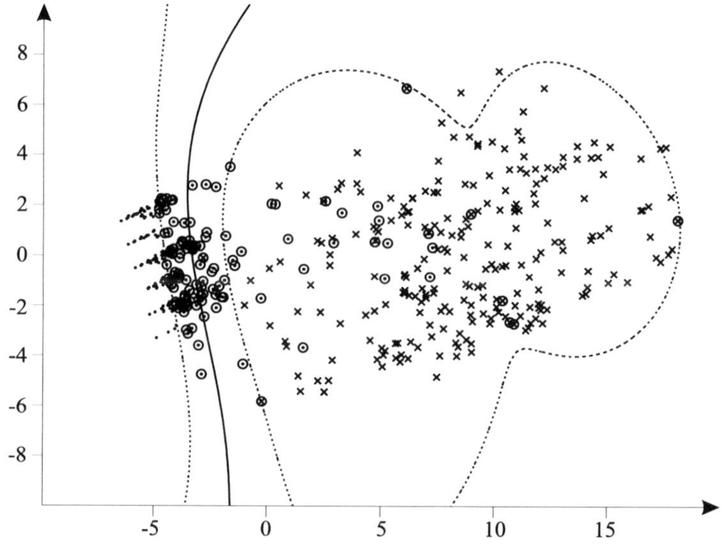

Rys. 4.11. Granice decyzyjne przy zastosowaniu różnych wartości parametru C dla poszczególnych klas ($\sigma = 7$)

5

Metoda najbliższego sąsiada

Jedną z najistotniejszych nieparametrycznych metod klasyfikacji jest metoda *J-najbliższych sąsiadów*, oznaczana przez J-NN[1]. Metoda ta została zaproponowana przez Fixa i Hodgesa (1951). W metodzie tej zaliczamy rozpoznawany obiekt do tej klasy, do której należy większość z jego J najbliższych sąsiadów. Metoda ta jest intuicyjnie bardzo łatwa w interpretacji. Estymator prawdopodobieństwa a posteriori przynależności obserwacji $\boldsymbol{x}$ do klasy k jest obliczany jako udział obserwacji z tej klasy wśród jej J najbliższych sąsiadów, czyli

$$\hat{p}(k|\boldsymbol{x}) = \frac{1}{J}\sum_{i=1}^{n} I(\rho(\boldsymbol{x},\ \boldsymbol{x}_i) \leq \rho(\boldsymbol{x},\ \boldsymbol{x}^{(J)}))I(y_i = k), \qquad k = 1,\ \ldots,\ K, \tag{5.1}$$

gdzie $I(A)$ jest funkcją indykatorową zdarzenia A, $\boldsymbol{x}^{(J)}$ jest J-tym co do odległości od $\boldsymbol{x}$ punktem z próby uczącej, natomiast ρ jest pewną odległością (lub ogólniej miarą niepodobieństwa obiektów). Miary niepodobieństwa obiektów zostaną szczegółowo omówione w podrozdz. 5.1.

Klasyfikator metody J-najbliższych sąsiadów ma zatem postać

$$\hat{d}_{NN}(\boldsymbol{x}) = \arg\max_{1\leq k\leq K} \hat{p}(k|\boldsymbol{x}),$$

gdzie $\hat{p}(k|\boldsymbol{x})$ jest dane wzorem (5.1).

Metoda ta ma bardzo wysoką efektywność, gdy liczba obserwacji rośnie nieskończenie; zostało to pokazane przez Covera i Harta (1967). Jednakże w wielu sytuacjach praktycznych liczba dostępnych obserwacji jest niewielka, co często prowadzi do drastycznego spadku efektywności metody

---

[1] Po ang. J-Nearest Neighbors (J-NN).

najbliższych sąsiadów. Metoda najbliższych sąsiadów nie wymaga estymacji warunkowych funkcji gęstości, jest więc zdecydowanie prostsza w implementacji. Dillon i Hart (1984) pokazali, że nawet jeśli mamy metodę, która jest zgodna (tzn. asymptotycznie daje optymalny błąd bez względu na rozkłady cech w klasach), np. taką jak metoda najbliższych sąsiadów, jej efektywność na zbiorze skończonym może być zupełnie niewystarczająca.

5.1. Miara niepodobieństwa

Szczególnie ważne w przypadku tej metody jest przyjęcie właściwej odległości, a w zasadzie miary niepodobieństwa obiektów[2]. Wiele informacji o miarach niepodobieństwa można znaleźć w książkach Krzanowskiego i Marriotta (1994, 1995). Miary takie są również podstawowym elementem analizy skupień (którą omówimy w rozdz. 12). Szczegółowe omówienie różnych miar niepodobieństwa, z punktu widzenia tej właśnie analizy, można znaleźć np. w monografii Kaufmana i Rousseeuwa (1990).

Określmy zatem dokładnie, co rozumiemy pod pojęciem miary niepodobieństwa między obiektami.

Niech $\boldsymbol{x}_r$ i $\boldsymbol{x}_s$ będą p-wymiarowymi wektorami obserwacji r-tego i s-tego obiektu ($r,\ s = 1,\ 2,\ \ldots,\ n$).

Definicja 5.1.
Funkcję $\rho : \mathcal{X} \times \mathcal{X} \to \mathbb{R}$ nazywamy *miarą niepodobieństwa*, jeśli:

1. $\rho(\boldsymbol{x}_r,\ \boldsymbol{x}_s) \geq 0$,
2. $\rho(\boldsymbol{x}_r,\ \boldsymbol{x}_s) = 0$ wtedy i tyko wtedy, gdy $\boldsymbol{x}_r = \boldsymbol{x}_s$,
3. $\rho(\boldsymbol{x}_r,\ \boldsymbol{x}_s) = \rho(\boldsymbol{x}_s,\ \boldsymbol{x}_r)$.

Określona w ten sposób miara jest semi-metryką na przestrzeni próby. Jak widać, nie musi ona być (choć często jest) metryką, tzn. nie musi spełniać warunku trójkąta: $\rho(\boldsymbol{x}_r,\ \boldsymbol{x}_s) \leq \rho(\boldsymbol{x}_r,\ \boldsymbol{x}_t) + \rho(\boldsymbol{x}_t,\ \boldsymbol{x}_s)$. Nierówność trójkąta nie jest nam potrzebna do określenia kolejności odległości punktów od $\boldsymbol{x}$, ponieważ nie interesują nas odległości między pozostałymi punktami. Wybór miary niepodobieństwa obiektów jest arbitralny i zależy głównie od charakteru danych. Poniżej podane są przykłady kilku najpopularniejszych tego typu miar.

[2]Po ang. dissimilarity measure.

Dla danych ilościowych, jako miarę niepodobieństwa między obiektami, stosuje się często zwykłą *odległość (metrykę) euklidesową*

$$\rho_1(\boldsymbol{x}_r,\ \boldsymbol{x}_s) = ((\boldsymbol{x}_r - \boldsymbol{x}_s)'(\boldsymbol{x}_r - \boldsymbol{x}_s))^{1/2} = \left(\sum_{i=1}^{p}(x_{ri} - x_{si})^2\right)^{1/2} \tag{5.2}$$

lub jej kwadrat

$$\rho_2(\boldsymbol{x}_r,\ \boldsymbol{x}_s) = (\boldsymbol{x}_r - \boldsymbol{x}_s)'(\boldsymbol{x}_r - \boldsymbol{x}_s) = \sum_{i=1}^{p}(x_{ri} - x_{si})^2. \tag{5.3}$$

Zwróćmy uwagę, że miara (5.3) nie jest metryką, ponieważ nie jest dla niej spełniony warunek trójkąta.

Jeżeli cechy opisujące obiekty są wyrażone w różnych jednostkach, to w celu zniwelowania ich wpływu możemy zastosować *ważoną odległość euklidesową*

$$\rho_3(\boldsymbol{x}_r,\ \boldsymbol{x}_s) = ((\boldsymbol{x}_r - \boldsymbol{x}_s)'\boldsymbol{W}^{-1}(\boldsymbol{x}_r - \boldsymbol{x}_s))^{1/2} = \left(\sum_{i=1}^{p}\frac{1}{w_i^2}(x_{ri} - x_{si})^2\right)^{1/2},$$

gdzie $\boldsymbol{W} = \mathrm{diag}\{w_1^2,\ \ldots,\ w_p^2\}$, a wagi w_i są odchyleniami standardowymi poszczególnych cech (lub rzadziej ich rozstępami). Aby miara uwzględniała również korelacje między cechami, stosujemy jako miarę niepodobieństwa *odległość Mahalanobisa*

$$\rho_4(\boldsymbol{x}_r,\ \boldsymbol{x}_s) = ((\boldsymbol{x}_r - \boldsymbol{x}_s)'\boldsymbol{S}^{-1}(\boldsymbol{x}_r - \boldsymbol{x}_s))^{1/2},$$

gdzie $\boldsymbol{S}$ jest estymatorem macierzy kowariancji.

Czasami, choć rzadko, stosuje się również inne miary niepodobieństwa obiektów. Przykładowo:
odległość miejską[3] (zwaną również *taksówkową* lub *manhatańską*)

$$\rho_5(\boldsymbol{x}_r,\ \boldsymbol{x}_s) = \sum_{i=1}^{p}|x_{ri} - x_{si}|.$$

Odległość ta, tak samo jak odległość euklidesowa, jest szczególnym przypadkiem *odległości Minkowskiego* w przestrzeni $\mathbb{R}^p$ danej wzorem:

$$\rho(\boldsymbol{x}_r,\ \boldsymbol{x}_s) = \left(\sum_{i=1}^{p}|x_{ri} - x_{si}|^q\right)^{1/q}$$

[3] Po ang. city-block distance.

odległość Czebyszewa

$$\rho_6(\boldsymbol{x}_r,\ \boldsymbol{x}_s) = \max_{1 \leq i \leq p} |x_{ri} - x_{si}|,$$

odległość Frechéta

$$\rho_7(\boldsymbol{x}_r,\ \boldsymbol{x}_s) = \sum_{i=1}^{p} \frac{|x_{ri} - x_{si}|}{1 + |x_{ri} - x_{si}|} \frac{1}{2^i},$$

odległość Canberry

$$\rho_8(\boldsymbol{x}_r,\ \boldsymbol{x}_s) = \sum_{i=1}^{p} \frac{|x_{ri} - x_{si}|}{|x_{ri} + x_{si}|},$$

odległość poczty

$$\rho_9(\boldsymbol{x}_r,\ \boldsymbol{x}_s) = \begin{cases} \rho_2(\boldsymbol{x}_r,\ 0) + \rho_2(0,\ \boldsymbol{x}_s), & \text{dla } \boldsymbol{x}_r \neq \boldsymbol{x}_s, \\ 0, & \text{dla } \boldsymbol{x}_r = \boldsymbol{x}_s, \end{cases}$$

odległość metra

$$\rho_{10}(\boldsymbol{x}_r,\ \boldsymbol{x}_s) = \begin{cases} 0, & \text{dla } \boldsymbol{x}_r = \boldsymbol{x}_s, \\ \rho_2(\boldsymbol{x}_r,\ \boldsymbol{x}_s), & \text{dla } \boldsymbol{x}_r \neq \boldsymbol{x}_s \text{ oraz } \boldsymbol{x}_r,\ \boldsymbol{x}_s \\ & \text{leżących na jednej prostej} \\ & \text{przechodzącej przez punkt } \mathbf{0}, \\ \rho_2(\boldsymbol{x}_r,\ 0) + \rho_2(0,\ \boldsymbol{x}_s), & \text{poza tym.} \end{cases}$$

W przypadku danych jakościowych, możemy w naturalny sposób zdefiniować miarę niepodobieństwa obiektów jako

$$\rho_{11}(\boldsymbol{x}_r,\ \boldsymbol{x}_s) = \frac{1}{p} \sum_{i=1}^{p} I(x_{ri} \neq x_{si}). \tag{5.4}$$

Miara ta jest nazywana *współczynnikiem podobieństwa Sneatha.*

Na szczególną uwagę zasługuje sytuacja danych binarnych, tzn. takich, gdzie każda cecha może przyjmować tylko dwie wartości (0 albo 1). Miara (5.4) przyjmuje wtedy postać

$$\rho_{12}(\boldsymbol{x}_r,\ \boldsymbol{x}_s) = \frac{b+c}{p} = 1 - \frac{a+d}{p},$$

gdzie a i d oznaczają liczbę cech zgodnych (1-1, 0-0 odpowiednio), b i c niezgodnych (1-0, 0-1 odpowiednio). Miara ta nosi nazwę *współczynnika dopasowania obiektów*.

Do innych, szczególnie często wykorzystywanych w tej sytuacji, miar niepodobieństwa obiektów należą, *statystyka* χ^2

$$\rho_{13}(\boldsymbol{x}_r,\ \boldsymbol{x}_s) = \frac{p(ad-bc)^2}{(a+b)(c+d)(a+c)(b+d)}$$

oraz *współczynnik Jaccarda*

$$\rho_{14}(\boldsymbol{x}_r,\ \boldsymbol{x}_s) = \frac{b+c}{a+b+c}$$

stosowany wtedy, gdy zerowa wartość cechy oznacza jej brak w danym obiekcie.

Rozważmy jeszcze następującą sytuację. Niech dane będą miary niepodobieństwa $\rho(x_{ri},\ x_{si})$, $(i = 1,\ \dots,\ p)$ oddzielnie dla każdej z p cech. Wtedy za *całkowitą miarę niepodobieństwa między obiektami* możemy przyjąć kombinację wypukłą miar brzegowych postaci

$$\rho(\boldsymbol{x}_r,\ \boldsymbol{x}_s) = \sum_{i=1}^{p} w_i^2 \rho(x_{ri},\ x_{si}), \qquad \sum_{i=1}^{p} w_i^2 = 1.$$

Wielkość w_i jest wagą uwzględniającą udział danej cechy. W szczególności, gdy $\rho(x_{ri},\ x_{si}) = (x_{ri} - x_{si})^2$, to całkowita miara niepodobieństwa $\rho(\boldsymbol{x}_r,\ \boldsymbol{x}_s)$ jest kwadratem ważonej odległości euklidesowej.

Warto podkreślić, że nawet racjonalny wybór miary niepodobieństwa nie gwarantuje dobrych wyników klasyfikacji. Dzieje się tak dlatego, że część obserwowanych cech może okazać się nieistotna z punktu widzenia klasyfikacji. Cechy te mają jedynie wpływ na kształt otoczeń punktów, nie mają natomiast wpływu na kształt granic decyzyjnych. W wyniku jedynie zakłócają proces decyzyjny.

5.2. Estymacja funkcji gęstości

Idea metody najbliższych sąsiadów została wykorzystana do nieparametrycznej estymacji gęstości prawdopodobieństwa. Niech $K(\boldsymbol{x})$ oznacza kulę o środku w punkcie $\boldsymbol{x}$ oraz objętości $V(\boldsymbol{x})$. Wówczas prawdopodobieństwo, że losowo wybrany punkt $\boldsymbol{x}_0$ leży w kuli $K(\boldsymbol{x})$ wynosi:

$$\theta = \int_{K(\boldsymbol{x})} f(\boldsymbol{x}) d\boldsymbol{x}. \tag{5.5}$$

Dla małych wartości $V(\boldsymbol{x})$ mamy:

$$\theta \approx f(\boldsymbol{x}) V(\boldsymbol{x}). \tag{5.6}$$

Ale również, jeżeli spośród n obserwacji dokładnie J leży w kuli $K(\boldsymbol{x})$, to

$$\theta \approx \frac{J}{n}.$$

Z zależności (5.5) oraz (5.6) otrzymujemy:

$$\hat{f}(\boldsymbol{x}) = \frac{J}{nV(\boldsymbol{x})}. \tag{5.7}$$

W metodzie J najbliższych sąsiadów, dla danego n, ustala się $J = J(n)$. Wówczas $V(\boldsymbol{x})$ jest objętością kuli o środku w punkcie $\boldsymbol{x}$ i promieniu $\rho(\boldsymbol{x},\ \boldsymbol{x}^{(J)})$, gdzie $\boldsymbol{x}^{(J)}$ jest J-tym co do odległości od $\boldsymbol{x}$ punktem z próby.

Estymator (5.7) można przedstawić w alternatywnej postaci:

$$\hat{f}(\boldsymbol{x}) = \frac{1}{n[\rho(\boldsymbol{x},\ \boldsymbol{x}^{(J)})]^p} \sum_{i=1}^{n} K\left(\frac{\boldsymbol{x} - \boldsymbol{x}_i}{\rho(\boldsymbol{x},\ \boldsymbol{x}^{(J)})}\right), \tag{5.8}$$

gdzie K jest funkcją jądrową i w tym przypadku jest to gęstość rozkładu jednostajnego na jednostkowej sferze w przestrzeni $\mathbb{R}^p$. Można więc metodę najbliższych sąsiadów rozpatrywać jako metodę jądrową, przy czym dodatkową zaletę stanowi fakt, że parametr gładkości $h = \rho(\boldsymbol{x},\ \boldsymbol{x}^{(J)})$ nie jest wielkością stałą, lecz różni się lokalnie (jest funkcją argumentu $\boldsymbol{x}$). Estymatory gęstości otrzymane w ten sposób mają często lokalnie dobre własności (dobrze oddają kształt gęstości w obszarze jej stosunkowo dużych wartości), ale można pokazać, że całka z takiego estymatora po całej przestrzeni $\mathbb{R}^p$ jest równa nieskończoności.

Przykład 5.1.
Wygenerowano 100 obserwacji z rozkładu $N(0,\ 1)$. Następnie za pomocą formuły (5.7) estymowano funkcję gęstości tego rozkładu. Wyniki przedstawiono na rys. 5.1. Jak widać, w problemie estymacji gęstości prawdopodobieństwa

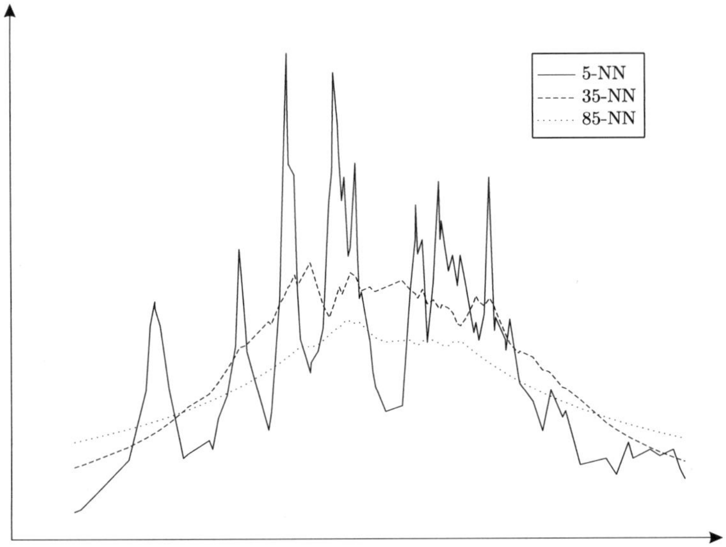

Rys. 5.1. Estymator gęstości prawdopodobieństwa dla różnych wartości parametru J

$f(x)$, parametr J odgrywa analogiczną rolę jak parametr gładkości h w metodzie jądrowej. Dla zbyt małych wartości J dochodzi do przeestymowania funkcji $f(x)$ na co wskazują „ostrza" w punktach próby. Natomiast duże wartości J prowadzą do zbyt „gładkiego" estymatora i utraty informacji o strukturze funkcji gęstości $f(x)$.

5.3. Własności graniczne

Okazuje się, że metoda najbliższych sąsiadów ma bardzo ciekawe własności asymptotyczne. Niech ε_n oznacza bezwarunkowe średnie prawdopodobieństwo błędu metody najbliższych sąsiadów na n-elementowym zbiorze i ε^* oznacza błąd bayesowski. Cover i Hart (1967) pokazali, że dla metody 1-NN (ogólna postać tego ograniczenia znajduje się w książce Dudy, Harta i Storka (2001)) zachodzi następująca nierówność:

$$\varepsilon^* \leq \varepsilon \leq \varepsilon^* \left(2 - \frac{K}{K-1}\varepsilon^*\right), \tag{5.9}$$

jeśli

$$\varepsilon = \lim_{n \to \infty} \varepsilon_n.$$

Można łatwo pokazać, że górne ograniczenie jest osiągane dla tzw. nieinformującego przypadku, gdzie gęstości we wszystkich klasach są takie same oraz prawdopodobieństwa a priori również są takie same. Kiedy błąd bayesowski jest mały, górne ograniczenie jest w przybliżeniu równe podwojonemu błędowi bayesowskiemu (rys. 5.2).

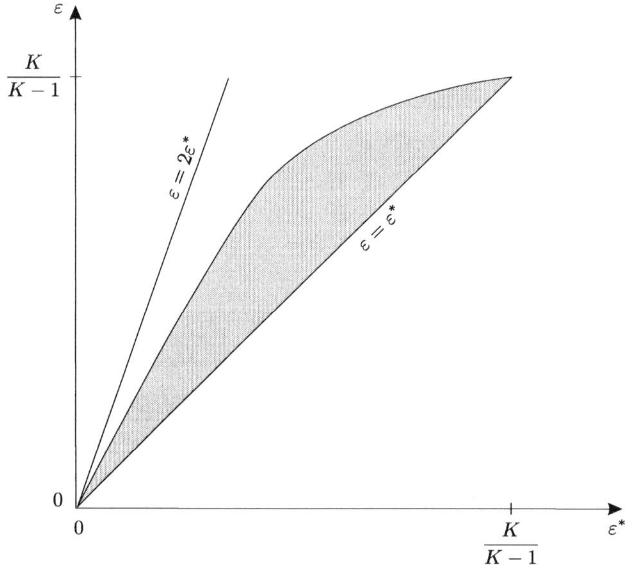

Rys. 5.2. Ograniczenie współczynnika błędu w metodzie J najbliższych sąsiadów (zaciemnione pole oznacza możliwe wartości poziomu błędu)

Ponieważ ε jest zawsze mniejszy bądź równy $2\varepsilon^*$, przynajmniej połowa informacji klasyfikacyjnej jest zawarta w najbliższym sąsiedzie. Co za tym idzie, żadna reguła klasyfikacyjna, oparta na nieskończonej próbie uczącej, typu najbliższy sąsiad nie poprawi błędu klasyfikacji o więcej niż połowę. Co ważne, zależność (5.9) jest prawdziwa bez względu na przyjętą miarę niepodobieństwa.

Przykład 5.2.
Rozważmy problem klasyfikacji do jednej z dwóch klas, zakładając, że prawdopodobieństwa a priori są równe, czyli $\pi_1 = \pi_0 = 0.5$. Gęstości w klasach są następujące:

$$f_1(x) = \begin{cases} 2x, & 0 \leq x \leq 1 \\ 0, & \text{w przeciwnym razie} \end{cases}$$

oraz

$$f_0(x) = \begin{cases} 2 - 2x, & 0 \le x \le 1 \\ 0, & \text{w przeciwnym razie.} \end{cases}$$

Skonstruujmy najpierw bayesowską regułę decyzyjną, która jest wyznaczona przez następującą nierówność:

$$P(Y = 1|X = x) > P(Y = 0|X = x)$$

lub równoważnie przez nierówność

$$f_1(x)\pi_1 > f_0(x)\pi_0.$$

Rozwiązując tę nierówność otrzymujemy, że obserwację zaliczamy do klasy pierwszej, jeśli $x > 0.5$. W przeciwnym razie obserwacja zaliczana jest do klasy zerowej. Możemy teraz znaleźć błąd bayesowski:

$$\varepsilon^* = \pi_1 \int_{L_0} f_1(x)dx + \pi_0 \int_{L_1} f_0(x)dx = 0.5 \int_0^{0.5} 2xdx + 0.5 \int_{0.5}^{1} (2 - 2x)dx = 0.25.$$

Obszary całkowania L_1 i L_0 to obszary klasyfikacji obserwacji x odpowiednio do klasy pierwszej i zerowej (rys. 5.3).

Z nierówności (5.9) wiemy, że błąd metody najbliższego sąsiada będzie się mieścił w przedziale $[0,\ 0.5]$. Można pokazać, że jeśli dysponujemy próbą

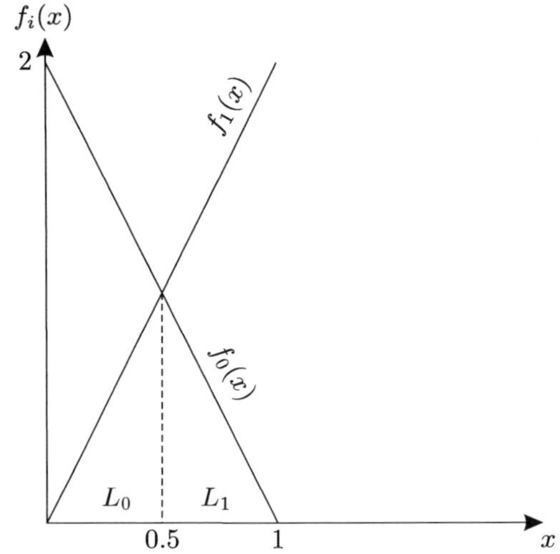

Rys. 5.3. Obszary klasyfikacji L_1 oraz L_0

uczącą o liczebności n to błąd klasyfikacji będzie na poziomie:

$$\varepsilon_n = \frac{1}{3} + \frac{1}{(n+1)(n+3)} + \frac{1}{2(n+2)(n+3)}. \tag{5.10}$$

Z tego otrzymujemy, że $\varepsilon = \frac{1}{3}$. Dla $n = 1$ otrzymujemy wartość 0.5, co odpowiada sytuacji, w której mamy jedną obserwację uczącą i zależnie, z której jest klasy, do tej zaliczamy wszystkie obserwacje testowe.

Metoda J-NN jest optymalna w sensie błędu bayesowskiego, gdyż jeśli $n \to \infty$, $J \to \infty$ oraz $J/n \to 0$, to prawdopodobieństwo błędu metody J-NN dąży do błędu bayesowskiego. Zauważmy, dodatkowo, że na powyższe fakty nie ma wpływu wybór miary niepodobieństwa.

5.4. Metody rozwiązywania sytuacji remisowych

Używając metody najbliższych sąsiadów, często nie jesteśmy pewni, do której klasy powinniśmy zaklasyfikować nową obserwację. Dzieje się tak dlatego, że bardzo często w najbliższym otoczeniu pewnych punktów znajdują się obserwacje ze wszystkich lub z kilku klas, czyli mamy takie same prawdopodobieństwa a posteriori dla dwóch lub więcej klas, tzn. dla j-tej klasy zachodzi

$$\max_{i \neq j} P(y_k = i|\mathbf{x}_k) = P(y_k = j|\mathbf{x}_k).$$

Mamy wtedy do czynienia z sytuacją remisową. Rozwiązywanie takich sytuacji stanowi jeden z głównych problemów związanych z metodą najbliższych sąsiadów. Musimy zatem wprowadzić jakieś metody rozwiązywania takich spornych sytuacji. Proponuje się dwie metody:

- W sytuacji remisowej dodajemy następnego sąsiada, czyli zamiast metody J najbliższych sąsiadów używamy metody $J+1$ najbliższych sąsiadów; jeśli nadal mamy remis, to dodajemy kolejnego sąsiada i postępujemy tak, aż do rozwiązania sytuacji remisowej.
- Wybieramy klasę, do której należy najbliższy sąsiad klasyfikowanej obserwacji, czyli w rzeczywistości stosujemy metodę 1 najbliższego sąsiada.

W przypadku pierwszej strategii rozwiązywania sytuacji remisowych pojawiają się pewne pytania. Jak wielu sąsiadów potrzeba, żeby zawsze dało się remis rozstrzygnąć, oraz jakie J możemy wybrać w zależności od liczebności próby n, aby zawsze sytuacja remisowa była rozstrzygalna. Odpowiedź na te pytanie dają dwa poniższe twierdzenia, których dowody można znaleźć w pracy Góreckiego (2005b).

Twierdzenie 5.1.

Aby rozwiązać sytuację remisową w metodzie J najbliższych sąsiadów dla K klas, potrzebujemy co najwyżej

$$\begin{cases} \frac{J}{2}(K-2)+1, & \text{dla parzystego } J, \\ \frac{J-1}{2}(K-2), & \text{dla nieparzystego } J \end{cases}$$

nowych sąsiadów.

Twierdzenie 5.2.

Używając pierwszej metody rozwiązywania sytuacji remisowych dla zbioru złożonego z n obserwacji oraz K klas możemy być pewni, że uda nam się rozwiązać taką sytuację za pomocą metody J najbliższych sąsiadów, jeśli

$$J \leq \left\lceil \frac{n}{K} \right\rceil * 2 - 1.$$

Przykład 5.3.

Rozważmy następujący przypadek. Niech $n = 50$ będzie liczbą obserwacji oraz niech będzie danych 7 klas. Rozpatrzymy następujące dwa przypadki:

a) $J = 6$,

b) $J = 7$.

Aby rozwiązać remis, potrzebujemy co najwyżej

a) $\frac{6}{2}(7-2)+1 = 16$,

b) $\frac{7-1}{2}(7-2) = 15$

nowych obserwacji. Jednakże jeśli chcemy być pewni, że ta metoda rozwiąże nam sytuację remisową, musimy wybrać J co najwyżej równe

$$J = \left\lceil \frac{50}{7} \right\rceil * 2 - 1 = 15.$$

5.5. Metody wyboru reprezentatywnego podzbioru obserwacji

Wadą metody najbliższego sąsiada jest duży nakład obliczeniowy niezbędny do klasyfikacji każdego obiektu. Jest to związane ze znaczną liczbą obliczeń niezbędną do wyznaczenia odległości między obiektami. W literaturze można znaleźć pewne techniki likwidujące te niedogodności. Z jednej strony są to techniki numeryczne poprawiające szybkość znajdowania odległości, z drugiej techniki dokonujące redukcji danych. Do tej pierwszej grupy należy *algorytm LAESA*[4] zaproponowany przez Mico, Oncina i Vidala (1994). Do drugiej grupy należą algorytmy edycyjne[5] oraz algorytmy kondensujące informację[6]. Celem tych pierwszych jest wybór niedużych i jednorodnych skupień obserwacji. Celem natomiast drugiego typu algorytmów jest wybranie jak najmniejszej liczby obserwacji, które dobrze reprezentują klasy (położone są one możliwie blisko brzegów). Podobne zadania spełnia metoda kwantyzacji wektorowej LVQ[7] (szczegóły w rozdz. 7). W wyniku tych procedur otrzymuje się pewien podzbiór $\mathbf{V}$ zbioru uczącego $\mathbf{Z}$. Do najczęściej wykorzystywanych technik tego rodzaju należy: *metoda Harta* (1968) oraz *metoda Wilsona* (1972). Ta pierwsza należy do technik kondensacji. Zaczynamy od zbioru $\mathbf{V}$ zawierającego jedną obserwację $\mathbf{z}_1$, która jest uznawana za uczącą. Do zbioru V przesuwamy wszystkie obserwacje, które są błędnie klasyfikowane przez metodę 1-NN uczoną na aktualnym zbiorze $\mathbf{V}$. Procedurę powtarzamy tak długo, aż wszystkie obserwacje ze zbioru $\mathbf{Z}$ są klasyfikowane poprawnie. Technika ta ma tendencję do zachowywania obserwacji leżących w pobliżu granic decyzyjnych oraz usuwania obserwacji leżących „głębiej". Metoda Wilsona należy natomiast do technik edycyjnych. W przypadku tej techniki jest wykonywana metoda J-NN (proponowane $J = 3$) na zbiorze $\mathbf{Z}$ i są usuwane wszystkie elementy, które zostały błędnie zaklasyfikowane. Pozostałe elementy tworzą zbiór $\mathbf{V}$, na którym powinna być używana metoda 1-NN. Zaskakująco dobre wyniki daje *metoda losowej edycji*. Polega ona na losowym podziale zbioru uczącego z góry ustaloną ilość razy, przy czym rozmiar zbioru $\mathbf{V}$ jest również ustalony. Wynikiem jest zbiór, na którym uzyskano najmniejszą ocenę błędu metodą resubstytucji.

---

[4] Po ang. Linear Approximating and Eliminating Search Algorithm (LAESA).

[5] Po ang. editing algorithms.

[6] Po ang. condensing algorithms.

[7] Po ang. Learning Vector Quantization (LVQ).

Przykład 5.4.
Wygenerowano następujące dane uczące:

Grupa I: 200 obserwacji z rozkładu $N\left(\begin{bmatrix}2\\1\end{bmatrix}, \begin{bmatrix}1 & 0\\0 & 2\end{bmatrix}\right)$.

Grupa II: mieszanina 200 obserwacji z rozkładu $N\left(\begin{bmatrix}5\\4\end{bmatrix}, \begin{bmatrix}2 & -1\\-1 & 4\end{bmatrix}\right)$

oraz 200 obserwacji z rozkładu $N\left(\begin{bmatrix}2\\8\end{bmatrix}, \begin{bmatrix}2 & 0\\0 & 3\end{bmatrix}\right)$.

Grupa III: 200 obserwacji z rozkładu $N\left(\begin{bmatrix}0\\5\end{bmatrix}, \begin{bmatrix}1 & 0\\0 & 1\end{bmatrix}\right)$.

Następnie w taki sam sposób wygenerowano zbiór testowy. Okazało się, że na takiej próbie otrzymano najmniejszą ocenę błędu klasyfikacji dla $J = 9$ i wyniosła ona 0.0862 (rys. 5.4). Następnie zbiór uczący poddano edycji (rys. 5.5a) oraz kondensacji (rys. 5.5b) – na rysunkach przedstawiono sytuację dla metody 1-NN.

Użycie metody edycji zredukowało próbę uczącą (tab. 5.1) o 133 obserwacje, natomiast użycie metody kondensacji o dalsze 657 obserwacji. W rezultacie otrzymujemy próbę uczącą wielkości 10. Jak widać zastosowanie algorytmu edycji i kondensacji nie tylko znacznie zredukowało liczebność próby uczącej

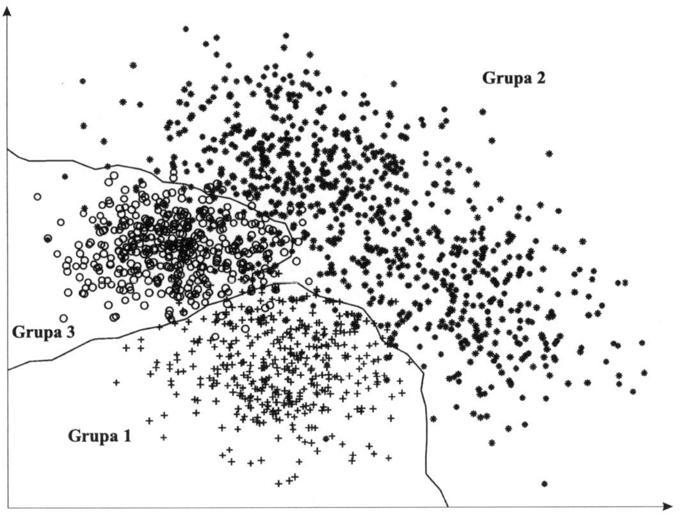

Rys. 5.4. Optymalne granice decyzyjne uzyskane dla metody 9 najbliższych sąsiadów

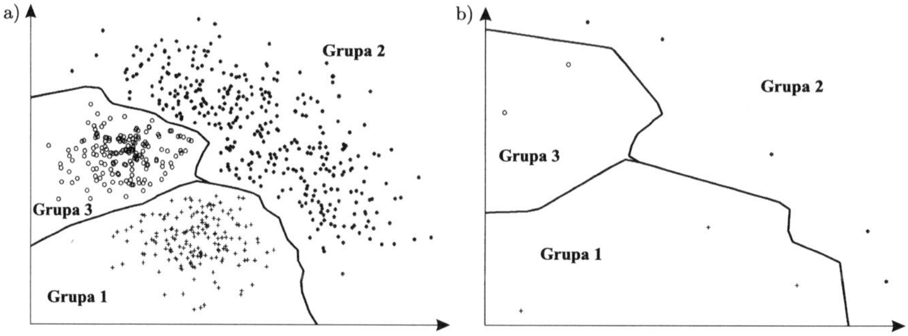

Rys. 5.5. a) Jednorodne skupienia obserwacji po edycji danych uczących, b) znacząca redukcja liczebności próby uczącej uzyskana w wyniku edycji i kondensacji nie wpływa istotnie na granice obszarów decyzyjnych

(a przez to zwiększyło szybkość obliczeń), ale również zmniejszyło błąd metody (tab. 5.2), dla 1-NN błąd zmalał z 11.87% do 8.38%. Natomiast w przypadku wybrania na początku metody 9-NN błąd zmalał z 8.62% do 8.38%.

Tab. 5.1. Redukcja próby po zastosowaniu algorytmu edycji i kondensacji

	Pierwotnie	Edycja	Kondensacja
Grupa 1	200	165	3
Grupa 2	400	339	5
Grupa 3	200	163	2
Razem	800	667	10

Tab. 5.2. Liczba błędnych decyzji klasyfikatorów przed i po zastosowaniu algorytmów edycji i kondensacji

	Próba pierwotna		Próba zredukowana	
	$J = 9$	$J = 1$	Edycja	Kondensacja
Grupa 1	13	25	16	20
Grupa 2	43	38	43	34
Grupa 3	13	32	10	13
Błąd	0.0862	0.1187	0.0862	0.0838

Przykład 5.5.
Przykład ten dotyczy tzw. problemu XOR. Jest to problem funkcji logicznej, dla której ciąg uczący przedstawiono w tab. 5.3. Mamy zatem dwie klasy, których nie można odseparować liniowo. Przykładowa granica decyzyjna jest przedstawiona na rys. 5.6.

Tab. 5.3. Ciąg uczący dla problemu XOR

x_1	x_2	$d = XOR(x_1, x_2)$
0	0	0
0	1	1
1	0	1
1	1	0

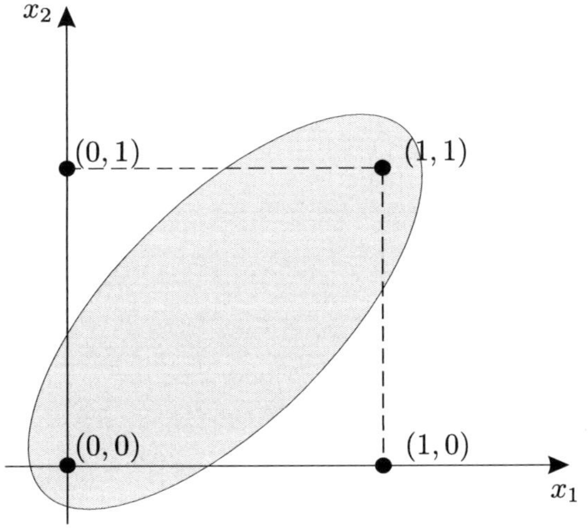

Rys. 5.6. Przykładowa granica decyzyjna dla problemu XOR

W celu pokazania własności techniki LVQ wygenerowano 2000 obserwacji dla tego problemu, z czego połowa została przeznaczona na zbiór uczący. Wybrano jednakowe liczebności klas. Wyniki działania metody najbliższego sąsiada przedstawiono w tab. 5.4.

Tab. 5.4. Błędy dla problemu XOR

	Błąd na zbiorze uczącym			Błąd na zbiorze testowym		
	Grupa 1	Grupa 0	Łączny	Grupa 1	Grupa 0	Łączny
1-NN bez LVQ	0.033	0.037	0.035	0.200	0.210	0.200
1-NN z LVQ	0.170	0.140	0.160	0.170	0.170	0.170

Widzimy, że technika LVQ prowadzi do ogromnej redukcji danych (w tym przypadku próba ucząca została zmniejszona z 1000 do 8, parametr określający wynikową liczebność próby uczącej może przybrać dowolną sensowną wartość, można więc zmniejszyć w tym przypadku próbę nawet do dwóch obserwacji).

Co ciekawe, taka redukcja nie powoduje spadku efektywności procesu klasyfikacji, wręcz przeciwnie, na zbiorze testowym widzimy zmniejszenie się błędu klasyfikacji. Interesujący jest fakt, że technika ta pozwala uniknąć zbytniego dopasowania się[8] metody do danych uczących (doskonałe wyniki metody bez LVQ na danych uczących i kiepskie na danych testowych). Dodatkowo widzimy (rys. 5.7a oraz 5.7b), że stosując metodę redukcji danych LVQ otrzymujemy dużo prostszą postać granic decyzyjnych. Bardzo ładnie granice decyzyjne wraz z pokazaniem, który punkt odpowiada za daną część całego obszaru, przedstawia diagram Voronoia (rys. 5.8). Więcej informacji na temat diagramów Voronoia znajduje się w podrozdz. 7.5.

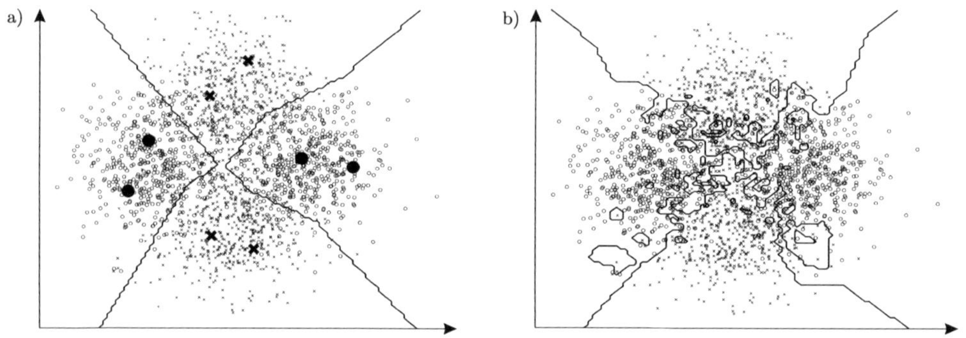

Rys. 5.7. Granice decyzyjne oraz punkty kluczowe dla metody 1-NN: a) z wykorzystaniem metody LVQ, b) bez wykorzystania metody LVQ

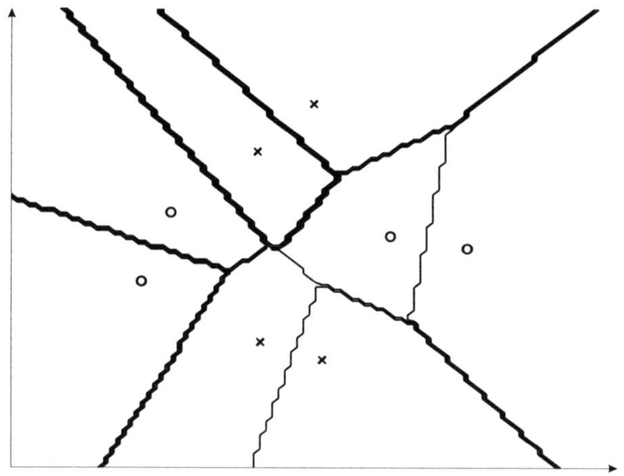

Rys. 5.8. Diagram Voronoia dla metody LVQ

[8]Po ang. overfitting.

5.6. Uwagi praktyczne

W praktyce wartość parametru J dobiera się eksperymentalnie, stosując bądź to próbę testową, bądź metodę sprawdzania krzyżowego (patrz Kohavi (1995)). Wiele wskazówek na temat doboru wartości parametru J znajduje się w pracy Friedmana (1997). Metoda J najbliższych sąsiadów jest coraz gładsza wraz ze wzrostem liczby sąsiadów i często okazuje się, że dla zbyt małych wartości parametru J otrzymujemy regułę, która jest dość odległa od reguły Bayesa. Metoda 1-NN ma tendencję do bycia zbyt czułą na pewną losowość (błędne sklasyfikowanie lub błędny pomiar cechy) ukrytą w danych. Kolejna ważna sprawa to dobór odpowiedniej miary niepodobieństwa obiektów (przeważnie z dobrym skutkiem stosuje się odległość euklidesową). Ponieważ pewne cechy, których wartości są duże mogą niwelować wpływ innych, dlatego niezbędna jest normalizacja wartości cech. Dla zmiennych ciągłych można użyć:

- normalizacja min-max[9]

$$x = \frac{x - \min(x)}{\max(x) - \min(x)},$$

- standaryzacja[10]

$$x = \frac{x - \bar{x}}{s(x)}.$$

Normalizacja jest niezbędna, ponieważ przeprowadza zmienne o różnych jednostkach w wielkości niemianowane i porównywalne. Metoda najbliższych sąsiadów bardzo często daje dobre wyniki, zwłaszcza przy rozpoznawaniu obrazów. Niestety, aby uzyskać satysfakcjonujące wyniki trzeba mieć duży zbiór uczący. Drugą wadą tej metody jest powolność na etapie klasyfikacji i związane z tym wymagania co do pamięci (należy przechowywać wszystkie dane uczące). Obliczenie odległości wymaga $O(n^2)$ operacji.

[9] Po ang. min-max normalization.
[10] Po ang. Z-score standardization.

6

Drzewa klasyfikacyjne

Konstrukcja *drzew klasyfikacyjnych*[1], a dokładnie *binarnych drzew klasyfikacyjnych* – bo tylko takie będziemy rozważali – polega na sekwencyjnym dzieleniu podzbiorów przestrzeni próby $\mathcal{X}$ na dwa rozłączne i dopełniające się podzbiory, rozpoczynając od całego zbioru $\mathcal{X}$. W każdym kroku podział dokonywany jest tak, aby uzyskane części były możliwie „jednorodne".

Podzbiory, które nie podlegają dalszemu podziałowi, będziemy nazywać końcowymi[2]. Do każdego podzbioru końcowego jest przypisana dokładnie jedna etykieta klasy. Sytuacja taka jest przedstawiona na rys. 6.1. Każdy

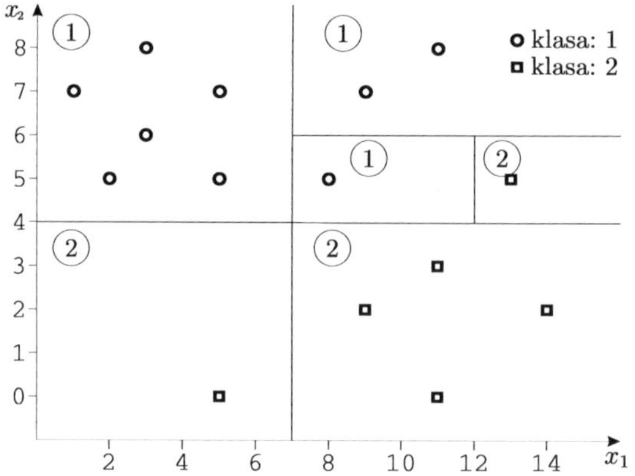

Rys. 6.1. Końcowy podział przestrzeni próby z wykorzystaniem jedynie podziałów jednowymiarowych. W każdym obszarze końcowym wewnątrz okręgów umieszczono etykietę przypisanej mu klasy

[1]Po ang. classification trees.
[2]Po ang. terminal.

element przestrzeni próby $\mathcal{X}$ jest opisany za pomocą dwóch cech X_1 i X_2, a w każdym obszarze końcowym wewnątrz okręgów umieszczono etykietę przypisanej mu klasy. Ponadto na rysunku umieszczono punkty odpowiadające obserwacjom uczącym, na bazie których był dokonywany sekwencyjny podział zbioru $\mathcal{X}$.

Zauważmy, że dwa lub więcej zbiorów końcowych może mieć przypisaną tę samą etykietę.

Idea konstrukcji drzew klasyfikacyjnych jest bardzo podobna do metody stosowanej w diagnostyce medycznej.

Lekarz pragnący zdiagnozować pacjenta zadaje mu szereg pytań. Kolejność pytań nie jest przypadkowa, a następne pytanie jest często uzależnione od uzyskanej odpowiedzi na pytanie poprzednie. Przykładowo, lekarz zaczyna od pytania najważniejszego ze względu na diagnozę, np. pytania o wiek pacjenta. Załóżmy, że istotne jest czy pacjent przekroczył 60 rok życia, czy nie. Sam wiek pacjenta jest niewystarczającą informacją potrzebną do podjęcia decyzji. Zatem lekarz zadaje następne pytanie. Zwróćmy uwagę, że być może w grupie osób starszych (wiek > 60) będzie ono dotyczyło poziomu cholesterolu we krwi, bo ta właśnie informacja jest najważniejsza w tej grupie wiekowej. Z kolei w grupie osób młodszych (wiek ≤ 60) pytanie będzie dotyczyło ciśnienia tętniczego krwi – bo ta właśnie informacja jest w tej grupie najważniejsza. Postępując w ten sposób, po kilku lub kilkunastu krokach, lekarz jest w stanie podjąć decyzję.

Opisany powyżej proces można przedstawić graficznie za pomocą grafu spójnego bez cykli, czyli drzewa (stąd nazwa metody).

Niech T oznacza drzewo klasyfikacyjne. Węzły[3] (wierzchołki) drzewa T utożsamiać będziemy z podzbiorami przestrzeni próby $\mathcal{X}$ i oznaczać literą t. Stąd $\mathcal{X}$ to węzeł, t_1 – korzeń drzewa T. Węzły odpowiadające podzbiorom końcowym będziemy nazywać liśćmi, a zbiór wszystkich liści drzewa T oznaczać będziemy przez $\widetilde{T}$. Przykładowo, sytuację przedstawioną na rys. 6.1. można zilustrować za pomocą drzewa o 11 węzłach (patrz rys. 6.2). Węzły wewnętrzne oznaczono okręgami, a liście kwadratami. Zauważmy, że węzłowi t_2 odpowiada podzbiór przestrzeni próby $\mathcal{X}$ opisany warunkiem $x_1 < 7$, a węzłowi końcowemu t_7 odpowiada podzbiór opisany warunkami $x_1 \geq 7$ i $x_2 \geq 6$.

Cechą charakterystyczną drzewa klasyfikacyjnego są związane z nim dwie funkcje. Określona na zbiorze liści $\widetilde{T}$ funkcja *ind* przyporządkowująca każdemu z nich etykietę klasy oraz określona dla każdego węzła wewnętrz-

[3]Po ang. node.

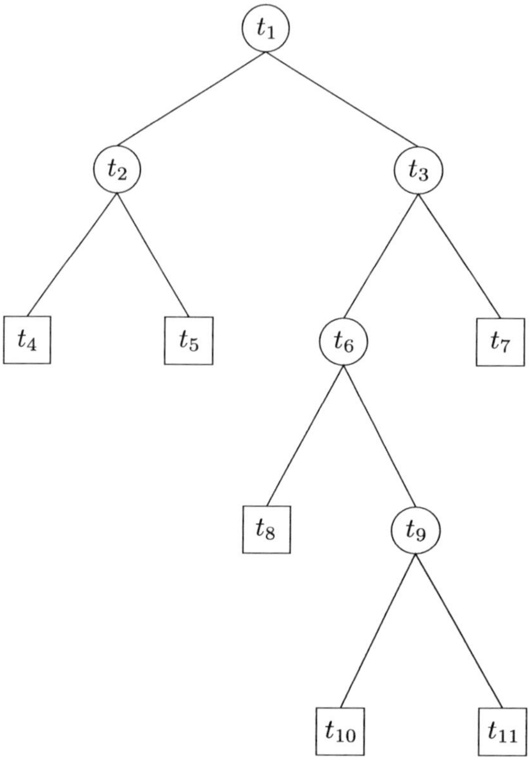

Rys. 6.2. Drzewo klasyfikacyjne o 11 węzłach. Węzły wewnętrzne oznaczono okręgami, a liście kwadratami

nego funkcja podziału[4] (podział) s. Funkcja ta każdemu elementowi węzła t przyporządkowuje jedną z dwóch wartości – prawda lub fałsz. Wszystkie elementy z węzła t, dla których podział s przyjmuje wartość prawda, są przyporządkowywane do węzła potomka t_L, a te elementy, dla których podział s przyjmuje wartość fałsz – do węzła potomka t_R. Zilustrowano to na rys. 6.3.

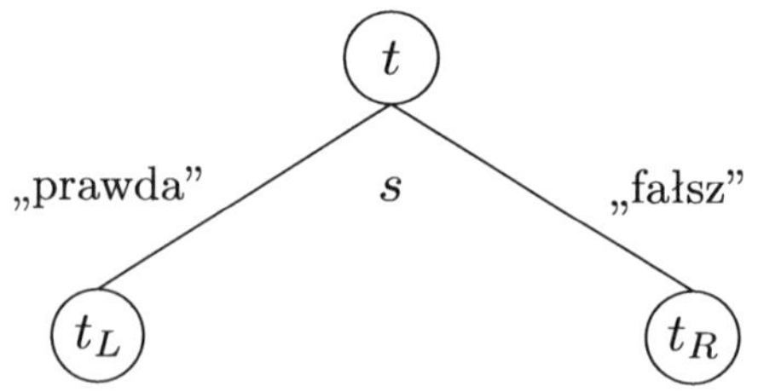

Rys. 6.3. Schematyczny podział węzła t drzewa klasyfikacyjnego

[4]Po ang. split.

Prezentując graficznie drzewo klasyfikacyjne, wygodnie jest nanieść w węzłach wewnętrznych drzewa kryteria podziału s, a liściom przyporządkować etykiety klas. Na rysunku 6.4 przedstawiono taką właśnie postać dla drzewa klasyfikacyjnego z rys. 6.2.

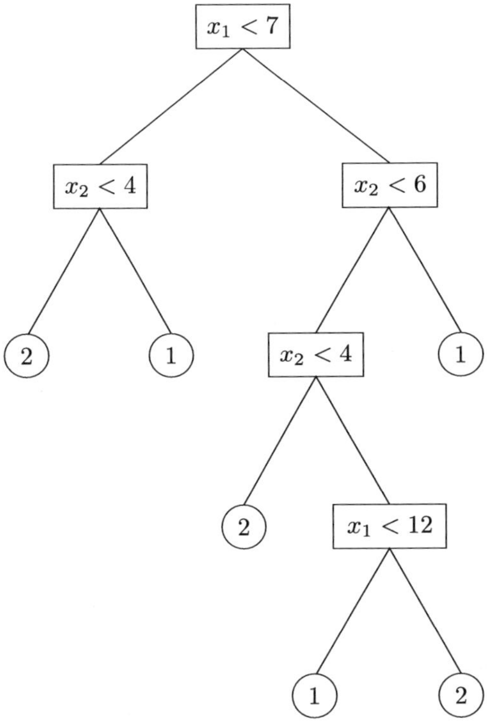

Rys. 6.4. Drzewo klasyfikacyjne. W węzłach wewnętrznych naniesiono kryteria podziału s, a liściom przyporządkowano etykiety klas

Podsumowując, każdy klasyfikator zbudowany na podstawie drzewa klasyfikacyjnego T ma następującą postać:

$$\hat{d}_T(\boldsymbol{x}) = \sum_{t \in \tilde{T}} ind(t) I(\boldsymbol{x} \in t).$$

Spośród wielu procedur konstrukcji drzew klasyfikacyjnych do najbardziej znanych i najczęściej używanych należą: opracowana przez Kassa metoda CHAID (Chi-squared Automatic Interaction Detector, 1980), opracowana przez Breimana, Friedmana, Olsena i Stona metoda CART (Classification and Regression Trees, 1984), opracowana przez Quinlana metoda C4.5 (1993) oraz opracowana przez Loha i Shiha metoda QUEST (Quick Unbiased Efficient Statistical Tree, 1997).

Kończąc wstępne rozważania dotyczące drzew klasyfikacyjnych, należy podkreślić, że do ich podstawowych zalet należą:

1. Bezproblemowy sposób wykorzystania do klasyfikacji zarówno cech ilościowych, jak i jakościowych.
2. Prosta forma końcowa drzewa klasyfikacyjnego umożliwiająca w szybki i efektywny sposób klasyfikację nowych obiektów.
3. Odporność na obserwacje odstające i sytuacje, w których niektóre zmienne są nieobserwowane.

6.1. Konstrukcja drzewa klasyfikacyjnego

Każda konstrukcja drzewa klasyfikacyjnego jest oparta na próbie uczącej $\mathcal{L}_n = \{(\boldsymbol{x}_1, y_1), \ldots, (\boldsymbol{x}_n, y_n)\}$, gdzie $\boldsymbol{x}_j$ jest wektorem obserwowanych cech, a y_j etykietą klasy ($j = 1, 2, \ldots, n$).

Składa się ona z trzech podstawowych elementów:

1. Wyboru metody „optymalnego" podziału węzłów.
2. Doboru optymalnej wielkości drzewa klasyfikacyjnego, czyli reguły decydującej o tym, czy dany węzeł ma być węzłem końcowym, czy ma podlegać dalszemu podziałowi.
3. Sposobu, w jaki każdemu węzłowi końcowemu przyporządkowujemy etykietę klasy.

Wprowadźmy następujące oznaczenia: n_i to liczba obserwacji w próbie uczącej $\mathcal{L}_n$ należących do i-tej klasy, $n_i(t)$ liczba obserwacji w węźle t należących do i-tej klasy ($i = 1, \ldots, K$). Wtedy $n = n_1 + \cdots + n_K$ jest liczbą obserwacji uczących, a $n(t) = n_1(t) + \cdots + n_K(t)$ jest liczbą obserwacji uczących w węźle t.

Dla dowolnego obiektu $\boldsymbol{Z} = (\boldsymbol{X}, Y)$ określmy następujące prawdopodobieństwa:

$$p(t) = \mathrm{P}(\boldsymbol{X} \in t), \qquad p(t|i) = \mathrm{P}(\boldsymbol{X} \in t | Y = i).$$

Ze wzoru na prawdopodobieństwo całkowite mamy

$$p(t) = \sum_{i=1}^{K} \pi_i p(t|i),$$

gdzie π_i jest prawdopodobieństwem a priori tego, że obiekt należy do i-tej klasy ($i = 1, \ldots, K$). Przyjmując estymator częstościowy prawdopodobieństwa $p(t|i)$ postaci

$$\hat{p}(t|i) = \frac{n_i(t)}{n_i},$$

otrzymujemy następujący estymator prawdopodobieństwa $p(t)$:

$$\hat{p}(t) = \sum_{i=1}^{K} \pi_i \frac{n_i(t)}{n_i}.$$

Ponadto, jeżeli prawdopodobieństwa a priori π_i oszacujemy przez n_i/n, to otrzymamy

$$\hat{p}(t) = \sum_{i=1}^{K} \frac{n_i}{n} \frac{n_i(t)}{n_i} = \frac{n(t)}{n}.$$

Co więcej, ze wzoru Bayesa mamy:

$$p(i|t) = \frac{\pi_i p(t|i)}{p(t)}, \qquad i = 1, \ldots, K.$$

Stąd

$$\hat{p}(i|t) = \frac{\pi_i \frac{n_i(t)}{n_i}}{\sum_{i=1}^{K} \pi_i \frac{n_i(t)}{n_i}}$$

lub po oszacowaniu prawdopodobieństw a priori

$$\hat{p}(i|t) = \frac{n_i(t)}{n(t)}.$$

Niech

$$p_L = \mathrm{P}(\boldsymbol{X} \in t_L | \boldsymbol{X} \in t), \qquad p_R = \mathrm{P}(\boldsymbol{X} \in t_R | \boldsymbol{X} \in t),$$

gdzie t_L i t_R są węzłami potomkami węzła t. Wtedy

$$p_L = \frac{p(t_L)}{p(t)}, \qquad p_R = \frac{p(t_R)}{p(t)}$$

oraz

$$\hat{p}_L = \frac{n(t_L)}{n(t)}, \qquad \hat{p}_R = \frac{n(t_R)}{n(t)}.$$

Na rysunku 6.3 przedstawiono schemat podziału elementów węzła wewnętrznego t na dwa węzły potomki t_L i t_R. Podział ten został dokonany przy użyciu kryterium (funkcji) podziału s. Ogólnie rzecz biorąc, każdy podział s jest generowany przez odpowiadające mu pytanie postaci:

$$\text{czy } \boldsymbol{x} \in \boldsymbol{A}, \quad \boldsymbol{A} \subset \mathcal{X}?$$

Postać powyższego pytania jest bardzo ogólna. Może ona dotyczyć jedynie jednej cechy (podziały jednowymiarowe) lub wielu cech (podziały wielowymiarowe). Poszukując optymalnego podziału s, powinniśmy rozważyć możliwie duży zbiór pytań. Im jednak będzie on większy, tym bardziej skomplikowana i długa będzie droga wyboru optymalnego podziału s. Kryterium wyboru takiego optymalnego podziału bazujące na wyczerpującym przeszukiwaniu wszystkich możliwych podziałów (metoda CART) zostanie omówione w podrozdz. 6.2, natomiast w podrozdz. 6.4 omówione zostanie kryterium wyboru oparte na kwadratowej analizie dyskryminacyjnej zastosowanej w metodzie QUEST.

Jednym z podstawowych elementów drzewa klasyfikacyjnego jest określona na liściach drzewa *funkcja ind*. Przyporządkowuje ona etykietę klasy każdemu z węzłów końcowych.

Zdefiniujmy ją następująco:

$$ind(t) = \arg \max_{1 \le i \le K} \hat{p}(i|t).$$

Jeżeli maksimum wypada dla dwóch lub więcej klas, wybór jednej z nich jest dowolny.

Wprowadzone powyżej wielkości dla drzewa klasyfikacyjnego przedstawionego na rys. 6.2 i 6.4 zawiera tab. 6.1.

Tab. 6.1. Parametry drzewa klasyfikacyjnego z rys. 6.4

t	s	$n(t)$	$n_1(t)$	$n_2(t)$	$\hat{p}(t)$	$\hat{p}(1\|t)$	$\hat{p}(2\|t)$	$\hat{p}_L$	$\hat{p}_R$	$ind(t)$
t_1	$x_1 < 7$	15	9	6	1	9/15	6/15	7/15	8/15	–
t_2	$x_2 < 4$	7	6	1	7/15	6/7	1/7	1/7	6/7	–
t_3	$x_2 < 6$	8	3	5	8/15	3/8	5/8	3/4	1/4	–
t_4	–	1	0	1	1/15	0	1	–	–	2
t_5	–	6	6	0	6/15	1	0	–	–	1
t_6	$x_2 < 4$	6	1	5	6/15	1/6	5/6	2/3	1/3	–
t_7	–	2	2	0	2/15	1	0	–	–	1
t_8	–	4	0	4	4/15	0	1	–	–	2
t_9	$x_1 < 12$	2	1	1	2/15	1/2	1/2	1/2	1/2	–
t_{10}	–	1	1	0	1/15	1	0	–	–	1
t_{11}	–	1	0	1	1/15	0	1	–	–	2

6.2. Kryteria podziału – metoda CART

Wspomnieliśmy poprzednio, że każdy podział s jest generowany przez odpowiadające mu pytanie postaci:

$$\text{czy } \boldsymbol{x} \in \boldsymbol{A}, \ \boldsymbol{A} \subset \mathcal{X}?$$

Załóżmy, że każde pytanie dotyczy tylko wartości jednej zmiennej. Idea poszukiwania optymalnego podziału s w metodzie CART bazuje na wyczerpującym przeszukiwaniu wszystkich możliwych podziałów. Oznacza to, że dla każdej zmiennej ilościowej X podziały s wyznaczają wszystkie pytania postaci:

$$\text{czy } x \leq c, \ c \in (-\infty, \ \infty)?$$

oraz dla każdej zmiennej jakościowej X o wartościach w zbiorze V, podziały wyznaczają pytania

$$\text{czy } x \in S,$$

gdzie S jest dowolnym niepustym podzbiorem zbioru V.

Ponieważ podział s jest dokonywany na podstawie n-elementowej próby uczącej $\mathcal{L}_n$, zatem zbiór istotnie różnych podziałów s jest zawsze skończony.

Niech zmienna X będzie ilościowa. Wtedy w próbie uczącej $\mathcal{L}_n$ może przyjmować ona co najwyżej n różnych wartości. Ustawmy wartości zmiennej X w szereg pozycyjny:

$$x_{(1)} \leq x_{(2)} \leq \cdots \leq x_{(n)}.$$

Widać, że możemy uzyskać co najwyżej $n-1$ różnych (dających niepuste węzły potomki) podziałów s. Generowane są one przez pytania postaci:

$$\text{czy } x \leq c_j,$$

gdzie c_j jest środkiem przedziału $[x_{(j)}, \ x_{(j+1)}]$ wyznaczonym przez różne wartości cechy X.

Załóżmy teraz, że cecha X jest jakościowa, a zbiór jej wartości V składa się z L elementów. W tej sytuacji istnieje $2^{L-1} - 1$ różnych podziałów s. Nawet w przypadku niewielkiej liczby L, liczba podziałów, które musimy przeszukać jest bardzo duża, co w konsekwencji prowadzi do czasochłonnych obliczeń.

Zajmijmy się teraz zagadnieniem optymalnego wyboru podziału s. Każdy podział s jest uwarunkowany przez obserwacje ze zbioru uczącego $\mathcal{L}_n$, należące do danego węzła t. Praktycznie trzeba zatem rozdzielić je na dwa podzbiory możliwie jednorodne ze względu na etykietę klasy. Gdyby zatem w węźle t znajdowały się obserwacje tylko z dwóch klas, to idealnym byłby taki podział, który przypisałby obserwacje uczące z jednej klasy do węzła t_L, a z drugiej klasy do węzła t_R.

Dla każdego węzła t określmy pewną miarę $i(t)$ niejednorodności elementów w tym węźle[5]. Stąd, dla każdego podziału s węzła t będziemy mogli zmierzyć niejednorodność elementów w tym węźle oraz w jego potomkach t_L i t_R.

Jeżeli zmianę niejednorodności przy podziale s węzła t określimy jako

$$\Delta i(s,\ t) = i(t) - p_L i(t_L) - p_R i(t_R),$$

to wówczas optymalnym będzie taki podział s^*, dla którego:

$$\Delta i(s^*,\ t) = \max_s \Delta i(s,\ t).$$

Optymalizacja ta dotyczy oddzielnie każdego węzła t drzewa decyzyjnego T. Ma zatem ona charakter lokalny. Rozwiązanie takie jest charakterystyczne dla wszystkich algorytmów konstrukcji drzew klasyfikacyjnych.

Przejdźmy teraz do omówienia różnych *miar niejednorodności* $i(t)$ elementów węzła t.

Niech ϕ oznacza funkcję określoną dla wszystkich K-elementowych ciągów prawdopodobieństw $(p_1,\ p_2,\ \ldots,\ p_K)$ takich, że $\sum_{i=1}^{K} p_i = 1$, $p_i \geq 0$ $(i = 1,\ 2,\ \ldots,\ K)$, spełniającą następujące warunki:

1. Funkcja ϕ osiąga maksimum tylko w punkcie $\left(\frac{1}{K},\ \frac{1}{K},\ \ldots,\ \frac{1}{K}\right)$.
2. Funkcja ϕ osiąga minimum tylko w punktach: $(1,\ 0,\ \ldots,\ 0)$, $(0,\ 1,\ \ldots,\ 0)$, $\ldots$, $(0,\ 0,\ \ldots,\ 1)$.
3. Funkcja ϕ jest symetryczną funkcją swoich argumentów.

Miarę niejednorodności $i(t)$ w węźle t definiujemy następująco:

$$i(t) = \phi(p(1|t),\ \ldots,\ p(K|t)).$$

Poniżej są podane trzy najpopularniejsze postaci funkcji ϕ

1. $\phi_1(p_1,\ \ldots,\ p_K) = 1 - \max\{p_1,\ \ldots,\ p_K\}$.

   Miara niejednorodności oparta na tej funkcji określa błąd klasyfikacji.

[5] Po ang. impurity function.

2. $\phi_2(p_1, \ldots, p_K) = -\sum_{i=1}^{K} p_i \log p_i.$

Funkcja ta jest nazywana *funkcją entropii.*

3. $\phi_3(p_1, \ldots, p_K) = 1 - \sum_{i=1}^{K} p_i^2.$

Funkcja ta jest nazywana *indeksem Giniego.*

Szczegółowe omówienie zalet i wad miar niejednorodności opartych na powyższych funkcjach oraz ich różne modyfikacje znaleźć można w książce Gatnara (2000).

Dla $K = 2$, ϕ jest funkcją jednej zmiennej postaci:

$$\phi_1(p) = 1 - \max\{p,\ 1-p\} = \min\{p,\ 1-p\},$$
$$\phi_2(p) = -p \log p - (1-p)\log(1-p),$$
$$\phi_3(p) = 2p(1-p),\ p \in [0,\ 1].$$

Wykresy tych funkcji przedstawiono na rys. 6.5.

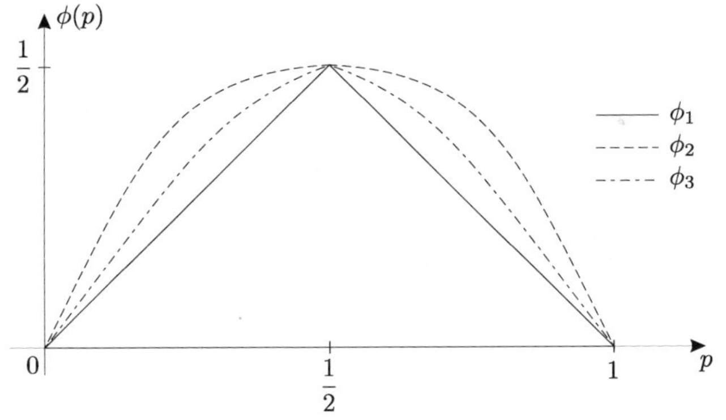

Rys. 6.5. Wykresy funkcji ϕ dla $K = 2$

Poszukiwanie optymalnego podziału s^* dla każdej cechy jakościowej o L różnych wartościach wymaga rozważenia $2^{L-1} - 1$ różnych podziałów. W przypadku używania jako kryterium podziału entropii lub indeksu Giniego, w zagadnieniach dwuklasowych można, bez utraty ogólności, zredukować liczbę przeszukiwanych podziałów do zaledwie $L - 1$.

Niech $b_1,\ b_2,\ \ldots,\ b_L$ będą różnymi wartościami cechy jakościowej X przyjmowanymi przez elementy próby uczącej $\mathcal{L}_n$ w węźle t. Ponadto, niech

$n_{il}(t)$ oznacza liczbę obserwacji uczących w węźle t należących do i-tej klasy ($i = 1,\ 2$), dla których cecha X przyjmuje wartość b_l ($l = 1,\ 2,\ \ldots,\ L$).

Wtedy jako oszacowanie prawdopodobieństwa

$$p(b_l|i) = \mathrm{P}(\boldsymbol{X} \in t,\ X = b_l|Y = i)$$

możemy przyjąć

$$\hat{p}(b_l|i) = \frac{n_{il}(t)}{n_i}, \qquad i = 1,\ 2.$$

Zatem

$$\hat{p}(i|b_l) = \frac{\pi_i \dfrac{n_{il}(t)}{n_i}}{\pi_1 \dfrac{n_{1l}(t)}{n_1} + \pi_2 \dfrac{n_{2l}(t)}{n_2}}, \qquad i = 1,\ 2.$$

Stąd, szacując prawdopodobieństwa a priori π_i przez n_i/n, otrzymujemy

$$\hat{p}(i|b_l) = \frac{n_{il}(t)}{n_{1l}(t) + n_{2l}(t)}, \qquad i = 1,\ 2.$$

Prawdziwe jest następujące twierdzenie.

TWIERDZENIE 6.1.

Niech $K = 2$ oraz $b_{(1)}, b_{(2)}, \ldots, b_{(L)}$ będą wartościami cechy jakościowej X przyjmowanymi w węźle t przez elementy próby $\mathcal{L}_n$ uporządkowanymi w taki sposób, że

$$\hat{p}(1|b_{(1)}) \leq \hat{p}(1|b_{(2)}) \leq \cdots \leq \hat{p}(1|b_{(L)}).$$

Wtedy dla entropii oraz indeksu Giniego podział s^*, dla którego

$$\Delta i(s^*,\ t) = \max_s \Delta i(s,\ t)$$

jest generowany przez jedno z pytań postaci:

czy $x \in S_L$,

gdzie $S_L = \{b_{(1)},\ b_{(2)},\ \ldots,\ b_{(l)}\}$, $l = 1,\ 2,\ \ldots,\ L-1$.

Rozważane przez nas dotychczas podziały węzła były oparte na jednej zmiennej. W przypadku cech o charakterze ilościowym oznacza to, że hiperpłaszczyzny wyznaczające podział były równoległe do osi współrzędnych. Dla niektórych zbiorów danych podejście to prowadzi do konieczności użycia wielu podziałów, aby uzyskać wystarczająco jednorodne podzbiory. Użycie hiperpłaszczyzn położonych ukośnie dałoby szybszy efekt. Przykładowo dla danych przedstawionych na rys. 6.1 do uzyskania jednorodnych podzbiorów potrzebowaliśmy 5 podziałów wyznaczonych zawsze przez wybraną jedną cechę. Gdybyśmy użyli obydwu cech jednocześnie do uzyskania jednorodności wystarczyłby jeden podział. Widać to wyraźnie na rys. 6.6.

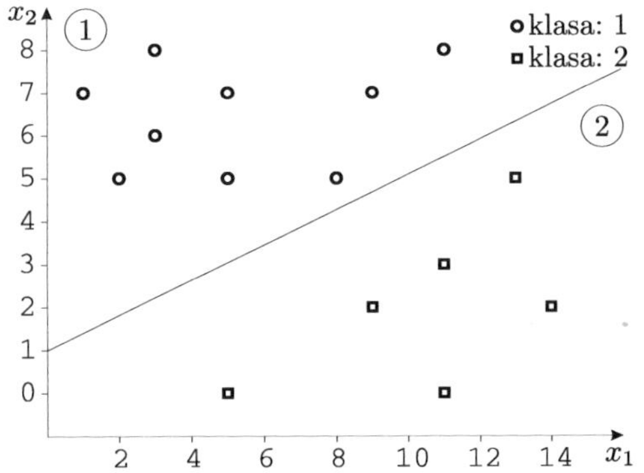

Rys. 6.6. Końcowy podział przestrzeni próby z wykorzystaniem podziału wielowymiarowego. W każdym obszarze końcowym wewnątrz okręgów umieszczono etykietę przypisanej mu klasy

Podział ten jest generowany przez pytanie:

czy $2x_1 - 5x_2 < -5$?

Odpowiadające temu (wielowymiarowemu) sposobowi podziału drzewo klasyfikacyjne jest przedstawione na rys. 6.7.

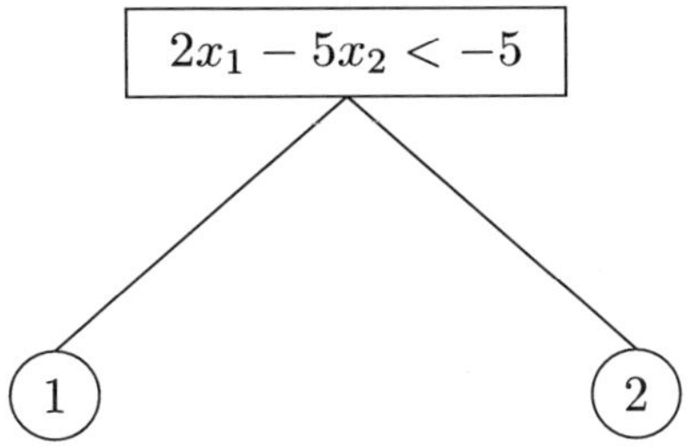

Rys. 6.7. Drzewo klasyfikacyjne z podziałem wielowymiarowym

Rozważmy dokładnie zagadnienie podziałów wielowymiarowych.

Niech $X_1,\ X_2,\ \ldots,\ X_p$ oznaczają tak jak dotychczas badane cechy, przy czym dodatkowo załóżmy, że są one uporządkowane tak, że pierwszych m z nich to cechy ilościowe, a pozostałe $p-m$ to cechy jakościowe.

Zajmijmy się najpierw cechami ilościowymi. Poszukując optymalnego podziału s węzła t, będziemy zakładać, że w zbiorze generujących je pytań, zamiast pytań dotyczących pojedynczej cechy X_j postaci:

$$\text{czy } x_j \le c,\ c \in (-\infty,\ \infty),\ j = 1,\ 2,\ \ldots,\ m,$$

będziemy rozważać pytania:

$$\text{czy } \sum_{j=1}^{m} a_j x_j \le c,\ \sum_{j=1}^{m} a_j^2 = 1,\ c \in (-\infty,\ \infty)?$$

Określone w ten sposób wielowymiarowe zagadnienie jest dużo bardziej skomplikowane, niż rozważane poprzednio zagadnienia jednowymiarowe. Poza doborem stałej c musimy jeszcze wyznaczyć współczynniki a_1, $a_2, \ldots,\ a_m$.

Zakładamy, że dla węzła t jest dana miara jego niejednorodności $i(t)$, np. indeks Giniego. Tak jak w przypadku podziałów jednowymiarowych szukamy podziału s^*, dla którego

$$\Delta i(s^*,\ t) = \max_s \Delta i(s,\ t).$$

Podanie algorytmu wyznaczającego podział s^* nie jest możliwe. Poniżej jest przedstawiony algorytm zaproponowany w metodzie CART. Podobnie jak inne tego typu algorytmy wyznacza on jedynie maksimum lokalne, które niekoniecznie musi być maksimum globalnym.

Rozpoczyna się on od przekształcenia wartości zmiennych $X_1,\ X_2, \ldots,$ X_m przez odjęcie od nich ich mediany i podzielenie ich wartości przez rozstęp międzykwartylowy. Oznaczmy uzyskane w ten sposób nowe zmienne przez $\tilde{X}_1,\ \tilde{X}_2,\ \ldots,\ \tilde{X}_m$. Startując z najlepszego podziału jednowymiarowego, w każdym następnym kroku, modyfikujemy ustawienie hiperpłaszczyzny tzn. zmieniamy wartości parametrów $a_1,\ a_2,\ \ldots a_m$ i c, poprawiając jakość podziału.

Niech s_l oznacza podział otrzymany w l-tym kroku tego algorytmu. Procedurę poprawiania ustawienia hiperpłaszczyzny kończymy, gdy jest spełniony warunek

$$|i_{s_l}(t) - i_{s_{l-1}}(t)| \le \varepsilon,$$

gdzie ε jest zadaną wartością progową.

Kończąc, wracamy do wyjściowych zmiennych, otrzymując podział s^* węzła t wyznaczony przez kryterium

$$v^* = \sum_{j=1}^{m} a_j^* x_j \leq c^*. \tag{6.1}$$

Jak wygląda pojedynczy (l-ty) krok algorytmu? Składa się on z m etapów odpowiadających liczbie rozważanych zmiennych. Niech

$$v = \sum_{j=1}^{m} a_j \tilde{x}_j \leq c \tag{6.2}$$

będzie kryterium podziału wyznaczonym w $l - 1$ kroku (krok zerowy to najlepszy podział jednowymiarowy). Modyfikujemy to kryterium, rozpoczynając od pierwszej zmiennej. Dla ustalonego γ ($\gamma = -0.25,\ 0,\ 0.25$), wyznaczamy najlepszy (ze względu na δ) podział postaci

$$v - \delta(\tilde{x}_1 + \gamma) \leq c,$$

gdzie stałą δ dobieramy w następujący sposób:

$$\delta \geq \frac{v - c}{\tilde{x}_1 + \gamma}, \quad \text{dla} \quad \tilde{x}_1 + \gamma \geq 0$$

oraz

$$\delta \leq \frac{v - c}{\tilde{x}_1 + \gamma}, \quad \text{dla} \quad \tilde{x}_1 + \gamma < 0.$$

Dalsza część procedury jest identyczna jak w przypadku jednowymiarowym.

Niech

$$u_j = \frac{v_j + c}{\tilde{x}_{1j} + \gamma}, \qquad j = 1,\ 2,\ \ldots,\ n(t).$$

Porządkujemy wartości u_j oraz dla u_j takich, że $\tilde{x}_{1j} + \gamma \geq 0$ znajdujemy najlepszy podział postaci $u \leq \delta_1$, a dla u_j takich, że $\tilde{x}_{1j} + \gamma < 0$ najlepszy podział postaci $u \geq \delta_2$. Wybierając lepszy z nich, otrzymujemy poszukiwaną wartość stałej δ. Modyfikujemy podział (6.2), otrzymując nową postać

$$v' = \sum_{j=1}^{m} a_j' \tilde{x}_j \leq c',$$

gdzie

$$\begin{aligned} a_1' &= a_1 - \delta, \\ a_j' &= a_j, \qquad j = 2,\ 3,\ \ldots,\ m, \\ c' &= c + \delta\gamma. \end{aligned}$$

Powtarzamy modyfikację kryterium dla wszystkich pozostałych $m - 1$ zmiennych.

W otrzymanym w ten sposób optymalnym kryterium podziału s^* zostały wykorzystane wszystkie zmienne. Zazwyczaj część z nich ma minimalny wpływ na jakość podziału i ich usunięcie powoduje jedynie jego minimalne pogorszenie, dając w zamian prostszy i łatwiejszy w interpretacji model.

Usuńmy zatem te niepotrzebne zmienne. Dla $k = 1,\ 2,\ \ldots,\ m$ oznaczmy przez s_k podział węzła t wyznaczony przez kryterium postaci

$$v_k = \sum_{j \neq k} a_j^* x_j \leq c_k,$$

gdzie współczynniki kombinacji liniowej są identyczne jak w optymalnym podziale s^* wyznaczony przez kryterium (6.1), a jedynie c_k jest zoptymalizowane. Niech

$$\Delta^* = \Delta i(s^*,\ t)$$

oraz

$$\Delta_k = \Delta i(s_k,\ t).$$

Zmienna X_j będzie najmniej istotna, gdy

$$\Delta_j = \max_k \Delta_k,$$

a najbardziej istotna, gdy

$$\Delta_j = \min_k \Delta_k.$$

Jeśli będzie spełniony warunek

$$\Delta^* - \max_j \Delta_j < \beta(\Delta^* - \min_j \Delta_j) \tag{6.3}$$

dla pewnego $\beta > 0$ (zazwyczaj $\beta = 0.1$ lub $\beta = 0.2$), to usuniemy tę najmniej istotną zmienną.

Powtarzamy operacje usuwania, wykorzystując pozostałe w modelu zmienne tak długo, aż dla najmniej istotnej z nich jest spełniony warunek (6.3).

Zajmijmy się teraz pozostałymi $p - m$ cechami jakościowymi.

Dla każdej takiej cechy X_j, jednowymiarowy podział s_j był generowany przez pytanie postaci:

$$\text{czy } x_j \in S_j,$$

gdzie S_j był dowolnym niepustym podzbiorem zbioru jej wartości V_j, $j = m+1, \ldots, p$.

W dalszym ciągu, dla tego typu zmiennych, będziemy poszukiwać podziałów wielowymiarowych generowanych przez pytania postaci:

$$\text{czy } x_{j_1} \in S_{j_1} \text{ i } x_{j_2} \in S_{j_2} \text{ i } \cdots \text{ i } x_{j_h} \in S_{j_h},$$

gdzie $\{j_1, j_2, \ldots, j_h\} \subset \{m+1, \ldots, p\}$.

Podział s generowany przez powyższe pytanie oznaczać będziemy przez $s_{j_1} \cap s_{j_2} \cap \ldots \cap s_{j_h}$. Jeżeli zatem dla węzła t jest dana jego miara niejednorodności $i(t)$, to optymalny podział s^* maksymalizuje $\Delta i(s, t)$ po wszystkich wielowymiarowych podziałach $s = s_{j_1} \cap s_{j_2} \cap \ldots \cap s_{j_h}$ takich, że $\{j_1, j_2, \ldots, j_h\} \subset \{m+1, \ldots, p\}$.

W metodzie CART zamiast bezpośredniej maksymalizacji $\Delta i(s, t)$, zaproponowano następującą procedurę krokową.

Niech s_j^* oznacza optymalny podział jednowymiarowy oparty na zmiennej X_j, a S^* rodzinę jednowymiarowych podziałów postaci:

$$S^* = \{s_{m+1}^*, \bar{s}_{m+1}^*, \ldots, s_p^*, \bar{s}_p^*\},$$

gdzie $\bar{s}_j^*$ jest podziałem dopełniającym podział s_j^*[6].

Procedurę rozpoczynamy od wyznaczenia najlepszego podziału jednowymiarowego ze zbioru S^*. Oznaczmy ten podział przez s_1^*. Następnie poszukujemy podziału $s \in S^*$ spełniającego warunek:

$$\max\{\Delta i(s_1^* \cap s, t), \Delta i(\bar{s}_1^* \cap s, t)\}.$$

Oznaczając go przez s_2^*, otrzymujemy podział $s_1^* \cap s_2^*$. Dodawanie następnych podziałów ze zbioru S^* kontynuujemy, aż do spełnienia warunku

$$\Delta i(s_1^* \cap \ldots \cap s_{l+1}^*, t) \leq (1+\beta)\Delta i(s_1^* \cap \ldots \cap s_l^*, t),$$

gdzie $\beta > 0$ jest ustaloną wartością progową.

Kończąc powyższą procedurę, otrzymujemy wielowymiarowy podział

$$s^* = s_1^* \cap \ldots \cap s_l^*.$$

[6]Jeżeli podział s jest generowany przez pytanie postaci: czy $\boldsymbol{x} \in \boldsymbol{A}$?, to podział dopełniający $\bar{s}$ jest generowany przez pytanie: czy $\boldsymbol{x} \notin \boldsymbol{A}$?

Konstrukcja drzewa klasyfikacyjnego przebiega sekwencyjnie. Na każdym etapie definiujemy dla aktualnego drzewa T odpowiadającą mu miarę niejednorodności jako

$$I(T) = \sum_{t \in \tilde{T}} i(t)p(t).$$

Weźmy dowolny węzeł $t \in \tilde{T}$ i dokonajmy jego podziału na dwa węzły t_L i t_R. Dla tak powstałego nowego drzewa T' mamy:

$$I(T') = \sum_{t \in \tilde{T}'} i(t)p(t).$$

Zatem

$$\begin{aligned}
\Delta I(s,\, t) &= I(T) - I(T') = \\
&= i(t)p(t) - i(t_L)p(t_L) - i(t_R)p(t_R) = \\
&= [i(t) - p_L i(t_L) - p_R i(t_R)]p(t) = \\
&= \Delta i(s,\, t)p(t).
\end{aligned}$$

Widać, że maksymalizacja $\Delta I(s,\, t)$ jest równoważna z wyborem optymalnego podziału s węzła t.

Mamy zatem następujące twierdzenie.

TWIERDZENIE 6.2.

Wybór optymalnego podziału s węzła t jest równoważny wyborowi podziału s minimalizującego miarę niejednorodności drzewa klasyfikacyjnego $I(T)$.

6.3. Optymalna wielkość drzewa klasyfikacyjnego

Wybór optymalnej wielkości drzewa klasyfikacyjnego jest równoważny z podaniem reguły decydującej o tym, czy dany węzeł ma podlegać podziałowi, czy ma być liściem drzewa.

Oznaczmy przez $e(t)$ prawdopodobieństwo błędnego zaklasyfikowania obiektu $\boldsymbol{Z} = (\boldsymbol{X},\, Y)$ należącego do węzła t, tzn.

$$e(t) = \mathrm{P}(\boldsymbol{X} \in t,\ ind(t) \neq Y).$$

Estymator tego prawdopodobieństwa uzyskany metodą ponownego podstawiania ma postać:

$$\hat{e}_R(t) = \hat{r}(t)\hat{p}(t),$$

gdzie

$$\hat{r}(t) = \sum_{i \neq ind(t)} \hat{p}(i|t) = 1 - \max_{1 \leq i \leq K} \hat{p}(i|t).$$

Widać, że $\hat{r}(t)$ jest estymatorem prawdopodobieństwa błędnego zaklasyfikowania obiektu $\boldsymbol{Z}$, pod warunkiem że wpada on do węzła t. Ponadto oczywista jest następująca nierówność:

$$\hat{e}_R(t) \geq \hat{e}_R(t_L) + \hat{e}_R(t_R).$$

Łączne prawdopodobieństwo błędnej klasyfikacji obiektu $\boldsymbol{Z}$ przez drzewo klasyfikacyjne T wyraża się wzorem

$$e(T) = \sum_{t \in \tilde{T}} e(t),$$

a odpowiadający mu estymator z metody ponownego podstawiania ma postać:

$$\hat{e}_R(T) = \sum_{t \in \tilde{T}} \hat{e}_R(t).$$

Zwróćmy uwagę, że stosując metodę ponownego podstawiania do szacowania poprawności klasyfikacji za pomocą drzewa klasyfikacyjnego T dochodzimy do następującego wniosku: jeżeli drzewo klasyfikacyjne T' powstało z drzewa klasyfikacyjnego T przez podział dowolnego węzła końcowego, to

$$\hat{e}_R(T') \leq \hat{e}_R(T).$$

Co więcej, kontynuując odpowiednio długo proces dzielenia węzłów, otrzymamy maksymalne drzewo klasyfikacyjne $T_{\max}$, dla którego w każdym węźle końcowym znajdować się będą tylko te obiekty z próby uczącej $\mathcal{L}_n$, które należą do tej samej klasy[7]. Dla takiego maksymalnego drzewa klasyfikacyjnego mamy:

$$\hat{e}_R(T_{\max}) = 0.$$

Budując jednak zbyt duże drzewo klasyfikacyjne, mamy do czynienia z tzw. efektem przetrenowania. Polega on na tym, że drzewo doskonale klasyfikuje

[7]Zakładamy, że w próbie uczącej $\mathcal{L}_n$ nie ma dwóch lub więcej obiektów o tym samym wektorze obserwacji, ale należących do różnych klas.

obiekty z próby uczącej $\mathcal{L}_n$ lecz coraz słabiej (w miarę zwiększania liczby liści) nowe elementy. Efekt ten widać wyraźnie na rys. 6.8. Przedstawiono tu typowe zachowanie się estymatorów z metody ponownego podstawiania oraz v-krokowej metody sprawdzania krzyżowego.

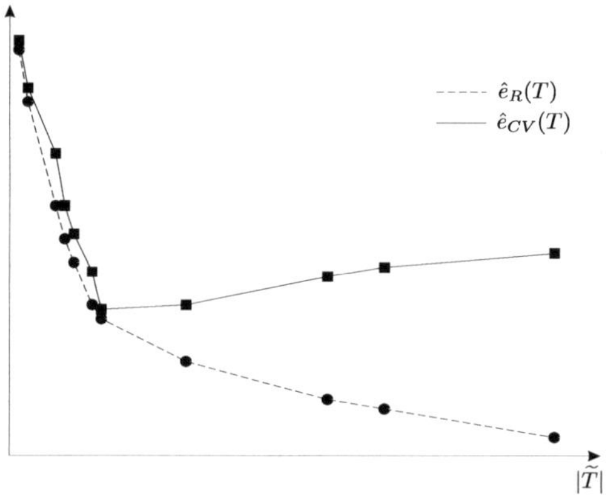

Rys. 6.8. Estymatory błędu e_R jako funkcje ilości liści drzewa klasyfikacyjnego

Pojawia się zatem problem optymalnego wyboru wielkości drzewa klasyfikacyjnego. Prostym rozwiązaniem (niestety, dającym zwykle słabe rezultaty) jest wprowadzenie tzw. *reguły stopu*. Zapewne najprostsza reguła tego typu, każe nie dzielić dalej węzła t, gdy liczba przyporządkowanych mu obserwacji ze zbioru uczącego $\mathcal{L}_n$ jest odpowiednio mała, tzn. węzeł t uznajemy za końcowy, gdy

$$n(t) < n_{\min},$$

gdzie $n_{\min}$ jest ustaloną wartością progową (zazwyczaj jest to mały procent liczby elementów próby uczącej).

Inna reguła stopu nawiązuje do wyniku z tw. 6.2. Każe ona uznać węzeł t za końcowy, po uzyskaniu odpowiedniej jednorodności drzewa klasyfikacyjnego T. Zgodnie z tą regułą węzeł t uznajemy za końcowy, gdy

$$\max_s \Delta I(s,\ t) < \beta,$$

gdzie $\beta > 0$ jest ustaloną wielkością progową.

Bardziej efektywnym podejściem (choć zwiększającym znacząco czasochłonność procesu konstrukcji optymalnego drzewa klasyfikacyjnego) jest procedura polegająca na wstępnym tworzeniu drzewa maksymalnego, a następnie na selektywnym *przycinaniu*[8] jego krawędzi.

[8]Po ang. pruning.

Proces przycinania ma na celu modyfikację postaci maksymalnego drzewa klasyfikacyjnego w ten sposób, aby uzyskać nowe drzewo, które daje możliwie minimalne prawdopodobieństwo błędnej klasyfikacji. Prowadzi on do redukcji wielkości drzewa – zostają wyeliminowane te podziały, które nie mają istotnego znaczenia dla poprawności klasyfikacji.

Przez T_t oznaczmy gałąź drzewa T o korzeniu $t \in T$. Składa się ona z węzła t i wszystkich jego węzłów potomków należących do drzewa T. $T - T_t$ będzie oznaczać drzewo T z przyciętą gałęzią T_t. Drzewo takie składa się ze wszystkich węzłów drzewa T oprócz węzłów gałęzi T_t oraz z węzła t.

Niech teraz drzewo T' powstanie z drzewa T przez przycięcie jego gałęzi. Drzewo T' będziemy nazywać *przyciętym poddrzewem drzewa* T i będziemy oznaczać je przez $T' \prec T$. Zauważmy, że drzewa T' i T mają ten sam korzeń. Ponadto, przez $|T|$ oznaczmy liczbę węzłów drzewa T.

Rozpoczynając procedurę przycinania, zakładamy, że mamy zbudowane maksymalne drzewo klasyfikacyjne $T_{\max}$ oraz że $|\tilde{T}_{\max}| = L$. W przypadku gdy liczba węzłów końcowych L jest bardzo duża, można zamiast drzewa maksymalnego użyć mniejszego drzewa, np. zakończyć proces podziału węzłów, gdy $n(t) < 5$.

Kontynuując procedurę przycinania drzewa $T_{\max}$, otrzymamy ciąg jego poddrzew postaci:

$$T_{\max} \succ T_1 \succ T_2 \succ \cdots \succ \{t_1\}.$$

Dla każdego takiego ciągu mamy

$$L = |\tilde{T}_{\max}| > |\tilde{T}_1| > |\tilde{T}_2| > \cdots > |\{t_1\}| = 1$$

oraz

$$0 = \hat{e}_R(T_{\max}) < \hat{e}_R(T_1) < \hat{e}_R(T_2) < \cdots < \hat{e}_R(\{t_1\}).$$

Ponadto chcielibyśmy, aby na każdym etapie przycinania wybierane było drzewo klasyfikacyjne o możliwie małym błędzie klasyfikacji, a zatem aby dla $|\tilde{T}_l| = l$ $(1 \le l \le L)$

$$\hat{e}_R(T_l) = \min_{T \in \mathcal{T}_l} \hat{e}_R(T), \tag{6.4}$$

gdzie $\mathcal{T}_l$ jest klasą wszystkich poddrzew drzewa klasyfikacyjnego $T_{\max}$ mających $L - l$ liści.

Zwróćmy uwagę, że prosta (lecz bardzo czasochłonna) rekurencyjna procedura polegająca na tym, aby dla każdego l (począwszy od $l = 1$) przeszukać całą klasę drzew T_l i wybrać to, dla którego jest spełniony warunek (6.4), prowadzi do ciągu

$$T_{\max},\ T_1,\ T_2,\ \ldots,\ \{t_1\},$$

w którym drzewo T_{l+1} wcale nie musi być poddrzewem drzewa T_l.

W procedurze CART zastosowano zatem inną metodę. Wprowadźmy następującą *miarę błędu i złożoności drzewa klasyfikacyjnego* T:

$$R_\alpha(T) = \hat{e}_R(T) + \alpha|\tilde{T}|,$$

gdzie $\alpha \geq 0$ jest *współczynnikiem złożoności.*

Miara ta ma charakter *funkcji kosztu.* Jest kombinacją błędu klasyfikacji (szacowanego metodą ponownego podstawiania) i złożoności drzewa klasyfikacyjnego (mierzonego liczbą liści). Współczynnik złożoności α pełni funkcję kary. Minimalizacja miary R_α, dla małych wartości α, będzie wskazywać na duże (w sensie liczby liści) drzewa. Zwiększając współczynnik złożoności α, będziemy wybierać coraz mniejsze drzewa, aż do momentu osiągnięcia drzewa minimalnego $\{t_1\}$ złożonego tylko z korzenia drzewa T.

Można pokazać, że dla każdego $\alpha \geq 0$ istnieje optymalnie przycięte poddrzewo $T(\alpha)$ drzewa $T_{\max}$ spełniające warunki:

1. $R_\alpha(T(\alpha)) = \min_{T \prec T_{\max}} R_\alpha(T)$.
2. Jeżeli $R_\alpha(T) = R_\alpha(T(\alpha))$, to $T(\alpha) \prec T$.

Zatem wzrost wartości współczynnika złożoności α powoduje wybieranie coraz to mniejszych, optymalnie przyciętych drzew $T(\alpha)$. Nie jest to proces ciągły. Drzewo $T(\alpha)$ będzie optymalnie przycięte tak długo, aż wartość współczynnika α nie przekroczy pewnego progu α', dla którego drzewem optymalnie przyciętym będzie inne mniejsze drzewo $T(\alpha')$. Zwiększając dalej wartość współczynnika α, zostanie osiągnięta następna wartość progowa α'' i następne odpowiadające jej mniejsze, optymalnie przycięte drzewo $T(\alpha'')$. Proces ten możemy kontynuować, aż do momentu, gdy optymalnie przycięte drzewo będzie się składać tylko z korzenia drzewa $T_{\max}$.

Opiszmy teraz dokładnie konstrukcję ciągu optymalnie przyciętych poddrzew drzewa $T_{\max}$.

W pierwszym kroku przyjmujemy $\alpha_1 = 0$ i wyznaczamy optymalnie przycięte drzewo $T_1 = T(0)$. Zwróćmy uwagę, że dla drzewa T_1 zachodzi

$$\hat{e}_R(T_1) = \hat{e}_R(T_{\max}).$$

Niech t_L i t_R będą liśćmi, potomkami węzła $t \in T_{\max}$. Jeżeli

$$\hat{e}_R(t) = \hat{e}_R(t_L) + \hat{e}_R(t_R), \tag{6.5}$$

to liście t_L i t_R obcinamy. Przypomnijmy, że dla każdego węzła t zachodzi

$$\hat{e}_R(t) \geq \hat{e}_R(t_L) + \hat{e}_R(t_R).$$

Kontynuując proces obcinania liści, dla których jest spełniony warunek (6.5), otrzymujemy drzewo T_1.

W drugim kroku algorytmu, bazując na drzewie T_1, wyznaczamy optymalnie przycięte drzewo T_2.

Niech $t \in T_1 \setminus \tilde{T}_1$. Dla gałęzi T_t mamy

$$\hat{e}_R(t) > \hat{e}_R(T_t) = \sum_{t' \in \tilde{T}_t} \hat{e}_R(t'). \tag{6.6}$$

Oznaczmy przez $\{t\}$ podgałąź gałęzi T_t składającą się tylko z węzła t. Rozważmy różnicę między

$$R_\alpha(\{t\}) = \hat{e}_R(t) + \alpha$$

oraz

$$R_\alpha(T_t) = \hat{e}_R(T_t) + \alpha|\tilde{T}_t|.$$

Dla α takich, że

$$R_\alpha(T_t) < R_\alpha(\{t\})$$

gałąź T_t ma mniejszą miarę błędu i złożoności niż pojedynczy węzeł $\{t\}$. Nie ma więc powodu, aby ją przyciąć. Dla

$$\alpha = \frac{\hat{e}_R(t) - \hat{e}_R(T_t)}{|\tilde{T}_t| - 1}, \tag{6.7}$$

$R_\alpha(T_t) = R_\alpha(\{t\})$. Zatem ze względu na mniejszą złożoność będziemy preferować $\{t\}$ w stosunku do całej gałęzi T_t. Biorąc pod uwagę nierówność (6.6), wyznaczoną ze wzoru (6.7), wartość α jest dodatnia. Dla każdego $t \in T_1$, określmy funkcję $g(t)$ wzorem

$$g(t) = \begin{cases} \dfrac{\hat{e}_R(t) - \hat{e}_R(T_t)}{|\tilde{T}_t| - 1}, & \text{dla } t \notin \tilde{T}_1, \\ \infty, & \text{dla } t \in \tilde{T}_1. \end{cases} \tag{6.8}$$

Niech teraz $t_0 \in T_1$ będzie takie, że

$$g(t_0) = \min_{t \in T_1} g(t).$$

Przyjmijmy

$$\alpha_2 = g(t_0)$$

oraz

$$T_2 = T_1 - T_{t_0}.$$

Jeżeli dla węzła t'_0 zachodzi

$$g(t'_0) = g(t_0),$$

to przycinamy także gałąź $T_{t'_0}$, tzn.

$$T_2 = T_1 - T_{t_0} - T_{t'_0}.$$

Po przycięciu wszystkich tego typu gałęzi otrzymujemy końcową postać drzewa T_2. Następnie powtarzamy drugi krok algorytmu, przyjmując jako bazowe drzewa T_2, T_3 itd. Postępując w ten sposób, po skończonej liczbie kroków, otrzymamy ciąg

$$T_1 \succ T_2 \succ \cdots \succ \{t_1\} \tag{6.9}$$

optymalnie przyciętych poddrzew drzewa klasyfikacyjnego $T_{\max}$.

Poniższe twierdzenie podaje własności ciągu poddrzew (6.9).

Twierdzenie 6.3.

Niech α_k $(k = 1, 2, \ldots)$ będą wartościami progowymi współczynnika złożoności α odpowiadającymi poddrzewom T_k drzewa maksymalnego $T_{\max}$ z ciągu (6.9).

Wtedy dla każdego $k \geq 1$ oraz $\alpha_k \leq \alpha < \alpha_{k+1}$ zachodzi

$$T(\alpha) = T(\alpha_k) = T_k.$$

Zilustrujmy podany powyżej algorytm optymalnego przycinania, wykorzystując dane przedstawione na rys. 6.1.

Łatwo zauważyć, że odpowiadające im drzewo klasyfikacyjne z rys. 6.4 jest drzewem maksymalnym. W tym przypadku procedura optymalnego przycinania składa się jedynie z trzech kroków i prowadzi do następującego ciągu

$$T_1 = T_{\max} \succ T_2 \succ T_3 = \{t_1\}.$$

Przebieg procesu przycinania jest przedstawiony w tab. 6.2, a drzewo T_2 na rys. 6.9.

Mamy za sobą wstępny etap przycinania drzewa klasyfikacyjnego $T_{\max}$ polegający na konstrukcji ciągu drzew optymalnie przyciętych. Z ciągu te-

Tab. 6.2. Przebieg przycinania drzewa klasyfikacyjnego z rys. 6.4

k	$\lvert\tilde{T}_k\rvert$	α_k	$R_\alpha(T_k)$
1	6	0	0
2	3	1/15	2/15
3	1	1/10	6/15

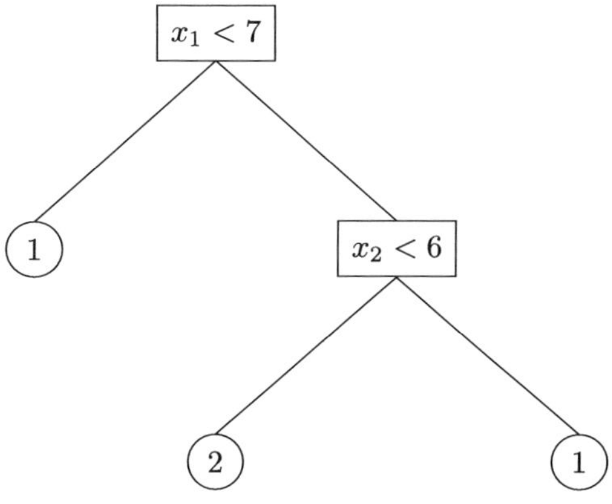

Rys. 6.9. Optymalnie przycięte drzewo klasyfikacyjne z rys. 6.4

go należy teraz wybrać drzewo o możliwie minimalnym błędzie klasyfikacji $e(T_k)$, $k = 1, 2, \ldots$.

Do oszacowania tego błędu użyjemy v-krokowej metody sprawdzania krzyżowego (standardowo przyjmujemy $v = 5$ lub $v = 10$). Inna możliwość, to wykorzystanie do szacowania błędu próby testowej – o ile taką mamy.

Niech zatem $\mathcal{L}_n = \tilde{\mathcal{L}}_n^{(1)} \cup \cdots \cup \tilde{\mathcal{L}}_n^{(v)}$. Jako estymator prawdopodobieństwa błędu dla drzewa klasyfikacyjnego T_k przyjmujemy

$$\hat{e}_{CV}(T_k) = \frac{1}{n} \sum_{l=1}^{v} \sum_{j=1}^{n} I(\boldsymbol{z}_j \in \tilde{\mathcal{L}}_n^{(l)}) I(\hat{d}_{T^l(\alpha'_k)}(\boldsymbol{x}_j) \neq y_j),$$

gdzie $T^l(\alpha'_k)$ jest elementem ciągu optymalnie przyciętych poddrzew drzewa $T^l_{\max}$, wyznaczonego na podstawie próby $\tilde{\mathcal{L}}_n \setminus \tilde{\mathcal{L}}_n^{(l)}$ oraz $\alpha'_k = \sqrt{\alpha_k \alpha_{k+1}}$, $k = 1, 2, \ldots$.

Za optymalne uznajemy drzewo T_0, dla którego

$$\hat{e}_{CV}(T_0) = \min_k \hat{e}_{CV}(T_k).$$

W metodzie CART zaproponowano dodatkowe wzmocnienie tej reguły przez wskazanie za optymalne nie drzewa T_0, lecz najmniejszego drzewa

T_k, dla którego $\hat{e}_{CV}(T_k)$ przekracza wartość $\hat{e}_{CV}(T_0)$ o nie więcej niż jedno odchylenie standardowe określone wzorem

$$s(T_0) = \sqrt{\frac{\hat{e}_{CV}(T_0)(1 - \hat{e}_{CV}(T_0))}{v}}.$$

Powyższa reguła wyboru nosi nazwę 1SE[9].

Przykład 6.1.
Proces konstrukcji optymalnego drzewa klasyfikacyjnego za pomocą metody CART zilustrujemy, wykorzystując rzeczywiste dane doświadczalne – Wisconsin Breast Cancer[10]. Analizowana próba obejmowała 699 kobiet z wykrytym guzem piersi – u 458 z nich stwierdzono guza łagodnego (klasa 1), a u pozostałych 241 guza złośliwego (klasa 2). W procesie klasyfikacji wykorzystamy pomiary 9 różnych cech o wartościach ze zbioru $\{1, 2, \ldots, 10\}$. Wszystkie cechy miały charakter ilościowy, a podczas konstrukcji drzewa klasyfikacyjnego wykorzystano jedynie podziały jednowymiarowe, przyjmując jako miarę niejednorodności indeks Giniego. Do wyznaczenia optymalnej wielkości drzewa klasyfikacyjnego zastosowano metodę przycinania. Ciąg optymalnie przyciętych poddrzew składał się z 11 elementów. Końcowego wyboru optymalnie przyciętego drzewa klasyfikacyjnego dokonano, stosując 10-krokową metodę sprawdzania krzyżowego z zastosowaniem reguły 1SE. Szczegóły procesu przycinania są zawarte w tab. 6.3.

Tab. 6.3. Szczegóły przebiegu przycinania drzewa klasyfikacyjnego

k	$\lvert\tilde{T}_k\rvert$	α_k	$\hat{e}_R(T_k)$	$\hat{e}_{CV}(T_k)$ $(\pm s(T_k))$
1	40	0.0000	0.0000	0.0601 (0.0090)
2	28	0.0005	0.0057	0.0544 (0.0086)
3	22	0.0007	0.0100	0.0486 (0.0081)
4	19	0.0010	0.0129	0.0444 (0.0078)
5	10	0.0014	0.0258	0.0515 (0.0084)
6	**8**	**0.0021**	**0.0300**	**0.0486 (0.0081)**
7	5	0.0043	0.0429	0.0587 (0.0089)
8	4	0.0057	0.0486	0.0587 (0.0089)
9	3	0.0086	0.0572	0.0587 (0.0089)
10	2	0.0186	0.0758	0.0830 (0.0104)
11	1	0.2690	0.3448	0.3448 (0.0179)

Optymalnej wielkości drzewo klasyfikacyjne ma 15 węzłów, z czego 8 to liście. Ponadto widać, że w procesie klasyfikacji bierze udział tylko 5 cech (pozostałe

[9] Po ang. one standard error (1SE).

[10] Dane pochodzą ze składnicy danych Uniwersytetu Kalifornijskiego w Irvine. http://www.ics.uci.edu/~mlearn/MLRepository.html.

4 nie mają jakiegokolwiek wpływu). Tabelaryczny oraz graficzny opis tego drzewa klasyfikacyjnego zawiera tab. 6.4 i rys. 6.10.

Tab. 6.4. Parametry drzewa klasyfikacyjnego z rys. 6.10

t	s	$n(t)$	$n_1(t)$	$n_2(t)$	$\hat{p}(t)$	$\hat{p}(1\|t)$	$\hat{p}(2\|t)$	ind(t)
t_1	$x_2 < 2.5$	699	458	241	1.000	0.655	0.345	–
t_2	$x_6 < 5.5$	429	417	12	0.614	0.972	0.028	–
t_3	$x_3 < 2.5$	270	41	229	0.386	0.152	0.848	–
t_4	–	421	416	5	0.602	0.988	0.012	1
t_5	–	8	1	7	0.011	0.125	0.875	2
t_6	$x_1 < 5.5$	23	18	5	0.033	0.783	0.217	–
t_7	$x_2 < 4.5$	247	23	224	0.353	0.093	0.907	–
t_8	–	19	18	1	0.027	0.947	0.053	1
t_9	–	4	0	4	0.006	0.000	1.000	2
t_{10}	$x_6 < 2.5$	70	18	52	0.100	0.257	0.743	–
t_{11}	–	177	5	172	0.253	0.028	0.972	2
t_{12}	$x_4 < 3.5$	14	10	4	0.020	0.714	0.286	–
t_{13}	–	56	8	48	0.080	0.143	0.857	2
t_{14}	–	11	10	1	0.016	0.909	0.091	1
t_{15}	–	3	0	3	0.004	0.000	1.000	2

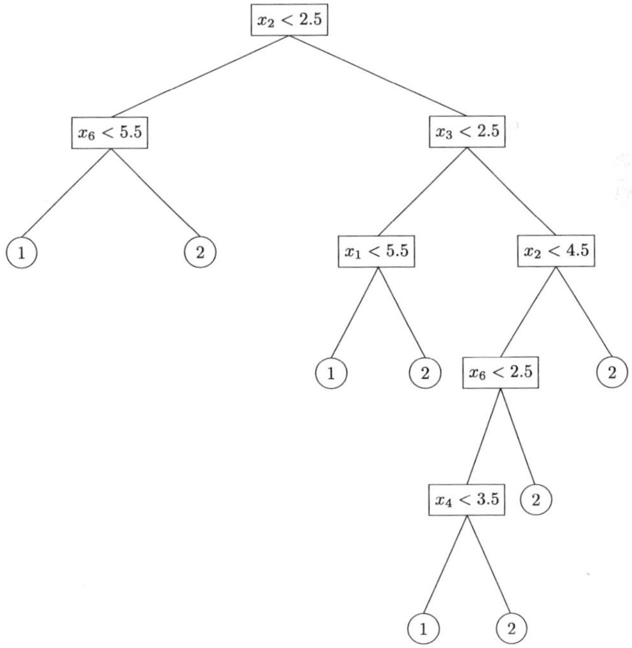

Rys. 6.10. Optymalne drzewo klasyfikacyjne uzyskane metodą CART

6.4. Kryteria podziału – metoda QUEST

W podrozdziale 6.2 został przedstawiony algorytm optymalnego podziału węzła t zaproponowany przez Breimana, Friedmana, Olsena i Stone'a (1984) w metodzie CART. Opiera się on na pomyśle wyczerpującego przeszukiwania wszystkich możliwych podziałów i wyborze optymalnego z wykorzystaniem miary niejednorodności węzła $i(t)$. Metoda ta ma jednak dwie podstawowe wady.

1. Dużą czasochłonność obliczeń, w szczególności gdy mamy do czynienia ze zmiennymi jakościowymi o wielu wartościach.
2. Obciążoność procesu wyboru zmiennych do podziału polegającą na skłonności do wybierania zmiennych, które prowadzą do większej liczby podziałów. Szczególnie widoczne jest to wtedy, gdy mamy do wyboru zarówno zmienne ilościowe, jak i jakościowe. Zmienne jakościowe są zdecydowanie częściej wybierane do podziału w stosunku do zmiennych ilościowych, a wśród zmiennych jakościowych częściej wybierane są te, które mają większą liczbę wartości.

W tym rozdziale zostanie przedstawiony szybki i nieobciążony algorytm podziału węzła zaproponowany przez Loha i Shiha (1997) w metodzie QUEST. Do wyznaczenia optymalnego podziału użyli oni procedury opartej na kwadratowej analizie dyskryminacyjnej omówionej w podrozdz. 1.4.

Analogicznie jak w podrozdz. 6.2 załóżmy, że badane cechy $X_1, \ldots, X_p$ są uporządkowane tak, że pierwszych m z nich to cechy ilościowe, a pozostałe $p - m$, to cechy jakościowe. Ponieważ użycie kwadratowej analizy dyskryminacyjnej do podziału węzła wymaga, aby wszystkie cechy były ilościowe, zatem rozpocznijmy od zastąpienia cech jakościowych przez odpowiadające im cechy ilościowe.

Niech X będzie cechą jakościową przyjmującą wartości ze zbioru $V = \{v_1, \ldots, v_L\}$. Zmiennej X przyporządkowujemy wektor $\boldsymbol{U} = (U_1, \ldots, U_L)'$ w taki sposób, że

$$U_l = \begin{cases} 1, & \text{jeżeli } X = v_l, \\ 0, & \text{w przeciwnym przypadku.} \end{cases}$$

Ostatecznie zmienną jakościową X zastępujemy zmienną ilościową $Y = \boldsymbol{a}'\boldsymbol{U}$, gdzie wektor $\boldsymbol{a}$ jest wyznaczony tak, aby Y była pierwszą zmienną dyskryminacyjną (patrz podrozdz. 1.7). Dodatkowo wartości zmiennej Y skalujemy tak, aby wszystkie znalazły się w przedziale $[-1, 1]$ – najmniejsza z nich przyjmowała wartość -1, a największa 1.

Rozważmy najpierw podziały jednowymiarowe, generowane przez pytanie postaci

czy $x < c$,

gdzie x jest wartością wybranej do podziału zmiennej, a c dobraną odpowiednio wartością progową.

W procedurze QUEST wyraźnie są wydzielone dwa etapy prowadzące do znalezienia optymalnego podziału węzła: pierwszy etap polega na wyborze zmiennej, a drugi na dobraniu dla wybranej w etapie pierwszym zmiennej wartości progowej c. Zacznijmy od wyboru jednej (optymalnej) zmiennej, na bazie której dokonamy podziału węzła t. Ponieważ podział ma prowadzić do uzyskania możliwie jednorodnych węzłów potomków, zatem do podziału powinniśmy wybrać tę z cech, której wartości w poszczególnych klasach są najbardziej zróżnicowane. Do wyboru tej zmiennej użyjemy testu analizy wariancji (ANOVA). W hipotezie zerowej tego testu zakładamy, że wartości oczekiwane danej cechy są sobie równe we wszystkich klasach. Do testowania tej hipotezy używamy statystyki F będącej ilorazem wariancji między i wewnątrz grupowej. Korzystając z tego testu, do podziału powinniśmy wybrać tę cechę, dla której wartość statystyki F jest maksymalna oraz zapewnia odrzucenie hipotezy zerowej na zadanym poziomie istotności.

Okazuje się jednak, że dla cech jakościowych i ich zastępczych zmiennych, wartości statystyki F testu ANOVA są nieadekwatnie duże w stosunku do wartości tej statystyki obliczonej dla pozostałych zmiennych ilościowych. W konsekwencji prowadziłoby to do nadmiernie częstego (obciążonego) wyboru do podziału zmiennych jakościowych. Aby tego uniknąć, dla cech jakościowych, zamiast testem ANOVA będziemy się posługiwać testem niezależności χ^2 (wykorzystując ich pierwotne wartości). Wartości statystyki testowej χ^2 obliczamy, korzystając z tablicy dwudzielczej utworzonej na podstawie etykiet klas i wartości danej cechy. Duże wartości tej statystyki przeczą hipotezie zerowej o niezależności (jednorodności) rozkładów badanej cechy w K klasach.

Procedura wyboru zmiennych jest zatem następująca. Dla każdej cechy ilościowej X_j, $j = 1, 2, \ldots, m$, obliczamy wartość F_j statystyki F w teście ANOVA. Niech j_1 oznacza numer pierwszej zmiennej, dla której

$$F_{j_1} = \max\{F_j : j = 1, 2, \ldots, m\}.$$

Dla tak wybranej zmiennej obliczamy p-wartość α_1 w teście ANOVA ze wzoru

$$\alpha_1 = \mathrm{P}(F > F_{j_1}),$$

gdzie $F \sim F_{K(t)-1,\ n(t)-K(t)}$.

Z kolei dla każdej cechy jakościowej X_j, $j = m+1, 2, \ldots, p$, obliczamy wartość χ_j^2 statystyki χ^2 w teście niezależności. Niech j_2 oznacza numer pierwszej zmiennej, dla której

$$\chi_{j_2}^2 = \max\{\chi_j^2 : j = m+1, 2, \ldots, p\}.$$

Dla tak wybranej zmiennej obliczamy p-wartość α_2 w teście niezależności χ^2 ze wzoru

$$\alpha_2 = \mathrm{P}(\chi^2 > \chi_{j_2}^2),$$

gdzie $\chi^2 \sim \chi^2_{((K(t)-1)(L(t)-1))}$ oraz $L(t)$ oznacza liczbę wartości danej cechy w węźle t.

Oznaczmy przez $\alpha_0 = \min\{\alpha_1, \alpha_2\}$ oraz dla $\alpha_1 \le \alpha_2$ przyjmijmy $j_0 = j_1$, a dla $\alpha_1 > \alpha_2$ przyjmijmy $j_0 = j_2$.

Wykonaliśmy zatem p-testów (po jednym dla każdej z p cech). Dla zapewnienia poziomu istotności całej wielokrotnej procedury testowej równej α (zazwyczaj $\alpha = 0.05$), przyjmujemy poziom istotności każdego z testów równy α/p. Z nierówności Bonferroniego wynika bowiem wtedy, że

$$\begin{aligned}
&\mathrm{P}(F > F_1 \cup \cdots \cup F > F_m \cup \chi^2 > \chi_{m+1}^2 \cup \cdots \cup \chi^2 > \chi_p^2) \le \\
&\le \mathrm{P}(F > F_1) + \cdots + \mathrm{P}(F > F_m) + \mathrm{P}(\chi^2 > \chi_{m+1}^2) + \cdots + \mathrm{P}(\chi^2 > \chi_p^2) \le \\
&\le \sum_{j=1}^{p} \frac{\alpha}{p} = \alpha.
\end{aligned}$$

Stąd, aby cecha X_{j_0} została wybrana do podziału węzła t, musi być spełniony warunek

$$\alpha_0 < \frac{\alpha}{p}. \tag{6.10}$$

Pozostała jeszcze do rozważenia sytuacja, gdy dla cechy X_{j_0} warunek (6.10) nie jest spełniony. Przekształćmy wtedy zmienne ilościowe $X_1, \ldots, X_m$ na nowe zmienne $Z_1, \ldots, Z_m$.

Niech $x_{jl}^{(i)}$ ($j = 1, \ldots, m$, $l = 1, \ldots, n_i(t)$, $i = 1, \ldots, K(t)$) oznacza l-tą obserwację j-tej cechy z klasy i-tej należącą do węzła t. Zdefiniujmy

$$z_{jl}^{(i)} = |x_{jl}^{(i)} - \bar{x}_j^{(i)}|,$$

gdzie

$$\bar{x}_j^{(i)} = \frac{1}{n_i(t)} \sum_{l=1}^{n_i(t)} x_{jl}^{(i)}.$$

Dla tych dodatkowych m zmiennych $Z_1, \ldots, Z_m$ wykonajmy również test ANOVA. Przez j_3 oznaczmy numer zmiennej o największej wartości statystyki F, a przez α_3 odpowiadającą jej p-wartość. W sumie wykonaliśmy teraz $p + m$ testów. Jeżeli

$$\alpha_3 < \frac{\alpha}{p+m}, \tag{6.11}$$

to do podziału wybieramy cechę X_{j_3}. Jeżeli jednak warunek (6.11) nie jest spełniony, to kończymy procedurę wyboru zmiennej, wybierając zmienną X_{j_0} (pomimo że nie jest spełniony warunek (6.10)).

Jak już wspomniano wcześniej, w metodzie QUEST wybór stałej c oparty zostanie na kwadratowej analizie dyskryminacyjnej. Niech zatem X będzie wybraną (z p możliwych) zmienną w opisany powyżej sposób. W przypadku gdy X jest zmienną jakościową, to w dalszym ciągu będziemy używać odpowiadającej jej zmiennej ilościowej Y.

Jeżeli w węźle t mamy $K(t) > 2$, to grupujemy obserwacje z tego węzła w dwie superklasy A i B, stosując metodę k-średnich ($k = 2$) omówioną w podrozdz. 10.2. Za punkt startowy w procedurze k-średnich przyjmujemy dwie najbardziej różniące się średnie z $K(t)$ klas. Gdyby wszystkie średnie były identyczne, to za superklasę A przyjmujemy tę klasę z $K(t)$ możliwych, która jest najliczniejsza, a superklasę B tworzą elementy z pozostałych klas.

Zakładając normalność rozkładów cechy X w obu superklasach, metoda QDA prowadzi do równania krzywej je rozdzielającej postaci:

$$\frac{p(A|t)}{s_A}\phi\left(\frac{x-\bar{x}_A}{s_A}\right) = \frac{p(B|t)}{s_B}\phi\left(\frac{x-\bar{x}_B}{s_B}\right), \tag{6.12}$$

gdzie ϕ jest funkcją gęstości rozkładu $N(0,\ 1)$, a $\bar{x}_A$, $\bar{x}_B$, s_A^2, s_B^2 są średnimi oraz wariancjami obliczonymi na podstawie prób uczących należących do A i B odpowiednio.

Równanie (6.12) można łatwo przekształcić do postaci

$$a_2x^2 + a_1x + a_0 = 0, \tag{6.13}$$

gdzie

$$\begin{aligned} a_2 &= s_A^2 - s_B^2 \\ a_1 &= 2(\bar{x}_A s_B^2 - \bar{x}_B s_A^2) \\ a_0 &= (\bar{x}_B s_A)^2 - (\bar{x}_A s_B)^2 + 2s_A^2 s_B^2 \log\{[p(A|t)s_B]/[p(B|t)s_A]\}. \end{aligned}$$

Dla $a_2 = 0$, tzn. gdy $s_A = s_B$ równanie (6.13) jest równaniem liniowym. Ponadto gdy $a_1 \neq 0$, tzn. gdy $\bar{x}_A \neq \bar{x}_B$, ma ono dokładnie jedno rozwiązanie, które przyjmujemy za wartość progową podziału c. Mamy wtedy

$$c = (\bar{x}_A + \bar{x}_B)/2 - (\bar{x}_A - \bar{x}_B)^{-1} s_A^2 \log[p(A|t)/p(B|t)].$$

Dla $a_2 = a_1 = 0$ równanie (6.13) może nie mieć żadnego lub mieć nieskończenie wiele rozwiązań. Wtedy przyjmujemy $c = \bar{x}_A = \bar{x}_B$.

Rozważmy teraz przypadek, gdy $a_2 \neq 0$. Gdy $\Delta < 0$ i równanie (6.13) nie ma pierwiastków, przyjmujemy $c = (\bar{x}_A + \bar{x}_B)/2$. W przypadku gdy $\Delta \geq 0$, za c przyjmujemy ten z pierwiastków równania, który jest bliższy $\bar{x}_A$ pod warunkiem jednak, że doprowadzi to do podziału na niepuste węzły potomki. Gdyby warunek ten nie miał być spełniony, to tak jak poprzednio przyjmujemy $c = (\bar{x}_A + \bar{x}_B)/2$.

Konstrukcja optymalnych podziałów wielowymiarowych opiera się na zastosowaniu zmiennych dyskryminacyjnych. Transformujemy wszystkie zmienne jakościowe do zmiennych ilościowych, a następnie, korzystając ze wszystkich p zmiennych, wyznaczamy pierwszą zmienną dyskryminacyjną. Dla tej nowej zmiennej szukamy optymalnej wartości progowej c, analogicznie jak w przypadku podziałów jednowymiarowych.

Ostateczną postać drzewa klasyfikacyjnego otrzymujemy, stosując bezpośrednią regułę stopu lub technikę przycinania opisaną w podrozdz. 6.3.

Przykład 6.2.

Porównajmy optymalne drzewa klasyfikacyjne utworzone za pomocą metody CART i QUEST, wykorzystując dane z przykładu 6.1. W algorytmie QUEST wykorzystamy również jedynie podziały jednowymiarowe, a do wyznaczenia optymalnej wielkości drzewa klasyfikacyjnego zastosujemy analogiczną technikę przycinania.

Tym razem ciąg optymalnie przyciętych poddrzew składa się tylko z 9 elementów, a optymalnej wielkości drzewo klasyfikacyjne ma tak jak poprzednio 15 węzłów, z czego 8 to liście – patrz tab. 6.5.

Z tabeli 6.3 oraz 6.5 widać również, że oszacowane metodą 10-krokowego sprawdzania krzyżowego prawdopodobieństwo błędnej klasyfikacji dla optymalnych drzew klasyfikacyjnych jest w obu metodach prawie identyczne (CART: 0.0486, QUEST: 0.0458). Istnieją jednak różnice w końcowej postaci drzew klasyfikacyjnych. Dla metody QUEST optymalne drzewo klasyfikacyjne przedstawiono w tab. 6.6 i na rys. 6.11. Na przykład do pierwszego podziału w metodzie CART została wybrana zmienna X_2, a w metodzie QUEST zmienna X_3. Zasadnicza jednak różnica to czas potrzebny do konstrukcji optymalnego drzewa klasyfikacyjnego. W metodzie QUEST jest on więcej niż o połowę krótszy

Tab. 6.5. Ciąg optymalnie przyciętych drzew klasyfikacyjnych

k	$\lvert\tilde{T}_k\rvert$	α_k	$\hat{e}_R(T_k)$	$\hat{e}_{CV}(T_k)$ $(\pm s(T_k))$
1	39	0.0000	0.0014	0.0529 (0.0085)
2	36	0.0005	0.0029	0.0501 (0.0082)
3	20	0.0007	0.0143	0.0486 (0.0081)
4	11	0.0014	0.0272	0.0415 (0.0075)
5	**8**	**0.0029**	**0.0358**	**0.0458 (0.0079)**
6	4	0.0039	0.0515	0.0501 (0.0082)
7	3	0.0072	0.0587	0.0558 (0.0087)
8	2	0.0186	0.0773	0.0829 (0.0104)
9	1	0.2675	0.3448	0.3448 (0.0179)

Tab. 6.6. Parametry drzewa klasyfikacyjnego z rys. 6.11

t	s	$n(t)$	$n_1(t)$	$n_2(t)$	$\hat{p}(t)$	$\hat{p}(1\|t)$	$\hat{p}(2\|t)$	ind(t)
t_1	$x_3 < 3.568$	699	458	241	1.000	0.655	0.345	–
t_2	$x_6 < 4.018$	468	436	32	0.670	0.932	0.068	–
t_3	$x_2 < 5.354$	231	22	209	0.330	0.095	0.905	–
t_4	–	435	426	9	0.622	0.979	0.021	1
t_5	$x_3 < 1.728$	33	10	23	0.047	0.303	0.697	–
t_6	$x_6 < 3.084$	89	17	72	0.127	0.191	0.809	–
t_7	–	142	5	137	0.203	0.035	0.965	2
t_8	–	9	7	2	0.013	0.778	0.222	1
t_9	–	24	3	21	0.034	0.125	0.875	2
t_{10}	$x_9 < 2.500$	22	13	9	0.031	0.591	0.409	–
t_{11}	–	67	4	63	0.096	0.060	0.940	2
t_{12}	$x_4 < 3.400$	19	13	6	0.027	0.684	0.316	–
t_{13}	–	3	0	3	0.004	0.000	1.000	2
t_{14}	–	15	13	2	0.021	0.867	0.133	1
t_{15}	–	4	0	4	0.006	0.000	1.000	2

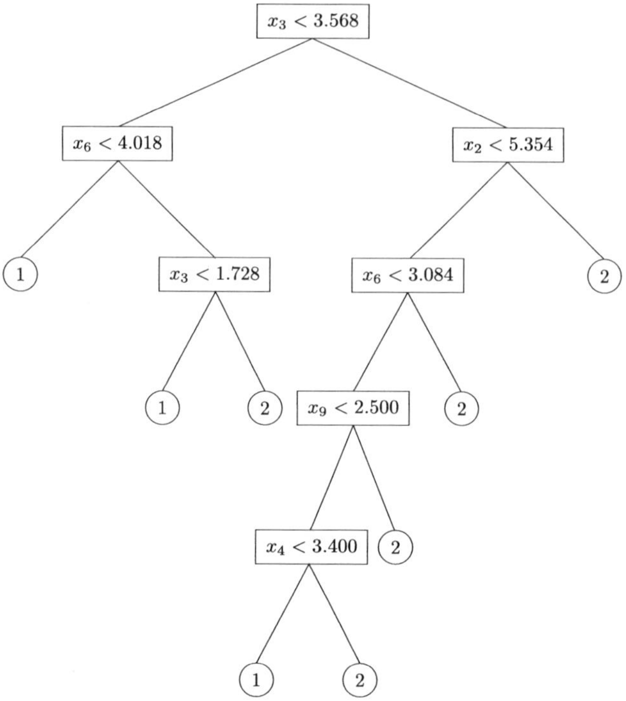

Rys. 6.11. Optymalne drzewo klasyfikacyjne uzyskane metodą QUEST

6.5. Brakujące wartości cech

Jedną z pozytywnych cech algorytmów drzew klasyfikacyjnych jest umiejętność radzenia sobie z obserwacjami (zarówno z próby uczącej, jak i nowymi, poddawanymi klasyfikacji), dla których istnieją cechy z brakującymi wartościami.

Sytuacja, w której brakująca wartość występuje w zbiorze uczącym $\mathcal{L}_n$ jest rozwiązywana w prosty sposób. W metodzie CART do wyznaczania optymalnego podziału są używane tylko niebrakujące wartości zmiennych (w przypadku podziałów wielowymiarowych są używane jedynie obserwacje z kompletnymi danymi dla wszystkich zmiennych ilościowych lub jakościowych odpowiednio). W metodzie QUEST brakujące wartości zmiennej są zastępowane średnimi obliczonymi z niebrakujących wartości tej zmiennej w danej klasie.

Bardziej skomplikowana sytuacja zachodzi wtedy, gdy brakujące wartości pewnych zmiennych występują w nowym, klasyfikowanym obiekcie.

Metoda CART wykorzystuje wtedy tzw. *podziały zastępcze*[11]. Idea podziału zastępczego jest następująca: jeżeli podział węzła opiera się na zmiennej X, a jej wartością nie dysponujemy, to zamiast tej zmiennej wykorzystamy zmienną X' (której wartością dysponujemy) wybraną w taki sposób, aby podział dokonany za pomocą zmiennej X' był możliwie zbliżony do podziału dokonanego za pomocą zmiennej X.

Niech s^* oznacza, jak zwykle, optymalny podział węzła t na dwa potomki t_L i t_R, ponadto niech podział s^* będzie wyznaczony przez zmienną X. Weźmy dowolną zmienną X' i oznaczmy przez S' zbiór wszystkich podziałów wyznaczonych przez tę zmienną, a przez $\bar{S}'$ zbiór wszystkich podziałów dopełniających podziały ze zbioru S'.

Dla dowolnego podziału $s' \in S' \cup \bar{S}'$ węzła t na dwa węzły potomki t'_L i t'_R oznaczmy przez $p_L(s^*, s')$ prawdopodobieństwo tego, że element z węzła t zarówno w wyniku podziału s^*, jak i s' znajdzie się w $t_L \cap t'_L$.

Stąd

$$p_L(s^*,\ s') = \frac{p(t_L \cap t'_L)}{p(t)}.$$

Analogicznie

$$p_R(s^*,\ s') = \frac{p(t_R \cap t'_R)}{p(t)}.$$

Podział $\tilde{s}$ węzła t wyznaczony przez zmienną X' będziemy nazywać zastępczym dla podziału s^* wyznaczonego przez zmienną X, gdy

$$p(s^*,\ \tilde{s}) = \max_{s' \in S' \cup \bar{S}'} [p_L(s^*,\ s') + p_R(s^*,\ s')].$$

W każdym węźle t drzewa klasyfikacyjnego możemy stworzyć listę podziałów zastępczych i przy braku wartości zmiennej wykorzystać najbardziej zgodny podział zastępczy.

Jako miarę zgodności podziałów s^* i $\tilde{s}$ przyjmijmy

$$\lambda = \frac{\min\{p_L,\ p_R\} - [1 - p(s^*,\ \tilde{s})]}{\min\{p_L,\ p_R\}}.$$

Zauważmy, że $\lambda \leq 1$, a wartość jeden odpowiada idealnej zgodności podziałów. Widać również, że λ może przyjmować wartości ujemne, co wskazuje na zupełny brak zgodności podziałów.

Technikę podziałów zastępczych możemy stosować zarówno do podziałów jedno-, jak i wielowymiarowych.

[11]Po ang. surrogate splits.

W algorytmie QUEST bardziej tradycyjnie zastępujemy brakujące dane dobranymi w odpowiedni sposób wartościami. W przypadku podziałów jednowymiarowych zastępowania takiego dokonujemy dopiero w tym węźle, w którym mamy brakującą wartość zmiennej wykorzystywanej do jego podziału. Natomiast przy podziałach wielowymiarowych zastępujemy wszystkie brakujące wartości zmiennych już w korzeniu drzewa klasyfikacyjnego.

Załóżmy, że dla nowego obiektu $\boldsymbol{Z}_0 = (\boldsymbol{X}_0, Y_0)$ brakuje nam wartość x_{j0} j-tej zmiennej i że zmienna ta jest ilościowa. Brakującą wartość zastąpimy j-tą składową środka ciężkości klasy, która leży najbliżej obiektu $\boldsymbol{Z}_0$. Odległość obiektu $\boldsymbol{Z}_0$ od i-tej klasy definiujemy następująco:

$$d_i = -\log[\hat{p}(i|t)] + \frac{1}{2}\sum_j \left[2\log(s_j^{(i)}) + \left(\frac{x_{j0} - \bar{x}_j^{(i)}}{s_j(i)} \right)^2 \right], \qquad (6.14)$$

gdzie $\bar{x}_j^{(i)}$ i $s_j^{(i)}$ oznaczają średnią oraz odchylenie standardowe j-tej zmiennej w i-tej klasie, a sumowanie dotyczy wyłącznie zmiennych o niebrakujących wartościach (zmienne jakościowe zastępujemy przez odpowiadające im zmienne ilościowe, tak jak przy tworzeniu podziałów).

W przypadku gdy brakującą wartość ma zmienna jakościowa, to jej wartość zastępujemy kategorią, której odpowiednik ilościowy jest najbliższy zastępującej ją wartości.

Rozważane przez nas algorytmy drzew klasyfikacyjnych prowadziły zawsze do drzew binarnych. Są one niewątpliwie najczęściej stosowane, co nie znaczy, że nie istnieją ciekawe i udane procedury drzew niebinarnych. Szczególnie ciekawą propozycją takiego algorytmu jest opracowana przez Kima i Loha metoda CRUISE (Classification Rule with Unbiased Interaction Selection and Estimation, 2001). W procedurze tej liczba potomków każdego węzła równa jest liczbie klas.

Kończąc rozważania dotyczące drzew klasyfikacyjnych, zwróćmy uwagę na ich następującą wadę. Otóż procedury drzew klasyfikacyjnych (tak samo jak sieci neuronowe omówione w rozdz. 7) charakteryzują się niestabilnością w tym sensie, że nawet niewielkie zmiany próby uczącej mogą powodować bardzo duże różnice w końcowej postaci optymalnego drzewa klasyfikacyjnego i co za tym idzie mogą w efekcie dawać duże różnice w klasyfikacji.

Aby zniwelować to zjawisko, stosuje się często różne techniki wzmacniania tego typu klasyfikatorów. Należą do nich algorytmy bagging i boosting, które zostaną omówione w rozdz. 9. Inną propozycją, dostosowaną wyłącznie do drzew klasyfikacyjnych, jest *algorytm lasów losowych* podany i rozwijany przez Breimana (2001).

7

Sieci neuronowe

Zainteresowanie sieciami neuronowymi zarówno wśród inżynierów, przedstawicieli nauk ścisłych – matematyki i fizyki, oraz biologów czy neurofizjologów wynika przede wszystkim z poszukiwań sposobów budowy bardziej efektywnych i niezawodnych urządzeń do przetwarzania informacji, a układ nerwowy człowieka jest tutaj niedościgłym wzorem. Mózg człowieka ciągle jest najpotężniejszym z istniejących obecnie urządzeń liczących do celów przetwarzania informacji w czasie rzeczywistym. Jego główną zaletą pozostaje możliwość pracy równoległej, a nie tak jak większości komputerów szeregowej oraz możliwość odrzucania przypadków bezsensownych.

W ogólności sieć neuronowa składa się z grupy albo grup połączonych ze sobą komórek nerwowych. Jedna komórka nerwowa może być łączona z wieloma innymi komórkami nerwowymi, co sprawia, że ogólna struktura sieci może być bardzo skomplikowana. *Sztuczna sieć neuronowa*[1], nazywana również *symulowaną siecią neuronową*[2] lub po prostu siecią neuronową[3] to połączona grupa sztucznych komórek nerwowych, która wykorzystuje model matematyczny w celu przetwarzania informacji. Pomysł tej techniki wywodzi się z badań bioelektrycznych sieci w mózgu utworzonych przez neurony i synapsy. W modelu sieci neuronowej węzły (zwane neuronami) są połączone między sobą, tworząc sieć węzłów – stąd określenie „sieć neuronowa".

Badania nad sztucznymi sieciami neuronowymi miały swoje początki w latach czterdziestych XX w. McCulloch i Pitts (1943) przedstawili pierwszy formalny model neuronu w postaci układu arytmetyczno-logicznego. Jego idea przetrwała lata i stanowi do dzisiaj podstawowe ogniwo większości

[1]Po ang. artificial neural network.
[2]Po ang. simulated neural network.
[3]Po ang. neural network.

używanych modeli. Istotnym elementem tego modelu jest sumowanie sygnałów wejściowych z odpowiednią wagą i poddanie otrzymanej sumy działaniu nieliniowej funkcji aktywacji. Sieci wykorzystujące ten model nie zostały praktycznie zrealizowane w ówczesnej technice, jednak wywarły wpływ na późniejszy rozwój tej dziedziny. W roku 1949 Hebb odkrył, że informacja może być przechowywana w strukturze połączeń między neuronami, i jako pierwszy zaproponował metodę uczenia sieci polegającą na zmianach wag połączeń między neuronami. Zaproponowana metoda jest dzisiaj nazywana regułą Hebba. W latach pięćdziesiątych powstały pierwsze sieci neuronowe. Szeroko znanym przykładem działającej sieci jest perceptron. Zbudowany został w roku 1958 przez Rosenblatta i Wightmana w Cornell Aeronautical Laboratory. Został on określony jako maszyna klasyfikująca obrazy i zdolna do uczenia się przez modyfikację połączeń prowadzących do układów progowych. Perceptron był wykorzystywany do rozpoznawania znaków alfanumerycznych, niestety, okazał się wrażliwy na transformacje znaków, pomimo tego działał poprawnie (jak na swoje możliwości) nawet po uszkodzeniu kilku elementów. W roku 1975 Fukushima zbudował *Cognitron*, a w 1978 jego rozbudowaną wersję *Neocognitron* potrafiącą rozpoznawać nawet pismo chińskie. Maszyna ta była dodatkowo odporna na zniekształcenia, przeskalowania, przesunięcia i obroty znaków. Cognitron był pierwszą siecią wielowarstwową. Przełom w wykorzystaniu sieci neuronowych nastąpił w roku 1986, kiedy to opracowano metodę uczenia sieci zwaną metodą wstecznej propagacji błędu[4].

Sieci neuronowe są z powodzeniem stosowane w bardzo wielu różnych dziedzinach, jak: finanse, medycyna, technika, geologia czy fizyka. Sieci neuronowe mogą być użyte wszędzie tam, gdzie pojawiają się zadania związane z predykcją, klasyfikacją czy sterowaniem. Olbrzymi sukces sieci neuronowych związany jest z dwoma czynnikami:

- Moc – jest bardzo zaawansowaną techniką modelowania, zdolną do odwzorowywania nadzwyczaj złożonych funkcji. W szczególności sieci neuronowe mają charakter nieliniowy, co istotnie wzbogaca możliwości ich zastosowań.
- Prostota – w praktyce same konstruują potrzebne użytkownikowi modele, ponieważ automatycznie uczą się na podanych przez niego przykładach. Odbywa się to w taki sposób, że użytkownik sieci gromadzi dane pokazujące, jak wyraża się interesująca go zależność, a następnie uruchamia algorytm uczenia, który ma na celu automatyczne wytworzenie

[4]Po ang. BackPropagation.

w pamięci sieci potrzebnej struktury danych. Opierając się na tej samodzielnie stworzonej strukturze danych, sieć realizuje potem wszystkie funkcje związane z eksploatacją utworzonego modelu.

7.1. Modele neuronów

Mózg i komputer, zastanówmy się, jakie mamy tutaj podobieństwa i jakie różnice. Wyobraźmy sobie komputer, który rozwiązując pewien problem, sam się uczy. Najpierw wprowadzamy do niego informacje o postawionym zadaniu, dane wejściowe problemu oraz wybrane przykłady wraz z poprawnymi ich rozwiązaniami. Następnie komputer analizuje wprowadzone informacje i ucząc się na swoich błędach, osiąga w końcu taki stan, w którym postawiony problem może być rozwiązany. W takiej działalności można zauważyć wiele podobieństw do działania człowieka. I tutaj pojawia się pytanie: czy potrafimy skonstruować urządzenie techniczne o podobnych właściwościach? Badania ostatnich lat sugerują odpowiedź twierdzącą – a właśnie sieci neuronowe wydają się być drogą prowadzącą do tego celu. Jak działa zatem ludzki mózg, gdy postawimy go w sytuacji, w której musi bardzo szybko dokonać wyboru jednej z wielu możliwości. W takiej sytuacji improwizujemy, tzn. z wielu czynników, które bierzemy pod uwagę, usuwamy te, których nie jesteśmy pewni, a następnie porównujemy sytuację ze wspomnieniami (są to wzorce podobnych sytuacji z przeszłości). Poszukujemy wzorca połączonego z reakcją, która okazała się dla nas korzystna. Jeśli taki wzorzec odnajdziemy, to postępujemy w taki właśnie sposób, mimo że sytuacja może się znacząco różnić od wzorca. Dlatego aby działać prawidłowo, należy uczyć się na wielu wzorcach, również tych negatywnych.

Z punktu widzenia informatyki interesujące jest porównanie własności komputera i mózgu. Sieci nerwowe mogą przeprowadzać niezwykłe obliczenia i działania, aczkolwiek jest rzeczą oczywistą, że w na przykład obliczeniach arytmetycznych mózg nie jest tak dobrym urządzeniem (szybkość, wydajność i dokładność) jak komputer. Z drugiej jednak strony, gdy ma się do czynienia z zadaniami takimi jak rozpoznawanie, skojarzenia czy klasyfikacja – mózg może pokonać nawet najszybszy superkomputer, pomimo że w tym procesie neurony jako jednostki przetwarzające są o wiele wolniejsze od swoich elektronicznych odpowiedników. Z punktu widzenia zasady działania, zarówno mózg, jak i konwencjonalne komputery realizują w zasadzie podobne funkcje – gromadzą i przetwarzają. Różnica nie leży więc w odmiennym działaniu, lecz na odmiennych zasadach gromadzenia i przetwarzania informacji. Jeżeli mamy wykonać proste działanie arytmetyczne, np. dodawanie liczb, to nie wykonujemy tego dodawania, lecz rozpoznajemy problem, a następnie przywołujemy z pamięci właściwą, skojarzoną z nim odpowiedź. Jedną z najciekawszych własności, najbardziej różniącą świat sieci neuronowych od świata komputerów, jest ich zdolność do poprawiania błędów. Mózg zdolny jest do rekonstrukcji, odtworzenia sygnału na podstawie informacji częściowej, a dodatkowo obarczonej błędami.

Jak wynika z badań anatomicznych i histologicznych na mózg człowieka składa się przede wszystkim bardzo duża liczba elementarnych komórek nerwowych, czyli *neuronów* (rys. 7.1). Ocenia się, że jest ich około 85 miliardów[5] (w mózgu o wadze 1350 g), połączonych ze sobą w formie skomplikowanej sieci. Ustalono, że średnio na jeden neuron przypada kilka tysięcy połączeń, których łącznie jest około 100 trylionów. Każdy z neuronów jest komórką organiczną mogącą przenosić i przetwarzać złożone sygnały elektrochemiczne. Neuron na ogół ma rozgałęzioną strukturę wielu wejść informacyjnych (*dendryty*), scalające sygnały ze wszystkich tych wejść ciało (*perikarion*) oraz opuszczający komórkę jako pojedyncze włókno nośnik informacji wyjściowej (*akson*), powielający potem przewodzony przez siebie wynik pracy neuronu i rozsyłający go do różnych neuronów odbiorczych przez rozgałęzioną strukturę wyjściową (*telodendron*). Akson jednej komórki łączy się z dendrytami innych komórek przez biochemiczne złącza, modyfikujące sygnały i stanowiące nośnik pamięci. Są to tak zwane *synapsy*, biologicznie bardzo skomplikowane, ale w sieciach neuronowych sprowadzane jedynie do operatorów przemnażania wejściowych sygnałów przez współczynniki ustalane w toku procesu uczenia. Pobudzony przez synapsy neuron przechodzi do stanu aktywnego, co objawia się tym, że wysyła on przez swój akson wyjściowy sygnał elektrochemiczny o charakterystycznym kształcie, amplitudzie i czasie trwania. Ogólna forma sygnału pobudzonego neuronu jest identyczna dla wszystkich żywych organizmów, chociaż jego czas trwania jest różny dla różnych organizmów. Najkrótszy jest on u ssaków i dlatego mówi się często, że pobudzony neuron ssaka wysyła impuls czynnościowy (iglicę). Przejście od stanu bezczynności do stanu generacji impulsów zachodzi w typowych neuronach bardzo gwałtownie, dlatego mówi się o „zapłonie" neuronu, który zaczyna generować na swoim wyjściu odpowiedni sygnał. Sygnał ten za pośrednictwem kolejnych synaps dociera do innych neuronów, które pod jego wpływem mogą również zostać postawione w stan zapłonu. Neuron przechodzi w stan zapłonu tylko wówczas, gdy łączny sygnał, który dotarł do ciała komórki przez dendryty przekroczy pewien poziom progowy.

[5]Lange W. (1975), Cell number and cell density in the cerebellar cortex of man and other mammals, *Cell Tissue Res.*, **157(1)**, s. 115–124.

Siła otrzymanego przez neuron sygnału (i związana z tym szansa na zapłon) zależy w największym stopniu od efektywności (współczynnika wagi) synapsy, do której dociera impuls. Ponieważ waga ta ma zasadnicze znaczenie z punktu widzenia objaśnienia działania sieci neuronowych, przyjrzymy się dokładniej temu, co się dzieje w synapsach. W każdej synapsie występuje szczelina wypełniona specjalną substancją, tak zwanym *neurotransmiterem* albo *neuromediatorem*. Mechanizm funkcjonowania neurotransmitera ma duże znaczenie w biologii układu nerwowego: niedobór niektórych neurotransmiterów powoduje groźne choroby (choroba Parkinsona), zaś wpływając na chemiczne przemiany neurotransmitera, można sztucznie wpływać na zachowanie człowieka (tak działa wiele leków i niektóre narkotyki). Neurotransmiter jest substancją chemiczną, która ma zdolność do przesłania sygnału przez występującą w każdej synapsie lukę między neuronem nadającym sygnał i neuronem, który ten sygnał odbiera.

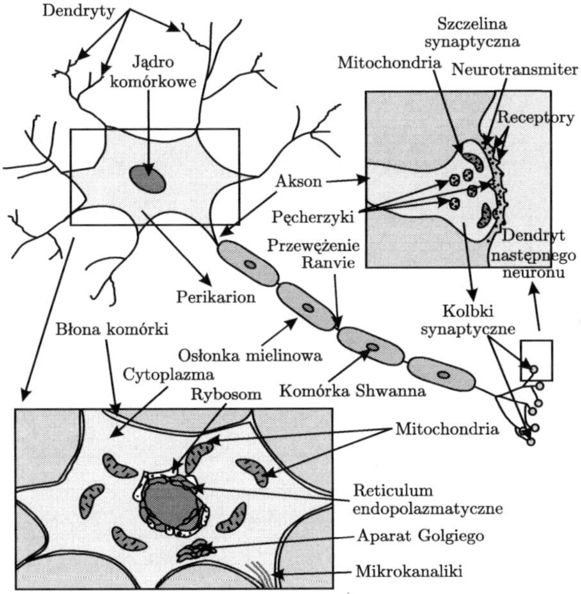

Rys. 7.1. Budowa neuronu

Jeden z najwybitniejszych badaczy systemów neurologicznych Donald Hebb głosił pogląd, że na proces uczenia składają się głównie zmiany „siły" połączeń synaptycznych. Na przykład, w klasycznym eksperymencie Pawłowa, dotyczącym odruchów warunkowych, w którym dźwięk dzwonka rozlega się przed podaniem psu obiadu, pies bardzo szybko uczy się łączyć dźwięk dzwonka z jedzeniem (wydzielanie śliny). Odbywa się to w ten sposób, że połączenia synaptyczne między właściwą częścią kory słuchowej a gruczołami ślinowymi są wzmacniane w kolejnych doświadczeniach, w których dźwięk dzwonka towarzyszy karmieniu, co w końcu powoduje, że kiedy kora słuchowa jest stymulowana przez dźwięk dzwonka, pies zaczyna się ślinić. Obecnie sądzi się, że korzystając z bardzo dużej liczby takich prostych mechanizmów uczenia oraz używając licznych, ale wyjątkowo prostych elementów przetwarzających informacje, jakimi są neurony, mózg jest zdolny do realizacji wszystkich tych wyjątkowo skomplikowanych zadań, jakie codziennie wykonuje. Oczywiście w rzeczywistym biologicznym mózgu występuje wiele bardziej złożonych mechanizmów przetwarzania informacji, angażujących wiele dodatkowych elementów (na przykład tak zwane komórki glejowe). To jednak nie stanowi tematu naszych rozważań.

Jest jednak rzeczą bardzo interesującą, a nawet intrygującą, że sztuczne sieci neuronowe mogą osiągać tak znaczące rezultaty praktyczne, korzystając z niesłychanie uproszczonego modelu neuronu, którego złożoność nie jest wiele większa od schematu polegającego na tym, że neuron jedynie wyznacza ważoną sumę swoich wejść i przechodzi do stanu zapłonu wtedy, gdy łączny sygnał wejściowy przekroczy pewien ustalony poziom progowy.

Prawdopodobnie największą różnicą między mózgiem a komputerem jest to, że komputery, w zdecydowanej większości, pracują sekwencyjnie, podczas gdy mózg potrafi wykonywać wiele operacji równolegle. Na koniec tej części dokonajmy jeszcze pewnego porównania ludzkiego mózgu i komputera. Szacuje się, że cały ludzki mózg może przetwarzać 10^{18} operacji logicznych na sekundę, podczas gdy 64-bitowy procesor PowerPC 970 wykonuje jedynie 10^{11} takich operacji[6]. Trzeba jednak pamiętać, że działanie biologicznych sieci neuronowych nie jest jeszcze dobrze zrozumiane i nie jest jasne, czy coś takiego jak „operacje logiczne" następują w ludzkim mózgu.

7.1.1. Perceptron

Zacznijmy od wprowadzenia pewnych oznaczeń (należy w tym miejscu zwrócić szczególną uwagę, gdyż ze względu na powszechnie przyjęte oznaczenia występujące w opisie sieci neuronowych następuje w tym miejscu zmiana pewnych oznaczeń): p – liczba wejść w neuronie, $\boldsymbol{x}$ – sygnał wejściowy, $\boldsymbol{w}$ – wagi synaptyczne, y – wartość wyjściowa neuronu, w_0 – wartość progowa oraz f – funkcja aktywacji. Działanie pojedynczego neuronu można opisać za pomocą formuły

$$y = f(s), \tag{7.1}$$

[6]Agre P.E. (1997), *Computation and human experience*, Cambridge University Press.

gdzie

$$s = \sum_{i=0}^{p} x_i w_i = \boldsymbol{x}'\boldsymbol{w} + w_0. \tag{7.2}$$

Neuron tej postaci przedstawiono na rys. 7.2. Jak widać działanie neuronu jest bardzo proste, sygnały wejściowe są mnożone przez odpowiednie wagi, a następnie iloczyny te są sumowane, w wyniku czego otrzymujemy sygnał wyjściowy s. Jeśli na tym kończymy i s jest sygnałem wyjściowym dla neuronu, to mówimy o neuronie liniowym, jeśli natomiast sygnał s jest poddawany jeszcze działaniu pewnej funkcji f zwanej funkcją aktywacji[7], to mówimy o neuronie nieliniowym. W tabeli 7.1 znajduje się zestawienie popularnych funkcji aktywacji.

Tab. 7.1. Popularne funkcje aktywacji

Nazwa	Oznaczenie	Działanie
liniowa	purelin	$f(x) = x$
unipolarna	hardlim	$f(x) = \begin{cases} 0, & \text{dla } x < 0 \\ 1, & \text{dla } x \geq 0 \end{cases}$
bipolarna	hardlims	$f(x) = \begin{cases} -1, & \text{dla } x < 0 \\ 1, & \text{dla } x \geq 0 \end{cases}$
niesymetryczna liniowa z nasyceniem	satlin	$f(x) = \begin{cases} 0, & \text{dla } x < 0 \\ x, & \text{dla } 0 \leq x \leq 1 \\ 1, & \text{dla } x > 1 \end{cases}$
symetryczna liniowa z nasyceniem	satlins	$f(x) = \begin{cases} -1, & \text{dla } x < 0 \\ x, & \text{dla } 0 \leq x \leq 1 \\ 1, & \text{dla } x > 1 \end{cases}$
logarytmiczno-sigmoidalna	logsig	$f(x) = \dfrac{1}{1 + e^{-x}}$
tangens hiperboliczny	tansig	$f(x) = \dfrac{e^x - e^{-x}}{e^x + e^{-x}}$
liniowa dodatnia	poslin	$f(x) = \begin{cases} 0, & \text{dla } x < 0 \\ x, & \text{dla } x \geq 0 \end{cases}$

[7]Po ang. activation function, transfer function.

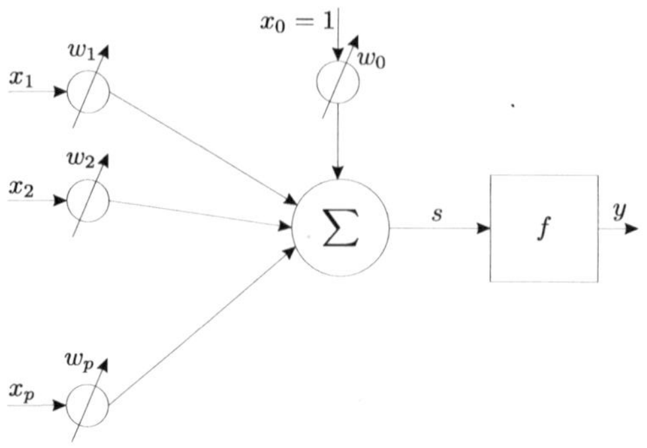

Rys. 7.2. Pojedynczy neuron

Zakładamy, że $x_0 = 1$, natomiast wagę w_0 nazywamy progiem[8]. W tak rozumianym neuronie cała wiedza (pamięć) kryje się w wagach, które łatwo możemy zmieniać, czyli uczyć neuron.

Formalnie działanie *perceptronu* (rys. 7.3) można zapisać następująco:

$$y = f(\boldsymbol{x}'\boldsymbol{w} + \theta), \tag{7.3}$$

gdzie funkcja f jest funkcją bipolarną (funkcja signum).

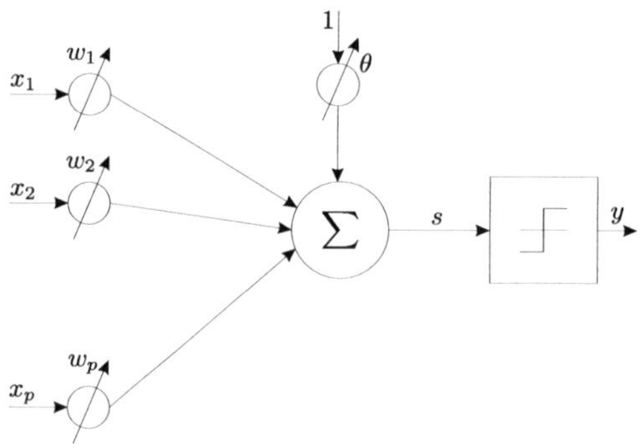

Rys. 7.3. Perceptron

Zauważmy, że model ten odpowiada modelowi (7.1) dla $w_0 = \theta$. Perceptron ze względu na postać swojej funkcji aktywacji może zostać łatwo użyty do zagadnień klasyfikacji obiektów do jednej z dwóch klas. W tym przypadku podział ten jest wyznaczony przez prostą daną równaniem

$$w_1 x_1 + w_2 x_2 + \theta = 0.$$

[8] Po ang. threshold, bias, offset.

Sieć taka odpowiada zatem liniowej funkcji dyskryminacyjnej. Geometryczne przedstawienie tego faktu znajduje się na rys. 7.4.

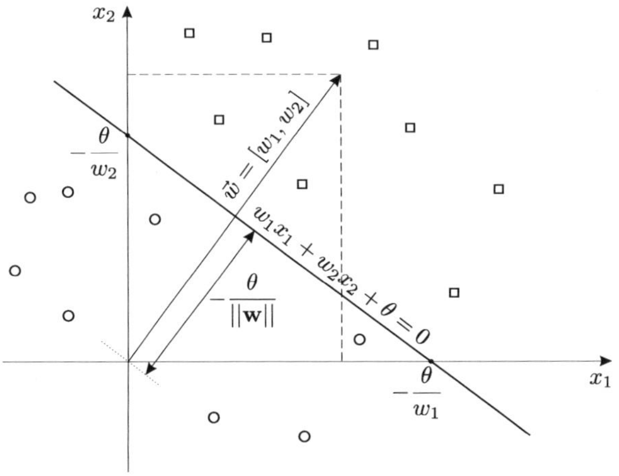

Rys. 7.4. Geometryczne przedstawienie funkcji dyskryminacyjnej i wag

Zauważmy, że wektor wag jest zawsze prostopadły do funkcji dyskryminacyjnej. W ogólnym przypadku, gdy perceptron ma p wejść, wówczas dzieli p-wymiarową przestrzeń na dwie półprzestrzenie, które są rozdzielone $(p-1)$-wymiarową hiperpłaszczyzną, zwaną *granicą decyzyjną*, określoną wzorem

$$\boldsymbol{x}'\boldsymbol{w} + \theta = 0.$$

Teraz postaramy się przedstawić metody uczenia perceptronu. Są to metody z grupy tzw. metod uczenia z nauczycielem (uczenie pod nadzorem). W tego typu zagadnieniach mamy zbiór uczący, dla którego znamy prawdziwe wartości klas, do których należą poszczególne obserwacje ze zbioru uczącego. Zbiór ten posłuży nam do znalezienia optymalnych wartości wag synaptycznych. W tego typu metodach, po podaniu obserwacji uczącej, obliczamy sygnał wyjściowy i modyfikujemy wagi tak, aby zminimalizować różnicę między wzorcem a sygnałem wyjściowym. Oznaczmy wzorce przez $d(\boldsymbol{x}_i)$, $i = 1, 2, \ldots, n$.

Algorytm uczenia perceptronu

1. Wybierz losowo początkowe wagi perceptronu.
2. Ucz perceptron, wykorzystując kolejne elementy $\boldsymbol{x}_i$ zbioru uczącego $\boldsymbol{Z}$.
3. Oblicz wartość wyjściową y_i perceptronu zgodnie z formułą (7.3).

4. Jeżeli wartość y_i różni się od wartości wzorcowej $d(\boldsymbol{x}_i)$, dokonaj modyfikacji wag według formuły

$$\boldsymbol{w}(t+1) = \boldsymbol{w}(t) + d(\boldsymbol{x}_i)\boldsymbol{x}_i. \tag{7.4}$$

5. Wróć do punktu 2.

Algorytm powtarza się tak długo, aż dla wszystkich obserwacji błąd na wyjściu będzie mniejszy od założonej tolerancji. Dany zbiór uczący może zostać użyty wielokrotnie, każde użycie zbioru uczącego jest zwane *epoką*. W trakcie każdej epoki wszystkie przypadki wchodzące w skład ciągu uczącego są po kolei prezentowane sieci, wartości wyjściowe są porównywane z wartościami zakładanymi i jest wyznaczana wartość błędu. Następnie cały proces jest powtarzany. W trakcie kolejnych epok sieci są pokazywane te same dane, jednak dla zachowania prawidłowego przebiegu procesu uczenia między jedną a drugą epoką przykłady powinny być mieszane, żeby nie były pokazywane wielokrotnie w tej samej kolejności. Zauważmy, że próg θ jest modyfikowany zgodnie ze wzorem

$$\theta(t+1) = \theta(t) + d(\boldsymbol{x}_i). \tag{7.5}$$

Okazuje się, że algorytm ten jest zbieżny.

TWIERDZENIE 7.1. (Novikoffa, 1962)

Jeżeli istnieje układ wag $\boldsymbol{w}^*$ klasyfikujący wszystkie obserwacje ze zbioru uczącego poprawnie, to algorytm uczący znajdzie rozwiązanie w skończonej liczbie kroków, bez względu na początkowy wybór wektora wag $\boldsymbol{w}$.

Dowód. Zakładamy, oczywiście, że dane można liniowo oddzielić (odseparować), ponieważ tylko wtedy można nauczyć perceptron poprawnej klasyfikacji wszystkich obserwacji ze zbioru uczącego. Należy pokazać, że w skończonej liczbie kroków perceptron będzie realizował odwzorowanie $y = d(\boldsymbol{x})$. Ponieważ funkcja aktywacji jest w perceptronie typu signum, nie musimy uwzględniać różnych długości wektora $\boldsymbol{w}^*$. Przyjmijmy zatem, że $\|\boldsymbol{w}^*\| = 1$. Sukces uczenia osiągamy wtedy, gdy cosinus kąta między wektorem $\boldsymbol{w}^*$ oraz wektorem $\boldsymbol{w}$ wyniesie 1, czyli wektory te się pokryją. Z faktu, że $|\boldsymbol{w}^* \cdot \boldsymbol{x}| > 0$ oraz że $\boldsymbol{w}^*$ jest rozwiązaniem (czyli jest prostopadły do funkcji dyskryminacyjnej) wynika, że istnieje taka stała $\delta > 0$, że $|\boldsymbol{w}^* \cdot \boldsymbol{x}| > \delta$ dla wszystkich elementów $\boldsymbol{x}$ zbioru uczącego. Z definicji iloczynu skalarnego mamy

$$\cos\alpha = \frac{\boldsymbol{w}^* \cdot \boldsymbol{w}}{\sqrt{\|\boldsymbol{w}^*\|^2 \|\boldsymbol{w}\|^2}} = \frac{\boldsymbol{w}^* \cdot \boldsymbol{w}}{\|\boldsymbol{w}\|}. \tag{7.6}$$

Zgodnie z algorytmem uczenia perceptronu, wagi są modyfikowane dla zadanego $\boldsymbol{x}$ zgodnie z formułą (7.4). Interesuje nas przypadek, gdy na wyjściu sieci pojawił się błąd i musimy zmodyfikować wagi. Mamy

$$\begin{aligned}\boldsymbol{w}^n \cdot \boldsymbol{w}^* &= \boldsymbol{w} \cdot \boldsymbol{w}^* + d(\boldsymbol{x})\boldsymbol{w}^* \cdot \boldsymbol{x} = \\ &= \boldsymbol{w} \cdot \boldsymbol{w}^* + \operatorname{sgn}(\boldsymbol{w}^* \cdot \boldsymbol{x})\boldsymbol{w}^* \cdot \boldsymbol{x} > \\ &> \boldsymbol{w}^* \cdot \boldsymbol{w} + |\boldsymbol{w}^* \cdot \boldsymbol{x}| > \boldsymbol{w}^* \cdot \boldsymbol{w} + \delta,\end{aligned}$$

$$\begin{aligned}\|\boldsymbol{w}^n\|^2 &= \|\boldsymbol{w} + d(\boldsymbol{x})\boldsymbol{x}\|^2 = \\ &= \|\boldsymbol{w}\|^2 + 2d(\boldsymbol{x})\boldsymbol{w} \cdot \boldsymbol{x} + \|\boldsymbol{x}\|^2 < \\ &< \|\boldsymbol{w}\|^2 + \|\boldsymbol{x}\|^2 \text{ (ponieważ } d(\boldsymbol{x}) = -\operatorname{sgn}(\boldsymbol{w} \cdot \boldsymbol{x})) = \\ &= \|\boldsymbol{w}\|^2 + M.\end{aligned}$$

Po t krokach modyfikacji wag sieci mamy

$$\boldsymbol{w}(t) \cdot \boldsymbol{w}^* > \boldsymbol{w}(t) \cdot \boldsymbol{w}^* + t\delta \tag{7.7}$$

oraz

$$\|\boldsymbol{w}(t)\|^2 < \|\boldsymbol{w}\|^2 + tM. \tag{7.8}$$

Ze wzorów (7.6)–(7.8) mamy

$$\cos\alpha(t) = \frac{\boldsymbol{w}^* \cdot \boldsymbol{w}(t)}{\|\boldsymbol{w}(t)\|} > \frac{\boldsymbol{w}^* \cdot \boldsymbol{w} + t\delta}{\sqrt{\|\boldsymbol{w}\|^2 + tM}}. \tag{7.9}$$

Zatem

$$\lim_{n\to\infty} \cos\alpha(t) = \lim_{n\to\infty} \frac{\delta}{\sqrt{M}}\sqrt{t} = \infty,$$

co oczywiście nie jest możliwe. Dlatego musi istnieć górna granica wartości $t = t_{\max}$. Istnieje zatem skończona liczba kroków modyfikacji wag, po których dowolny wektor wag początkowych będzie realizował odwzorowanie $y = d(\boldsymbol{x})$. Jeśli zaczniemy od $\boldsymbol{w} = \boldsymbol{0}$, to

$$t_{\max} = \frac{M}{\delta^2}.$$

■

Przykład 7.1.
Niech wagi początkowe perceptronu będą równe: $w_1 = 1$, $w_2 = 2$ oraz $\theta = -2$. Mamy do rozwiązania problem klasyfikacji przedstawiony na rys. 7.5.

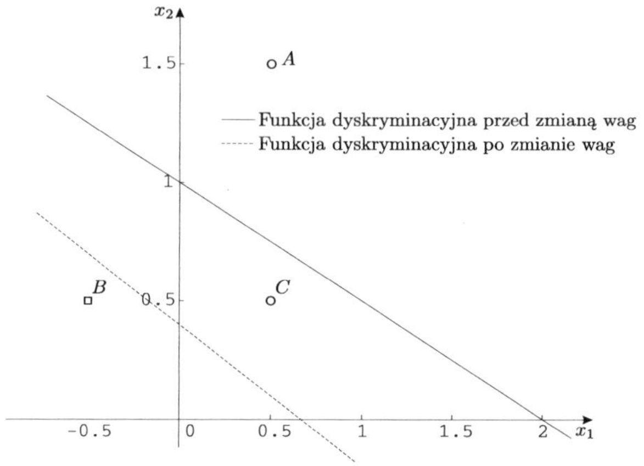

Rys. 7.5. Liniowa funkcja dyskryminacyjna przed zmianą i po zmianie wag

Pierwsza obserwacja $A(0.5;\ 1.5)$ z pierwszej klasy ($d(\boldsymbol{x}) = +1$) zostaje pokazana sieci. Ponieważ obserwacja ta jest poprawnie klasyfikowana, nie trzeba modyfikować wag. To samo dzieje się z obserwacją $B(-0.5;\ 0.5)$ z drugiej klasy ($d(\boldsymbol{x}) = -1$). Kiedy sieć dostaje obserwację $C(0.5;\ 0.5)$ z pierwszej klasy ($d(\boldsymbol{x}) = +1$), sieć jej nie rozpoznaje poprawnie (na wyjściu pojawia się -1). Musimy zatem zmodyfikować wagi zgodnie z formułami (7.4) oraz (7.5). Mamy $\Delta(w_1) = 0.5$, $\Delta(w_2) = 0.5$ oraz $\Delta(\theta) = 1$. Wagi wynoszą teraz: $w_1 = 1.5$, $w_2 = 2.5$, $\theta = -1$. Po tych zmianach obserwacja C jest klasyfikowana poprawnie.

Przykład 7.2.
Spróbujmy rozwiązać, wykorzystując pojedynczy perceptron, następujący problem klasyfikacyjny:

$$\boldsymbol{x}_1 = \begin{bmatrix} 1 \\ 2 \end{bmatrix}, \quad d(\boldsymbol{x}_1) = 1$$

$$\boldsymbol{x}_2 = \begin{bmatrix} -1 \\ 2 \end{bmatrix}, \quad d(\boldsymbol{x}_2) = -1$$

$$\boldsymbol{x}_3 = \begin{bmatrix} 0 \\ -1 \end{bmatrix}, \quad d(\boldsymbol{x}_3) = -1.$$

Załóżmy, że początkowe wagi wynoszą

$$\boldsymbol{w}(0) = [1.0,\ -0.8]'.$$

Przyjmijmy dodatkowo założenie, że granica decyzyjna przechodzi przez początek układu współrzędnych, czyli $\theta = 0$. Kolejne kroki działania sieci wyglądają następująco:

$$y_1 = \mathrm{sgn}(\boldsymbol{x}_1'\boldsymbol{w}(0)) = \mathrm{sgn}\left([1,\ 2][1,\ -0.8]'\right) = \mathrm{sgn}(-0.6) = -1,$$

$$\boldsymbol{w}(1) = \boldsymbol{w}(0) + d(\boldsymbol{x}_1)\boldsymbol{x}_1 = \begin{bmatrix} 1.0 \\ -0.8 \end{bmatrix} + \begin{bmatrix} 1 \\ 2 \end{bmatrix} = \begin{bmatrix} 2 \\ 1.2 \end{bmatrix},$$

$$y_2 = \mathrm{sgn}(\boldsymbol{x}_2'\boldsymbol{w}(1)) = \mathrm{sgn}\left([-1,\ 2][2,\ 1.2]'\right) = \mathrm{sgn}(0.4) = 1,$$

$$\boldsymbol{w}(2) = \boldsymbol{w}(1) + d(\boldsymbol{x}_2)\boldsymbol{x}_2 = \begin{bmatrix} 2.0 \\ 1.2 \end{bmatrix} - \begin{bmatrix} -1 \\ 2 \end{bmatrix} = \begin{bmatrix} 3 \\ -0.8 \end{bmatrix},$$

$$y_3 = \mathrm{sgn}(\boldsymbol{x}_3'\boldsymbol{w}(2)) = \mathrm{sgn}\left([0,\ -1][3,\ -0.8]'\right) = \mathrm{sgn}(0.8) = 1,$$

$$\boldsymbol{w}(3) = \boldsymbol{w}(2) + d(\boldsymbol{x}_3)\boldsymbol{x}_3 = \begin{bmatrix} 3.0 \\ -0.8 \end{bmatrix} - \begin{bmatrix} 0 \\ -1 \end{bmatrix} = \begin{bmatrix} 3 \\ 0.2 \end{bmatrix}.$$

Tak nauczona sieć potrafi już rozpoznawać wszystkie obserwacje. Zmiany zachodzące podczas uczenia sieci można zaobserwować na rys. 7.6.

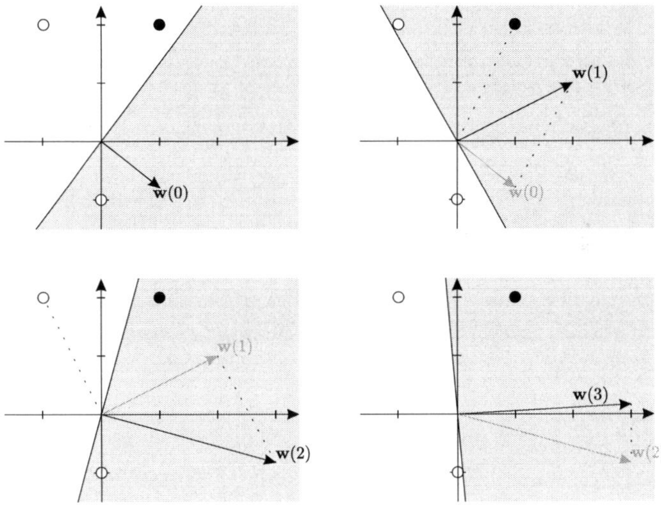

Rys. 7.6. Proces uczenia sieci z przykładu 7.2

Przykład 7.3.
Rozważmy następujący przykład praktyczny zastosowania sieci neuronowej. Handlarz owoców posiada magazyn, w którym przechowuje różne rodzaje owoców oraz warzyw. Różne typy owoców mogą być pomieszane w magazynie. Handlarz potrzebuje maszyny, która będzie sortować owoce. Istnieje taśma przesyłowa, na której są kładzione owoce. Na tej taśmie są umieszczone czujniki, które mierzą trzy własności owoców: kształt, fakturę oraz wagę. Czujnik

kształtu zwraca 1, jeśli kształt owocu jest zbliżony do kuli, w przeciwnym razie zwraca wynik -1. Czujnik faktury zwraca 1, jeżeli powierzchnia owocu jest gładka, w przeciwnym razie tzn. jeśli powierzchnia jest chropowata otrzymujemy na wyjściu czujnika -1. Ostatni czujnik, czujnik wagi daje w wyniku 1, jeżeli waga owocu jest większa od 0,5 kg, jeśli natomiast waga jest mniejsza od 0,5 kg, otrzymujemy -1. Te trzy wskazania czujników stanowią dane wejściowe dla sieci neuronowej, której zadaniem jest rozpoznanie owocu, który w zależności od tego rozpoznania zostanie skierowany do odpowiedniego magazynu końcowego. W celu uproszczenia obliczeń rozważać będziemy jedynie problem rozpoznawania dwóch typów owoców: pomarańcz oraz jabłek. Odpowiedni schemat doświadczenia jest przedstawiony na rys. 7.7.

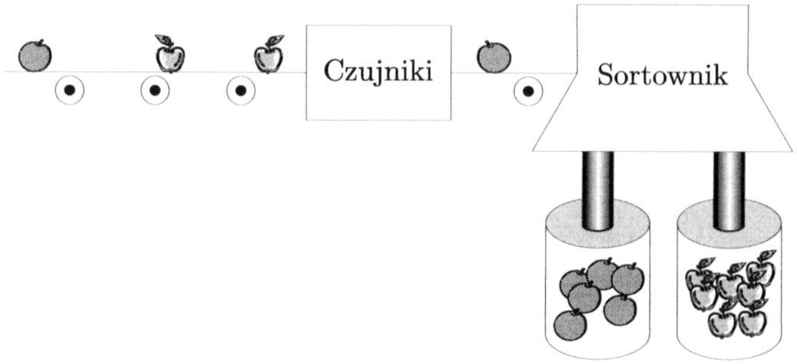

Rys. 7.7. Schemat hurtowni do przykładu 7.3

Kiedy zatem owoc przejdzie przez system czujników jest reprezentowany przez trójwymiarowy wektor. Pierwszy element tego wektora dotyczy kształtu, drugi faktury, a trzeci wagi. Prototyp pomarańczy wygląda zatem następująco:

$$\boldsymbol{x}_1 = [1,\ -1,\ -1]',$$

a jabłka

$$\boldsymbol{x}_2 = [1,\ 1,\ -1]'.$$

Przykład można oczywiście skomplikować, dodając kolejne owoce, np. banana ($[-1,\ 1,\ -1]'$) czy ananasa ($[-1,\ -1,\ 1]'$). Powoduje to jednak jedynie komplikację obliczeń i sieci, a nie wnosi nic nowego do samej konstrukcji.

Sieć neuronowa otrzymuje te dwa trójwymiarowe wektory jako wzorce uczące. Na ich podstawie musi podjąć decyzję, co do każdego owocu, który pojawi się na taśmie. Do rozwiązania tego problemu zastosujemy perceptron. Ponieważ są jedynie dwie klasy, wystarczy nam jeden neuron. Musimy znaleźć wagi oraz obciążenie w taki sposób, aby móc rozpoznawać pomarańcze i jabłka. Przyjmijmy, że jeśli na wyjściu perceptronu pojawi się 1, to mamy do czynienia z jabłkiem, w przeciwnym razie sieć rozpoznała pomarańczę. Musimy znaleźć granicę decyzyjną, która rozdziela pomarańcze i jabłka. Na rysunku 7.8 są przedstawione wzorce w przestrzeni $\mathbb{R}^3$.

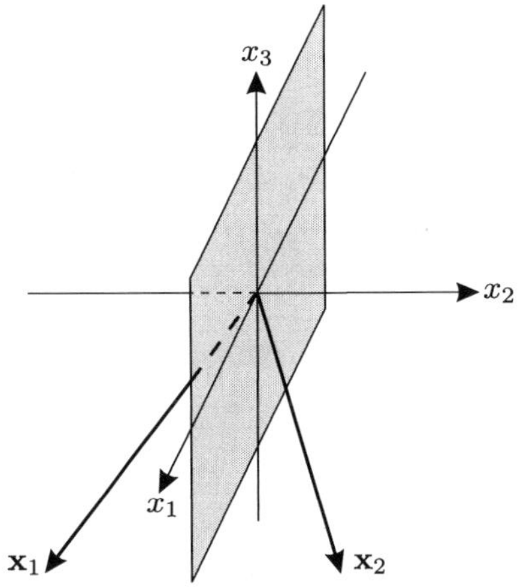

Rys. 7.8. Wektory wzorców

Z rysunku widać, że granicą liniową, która rozdziela te dwa sygnały symetrycznie, jest płaszczyzna x_1x_2, która może być opisana równaniem

$$x_2 = 0,$$

lub alternatywnie

$$[x_1,\ x_2,\ x_3][0,\ 1,\ 0]' + 0 = 0.$$

Stąd

$$\boldsymbol{w} = [0,\ 1,\ 0]',$$
$$\theta = 0.$$

Wektor wag jest ortogonalny do granicy decyzyjnej. Obciążenie wynosi 0, gdyż granica przechodzi przez początek układu współrzędnych. Sprawdźmy, czy nasza sieć poprawnie klasyfikuje sygnały wzorcowe.

$$y_1 = \operatorname{sgn}([1,\ -1,\ -1][0,\ 1,\ 0]' + 0) = -1 \text{ (pomarańcza)},$$
$$y_2 = \operatorname{sgn}([1,\ 1,\ -1][0,\ 1,\ 0]' + 0) = 1 \text{ (jabłko)}.$$

Widać, że sieć nie ma najmniejszych problemów z rozpoznaniem wzorców. Co się jednak stanie, gdy na taśmę trafi pomarańcza o kształcie bardziej eliptycznym niż kulistym, czyli obiekt scharakteryzowany przez wektor $\boldsymbol{x}_3 = [-1,\ -1,\ -1]'$. Odpowiedź sieci wynosi

$$y_3 = \operatorname{sgn}([-1,\ -1,\ -1][0,\ 1,\ 0]' + 0) = -1 \text{ (pomarańcza)}.$$

Sieć zachowała się zatem właściwie. Jeśli dokładniej przyjrzymy się zagadnieniu dostrzeżemy, że w tym przypadku znaczenie ma jedynie faktura owocu,

gdyż wagi przy pozostałych czujnikach są wyzerowane. Wszystko zatem co ma porowatą fakturę skórki zostanie sklasyfikowane jako pomarańcza, a to co gładką jako jabłko.
Spróbujmy teraz znaleźć optymalne wagi za pomocą opisanego algorytmu perceptronowego. Załóżmy, że początkowe wagi wynoszą

$$\boldsymbol{w}(0) = [0.5,\ -1,\ -0.5]',$$

a obciążenie wynosi $\theta(0) = 0.5$. Pokazujemy teraz sieci obserwację $\boldsymbol{x}_1$. Sprawdźmy, jak klasyfikuje ją sieć:

$$\begin{aligned} y_1 &= \operatorname{sgn}(\boldsymbol{x}_1'\boldsymbol{w}(0) + \theta(0)) = \operatorname{sgn}([1,\ -1,\ -1][0.5,\ -1,\ -0.5]' + 0.5) = \\ &= \operatorname{sgn}(1.5) = 1. \end{aligned}$$

Obserwacja została zaklasyfikowana błędnie. Niezbędna jest więc modyfikacja wag zgodnie ze wzorem (7.4). Mamy

$$\boldsymbol{w}(1) = \boldsymbol{w}(0) + d(\boldsymbol{x}_1)\boldsymbol{x}_1 = \begin{bmatrix} 0.5 \\ -1 \\ -0.5 \end{bmatrix} - \begin{bmatrix} 1 \\ -1 \\ -1 \end{bmatrix} = \begin{bmatrix} -0.5 \\ 0 \\ 0.5 \end{bmatrix}$$

oraz

$$\theta(1) = \theta(0) + d(\boldsymbol{x}_1) = 0.5 - 1 = -0.5.$$

W drugim kroku analizie podlega druga obserwacja. Jak wyżej najpierw sprawdzamy, jak rozpoznaje ją sieć

$$\begin{aligned} y_2 &= \operatorname{sgn}(\boldsymbol{x}_2'\boldsymbol{w}(1) + \theta(1)) = \operatorname{sgn}([1,\ 1,\ -1][-0.5,\ 0,\ 0.5]' - 0.5) = \\ &= \operatorname{sgn}(-1.5) = -1. \end{aligned}$$

Jest ona zatem również błędnie klasyfikowana. Przystępujemy do modyfikacji wag

$$\boldsymbol{w}(2) = \boldsymbol{w}(1) + d(\boldsymbol{x}_2)\boldsymbol{x}_2 = \begin{bmatrix} -0.5 \\ 0 \\ 0.5 \end{bmatrix} + \begin{bmatrix} 1 \\ 1 \\ -1 \end{bmatrix} = \begin{bmatrix} 0.5 \\ 1 \\ -0.5 \end{bmatrix}$$

oraz

$$\theta(2) = \theta(1) + d(\boldsymbol{x}_2) = -0.5 + 1 = 0.5.$$

Z elementarnych rachunków wynika, że taka sieć bezbłędnie klasyfikuje oba owoce. Spróbujmy jeszcze zaklasyfikować obiekt $\boldsymbol{x}_3$. Otrzymujemy

$$\begin{aligned} y_3 = \operatorname{sgn}(\boldsymbol{x}_3'\boldsymbol{w}(2) + \theta(2)) &= \operatorname{sgn}([-1,\ -1,\ -1][0.5,\ 1,\ -0.5]' + 0.5) = \\ &= \operatorname{sgn}(-0.5) = -1. \end{aligned}$$

Sieć zachowuje się zatem poprawnie również w tej sytuacji.

7.1.2. ADALINE (ADAptive LInear NEuron)

Budowa neuronu typu *ADALINE* jest przedstawiona na rys. 7.9. Model neuronu tego typu został opracowany przez Widrowa i Hoffa (1960). Neuron ten różni się od perceptronu jedynie algorytmem uczenia. W analogiczny jak w perceptronie sposób, wyznaczamy sygnał wyjściowy. Następuje jednak porównanie wartości otrzymanej z części liniowej neuronu (sumator) z wzorcem. W ten sposób, jeżeli neuron nie jest nauczony, otrzymujemy błąd

$$\varepsilon(\boldsymbol{x}) = d(\boldsymbol{x}) - s(\boldsymbol{x}).$$

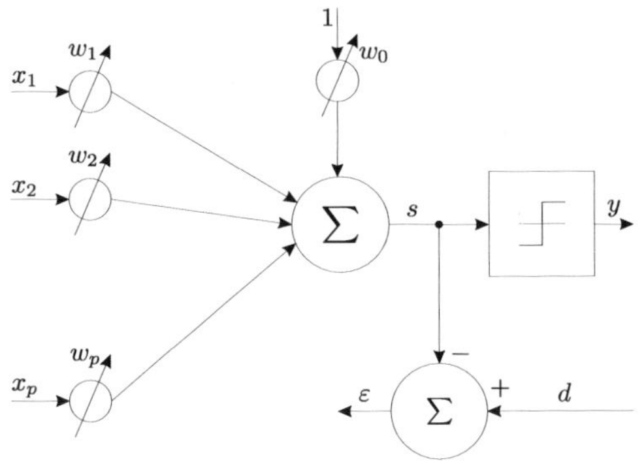

Rys. 7.9. Neuron ADALINE

Proces uczenia sprowadza się do minimalizacji następującej funkcji:

$$Q(\boldsymbol{w}) = \frac{1}{2}\varepsilon^2(\boldsymbol{x}) = \frac{1}{2}\left[d(\boldsymbol{x}) - \left(\boldsymbol{x}'\boldsymbol{w} + \theta\right)\right]^2 \tag{7.10}$$

zwanej błędem średniokwadratowym. Ponieważ funkcja (7.10) jest różniczkowalna względem $\boldsymbol{w}$, możemy zatem do modyfikacji wag użyć algorytmu gradientowego. Do minimalizacji funkcji (7.10) użyjemy metody największego spadku, która zostanie omówiona w ogólnym przypadku przy okazji opisu algorytmu wstecznej propagacji błędu. Wagi modyfikujemy zatem zgodnie z formułą

$$\boldsymbol{w}(t+1) = \boldsymbol{w}(t) - \eta\frac{\partial Q(\boldsymbol{w})}{\partial \boldsymbol{w}}, \tag{7.11}$$

gdzie η jest *współczynnikiem uczenia*[9], który decyduje o szybkości uczenia. Współczynnik ten może być w zasadzie dobierany dowolnie. W praktyce

[9] Po ang. learning rate.

stosuje się jednak pewne ograniczenia. Przede wszystkim parametr ten powinien mieć charakter stochastyczny, czyli zależeć od sygnału wejściowego. Tadeusiewicz (1993), opierając się na teorii aproksymacji stochastycznej, podaje następujące warunki, jakie powinien spełniać ciąg (η_n):

$$\sum_{n=1}^{\infty} \eta_n = \infty, \quad \sum_{n=1}^{\infty} (\eta_n)^2 < \infty.$$

Najprostszym ciągiem spełniającym te warunki jest ciąg $\eta_n = \frac{\eta_0}{n}$. Ciąg ten jednak bardzo szybko maleje i prowadzi do szybkiego zatrzymania procesu uczenia. Z tego względu używa się raczej ciągu

$$\eta_n = \left(\sum_{i=1}^{n} ||\boldsymbol{x}_i||^2 \right)^{-1}. \tag{7.12}$$

W praktyce najczęściej rezygnuje się z tych propozycji na rzecz prostoty, przyjmując pewną ustaloną wartość η niezależną od kroku uczenia. Najczęściej przyjmuje się wartość $\eta = 0.6$ (zobacz: Tadeusiewicz (1993)) i obserwując zachowanie sieci, odpowiednio modyfikuje.

Definicja 7.1.
Optymalnym współczynnikiem uczenia nazywamy taki współczynnik uczenia, który w jednym kroku prowadzi do minimalizacji funkcji błędu.

Korzystając z formuły (7.11), możemy wyrazić ten optymalny współczynnik uczenia następująco:

$$\eta_{opt} = \left(\frac{\partial^2 Q(\boldsymbol{w})}{\partial \boldsymbol{w}^2} \right)^{-1}.$$

Współczynnik uczenia nie może być większy niż $2\eta_{opt}$. Najlepszy efekt osiągniemy, jeśli współczynnik ten będzie obliczany osobno dla każdej wagi (zobacz: Duda, Hart i Stork (2001)). Pamiętać jednak należy, że dobór zbyt małego współczynnika uczenia η powoduje, że proces uczenia jest nieefektywny (trwa bardzo długo). Wybór natomiast zbyt dużego parametru sprawia, że możemy „przeskoczyć" właściwe rozwiązanie i uczenie nie skończy się (proces nie będzie zbieżny). Zachowanie gradientu, w przypadku jednowymiarowym, dla różnych wartości współczynnika η jest przedstawione na rys. 7.10.

Wróćmy do modyfikacji wag w neuronie typu ADALINE. Wzór (7.11) możemy interpretować następująco: poprawka wag musi być proporcjonalna

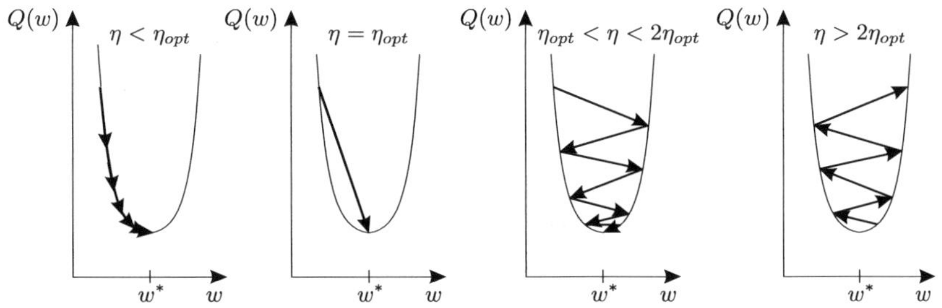

Rys. 7.10. Zachowanie gradientu w zależności od współczynnika uczenia

do wektora gradientu funkcji Q. Znak – wynika z tego, że gradient wskazuje kierunek najszybszego wzrostu, podczas gdy my chcemy funkcję zminimalizować. Korzystając z reguły łańcuchowej, mamy

$$\frac{\partial Q(\boldsymbol{w})}{\partial \boldsymbol{w}} = \frac{\partial Q(\boldsymbol{w})}{\partial s(\boldsymbol{x})} \cdot \frac{\partial s(\boldsymbol{x})}{\partial \boldsymbol{w}}.$$

Dodatkowo, ponieważ $s(\boldsymbol{x})$ jest funkcją liniową względem wektora $\boldsymbol{w}$, mamy

$$\frac{\partial s(\boldsymbol{x})}{\partial \boldsymbol{w}} = \boldsymbol{x}.$$

Ponadto

$$\frac{\partial Q(\boldsymbol{w})}{\partial s(\boldsymbol{x})} = -(d(\boldsymbol{x}) - s(\boldsymbol{x})).$$

Reasumując, wektor wag możemy modyfikować według wzoru

$$\boldsymbol{w}(t+1) = \boldsymbol{w}(t) + \eta\delta(\boldsymbol{x})\boldsymbol{x},$$

gdzie $\delta(\boldsymbol{x}) = d(\boldsymbol{x}) - s(\boldsymbol{x})$. Reguła ta jest zwana *regułą delta* (dokładnie jest to jej szczególny przypadek dla liniowej funkcji aktywacji neuronu). Z regułą delta mamy do czynienia, gdy uaktualnianie wag następuje każdorazowo po prezentacji jednej pary uczącej. Gdy uaktualnianie zachodzi po prezentacji wszystkich par uczących, mamy wtedy do czynienia z tzw. skumulowaną regułą delta. Reguła delta jest przeznaczona dla neuronów z ciągłymi funkcjami aktywacji i stanowi odpowiednik reguły perceptronowej. Nazywana jest ona również ciągłą regułą perceptronową.

Neuron typu ADALINE można również uczyć za pomocą *rekurencyjnej metody najmniejszych kwadratów*[10]. Szczegółowy opis metody znajduje się

[10] Po ang. Recursive Least Squares (RLS).

np. w książce Rutkowskiego (2005). W tym przypadku jako miarę błędu przyjmujemy

$$Q(t) = \sum_{k=1}^{t} \gamma^{t-k} \varepsilon^2(\boldsymbol{x}_k) = \sum_{k=1}^{t} \gamma^{t-k} [d(\boldsymbol{x}_k) - \boldsymbol{x}_k' \boldsymbol{w}(t)]^2,$$

gdzie γ jest współczynnikiem zapominania[11] wybieranym z przedziału [0, 1]. Użycie współczynnika zapominania powoduje, że wagi w sieci nie są zbyt duże. Stanowi on zazwyczaj niewielki procent współczynnika uczenia η. Przyjęcie zbyt dużej jego wartości powoduje, że neuron zapomina to, czego się wcześniej nauczył.

Ostatecznie algorytm RLS zastosowany do uczenia neuronów typu ADALINE przybiera następującą postać:

$$\begin{aligned}
\varepsilon(\boldsymbol{x}_t) &= d(\boldsymbol{x}_t) - y(\boldsymbol{x}_t), \\
\boldsymbol{g}(\boldsymbol{x}_t) &= \frac{\boldsymbol{P}(\boldsymbol{x}_{t-1})(\boldsymbol{x}_t)}{\gamma + \boldsymbol{x}_t' \boldsymbol{P}(\boldsymbol{x}_{t-1})(\boldsymbol{x}_t)}, \\
\boldsymbol{P}(\boldsymbol{x}_t) &= \gamma^{-1}[\boldsymbol{I} - \boldsymbol{g}(\boldsymbol{x}_t)\boldsymbol{x}_t'] \boldsymbol{P}(\boldsymbol{x}_{t-1}), \\
\boldsymbol{w}(t) &= \boldsymbol{w}(t-1) + \boldsymbol{g}(\boldsymbol{x}_t)\varepsilon(\boldsymbol{x}_t),
\end{aligned}$$

gdzie jako wartości początkowe przyjmujemy

$$\boldsymbol{P}(\boldsymbol{x}_0) = \lambda \boldsymbol{I}, \tag{7.13}$$

oraz $\lambda > 0$ jest pewną stałą.

Przykład 7.4.
Wróćmy do przykładu 7.3. Wykorzystamy neuron ADALINE do klasyfikacji owoców. Niech

$$\begin{aligned}
\eta &= 0.4, \\
\theta &= 0, \\
\boldsymbol{w}(0) &= \begin{bmatrix} 0 \\ 0 \\ 0 \end{bmatrix}.
\end{aligned}$$

[11] Po ang. forgetting factor.

W kolejnych iteracjach otrzymujemy

$$s(0) = \boldsymbol{x}_1'\boldsymbol{w}(0) = [1,\ -1,\ -1]\begin{bmatrix}0\\0\\0\end{bmatrix} = 0,$$

$$\delta(0) = -1 - 0 = -1,$$

$$\boldsymbol{w}(1) = \boldsymbol{w}(0) + \eta\delta(0)\boldsymbol{x}_1' = \begin{bmatrix}0\\0\\0\end{bmatrix} + 0.4(-1)\begin{bmatrix}1\\-1\\-1\end{bmatrix} = \begin{bmatrix}-0.4\\0.4\\0.4\end{bmatrix},$$

$$s(1) = \boldsymbol{x}_2'\boldsymbol{w}(1) = [1,\ 1,\ -1]\begin{bmatrix}-0.4\\0.4\\0.4\end{bmatrix} = -0.4,$$

$$\delta(0) = -1 - (-0.4) = 1.4,$$

$$\boldsymbol{w}(2) = \boldsymbol{w}(1) + \eta\delta(1)\boldsymbol{x}_2' = \begin{bmatrix}-0.4\\0.4\\0.4\end{bmatrix} + 0.4(1.4)\begin{bmatrix}1\\1\\-1\end{bmatrix} = \begin{bmatrix}0.16\\0.96\\-0.16\end{bmatrix},$$

$$s(2) = \boldsymbol{x}_1'\boldsymbol{w}(2) = [1,\ -1,\ -1]\begin{bmatrix}0.16\\0.96\\-0.16\end{bmatrix} = -0.64,$$

$$\delta(2) = -1 - (-0.64) = -0.36,$$

$$\boldsymbol{w}(3) = \boldsymbol{w}(2) + \eta\delta(2)\boldsymbol{x}_1' = \begin{bmatrix}0.16\\0.96\\-0.16\end{bmatrix} + 0.4(-1)\begin{bmatrix}1\\-1\\-1\end{bmatrix} = \begin{bmatrix}0.016\\1.104\\-0.016\end{bmatrix},$$

$$\boldsymbol{w}(\infty) = \begin{bmatrix}0\\1\\0\end{bmatrix}.$$

Jak widać, w tym przypadku o wszystkim decyduje druga składowa, gdyż tylko dla niej waga jest niezerowa i odpowiada dokładnie klasie. Oczywiście obydwa owoce są klasyfikowane poprawnie.

7.1.3. Neuron sigmoidalny

Neuron typu *sigmoidalnego* (rys. 7.11) ma strukturę podobną do modelu perceptronu i modelu ADALINE, z tą różnicą, że funkcja aktywacji jest ciągła i przyjmuje postać funkcji sigmoidalnej unipolarnej

$$f_1(x) = \frac{1}{1 + e^{-\beta x}}$$

lub bipolarnej

$$f_2(x) = \tanh(\beta x) = \frac{1 - e^{\beta x}}{1 + e^{-\beta x}}.$$

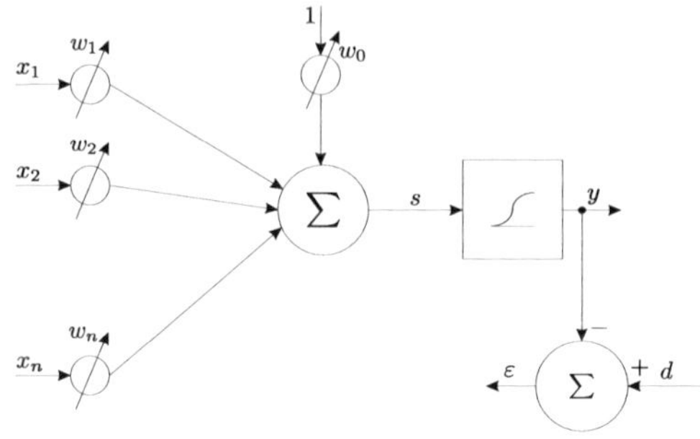

Rys. 7.11. Neuron sigmoidalny

Na rysunku 7.12 jest przedstawiony przebieg funkcji unipolarnej dla różnych wartości parametru β. Parametr β jest dobierany przez użytkow-

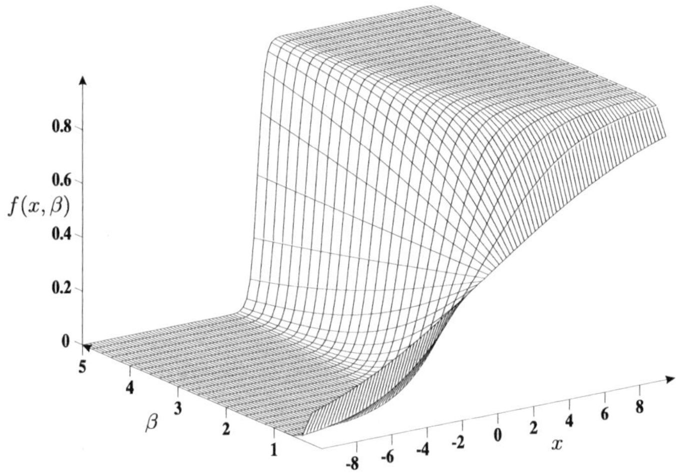

Rys. 7.12. Wykres funkcji unipolarnej dla różnych wartości parametru β

nika i jego wartość wpływa na kształt funkcji aktywacji. Przy jego małych wartościach funkcja ma mały kąt nachylenia, w miarę wzrostu jego wartości przebieg staje się bardziej stromy. Przy β dążącym do nieskończoności funkcja sigmoidalna przechodzi w funkcję progową, identyczną z funkcją aktywacji perceptronu. Ważną cechą funkcji sigmoidalnej jest różniczkowalność. Ponadto pochodne tych funkcji można łatwo obliczyć i przybierają one postać

$$f_1'(x) = \beta f_1(x)(1 - f_1(x)),$$
$$f_2'(x) = \beta(1 - f_2^2(x)).$$

Uczenie neuronu sigmoidalnego odbywa się zwykle w trybie z nauczycielem, przez minimalizację funkcji celu, podobnie jak w przypadku perceptronu. Założenie ciągłej funkcji aktywacji umożliwia zastosowanie w uczeniu metody gradientowej. Najprościej jest przyjąć metodę największego spadku, zgodnie z którą aktualizacja wektora wag odbywa się w kierunku ujemnego gradientu funkcji celu. Miarę błędu Q definiujemy następująco:

$$Q(\boldsymbol{w}) = \frac{1}{2}\left[d(\boldsymbol{x}) - f(\boldsymbol{w}'\boldsymbol{x})\right]^2. \tag{7.14}$$

Jeśli uwzględnimy w rachunkach w metodzie największego spadku postać funkcji aktywacji, otrzymamy następującą formułę na uaktualnione wagi neuronu:

$$\boldsymbol{w}(t+1) = \boldsymbol{w}(t) - \eta\frac{\partial Q(\boldsymbol{w})}{\partial \boldsymbol{w}} = \boldsymbol{w}(t) + \eta(d - f(s(\boldsymbol{x})))f'(s(\boldsymbol{x}))\boldsymbol{x}, \tag{7.15}$$

gdzie

$$s(\boldsymbol{x}) = \boldsymbol{w}'\boldsymbol{x}.$$

Istnieje alternatywna metoda uczenia neuronu sigmoidalnego wykorzystująca algorytm RLS. Szczegóły tego podejścia można znaleźć np. w książce Rutkowskiego (2005). My natomiast podamy jedynie finalną postać algorytmu RLS do uczenia neuronu sigmoidalnego

$$\begin{aligned}
\varepsilon(\boldsymbol{x}_t) &\approx d(\boldsymbol{x}_t) - y(\boldsymbol{x}_t),\\
\boldsymbol{g}(\boldsymbol{x}_t) &= \frac{f'(s(\boldsymbol{x}_t))\boldsymbol{P}(\boldsymbol{x}_{t-1})(\boldsymbol{x}_t)}{\gamma + f'^2(s(\boldsymbol{x}_t))\boldsymbol{x}_t'\boldsymbol{P}(\boldsymbol{x}_{t-1})(\boldsymbol{x}_t)},\\
\boldsymbol{P}(\boldsymbol{x}_t) &= \gamma^{-1}[\boldsymbol{I} - f'(s(\boldsymbol{x}_t))\boldsymbol{g}(\boldsymbol{x}_t)\boldsymbol{x}_t']\boldsymbol{P}(\boldsymbol{x}_{t-1}),\\
\boldsymbol{w}(t) &= \boldsymbol{w}(t-1) + \boldsymbol{g}(\boldsymbol{x}_t)\varepsilon(\boldsymbol{x}_t),
\end{aligned}$$

gdzie wartości początkowe określa zależność (7.13).

7.1.4. Neuron Hebba

Hebb (1949) zauważył podczas swoich badań, że powiązanie dwóch komórek jest wzmacniane, jeśli obie komórki są pobudzone (stają się aktywne) w tym samym czasie. Spostrzeżenie to jest obecnie znane pod nazwą *reguły Hebba*. Model neuronu Hebba ma identyczną strukturę jak model typu sigmoidalnego, ale charakteryzuje się odmienną metodą uczenia. Hebb zaproponował algorytm, w którym wagi modyfikujemy następująco:

$$\boldsymbol{w}(t+1) = \boldsymbol{w}(t) + \eta y\boldsymbol{x}. \tag{7.16}$$

Ucząc zatem pojedynczy neuron, modyfikujemy wagi $\boldsymbol{w}$ proporcjonalnie do sygnału wejściowego $\boldsymbol{x}$, do sygnału wyjściowego y oraz z uwzględnieniem współczynnika uczenia η. Formuła (7.16) odpowiada uczeniu bez nauczyciela, gdyż nie podajemy wzorca. Łatwo można ją zmodyfikować, w celu otrzymania metody uczenia z nauczycielem. W takim przypadku otrzymujemy formułę

$$\boldsymbol{w}(t+1) = \boldsymbol{w}(t) + \eta \boldsymbol{x} d(\boldsymbol{x}). \tag{7.17}$$

Istnieje niebezpieczeństwo zbytniego zwiększenia się wag synaptycznych, które wzrastają wykładniczo. Aby tego uniknąć, stosuje się współczynnik zapominania

$$\boldsymbol{w}(t+1) = \boldsymbol{w}(t)(1-\gamma) + \Delta \boldsymbol{w}, \tag{7.18}$$

który jednak zawodzi w przypadku liniowej funkcji aktywacji neuronu Hebba. Niestabilność reguły Hebba można wyeliminować, ograniczając w trakcie uczenia wektor wag. Uzyskuje się to przez ponowną normalizację oraz odpowiedni dobór w trakcie uczenia współczynnika skali. Są to jednak techniki skomplikowane obliczeniowo

$$\boldsymbol{w}^* = \alpha \boldsymbol{w}, \quad ||\boldsymbol{w}^*|| = 1. \tag{7.19}$$

Innym podejściem do stabilizacji procesu uczenia neuronu Hebba jest zastosowanie *reguły Oji*, która nie wymaga renormalizacji wag, lecz sprawia, że w trakcie uczenia wektor wag dąży samoczynnie do $||\boldsymbol{w}|| = 1$. Zmiana wag w tej metodzie jest wykonywana według formuły (wersja bez nauczyciela)

$$\Delta \boldsymbol{w} = \eta y (\boldsymbol{x} - y \boldsymbol{w}).$$

Oja (1982) udowodnił następujące twierdzenie:

TWIERDZENIE 7.2.

Jeżeli zbiór danych uczących ma macierz kowariancji $\boldsymbol{C}$, to wektory wag sieci o s neuronach spełniają następujące równanie macierzowe:

$$\boldsymbol{C}\boldsymbol{w}_i = \lambda_i \boldsymbol{w}_i,$$

gdzie

$$\lambda_i = \boldsymbol{w}_i' \boldsymbol{C} \boldsymbol{w}_i.$$

Można zatem powiedzieć, że wektor $\boldsymbol{w}_i$ jest wektorem własnym macierzy kowariancji $\boldsymbol{C}$.

Dowód. Po wielokrotnej prezentacji danych uczących wartości oczekiwane przyrostów wag powinny wynosić zero (co oznacza, że wagi od pewnego momentu nie ulegają zmianie)

$$E_{\boldsymbol{X}}(\Delta \boldsymbol{w}_i) = E_{\boldsymbol{X}}(\eta Y_i(\boldsymbol{X} - Y_i \boldsymbol{w}_i)) = \boldsymbol{0}.$$

Dalej mamy

$$E_{\boldsymbol{X}}(Y_i \boldsymbol{X}) - E_{\boldsymbol{X}}(Y_i Y_i \boldsymbol{w}_i) = \boldsymbol{0}.$$

Korzystając z tego, że $Y_i = \boldsymbol{w}_i' \boldsymbol{X} = \boldsymbol{X}' \boldsymbol{w}_i$, otrzymujemy

$$E_{\boldsymbol{X}}(\boldsymbol{X}\boldsymbol{X}'\boldsymbol{w}_i) - E_{\boldsymbol{X}}(\boldsymbol{w}_i'\boldsymbol{X}\boldsymbol{X}'\boldsymbol{w}_i\boldsymbol{w}_i) = \boldsymbol{0}.$$

Oznaczmy

$$E_{\boldsymbol{X}}(\boldsymbol{X}\boldsymbol{X}') = \boldsymbol{C}, \tag{7.20}$$

$$\boldsymbol{w}_i'\boldsymbol{C}\boldsymbol{w}_i = \lambda_i. \tag{7.21}$$

Stąd otrzymujemy

$$\boldsymbol{C}\boldsymbol{w}_i = \lambda_i \boldsymbol{w}_i, \tag{7.22}$$

czyli

$$(\boldsymbol{C} - \lambda_i \boldsymbol{I})\boldsymbol{w}_i = \boldsymbol{0}.$$

Mnożąc teraz lewostronnie równanie (7.22) przez $\boldsymbol{w}_i'$ oraz korzystając z (7.21), mamy

$$\boldsymbol{w}_i'\boldsymbol{C}\boldsymbol{w}_i = \lambda_i \boldsymbol{w}_i'\boldsymbol{w}_i,$$

$$\lambda_i = \lambda_i ||\boldsymbol{w}_i||^2.$$

Ostatnia równość jest prawdziwa wtedy i tylko wtedy, gdy $||\boldsymbol{w}_i|| = 1$. ∎

Oja (1982) podał efektywne wzory, jak znajdywać metodą sekwencyjną kolejne wektory wag. Jednak nauka taka trwa długo.

Definicja 7.2.
Sygnały napływające do neuronu noszą nazwę *sygnałów presynaptycznych*. Sygnały wytwarzane przez neuron to *sygnały postsynaptyczne*.

Algorytm uczący Hebba można zaliczyć do uczenia typu korelacyjnego, w którym siła połączenia międzyneuronowego wzrasta przy istnieniu korelacji między sygnałami presynaptycznymi i postsynaptycznymi neuronu.

Przykład 7.5.
Zastosujemy algorytm Hebba do przykładu 7.3. Dokonamy najpierw normalizacji danych w taki sposób, aby długość każdego wektora uczącego wynosiła 1. Tak zmodyfikowane sygnały mają postać

$$\boldsymbol{x}_1 = [0.5774,\ -0.5774,\ -0.5774]', \qquad \boldsymbol{x}_2 = [0.5774,\ 0.5774,\ -0.5774]'.$$

Przyjmijmy współczynnik uczenia 0.2 oraz funkcję aktywacji typu signum. Niech początkowo wektor wag będzie równy $\mathbf{0}$. Nie wprowadzimy również obciążenia do naszej sieci. Kolejne kroki algorytmu wyglądają następująco:

$$\boldsymbol{w}(1) = \boldsymbol{w}(0) + 0.2d(\boldsymbol{x}_1)\boldsymbol{x}_1 = -0.2\begin{bmatrix} 0.5774 \\ -0.5774 \\ -0.5774 \end{bmatrix} = \begin{bmatrix} -0.1155 \\ 0.1155 \\ 0.1155 \end{bmatrix},$$

$$\boldsymbol{w}(2) = \boldsymbol{w}(0) + 0.2d(\boldsymbol{x}_2)\boldsymbol{x}_2 = \begin{bmatrix} -0.1155 \\ 0.1155 \\ 0.1155 \end{bmatrix} + 0.2\begin{bmatrix} 0.5774 \\ 0.5774 \\ -0.5774 \end{bmatrix} = \begin{bmatrix} 0 \\ 0.2310 \\ 0 \end{bmatrix}.$$

Sprawdźmy teraz, czy już ta jedna epoka wystarczy, aby sieć poprawnie rozpoznawała oba wzorce

$$\begin{aligned} y_1 = \operatorname{sgn}(\boldsymbol{x}_1'\boldsymbol{w}(2)) &= \operatorname{sgn}([0.5774,\ -0.5774,\ -0.5774][0,\ 0.2300,\ 0]') = \\ &= \operatorname{sgn}(-0.1334) = -1, \\ y_2 = \operatorname{sgn}(\boldsymbol{x}_2'\boldsymbol{w}(2)) &= \operatorname{sgn}([0.5774,\ 0.5774,\ -0.5774][0,\ 0.2300,\ 0]') = \\ &= \operatorname{sgn}(0.1334) = 1. \end{aligned}$$

Pozostaje jeszcze sprawdzić „eliptyczną” pomarańczę

$$\begin{aligned} y_3 = \operatorname{sgn}(\boldsymbol{x}_3'\boldsymbol{w}(2)) &= \operatorname{sgn}([-0.5774,\ -0.5774,\ -0.5774][0,\ 0.2300,\ 0]') = \\ &= \operatorname{sgn}(-0.1334) = -1. \end{aligned}$$

Jak widać sieć działa poprawnie na wszystkich obserwacjach uczących, jak i na danych zaburzonych.

7.2. Sieci wielowarstwowe

Jeżeli sieć ma mieć wartość praktyczną, to musi mieć wejścia (służące do wprowadzania wartości zmiennych obserwowanych do sieci) oraz wyjścia (które wyznaczają wynik obliczeń). Dodatkowo w większości sieci istnieją *neurony ukryte*. Są to po prostu te elementy sieci, których nie daje się bezpośrednio obserwować ani od strony wejścia, ani od strony wyjścia. Najczęściej na każdy element sieci, tzn. wejście, warstwa ukryta oraz wyjście, składa się wiele neuronów, które tworzą warstwę – rys. 7.13. *Warstwa ukryta*[12] składa się z neuronów, które przetwarzają informacje napływające z warstwy wejściowej lub z poprzedniej warstwy ukrytej i przekazują do następnej warstwy ukrytej lub do warstwy wyjściowej. Wprowadźmy oznaczenia, które pozwolą nam na opis sieci wielowarstwowych. W każdej warstwie znajduje się N_k, $k = 1, \ldots, L$ neuronów, które oznaczamy jako N_i^k, $i = 1, \ldots, N_k$. Sieć ma N_0 wejść, na które są podawane sygnały $\boldsymbol{x}(t) = [x_1(t), \ldots, x_{N_0}(t)]'$. Sygnał wyjściowy i-tego neuronu w k-tej warstwie oznaczamy $y_i^{(k)}(t)$. Zauważmy, że sygnał wejściowy neuronu N_i^k jest powiązany z sygnałem wyjściowym $k-1$-warstwy następująco:

$$x_i^{(k)}(t) = \begin{cases} x_i(t), & \text{dla } k = 1,\ i \neq 0, \\ y_i^{(k-1)}(t), & \text{dla } k = 1, \ldots, L, \quad i \neq 0, \\ 1, & \text{dla } i = 0, \quad k = 1, \ldots, L. \end{cases}$$

Przez $w_{ij}^{(k)}(t)$ oznaczymy wagę i-tego neuronu k-warstwy łączącą ten neuron z j-tym sygnałem wejściowym. Wektor wag oznaczymy przez $\boldsymbol{w}_i^{(k)}(t) = [w_{i0}^{(k)}(t), \ldots, w_{iN_{k-1}}^{(k)}(t)]'$. Sygnał wyjściowy neuronu N_i^k w kroku t jest określony następująco:

$$y_i^{(k)}(t) = f\left(s_i^{(k)}(t)\right), \tag{7.23}$$

gdzie

$$s_i^{(k)}(t) = \sum_{j=0}^{N_{k-1}} w_{ij}^{(k)}(t) x_j^{(k)}(t).$$

Zauważmy, że sygnały wyjściowe ostatniej warstwy są równocześnie sygnałami wyjściowymi całej sieci i są one porównywane z sygnałami wzorcowymi $d_1^L(t), \ldots, d_{N_L}^L(t)$.

[12]Po ang. hidden layer.

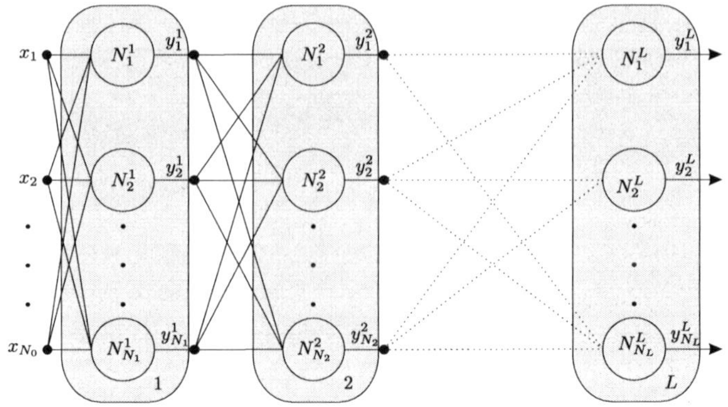

Rys. 7.13. Wielowarstwowa sieć neuronowa

Przykład 7.6.

Pojedynczy perceptron nie jest w stanie rozwiązać problemu XOR, ale już dodanie jednej warstwy ukrytej powoduje, że problem jest rozwiązywalny przez perceptron. Na rysunku 7.14a jest pokazany schemat perceptronu z dwoma wejściami i jedną warstwą ukrytą (wagi obok linii łączących, próg w kółku). Na rysunku 7.14b te cztery punkty są teraz liniowo separowalne przez płaszczyznę.

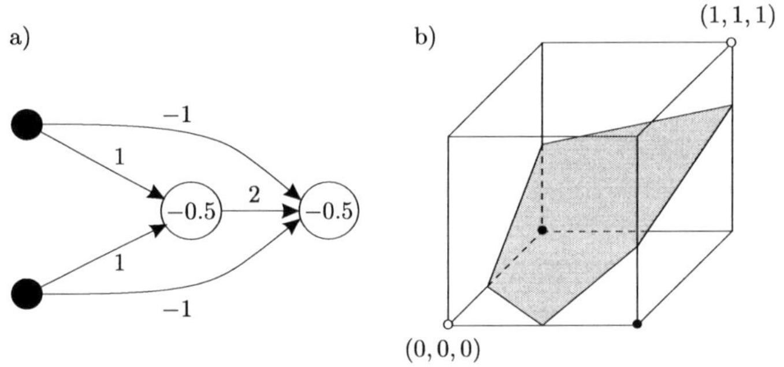

Rys. 7.14. Graficzna prezentacja rozwiązania problemu XOR

Sieć bez warstw ukrytych pozwala rozwiązać jedynie problemy, które są liniowo separowalne (czyli klasy można rozdzielić za pomocą hiperpłaszczyzn). W praktyce większość problemów wymaga użycia sieci z co najmniej jedną warstwą ukrytą. Okazuje się, że taka właśnie sieć w zupełności wystarcza w większości przypadków, jedynie w pewnych bardzo specyficznych sytuacjach należy zastosować sieć, która ma dwie warstwy ukryte. Sieci z dwoma warstwami ukrytymi wymagane są tam, gdzie niezbędna jest niezmienniczość sieci ze względu na przesunięcia i obroty (oraz inne deformacje), np. w rozpoznawaniu zdjęć chcemy, aby sieć potrafiła rozpoznać zdjęcia również lekko zdeformowane przez przekształcenia geometryczne. Część problemów może zostać również rozwiązana przez sieć z większą liczbą warstw ukrytych, ale z mniejszą całkowitą liczbą neuronów w tych warstwach. Większość specjalistów sądzi, że potrzeba zastosowania trzech warstw ukrytych nie pojawia się praktycznie nigdy.

Uwaga 7.1. Sieć z jedną warstwą ukrytą z liniową funkcją aktywacji w tej warstwie jest równoważna sieci bez warstw ukrytych.

Neurony w sieci muszą zostać wzajemnie połączone. Musi zatem zostać wybrana odpowiednia struktura sieci. Pierwszą ważną kwestią przy wyborze struktury sieci jest występowanie lub brak w tej strukturze sprzężenia zwrotnego. Proste sieci mają *strukturę jednokierunkową*[13]: sygnał przepływa w nich tylko w jednym kierunku – od wejść, przez kolejne warstwy ukryte, osiągając ostatecznie neurony wyjściowe. Strukturę taką charakteryzuje zawsze stabilne zachowanie, co jest jej zaletą. Jednak jeśli sieć ma wbudowane *sprzężenie zwrotne* (tzn. zawiera połączenia powrotne od późniejszych do wcześniejszych neuronów), to wówczas może wykonywać bardziej skomplikowane obliczenia. Dodatkowo sieć ze sprzężeniami zwrotnymi o mniejszej liczbie neuronów może wykonywać równie złożone obliczenia jak sieć bez sprzężeń, zawierająca znacznie większą liczbę neuronów. Sieć ze sprzężeniami płaci jednak za to niestabilnością i bardzo złożoną dynamiką, prowadzącą nawet do chaosu deterministycznego[14].

Uwaga 7.2. Jednokierunkową sieć neuronową można zapisać w następującej postaci:

$$y(\boldsymbol{x}) = \sum_{j=1}^{m} \lambda_j f(\theta_j + \boldsymbol{x}'\boldsymbol{w}_j), \tag{7.24}$$

która przypomina parametryczne zagadnienie regresji nieliniowej. Takie sformułowanie problemu jest bardzo zbliżone do metody projection pursuit, o której szczegółowo można przeczytać w monografii Webba (2003). Podobnie formułowany jest również uogólniony model addytywny, o którym można przeczytać w książce Devroye'a, Györfiego oraz Lugosiego (1996).

---

[13] Po ang. feedforward.

[14] Zachowanie układu deterministycznego jest niezwykle wrażliwe na warunki początkowe, do tego stopnia, że w praktyce nieprzewidywalne, przypominające proces przypadkowy, mimo że jest ono deterministyczne w sensie matematycznym, tzn. jest opisywane równaniami o charakterze deterministycznym, a nie probabilistycznym.

Gdy znamy już strukturę sieci, musimy zastanowić się, jak te neurony ze sobą połączyć. Jak wiadomo struktura połączeń ma istotny wpływ na działanie sieci, więc dobry dobór połączeń może znacznie przyspieszyć uczenie sieci. Nie potrafimy a priori powiedzieć, które połączenia w sieci są potrzebne, a które nie. Dobrym rozwiązaniem jest połączenie typu „każdy z każdym". Przy przyjęciu tej zasady każdy neuron ukryty i każdy neuron wyjściowy połączony jest z każdym neuronem z warstwy poprzedniej. Taka reguła łączenia jest obliczeniowo dość kosztowna – przy n łączonych neuronach mamy $(n-1)n/2$ połączeń. Dzięki jednak zdefiniowaniu połączeń według zasady „każdy z każdym" niczego się z góry nie wyklucza i dopiero proces uczenia kształtuje ostateczną strukturę sieci, ustalając niezerowe wartości współczynników wagowych tylko na niektórych drogach przepływu sygnałów. Te połączenia, dla których proces uczenia ustali zerowe (lub bardzo małe) bezwzględne wartości współczynników wag, traktować można jako nieistniejące w sieci. Połączenia takie można potem z sieci całkowicie usunąć.

Liczba neuronów w warstwach wejściowych i wyjściowych jest uzależniona od rozwiązywanego problemu. Określenie liczby warstw ukrytych i liczby neuronów znajdujących się w tych warstwach nie jest jednak już takie proste. Jako punkt wyjścia można przyjąć sieć z jedną warstwą ukrytą, zawierającą taką liczbę neuronów, która jest równa połowie sumy liczby neuronów wejściowych i liczby neuronów wyjściowych. Duża liczba neuronów w warstwach ukrytych zmniejsza poziom błędu na zbiorze uczącym, jednak niekoniecznie również na zbiorze testowym. Dodawanie dodatkowych neuronów w warstwach ukrytych zawsze prowadzi do spadku poziomu tego pierwszego błędu. W przypadku błędu szacowanego dla zbioru testowego początkowo dodawanie daje jego redukcję, jednak dalsze dodawanie neuronów prowadzi do wzrostu poziomu tego błędu (rys. 7.15). Efekt ten jest nazywany *efektem szczytu*[15]. Duda, Hart i Stork (2001) proponują wybór takiej liczby neuronów w warstwach ukrytych, aby całkowita liczba wag w sieci była w przybliżeniu równa $N/10$.

Po określeniu liczby warstw oraz liczby neuronów w każdej warstwie należy dobrać wartości wag oraz wartości progowe wszystkich neuronów. Parametry te wybiera się w sposób zapewniający minimalizację błędu działania sieci. Zadanie to jest realizowane przez algorytm uczenia.

Sieci neuronowe wnoszą w modelowaniu nową jakość. Sieć w trakcie procesu uczenia może całkiem sama znaleźć nieliniowy model rozważanego problemu, przy czym twórca sieci nie musi jej podawać żadnych założeń doty-

[15]Po ang. peaking effect.

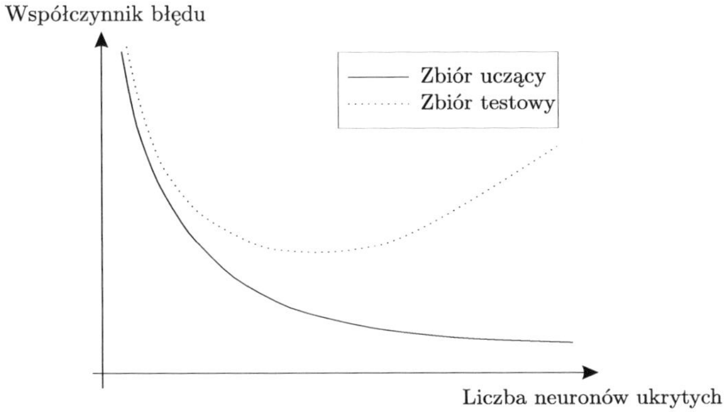

Rys. 7.15. Efekt szczytu

cząców kształtu modelu. Do zrozumienia istoty trudności, na jakie natrafia się przy modelowaniu nieliniowych zależności przy użyciu sieci neuronowych, pomocna może być koncepcja *powierzchni błędu*. Każda z wag i wartości progowych sieci jest traktowana jako jeden z wymiarów przestrzeni. W ten sposób każdy stan sieci, wyznaczony przez aktualne wartości jej N parametrów, może być traktowany jako punkt na N-wymiarowej hiperpłaszczyźnie. $(N + 1)$-wymiarem (zaznaczanym jako wysokość ponad wspomnianą wyżej hiperpowierzchnią) jest błąd, jaki popełnia sieć. Dla każdego możliwego zestawu wag i progów może więc zostać narysowany punkt w przestrzeni $(N + 1)$-wymiarowej, w taki sposób, że stan sieci wynikający z aktualnego zestawu jej parametrów lokuje ten punkt na wspomnianej wyżej N-wymiarowej hiperpłaszczyźnie, zaś wartość błędu, jaki popełnia sieć dla tych właśnie wartości parametrów stanowi wysokość umieszczenia punktu ponad tą płaszczyzną. Gdybyśmy opisaną procedurę powtórzyli dla wszystkich możliwych wartości kombinacji wag i progów sieci, wówczas otrzymalibyśmy „chmurę" punktów rozciągających się ponad wszystkimi punktami N-wymiarowej hiperpłaszczyzny parametrów sieci, tworzącą właśnie rozważaną powierzchnię błędu. Celem uczenia sieci jest znalezienie na tej wielowymiarowej powierzchni punktu o najmniejszej wysokości, czyli ustalenie takiego zestawu wag i progów, który odpowiada najmniejszej wartości błędu. Przy stosowaniu modeli liniowych z funkcją błędu opartą na sumie kwadratów powierzchnia błędu ma kształt paraboloidy (rys. 7.16).

Z tego powodu wyznaczenie w tym przypadku wartości minimalnej nie stwarza większych problemów. Powierzchnia błędu dla sieci neuronowej budującej model nieliniowy jest, niestety, o wiele bardziej złożona. Ma ona pewne cechy mogące być źródłem wielu problemów: minima lokalne (im większa liczba warstw ukrytych, tym jest to bardziej prawdopodobne), pła-

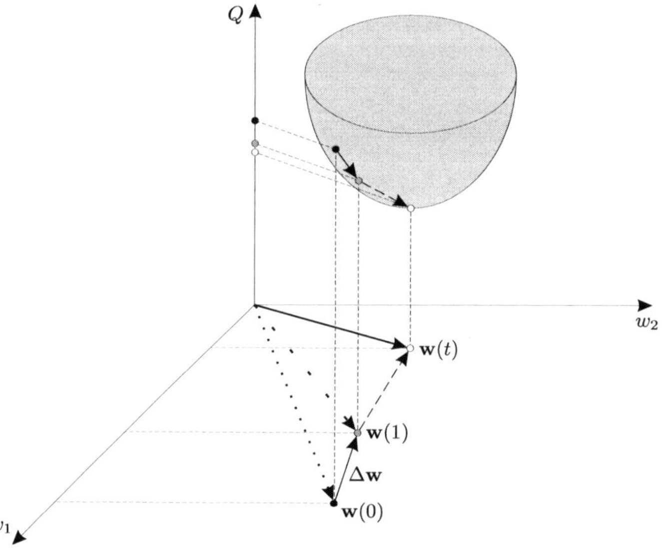

Rys. 7.16. Powierzchnia błędu

skie powierzchnie[16], punkty siodłowe i długie, wąskie wąwozy. Określenie minimum globalnego powierzchni błędu w sposób analityczny nie jest możliwe i dlatego uczenie sieci neuronowej można uznać zasadniczo za eksploracyjne badanie powierzchni błędu. Począwszy od początkowego, losowego układu wag i wartości progowych (układ ten jest reprezentowany przez pewien punkt położony w losowo wybranym miejscu na powierzchni błędu), algorytm uczenia realizuje kolejne kroki w taki sposób, by zmierzać ku poszukiwaniu minimum globalnego. W typowych przypadkach jest to realizowane przez obliczenie gradientu (nachylenia) powierzchni błędu w aktualnym punkcie. Wykorzystanie informacji o kierunku spadku funkcji błędu umożliwia wykonanie kroku (tzn. na zmianę wartości wag neuronów w sieci) w taki sposób, by przemieścić się po powierzchni błędu w kierunku jej spadku. Ostatecznie algorytm uczenia zatrzymuje się w jakimś minimum (gdy gradient wskazuje, że we wszystkich kierunkach dookoła możliwe jest tylko pogorszenie funkcji błędu). Osiągnięte przez proces uczenia minimum może mieć charakter lokalny lub (w sprzyjającym przypadku) może być minimum globalnym. Podczas uczenia wskazana powierzchnia zmienia się w następujący sposób. W początkowej fazie uczenia sieci, wagom oraz wartościom progowym nadaje się wartości będące niewielkimi liczbami losowymi. Powoduje to, że powierzchnia odpowiedzi neuronów ma kształt losowo zaburzonej płaskiej powierzchni zawierającej jedynie bardzo niewielkie na-

[16]Po ang. plateau.

chylenia. W rzeczywistości powierzchnie tego typu z reguły nie są przydatne do rozwiązywania stawianych przed sieciami problemów. W trakcie trwania uczenia powierzchnie odpowiedzi dla poszczególnych neuronów ulegają obróceniu i przesunięciu w sposób zapewniający docelowo przyjęcie przez nie prawidłowej pozycji, wynikających z pożądanej (oczekiwanej) funkcji decyzyjnej. Towarzyszy temu wzrost wartości wag, prowadzący do tego, że powierzchnie te stają się znacznie bardziej wydatne. Takie silnie uformowane funkcje są w stanie modelować poszczególne części zadanej powierzchni odpowiedzi, a ich odpowiednie złożenie stanowić może podstawę do tworzenia właściwej funkcji odpowiedzi sieci.

7.2.1. MADALINE (Many ADALINE's)

Jest to sieć złożona z wielu neuronów typu ADALINE. Sieć taka została zaproponowana przez Widrowa i Hoffa (1960) i stała się praktycznie pierwszą powszechnie stosowaną siecią neuronową. W sposób analogiczny do algorytmu uczenia pojedynczego uczy się także całą sieć elementów liniowych. W swej pierwotnej postaci sieć taka ma wszystkie neurony ukryte typu ADALINE, natomiast wyjście sieci może być różne:

- MRI[17] – wynik otrzymujemy przez operację sumy logicznej,
- MRII – wynik otrzymujemy również za pomocą neuronu typu ADALINE.

Później wprowadzono jeszcze MRIII, która działa podobnie jak MRII, z tą jednak różnicą, że neuron wyjściowy jest typu sigmoidalnego.

Uwaga 7.3. Sieci neuronowe typu MADALINE są chętnie stosowane jako filtry adaptacyjne (np. do eliminacji efektu echa w liniach telefonicznych). Filtry te mogą być wykorzystywane do typowych zadań (dolno- lub górnoprzepustowa filtracja sygnału, eliminacja zakłóceń, analiza widmowa itp.).

7.3. Metody uczenia sieci neuronowych

Uczenie (trenowanie) sieci jest najważniejszym z etapów działania sieci neuronowych. Od nauczenia sieci zależy jakość wyników, jakie ona generuje. Są dwa warianty uczenia sieci neuronowej: z nauczycielem[18] i bez nauczycie-

[17] MADALINE Rule I.
[18] Po ang. supervised learning.

la[19]. Pierwszy polega na tym, że sieci podaje się prawidłowe odpowiedzi na konkretne sygnały wejściowe i porównuje się je z wyjściem sieci. Różnica jest traktowana jako błąd i wpływa na proces uczenia sieci. Jest to tzw. reguła „delta". Na niej opierają się różne strategie uczenia z nauczycielem, z których najpopularniejsza jest metoda wstecznej propagacji błędu. Drugi wariant – uczenie bez nauczyciela, nie podaje prawidłowej odpowiedzi. Sieć sama musi wydobyć pewne kategorie lub cechy w zbiorze danych wejściowych. Jest to analogia do działania ludzkiego mózgu. Metoda ta jest oparta na regule Hebba, która zakłada wzmacnianie połączeń między źródłami silnych sygnałów a neuronami, które na nie silnie reagują. W wyniku uczenia sygnały podobne do siebie będą rozpoznawane przez pewne neurony, inne zaś będą ignorowane. Celem procesu uczenia jest uzyskanie zgodności między wyjściem neuronu a wymaganą odpowiedzią.

7.3.1. Funkcja błędu

Funkcja błędu $Q = Q(\boldsymbol{w})$ jest w ogólnym przypadku nieliniową funkcją wektora wszystkich wag w sieci. W przypadku zagadnień regresyjnych (tj. aproksymacji lub predykcji funkcji wielu zmiennych) funkcja błędu jest minimalizowana najczęściej według zasady najmniejszych kwadratów, tj. minimalizacji podlega funkcja

$$Q = \frac{1}{2}\sum_{i=1}^{N}[\boldsymbol{d}_i - \boldsymbol{y}(\boldsymbol{x}_i,\ \boldsymbol{w})]'[\boldsymbol{d}_i - \boldsymbol{y}(\boldsymbol{x}_i,\ \boldsymbol{w})], \tag{7.25}$$

gdzie $\boldsymbol{y}(\boldsymbol{x}_i,\ \boldsymbol{w})$ jest odpowiedzią sieci. Przy takim kryterium optymalny wektor wag otrzymujemy jako rozwiązanie układu równań normalnych. Czasami układ równań normalnych jest źle uwarunkowany, co prowadzi do błędnych rozwiązań. Aby temu zapobiec, wprowadza się tzw. czynnik regularyzujący[20] $\alpha > 0$. Minimalizuje się wtedy

$$Q_r = Q + \alpha \sum_i w_i^2.$$

W przypadku zagadnień klasyfikacyjnych minimalizacji podlega funkcja

$$Q = -\sum_{i=1}^{N} \boldsymbol{d}_i' \ln \boldsymbol{y}(\boldsymbol{x}_i,\ \boldsymbol{w})$$

[19]Po ang. unsupervised learning.

[20]Po ang. weight decay penalty term.

nazywana *entropią krzyżową*[21]. W tym przypadku $\boldsymbol{y}(\boldsymbol{x}_i, \boldsymbol{w})$ jest zwracanym przez sieć wektorem prawdopodobieństw a posteriori.

7.3.2. Ogólna idea algorytmów iteracyjnych

Naszym celem jest znalezienie rozwiązania optymalnego, wychodząc z pewnego wektora początkowego $\boldsymbol{w}(0)$. W kolejnych krokach (iteracjach) konstruujemy ciąg kolejnych przybliżeń wektora $\boldsymbol{w}$, który ma dać minimum funkcji błędu Q. Kolejne przybliżenia konstruujemy według zasady

$$\boldsymbol{w}(t+1) = \boldsymbol{w}(t) + \Delta\boldsymbol{w}(t). \tag{7.26}$$

Problem polega na tym, aby znaleźć poprawkę $\Delta\boldsymbol{w}(t)$. Poprawka ta polega najczęściej na znalezieniu pewnego wektora $\boldsymbol{p}(t)$ i dodaniu jego części do aktualnie posiadanego wektora wag $\boldsymbol{w}(t)$. Tak więc mamy

$$\Delta\boldsymbol{w}(t) = \eta\boldsymbol{p}(t),$$

gdzie η jest omawianym wcześniej współczynnikiem uczenia. Najczęściej wykorzystywane w praktyce są tzw. *algorytmy gradientowe*, które jako kierunek zmian wybierają ujemny gradient funkcji Q w punkcie $\boldsymbol{w}(t)$. W takim przypadku mamy

$$\Delta\boldsymbol{w}(t) = -\eta\nabla Q(\boldsymbol{w}(t)). \tag{7.27}$$

Aby przeciwdziałać zbyt chaotycznym zmianom, wprowadza się do wzoru (7.27) dodatkowy człon zwany *momentum* lub pędem. W takim przypadku wzór (7.27) przyjmuje postać

$$\Delta\boldsymbol{w}(t) = -\eta\nabla Q(\boldsymbol{w}(t)) + \alpha(\boldsymbol{w}(t) - \boldsymbol{w}(t-1)), \tag{7.28}$$

gdzie $\alpha > 0$. Najczęściej zakłada się, że $0 < \alpha < 1$, w takim przypadku wzór (7.28) ma postać

$$\Delta\boldsymbol{w}(t) = -(1-\alpha)\eta\nabla Q(\boldsymbol{w}(t)) + \alpha(\boldsymbol{w}(t) - \boldsymbol{w}(t-1)). \tag{7.29}$$

Na rysunku 7.17 została przedstawiona idea momentum, po lewej stronie uczenie bez członu momentum, a po prawej z momentum. Dodatkowy człon momentum uzależnia wartość wagi w kolejnym kroku nie tylko od jej wartości w kroku obecnym, ale również od wartości w kroku poprzed-

[21] Po ang. cross-entropy.

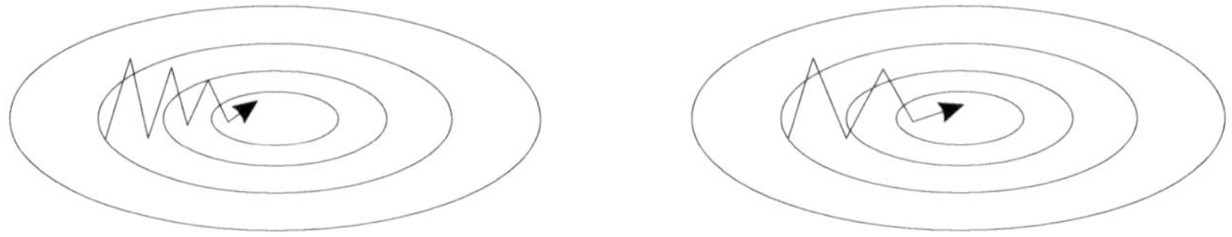

Rys. 7.17. Idea momentum

nim. Jeżeli w kolejnych iteracjach kierunek był taki sam, to następuje przyspieszenie procesu uczenia przez szybsze przesunięcie wagi w tym kierunku. W przypadku odwrotnym następuje zahamowanie procesu uczenia. Pozwala to uniknąć lokalnych minimów. Jednoczesny dobór współczynnika uczenia η oraz *współczynnika momentum* α jest dokonywany za pomocą techniki opisanej przez Bishopa (1996), zwanej *odważnym kierowcą*[22]. Polega ona na tym, że po każdej iteracji, jeśli wartość funkcji błędu się zmniejszyła, zwiększamy współczynniki o $1\% - 5\%$. Jeśli natomiast wartość funkcji błędu wzrosła, to cofamy ostatnią zmianę wag i drastycznie zmniejszamy współczynniki, najczęściej o 50%.

Rozwińmy funkcję błędu w szereg Taylora w punkcie $\boldsymbol{w}$ w kierunku $\boldsymbol{p}$. Mamy

$$Q(\boldsymbol{w}+\boldsymbol{p}) = Q(\boldsymbol{w}) + [\boldsymbol{g}(\boldsymbol{w})]'\boldsymbol{p} + \frac{1}{2}\boldsymbol{p}'\boldsymbol{H}(\boldsymbol{w})\boldsymbol{p} + \ldots, \tag{7.30}$$

gdzie:

$$\boldsymbol{g}(\boldsymbol{w}) = \nabla Q(\boldsymbol{w}) = \left(\frac{\partial Q}{\partial w_i}\right)(\boldsymbol{w}),$$

$$\boldsymbol{H}(\boldsymbol{w}) = \left(\frac{\partial^2 Q}{\partial w_i \partial w_j}\right)(\boldsymbol{w}).$$

Macierz $\boldsymbol{H}$ nazywa się *hesjanem* i zawiera w sobie informacje o krzywiźnie powierzchni błędu.

Różne algorytmy minimalizacji korzystają z rozwinięcia (7.30). Niektóre odmiany tych algorytmów biorą tylko człony liniowe, a inne zarówno człony liniowe, jak i kwadratowe.

Ważnym elementem metod iteracyjnych jest kryterium stopu, czyli warunki powodujące zakończenie algorytmu. W przypadku algorytmów gradientowych stosuje się następujące kryteria:

1. $||\nabla Q(\boldsymbol{w}(t))|| < \varepsilon$ – test stacjonarności.

---

[22] Po ang. bold driver.

2. $||\boldsymbol{w}(t) - \boldsymbol{w}(t-1)|| < \varepsilon$ lub $||Q(\boldsymbol{w}(t)) - Q(\boldsymbol{w}(t-1))|| < \varepsilon$ – test szybkości zbieżności.
3. $t > T_0$ – test liczby iteracji.

W praktyce najczęściej będziemy poszukiwać minimum skalarnej funkcji wielowymiarowego wektora wag. Wymiar ten może być dość duży, toteż rozwiązanie problemu jest zdecydowanie trudniejsze niż w przypadku jednowymiarowym. Możliwa jest bowiem sytuacja, w której lokalnych minimów jest o wiele więcej niż w przestrzeniach niskowymiarowych. Pojawia się tutaj dobrze znany problem zwany przekleństwem wymiarowości[23]. Przypuśćmy, że chcemy znaleźć n-wymiarowy wektor wag, w którym każda składowa leży w przedziale [0, 1]. Wówczas wektor ten jest określony na jednostkowej kostce. Przyjmijmy teraz, że wiemy, iż optymalna wartość każdej składowej nie leży w przedziale [0, 0.5]. Gdy $n = 1$, oznacza to, że połowa przestrzeni została wyeliminowana, gdy $n = 10$ jedynie $\frac{1}{1024} \approx \frac{1}{1000}$, a gdy $n = 20$, jedynie $\frac{1}{1000000}$. Z tego widzimy, na czym polega problem z przeoczeniem globalnego minimum.

Spełnienie warunku koniecznego istnienia minimum funkcji w punkcie nie daje oszacowania wartości funkcji błędu w tym punkcie. Powstaje zatem pytanie – czy dla dowolnej aproksymowanej funkcji istnieje sieć neuronowa, której błąd aproksymacji będzie nie większy niż pewien założony limit. Odpowiedzi na to pytanie udziela poniższe twierdzenie.

TWIERDZENIE 7.3. (O uniwersalnych właściwościach aproksymacyjnych)

Niech $D \subset \mathbb{R}^n$ będzie zbiorem otwartym o skończonej mierze; $f : \mathbb{R}^n \to \mathbb{R}$ będzie funkcją ciągłą, taką że $|f(\boldsymbol{x})| < \infty$ na zbiorze D. Dodatkowo niech $\hat{y}(\boldsymbol{x}) = \sum_i g(\boldsymbol{w}_i'\boldsymbol{x})$, gdzie g jest ograniczoną i monotoniczną funkcją ciągłą. Wówczas dla każdego $\varepsilon > 0$ istnieje taki zbiór wektorów wag $\{\boldsymbol{w}_i\}$, że dla każdego $\boldsymbol{x} \in D$ zachodzi

$$|\hat{y}(\boldsymbol{x}) - f(\boldsymbol{x})| < \varepsilon.$$

Tak więc na mocy powyższego twierdzenia wiemy, że istnieje taka struktura sieci neuronowej, w której da się zrealizować aproksymację z dowolnie małym błędem.

[23] Po ang. curse of dimensionality.

W tym miejscu warto wspomnieć, że istnieje interesująca reprezentacja dla każdej funkcji ciągłej na $[0,\ 1]^d$.

TWIERDZENIE 7.4. (Kołmogorowa, 1957, Lorentza, 1976)

Niech f będzie funkcją ciągłą na $[0,\ 1]^d$. Wtedy f może zostać zapisana w następującej postaci: niech $\delta > 0$ będzie stałą i wybierzmy wymierne $0 < \varepsilon < \delta$. Wtedy

$$f = \sum_{k=1}^{2d+1} g(z_k), \tag{7.31}$$

gdzie $g : \mathbb{R} \to \mathbb{R}$ jest funkcją ciągłą (zależy od f oraz ε), a każde z_k ma postać

$$z_k = \sum_{j=1}^{d} \lambda^k \varphi(x_j + \varepsilon k) + k, \tag{7.32}$$

gdzie $\lambda \in \mathbb{R}$, a φ jest rosnącą funkcją monotoniczną. Dodatkowo λ oraz φ nie zależą od f i φ jest funkcją lipschitzowską: $|\varphi(x) - \varphi(y) \leq c|x - y|$ dla pewnego $c > 0$.

Twierdzenie to orzeka zatem, że funkcja f może być reprezentowana przez bardzo prostą sieć zwaną siecią Kołmogorowa–Lipschitza. W sieci tej pierwsza warstwa jest ustalona i znana z góry, gdyż jedynie funkcja g zależy od f. Podejście takie prowadzi do ciekawego rozwiązania: nie musimy mieszać zmiennych wejściowych, gdyż za pomocą prostych funkcji jednej zmiennej możemy przedstawić wszystkie wielowymiarowe funkcje ciągłe.

Analizując gradient, umiemy stwierdzić, czy zestaw wag sieci jest minimum lokalnym funkcji błędu. Pozostaje rzecz najtrudniejsza – wyznaczyć taki zestaw wag, który jest nie tylko minimum lokalnym, lecz globalnym funkcji błędu. Omawiane algorytmy iteracyjne uczenia sieci korzystają z warunku koniecznego istnienia minimum funkcji w punkcie. Oznacza to, że wektor wag będzie zdążać do takiego wektora, dla którego przypada minimum lokalne funkcji błędu, niekoniecznie będące minimum globalnym. Nie jest znany algorytm dający gwarancję znalezienia minimum globalnego dowolnej funkcji w $\mathbb{R}^n$, a zatem nie możemy liczyć na gwarancję wyznaczenia optymalnego wektora wag. Można jednak zwiększyć prawdopodobieństwo wyznaczenia minimum globalnego, stosując np. metody przeszukiwania wykorzystujące losowość (np. algorytm symulowanego wyżarzania).

7.3.3. Algorytm największego spadku

W algorytmie największego spadku[24] ograniczamy się w rozwinięciu (7.30) jedynie do członu liniowego

$$Q(\boldsymbol{w}(t) + \boldsymbol{p}(t)) \approx Q(\boldsymbol{w}(t)) + [\boldsymbol{g}(\boldsymbol{w}(t))]'\boldsymbol{p}(t).$$

Chcemy, aby funkcja błędu w każdym kolejnym kroku malała, czyli aby

$$Q(\boldsymbol{w}(t+1)) < Q(\boldsymbol{w}(t)).$$

Wystarczy zatem tak dobrać wagi, aby zachodziło

$$[\boldsymbol{g}(\boldsymbol{w}(t))]'\boldsymbol{p}(t) < 0.$$

Zauważmy, że warunek ten jest spełniony, jeśli przyjmiemy

$$\boldsymbol{p}(t) = -\boldsymbol{g}(\boldsymbol{w}(t)).$$

Wstawiając tę zależność do wzoru (7.26), otrzymujemy

$$\boldsymbol{w}(t+1) = \boldsymbol{w}(t) - \eta \boldsymbol{g}(\boldsymbol{w}(t)). \tag{7.33}$$

Wzór ten jest znany pod nazwą *reguły największego spadku*[25] i został również wykorzystany w formule (7.11). Wadą tej metody jest niewykorzystywanie informacji o krzywiźnie powierzchni błędu oraz bardzo powolne zbliżanie się do minimum (zastosowanie momentum znacznie skraca ten czas). Największymi zaletami są natomiast prostota oraz mała złożoność obliczeniowa.

Przykład 7.7.
Korzystając z metody największego spadku, znajdziemy minimum funkcji $F(\boldsymbol{x}) = x_1^2 + 2x_1x_2 + 2x_2^2 + x_1$. Niech

$$\boldsymbol{x}(0) = \begin{bmatrix} 0.5 \\ 0.5 \end{bmatrix}, \quad \eta = 0.1.$$

---

[24] Cauchy zaproponował w 1847 r. algorytm wykorzystujący w podobny sposób gradient do rozwiązywania układu równań nieliniowych.

[25] Po ang. gradient descent.

Otrzymujemy

$$\nabla F(\boldsymbol{x}) = \begin{bmatrix} 2x_1 + 2x_2 + 1 \\ 2x_1 + 4x_2 \end{bmatrix},$$

$$\boldsymbol{g}(0) = \nabla F(\boldsymbol{x})|_{\boldsymbol{x}(0)} = \begin{bmatrix} 3 \\ 3 \end{bmatrix},$$

$$\boldsymbol{x}(1) = \boldsymbol{x}(0) - \eta \boldsymbol{g}(0) = \begin{bmatrix} 0.5 \\ 0.5 \end{bmatrix} - 0.1 \begin{bmatrix} 3 \\ 3 \end{bmatrix} = \begin{bmatrix} 0.2 \\ 0.2 \end{bmatrix},$$

$$\boldsymbol{g}(1) = \nabla F(\boldsymbol{x})|_{\boldsymbol{x}=\boldsymbol{x}(1)} = \begin{bmatrix} 1.8 \\ 1.2 \end{bmatrix},$$

$$\boldsymbol{x}(2) = \boldsymbol{x}(1) - \eta \boldsymbol{g}(1) = \begin{bmatrix} 0.2 \\ 0.2 \end{bmatrix} - 0.1 \begin{bmatrix} 1.8 \\ 1.2 \end{bmatrix} = \begin{bmatrix} 0.02 \\ 0.08 \end{bmatrix}$$

itd.

Pełny przebieg algorytmu znajduje się na rys. 7.18. Minimum funkcji osiągamy w 127 kroku algorytmu.

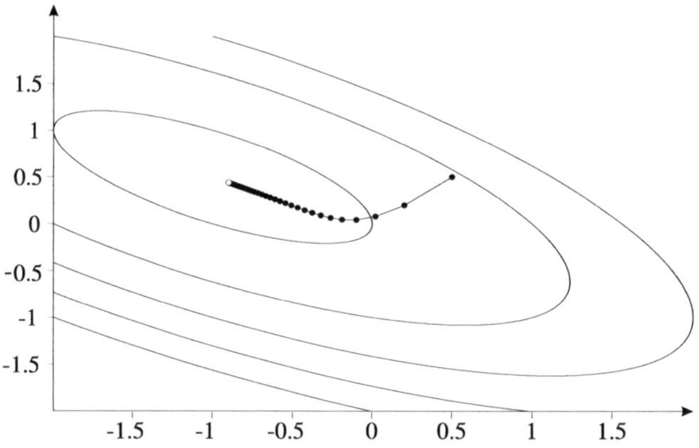

Rys. 7.18. Przebieg algorytmu największego spadku

7.3.4. Algorytm wstecznej propagacji błędu

Najlepiej znanym algorytmem uczenia sieci neuronowej jest *metoda wstecznej propagacji błędów*[26]. Została ona opisana przez Werbosa (1974), następnie rozwinięta przez Rumelharta, Hintona i Williamsa (1986). Algorytm rozpoczyna się od wyznaczenia przez kolejne warstwy sygnałów wyjściowych

[26]Po ang. BackPropagation (BP).

(wzór (7.23)). Tak otrzymane sygnały są wejściami dla neuronów następnej warstwy. Proces powtarzamy aż do ostatniej warstwy. Znając sygnał wyjściowy, porównujemy go ze wzorcem i wyznaczamy błąd sieci zgodnie ze wzorem

$$Q(t) = \sum_{i=1}^{N_L} \varepsilon_i^{(L)^2}(t) = \sum_{i=1}^{N_L} \left(d_i^{(L)}(t) - y_i^{(L)}(t) \right)^2 . \tag{7.34}$$

Wykorzystując teraz regułę delta, możemy (podobnie jak dla neuronów sigmoidalnych) zmodyfikować wagi neuronów ostatniej warstwy w następujący sposób:

$$w_{ij}^{(k)}(t+1) = w_{ij}^{(k)}(t) + 2\eta\delta_i^{(k)}(t)x_j^{(k)}(t), \tag{7.35}$$

$$\delta_i^{(k)}(t) = \varepsilon_i^{(k)}(t)f'\left(s_i^{(k)}(t)\right). \tag{7.36}$$

W ten sposób nie zmodyfikujemy jednak wag w warstwach ukrytych (nie znamy bowiem wartości błędu w poszczególnych miejscach sieci). Dlatego błąd wyjściowy jest propagowany od tyłu (od warstwy wyjściowej do wejściowej) zgodnie z połączeniami neuronów i z uwzględnieniem funkcji aktywacji. Błąd w pojedynczym neuronie warstw ukrytych możemy obliczyć jako

$$\varepsilon_i^{(k)}(t) = \sum_{m=1}^{N_{k+1}} \delta_m^{(k+1)}(t)w_{mi}^{(k+1)}(t), \qquad k = 1, \ldots, L-1. \tag{7.37}$$

Nowoczesne algorytmy drugiego rzędu, takie jak: *metoda gradientów sprzężonych* i *metoda Levenberga–Marquardta* są w większości zastosowań istotnie szybsze (nawet o rząd wielkości), co mogłoby skłaniać do ich częstszego stosowania. Klasyczna metoda BP też ma jednak wiele istotnych zalet, co powoduje, że jest zdecydowanie najchętniej stosowana przez większość użytkowników sieci neuronowych. Jest wiele powodów tego, że BP jest najchętniej stosowanym algorytmem uczenia sieci:

- jest algorytmem najprostszym do zrozumienia, co jest nie bez znaczenia, gdy idzie o zdobycie zaufania użytkownika,
- pomimo rozwoju alternatywnych algorytmów, algorytm BP ciągle ma przewagę w pewnych sytuacjach. Dotyczy to głównie przypadku, kiedy założenia niezbędne do wykorzystania innych algorytmów są trudne do spełnienia lub sprawdzenia,

- jest wprawdzie metodą powolną, ale dość pewną – to znaczy za pomocą tej metody można wyjątkowo łatwo (niestety, niekoniecznie bardzo sprawnie) rozwiązywać praktycznie każde zadanie, podczas gdy inne metody w pewnych przypadkach gwarantują bardzo dużą efektywność przetwarzania, w innych jednak zawodzą całkowicie.

Niestety, algorytm wstecznej propagacji błędów nie jest doskonały. Nie można zapewnić, że w trakcie procesu uczenia zostanie odnalezione minimum globalne, a nie, jak to się często zdarza, minimum lokalne. Wybranie niewłaściwego punktu startowego, czyli niewłaściwy dobór wartości początkowych wag może spowodować wejście w minimum lokalne, którego algorytm nie będzie w stanie opuścić. Na poprawę efektywności procesu uczenia może mieć wpływ wiele czynników:

- wagi początkowe nie mogą być zbyt duże, zaleca się losowy wybór wag, w taki sposób, by sygnał wyjściowy nieliniowej części neuronu był nieco mniejszy od jedności,
- korzystnie wpływa kilkukrotne powtarzanie uczenia, rozpoczynające się od różnych wartości wag,
- zaleca się podawanie wektorów wzorcowych w losowej kolejności.

Przedstawione powyżej propozycje mogą w pewnym stopniu zapobiec problemom pojawiającym się w trakcie uczenia sieci. Najlepszym rozwiązaniem jest jednak dokonanie modyfikacji samego algorytmu wstecznej propagacji błędów tak, aby przyspieszyć zbieżność procesu uczenia oraz uniknąć minimów lokalnych.

7.3.5. Modyfikacje algorytmu BP

Algorytm BP z członem momentum

Klasyczna metoda wstecznej propagacji błędów wymaga dużej liczby iteracji, by osiągnąć zbieżność oraz jest wrażliwa na występowanie minimów lokalnych. Podstawowy algorytm BP może się okazać zbyt wolny, jeżeli przyjmie się za mały współczynnik uczenia, z kolei zbyt duża jego wartość grozi wystąpieniem oscylacji. Jednym z rozwiązań umożliwiających bezpieczne zwiększenie efektywnego tempa uczenia bez pogarszania stabilności procesu jest zastosowanie momentowej metody wstecznej propagacji błędu[27]. W takim przypadku modyfikacja wag odbywa się zgodnie z wyrażeniem

[27] Po ang. Momentum BackPropagation (MBP).

$$w_{ij}^{(k)}(t+1) = w_{ij}^{(k)}(t) + 2\eta\delta_i^{(k)}(t)x_j^{(k)}(t) + \alpha\left[w_{ij}^{(k)}(t) - w_{ij}^{(k)}(t-1)\right]. \quad (7.38)$$

Znaczenie członu momentum widoczne jest wtedy, gdy kierunek, w którym są wykonywane kolejne kroki, przez pewien czas się nie zmienia. Następuje wówczas zwiększanie się prędkości działania algorytmu (przy zachowaniu stałego gradientu dla $\alpha = 0.9$ następuje 10-krotne przyspieszenie procesu uczenia), co (czasami) stwarza możliwość ucieczki z minimum lokalnego oraz szybkiego przejścia przez obszary płaskie. Natomiast w fazach uczenia związanych z pokonywaniem obszarów, w których funkcja błędu ma liczne drobne zafalowania, dodanie składnika momentum powoduje uodpornienie algorytmu uczenia na efekty „myszkowania" w poszukiwaniu drogi zmian wag wiodącej najdokładniej wzdłuż linii najszybszego spadku błędu (ciągle zmieniającej kierunek). Modyfikacja momentowa algorytmu propagacji wstecznej powoduje zatem przyspieszenie procesu uczenia przy pewnych stałych tendencjach zmian w przestrzeni wag jednocześnie, przynajmniej częściowo, chroniąc przed wystąpieniem oscylacji i utknięciem w minimum lokalnym. Wprowadzenie składnika momentum do algorytmu wstecznej propagacji błędu sprawia, że krok uczenia jest średnią wykładniczą wszystkich poprzednich kroków uczenia, co zostało pokazane przez Reeda i Marksa (1999).

Metody doboru współczynnika uczenia

Aby przyspieszyć działanie metody BP, stosuje się specjalne metody doboru współczynnika uczenia. Podstawowy pomysł polega na tym, że gdy w kolejnym kroku algorytmu wartość funkcji błędu maleje, to współczynnik uczenia η jest nieznacznie zwiększany, natomiast gdy wartość błędu wzrasta w znacznym stopniu w stosunku do poprzedniego kroku algorytmu (prawdopodobnie następuje oddalanie się od minimum funkcji błędu), wartość współczynnika η jest zmniejszana. Sposoby zwiększania i zmniejszania współczynnika η najczęściej są arbitralne. Istnieje kilka metod zmian współczynnika:

- metoda Silvy i Almeidy (1990),
- metoda Jacobsa (1988),
- metoda delta-delta,
- metoda delta-bar-delta.

Czasem dla polepszenia jakości działania algorytmu BP stosuje się też mechanizm zanikania wag[28] zaproponowany przez Werbosa (1988). W tym

[28]Po ang. weight decay.

przypadku dodatkowo w każdym kroku wagi są zmniejszane o pewną część β swojej poprzedniej wartości.

W celu poprawy działania metody BP jest stosowana również technika kumulowania poprawek[29] polegająca na tym, że po prezentacji każdego wzorca jest obliczany wektor zmiany wag $\Delta\boldsymbol{w}(i)$, ale modyfikacja wartości wag jest dokonywana po prezentacji wszystkich elementów ciągu uczącego (raz w ciągu całej epoki). Wagi są wtedy zmieniane o wartość

$$\Delta\boldsymbol{w} = \sum_{i=1}^{n} \Delta\boldsymbol{w}(i).$$

Algorytm Quickprop

Wszystkie te pomysłowe metody w naprawdę trudnych zadaniach dają raczej kiepskie wyniki. Dlatego wymyślono metodę przyspieszonej wstecznej propagacji, zwaną Quickprop. Algorytm został opracowany przez Fahlmana (1988). W metodzie tej zakłada się, że funkcja błędu jest lokalnie paraboloidalna. Korzystając z tego założenia, wartość wektora poprawki $\Delta\boldsymbol{w}(t)$ oblicza się w taki sposób, aby w trakcie jednej iteracji osiągnąć minimum funkcji błędu (minimum paraboloidy). Wartość poprawki jest dana wzorem

$$\Delta\boldsymbol{w}(t) = \frac{\nabla Q(\boldsymbol{w}(t))}{\nabla Q(\boldsymbol{w}(t-1)) - \nabla Q(\boldsymbol{w}(t))} \Delta\boldsymbol{w}(t-1).$$

Chociaż zazwyczaj przytoczone wyżej założenie nie jest dokładnie spełnione, to jednak algorytm Quickprop powoduje bardzo duże przyspieszenie procesu uczenia. Ponadto zabezpiecza on przed utknięciem w płytkim minimum lokalnym.

Algorytm Rprop

Odporny algorytm wstecznej propagacji błędu[30] został opracowany przez Riedmillera i Brauna (1992). Istotą metody jest uwzględnienie w procesie aktualizacji wagi tylko znaku składowej gradientu (jej wartość jest pomijana). Minimalizacji podlega zmodyfikowana funkcja błędu postaci

$$Q = \sum (d_i - y_i)^2 + 10^{-\gamma} \sum w_{ij}^2,$$

gdzie parametr β (najczęściej $\beta = 4$) określa tempo zanikania wag.
Modyfikacja wartości wag w algorytmie Rprop jest dokonywana po prezentacji wszystkich elementów ciągu uczącego (uczenie skumulowane). W meto-

[29] Po ang. cumulative weight adjustment.

[30] Po ang. resilient backpropagation.

dzie tej każda z modyfikowanych wag występujących w sieci jest zmieniana w następujący sposób:

$$\Delta \boldsymbol{w}_{ij}(t) = \begin{cases} -\Delta_{ij}(t), & \text{jeśli } \dfrac{\partial Q(\boldsymbol{w}(t))}{\partial w_{ij}} > 0 \\ \Delta_{ij}(t), & \text{jeśli } \dfrac{\partial Q(\boldsymbol{w}(t))}{\partial w_{ij}} < 0 \\ 0, & \text{jeśli } \dfrac{\partial Q(\boldsymbol{w}(t))}{\partial w_{ij}} = 0 \end{cases},$$

gdzie $\Delta_{ij}(t)$ jest wielkością, o którą zmienia się wartość wagi w_{ij} w epoce t. Każdorazowo po dokonaniu zmiany wag następuje proces modyfikowania wielkości $\Delta_{ij}(t)$ według wzoru

$$\Delta_{ij}(t+1) = \begin{cases} \theta^{+}\Delta_{ij}(t), & \text{jeśli } \dfrac{\partial Q(\boldsymbol{w}(t))}{\partial w_{ij}}\dfrac{\partial Q(\boldsymbol{w}(t+1))}{\partial w_{ij}} > 0 \\ \theta^{-}\Delta_{ij}(t), & \text{jeśli } \dfrac{\partial Q(\boldsymbol{w}(t))}{\partial w_{ij}}\dfrac{\partial Q(\boldsymbol{w}(t+1))}{\partial w_{ij}} < 0 \\ \Delta_{ij}(t), & \text{jeśli } \dfrac{\partial Q(\boldsymbol{w}(t))}{\partial w_{ij}}\dfrac{\partial Q(\boldsymbol{w}(t+1))}{\partial w_{ij}} = 0 \end{cases},$$

gdzie $0 < \theta^{-} < 1 < \theta^{+}$. Najczęściej przyjmuje się $\theta^{-} = 0.5$ oraz $\theta^{+} = 1.2$. Powyższa reguła powoduje, że w przypadku, gdy odpowiednia pochodna cząstkowa zmieniła swój znak, następuje zmniejszenie wartości $\Delta_{ij}(t)$. Jeżeli znak pochodnej nie uległ zmianie, to wartość $\Delta_{ij}(t)$ jest nieznacznie zwiększana. Parametrami metody Rprop oprócz współczynnika zapominania wag γ są: wartość początkowa Δ_0 (najczęściej 0.1) oraz wartość maksymalna $\Delta_{\max}$ (najczęściej 50) i wartość minimalna $\Delta_{\min}$ (najczęściej 10^{-6}). Po osiągnięciu jednej z podanych wartości granicznych wielkość Δ_{ij} nie jest już więcej modyfikowana.

Algorytm Rprop powoduje znaczne przyspieszenie procesu uczenia zwłaszcza w obszarach o niewielkim nachyleniu funkcji błędu.

Algorytm QRprop

Algorytm ten łączy cechy metody Quickprop oraz Rprop. Jego autorami są Pfister i Rojas (1994). Wagi w tym algorytmie są modyfikowane w następujący sposób:

- jeżeli $\nabla_i Q(\boldsymbol{w}(t))$ oraz $\nabla_i Q(\boldsymbol{w}(t-1))$ nie zmieniły znaku (ich iloczyn jest dodatni), to jest stosowana modyfikacja wag zgodna z metodą Rprop,

- jeżeli $\nabla_i Q(\boldsymbol{w}(t))$ oraz $\nabla_i Q(\boldsymbol{w}(t-1))$ zmieniły znak (ich iloczyn jest ujemny i oznacza to, że nastąpił przeskok nad minimum lokalnym), to jest stosowana modyfikacja wag zgodna z metodą Quickprop, zaś wartości składowej gradientu $\nabla_i Q(\boldsymbol{w}(t))$ przypisuje się wartość 0,
- w przypadku, gdy iloczyn jest równy zeru (co następuje wówczas, gdy w kroku poprzednim nadano składowej gradientu wartość 0 lub wtedy, gdy osiągnięto minimum lokalne), wykonuje się krok zgodnie z metodą Quickprop, wykorzystując składowe wektorów gradientu $\nabla_i Q(\boldsymbol{w}(t))$ oraz $\nabla_i Q(\boldsymbol{w}(t-2))$.

Uwaga 7.4. Wszystkie te „sztuczki" nie zmieniają faktu, że wszystkie algorytmy bazujące na koncepcji wstecznej propagacji błędu działają jednak raczej wolno.

7.3.6. Algorytm zmiennej metryki

Kolejną dość powszechnie stosowaną modyfikacją algorytmu wstecznej propagacji błędu jest algorytm zmiennej metryki zwany też quasi-Newtona. W algorytmie tym, wyznaczając wektor poszukiwań, wykorzystujemy kwadratowe przybliżenie funkcji celu, a wektor poszukiwań wyznaczamy, korzystając z formuły

$$\boldsymbol{p}(t) = -[\boldsymbol{H}(\boldsymbol{w}(t))]^{-1}\boldsymbol{g}(\boldsymbol{w}(t)). \tag{7.39}$$

Niestety, aby móc skorzystać z tej zależności, niezbędne jest wyznaczenie hesjanu w każdym kroku, a co gorsza jego odwrócenie (musi być dodatnio określony). W tym stanie rzeczy w praktycznych implementacjach rezygnuje się z wyznaczania dokładnej wartości hesjanu (jeśli wyznaczamy hesjan dokładnie, to mamy do czynienia z metodą Newtona zwaną w przypadku jednowymiarowym metodą Newtona–Raphsona), a zamiast tego korzysta się z jego wartości przybliżonej $\boldsymbol{G}(\boldsymbol{w}(t))$. W metodzie zmiennej metryki w każdym kroku iteracji modyfikuje się $\boldsymbol{G}$ z kroku poprzedniego o pewną poprawkę. Poprawka jest tak dobrana, aby aktualna wartość $\boldsymbol{G}(\boldsymbol{w}(t))$ przybliżała funkcję celu zgodnie z zależnością

$$\boldsymbol{G}(\boldsymbol{w}(t))(\boldsymbol{w}(t) - \boldsymbol{w}(t-1)) = \boldsymbol{g}(\boldsymbol{w}(t)) - \boldsymbol{g}(\boldsymbol{w}(t-1)). \tag{7.40}$$

Przybliżenie hesjanu $\boldsymbol{G}(\boldsymbol{w}(t))$ powinno być tak skonstruowane, aby zapewnić istnienie odwrotności $\boldsymbol{V}(t) = [\boldsymbol{G}(\boldsymbol{w}(t))]^{-1}$. Powszechnie są stosowane dwie metody mające tę własność. W celu ich podania przyjmijmy następujące oznaczenia:

$$\boldsymbol{c}(t) = \boldsymbol{w}(t) - \boldsymbol{w}(t-1),$$
$$\boldsymbol{r}(t) = \boldsymbol{g}(\boldsymbol{w}(t)) - \boldsymbol{g}(\boldsymbol{w}(t-1)).$$

Przy takich oznaczeniach mamy:

- *metoda Davidona–Fletchera–Powella*

$$\boldsymbol{V}(t) = \boldsymbol{V}(t-1) + \frac{\boldsymbol{c}(t)\boldsymbol{c}'(t)}{\boldsymbol{c}'(t)\boldsymbol{r}(t)} - \frac{\boldsymbol{V}(t-1)\boldsymbol{r}(t)\boldsymbol{r}'(t)\boldsymbol{V}(t-1)}{\boldsymbol{r}'(t)\boldsymbol{V}(t-1)\boldsymbol{r}(t)},$$

- *metoda Broydena–Fletchera–Goldfarba–Shanno*

$$\boldsymbol{V}(t) = \boldsymbol{V}(t-1) + \left[1 + \frac{\boldsymbol{r}'(t)\boldsymbol{V}(t-1)\boldsymbol{r}(t)}{\boldsymbol{c}'(t)\boldsymbol{r}(t)}\right]\frac{\boldsymbol{c}(t)\boldsymbol{c}'(t)}{\boldsymbol{c}'(t)\boldsymbol{r}(t)} - \\ - \frac{\boldsymbol{c}(t)\boldsymbol{r}'(t)\boldsymbol{V}(t-1) + \boldsymbol{V}(t-1)\boldsymbol{r}(t)\boldsymbol{c}'(t)}{\boldsymbol{c}'(t)\boldsymbol{r}(t)}.$$

Za wartość startową przyjmuje się zwykle $\boldsymbol{V}(0) = \boldsymbol{I}$ oraz pierwszy krok algorytmu wykonuje się zgodnie z algorytmem największego spadku. Istotną cechą algorytmów zmiennej metryki jest tak zwana odnowa, polegająca na wykonaniu podstawienia $\boldsymbol{V}(t) = \boldsymbol{I}$, dla pewnych wartości t. Odnowę przeprowadza się wówczas, gdy wyznaczony kierunek zmian nie jest kierunkiem poprawy, jak również wtedy, gdy liczba iteracji od czasu ostatniej odnowy przekroczyła wartość $2n + 1$, gdzie n jest liczbą szacowanych parametrów sieci.

Metoda zmiennej metryki charakteryzuje się zbieżnością superliniową[31], jest zatem znacznie lepsza od liniowo zbieżnej metody największego spadku.

[31] Zbieżny ciąg $(x_k)_{k=0}^{\infty}$ otrzymany w kolejnych krokach algorytmu, jest zbieżny:

- liniowo, gdy dodatkowo

$$\lim_{k\to\infty} \frac{||x_{k+1} - x_0||}{||x_k - x_0||} = \rho > 0,$$

- superliniowo, gdy dodatkowo

$$\lim_{k\to\infty} \frac{||x_{k+1} - x_0||}{||x_k - x_0||} = 0,$$

- kwadratowo, gdy dodatkowo

$$\lim_{k\to\infty} \frac{||x_{k+1} - x_0||}{||x_k - x_0||^2} = \rho \geq 0.$$

Zbieżność liniowa jest najwolniejsza, a kwadratowa najszybsza. Z różnych eksperymentów wynika, że w zastosowaniach praktycznych szybkość liniowa jest do zaakceptowania, gdy granica ρ jest dostatecznie mała, nie większa niż $\frac{1}{4}$.

Jest obecnie uważana za jedną z najlepszych metod optymalizacji funkcji wielu zmiennych. Jednakże konieczność aproksymowania macierzy hesjanu ogranicza jej zastosowanie do problemów z niewielką ilością zmiennych (w praktyce dla mniej niż 1000).

Przykład 7.8.
Do minimalizacji funkcji z przykładu 7.7 zastosujemy teraz metodę Newtona. Aby z niej skorzystać, musimy mieć najpierw hesjan tej funkcji. Mamy

$$\boldsymbol{H} = \begin{bmatrix} 2 & 2 \\ 2 & 4 \end{bmatrix}.$$

Teraz

$$\boldsymbol{x}(1) = \boldsymbol{x}(0) - \boldsymbol{H}^{-1}\boldsymbol{g}(0) = \begin{bmatrix} 0.5 \\ 0.5 \end{bmatrix} - \begin{bmatrix} 2 & 2 \\ 2 & 4 \end{bmatrix}^{-1} \begin{bmatrix} 3 \\ 3 \end{bmatrix} = \begin{bmatrix} -1 \\ 0.5 \end{bmatrix}.$$

Jak widać, w tym przypadku jest potrzebny jedynie jeden krok (rys. 7.19), aby zminimalizować funkcję $F(\boldsymbol{x})$.

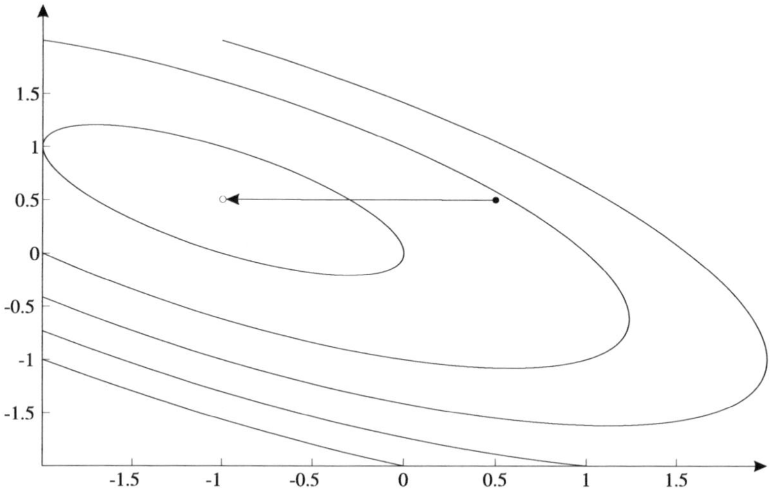

Rys. 7.19. Przebieg algorytmu Newtona

7.3.7. Metoda gradientów sprzężonych

Metoda gradientów sprzężonych[32] bazuje na założeniu, że powierzchnia błędu ma kształt paraboliczny. Została zaproponowana przez Fletchera i Re-

[32]Po ang. conjugate gradient descent.

evesa (1964). W rzeczywistych przypadkach kształt funkcji błędu zazwyczaj odbiega od tego założonego, jednakże w pobliżu minimum w dużym stopniu przypomina on paraboloidę. Obliczenia rozpoczyna się od pewnego punktu startowego $\boldsymbol{w}(0)$. W tym punkcie wyznacza się kierunek przeciwny do gradientu funkcji błędu. Następnie wzdłuż tak wyznaczonego kierunku przeprowadza się proces poszukiwania minimum funkcji błędu. Wyznaczanie minimum funkcji błędu na danym kierunku przeprowadza się, wykorzystując aproksymację Taylora postaci

$$Q(\boldsymbol{w}(t) + \alpha(t)\boldsymbol{p}(t)) \approx Q(\boldsymbol{w}(t)) + \alpha(t)\boldsymbol{g}'(\boldsymbol{w}(t))\boldsymbol{p}(t) + \frac{1}{2}\alpha^2(t)\boldsymbol{p}(t)'\boldsymbol{H}\boldsymbol{p}(t). \tag{7.41}$$

Szukamy zatem takiej wartości α, która minimalizuje wyrażenie (7.41). Po obliczeniu pochodnej cząstkowej względem α i przyrównaniu jej do zera otrzymujemy

$$\alpha_{\min}(t) = -\frac{\boldsymbol{p}'(t)\boldsymbol{g}(\boldsymbol{w}(t))}{\boldsymbol{p}'(t)\boldsymbol{H}\boldsymbol{p}(t)}.$$

Zauważmy, że do wyznaczenia tej wartości wystarczy znajomość $\boldsymbol{H}\boldsymbol{p}(t)$, a tę wartość można obliczyć pośrednio bez obliczania całego hesjanu. Powyższa technika nosi nazwę *przeszukiwania liniowego*[33].

Teraz rozpoczyna się najważniejszy krok działania algorytmu. Należy mianowicie wyznaczyć nowy kierunek poszukiwań. W metodzie gradientów sprzężonych przyjmuje się, że nowy kierunek poszukiwań jest wyznaczony przez wektor sprzężony z wektorem określającym poprzedni kierunek.

Definicja 7.3.
Niezerowe wektory $\boldsymbol{a}$ i $\boldsymbol{b}$ są *sprzężone* względem pewnej funkcji paraboloidalnej, jeśli $\boldsymbol{a}'\boldsymbol{H}\boldsymbol{b} = 0$, gdzie $\boldsymbol{H}$ jest hesjanem tej funkcji. Ogólniej można powiedzieć, że niezerowe wektory $\boldsymbol{a}$ i $\boldsymbol{b}$ są *sprzężone* względem dodatnio określonej macierzy $\boldsymbol{Q}$, jeśli $\boldsymbol{a}'\boldsymbol{Q}\boldsymbol{b} = 0$.

Uwaga 7.5. r-elementowy zbiór niezerowych wektorów $\boldsymbol{a}_1$, $\boldsymbol{a}_n$, ..., $\boldsymbol{a}_r$ w przestrzeni n-wymiarowej ($r \in [2, n]$) nazywamy układem $\boldsymbol{Q}$-sprzężonym, jeżeli

$$\forall i \neq j \qquad \boldsymbol{a}_i'\boldsymbol{Q}\boldsymbol{a}_j = 0.$$

[33]Po ang. line search.

Okazuje się, że taki układ jest układem wektorów liniowo niezależnych. Najprostszym przykładem dwu wektorów sprzężonych są wektory prostopadłe, które są sprzężone względem macierzy jednostkowej. Trzy wektory w przestrzeni trójwymiarowej są sprzężone względem macierzy jednostkowej, jeżeli każdy z nich jest prostopadły do płaszczyzny wyznaczonej przez dwa pozostałe. Zatem już w przestrzeni trójwymiarowej sprzężenie względem macierzy jednostkowej to inna własność niż prostopadłość.

Jeśli jakieś wektory są sprzężone, to oznacza to ich specjalne położenie względem pewnej funkcji kwadratowej. W naszym przypadku tą funkcją będzie funkcja błędu. Warunek z powyższej definicji w przypadku sieci neuronowych wygląda następująco:

$$\boldsymbol{p}'(t-1)\boldsymbol{H}\boldsymbol{p}(t) = 0.$$

Po uwzględnieniu warunków ortogonalności i warunków sprzężenia względem hesjanu $\boldsymbol{H}$ otrzymujemy

$$\boldsymbol{p}(t) = -\boldsymbol{g}(\boldsymbol{w}(t)) + \beta_t \boldsymbol{p}(t-1),$$

gdzie współczynnik β_t może być liczony według jednej z trzech formuł. Pierwsza z nich zwana *formułą Fletchera–Reevesa* ma postać

$$\beta_t = \frac{\boldsymbol{g}'(\boldsymbol{w}(t))\boldsymbol{g}(\boldsymbol{w}(t))}{\boldsymbol{g}'(\boldsymbol{w}(t-1))\boldsymbol{g}(\boldsymbol{w}(t-1))}.$$

Druga zwana *formułą Hestenesa–Stiefeljesta* ma postać

$$\beta_t = \frac{\boldsymbol{g}'(\boldsymbol{w}(t))(\boldsymbol{g}(\boldsymbol{w}(t)) - \boldsymbol{g}(\boldsymbol{w}(t-1))}{\boldsymbol{p}'(\boldsymbol{w}(t-1))(\boldsymbol{g}(\boldsymbol{w}(t)) - \boldsymbol{g}(\boldsymbol{w}(t-1))}.$$

Najczęściej jest jednak wykorzystywana *formuła Polaka–Ribiere'a*, która jest najbardziej odporna na niekwadratowe funkcje błędu oraz znajduje najczęściej rozwiązanie w mniejszej liczbie kroków niż dwie poprzednie metody (wykazały to badania praktyczne, niestety, nie ma żadnego formalnego uzasadnienia obu faktów, Powell (1977) sugeruje, że algorytm ten ma właściwość samoodnowy, w sytuacji gdy nastąpiło ugrzęźnięcie) i ma postać

$$\beta_t = \frac{\boldsymbol{g}'(\boldsymbol{w}(t))(\boldsymbol{g}(\boldsymbol{w}(t)) - \boldsymbol{g}(\boldsymbol{w}(t-1))}{\boldsymbol{g}'(\boldsymbol{w}(t-1))\boldsymbol{g}(\boldsymbol{w}(t-1))}.$$

Wybierając nowy kierunek według reguły gradientów sprzężonych, zapewniamy sobie to, że dla punktów leżących wzdłuż nowego kierunku, kierunek najszybszego spadku wartości funkcji będzie zawsze prostopadły do kierunku poprzedniego. Dzięki temu przy dalszych poszukiwaniach nie zagubimy minimum osiągniętego względem poprzedniego kierunku. Po wyznaczeniu nowego kierunku cały proces się powtarza: wyznaczamy minimum funkcji wzdłuż tego kierunku i określamy nowy kierunek poszukiwań, sprzężony z poprzednim.

Jeśli minimalizowana funkcja błędu jest wyznaczona w przestrzeni n-wymiarowej, to wystarczy n powtórzeń, aby osiągnąć minimum. Znalezienie minimum jest gwarantowane przy założeniu, że funkcja jest paraboloidalna.

Jak widać, sprzężony algorytm gradientowy jest analogiczny do zwykłych metod gradientowych z członem momentum, gdzie rolę momentum odgrywa β_t.

W rzeczywistości na skomplikowanej powierzchni funkcji błędu, dokładne sprzężenie kierunków poszukiwań zdecydowanie zawodzi, co powoduje, że dla odnalezienia minimum funkcji n zmiennych potrzeba więcej niż n iteracji. Jednak i tak w typowych przypadkach metoda gradientów sprzężonych wymaga znacznie mniejszej liczby powtórzeń niż typowo stosowana metoda wstecznej propagacji błędów. Ze względu na kumulację błędów zaokrągleń, metoda zatraca własność ortogonalności między wektorami, zatem co określoną ilość iteracji należy przeprowadzać jej ponowny start (proces ten jest nazywany odnową). Efektywność tej metody jest słabsza niż efektywność metody zmiennej metryki, jednakże jej prostota i niewielkie wymagania pamięciowe czynią ją najlepszą metodą optymalizacji funkcji o dużej ilości zmiennych.

Przykład 7.9.

Ponownie rozpatrzmy zagadnienie minimalizacji funkcji z przykładu 7.7. W pierwszym kroku metody gradientów sprzężonych należy poszukać współczynnika uczenia w celu minimalizacji wzdłuż kierunku gradientu. Otrzymujemy

$$\alpha(0) = -\frac{\boldsymbol{p}'(0)\boldsymbol{g}(0)}{\boldsymbol{p}'(0)\boldsymbol{H}\boldsymbol{p}(0)} = -\frac{\begin{bmatrix} -3 & -3 \end{bmatrix}\begin{bmatrix} 3 \\ 3 \end{bmatrix}}{\begin{bmatrix} -3 & -3 \end{bmatrix}\begin{bmatrix} 2 & 2 \\ 2 & 4 \end{bmatrix}\begin{bmatrix} -3 \\ -3 \end{bmatrix}} = 0.2,$$

$$\boldsymbol{x}(1) = \boldsymbol{x}(0) + \alpha_0\boldsymbol{p}(0) = \begin{bmatrix} 0.5 \\ 0.5 \end{bmatrix} + 0.2\begin{bmatrix} -3 \\ -3 \end{bmatrix} = \begin{bmatrix} -0.1 \\ -0.1 \end{bmatrix}.$$

Teraz możemy poszukać wektora sprzężonego. Mamy

$$\boldsymbol{g}(1) = \nabla F(\boldsymbol{x})|_{\boldsymbol{x}=\boldsymbol{x}(1)} = \begin{bmatrix} 0.6 \\ -0.6 \end{bmatrix},$$

$$\beta_1 = \frac{\boldsymbol{g}'(1)\boldsymbol{g}(1)}{\boldsymbol{g}'(0)\boldsymbol{g}(0)} = 0.04,$$

$$\boldsymbol{p}(1) = -\boldsymbol{g}(1) + \beta_1\boldsymbol{p}(0) = \begin{bmatrix} -0.72 \\ 0.48 \end{bmatrix},$$

$$\alpha(1) = -\frac{\boldsymbol{p}'(1)\boldsymbol{g}(1)}{\boldsymbol{p}'(1)\boldsymbol{H}\boldsymbol{p}(1)} = -\frac{\begin{bmatrix} -0.72 & 0.48 \end{bmatrix}\begin{bmatrix} 0.6 \\ -0.6 \end{bmatrix}}{\begin{bmatrix} -0.72 & 0.48 \end{bmatrix}\begin{bmatrix} 2 & 2 \\ 2 & 4 \end{bmatrix}\begin{bmatrix} -0.72 \\ 0.48 \end{bmatrix}} = 1.25,$$

$$\boldsymbol{x}(2) = \boldsymbol{x}(1) + \alpha_1\boldsymbol{p}(1) = \begin{bmatrix} -0.1 \\ -0.1 \end{bmatrix} + 1.25\begin{bmatrix} -0.72 \\ 0.48 \end{bmatrix} = \begin{bmatrix} -1 \\ 0.5 \end{bmatrix}.$$

Zgodnie z oczekiwaniami wynik otrzymaliśmy w dwóch krokach. Warto jeszcze zauważyć, że w przypadku stosowania cały czas metody przeszukiwania liniowego wynik otrzymujemy w 14 krokach (przy czym bardzo blisko rozwiązania jesteśmy już po pięciu krokach) (rys. 7.20).

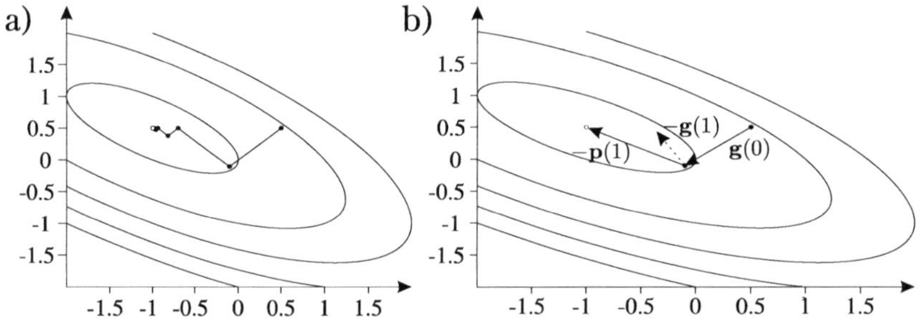

Rys. 7.20. a) Przebieg algorytmu przeszukiwania liniowego, b) oraz gradientów sprzężonych

7.3.8. Algorytm Levenberga–Marquardta

Rozwiązaniem problemów występujących w metodzie BP może być zastosowanie metody Levenberga–Marquardta (Levenberg (1944) oraz Marquardt (1963)). W pobliżu minimum ma ona cechy metody Newtona, zaś wraz z oddalaniem się punktu startowego od minimum sposób jej działania zbliża się do metody najszybszego spadku. W zapisie formalnym różnica między formułą Newtona i Levenberga–Marquardta jest niewielka – polega wyłącznie na tym, że w tej drugiej do hesjanu jest dodawana macierz jednostkowa pomnożona przez pewien dodatni współczynnik μ zwany *parametrem Levenberga–Marquardta* (parametr regularyzacji), którego wartość jest modyfikowana w trakcie uczenia. Jeśli wartość tego współczynnika jest bliska zeru, to wektor wag będzie bardzo zbliżony do wartości wyznaczonych przy zastosowaniu metody Newtona. Zwiększaniu wartości parametru μ towarzyszy wzrost znaczenia kierunku poprawy wyznaczonego na podstawie gradientu funkcji błędu. Zauważmy, że parametr ten wyznacza dodatkowo wielkość kroku w kierunku poprawy bazującym na gradiencie – im większa wartość μ, tym mniejszy krok.

W metodzie tej wykorzystuje się rozwinięcie funkcji błędu do trzeciego składnika. Kierunek minimalizacji jest identyczny jak w metodzie zmiennej metryki, czyli ma postać (7.39). Jeśli przyjmiemy, że wektor wyjściowy ma M składowych, to miarę błędu wyraża

$$Q(\boldsymbol{w}(t)) = \frac{1}{2}\sum_{i=1}^{M} e_i^2(\boldsymbol{w}(t)), \tag{7.42}$$

gdzie $e_i(t) = d_i(t) - y_i(t)$.

Zarówno gradient, jak i przybliżoną macierz hesjanu $\boldsymbol{G}(\boldsymbol{w}(t))$ wyznaczamy na podstawie jakobianu $\boldsymbol{J}(\boldsymbol{w}(t)) = \left(\frac{\partial e}{\partial w_i}\right)(\boldsymbol{w})(t)$. Otrzymujemy

$$\boldsymbol{g}(\boldsymbol{w}(t)) = \boldsymbol{J}'(\boldsymbol{w}(t))\boldsymbol{e}(\boldsymbol{w}(t)), \tag{7.43}$$

$$\boldsymbol{G}(\boldsymbol{w}(t)) = \boldsymbol{J}'(\boldsymbol{w}(t))\boldsymbol{J}(\boldsymbol{w})(t) + \boldsymbol{S}(\boldsymbol{w}(t)). \tag{7.44}$$

Składnik $\boldsymbol{S}(\boldsymbol{w}(t))$ odpowiada składnikom hesjanu, zawierającym pochodne wyższych rzędów. Składnik ten jest aproksymowany jako

$$\boldsymbol{S}(\boldsymbol{w}(t)) = \mu \boldsymbol{I}, \tag{7.45}$$

gdzie μ jest omawianym wcześniej parametrem regularyzacji. Jego wartość jest dobierana arbitralnie i zmniejszana w trakcie uczenia do zera w okolicy rozwiązania minimalnego. Na początku działania algorytmu, gdy wartość μ jest duża, poszukiwanie odbywa się zgodnie z metodą największego spadku

$$\boldsymbol{p}(t) = -\frac{\boldsymbol{g}(\boldsymbol{w}(t))}{\mu}.$$

W miarę zbliżania się do rozwiązania wartość parametru Levenberga–Marquardta jest zmniejszana i pierwszy składnik wzoru na przybliżony hesjan zaczyna odgrywać coraz większą rolę.

W typowych przypadkach algorytm Levenberga–Marquardta jest najszybszym algorytmem uczenia sieci jednokierunkowych. Niestety, ma on trzy istotne ograniczenia:

- algorytm ten może być użyty wyłącznie dla sieci z pojedynczym neuronem wyjściowym,
- funkcja błędu musi być w postaci sumy kwadratów odchyleń (z tego względu zazwyczaj zalecany jest tylko do zagadnień regresyjnych),
- jego wymagania w zakresie pamięci są proporcjonalne do kwadratu liczby parametrów, co powoduje, że jest on niepraktyczny w przypadku dużych sieci.

Powyższe ograniczenia powodują, że metoda Levenberga–Marquardta nie jest używana aż tak często, jak by na to (pozornie) zasługiwała. Metoda gradientów sprzężonych jest prawie tak samo dobra, a nie ma wskazanych ograniczeń.

7.3.9. Rekurencyjna metoda najmniejszych kwadratów

Opisywane wyżej metody każdorazowo odwoływały się do pojęcia wyboru kierunku zmian wag gwarantującego szybki spadek funkcji błędu. Czasami warto jednak sięgnąć do metod, które tego elementu w sposób jawny nie zawierają, a mimo to (a często właśnie dlatego) umożliwiają w miarę szybki postęp procesu uczenia. Wśród tych metod na uwagę z pewnością zasługuje rekurencyjna metoda najmniejszych kwadratów[34]. W przypadku tej metody jako miarę błędu przyjmuje się

$$Q(t) = \sum_{l=1}^{t} \gamma^{t-l} \sum_{j=1}^{N_L} \varepsilon_j^{(L)^2}(l), \tag{7.46}$$

gdzie $\gamma \in (0,\ 1]$ jest współczynnikiem zapominania. Zauważmy, że występowanie czynnika γ^{t-l} sprawia, że na wartość funkcji $Q(t)$ większy wpływ mają błędy uzyskane w późniejszych krokach algorytmu. Warunkiem minimalizacji formuły (7.46) jest zerowanie się pierwszych pochodnych funkcji $Q(t)$ liczonych względem wag neuronów poszczególnych warstw, czyli

$$\frac{\partial Q(t)}{\partial \boldsymbol{w}_i^{(k)}(t)} = \boldsymbol{0}. \tag{7.47}$$

Po dokonaniu odpowiednich przekształceń równanie normalne zapisać można w postaci

$$\boldsymbol{r}_i^{(k)}(t) = \boldsymbol{R}_i^{(k)}(t)\boldsymbol{w}_i^{(k)}(t), \tag{7.48}$$

[34] Po ang. Recursive Least Squares (RLS).

gdzie

$$\boldsymbol{R}_i^{(k)}(t) = \sum_{l=1}^{t} \gamma^{t-l} f'^2 \left(s_i^{(k)}(l) \right) \boldsymbol{x}^{(k)}(l)(\boldsymbol{x}^{(k)}(l))'.$$

Równanie (7.48) można rozwiązać w sposób rekurencyjny, bez konieczności odwracania macierzy $\boldsymbol{R}_i^{(k)}(t)$. Wymaga to zastosowania algorytmu RLS. W rezultacie adaptacyjna korekta jest przeprowadzana następująco:

$$\boldsymbol{w}_i^{(k)}(t) = \boldsymbol{w}_i^{(k)}(t-1) + \boldsymbol{g}_i^{(k)}(t)\varepsilon_i^{(k)}(t), \quad i = 1, \ldots, N_k, \quad k = 1, \ldots, L, \tag{7.49}$$

$$\varepsilon_i^{(k)}(t) = \begin{cases} d_i^{(L)}(t) - y_i^{(L)}(t), & \text{dla } k = L, \\ \sum\limits_{j=1}^{N_{k+1}} f'\left(s_j^{(k+1)}(t) \right) w_{ji}^{(k+1)}(t)\varepsilon_j^{(k+1)}(t), & \text{dla } k = 1, \ldots, L-1, \end{cases} \tag{7.50}$$

$$\boldsymbol{g}_i^{(k)}(t) = \frac{f'\left(s_i^{(k)}(t) \right) \boldsymbol{P}_i^{(k)}(t-1)\boldsymbol{x}^{(k)}(t)}{\gamma + f'^2\left(s_i^{(k)}(t) \right) (\boldsymbol{x}^{(k)}(t))'\boldsymbol{P}_i^{(k)}(t-1)\boldsymbol{x}^{(k)}(t)}, \tag{7.51}$$

$$\boldsymbol{P}_i^{(k)}(t) = \gamma^{-1} \left[\boldsymbol{I} - f'\left(s_i^{(k)}(t) \right) \boldsymbol{g}_i^{(k)}(t)(\boldsymbol{x}^{(k)}(t))' \right] \boldsymbol{P}_i^{(k)}(t-1). \tag{7.52}$$

Przy prowadzeniu iteracyjnych obliczeń zmierzających do ustalenia wartości wag przyjmuje się następujące warunki początkowe:

$$\boldsymbol{P}^{(k)}(0) = \delta \boldsymbol{I}, \qquad \delta \gg 0, \qquad \boldsymbol{w}_i^{(k)}(0) = 0.$$

Można również wyznaczyć początkowe wagi sieci w sposób losowy.

7.3.10. Algorytm symulowanego wyżarzania

Stosując powyższe metody uczenia sieci neuronowych, nie unikamy, niestety, problemu minimów lokalnych, które są powiązane z punktami startowymi algorytmu. Dzieje się tak zwłaszcza wtedy, gdy rośnie wymiar danych lub złożoność metody. W takich przypadkach przeszukiwanie zbioru wszystkich możliwych konfiguracji jest zupełnie nieefektywne i nierealne z obliczeniowego punktu widzenia. W takich przypadkach na pomoc przychodzą nam metody, w których istotną rolę odgrywa losowość. Jedną z metod tego typu jest *algorytm symulowanego wyżarzania*[35]. Termin pochodzi z pracy Metropolisa i in. (1953). Wyżarzanie jest pojęciem metalurgicznym. W sytuacji

[35] Po ang. simulated annealing.

gdy atomy w kawałku metalu są rozmieszczone przypadkowo, metal jest kruchy i łamliwy. W procesie wyżarzania metal podgrzewa się do wysokiej temperatury, powodując gwałtowne wstrząsanie atomami. Jeżeli będzie on nagle ochłodzony, to mikrostruktura zostanie uwięziona w niestabilnym stanie przypadkowym. Jeśli jednak ochładzanie będzie powolne, to wraz ze spadkiem temperatury atomy będą tworzyły regularne układy, dość stabilne w danej temperaturze. Jeśli spadek temperatury jest dostatecznie wolny, to w rezultacie metal osiągnie stabilną, uporządkowaną strukturę. Ten stan odpowiada globalnemu minimum energii. W procesie może wystąpić chwilowy wzrost energii, który umożliwia uniknięcie minimów lokalnych (w algorytmie reprezentowany przez losowy szum). Algorytm symulowanego wyżarzania przedstawia się następująco (k oznacza numer iteracji):

Algorytm symulowanego wyżarzania

1. Losujemy wagi sieci, ustalamy wysoką temperaturę początkową T_0 (oczywiście jest to tylko parametr procedury i nie ma nic wspólnego z prawdziwą fizyczną temperaturą) oraz końcową (może to być nawet zero absolutne).
2. Losujemy węzeł sieci i dla niego nową wagę w^*.
3. Jeżeli $E(w^*) < E(w)$, to $w = w^*$, w przeciwnym razie $w = w^*$ z prawdopodobieństwem $\exp(-\frac{\Delta E}{T_k})$ (losowy szum).
4. Zmniejszamy temperaturę zgodnie z jednym z trzech schematów[36]:
   - $T_k := \dfrac{T_0}{1 + \ln k}$ – schemat logarytmiczny (Boltzmanna),
   - $T_k := \dfrac{T_0}{k}$ – schemat liniowy (Cauchy'ego),
   - $T_k := T_0 a^k$, $0 < a < 1$ – schemat geometryczny.
5. Powtarzamy kroki 2–4, aż do osiągnięcia temperatury końcowej.

Losowanie wag może się odbywać pojedynczo lub jest losowany cały zestaw wag dla sieci i jest obliczona zmiana energii. Algorytm ten napotyka jednak również wiele problemów praktycznych związanych z doborem parametrów startowych i schematu redukcji temperatury. Użyteczne w wyborze schematu chłodzenia, z punktu widzenia teoretycznego, może się okazać twierdzenie:

[36] T_k jest nazywane schematem wyżarzania lub chłodzenia (po ang. annealing schedule lub cooling schedule). Ogólnie za ten schemat często przyjmuje się $T_{k+1} = cT_k$, gdzie $0 < c < 1$. Oznacza to oczywiście stałą wartość kroku chłodzenia, najczęściej wybiera się wartość $0.8 < c < 0.99$.

TWIERDZENIE 7.5. (Gemana i Gemana, 1984)

Jeśli temperatura początkowa jest dostatecznie wysoka oraz $T_k \geq \frac{T_0}{1 + \ln k}$ *dla każdego* k*, to algorytm symulowanego wyżarzania minimalizuje funkcję energii z prawdopodobieństwem* 1.

Powyższe twierdzenie sugeruje schemat logarytmiczny, nie ma on jednak zastosowania praktycznego, gdyż zbyt wolno prowadzi do zadowalających wyników. Z badań empirycznych wynika, że największą przydatność praktyczną ma schemat geometryczny, który jest najszybszy.

7.3.11. Algorytmy ewolucyjne i genetyczne

Inną metodą uniknięcia minimów lokalnych jest wykorzystanie algorytmów genetycznych i ewolucyjnych[37]. *Algorytmy ewolucyjne* próbują naśladować proces ewolucji, ich szczególnym przypadkiem są *algorytmy genetyczne*. Idea algorytmów genetycznych jest inspirowana procesami, obserwowanymi w przyrodzie, którymi są selekcja osobników i ewolucja gatunków, mechanizmy rozmnażania i związane z nimi dziedziczenie cech. W wyniku działania mechanizmów naturalnych powstają nowe gatunki, lepiej dostosowane do środowiska, w którym żyją, wypierając gatunki gorzej dostosowane. Co więcej, mechanizmy te umożliwiają dostosowanie się do nowych, zmiennych warunków życia, a więc są adaptacyjne. Algorytmy genetyczne, realizowane w postaci programów, polegają na generowaniu populacji pewnych obiektów, które dopasowują się do danych wymagań. Postępując w ten sposób, można na przykład znaleźć ekstremum globalne funkcji. Są one algorytmami łatwymi do implementacji i uniwersalnymi. Algorytm genetyczny charakteryzuje się uniwersalnym podejściem do wszystkich rozwiązywanych problemów. W celu właściwego ukierunkowania poszukiwania, realizowanego przez algorytm genetyczny, wprowadzono trzy operacje genetyczne: *reprodukcję*, *krzyżowanie* i *mutację*. Każdy problem rozwiązuje się za pomocą algorytmu genetycznego w taki sam sposób, używając tych samych operatorów. Operator reprodukcji zapewnia ciągłość informacji opisującej dane zagadnienie, przekazywanej z pokolenia na pokolenie. Operator krzyżowania ma za zadanie łączyć w różnych kombinacjach cechy pochodzące z różnych osobników populacji, zaś operator mutacji ma za zadanie zwiększać różno-

[37]Przez algorytmy ewolucyjne będziemy rozumieć rodzinę probabilistycznych algorytmów optymalizacyjnych wzorowanych na zasadach naturalnej ewolucji.

rodność tych osobników, czyli zapobiec przedwczesnej zbieżności algorytmu. Tak zdefiniowany algorytm genetyczny, w połączeniu z odpowiednim sposobem kodowania ma wiele wspólnych cech z procesem naturalnej ewolucji: wynikiem jego działania jest populacja osobników coraz lepiej dostosowanych do „środowiska" (tzn. mających coraz wyższą funkcję celu), mimo że modyfikacje osobników mają charakter losowy. Metoda ta przy tym okazuje się zaskakująco skuteczna w wielu praktycznych zadaniach, dla których nie znamy rozwiązań dokładnych (chociaż, co należy podkreślić, daje ona zwykle jedynie rozwiązania przybliżone).

Przykład 7.10.
Najbardziej znanym algorytmem genetycznym jest jego „klasyczna postać binarna". Kodowanie przeprowadza daną przestrzeń S na zbiór $\Omega = \{0,1\}^n$, a więc chromosomy są n-elementowymi ciągami binarnymi (pojedynczy element chromosomu jest nazywany genem). Operator mutacji polega na zastosowaniu na osobnikach populacji pewnej liczby operatorów v_{mut} zmieniających losowy gen losowego osobnika na przeciwny (w poniższym przykładzie mutowany gen oznaczono kreską)

$$v_{mut}(101\bar{1}1001) = 101\bar{0}1001.$$

Mutacja poszczególnych genów w populacji jest niezależna i zachodzi z pewnym, zwykle niewielkim (np. 0.01) prawdopodobieństwem. Krzyżowanie jest operatorem polegającym na zastosowaniu na pewnej liczbie par osobników operatora v_{cross} wymieniającego odcinek chromosomu między jednym osobnikiem a drugim

$$v_{cross}\begin{pmatrix} 101|11001 \\ 001|10100 \end{pmatrix} = \begin{pmatrix} 10110100 \\ 00111001 \end{pmatrix}.$$

W powyższym przykładzie wartości genów dwóch osobników rodzicielskich i dwóch potomnych są wypisane w poziomych wierszach. Początek wymienianego odcinka (oznaczony pionową kreską) jest ustalany drogą losowania. Prawdopodobieństwo, że dany osobnik zostanie skrzyżowany z innym jest ustalone na pewnym poziomie, najczęściej wynosi ono 0.5. Nowa populacja jest tworzona za pomocą operatora selekcji, faworyzującego osobniki o wysokiej wartości funkcji celu. W tym przypadku jest używany algorytm „koła ruletki" – wartości funkcji celu osobników są normalizowane i traktowane jako rozkład prawdopodobieństwa na zbiorze osobników. Następnie jest losowana populacja nowych osobników (ze zwracaniem), zgodnie z tym rozkładem.

Algorytmy ewolucyjne różnią się od algorytmów genetycznych przede wszystkim inną koncepcją mutacji (parametr charakteryzujący mutację podlega również ewolucji) oraz tym, że operator krzyżowania jest operatorem ubocznym i często nie jest wykorzystywany. Zainteresowanych tematyką algorytmów ewolucyjnych, a w szczególności genetycznych do uczenia sieci neuronowych odsyłamy do książki Rutkowskiej, Pilińskiego i Rutkowskiego (1999).

7.4. Sieci rekurencyjne

Dotychczas opisywane sieci charakteryzowały się jednokierunkowym przepływem sygnałów od wejścia do wyjścia. W sieciach rekurencyjnych istnieją sprzężenia zwrotne między wejściem a wyjściem. Zależności dynamiczne jakie panują w sieci są widoczne na każdym etapie działania. Zmiana stanu jednego neuronu przenosi się przez masowe sprzężenie zwrotne na całą sieć, wywołując stan przejściowy, kończący się określonym stanem ustalonym, na ogół innym niż stan poprzedni. Przy masowym sprzężeniu zwrotnym pobudzeniami dla neuronów są sygnały wejściowe innych neuronów. Proces ustalania się sygnałów wyjściowych sieci rekurencyjnych jest procesem dynamicznym. Biorąc pod uwagę nieliniowość funkcji aktywacji neuronów, jest to dynamika nieliniowa, stanowiąca istotną cechę tego rodzaju sieci.

7.4.1. Sieci Hopfielda

Jednym z podstawowych przedstawicieli sieci rekurencyjnych jest *sieć Hopfielda* (ze względu na pełnioną funkcję często nazywana *pamięcią autoasocjacyjną*). Została ona zaprezentowana w pracy Hopfielda z 1982 roku i była ona punktem zwrotnym w badaniach nad sieciami neuronowymi i przyczyniła się do intensywnych prac badawczych w tej dziedzinie. Sieć Hopfielda jest często podawana jako przykład pamięci skojarzeniowej lub jako układ do rozwiązywania zadań z zakresu optymalizacji. Strukturę sieci Hopfielda można opisać bardzo prosto – jest to układ wielu identycznych elementów połączonych metodą każdy z każdym. Jest zatem najczęściej rozpatrywana jako struktura jednowarstwowa. W odróżnieniu od sieci warstwowych typu perceptron sieć Hopfielda jest siecią rekurencyjną, gdzie neurony są wielokrotnie pobudzane w jednym cyklu rozpoznawania, co uzyskuje się przez pętle sprzężenia zwrotnego (charakterystyczne jest, że nie występuje sprzężenie zwrotne neuronu z samym sobą – waga $w_{ii} = 0$). Wagi w tego typu sieci są symetryczne, czyli $w_{ij} = w_{ji}$. Założenie o symetrii wag sprawia, że sieć ta zawsze się stabilizuje (czyli dla pewnego t zachodzi warunek $y_k(t+1) = y_k(t)$, dla każdego neuronu). Symetria jest wygodna z teoretycznego punktu widzenia, pozwala wprowadzić funkcję energii, jest natomiast nierealistyczna z punktu widzenia biologicznego. Jeśli założymy, że funkcja aktywacji ma postać symetrycznej funkcji liniowej z nasyceniem, to działanie tego typu sieci w kroku t można opisać wzorem

$$y_k(t+1) = \text{satlins}\left(\sum_{j=1,\ j\neq k}^{p} w_{kj} y_j(t) + \theta_k\right),$$

przy czym $y_j(0) = x_j$. Sygnał na wyjściu będzie się zmieniał tak długo, aż $y_k(t) = y_k(t-1)$ dla wszystkich neuronów sieci. Stan energetyczny sieci można wyrazić za pomocą funkcji Lapunowa (w fizyce statystycznej funkcja Hamiltona, w teorii optymalizacji funkcja celu lub kosztu, w obliczeniach ewolucyjnych funkcja przystosowania) postaci

$$E(\boldsymbol{W}) = -\frac{1}{2}\sum_{j}\sum_{k\neq j} y_j y_k w_{kj} - \sum_{k} \theta_k y_k. \tag{7.53}$$

Funkcja energii jest ograniczona od dołu. Można pokazać, że zmiana energii

$$\Delta E(\boldsymbol{W}) = -\Delta y_k \left(\sum_{j\neq k} y_j w_{kj} + \theta_k\right)$$

jest zawsze ujemna. Stąd funkcja energii jest funkcją malejącą w kolejnych krokach. Dlatego jesteśmy pewni, że osiągnie ona minimum, w którym sieć się ustabilizuje. Na rysunku 7.21 są przedstawione obszary przyciągania sieci Hopfielda.

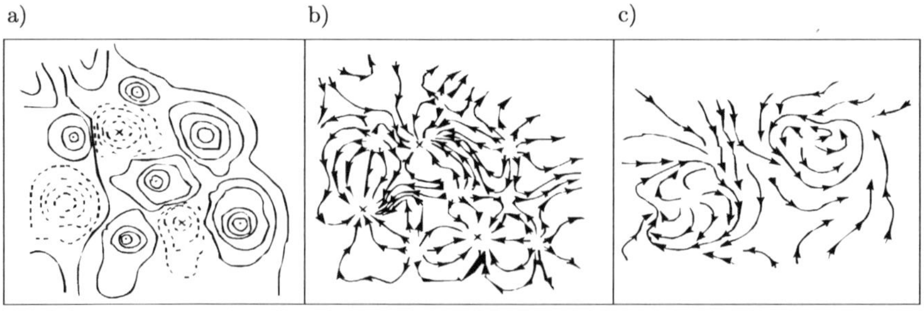

Rys. 7.21. Obszary przyciągania sieci Hopfielda: a) krajobraz energetyczny sieci, linie ciągłe oznaczają wzniesienie, przerywane dolinę, b) możliwe linie trajektorii sieci, c) bardziej skomplikowana dynamika działania sieci

Uczenie sieci tego typu odbywa się najczęściej za pomocą uogólnionej reguły Hebba postaci

$$w_{kj} = \frac{1}{n} \sum_{i=1}^{K} x_{ik} x_{ij}, \tag{7.54}$$

która nie daje, niestety, oczekiwanych rezultatów (pojemność sieci – maksymalna liczba wzorców zapamiętanych i odtworzonych z pewnym akceptowalnym błędem 1% – wynosi jedynie 13.8% liczby neuronów). Z tego powodu w praktyce używa się metody *pseudoinwersji*. W metodzie tej zakłada się, że celem uczenia jest taki dobór wag, aby na wyjściu otrzymać taki sam sygnał jak na wejściu, czyli

$$\boldsymbol{W}\boldsymbol{X} = \boldsymbol{X},$$

co prowadzi do rozwiązania

$$\boldsymbol{W} = \boldsymbol{X}\boldsymbol{X}^{+},$$

gdzie $\boldsymbol{X}^{+}$ oznacza odwrotność Moore'a–Penrose'a.

Definicja 7.4.
Uogólniona odwrotność $\boldsymbol{A}^{+}$ macierzy $\boldsymbol{A}$ nazywa się *MP-odwrotnością* (*Moore'a–Penrose'a odwrotnością*) wtedy i tylko wtedy, gdy zachodzą następujące warunki:

$$\begin{aligned} \boldsymbol{A}\boldsymbol{A}^{+}\boldsymbol{A} &= \boldsymbol{A}, \\ \boldsymbol{A}^{+}\boldsymbol{A}\boldsymbol{A}^{+} &= \boldsymbol{A}^{+}, \\ (\boldsymbol{A}\boldsymbol{A}^{+})' &= \boldsymbol{A}\boldsymbol{A}^{+}, \\ (\boldsymbol{A}^{+}\boldsymbol{A})' &= \boldsymbol{A}^{+}\boldsymbol{A}. \end{aligned}$$

Sieć uczona tą metodą ma pojemność n.

Przykład 7.11.
Przeprowadzimy teraz próbę nauczenia sieci Hopfielda rozpoznawania trzech symboli. Za pomocą sieci złożonej z 9 neuronów rozpoznajemy obrazek złożony z 9 pikseli. Czarny piksel jest reprezentowany jako 1, natomiast biały jako -1. W ten sposób otrzymaliśmy 9-elementowy wektor testowy. Za pomocą reguły (7.54) nauczyliśmy sieć 3 kształtów, z których 2 przypominają kółko i krzyżyk, a jako trzeci wektor wzorcowy podaliśmy same -1, co odpowiada pustemu kwadratowi. Wzorce są zobrazowane na rys. 7.22.

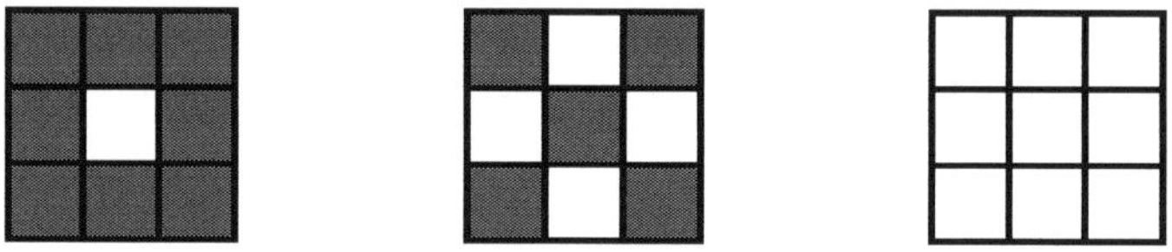

Rys. 7.22. Wzorce uczące do przykładu 7.11

W ten sposób otrzymujemy następujące dane uczące:

$$\boldsymbol{x}_1 = [1,\ 1,\ 1,\ 1,\ -1,\ 1,\ 1,\ 1,\ 1]',$$

$$\boldsymbol{x}_2 = [1,\ -1,\ 1,\ -1,\ 1,\ -1,\ 1,\ -1,\ 1]',$$

$$\boldsymbol{x}_3 = [-1,\ -1,\ -1,\ -1,\ -1,\ -1,\ -1,\ -1,\ -1]'.$$

Następnie uczymy sieć Hopfielda za pomocą powyższych wzorców. Okazuje się, że gdy na wejście sieci podamy wektor identyczny z wzorcem, wówczas sieć nie zmieni swojego stanu. Oznacza to, że sieć idealnie rozpoznaje dane niezaburzone. Teraz zaburzamy dane wejściowe (rys. 7.23).

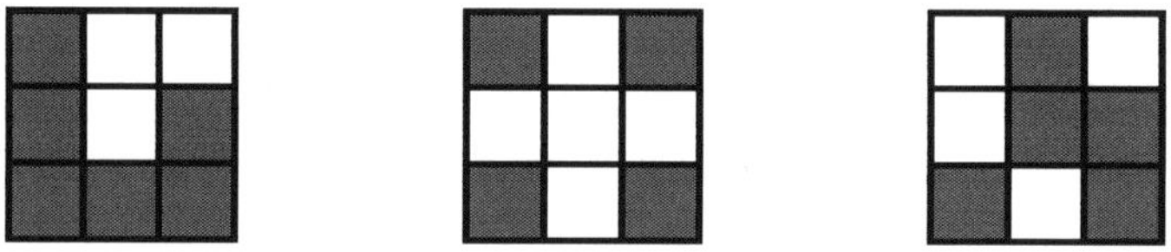

Rys. 7.23. Zaburzone dane do przykładu 7.11

Zaburzone dane mają więc postać:

$$\tilde{\boldsymbol{x}}_1 = [1,\ -1,\ -1,\ 1,\ -1,\ 1,\ 1,\ 1,\ 1]',$$

$$\tilde{\boldsymbol{x}}_2 = [1,\ -1,\ 1,\ -1,\ -1,\ -1,\ 1,\ -1,\ 1]',$$

$$\tilde{\boldsymbol{x}}_3 = [-1,\ 1,\ -1,\ -1,\ 1,\ 1,\ 1,\ -1,\ 1]'.$$

Dwa pierwsze zaburzone sygnały zostały poprawnie rozpoznane po 2 iteracjach, natomiast dla sygnału trzeciego wyjście sieci wciąż było nieustabilizowane po 20 iteracjach, co świadczy, że sieć nie mogła rozpoznać zniekształconego sygnału.

Przykład 7.12.
Wróćmy teraz do przykładu 7.3. Spróbujemy go rozwiązać za pomocą sieci Hopfielda (lekko zmodyfikowanej, bo ze sprzężeniem zwrotnym obejmującym ten sam neuron – bardzo podobna sieć do omówionej później sieci MAXNET, przy czym waga sprzężenia ze samym sobą nie musi być równa jeden). Sieć

Hopfielda składa się w tym przykładzie z 3 neuronów i posłuży do rozpoznawania dwóch typów obiektów. W tym przypadku po zastosowaniu metody pseudoinwersji dla macierzy

$$\boldsymbol{X} = \begin{bmatrix} 1 & 1 \\ -1 & 1 \\ -1 & -1 \end{bmatrix}$$

otrzymano macierz wag postaci

$$\boldsymbol{W} = \begin{bmatrix} 0.5 & 0 & -0.5 \\ 0 & 1 & 0 \\ -0.5 & 0 & 0.5 \end{bmatrix}.$$

Wektor progowy ustalamy arbitralnie, sensowny wydaje się np. wektor

$$\boldsymbol{\theta} = \begin{bmatrix} 0.9 \\ 0 \\ -0.9 \end{bmatrix}.$$

Tak dobrany próg spowoduje, że pierwsza składowa wektora wynikowego będzie szybko dążyć do wartości 1, a trzecia do wartości -1 (są one wspólne dla jabłek i pomarańczy). Natomiast druga składowa wektora wyjściowego będzie jedynie zależna od wejścia i macierzy wag, która jednak nie zmienia wejścia. Poszczególne składowe wektora wynikowego zmieniają się zatem według następujących formuł:

$$\begin{aligned} y_1(t+1) &= \operatorname{satlins}(0.5y_1(t) - 0.5y_3(t) + 0.9), \\ y_2(t+1) &= \operatorname{satlins}(y_2(t)), \\ y_3(t+1) &= \operatorname{satlins}(-0.5y_1(t) + 0.5y_3(t) - 0.9). \end{aligned}$$

Wektory wzorcowe sieć klasyfikuje poprawnie. Natomiast dla wektora $[-1,\ -1,\ -1]'$ tak skonstruowana sieć zwraca następujące elementy na wyjściu w kolejnych krokach

$$\boldsymbol{y}(0) = \begin{bmatrix} -1 \\ -1 \\ -1 \end{bmatrix}, \quad \boldsymbol{y}(1) = \begin{bmatrix} 0.9 \\ -1 \\ -0.9 \end{bmatrix}, \quad \boldsymbol{y}(2) = \begin{bmatrix} 1 \\ -1 \\ -1 \end{bmatrix}, \quad \boldsymbol{y}(3) = \begin{bmatrix} 1 \\ -1 \\ -1 \end{bmatrix}.$$

Czyli podobnie jak perceptron sieć Hopfielda zaklasyfikowała obiekt jako pomarańczę. W przeciwieństwie jednak do perceptronu, który jako wynik podawał etykietę klasy, sieć Hopfielda dostarcza wzorzec wynikowy, który jest najbardziej podobny do klasyfikowanego obiektu.

7.4.2. Sieci Hamminga

Sieć Hamminga jest trójwarstwową strukturą rekurencyjną, stanowiącą rozwinięcie sieci Hopfielda i jest wyspecjalizowaną strukturą pamięci heteroasocjacyjnej (kojarzenie dwóch różnych wektorów). Została ona zaproponowana przez Lippmanna w 1987 roku. Jej działanie polega na wykorzystaniu minimalizacji odległości Hamminga wektora testującego, położonego na wejściu sieci od wektorów reprezentujących wzorce uczące, zakodowane w strukturze sieci. Liczba neuronów w warstwie pierwszej oraz drugiej jest taka sama. *Odległość Hamminga* w przypadku wielkości binarnych można wyrazić następująco:

$$\rho(\boldsymbol{x},\ \boldsymbol{y}) = \sum_{i=1}^{p} \left[x_i(1 - y_i) + y_i(1 - x_i) \right]. \tag{7.55}$$

Natomiast w przypadku wektorów bipolarnych

$$\rho(\boldsymbol{x},\ \boldsymbol{y}) = \frac{1}{2} \left[p - \sum_{i=1}^{p} x_i y_i \right]. \tag{7.56}$$

Odległość Hamminga jest równa zeru, jeśli $\boldsymbol{x} = \boldsymbol{y}$, w przeciwnym razie jest ona równa liczbie bitów, którymi różnią się oba wektory.

Warstwa pierwsza jest jednokierunkowa z przepływem sygnałów od wejścia do wyjścia i z ustalonymi wartościami wag. Warstwę drugą tworzy tzw. MAXNET, czyli neurony połączone sprzężeniem zwrotnym każdy z każdym i, co należy zauważyć, istnieje sprzężenie neuronu z własnym wyjściem (odwrotnie niż w sieci Hopfielda). Sprzężenie neuronu z własnym wyjściem jest pobudzające, o wartości wagowej równej $+1$. Wagi polaryzacji przyjmują wartości odpowiadające zeru. Jest to warstwa odpowiadająca sieci Hopfielda. Wartości wag w tej warstwie dobiera się hamująco (najczęściej są one postaci $-1/S$, gdzie S oznacza liczbę neuronów w warstwie rekurencyjnej), w ten sposób następuje wygaszanie wszystkich wyjść poza tym, które było najsilniejsze w warstwie pierwszej. Warstwa wyjściowa jednokierunkowa reprezentuje wektor wyjściowy skojarzony z wektorem wejściowym. Wagi neuronów tej warstwy ustala się niezależnie od tego wektora. W działaniu omawianej sieci można wyróżnić trzy etapy:

Etap I – podawany jest na wejście sieci wektor $\boldsymbol{x}$. Po jego prezentacji na wyjściu neuronów warstwy pierwszej generowane są sygnały inicjujące stan początkowy neuronów warstwy drugiej (MAXNET-u).

Etap II – sygnały inicjujące działanie MAXNET-u są usuwane i rozpoczyna się proces iteracyjny w warstwie MAXNET-u, inicjowany stanem początkowym, wynikającym z fazy pierwszej. Proces iteracyjny trwa do chwili, aż wszystkie neurony z wyjątkiem jednego osiągną stan zerowy. Neuron o niezerowym sygnale wyjściowym (zwycięzca) reprezentuje klasę, do której należy wektor wejściowy.

Etap III – neuron zwycięzca za pośrednictwem wag łączących go z neuronem warstwy wyjściowej wytwarza na wyjściu sieci odpowiedź stanowiącą wektor $\boldsymbol{y}$, skojarzony z wektorem pobudzającym $\boldsymbol{x}$.

Ważną zaletą sieci Hamminga jest duża oszczędność połączeń wagowych. Przykładowo dla sieci o 100 wejściach, kodujących 10 różnych klas wektorowych (czyli jest po 10 neuronów w każdej z warstw):

- sieć Hopfielda – 10 000 połączeń wagowych o dobieranych wartościach wag,
- sieć Hamminga – 1000 wag w warstwie pierwszej oraz 100 wag w warstwie MAXNET-u.

Na podstawie licznych eksperymentów wykazano, że sieć rekurencyjna Hamminga daje lepsze rezultaty niż sieć Hopfielda. Jedyny problem dotyczący sieci Hamminga występuje wówczas, gdy wzorce zaburzone są jednakowo odległe w sensie odległości Hamminga od dwóch lub większej liczby wzorców oryginalnych. Wtedy wskazanie sieci na jeden z tych wzorców jest losowe.

Uwaga 7.6. Sieć Hamminga ma pojemność $n/2$.

Przykład 7.13.
Ponownie rozważmy przykład 7.3. Tym razem jednak do rozwiązania problemu posłuży nam sieć Hamminga. Klasyfikacji poddamy wektor $\boldsymbol{x} = [-1,\ -1,\ -1]'$. W warstwie pierwszej mamy 2 neurony (ponieważ tyle jest wzorców), a wagi są ustalone jako zbiór wzorców, czyli

$$\boldsymbol{W}^1 = \begin{bmatrix} 1 & -1 & -1 \\ 1 & 1 & -1 \end{bmatrix}.$$

Każda składowa wektora progowego jest równa wymiarowi wektora wzorcowego, czyli

$$\boldsymbol{\theta}^1 = \begin{bmatrix} 3 \\ 3 \end{bmatrix}.$$

Teraz możemy już obliczyć wyjście pierwszej części sieci Hamminga

$$\boldsymbol{y}^1 = \boldsymbol{W}^1\boldsymbol{x} + \boldsymbol{\theta}^1 = \begin{bmatrix} 1 & -1 & -1 \\ 1 & 1 & -1 \end{bmatrix} \begin{bmatrix} -1 \\ -1 \\ -1 \end{bmatrix} + \begin{bmatrix} 3 \\ 3 \end{bmatrix} = \begin{bmatrix} 4 \\ 2 \end{bmatrix}.$$

Wektor ten staje się wejściem dla warstwy rekurencyjnej, która składa się również z dwóch neuronów. Mamy zatem

$$\boldsymbol{y}^2(0) = \begin{bmatrix} 4 \\ 2 \end{bmatrix}.$$

Macierz wag dla sieci rekurencyjnej ma postać

$$W^2 = \begin{bmatrix} 1 & -0.5 \\ -0.5 & 1 \end{bmatrix}.$$

Kolejne iteracje w warstwie rekurencyjnej wyglądają następująco:

$$\begin{aligned} y^2(1) &= \text{poslin}(W^2 y^2(0)) = \text{poslin}\left(\begin{bmatrix} 1 & -0.5 \\ -0.5 & 1 \end{bmatrix}\begin{bmatrix} 4 \\ 2 \end{bmatrix}\right) = \\ &= \text{poslin}\left(\begin{bmatrix} 3 \\ 0 \end{bmatrix}\right) = \begin{bmatrix} 3 \\ 0 \end{bmatrix}. \\ y^2(2) &= \text{poslin}(W^2 y^2(1)) = \text{poslin}\left(\begin{bmatrix} 1 & -0.5 \\ -0.5 & 1 \end{bmatrix}\begin{bmatrix} 3 \\ 0 \end{bmatrix}\right) = \\ &= \text{poslin}\left(\begin{bmatrix} 3 \\ -1.5 \end{bmatrix}\right) = \begin{bmatrix} 3 \\ 0 \end{bmatrix}. \end{aligned}$$

Otrzymaliśmy zatem rozwiązanie. Został wybrany wzorzec pomarańczy, gdyż tylko neuron wyjściowy numer jeden ma niezerową wartość (a jako pierwszy do macierzy wag został dodany wzorzec pomarańczy). Jest to oczywiście wynik poprawny, gdyż odległość Hamminga wektora x od wzorca jabłka wynosi 2, a od wzorca pomarańczy tylko 1.

Uwaga 7.7. Istnieją również inne sieci rekurencyjne. Przykładem takiej sieci jest *sieć typu BAM*[38], która jest uogólnieniem sieci Hopfielda, na dwuwarstwową strukturę rekurencyjną. Koncepcja tej sieci wywodzi się od Grossberga (1982), ale najważniejsze idee na temat BAM przedstawił Kosko (1988). Przepływ sygnałów jest dwukierunkowy: od wejścia do wyjścia i z powrotem. W jednym cyklu następuje przepływ informacji w jednym kierunku, są określane stany neuronów odbiorczych, po czym w następnym cyklu neurony te stanowią źródło, które wysyła sygnały w stronę przeciwną i cykl się powtarza, aż do ustalenia stanu równowagi. Funkcje aktywacji neuronów są unipolarne lub bipolarne. Na etapie uczenia używa się wyłącznie wersji bipolarnej sygnałów uczących, gdyż można zauważyć lepsze własności sieci w trybie odtworzeniowym. Przy jednym kierunku przepływu sygnałów wagi są opisane macierzą W, przy kierunku przeciwnym macierzą W'. Macierz wag W określa się jako macierz korelacji postaci

$$W = \sum_{i=1}^{n} x_i d'(x_i).$$

Czyli jest regułą Hebba ze współczynnikiem η równym 1. Kosko (1988) wykazał, że sieci tego typu są bezwarunkowo stabilne bez względu na to,

[38]Po ang. Bidirectional Associative Memory (BAM).

jaka jest macierz wag $\boldsymbol{W}$. Sieć BAM sprowadza się do sieci Hopfielda, kiedy macierz wag $\boldsymbol{W}$ jest kwadratowa i symetryczna. W literaturze można znaleźć wiele oszacowań pojemności pamięci sieci BAM, jednak najbardziej realistyczne wydaje się oszacowanie zaproponowane przez McEliece'a i in. (1987), które wynosi $\dfrac{l}{4\log_2 l}$, gdzie l oznacza liczbę neuronów w mniejszej z warstw ukrytych.

Oprócz sieci BAM często są wykorzystywane *sieci Elmana* (zwane również SRN[39]) oraz *sieci RTRN*[40]. Sieć Elmana jest modyfikacją warstwowej sieci perceptronowej. Charakteryzuje się częściową rekurencją w postaci sprzężenia zwrotnego między warstwą ukrytą a warstwą wejściową. Każdy neuron warstwy ukrytej ma swój odpowiednik w tzw. warstwie kontekstowej, stanowiącej wspólnie z wejściami sieci wektor wejściowy. Warstwa kontekstowa stanowi odpowiedni zestaw opóźnień jednostkowych względem wyjść neuronów warstwy ukrytej. Sieci Elmana uczy się za pomocą metody największego spadku. W sieci RTRN, w przeciwieństwie do sieci Elmana, sprzężenie zwrotne obejmuje również neurony z warstwy wyjściowej. Sieci tego typu uczy się za pomocą algorytmu uczenia rekurencyjnego w czasie rzeczywistym (RTRL)[41].

Rekurencyjne sieci wielowarstwowe są wykorzystywane do modelowania ciągów czasowych oraz do identyfikacji obiektów dynamicznych.

Przykład 7.14.

Wróćmy do przykładu 7.3. Weźmy pod uwagę cztery typy owoców: pomarańcze, jabłka, banany oraz ananasy. Każdemu z owoców będzie odpowiadał wektor $\boldsymbol{y}$, którego poszczególne składowe będą oznaczały: pochodzenie owocu (-1 – krajowy, 1 – zagraniczny), smak owocu (-1 – kwaśny, 1 – słodki) oraz popularność w Polsce (-1 – mała lub średnia, 1 – duża). Odpowiednie wektory wejściowe mają zatem postać

$$\boldsymbol{x}_1 = \begin{bmatrix} 1 \\ -1 \\ -1 \end{bmatrix}, \quad \boldsymbol{y}_1 = \begin{bmatrix} 1 \\ -1 \\ 1 \end{bmatrix},$$

$$\boldsymbol{x}_2 = \begin{bmatrix} 1 \\ 1 \\ -1 \end{bmatrix}, \quad \boldsymbol{y}_2 = \begin{bmatrix} -1 \\ -1 \\ 1 \end{bmatrix},$$

[39]Po ang. Simple Recurrent Network (SRN).
[40]Po ang. Real Time Recurrent Network (RTRN).
[41]Po ang. Real Time Recurrent Learning (RTRL).

$$\boldsymbol{x}_3 = \begin{bmatrix} -1 \\ 1 \\ -1 \end{bmatrix}, \quad \boldsymbol{y}_3 = \begin{bmatrix} 1 \\ 1 \\ 1 \end{bmatrix},$$

$$\boldsymbol{x}_4 = \begin{bmatrix} -1 \\ -1 \\ 1 \end{bmatrix}, \quad \boldsymbol{y}_4 = \begin{bmatrix} 1 \\ 1 \\ -1 \end{bmatrix}.$$

Teraz wykorzystując sieć BAM, nauczymy ją tych skojarzeń. Założymy, że początkowo wszystkie wagi były równe 0. W kolejnych krokach zmiany wag wyglądają następująco:

$$\Delta\boldsymbol{W}(1) = \boldsymbol{x}_1\boldsymbol{y}_1' = \begin{bmatrix} 1 \\ -1 \\ -1 \end{bmatrix} [1,\ -1,\ 1] = \begin{bmatrix} 1 & -1 & 1 \\ -1 & 1 & -1 \\ -1 & 1 & -1 \end{bmatrix},$$

$$\Delta\boldsymbol{W}(2) = \boldsymbol{x}_2\boldsymbol{y}_2' = \begin{bmatrix} 1 \\ 1 \\ -1 \end{bmatrix} [-1,\ -1,\ 1] = \begin{bmatrix} -1 & -1 & 1 \\ -1 & -1 & 1 \\ 1 & 1 & -1 \end{bmatrix},$$

$$\Delta\boldsymbol{W}(3) = \boldsymbol{x}_3\boldsymbol{y}_3' = \begin{bmatrix} -1 \\ 1 \\ -1 \end{bmatrix} [1,\ 1,\ 1] = \begin{bmatrix} -1 & -1 & -1 \\ 1 & 1 & 1 \\ -1 & -1 & -1 \end{bmatrix},$$

$$\Delta\boldsymbol{W}(4) = \boldsymbol{x}_4\boldsymbol{y}_4' = \begin{bmatrix} -1 \\ -1 \\ 1 \end{bmatrix} [1,\ 1,\ -1] = \begin{bmatrix} -1 & -1 & 1 \\ -1 & -1 & 1 \\ 1 & 1 & -1 \end{bmatrix}.$$

Końcowa postać macierzy wag (zawartość pamięci sieci BAM) jest następująca:

$$\boldsymbol{W} = \Delta\boldsymbol{W}(1) + \Delta\boldsymbol{W}(2) + \Delta\boldsymbol{W}(3) + \Delta\boldsymbol{W}(4) = \begin{bmatrix} -2 & -4 & 2 \\ -2 & 0 & 2 \\ 0 & 2 & -4 \end{bmatrix}.$$

Sprawdźmy teraz działanie naszej sieci. Podajemy na wejście sygnał $\boldsymbol{x}_2$. Sieć odpowiada następująco:

$$\boldsymbol{y} = \operatorname{sgn}(\boldsymbol{W}\boldsymbol{x}_2) = \operatorname{sgn}\left(\begin{bmatrix} -2 & -4 & 2 \\ -2 & 0 & 2 \\ 0 & 2 & -4 \end{bmatrix} \begin{bmatrix} 1 \\ 1 \\ -1 \end{bmatrix}\right) = \operatorname{sgn}\left(\begin{bmatrix} -8 \\ -4 \\ 6 \end{bmatrix}\right) = \begin{bmatrix} -1 \\ -1 \\ 1 \end{bmatrix}.$$

Otrzymany wynik odpowiada skojarzonemu z wektorem $\boldsymbol{x}_2$ wektorowi $\boldsymbol{y}_2$. Teraz na wejście sieci podajmy właśnie wektor $\boldsymbol{y}_2$. Odpowiedź sieci będzie następująca:

$$\boldsymbol{x} = \operatorname{sgn}(\boldsymbol{W}'\boldsymbol{y}_2) = \operatorname{sgn}\left(\begin{bmatrix} -2 & -2 & 0 \\ -4 & 0 & 2 \\ 2 & 2 & -4 \end{bmatrix} \begin{bmatrix} -1 \\ -1 \\ 1 \end{bmatrix}\right) = \operatorname{sgn}\left(\begin{bmatrix} 4 \\ 6 \\ -8 \end{bmatrix}\right) = \begin{bmatrix} 1 \\ 1 \\ -1 \end{bmatrix}.$$

Ponownie sieć odtworzyła poprawnie skojarzony sygnał. Podobnie można stwierdzić poprawne działanie sieci dla pozostałych uprzednio zapamiętanych wektorów.

7.5. Sieci samoorganizujące się z konkurencją (współzawodnictwem)

Sieci samoorganizujące się są stosowane całkowicie odmiennie niż wszystkie dotychczas omawiane sieci. Podczas gdy wszystkie inne sieci są uczone w trybie „z nauczycielem", to sieć samoorganizująca się jest uczona bez „nauczyciela". Podczas gdy w uczeniu z nauczycielem dane uczące zawierają przypadki opisujące zarówno zmienne wejściowe, jak i zmienne wyjściowe, to w uczeniu bez nauczyciela pojawiają się wyłącznie przypadki obejmujące same tylko zmienne wejściowe.

W sieciach samoorganizujących się na podstawie prezentowanych danych, neurony kierują się do obszarów zajmowanych przez te dane i uczą się je rozpoznawać. Po każdej prezentacji wzorca zwycięzcą zostaje tylko jeden neuron. Neuron-zwycięzca ma prawo uaktualnić swoje wagi według jednej z dwóch zasad: *WTA* oraz *WTM*. W efekcie współzawodnictwa wyrażonego zasadami WTA i WTM następuje samoorganizacja procesu uczenia. Neurony dopasowują swoje wagi w ten sposób, że przy prezentacji grup wektorów wejściowych zbliżonych do siebie zwycięża zawsze ten sam neuron. Może się zdarzyć, że niektóre neurony nigdy nie będą wygrywać lub też nigdy nie znajdą się w sąsiedztwie wygrywających neuronów. Takie neurony nazywamy *martwymi*.

W efekcie tych procedur cała przestrzeń zostaje podzielona na rozłączne obszary (skwantowana) odzwierciedlające strefy wpływów poszczególnych neuronów. W przypadku używania metryki euklidesowej są to obszary wypukłe, które nazywamy wielobokami Voronoia. Reprezentujące te obszary neurony-prototypy nazywa się *wektorami Voronoia* lub *wektorami kodowymi* i oznacza przez $\boldsymbol{v}_i$.

Definicja 7.5.
Obszarem Voronoia V_i nazywamy miejsce geometryczne punktów $\boldsymbol{x}$ spełniających następujący warunek:

$$V_i = \{\boldsymbol{x} \colon ||\boldsymbol{v}_i - \boldsymbol{x}|| < ||\boldsymbol{v}_j - \boldsymbol{x}||,\ j \neq i\}.$$

Łącząc neurony, których obszary Voronoia mają wspólną krawędź, otrzymujemy *triangulację Delaunaya*. Przykładowy diagram Voronoia jest przedstawiony na rys. 5.5. Kohonen nazwał proces tworzenia reprezentantów danych *kwantowaniem wektorowym*[42] lub dokładniej *adaptacyjnym kwantowaniem wektorowym* (LVQ)[43]. Można w przypadku takiej reprezentacji danych mówić o dwóch wskaźnikach jakości podziału:

- *błąd kwantyzacji*[44] – średnia odległość danych od ich reprezentantów, czyli prototypów,
- *błąd topologicznej reprezentacji*[45] – procent punktów danych, dla których pierwsi dwaj najbliżsi reprezentanci nie są sąsiadami na mapie.

Uwaga 7.8. Sieci samoorganizujące się są wykorzystywane do kompresji obrazów, wykrywania typu uszkodzenia oraz prognozowania obciążeń systemu elektroenergetycznego.

7.5.1. Sieci WTA (Winner Takes All)

W tego rodzaju sieci początkowy sygnał $\boldsymbol{x}$ trafia do wszystkich neuronów. Następnie obliczana jest pewna miara podobieństwa (najczęściej odległość euklidesowa, popularne są również: iloczyn skalarny, norma L_1 oraz norma L_∞) tego sygnału do wektora wag (początkowe wartości wag są losowe). Neuron, który zwycięża (jest najbliżej sygnału), jako wyjście przyjmuje war-

---

[42] Po ang. Vector Quantization.

[43] LVQ jest nadzorowaną wersją algorytmu Kohonena. Standardowy algorytm Kohonena próbuje przemieścić wektory wzorcowe na pozycje odpowiadające centrom skupień występujących w danych. Nie są przy tym brane pod uwagę etykiety klas wektorów uczących. Dla osiągnięcia najlepszej jakości klasyfikacji pożądane jest, aby wektory wzorcowe były rozmieszczane, w pewnym stopniu w zakresie klas. Tak, by reprezentowały naturalne skupienia wewnątrz każdej klasy. Istnieje kilka wariantów algorytmu LVQ. Podstawowa wersja, LVQ1, jest bardzo podobna do algorytmu Kohonena, gdzie jest wyszukiwany wektor najbliższy danego wektora uczącego i odpowiednio modyfikowane jest jego położenie. O ile jednak algorytm Kohonena przesuwa wektor w stronę wektora uczącego, to LVQ1 sprawdza zgodność klas obu wektorów. Jeśli są zgodne, wektor wzorcowy jest przesuwany w stronę wektora uczącego, a w przeciwnym przypadku jest od niego odsuwany. Bardziej wyszukane są algorytmy LVQ2.1 i LVQ3. Wyszukują one dwa wektory wzorcowe najbliższe danemu wektorowi uczącemu. Jeżeli jeden z wektorów wzorcowych ma klasę zgodną z wektorem uczącym, a drugi niezgodną, to zgodny wektor jest przesuwany do wektora uczącego, a niezgodny jest od niego odsuwany. LVQ3 przybliża oba wektory wzorcowe do wektora uczącego, jeżeli oba mają zgodne z przypadkiem klasy. W obydwu algorytmach, LVQ2.1 i LVQ3, zasadą jest odsuwanie wektorów wzorcowych od miejsc zagrożonych błędnymi klasyfikacjami.

[44] Po ang. average quantization error.

[45] Po ang. topographic error.

tość 1, a pozostałe 0. Modyfikowane są jedynie wagi zwycięskiego neuronu według formuły

$$\boldsymbol{w}_j(t+1) = \boldsymbol{w}_j(t) + \eta(t)[\boldsymbol{x}(t) - \boldsymbol{w}_j(t)],$$

gdzie η jest współczynnikiem uczenia.

Uwaga 7.9. W tego rodzaju sieciach opłaca się normalizacja sygnałów wejściowych np. w następujący sposób:

$$x_i^* = \frac{x_i}{\sqrt{\sum_{i=1}^{p} x_i^2}}.$$

Wykazano, że proces samoorganizacji prowadzi zawsze do podziału na spójne obszary przestrzeni cech, gdy choć jeden z wektorów $\boldsymbol{x}$ lub $\boldsymbol{w}$ podlega normalizacji. Normalizacja sygnału powoduje, że wektor wag w procesie normalizacji również będzie znormalizowany i na działanie neuronu będzie miał wpływ jedynie kąt między wektorem wag a sygnałem. W przypadku danych o wymiarowości większej od 200 normalizacja nie ma znaczenia.

Uwaga 7.10. Początkowy współczynnik uczenia η dobiera się eksperymentalnie z przedziału (0.1, 1). Najczęściej jest on malejącą funkcją czasu. Można dla każdego neuronu ustalić inny współczynnik uczenia η_i.

Na koniec procesu uczenia cała przestrzeń cech jest podzielona na rozdzielne strefy wpływów poszczególnych neuronów. Jednocześnie można zaobserwować pogrupowanie danych. Cały zbiór danych został podzielony na skupiska reprezentowane przez poszczególne neurony. Zastosowanie różnych miar odległości między wektorami kształtuje podział stref wpływów inaczej. W szczególności zastosowanie iloczynu skalarnego bez normalizacji wektorów może prowadzić do niespójnego podziału przestrzeni cech, w którym występuje kilka neuronów w jednym obszarze, a w innym nie ma żadnego. Na rysunku 7.24 są przedstawione różne strefy, w zależności do wybranej miary podobieństwa.

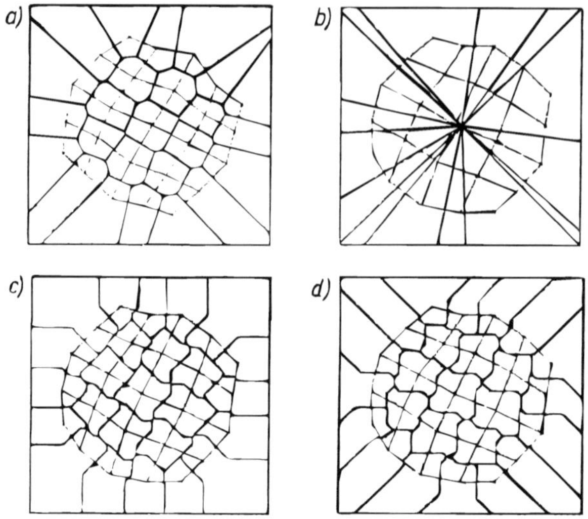

Rys. 7.24. Strefy wpływów w zależności od miary podobieństwa: a) miara euklidesowa, b) iloczyn skalarny, c) norma L_1, d) norma L_∞

7.5.2. Sieci WTM (Winner Takes Most)

Algorytmy WTA, w których tylko jeden neuron może podlegać adaptacji w każdej iteracji, są algorytmami słabo zbieżnymi, zwłaszcza przy dużej liczbie neuronów. W praktyce zostały one zastąpione algorytmami WTM, w których oprócz zwycięzcy uaktualniają swoje wagi również neurony z jego sąsiedztwa. Wybór zwycięskiego neuronu jest analogiczny jak w sieci WTA, natomiast wagi neuronów modyfikuje się następująco:

$$\boldsymbol{w}_j(t+1) = \boldsymbol{w}_j(t) + \eta(t)G(j,\ \boldsymbol{x})[\boldsymbol{x}(t) - \boldsymbol{w}_j(t)],$$

gdzie $G(j,\ \boldsymbol{x})$ jest *funkcją sąsiedztwa* dla j-tego neuronu. Definiując

$$G(j,\ \boldsymbol{x}) = \begin{cases} 1, & \text{dla zwycięskiego neuronu} \\ 0, & \text{dla pozostałych} \end{cases},$$

otrzymujemy sieć WTA. Funkcja sąsiedztwa może mieć bardzo zróżnicowaną postać. Często przyjmuje się, że wszystkie neurony z sąsiedztwa (trzeba jeszcze przyjąć jakiś próg, określający, jak daleko sięga sąsiedztwo) mają modyfikowane wagi z wartością funkcji sąsiedztwa równą 1. Inny pomysł to zmniejszanie wartości funkcji sąsiedztwa wraz z oddalaniem się od zwycięskiego neuronu np. w następujący sposób:

$$G(j,\ \boldsymbol{x}) = e^{-r^2},$$

gdzie r jest odległością między zwycięskim neuronem a jego sąsiadem.

Uwaga 7.11. Często uczenie sieci WTM jest celowo przeprowadzane w dwóch wyraźnie oddzielonych od siebie fazach: na początku stosunkowo krótka faza z dużym współczynnikiem uczenia i rozległym sąsiedztwem (początkowo powinno obejmować większą niż połowa z ogólnej liczby neuronów) oraz długa faza „douczania" z niską wartością współczynnika uczenia i zerowym lub prawie zerowym sąsiedztwem.

Uwaga 7.12. Niestety, w przypadku tego typu sieci nie są znane żadne fakty dotyczące zbieżności lub chociażby punktów stacjonarnych sieci. Można znaleźć jedynie pewne fakty w szczególnych przypadkach (np. dla procesu Markowa z czasem ciągłym na prostej zachodzi powolna zbieżność wymagająca około 10^4–10^6 iteracji).

Algorytm sieci WTM

1. Inicjalizacja: przypadkowe $\boldsymbol{w}_j(0)$. Definiuj wokół neuronu położonego w miejscu r_c siatki obszar sąsiedztwa $Os(r_c,\ \sigma(t))$ o promieniu $\sigma(t)$.
2. Oblicz odległości

$$d(\boldsymbol{x},\ \boldsymbol{w}_j) = \sqrt{\sum_i (x_i - w_{ji})^2}.$$

3. Znajdź neuron

$$c = \arg\min_i ||\boldsymbol{x} - \boldsymbol{w}_j||$$

z wagami $\boldsymbol{w}_c$ najbardziej podobnymi do $\boldsymbol{x}$ (neuron zwycięzca).

4. Zmień wagi wszystkich neuronów r_i w sąsiedztwie $Os(r_c,\ \sigma(t))$

$$\boldsymbol{w}_j(t+1) = \boldsymbol{w}_j(t) + h(r_j,\ r_c)(\boldsymbol{x} - \boldsymbol{w}_j(t)),$$

gdzie

$$h(r_j,\ r_c) = \eta(t) e^{-||r_j - r_c||^2/2\sigma_c^2(t)}.$$

5. Powoli zmniejszaj siłę $\eta(t)$ i promień $\sigma(t)$.
6. Powtarzaj aż ustaną zmiany.

Uwaga 7.13. Jeśli przyjmiemy $\sigma(t) = 0$, to sieć WTM zachowuje się analogicznie do analizy skupień za pomocą metody k-średnich (patrz podrozdz. 12.2).

Uwaga 7.14. W sieci typu WTM część neuronów może nie znaleźć się w sąsiedztwie neuronów wygrywających i tym samym może nie zmienić swoich początkowych wag (nie uczestniczy w procesie uczenia). Aby temu zapobiec, stosuje się tzw. mechanizm zmęczenia lub mechanizm sumienia[46], który polega na tym, że wygrywający neuron pauzuje przez pewien czas – nie uczestniczy w procesie uczenia – dając szansę mniej aktywnym neuronom.

Uwaga 7.15. Sieć WTM jest inspirowana przez pewne znane własności mózgu. Kora mózgowa jest w rzeczywistości dużym, płaskim płatem komórek nerwowych. Powierzchnia tego płata wynosi około 0.5 metra kwadratowego i ma pofałdowaną strukturę tylko po to, aby dostosować się do kształtu czaszki. Działanie tego płata jest związane ze znanymi własnościami topologicznymi – na przykład obszar odpowiadający dłoni znajduje się w nim obok obszaru przedramienia, a ten z kolei sąsiaduje z obszarem ramienia itd. Tak więc powierzchnia kory mózgowej (zarówno w obszarach czuciowych, jak i w obszarach sterujących ruchem) jest swoistą mapą topologiczną, odpowiadającą całej złożonej konstrukcji ciała człowieka.

Uwaga 7.16. Dotychczas omawiane uczenie sekwencyjne polega na tym, że po kolei są prezentowane sieci kolejne obserwacje. Aktualizacja wag następuje po każdej prezentacji według zasady WTA lub WTM. W przypadku konstrukcji map szybsze i stabilniejsze okazuje się być uczenie wsadowe[47]. Polega ono na wykonywaniu aktualizacji wag tylko na zakończenie każdej epoki, czyli gdy zostały zaprezentowane wszystkie dane uczące. Szczegóły uczenia wsadowego można znaleźć np. w monografii Kohonena (1995).

Uwaga 7.17. Najbardziej znaną siecią WTM jest *samoorganizująca się mapa* (SOM)[48] (zwana też od jej twórcy siecią Kohonena). W tego rodzaju sieci należy zdefiniować sąsiedztwo na stałe, tzn. definiując, które neurony się ze sobą łączą, tworząc siatkę, co prowadzi do kilku powszechnych (prostokątne i heksagonalne) możliwości zobrazowanych na rys. 7.25.

---

[46] Po ang. conscience mechanism.

[47] Po ang. batch training.

[48] Po ang. Self Organizing Map (SOM).

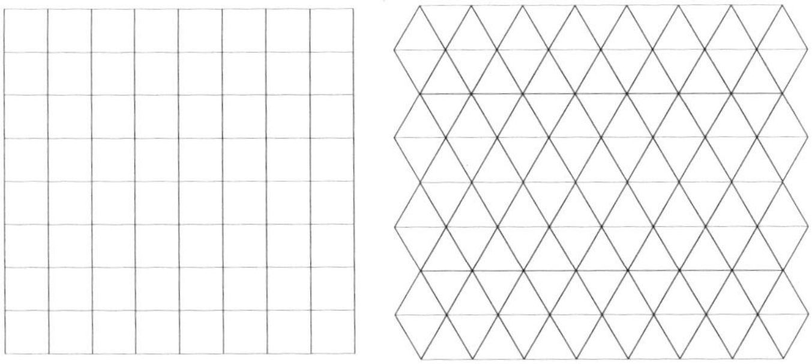

Rys. 7.25. Topologie sąsiedztwa

Tak zdefiniowane sąsiedztwa są nazywane *topologicznymi uporządkowaniami*. Neurony w warstwie tworzącej mapę topologiczną są rozważane w taki sposób, jakby były rozmieszczone w przestrzeni według jakiegoś ustalonego wzoru. Adaptacja wag przebiega analogicznie jak w przypadku klasycznej sieci WTM, jednak adaptacji podlegają jedynie neurony z topologicznego sąsiedztwa. Kohonen stwierdził, że można znacznie przyspieszyć proces uczenia, jeśli zamiast czysto losowego inicjowania wektorów wag rozstawi się je na płaszczyźnie pierwszych dwóch składowych głównych analizowanych danych. To spostrzeżenie pozwala zaoszczędzić tysiące iteracji.

Przykład 7.15.
W przykładzie tym skonstruujemy samoorganizującą się mapę województw na podstawie 17 zmiennych socjoekonomicznych (dane pochodzą z roku 2000[49]). Wszystkie obliczenia i rysunki zostały wykonane za pomocą pakietu SOM Toolbox 2.0[50]. Na rysunku 7.26 jest przedstawiona mapa Kohonena o wymiarach[51] 6×6 (czyli 36 neuronów) oraz 10×10 (100 neuronów).

To, co widzimy na mapie, jest obrazem wielowymiarowej przestrzeni. Mapa składa się z obszarów heksagonalnych, w których środkach znajdują się wektory referencyjne (współrzędne siatki, przy czym numeracja z góry na dół i od lewa do prawa) odpowiadające wektorom kodowym. Są to tzw. *U-macierze* od nazwiska niemieckiego informatyka Ultscha (1993). Na czym polega ten sposób wizualizacji? Przypuśćmy, że rozważamy mapę o rozmiarze $M = m_1 \times m_2$. Mapę tę poszerzamy w kierunku wschodnim o $m_2 - 1$ pustych jednostek ulokowanych między oryginalnymi jednostkami mapy (zawierającymi wektory referencyjne). Otrzymaną w ten sposób strukturę poszerzamy jeszcze raz w kierunku południowym, lokując między każdą parą wierszy dodatkowy wiersz

[49] http://www.stat.gov.pl/.
[50] http://www.cis.hut.fi/projects/somtoolbox/.
[51] Jeśli nie wyspecyfikujemy rozmiaru mapy, to liczba neuronów jest najczęściej ustalana jako $5\sqrt{N}$.

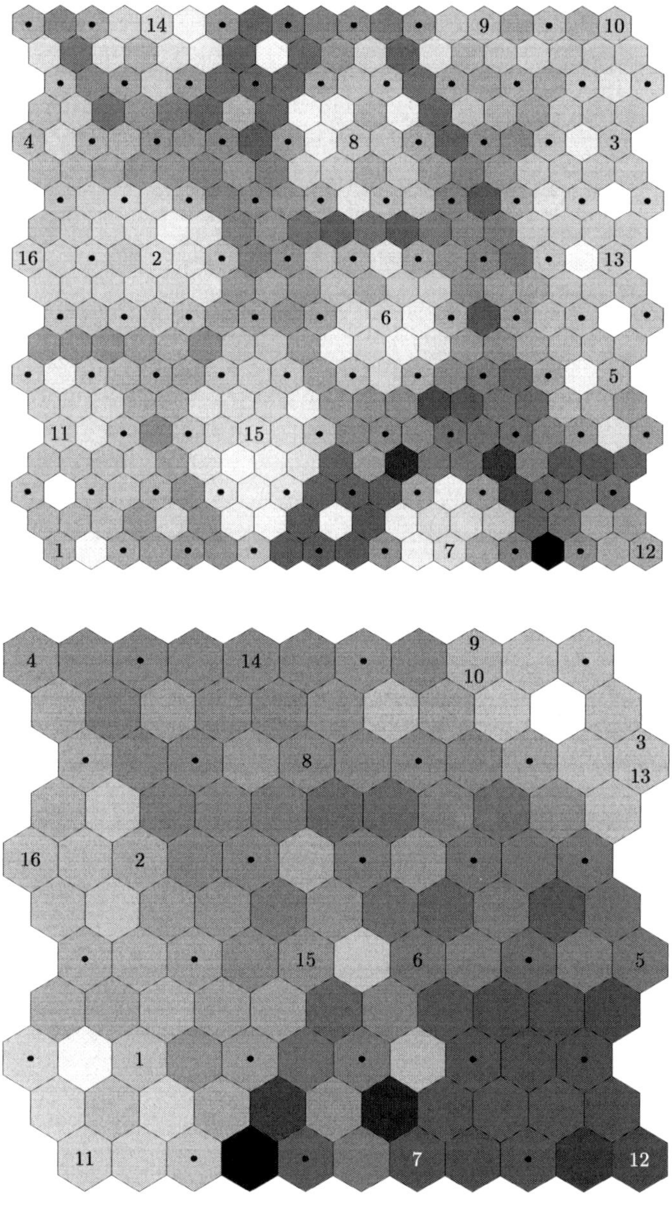

Rys. 7.26. Mapy Kohonena

zawierający $2m_2 - 1$ pustych jednostek. Ostatecznie otrzymujemy strukturę zawierającą $(2m_1 - 1) \times (2m_2 - 1)$ jednostek. Mapa kolorowana jest w taki sposób, że w punktach referencyjnych odcień koloru jest proporcjonalny do mediany odległości tego punktu od wszystkich jego sąsiadów. Natomiast w punktach pustych odcień jest proporcjonalny do mediany między dwoma sąsiednimi punktami. Obszary bliskie są jasne, ciemny kolor oznacza duże odległości, a więc może oznaczać granice skupień w przestrzeni $\mathbb{R}^{17}$. Województwa zostały ponumerowane alfabetycznie, czyli: 1 – dolnośląskie, 2 – kujawsko–pomorskie, 3 – lubelskie, 4 – lubuskie, 5 – łódzkie, 6 – małopolskie, 7 – mazowieckie, 8 – opolskie, 9 – podkarpackie, 10 – podlaskie, 11 – pomorskie, 12 – śląskie, 13 – świętokrzyskie, 14 – warmińsko–mazurskie, 15 – wielkopolskie, 16 – zachodniopomorskie. W obu mapach początkowe wagi były inicjowane losowo w przestrzeni dwóch pierwszych składowych głównych (przyspiesza uczenie i mapy dla różnych rozmiarów nie różnią się znacząco). W rezultacie powstałe mapy są do pewnego stopnia podobne, mówimy, że jest to topologiczne podobieństwo (w obu przypadkach błąd topologicznej reprezentacji wynosi 0). Przypatrując się położeniu „geograficznemu" punktów na mapie, można zauważyć pewną odmienność rozłożenia województw na obu mapach (dla pierwszej błąd kwantyzacji wynosi 0.467, a dla drugiej 1.680). Można zauważyć, że województwa o numerach 3, 9, 10, 13 wyraźnie tworzą jeden obszar. Pozostałe obszary nie są już tak wyraźne i różnią się nieco na obu mapach. Wyraźnie natomiast odstają od pozostałych województwo mazowieckie i śląskie, które między sobą również różnią się dość znacząco.

7.6. Sieci rezonansowe

Sieci tego typu dotyczą uczenia bez nauczyciela. Celem działania takiej sieci jest rozpoznawanie zapamiętanych wcześniej obrazów binarnych oraz rozróżnianie kształtów dotychczas nieznanych. Nowe wzorce są zapamiętywane w sieci, jeśli ich stopień podobieństwa do dotychczas znanych jest mały, jak mały określa tzw. *próg czujności* sieci τ. Autorami prac dotyczących *adaptacyjnej teorii rezonansu*[52], które są podstawą działania sieci rezonansowych, są Carpenter i Grossberg (1987a, 1987b, 1988). Wymienieni twórcy przedstawili dwie struktury: ART1 – służącą do rozpoznawania sygnałów binarnych oraz ART2 – służącą do klasyfikacji sygnałów ciągłych (później została zmodyfikowana do ART3). Sieci typu ART2 oraz ART3 nadają się również do rozpoznawania obrazów binarnych. Sieci rezonansowe składają się z dwóch zasadniczych warstw: dolnej i górnej oraz układów: kontrolnego i orientującego. W pierwszej, warstwie wejściowej (dolnej, porównującej), jest przechowywany kształt wejściowy. Druga warstwa, wyjściowa (górna, rozpoznająca), odpowiada za wskazanie klasy, do której klasyfikuje się kształt wejściowy. W warstwie tej tylko jeden z neuronów ma stan jedynki, natomiast wszystkie pozostałe są wyzerowane. Jednocześnie jest weryfikowany stopień podobieństwa kształtu wejściowego do przechowywa-

[52]Po ang. Adaptive Resonance Theory (ART).

nych wzorców i w zależności od oceny wyniku weryfikacji, odpowiedź górnej warstwy (czyli kształt wybranego wzorca) wraca do warstwy dolnej. Jeśli zwrócony sygnał dokładnie odpowiada wzorcowi, to klasyfikacja jest kończona. Jeśli natomiast sygnał odpowiedzi w znaczący sposób różni się od sygnału wejściowego, wówczas dokonują się odpowiednie modyfikacje wag połączeń. Poszczególne warstwy komunikują się w górę i w dół, czyli rezonują, dopasowując wagi, a proces ten powtarza się aż do osiągnięcia maksymalnego podobieństwa.

Sieć ART jest siecią uczącą się w czasie rzeczywistym. Uczenie to polega głównie na rozpoznawaniu kształtów, które znacznie różnią się od przechowywanych wzorców oraz ich zapamiętywaniu jako nowych wzorców. Ponadto, przy niewielkich różnicach stwierdzonych przy porównywaniu wejść z zapamiętanymi wzorcami dokonuje się „douczenia", które polega na poprawieniu wzorców zgodnie z obserwowanymi odchyleniami. Nad uaktywnieniem lub wygaszeniem odpowiednich decyzji w sieci czuwają *systemy kontrolny* i *orientujący*.

Układ kontrolny ma dwa zasadnicze zadania:

- zablokowanie reakcji warstwy dolnej na bodźce pochodzące z warstwy górnej, jeżeli brak jest sygnałów wejściowych,
- dodatkowe wzmocnienie czułości neuronów warstwy dolnej.

Układ kontrolny podlega działaniu trzech sygnałów: wejściowego, zwróconego z warstwy górnej oraz zewnętrznego sygnału blokującego. Pierwszy z wymienionych sygnałów jest logiczną jedynką, jeśli którekolwiek z wejść ma stan równy jeden. Podobnie sprawdzany jest stan wyjść warstwy górnej i w wypadku jej pobudzenia sygnał z tej warstwy przyjmuje wartość jeden. Zewnętrzny sygnał blokujący pełni funkcję nadrzędną w stosunku do obu sygnałów detekcji aktywności warstw. Każdy z neuronów znajdujących się w warstwie dolnej sieci rezonansowej podlega działaniu trzech sygnałów: wejściowego, z warstwy górnej oraz sterującego (wyjścia układu kontrolnego).

Celem układu orientującego jest sterowanie precyzją odwzorowania poszczególnych kategorii w sieci ART. System orientujący jest pobudzany przez sygnał wejściowy i hamowany przez sygnał globalny aktywności dolnej warstwy sieci. System jest nieaktywny, gdy sygnałowi wejściowemu towarzyszy reakcja w postaci pobudzenia dolnej warstwy, gdyż wtedy sieć rozpoznaje jeden z wcześniej zapamiętanych wzorców wejściowego sygnału i żadne dodatkowe działanie nie jest potrzebne. W przypadku gdy pojawia się nowy bodziec, sieć nie zna go, zatem nie dochodzi do „rezonansu" i pobudzenie dolnej warstwy jest niewielkie. Jest to przedstawiciel nowej kategorii sygnałów, które powinny być zapamiętane. Nowy sygnał wejściowy pobudza pewne neurony w warstwie górnej, a te poprzez sprzężenie zwrotne będą oddziaływać na neurony warstwy dolnej. Brak korelacji między uprzednio zapamiętanymi wzorcami a nowym sygnałem jest powodem, że wynikowe pobudzenie warstwy dolnej maleje i efekt rezonansu nie działa. Dlatego też stosuje się system orientujący, który powoduje, że wszystkie aktywne neurony warstwy górnej zostają zablokowane, co powoduje, że rozkład pobudzeń w warstwie dolnej może swobodnie się kształtować. W efekcie tych zdarzeń zostaje wyłoniony neuron zwycięzca, który staje się detektorem nowej klasy. System orientujący odgrywa rolę „zastępczego nauczyciela", który klasyfikuje pojawiające się sygnały na wejściach na znane i nowe, dzięki czemu poszczególne neurony nabywają zdolności do rozpoznawania i sygnalizowania różnych wzorców.

Oznaczmy przez v_{ij}, $i = 1, \ldots, p$, $j = 1, \ldots, K^*$, wagi połączeń między wyjściami neuronów górnej warstwy a wejściami neuronów dolnej warstwy. Natomiast wagi połączeń między wyjściami neuronów dolnej warstwy a wejściami neuronów górnej warstwy oznaczmy przez w_{ij}, $i = 1, \ldots, p$, $j = 1, \ldots, K^*$. Jak widzimy charakterystyczne dla tego typu sieci jest istnienie dwóch rodzajów połączeń międzyneuronowych, „z góry na dół" (pamięć krótkotrwała) – v_{ij} oraz „z dołu do góry" (pamięć długotrwała) – w_{ij}. Liczba p oznacza liczbę neuronów warstwy dolnej i jest wymiarem wektora sygnału wejściowego. Liczba K^* jest liczbą neuronów warstwy górnej, klasyfikującej, a zatem liczba ta jest zwykle nieco większa niż liczba rozpoznawanych klas, gdyż przewiduje się możliwość poszerzenia zbioru rozpoznawanych klas w wyniku uczenia sieci. Podczas działania sieci ART możemy wyróżnić następujące podstawowe etapy:

1. Inicjalizacja. Polega na wstępnym ustawieniu wag połączeń oraz progu zadziałania układu orientującego. Zazwyczaj dla wag połączeń z góry na dół przyjmujemy $v_{ij}(0) = 1$, a dla połączeń z dołu do góry $w_{ij}(0) = \frac{1}{1+p}$. Jako próg czujności przyjmuje się liczbę z przedziału $(0,\ 1)$.
2. Rozpoznawanie. W tej części działania sieci następuje uaktywnienie jednego z neuronów warstwy wyjściowej. W przypadku gdy dzieje się to po raz pierwszy po inicjalizacji działania sieci, to zwycięzcą może być dowolny z neuronów. Temu neuronowi przypisuje się jako kształt zapamiętany, kształt podany na wejście. Natomiast podczas normalnej pracy sieci w tej fazie jest obliczana suma $\sum_{i=1}^{p} w_{ij} x_i$, dla każdego neuronu górnej warstwy. Ten, dla którego wartość sumy jest największa zwycięża

i jego stan jest równy 1 (oznaczmy jego numer przez j^*). Stany pozostałych neuronów są wyzerowane.

3. Uczenie nowych wzorców. Jeżeli nie jest spełniony warunek wystarczającego podobieństwa kształtu wejściowego i wzorców dotychczas zapamiętanych, następuje uczenie się przez sieć nowego wzorca. Po wskazaniu klasy, do której badany obiekt najlepiej pasuje, należy sprawdzić, jak bardzo jest on podobny do obiektów wcześniej przypisanych do tej klasy. Obliczamy

$$D = \frac{\sum_{i=1}^{p} v_{ij^*} x_i}{\sum_{i=1}^{p} x_i}. \tag{7.57}$$

Liczba D jest porównywana z przyjętym wcześniej parametrem czułości τ układu orientującego. Jeżeli $\tau > D$, następuje blokada aktywnego neuronu warstwy górnej i rozpoczyna się powtórny proces porównania wejścia z pozostałymi zapamiętanymi wzorcami (za neuron zwycięski zostaje przyjęty najbardziej pobudzony spośród pozostałych). Jeżeli w wyniku porównania ze wszystkimi wzorcami nie uda się zaliczyć nowego obiektu do żadnej znanej klasy, to tworzymy nową klasę, przypisując jej pierwszy wolny neuron warstwy wyjściowej. Jeżeli „test podobieństwa" wypadnie pozytywnie, to mówimy, że sieć oraz sygnał wejściowy są w rezonansie (stąd nazwa tego typu sieci). Wybór małych wartości progowych powoduje, że bardzo łatwo uzyskać podobieństwo do wzorca nawet dla stosunkowo różnych obiektów. Sieć klasyfikuje wówczas obserwacje do niewielu klas, mówi się wtedy, że sieć klasyfikuje obrazy „z grubsza". Gdy próg czułości jest wysoki, otrzymujemy w wyniku działania sieci ART wiele klas, często o jedynie subtelnych różnicach i rozpoznawanie jest bardzo precyzyjne. Z powodu tak prostego obliczania podobieństwa sieć nie radzi sobie dobrze z rozpoznawaniem zaburzonych danych (zniekształcone obrazy). Wtedy nie mamy wyboru i możemy rozpoznawać jedynie „z grubsza".

4. Uczenie podczas rezonowania. W tej fazie są modyfikowane wagi połączeń między neuronami. Następuje korekta wag według wzorów

$$v_{ij^*}(t+1) = v_{ij^*}(t) x_i,$$

$$w_{ij^*}(t+1) = \frac{v_{ij^*}(t+1)}{0.5 + \sum_{i=1}^{p} v_{ij^*}(t+1) x_i}.$$

Jest to element rezonowania obu warstw i może być nazywany raczej „douczaniem" sieci. Jest ono rozumiane jako adaptacja przechowywanego wzorca do aktualnie rozpoznanego kształtu. Osiągane jest to przez nałożenie obu sygnałów na siebie. Wagi połączeń dla zgodnych bitów są wzmacniane, a dla niezgodnych osłabiane lub pozostawiane bez zmian.

Sieć kończy swoje działanie, gdy wartości wag nie ulegają zmianie lub też nie ma już wolnych neuronów w warstwie wyjściowej, a co za tym idzie nie ma gdzie przypisać nowej klasy.

7.7. Sieci o radialnych funkcjach bazowych

Odmienną koncepcją w podejściu do sieci neuronowych jest zastosowanie nieliniowego przekształcenia przestrzeni danych wejściowych przez wprowadzenie warstwy neuronów ukrytych realizujących funkcje zmieniające się radialnie wokół wybranego centrum. Neuron radialny stanowi naturalne uzupełnienie neuronu sigmoidalnego, umożliwiające w przypadku wystąpienia symetrii kołowej danych wydatne zmniejszenie liczby neuronów oraz liczby warstw ukrytych potrzebnych do realizacji zadania klasyfikacyjnego. Neuron sigmoidalny reprezentował w przestrzeni wielowymiarowej hiperpłaszczyznę separującą tę przestrzeń na dwie klasy, neuron radialny z kolei reprezentuje hipersferę, dokonującą podziału kołowego wokół punktu centralnego.

Sieci radialne (RBF)[53] służą najczęściej do nieliniowej aproksymacji funkcji. Używane bywają również w zagadnieniach klasyfikacyjnych, gdzie odtwarzają funkcję rozkładu. Okazuje się, że sieć taka może służyć jako uniwersalny aproksymator i przybliżać z dowolną dokładnością każdą funkcję ciągłą. Podstawą teoretyczną działania sieci RBF jest twierdzenie Covera o separowalności wzorców.

TWIERDZENIE 7.6. (Covera, 1965)

Złożony problem klasyfikacyjny zrzutowany nieliniowo na przestrzeń wielowymiarową jest liniowo separowalny z większym prawdopodobieństwem niż przy rzutowaniu na przestrzeń o mniejszej liczbie wymiarów. Prawdopodobieństwo $P(N, m_1)$, że dwuklasowy problem klasyfikacyjny (zbiór uczą-

[53] Po ang. Radial Basis Functions (RBF).

cy ma rozmiar N) jest liniowo separowalny w przestrzeni o wymiarze m_1 wynosi

$$P(N,\ m_1) = \left(\frac{1}{2}\right)^{N-1} \sum_{m=0}^{m_1} \binom{N-1}{m}. \tag{7.58}$$

Wynika z tego (patrz Haykin (1994)), że każdy zbiór wzorców losowo rozmieszczony w przestrzeni wielowymiarowej jest φ-separowalny z prawdopodobieństwem równym jeden, pod warunkiem zastosowania odpowiednio dużego wymiaru, na który jest rzutowana ta przestrzeń. W praktyce oznacza to, że zastosowanie dwu warstw sieci, jednej zawierającej funkcje radialne i wyjściowej warstwy liniowej zapewnia rozwiązanie problemu klasyfikacji nieliniowej. Dzieje się tak dlatego, iż wystarczy jedynie przyjąć dostatecznie dużą liczbę neuronów w warstwie ukrytej.

Definicja 7.6.
Radialną funkcją bazową nazywamy funkcję φ postaci

$$\varphi(\boldsymbol{x},\ \boldsymbol{c}) = \varphi(r(\boldsymbol{x},\ \boldsymbol{c})),$$

gdzie $r(\boldsymbol{x},\ \boldsymbol{c})$ jest odległością między punktami $\boldsymbol{x}$ i $\boldsymbol{c}$. Punkt $\boldsymbol{c}$ nazywamy *centrum*.

Funkcje radialne są nazywane czasami *funkcjami jądrowymi*. Formuła określająca funkcję radialną może zawierać dodatkowe parametry, np. parametr σ określający szerokość jądra. Poniżej są przedstawione typowe przykłady radialnych funkcji bazowych

$$\varphi(r) = \exp\left(-\frac{r^2}{2\sigma^2}\right) \text{ – funkcja gaussowska,}$$

$$\varphi(r) = \frac{r}{\sigma^2} \text{ – funkcja liniowa,}$$

$$\varphi(r) = (\sigma^2 + r^2)^{-\alpha}, \quad \alpha > 0 \text{ – odwrotna funkcja wielokwadratowa,}$$

$$\varphi(r) = (\sigma^2 + r^2)^{\beta}, \quad 0 < \beta < 1 \text{ – funkcja wielokwadratowa,}$$

$$\varphi(r) = (r\sigma)^2 \ln(r\sigma) \text{ – funkcja TPS}^{54},$$

[54]Po ang. Thin Plate Spline (TPS).

$\varphi(r) = (r\sigma)^4 \ln(r\sigma)$ – radykalna funkcja TPS,

przy czym $\sigma > 0$. Na rysunku 7.27 przedstawiono graficznie powyższe funkcje.

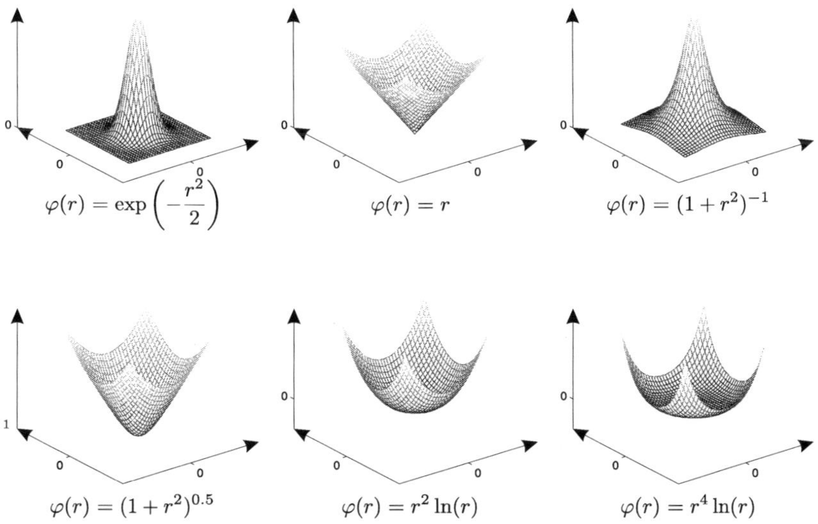

Rys. 7.27. Wybrane funkcje radialne

Przykład 7.16.
Nie jest możliwe rozdzielenie za pomocą separatora liniowego danych wejściowych należących do dwóch klas (klasa pierwsza – punkty A_1 i A_2, klasa druga – punkt B_1) przedstawionych na rys. 7.28a, gdzie A_1, A_2 i B_1 leżą na jednej prostej. Za pomocą funkcji

$$\varphi(\boldsymbol{x},\ \boldsymbol{c}) = \exp\left(-\frac{||\boldsymbol{x} - \boldsymbol{c}||^2}{2\sigma^2}\right),$$

gdzie $\boldsymbol{c}' = [2.5,\ 2]$ jest centrum oraz $\sigma = 0.8$ przestrzeń danych wejściowych dwuwymiarowych zostaje przekształcona w przestrzeń jednowymiarową separowalną liniowo (rys. 7.28b). Podobnego przekształcenia możemy dokonać za pomocą dwóch funkcji gaussowskich postaci

$$\varphi_1(\boldsymbol{x},\ \boldsymbol{c}_1) = \exp\left(-\frac{||\boldsymbol{x} - \boldsymbol{c}_1||^2}{2\sigma^2}\right),$$

$$\varphi_2(\boldsymbol{x},\ \boldsymbol{c}_2) = \exp\left(-\frac{||\boldsymbol{x} - \boldsymbol{c}_2||^2}{2\sigma^2}\right),$$

gdzie $\boldsymbol{c}_1' = [1.5\ 1]$ oraz $\boldsymbol{c}_2' = [2.5\ 3]$ są centrami. W tym przypadku przestrzeń dwuwymiarowa zostaje przekształcona w przestrzeń dwuwymiarową, dla której dane przedstawione mogą być rozdzielone za pomocą prostej. Przestrzeń dwuwymiarowa po przekształceniu jest przedstawiona na rys. 7.28c.

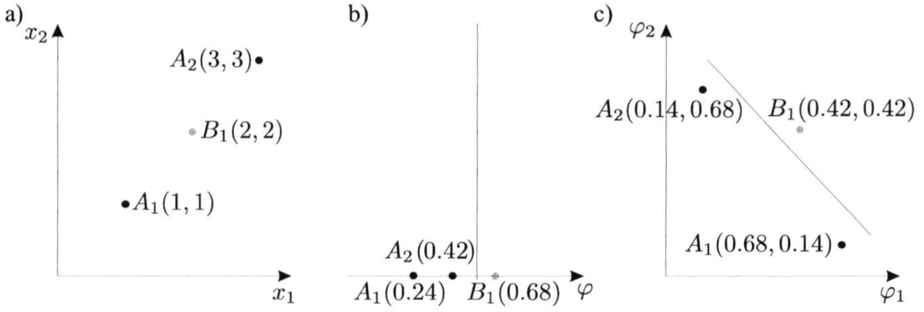

Rys. 7.28. a) Przestrzeń danych wejściowych przed przekształceniem, b) przestrzeń danych wejściowych po przekształceniu (1 wymiar), c) przestrzeń danych wejściowych po przekształceniu (2 wymiary)

Sieci RBF składają się z trzech warstw: ukrytej warstwy z neuronami o radialnych funkcjach aktywacji, warstwy wyjściowej zawierającej neurony liniowe oraz warstwy wejściowej. Warstwa ukryta składa się z $H < N$ neuronów (przy liczbie funkcji radialnych równej liczbie danych uczących sieć będzie się dopasowywała do różnego rodzaju szumów i nieregularności występujących w danych – straci zdolność uogólniania). Każdy z nich ma swoje centrum $\boldsymbol{c}_h$. Centra te są inicjowane podczas tworzenia struktury sieci, a następnie podczas procesu uczenia sieci są dostosowywane do danych uczących. Neuron warstwy wyjściowej ma przypisany wektor wag $\boldsymbol{w}$ oraz wartość progową w_0. Neuron ten wykonuje sumowanie dochodzących do niego z warstwy ukrytej sygnałów i przekazuje je na zewnątrz jako wynik odpowiadający wektorowi wejściowemu. Schemat tego rodzaju sieci znajduje się na rys. 7.29.

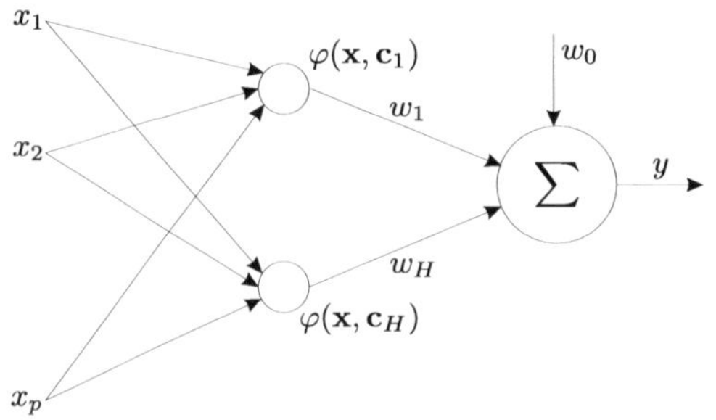

Rys. 7.29. Schemat sieci RBF

Algorytm uczenia sieci radialnej składa się z dwóch faz. W fazie pierwszej dobiera się położenia oraz kształt funkcji bazowych za pomocą:

- doboru losowego,
- metody k-średnich (samoorganizacja),
- metody wstecznej propagacji błędu.

Parametr określający szerokość jądra wyznacza się najczęściej jako średnią odległość centrów od ich najbliższych sąsiadów. Faza druga polega na doborze macierzy wag.

Definicja 7.7.
Macierz

$$\boldsymbol{G} = \begin{bmatrix} \varphi(\boldsymbol{x}_1, \boldsymbol{c}_1) & \varphi(\boldsymbol{x}_1, \boldsymbol{c}_2) & \ldots & \varphi(\boldsymbol{x}_1, \boldsymbol{c}_H) \\ \varphi(\boldsymbol{x}_2, \boldsymbol{c}_1) & \varphi(\boldsymbol{x}_2, \boldsymbol{c}_2) & \ldots & \varphi(\boldsymbol{x}_2, \boldsymbol{c}_H) \\ \ldots & \ldots & \ldots & \ldots \\ \varphi(\boldsymbol{x}_N, \boldsymbol{c}_1) & \varphi(\boldsymbol{x}_N, \boldsymbol{c}_2) & \ldots & \varphi(\boldsymbol{x}_N, \boldsymbol{c}_H) \end{bmatrix}$$

nazywamy *macierzą Greena.*

Macierz wag jest wyznaczana za pomocą *metody pseudoinwersji* macierzy Greena, tzn.

$$\boldsymbol{w} = \boldsymbol{G}^{+}\boldsymbol{d}(\boldsymbol{X}).$$

Sieci radialne wymagają zazwyczaj użycia większej liczby neuronów niż sieci jednokierunkowe o sigmoidalnej funkcji aktywacji. Pomimo to, uczenie ich trwa o wiele szybciej niż w przypadku sieci perceptronowych. Sieci radialne bazują na funkcjach mających wartość niezerową jedynie w określonej przestrzeni wokół centrów, realizują zatem aproksymację typu lokalnego, której zasięg działania jest bardzo ograniczony. Można się spodziewać, że zdolności do uogólniania takiej sieci są gorsze niż dla sieci sigmoidalnych. Na ogół uważa się, że sieci radialne lepiej niż sieci sigmoidalne nadają się do takich zadań klasyfikacyjnych, jak wykrywanie uszkodzeń w różnego rodzaju systemach, rozpoznawanie wzorców itp.

Przykład 7.17.
W przykładzie tym spróbujemy dokonać aproksymacji funkcji $f(x) = \sin x$ za pomocą radialnej sieci neuronowej. Wszystkie obliczenia i rysunki zostały

wykonane za pomocą pakietu Netlab[55]. W warstwie wyjściowej będzie oczywiście jeden neuron, którego zadaniem będzie pokazywanie wartości funkcji sinus dla konkretnego argumentu. Jako zbiór uczący posłużył zbiór 20 punktów wygenerowanych w następujący sposób: $(x_i,\ y_i = \sin(x_i) + 0.2\varepsilon)$, gdzie $\varepsilon \sim N(0,\ 1)$. Punkty x_i zostały równomiernie rozłożone na przedziale $[0,\ 2\pi]$. Maksymalna liczba epok została ustalona na 10. Za funkcję błędu przyjęto błąd średniokwadratowy. Wykonano doświadczenie dla $H = 2,\ 5,\ 8,\ 18$ neuronów w warstwie ukrytej. Jako funkcje radialne zostały wybrane funkcja gaussowska oraz funkcja TPS. Wynik działania sieci został przedstawiony na rys. 7.30.

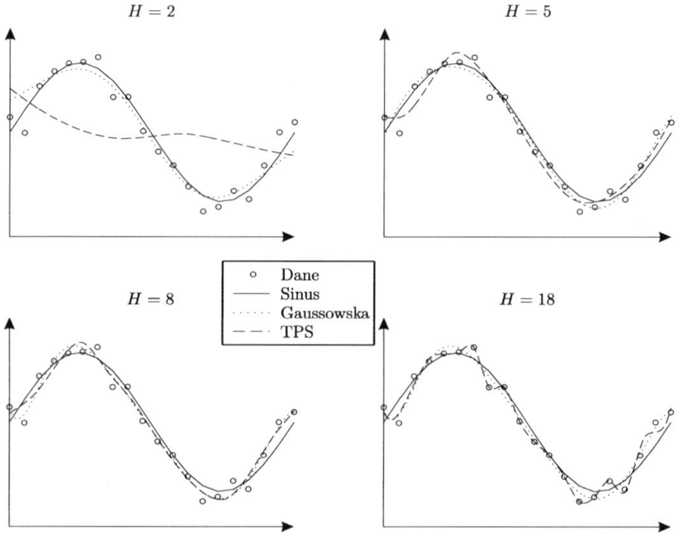

Rys. 7.30. Aproksymacja funkcji sinus za pomocą sieci RBF

Widzimy wyraźnie, że jedynie dwa neurony w warstwie ukrytej to zdecydowanie zbyt mało. Sieć w takim przypadku jest niedouczona. Optymalna sytuacja wydaje się zachodzić dla ośmiu neuronów w warstwie ukrytej. Taka sieć dobrze już rozpoznaje kształt funkcji sinus, a nie jest jeszcze przeuczona (jak w przypadku 18 neuronów ukrytych). Przeuczenie sieci skutkuje utratą zdolności generalizacji. Sieć taka nadmiernie dopasowuje się do danych uczących, co wyraźnie widać na rysunku. Sieć z funkcją gaussowską jako funkcją aktywacji wydaje się lepiej funkcjonować: szybciej się uczy i nie popada aż tak w przeuczenie.

Uwaga 7.18. Okazuje się, że jeśli w metodzie wektorów nośnych używa się jądra sigmoidalnego, to metoda ta sprowadza się do dwuwarstwowej sieci

[55] http://www.ncrg.aston.ac.uk/netlab/index.php.

RBF, gdzie liczba centrów równa się liczbie wektorów nośnych. Okazuje się, że taka metoda zachowuje się lepiej od klasycznej sieci RBF (Schölkopf i inni (1997)).

7.8. Probabilistyczne sieci neuronowe

W przypadku problemów klasyfikacyjnych użyteczną interpretacją wartości wyjściowych sieci jest ich traktowanie jako prawdopodobieństw przynależności do klas. W takiej sytuacji sieć w trakcie uczenia w rzeczywistości uczy się estymować funkcję gęstości prawdopodobieństwa reprezentowaną przez zgromadzone dane. Nauka rozwiązywania przez sieć problemu klasyfikacyjnego polega w rzeczywistości na próbie nauczenia się przez nią aproksymacji funkcji gęstości prawdopodobieństwa. W przypadku wykorzystania sieci neuronowych do aproksymacji funkcji gęstości wykorzystuje się podejście jądrowe. Działające w ten sposób sieci neuronowe nazywamy *probabilistycznymi sieciami neuronowymi* (PNN)[56].

W sieci tego typu występują przynajmniej trzy warstwy: wejściowa, ukryta radialna i wyjściowa. Neurony w warstwie ukrytej mają parametry kopiowane bezpośrednio z danych uczących; każdy z nich odpowiada jednemu przypadkowi. W sumie jest zatem w takiej sieci $2n + K$ neuronów. Każdy z neuronów ukrytych modeluje rozkład normalny (można oczywiście modelować inny rozkład, biorąc np. jądro Parzena) wycentrowany nad „swoim" przypadkiem uczącym. Natomiast w warstwie wyjściowej każdej klasie odpowiada jeden neuron. Do każdego z tych neuronów wyjściowych docierają połączenia od tych neuronów ukrytych, które zostały ustawione nad punktami należącymi do danej klasy. Nie ma połączeń neuronów wyjściowych z innymi neuronami ukrytymi. Neurony wyjściowe sumują więc wartości wyjściowe pojawiające się na wyjściach neuronów ukrytych należących do klasy odpowiadającej danemu neuronowi wyjściowemu. Wartości wyjściowe neuronów wyjściowych są więc proporcjonalne do estymatorów jądrowych funkcji gęstości prawdopodobieństwa dla różnych klas. Po zastosowaniu normalizacji zapewniającej ich sumowanie do jedności stanowią wprost oszacowania prawdopodobieństw a posteriori.

Powyższa cecha probabilistycznych sieci neuronowych jest niewątpliwie ich największą zaletą. Drugi kluczowy czynnik to duża szybkość ich uczenia (w zasadzie go nie ma, funkcję wag pełnią kolejne składowe wektorów obserwacji). Największą wadą sieci tego typu jest ich rozmiar. Musi ona bowiem

[56] Po ang. Probabilistic Neural Networks (PNN).

zawierać w swojej strukturze neurony ukryte odpowiadające wszystkim elementom zbioru uczącego, co powoduje, że w rzeczywistości w strukturze sieci jest odwzorowany cały zbiór uczący.

7.9. Uwagi praktyczne

Sieć neuronowa jest w swojej istocie pewnym wzorem (aproksymatorem regresyjnym), który da się przedstawić w formie graficznej. Wielokrotne ważone sumowanie i przekształcanie funkcją aktywacji występuje tyle razy, ile jest warstw neuronów w sieci. Jak więc widać, wyjście sieci neuronowej jest funkcją jej wejścia, przy czym funkcja ta jest parametryzowana zestawem parametrów $\boldsymbol{w}$.

W zagadnieniach klasyfikacyjnych celem stawianym sieci jest przypisanie każdego przypadku reprezentowanego przez odpowiedni zestaw danych wejściowych do jednej z wybranych klas. Mówiąc bardziej ogólnie, można zakładać, że zadaniem sieci jest estymacja prawdopodobieństwa przynależności przypadku do danej klasy.

Oprócz niskiego poziomu błędu uzyskiwanego przez sieć, kluczową własnością sieci neuronowej jest zdolność generalizacji swojej wiedzy na nowe przypadki. Niestety, zasadniczym problemem jest to, że w rzeczywistości w trakcie uczenia nie jest minimalizowany oczekiwany błąd sieci, lecz błąd wyznaczony dla zbioru uczącego. Najistotniejszym przejawem przedstawionego rozróżnienia między tym, czego chcielibyśmy uczyć sieć, a tym, czego ją naprawdę uczymy jest problem tak zwanego *przeuczenia sieci* (ang. overfitting). Najłatwiej jest to wyjaśnić na przykładzie regresji wielomianowej. Jeśli mamy pewne dane uczące, to zawsze możemy dobrać tak wysoki stopień wielomianu, że przejdzie on przez wszystkie punkty uczące. Dane te są jednak najprawdopodobniej zniekształcone przez szumy, dlatego też nie należy oczekiwać, aby najlepszy model przechodził dokładnie przez każdy z punktów. Przeciwnie, dobry model to taki, który odwzorowuje ogólną postać poszukiwanej zależności, ale abstrahuje od drobnych, zwykle nieistotnych „zafalowań" wejściowych danych. Skutek nadmiernego dopasowania łatwo zauważyć na wykresie wielomianu, który ma bardzo pofałdowany kształt, nieprzedstawiający istoty zależności. W przypadku sieci neuronowych pojawia się dokładnie ten sam problem. Sieć z większą liczbą wag może modelować bardziej złożone funkcje i z tego powodu ma większą skłonność do zbytniego dopasowywania się do danych. Sieć z mniejszą liczbą wag może z kolei nie być dostatecznie mocnym narzędziem do opisu występującej w rzeczywistości zależności. Na przykład sieć, która nie ma warstw ukrytych, może w rzeczywistości modelować wyłącznie proste zależności liniowe. Natomiast sieć o zbyt dużej liczbie warstw i zbyt dużej liczbie neuronów w warstwach ukrytych będzie miała skłonność do uczenia się „na pamięć" całego zbioru uczącego.

Sieć ma zdolność do *generalizacji*, jeśli jest w stanie prawidłowo zaklasyfikować dane należące do zbioru T (zbiór testowy) po nauczeniu jej za pomocą danych ze zbioru L (zbiór uczący) i zweryfikowaniu jej działania za pomocą danych ze zbioru V (zbiór *walidacyjny*) – rys. 7.31.

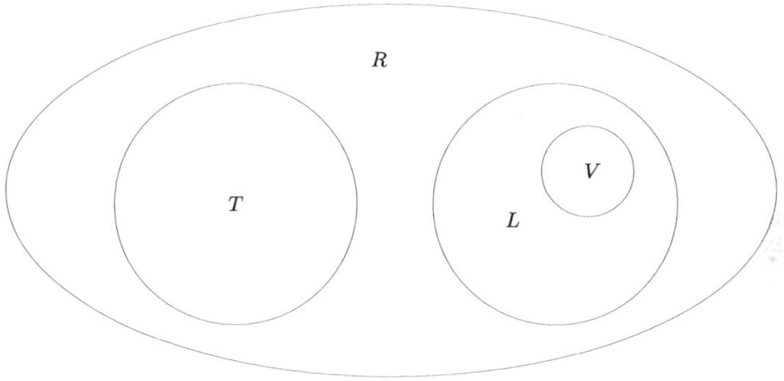

Rys. 7.31. Podział danych w przypadku uczenia sieci neuronowych

Dane należące do zbioru V nie są bezpośrednio stosowane w trakcie uczenia sieci. Są one wykorzystywane do przeprowadzenia niezależnej kontroli postępów algorytmu uczenia. W trakcie uczenia błąd popełniany przez sieć zmniejsza się, jak również zmniejsza się błąd walidacyjny. Na ogół po początkowym szybkim spadku obu rodzajów błędów obserwujemy szybszy spadek błędu dla zbioru L i wolniejszy dla zbioru V. Jest to zjawisko normalne, ponieważ sieć poprawia swoje działanie wyłącznie na podstawie danych ze zbioru uczącego, więc dopasowanie jej zachowania do tego właśnie zbioru musi być szybsze i dokładniejsze. Jeśli jednak w trakcie uczenia zaobserwujemy, że spadek błędu walidacyjnego zatrzymał się lub też błąd ten zaczyna rosnąć, to świadczy to, że sieć zaczęła zbytnio dopasowywać się do danych uczących i traci zdolność do generalizacji wyników uczenia. W takim przypadku proces uczenia powinien zostać zatrzymany, a nawet cofnięty do takiego punktu, w którym błąd walidacji miał najmniejszą wartość. Przeuczenie sieci polega więc na pojawiającym się w trakcie uczenia nadmiernym dopasowaniu się sieci do danych uczących. W takim przypadku wskazane jest, aby zmniejszyć liczbę neuronów ukrytych lub liczbę warstw ukrytych, ponieważ pojawienie się przeuczenia sugeruje, że rozwiązując istniejący problem, zastosowano sieć o zbyt dużych możliwościach (a co za tym idzie – komplikacji). W sytuacji przeciwnej, gdy sieć nie ma dostatecznych możliwości do modelowania rzeczywistej funkcji, przeuczenie się nie pojawi, ale wtedy mimo długiego uczenia, ani błąd uczenia, ani błąd walidacyjny nie spadnie do satysfakcjonującego poziomu.

Dane powinny być standaryzowane w taki sposób, aby średnia każdej zmiennej wynosiła 0, a jej wariancja była równa 1.

Wagi początkowe są generowane losowo z rozkładu $U\left(\frac{-1}{\sqrt{d}}, \frac{1}{\sqrt{d}}\right)$ (przy założeniu, że dane zostały poddane standaryzacji), gdzie d oznacza liczbę wejść do neuronu.

Niewskazane jest występowanie w zbiorze uczącym elementów powtarzających się. Jest to zjawisko „nadreprezentacji" – nadmiernej reprezentacji pewnej części dziedziny aproksymowanej funkcji. Jest ono niepożądane, ponieważ algorytm doboru parametrów może mieć tendencję do „skupiania się" na tych punktach – ewentualny błąd w owych punktach będzie miał większą wagę i zostanie zminimalizowany w pierwszej kolejności. Wynik może być podobny, jak przy braku zbioru testującego – w części dziedziny sieć będzie miała mniejszy błąd, co zapewne spowoduje (dużo) większy błąd w innych jej częściach. Jeśli nie zależy nam na szczególnie dobrej aproksymacji w wybranych elementach (w wybranej części dziedziny aproksymowanej funkcji), powinniśmy usunąć wszystkie powtórzenia z danych.

Należy pamiętać, że wraz ze wzrostem liczby neuronów w sieci rośnie czas jej uczenia. Z tego względu powstało wiele metod służących do redukcji struktury sieci. W metodach tych usuwa się najczęściej neurony lub połączenia, które nie mają znaczenia. Ingerencja w strukturę sieci następuje po jej nauczeniu. Najprostszą metodą jest usuwanie wag mniejszych od pewnego ustalonego progu. Niestety, nie zawsze jest to działanie, które prowadzi do spadku błędu sieci. Należy zatem pamiętać, aby zawsze po usunięciu połączeń jeszcze raz uczyć sieć. Jedną z metod upraszczania sieci i uniknięcia zjawiska nadmiernego dopasowania się sieci do danych uczących jest metoda zwana *rozpadem wag*[57]. Przyjmujemy w tym przypadku założenie, że wagi powinny być małe (niestety, nie zawsze prowadzi to do poprawy działania sieci, a wręcz może spowodować spadek jej efektywności). Po każdej zmianie wag następuje jeszcze przemnożenie nowej wagi przez czynnik $1 - \varepsilon$, gdzie $\varepsilon \in (0,\ 1)$. Jest to podejście bardzo zbliżone do wykorzystania współczynnika zapominania w regule Hebba. Można pokazać, że metoda ta jest równoważna zastosowaniu w metodzie największego spadku następują-

[57]Po ang. weight decay.

cej funkcji błędu

$$Q_{wd}(\boldsymbol{w}) = Q(\boldsymbol{w}) + \frac{2\varepsilon}{\eta}\boldsymbol{w}'\boldsymbol{w}. \tag{7.59}$$

Drugi człon po prawej stronie wzoru (7.59) jest zwany czynnikiem regularyzującym. Inną metodą redukcji struktury sieci jest *metoda OBD*[58] zaproponowana przez Le Cuna, Denkera i Sollę (1990). Ponieważ naszym celem jest usunięcie z już nauczonej sieci pewnych elementów, musimy zatem zbadać ich wpływ na działanie sieci. Ponieważ redukcji wag dokonujemy, gdy sieć jest już nauczona, możemy pominąć wektor gradientu (jesteśmy w minimum). Obcinając rozwinięcie szeregu Taylora do trzeciego wyrazu, mamy

$$\Delta Q(\boldsymbol{w}) = \frac{1}{2}(\boldsymbol{w} - \boldsymbol{w}^*)'\boldsymbol{H}(\boldsymbol{w}^*)(\boldsymbol{w} - \boldsymbol{w}^*), \tag{7.60}$$

gdzie $\boldsymbol{w}$ jest zbiorem wag po operacji usuwania, natomiast $\boldsymbol{w}^*$ jest zbiorem wag sieci po skończeniu procesu uczenia, ale przed usuwaniem. Ponieważ wyznaczenie hesjanu we wzorze (7.60) jest uciążliwe ze względu na dużą liczbę wag, przyjmuje się, że największy wpływ na wartość hesjanu mają jego elementy diagonalne. Stąd miarę wrażliwości[59] dla każdej wagi obliczamy następująco:

$$s_{ij} = \frac{1}{2}\frac{\partial^2 Q}{\partial w_{ij}^2} w_{ij}^2. \tag{7.61}$$

Po obliczeniu dla wszystkich wag współczynnika wrażliwości są usuwane te wagi, dla których wartość tego współczynnika jest niewielka i sieć jest ponownie uczona. Ponieważ hesjan jest macierzą niediagonalną, algorytm OBD może usunąć z sieci istotne wagi. Z tego powodu Hassibi, Stork i Wolf (1993) zaproponowali *metodę OBS*[60], w której uwzględnia się wszystkie składowe hesjanu. Najpierw obliczymy współczynniki wrażliwości

$$s_j = \frac{1}{2}\frac{w_j^2}{\boldsymbol{H}_{jj}^{-1}}. \tag{7.62}$$

Teraz zmniejszeniu ulegają wagi, dla których wartość wyrażenia (7.62) jest mniejsza od pewnego ustalonego progu. Poprawka ma wartość

$$\Delta \boldsymbol{w} = -\frac{w_j}{\boldsymbol{H}_{jj}^{-1}}\boldsymbol{H}^{-1}\boldsymbol{i}_j, \tag{7.63}$$

[58] Po ang. Optimal Brain Damage (OBD).

[59] Po ang. saliency.

[60] Po ang. Optimal Brain Surgeon (OBS).

gdzie $\boldsymbol{i}_j$ jest wektorem składającym się z samych zer poza j-tą pozycją, na której znajduje się jedynka. Omówione strategie prowadzące do prostszej struktury sieci działają w sposób zstępujący, tzn. uczymy dużą sieć, a potem redukujemy. Można sobie wyobrazić sytuację odwrotną, w której zaczynamy od jak najprostszej sieci, którą następnie rozbudowujemy, aż do uzyskania satysfakcjonującego poziomu błędu. Jedną z najbardziej znanych metod tego typu jest *metoda korelacji kaskadowej*[61] zaproponowana przez Fahlmana i Lebiere'a (1989).

Algorytm korelacji kaskadowej

1. Zaczynamy od jednej warstwy wejściowej i jednej wyjściowej.
2. Uczymy sieć.
3. Dodajemy jeden neuron i podłączamy go do warstwy wejściowej i do warstwy ukrytej.
4. Uczymy tylko wagi tego nowego neuronu, maksymalizując korelację między jego wyjściem a błędem sieci.
5. Zamieniamy nowo wprowadzony neuron na neuron ukryty.
6. Uczymy wagi wyjściowe.
7. Obliczamy błąd sieci i jeśli jest zadowalający, to kończymy, w przeciwnym razie wracamy do punktu 3.

Graficzna prezentacja algorytmu korelacji kaskadowej znajduje się na rys. 7.32.

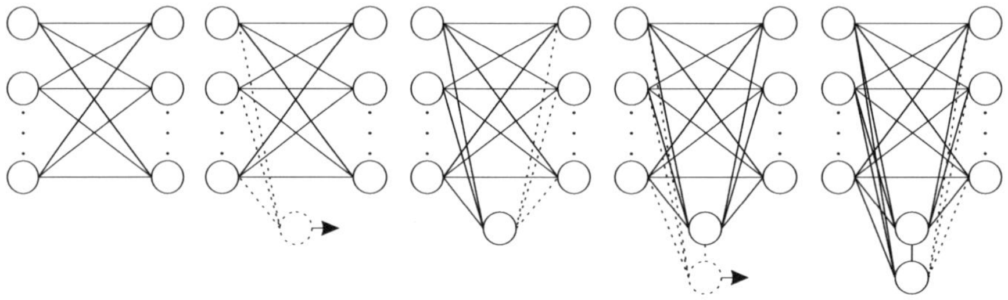

Rys. 7.32. Algorytm korelacji kaskadowej

Przy wyborze sieci należy postępować zgodnie z regułą głoszącą, że w przypadku gdy wszystkie parametry dwóch modeli są identyczne, to należy wybrać model prostszy, odrzucając model o większej złożoności.

---

[61] Po ang. cascade correlation.

Jedną z niewątpliwych wad sieci neuronowych jest ich nieprzezroczystość. W przeciwieństwie do drzew decyzyjnych nie istnieją bezpośrednie procedury przetwarzające wagi sieci neuronowych w zbiór reguł decyzyjnych. Istnieje jednak pewna technika zwana *analizą czułości*[62], która umożliwia względny pomiar wpływu każdej z cech na wynik działania sieci. Larose (2006) proponuje następujący algorytm analizy czułości.

Algorytm analizy czułości

1. Stwórz nową obserwację $\bar{\boldsymbol{x}}$, z wartością każdej cechy równą średniej ze wszystkich obserwacji zbioru testowego.
2. Oblicz wyjście sieci $\bar{y}$ dla $\bar{\boldsymbol{x}}$.
3. Cecha po cesze zmieniaj $\bar{\boldsymbol{x}}$ tak, aby odzwierciedlić wartości minimalną i maksymalną cechy. Oblicz wyjście sieci dla każdej zmiany i porównaj z $\bar{y}$.

Z analizy czułości wynika, że zmiana pewnych wartości cech z wartości minimalnych na maksymalne będzie miała większy wpływ na wynik działania sieci niż w przypadku innych cech.

[62] Po ang. sensitivity analysis.

8

Dekompozycja zagadnień wieloklasowych

Rozważmy zagadnienie klasyfikacyjne, dla którego liczba klas K jest większa od dwóch. Zagadnienie takie będziemy nazywali *wieloklasowym*. Jak jednak rozwiązać problem klasyfikacji wieloklasowej w sytuacji, gdy dana metoda daje się zastosować jedynie w przypadku *dwuklasowym* (*binarnym*). Przykładami takich metod są omówione w podrozdz. 1.8 klasyfikatory liniowe Andersona–Bahadura. Co więcej, nawet gdy istnieje „naturalne" rozwiązanie wieloklasowe, to zazwyczaj przypadek $K = 2$ jest najprostszy obliczeniowo. W tej sytuacji sensowne wydaje się „sprowadzenie" jednego dużego zagadnienia wieloklasowego do być może wielu, ale jednak o wiele prostszych, zagadnień binarnych.

W rozdziale tym omówimy zagadnienie dekompozycji wieloklasowego zagadnienia klasyfikacyjnego oraz związane z tym metody łączenia klasyfikatorów binarnych w jeden klasyfikator wieloklasowy.

8.1. Metoda OPC

Rozpocznijmy od najprostszej metody dekompozycji zwanej *jeden przeciwko wszystkim* (OPC)[1]. Zagadnienie z $K > 2$ klasami rozkładamy na K zagadnień binarnych. W każdym z nich ustalamy jedną z klas (obserwacjom z próby uczącej $\mathcal{L}_n$ z tej klasy przypisujemy nową etykietę 1), a drugą „super klasę" tworzą pozostałe $K - 1$ klasy (obserwacjom uczącym z tych klas nadajemy wspólną etykietę 0). Ten schemat dekompozycyjny można przedstawić w postaci *macierzy dekompozycji* $\boldsymbol{D}$, która dla $K = 4$ ma postać

[1] Po ang. One Per Class (OPC).

$$
\boldsymbol{D} = \begin{pmatrix} 1\,0\,0\,0 \\ 0\,1\,0\,0 \\ 0\,0\,1\,0 \\ 0\,0\,0\,1 \end{pmatrix}.
$$

Wiersze macierzy dekompozycji $\boldsymbol{D}$ odpowiadają poszczególnym klasom, natomiast kolumny klasyfikatorom binarnym. Każdej klasie możemy zatem przypisać kod złożony z K zer i jedynek, np. dla $K = 4$ kod klasy drugiej ma postać: 0100. Ogólnie kod przypisany klasie k-tej ma postać $w_k = w_{k1}w_{k2}\cdots w_{kK}$, gdzie $w_{ki} = 1$, gdy $k = i$ oraz $w_{ki} = 0$, gdy $k \neq i$, $(k,\ i\ = 1,\ 2,\ \ldots,\ K)$.

Aby zaklasyfikować nowy obiekt $\boldsymbol{x}$, wyznaczamy odpowiadający mu kod

$$
w(\boldsymbol{x}) = \hat{d}_1(\boldsymbol{x})\hat{d}_2(\boldsymbol{x})\cdots\hat{d}_K(\boldsymbol{x}),
$$

gdzie $\hat{d}_i$ $(i = 1,\ 2,\ \ldots,\ K)$ oznacza odpowiedni klasyfikator binarny. Następnie, korzystając z tego kodu, obliczamy odległość obiektu $\boldsymbol{x}$ od każdej z K klas. W tym celu wykorzystujemy odległość postaci

$$
\rho_1(\boldsymbol{x},\ G_k) = \sum_{i=1}^{K} |\hat{d}_i(\boldsymbol{x}) - w_{ki}|, \qquad k = 1,\ 2,\ \ldots,\ K, \tag{8.1}
$$

która jest w tym przypadku równa liczbie niezgodnych znaków w kodach odpowiadających klasie k i obserwacji $\boldsymbol{x}$. Odległość tę nazywamy *odległością Hamminga*.

Klasyfikator metody OPC ma wtedy postać

$$
\hat{d}_{OPC}(\boldsymbol{x}) = \arg \min_{1\leq k\leq K} \rho_1(\boldsymbol{x},\ G_k). \tag{8.2}
$$

Zwróćmy uwagę, że klasyfikator określony wzorem (8.2) prowadzi często do sytuacji remisowych (minimalna odległość jest osiągana dla więcej niż jednej klasy). Wyjściem z tej sytuacji może być zastąpienie etykiety $\hat{d}_i(\boldsymbol{x})$ estymatorem prawdopodobieństwa a posteriori $\hat{p}_i(\boldsymbol{x})$ przynależności obserwacji $\boldsymbol{x}$ do klasy z etykietą 1, uzyskanym w i-tym zagadnieniu binarnym $(i = 1,\ 2,\ \ldots,\ K)$.

Idea konstrukcji klasyfikatora (8.2) nie ulega zmianie. Różnica dotyczy jedynie sposobu obliczania odległości obiektu $\boldsymbol{x}$ od k-tej klasy. Odległość daną wzorem (8.1) zastępujemy odległością postaci

$$
\rho_2(\boldsymbol{x},\ G_k) = \sum_{i=1}^{K} |\hat{p}_i(\boldsymbol{x}) - w_{ki}|, \qquad k = 1,\ 2,\ \ldots,\ K. \tag{8.3}
$$

W algorytmach wykorzystujących metodę dekompozycji OPC zamiast odległością (8.3) posługujemy się często odległością

$$\rho_3(\boldsymbol{x},\ G_k) = \max_{1\le i\le K} |\hat{p}_i(\boldsymbol{x}) - w_{ki}|, \qquad k = 1,\ 2,\ \ldots,\ K. \tag{8.4}$$

Reguła klasyfikacyjna oparta na odległości (8.4) przyporządkowuje obserwacje do tej klasy, dla której oszacowane na bazie klasyfikatora binarnego $\hat{d}_i$ prawdopodobieństwo a posteriori $\hat{p}_i(\boldsymbol{x})$ jest maksymalne.

8.2. Metoda ECOC

Opisana w poprzednim podrozdziale metoda dekompozycji OPC, choć bardzo prosta, jest jednak mało efektywna. Dla K klas macierz dekompozycji $\boldsymbol{D}$ jest równa macierzy jednostkowej I_K, co między innymi oznacza, że w tym schemacie liczba używanych klasyfikatorów binarnych L jest zawsze równa liczbie klas K, a w kodzie odpowiadającym każdej z klas znajduje się dokładnie jedna jedynka.

W dalszym ciągu zajmiemy się bardziej skomplikowanymi schematami dekompozycyjnymi. Rozpocznijmy od następującego przykładu.

Przykład 8.1.
Rozważamy zagadnienie rozpoznawania ręcznie napisanych cyfr. W tym przypadku $K = 10$. Dokonajmy dekompozycji tego 10-klasowego zagadnienia na $L = 6$ zagadnień binarnych. Macierz dekompozycji $\boldsymbol{D}$ ma postać

$$\boldsymbol{D} = \begin{pmatrix} 0\ 0\ 0\ 1\ 0\ 0 \\ 1\ 0\ 0\ 0\ 0\ 0 \\ 0\ 1\ 1\ 0\ 1\ 0 \\ 0\ 0\ 0\ 0\ 1\ 0 \\ 1\ 1\ 0\ 0\ 0\ 0 \\ 1\ 1\ 0\ 0\ 1\ 0 \\ 0\ 0\ 1\ 1\ 0\ 1 \\ 0\ 0\ 1\ 0\ 0\ 0 \\ 0\ 0\ 0\ 1\ 0\ 0 \\ 0\ 0\ 1\ 1\ 0\ 0 \end{pmatrix}.$$

Wiersze macierzy dekompozycji $\boldsymbol{D}$ odpowiadają, po kolei, cyfrom: 0, 1, 2, 3, 4, 5, 6, 7, 8, 9, kolumny zaś (odpowiadające klasyfikatorom binarnym) mają następującą interpretację:

- kolumna 1: cyfra zawiera linię pionową,
- kolumna 2: cyfra zawiera linię poziomą,
- kolumna 3: cyfra zawiera linię ukośną,
- kolumna 4: cyfra zawiera linię zamkniętą,
- kolumna 5: cyfra zawiera linię otwartą z lewej strony,
- kolumna 6: cyfra zawiera linię otwartą z prawej strony.

Każdej cyfrze (klasie) odpowiada zatem 6-znakowy kod złożony z zer i jedynek, np. dla cyfry 5 kod ten ma postać: 110010.
Aby zaklasyfikować nową, ręcznie zapisaną cyfrę, wyznaczamy odpowiadający jej kod, rozwiązując 6 zagadnień binarnych (opisanych przez kolumny macierzy $\boldsymbol{D}$). Powiedzmy, że uzyskany w ten sposób kod to: 110001. Analogicznie jak w metodzie OPC, cyfrę klasyfikujemy do klasy najbliższej jej w sensie odległości Hamminga – w naszym przypadku jest to klasa odpowiadająca cyfrze 4 z kodem 110000.

Zwróćmy uwagę na to, że w powyższym przykładzie minimalna odległość Hamminga między kodami różnych klas wynosi 1, np. między kodami odpowiadającymi cyfrom 0 i 9. Stąd pojedynczy błąd jednego z klasyfikatorów binarnych może zadecydować o końcowej błędnej klasyfikacji.

Dietterich i Bakiri (1991, 1995) zaproponowali schemat dekompozycyjny umożliwiający korekcję co najmniej kilku błędnych decyzji klasyfikatorów binarnych. Schemat ten będziemy oznaczać ECOC[2]. Formalnie klasyfikator metody ECOC różni się od klasyfikatora metody OPC jedynie liczbą klasyfikatorów binarnych. W metodzie OPC ich liczba była zawsze równa liczbie klas ($L = K$). W schematach pozwalających na korekcję błędów liczba klasyfikatorów binarnych L jest z reguły dużo większa od liczby klas K. Ogólna postać takiego klasyfikatora, to

$$\hat{d}_{ECOC}(\boldsymbol{x}) = \arg \min_{1 \le k \le K} \rho_i(\boldsymbol{x},\ G_k), \qquad i = 1,\ 2, \tag{8.5}$$

a wykorzystywane do jego konstrukcji odległości różnią się od tych danych wzorami (8.1) i (8.3) jedynie długością kodów.

Możliwość korekcji błędnych decyzji klasyfikatorów binarnych jest zależna od odległości Hamminga między kodami klas. Jeżeli minimalna odległość jest równa h, to skorygować możemy co najmniej $\lfloor (h-1)/2 \rfloor$ błędnych decyzji pojedynczych klasyfikatorów binarnych. Przykładowo poniższa macierz dekompozycji $\boldsymbol{D}$ zastosowana do sytuacji rozpoznawania cyfr z przykładu 8.1 ma minimalną wartość odległości Hamminga równą 7, co umożliwia korekcję co najmniej trzech błędnych decyzji

[2]Po ang. Error-Correcting Output Codes (ECOC).

$$
\boldsymbol{D} = \begin{pmatrix}
1&1&0&0&0&0&1&0&1&0&0&1&1&0&1\\
0&0&1&1&1&1&0&1&0&1&1&0&0&1&0\\
1&0&0&1&0&0&0&1&1&1&1&0&1&0&1\\
0&0&1&1&0&1&1&1&0&0&0&0&1&0&1\\
1&1&1&0&1&0&1&1&0&0&1&0&0&0&1\\
0&1&0&0&1&1&0&1&1&1&0&0&0&0&1\\
1&0&1&1&1&0&0&0&0&1&0&1&0&0&1\\
0&0&0&1&1&1&1&0&1&0&1&1&0&0&1\\
1&1&0&1&0&1&1&0&0&1&0&0&0&1&1\\
0&1&1&1&0&0&0&0&1&0&1&0&0&1&1
\end{pmatrix}.
$$

Jak zatem powinna wyglądać macierz dekompozycji $\boldsymbol{D}$ dobrze korygująca błędy pojedynczych klasyfikatorów binarnych?

- Przede wszystkim macierz $\boldsymbol{D}$ powinna mieć możliwie duże odległości Hamminga między wierszami – kodami klas.
- Żadna z kolumn macierzy $\boldsymbol{D}$ nie może być złożona wyłącznie z samych zer albo jedynek.
- Żadne dwie kolumny macierzy $\boldsymbol{D}$ nie mogą być uzupełniające się, tzn. różnić się jedynie tym, że w miejscu zera występuje jedynka, a w miejscu jedynki występuje zero. Gdyby tak było, to decyzje podejmowane przez klasyfikatory odpowiadające tym kolumnom byłyby identyczne.
- Kolumny macierzy $\boldsymbol{D}$ powinny być możliwie mocno odległe od siebie. Klasyfikatory binarne odpowiadające bliskim sobie kolumnom będą dawały podobne (skorelowane) błędy.

Dla K klas macierz dekompozycji $\boldsymbol{D}$ może mieć maksymalnie 2^K różnych kolumn. Usuwając kolumny złożone z samych zer albo jedynek oraz kolumny uzupełniające się, stwierdzamy, że liczba „użytecznych" kolumn wynosi co najwyżej $2^{K-1} - 1$.

Dla niewielkiej liczby klas K ($3 \leq K \leq 7$) można jako macierz dekompozycji przyjąć macierz składającą się z wszystkich użytecznych kolumn. Jej konstrukcja jest następująca: pierwszy wiersz składa się tylko z jedynek, drugi wiersz składa się z następujących po sobie ciągów 2^{K-2} zer i $2^{K-2} - 1$ jedynek, trzeci wiersz tworzą ciągi 2^{K-3} zer, 2^{K-3} jedynek, 2^{K-3} zer i $2^{K-3} - 1$ jedynek itd. Ogólnie, każdy wiersz tej macierzy (poza pierwszym) to naprzemiennie ułożone ciągi 2^{K-i} zer i 2^{K-i} jedynek.

Przykładowo, dla $K = 5$ uzyskana w powyższy sposób macierz dekompozycji $\boldsymbol{D}$ ma postać

$$\boldsymbol{D} = \begin{pmatrix} 1\,1\,1\,1\,1\,1\,1\,1\,1\,1\,1\,1\,1\,1\,1 \\ 0\,0\,0\,0\,0\,0\,0\,1\,1\,1\,1\,1\,1\,1\,1 \\ 0\,0\,0\,0\,1\,1\,1\,1\,0\,0\,0\,0\,1\,1\,1 \\ 0\,0\,1\,1\,0\,0\,1\,1\,0\,0\,1\,1\,0\,0\,1 \\ 0\,1\,0\,1\,0\,1\,0\,1\,0\,1\,0\,1\,0\,1\,0 \end{pmatrix}.$$

Minimalna odległość Hamminga między wierszami tej macierzy wynosi 8. Pozwala to nam na korekcję co najmniej trzech błędnych decyzji klasyfikatorów binarnych.

Przy dużej liczbie klas K liczba użytecznych klasyfikatorów binarnych $L = 2^{K-1} - 1$ jest na tyle duża, że nie ma praktycznej możliwości użycia ich wszystkich. Przykładowo, dla zagadnienia z 10 klasami, rozważanego w przykładzie 8.1, wynosi ona $2^9 - 1 = 511$. Posługujemy się wtedy mniejszą liczbą klasyfikatorów binarnych mających jednak tę własność, że odległość Hamminga między kodami klas jest możliwie duża (co zapewnia możliwość korekcji wielu błędnych decyzji).

Dietterich i Bakiri proponują wykorzystanie do konstrukcji takich kodów algorytm BCH (Bose i Ray-Chaudhuri (1960), Hocquenghem (1950)). Opis tego algorytmu, wykorzystującego algebraiczne własności ciał Galois, znaleźć można w wielu monografiach, np. MacWilliams i Sloane (2006).

Metoda ECOC może być również traktowana jako technika wzmacniania klasyfikatorów. Inne techniki wzmacniania klasyfikatorów zostaną omówione w rozdziale 9. Zastosowanie metody ECOC powoduje jednoczesną redukcję wariancji i obciążenia klasyfikatora wieloklasowego. Ma to szczególne znaczenie w przypadku takich niestabilnych algorytmów, jak drzewa klasyfikacyjne czy sieci neuronowe. Dokładne informacje na ten temat znaleźć można w pracy Konga i Dietterícha (1995).

8.3. Metoda PWC

Zajmiemy się teraz innym schematem dekompozycyjnym, który będziemy oznaczać PWC[3].

Tym razem zagadnienie z $K > 2$ klasami rozkładamy na $L = K \times (K-1)/2$ zagadnień binarnych. Każde z nich dotyczy jednej z możliwych

[3] Po ang. pairwise coupling (PWC).

par klas (i, j), gdzie $i, j = 1, 2, \ldots, K, i < j$. Obserwacjom z próby uczącej z klasy i-tej przypisujemy etykietę 1, a obserwacjom uczącym z klasy j-tej nadajemy etykietę 0. Powyższemu schematowi dekompozycyjnemu odpowiada macierz dekompozycji D, która dla $K = 4$ ma postać

$$D = \begin{pmatrix} 1 & 1 & 1 & - & - & - \\ 0 & - & - & 1 & 1 & - \\ - & 0 & - & 0 & - & 1 \\ - & - & 0 & - & 0 & 0 \end{pmatrix}.$$

Symbol „–” w macierzy dekompozycyjnej oznacza, że elementy próby uczącej z danej klasy nie uczestniczą w budowie klasyfikatora binarnego.

Najprostszym sposobem na połączenie informacji z L klasyfikatorów binarnych jest zwykła procedura głosowania. Klasyfikator taki ma postać

$$\hat{d}_V(\boldsymbol{x}) = \arg \max_{1 \leq k \leq K} \sum_{l=1}^{L} \mathrm{I}(\hat{d}_l(\boldsymbol{x}) = w_{kl}), \tag{8.6}$$

gdzie $\hat{d}_l$ oznaczają klasyfikatory binarne, a w_k jest kodem odpowiadającym k-tej klasie (tym razem kod zawiera znaki: 0, 1, –).

Podstawową wadą tej procedury jest brak sensownego sposobu na rozwiązywanie pojawiających się często sytuacji remisowych. Taki „bezdecyzyjny” obszar widać wyraźnie na rys. 8.1.

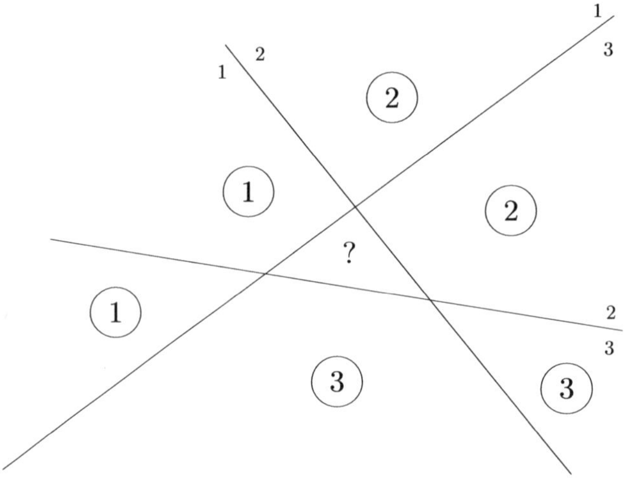

Rys. 8.1. Obszary klasyfikacyjne w procedurze głosowania. Obszar bezdecyzyjny oznaczono znakiem zapytania

Wady tej nie ma procedura zaproponowana w 1998 roku przez Hastiego i Tibshiraniego.

Niech $p_i(\boldsymbol{x}) = \mathrm{P}(Y = i|\boldsymbol{X} = \boldsymbol{x})$, $i = 1, 2, \ldots, K$, oznacza prawdopodobieństwo a posteriori przynależności obserwacji do i-tej klasy.

W klasyczny sposób regułę klasyfikacyjną definiujemy następująco:

$$\hat{d}_{HT}(\boldsymbol{x}) = \arg\max_{1 \leq i \leq K} \hat{p}_i(\boldsymbol{x}). \tag{8.7}$$

W jaki sposób jednak oszacować prawdopodobieństwa a posteriori $p_i(\boldsymbol{x})$, wykorzystując jedynie klasyfikatory binarne?

Niech $p_{ij}(\boldsymbol{x}) = \mathrm{P}(Y = i|Y = i \text{ lub } Y = j,\ \boldsymbol{X} = \boldsymbol{x})$, $i, j = 1, 2, \ldots, K$, $j \neq i$, oznaczają prawdopodobieństwa a posteriori przynależności obserwacji do i-tej klasy, pod warunkiem że rozpatrujemy tylko parę klas (i, j).

Wtedy

$$p_{ij}(\boldsymbol{x}) = \frac{p_i(\boldsymbol{x})}{p_j(\boldsymbol{x}) + p_j(\boldsymbol{x})}, \quad i < j, \quad (p_{ji}(\boldsymbol{x}) = 1 - p_{ij}(\boldsymbol{x})). \tag{8.8}$$

Prawdopodobieństwa $p_{ij}(\boldsymbol{x})$ oszacować możemy, wykorzystując klasyfikatory binarne. Bezpośrednie wykorzystanie zależności (8.8) do oszacowania prawdopodobieństw $p_i(\boldsymbol{x})$ jest, niestety, niemożliwe. Prowadzi ono bowiem do rozwiązania, zazwyczaj sprzecznego, układu $L = K(K-1)/2$ równań z K niewiadomymi.

W swojej procedurze Hastie i Tibshirani zaproponowali oszacowanie prawdopodobieństw a posteriori $p_i(\boldsymbol{x})$ poprzez minimalizację odległości Kullbacka–Leiblera postaci

$$\rho_{KL}(\hat{p}_1(\boldsymbol{x}), \ldots, \hat{p}_K(\boldsymbol{x})) = \sum_{j \neq i} n_{ij} \left[\hat{p}_{ij}(\boldsymbol{x}) \log \frac{\hat{p}_{ij}(\boldsymbol{x})}{p_{ij}(\boldsymbol{x})}\right], \tag{8.9}$$

gdzie liczebności $n_{ij} = n_i + n_j$ pełnią funkcję wag.

Do wyznaczenia tych oszacowań wykorzystujemy algorytm Bradleya–Terry'ego.

Algorytm Bradleya–Terry'ego

1. Przyjmujemy wartości początkowe prawdopodobieństw $\hat{p}_1(\boldsymbol{x}), \ldots$ $\ldots, \hat{p}_K(\boldsymbol{x})$ tak, aby $\sum_{k=1}^{K} \hat{p}_k(\boldsymbol{x}) = 1$.

2. Powtarzamy następujące kroki, aż do uzyskania zbieżności:
   (a) dla każdego i, $j = 1, 2, \ldots, K$, $j \neq i$ obliczamy prawdopodobieństwa p_{ij}, korzystając z zależności (8.8),
   (b) dla każdego $i = 1, 2, \ldots, K$

$$\hat{p}_i(\boldsymbol{x}) \leftarrow \hat{p}_i(\boldsymbol{x}) \frac{\sum\limits_{j \neq i} n_{ij} \hat{p}_{ij}(\boldsymbol{x})}{\sum\limits_{j \neq i} n_{ij} p_{ij}(\boldsymbol{x})},$$

   (c) dla każdego $i = 1, 2, \ldots, K$

$$\hat{p}_i(\boldsymbol{x}) \leftarrow \frac{\hat{p}_i(\boldsymbol{x})}{\sum\limits_{k=1}^{K} \hat{p}_k(\boldsymbol{x})}.$$

Zwróćmy jeszcze uwagę na dodatkową własność procedury Hastiego i Tibshiraniego. Dla każdego $i = 1, 2, \ldots, K$ jest prawdziwa następująca tożsamość:

$$p_i(\boldsymbol{x}) = \sum_{j \neq i} \left(\frac{p_i(\boldsymbol{x}) + p_j(\boldsymbol{x})}{K-1} \right) \left(\frac{p_i(\boldsymbol{x})}{p_i(\boldsymbol{x}) + p_j(\boldsymbol{x})} \right). \tag{8.10}$$

Korzystając z zależności (8.8) oraz przyjmując, że $p_1(\boldsymbol{x}) = \cdots = p_K(\boldsymbol{x}) = 1/K$, otrzymujemy oszacowanie prawdopodobieństwa $p_i(\boldsymbol{x})$ postaci

$$\tilde{p}_i(\boldsymbol{x}) = \frac{2}{K(K-1)} \sum_{j \neq i} \hat{p}_{ij}(\boldsymbol{x}). \tag{8.11}$$

To proste oszacowanie jest w szczególnym (lecz bardzo istotnym) przypadku równoważne (w sensie klasyfikacji) oszacowaniu otrzymanemu poprzez minimalizację odległości Kullbacka–Leiblera. Mówi o tym poniższe twierdzenie.

TWIERDZENIE 8.1.

Dla $n_1 = n_2 = \cdots = n_K$ zachodzi: $\tilde{p}_i(\boldsymbol{x}) > \tilde{p}_j(\boldsymbol{x})$ wtedy i tylko wtedy, gdy $\hat{p}_i(\boldsymbol{x}) > \hat{p}_j(\boldsymbol{x})$.

Widać zatem, że w przypadku równych liczebności reguła klasyfikacyjna (8.7) przyjmuje postać

$$\hat{d}_{HT}(\boldsymbol{x}) = \arg \max_{1 \leq i \leq K} \sum_{j \neq i} \hat{p}_{ij}(\boldsymbol{x}). \tag{8.12}$$

Co więcej, oszacowania prawdopodobieństw $p_i(\boldsymbol{x})$ dane wzorem (8.11) są dobrym punktem startowym algorytmu Bradleya–Terry'ego.

Przykład 8.2.
Rozważmy zagadnienie automatycznego rozpoznawania samogłosek języka angielskiego (jest ich 11). Rozpoznawanie oparto na wartości 10 różnych cech, a w przygotowaniu danych uczących wzięło udział 15 osób (8 mężczyzn i 7 kobiet). Dane uczące (990 obserwacji) wraz z opisem znajdują się w składnicy danych Uniwersytetu Kalifornijskiego w Irvin (http://www.ics.uci.edu/~mlearn/ /MLRepository.html).
Jeżeli do klasyfikacji wykorzystamy omówioną w podrozdz. 1.4 wieloklasową procedurę liniowej analizy dyskryminacyjnej (LDA), to oszacowany 10-krokową metodą sprawdzania krzyżowego aktualny poziom błędu wynosi 0.3974.
Jeżeli natomiast dokonamy dekompozycji tego 11-klasowego zagadnienia na 55 zagadnień binarnych, do rozwiązania każdego z nich zastosujemy również procedurę liniowej analizy dyskryminacyjnej, a następnie końcową decyzję podejmiemy, wykorzystując klasyfikator Hastiego i Tibshiraniego dany wzorem (8.12), to aktualny poziom błędu zostanie zredukowany do wartości 0.2532.

Procedura Hastiego i Tibshiraniego ma jedną zasadniczą wadę. Klasyfikując obserwację $\boldsymbol{x}$, musimy, za pomocą klasyfikatorów binarnych, oszacować $K(K-1)/2$ prawdopodobieństw $p_{ij}(\boldsymbol{x})$. Zauważmy jednak, że jeżeli klasyfikowana obserwacja w rzeczywistości pochodzi z i-tej klasy, to w oszacowaniu tych prawdopodobieństw obserwacje uczące z i-tej klasy biorą udział tylko $K-1$ razy. Do pozostałych $(K-1)(K-2)/2$ oszacowań obserwacje uczące z i-tej klasy nie są w ogóle wykorzystywane, co oznacza, że ich wynik może być w zasadzie zupełnie dowolny.

Zilustrujmy ten fakt na prostym przykładzie z trzema klasami. Załóżmy, że obserwacja $\boldsymbol{x}$ pochodzi z pierwszej klasy (tzn. $Y = 1$) oraz że na podstawie próby uczącej oszacowano $\hat{p}_{12}(\boldsymbol{x}) = 0.58$, $\hat{p}_{13}(\boldsymbol{x}) = 0.69$. Ponieważ do oszacowania $\hat{p}_{23}(\boldsymbol{x})$ nie wykorzystujemy obserwacji uczących z pierwszej klasy, zatem oznaczmy dowolny wynik tego oszacowania przez p, $0 \leq p \leq 1$. Obliczone za pomocą wzoru (8.11) oszacowania prawdopodobieństw a posteriori przynależności obserwacji $\boldsymbol{x}$ do poszczególnych klas wynoszą: $\tilde{p}_1(\boldsymbol{x}) = 0.42$, $\tilde{p}_2(\boldsymbol{x}) = 0.14 + 0.33p$, $\tilde{p}_3(\boldsymbol{x}) = 0.44 - 0.33p$. Z poniższej tabeli widać wyraźnie, jak wartość p wpływa na końcowy wynik klasyfikacji. Pomimo że oba oszacowania $\hat{p}_{12}(\boldsymbol{x})$ i $\hat{p}_{13}(\boldsymbol{x})$ wskazują na klasę pierwszą, aż w 19% źle dobrana wartość p powoduje błędną decyzję.

p	$\hat{d}_{HT}(\boldsymbol{x})$
0.00–0.04	3
0.04–0.85	1
0.85–1.00	2

Wprowadźmy modyfikację procedury Hastiego i Tibshiraniego polegającą na wprowadzeniu w procesie estymacji prawdopodobieństw $p_{ij}(\boldsymbol{x})$ wag uzależnionych od tego, czy obserwacja $\boldsymbol{x}$ należy od klasy i-tej lub j-tej, czy raczej do jednej z pozostałych klas.

Niech $q_{ij}(\boldsymbol{x})$ oznacza następujące prawdopodobieństwo:

$$q_{ij}(\boldsymbol{x}) = \mathrm{P}(Y = i \quad \text{lub} \quad Y = j | \boldsymbol{X} = \boldsymbol{x}).$$

Zatem $q_{ij}(\boldsymbol{x}) = p_i(\boldsymbol{x}) + p_j(\boldsymbol{x})$, a tożsamość (8.10) przyjmuje postać

$$p_i(\boldsymbol{x}) = \sum_{j \neq i} \left(\frac{q_{ij}(\boldsymbol{x})}{K-1} \right) p_{ij}(\boldsymbol{x}). \tag{8.13}$$

Do oszacowania prawdopodobieństwa $q_{ij}(\boldsymbol{x})$ wykorzystujemy klasyfikatory binarne uczone na próbie zawierającej tylko dwie klasy. Do pierwszej z nich zaliczamy obserwacje z klas i-tej i j-tej, a do drugiej pozostałe obserwacje.

Wykorzystując tożsamość (8.13), poprawiona (ważona) reguła klasyfikacyjna (8.7) przyjmie postać

$$\hat{d}_{WHT}(\boldsymbol{x}) = \arg \max_{1 \leq i \leq K} \sum_{j \neq i} \hat{q}_{ij}(\boldsymbol{x}) \hat{p}_{ij}(\boldsymbol{x}). \tag{8.14}$$

Inną metodę wprowadzenia wag zaproponowali Jelonek i Stefanowski (1998). Wprowadzili oni skojarzony z każdym klasyfikatorem binarnym współczynnik wiarogodności t_{ij} postaci

$$t_{ij} = \frac{v_i}{v_i + e_j},$$

gdzie v_i jest liczbą poprawnie zaklasyfikowanych obiektów z klasy i-tej, natomiast e_j jest liczbą błędnie zaklasyfikowanych obiektów z klasy j-tej. Zwróćmy uwagę na to, że w metodzie tej obliczanie współczynników wiarogodności jest przeprowadzane w fazie uczenia klasyfikatorów binarnych, tzn. współczynniki te są takie same dla każdego klasyfikowanego obiektu.

Reguła klasyfikacyjna Jelonka i Stefanowskiego ma postać

$$\hat{d}_{JS}(\boldsymbol{x}) = \arg \max_{1 \leq i \leq K} \sum_{j \neq i} t_{ij} \hat{p}_{ij}(\boldsymbol{x}). \tag{8.15}$$

Przykład 8.3.
Powróćmy jeszcze raz do zagadnienia automatycznego rozpoznawania samogłosek języka angielskiego rozważanego w przykładzie 8.2. Tym razem do klasyfikacji wykorzystamy omówioną w rozdz. 6 procedurę drzew klasyfikacyjnych (algorytm CART). Oszacowany 10-krokową metodą sprawdzania krzyżowego aktualny poziom błędu dla metody wieloklasowej wynosi 0.2991.
Po dokonaniu dekompozycji tego zagadnienia na zagadnienia binarne, wykorzystując klasyfikator (8.12), możemy zredukować aktualny poziom błędu do wartości 0.1861.
Wykorzystując ważoną regułę klasyfikacyjną (8.14), redukcja tego błędu będzie znacząco większa. Oszacowany aktualny poziom błędu wynosi bowiem 0.1602.

9

Wzmacnianie klasyfikatorów

W rozdziale tym zostaną przedstawione dwie metody wzmacniania klasyfikatorów. Są to – zaproponowane w latach dziewięćdziesiątych XX wieku – metody znane w literaturze pod nazwami bagging i boosting. Owo wzmacnianie odbywa się na drodze uczenia ciągu klasyfikatorów na obserwacjach kolejno pobieranych prób bootstrapowych (patrz podrozdz. 1.10).

Metoda *bagging*[1] została zaproponowana przez Breimana (1996b) i w rzeczywistości jest metodą redukcji zmienności (wariancji) klasyfikatora. Krytycznym czynnikiem jest tutaj niestabilność danej rodziny klasyfikatorów: jeżeli małe zmiany próby uczącej prowadzą do klasyfikatorów różniących się w sposób znaczący, wówczas metoda bagging prowadzi do klasyfikatora o mniejszej zmienności (bardziej stabilnego). Z tej właśnie przyczyny algorytm bagging stosuje się najczęściej dla drzew decyzyjnych czy sieci neuronowych, które należą do metod niestabilnych. Bagging nie gwarantuje jednak równoczesnego zmniejszenia obciążenia. Można pokazać, że w pewnych patologicznych przypadkach obciążenie wynikowego klasyfikatora może nawet wzrosnąć (Kuncheva, 2004).

Pewnego rodzaju uogólnieniem baggingu jest metoda *boosting*[2]. W swej pierwotnej postaci została ona zaproponowana (niezależnie od baggingu Breimana) przez Schapire'a (1990), a następnie rozwinięta przez Freunda i Schapire'a (1997). W przeciwieństwie do baggingu, boosting redukuje nie tylko wariancję, ale i obciążenie. Zazwyczaj startujemy z prostym (stąd – bardzo obciążonym) klasyfikatorem, a następnie stopniowo redukujemy obciążenie przez nadawanie obserwacjom próby uczącej odpowiednio dobranych wag. Niedogodnością tej metody jest to, że klasyfikator końcowy ma często dość skomplikowaną postać.

---

[1] Akronim słów: ang. **b**ootstrap **agg**regat**ing**.

[2] Po ang. boost – ulepszać, wzmacniać.

9.1. Algorytm bagging

Jak już wspomnieliśmy, bagging jest metodą wzmacniania klasyfikatorów opartą na repróbkowaniu, a dokładniej na technice bootstrapu[3]. Przypomnijmy, że przez n-elementową próbę bootstrapową rozumiemy próbę powstałą w wyniku n-krotnego losowania pojedynczej obserwacji z próby uczącej ze zwracaniem.

Niech $\mathcal{L}_n = \{\boldsymbol{z}_1, \ldots, \boldsymbol{z}_n\}$ będzie n-elementową próbą uczącą, gdzie $\boldsymbol{z}_j = (\boldsymbol{x}_j, y_j)$ oraz $\boldsymbol{x}_j$ jest wektorem obserwowanych cech, a y_j etykietą klasy ($j = 1, 2, \ldots, n$).

Algorytm bagging

1. Dla $b = 1, 2, \ldots, B$:
   1a) pobierz próbę bootstrapową $\mathcal{L}_n^{*b}$ z próby $\mathcal{L}_n$,
   1b) konstruuj klasyfikator $\hat{d}_b$ dla próby bootstrapowej $\mathcal{L}_n^{*b}$.
2. Klasyfikuj obserwację $\boldsymbol{x}$ wg reguły

$$\hat{d}_{Bag1}(\boldsymbol{x}) = \arg \max_{1 \le k \le K} \sum_{b=1}^{B} I(\hat{d}_b(\boldsymbol{x}) = k). \tag{9.1}$$

Widzimy, że łączona (zagregowana) reguła klasyfikacyjna klasyfikuje obserwację według schematu głosowania przez zwykłą większość[4], tzn. przydziela ją do grupy k-tej, jeżeli do tej grupy zaklasyfikowało ją najwięcej z klasyfikatorów składowych.

W przypadku gdy wynikami klasyfikatora składowego $\hat{d}_b$ ($b = 1, 2, \ldots$ $\ldots, B$) są estymatory prawdopodobieństw a posteriori $\hat{p}^{(b)}(k|\boldsymbol{x})$, można zastosować metodę sumacyjną (patrz podrozdz. 8.1). Klasyfikator (9.1) przyjmuje wówczas postać

$$\hat{d}_{Bag1}(\boldsymbol{x}) = \arg \max_{1 \le k \le K} \sum_{b=1}^{B} \hat{p}^{(b)}(k|\boldsymbol{x}). \tag{9.2}$$

Oczywiście równoprawnie można użyć w tym przypadku metody iloczynowej (patrz podrozdz. 8.2).

Ponieważ każdorazowo do próby bootstrapowej nie zostaje wybranych blisko 37% wszystkich elementów próby uczącej (patrz podrozdz. 1.10), ele-

[3]Technika ta została omówiona w podrozdz. 1.10 jako metoda oceny aktualnego poziomu błędnej klasyfikacji.

[4]Po ang. plurality voting.

menty te można potraktować jako zbiór testowy dla danego klasyfikatora składowego. Poziom poprawnej predykcji może wtedy posłużyć do określenia wielkości wag, z jakimi poszczególne klasyfikatory składowe zostaną włączone przy konstrukcji klasyfikatora łączonego: im większy odsetek poprawnych zaklasyfikowań, tym większy udział klasyfikatora w ostatecznej regule łączonej.

Algorytm bagging (wersja ważona)

1. Dla $b = 1, 2, \ldots, B$:

   1a) pobierz próbę bootstrapową $\mathcal{L}_n^{*b}$ z próby $\mathcal{L}_n$,

   1b) konstruuj klasyfikator $\hat{d}_b$ dla próby bootstrapowej $\mathcal{L}_n^{*b}$,

   1c) oblicz $w^{(b)} := \dfrac{\sum_{j=1}^{n} \left[I(\boldsymbol{z}_j \notin \mathcal{L}_n^{*b}) I(\hat{d}_b(\boldsymbol{x}_j) = y_j) \right]}{\sum_{j=1}^{n} I(\boldsymbol{z}_j \notin \mathcal{L}_n^{*b})}$.

2. Normalizuj $w^{(b)} := w^{(b)} / \sum_{b=1}^{B} w^{(b)}$.

3. Klasyfikuj obserwację $\boldsymbol{x}$ wg reguły

$$\hat{d}_{Bag2}(\boldsymbol{x}) = \arg \max_{1 \le k \le K} \sum_{b=1}^{B} w^{(b)} I(\hat{d}_b(\boldsymbol{x}) = k). \tag{9.3}$$

Algorytm bagging znajduje zastosowanie zwłaszcza dla niestabilnych metod klasyfikacyjnych (z tej przyczyny nie zaleca się np. przycinania drzew decyzyjnych przed jego zastosowaniem). Wiemy, że w każdej z prób bootstrapowych $\mathcal{L}_n^{*b}$ nie znajdzie się około 1/3 elementów próby uczącej $\mathcal{L}_n$. Można zatem oczekiwać, iż dla metod niestabilnych, postaci uzyskanych B klasyfikatorów składowych będą znacząco różne. Z uwagi na fakt, że technika głosowania jest formą uśredniania, algorytm bagging może znacząco zredukować poziom wariancji takich niestabilnych klasyfikatorów. Zauważmy jednak, iż po jego zastosowaniu nie zostaje zachowana struktura modelu – zagregowany klasyfikator drzew decyzyjnych nie jest już drzewem. W ten sposób utracona zostaje kluczowa zaleta tejże rodziny klasyfikatorów, jaką jest prostota interpretacji.

9.2. Algorytmy typu boosting

Jak zaznaczyliśmy we wstępie do rozdziału, boosting powstał niezależnie od baggingu. Jednakże metoda ta może być rozpatrywana jako jego uogólnienie. Co zatem różni te dwie metody? Zasadnicza odmienność tkwi w sposobie generowania zbiorów uczących dla kolejnych klasyfikatorów składowych. W przypadku baggingu każdy element losowany był do danej próby bootstrapowej według rozkładu jednostajnego nałożonego na elementy próby uczącej. Innymi słowy, każda obserwacja próby miała niejako przypisaną tę samą wagę równą $1/n$ w każdym kroku procedury. Innowacją w boostingu jest natomiast fakt, że w każdej kolejnej iteracji (tj. dla każdego kolejnego klasyfikatora składowego) rozkład prawdopodobieństwa nałożony na elementy próby uczącej jest różny.

Należy w tym miejscu postawić dwa podstawowe pytania:

1. W jaki sposób nadawać elementom próby uczącej $\mathcal{L}_n$ wagi w poszczególnych iteracjach?
2. Jak łączyć decyzje klasyfikatorów składowych?

Pierwszą procedurę opartą na boostingu zaproponował Schapire (1990). Pokazał on, że w problemie dwóch klas każdy słaby klasyfikator[5] może zostać ulepszony przez wyuczenie dwóch dodatkowych klasyfikatorów na przefiltrowanym zbiorze danych uczących. Połowę próby uczącej dla pierwszego z nich stanowiły obserwacje błędnie zaklasyfikowane przez klasyfikator wyjściowy, natomiast drugi z nich był uczony na tych obserwacjach próby, które przez dwa poprzednie klasyfikatory zostały rozpoznane jako elementy różnych grup. Ostatecznie, decyzja klasyfikatora łączonego była wyznaczana jako zwykła (nieważona) większość z decyzji trzech klasyfikatorów składowych. Pięć lat później, Freund (1995) zaproponował odmianę łączenia (przez większość) wielu słabych klasyfikatorów jednocześnie, ulepszając pierwotny algorytm Schapire'a. Następnie Freund i Schapire (1997) przedstawili wspólnie nowy algorytm zwany *AdaBoost*[6], który rozwiązał wiele praktycznych trudności napotykanych w przypadku wcześniejszych procedur. Pierwotnie *AdaBoost* był algorytmem dla problemu 2 klas i dopiero później został rozszerzony na przypadek K klas.

[5]Przez *słaby klasyfikator* (ang. weak classifier) będziemy rozumieć klasyfikator o poziomie błędu nieznacznie mniejszym od 1/2, tj. nieznacznie mniejszym od oceny błędu klasyfikatora, którego decyzja jest determinowana przez rzut symetryczną monetą.

[6]Akronim słów: ang. **Ada**ptive **Boost**ing.

Algorytm AdaBoost

1. Ustal wektor wag $\boldsymbol{w}^{(1)} = (w_1^{(1)}, w_2^{(1)}, \dots, w_n^{(1)})'$, przy czym

$$w_j^{(1)} \in [0, 1], \quad \sum_{j=1}^{n} w_j^{(1)} = 1.$$

(Zazwyczaj: $w_j^{(1)} = \frac{1}{n}$).

2. Dla $b = 1, 2, \dots, B$:

2a) pobierz próbę bootstrapową $\mathcal{L}_n^{*b}$ z próby $\mathcal{L}_n$ według rozkładu

$$\mathrm{P}(\boldsymbol{z}_j \in \mathcal{L}_n^{*b}) = w_j^{(b)}, \quad j = 1, 2, \dots, n,$$

2b) konstruuj klasyfikator $\hat{d}_b$ dla próby bootstrapowej $\mathcal{L}_n^{*b}$,

2c) oblicz

$$\hat{e}_b = \sum_{j=1}^{n} w_j^{(b)} l_j^{(b)}, \tag{9.4}$$

gdzie

$$l_j^{(b)} = \begin{cases} 1, & \text{jeżeli } \hat{d}_b(\boldsymbol{x}_j) \neq y_j \\ 0, & \text{jeżeli } \hat{d}_b(\boldsymbol{x}_j) = y_j \end{cases},$$

2d) jeżeli $\hat{e}_b \in (0; 1/2)$, oblicz

$$\beta_b = \frac{\hat{e}_b}{1 - \hat{e}_b}. \tag{9.5}$$

W przeciwnym przypadku podstaw $w_j^{(b)} = 1/n$ $(j = 1, 2, \dots, n)$ i krok 2a,

2e) aktualizuj wagi

$$w_j^{(b+1)} = \frac{w_j^{(b)} \beta_b^{1 - l_j^{(b)}}}{\sum_{i=1}^{n} w_i^{(b)} \beta_b^{1 - l_i^{(b)}}}, \quad j = 1, 2, \dots, n. \tag{9.6}$$

3. Klasyfikuj obserwację $\boldsymbol{x}$ wg reguły

$$\hat{d}_{AdaB}(\boldsymbol{x}) = \arg \max_{1 \le k \le K} \sum_{b=1}^{B} \left[\ln\left(\frac{1}{\beta_b}\right) I(\hat{d}_b(\boldsymbol{x}) = k) \right]. \tag{9.7}$$

Jak widzimy – w przeciwieństwie do baggingu, który był przykładem algorytmu równoległego (każdy klasyfikator składowy jest uczony oddzielnie, niezależnie od pozostałych) – w algorytmie *AdaBoost* każdy kolejny klasyfikator zależy od uprzednio skonstruowanego. W b-tym przebiegu algorytmu, w kroku 2c estymowany jest poziom błędu klasyfikatora $\hat{d}_b$, a następnie są wyznaczane wielkości β_b, które wykorzystywane są w kroku 2e do aktualizacji wag. Wagi obserwacji błędnie zaklasyfikowanych przez $\hat{d}_b$ są zwiększane (tym samym, wagi obserwacji klasyfikowanych poprawnie są zmniejszane). W ten sposób wzrasta prawdopodobieństwo, z jakim obserwacja klasyfikowana błędnie przez klasyfikator $\hat{d}_b$ zostanie wylosowana jako element próby uczącej klasyfikatora $\hat{d}_{b+1}$. Również klasyfikatory składowe otrzymują wagi – równe $\ln(1/\beta_b)$, tj. klasyfikatory o większej poprawności predykcji mają większy udział w ostatecznej decyzji klasyfikatora łączonego. Nowo zaobserwowany obiekt zostaje przydzielony do grupy k, jeżeli suma wag klasyfikatorów składowych identyfikujących go jako obiekt grupy k jest największa.

Zwróćmy uwagę, że obecność wag $w_j^{(b)}$ istotnie zmienia sposób wyznaczania poziomu błędu. Zgodnie ze wzorem (9.4) jest to ważona suma, a nie frakcja obserwacji błędnie zaklasyfikowanych. Co więcej, z uwagi na fakt, że przez nadanie wyraźnie większych wag niewielkiej liczbie obserwacji możliwe jest dokonanie znaczących zmian w próbie uczącej, kluczową własnością (w przeciwieństwie do baggingu) przestaje być niestabilność klasyfikatora bazowego.

Powyżej przedstawiliśmy odmianę algorytmu *AdaBoost* opartą na repróbkowaniu[7], czyli generowaniu pseudoprób. W każdej z B iteracji jest losowanych n obserwacji ze zwracaniem i z prawdopodobieństwem proporcjonalnym do ich bieżących wag. Warto zaznaczyć, że istnieje też inna, alternatywna wersja[8]. Stosowana jest ona w przypadku klasyfikatorów, które w swej implementacji dopuszczają bezpośrednio jako dane wejściowe ważone obserwacje próby. W ten sposób każdy z klasyfikatorów składowych wykorzystuje informację (odpowiednio ważoną) o każdym elemencie próby uczącej. Ponieważ w tym przypadku próbkowanie jest zbędne, algorytm jest całkowicie deterministyczny.

Jednym z uzasadnień sukcesu, jaki odniósł algorytm *AdaBoost* może być fakt, że estymator błędu wyznaczony za pomocą elementów próby uczącej bardzo szybko zbiega do zera[9]. Freund i Schapire (1997) udowodnili twier-

[7]Po ang. boosting by resampling.

[8]Po ang. boosting by reweighting.

[9]Badania wskazują, że najczęściej wystarczy kilka pierwszych iteracji algorytmu (Kuncheva, 2004).

dzenie podające ograniczenie górne na poziom błędu (estymowany za pomocą obserwacji wyjściowej próby uczącej) klasyfikatora uzyskanego w wyniku zastosowania powyższego algorytmu dla problemu dyskryminacji dwóch klas.

TWIERDZENIE 9.1.

Niech będzie dane zagadnienie dyskryminacji dwóch klas. Niech ponadto $\hat{e}_R$ *oznacza estymator błędu klasyfikatora łączonego (9.7) uzyskany metodą ponownego podstawienia. Wówczas*

$$\hat{e}_R < 2^B \prod_{b=1}^{B} \sqrt{\hat{e}_b(1-\hat{e}_b)}, \tag{9.8}$$

gdzie $\hat{e}_b$ *są ważonymi estymatorami błędu klasyfikatorów składowych danymi wzorem (9.4).*

Dowód. Dowód twierdzenia będzie składał się z dwóch części. Najpierw wyprowadzimy zależność między estymatorem $\hat{e}_R$ i – występującymi w algorytmie AdaBoost – wielkościami β_b, a następnie wykażemy, że dla β_b danych wzorem (9.5) udowodnione uprzednio ograniczenie górne na $\hat{e}_R$ osiąga minimum. Minimum to będzie właśnie równe prawej stronie tezy twierdzenia. W dowodzie twierdzenia będzie pomocny następujący lemat:

LEMAT 9.1.
Niech $a \geq 0$ *oraz* $r \in [0;\ 1]$. *Wówczas*

$$a^r \leq 1 - (1-a)r \tag{9.9}$$

Dowód. Ponieważ

$$\frac{\partial^2(a^r)}{\partial r^2} = a^r \ln^2 a$$

przyjmuje wartości nieujemne dla każdego r, więc odwzorowanie $r \mapsto a^r$ jest funkcją wypukłą.
Dla każdego $t \in [0;\ 1]$ mamy zatem

$$a^{(1-t)p+tq} \leq (1-t)a^p + ta^q, \tag{9.10}$$

gdzie p, q są dowolnymi punktami prostej rzeczywistej.
Obierając $t = r$, $p = 0$, $q = 1$, otrzymujemy tezę. ■

W kroku drugim algorytmu, po skonstruowaniu klasyfikatora $\hat{d}_b$, obliczamy – zgodnie ze wzorem (9.4) – jego ważony estymator błędu $\hat{e}_b$. W kroku 2d z $\hat{e}_b$ wyznaczamy wielkość β_b. Na tym etapie dowodu nie zakładamy jednak żadnej konkretnej zależności między nimi. Celem naszym jest dopiero wyrażenie wielkości β_b jako funkcji $\hat{e}_b$, i to minimalizującej poziom błędu $\hat{e}_R$.

Kolejno, w kroku 2e są aktualizowane wagi $w_j^{(b)}$. Oznaczmy przez D_b wielkość normującą w zależności (9.6) b-tej iteracji algorytmu, to jest

$$D_b = \sum_{j=1}^{n} w_j^{(b)} \beta_b^{1-l_j^{(b)}}. \tag{9.11}$$

Wówczas równanie (9.6) przyjmuje postać

$$w_j^{(b+1)} = w_j^{(1)} \prod_{i=1}^{b} \frac{\beta_i^{1-l_j^{(i)}}}{D_i}. \tag{9.12}$$

Niech

$$\mathcal{L}_n^{(-)} := \{\boldsymbol{z}_j \colon \hat{d}_{AdaB}(\boldsymbol{x}_j) \neq y_j, \quad j = 1,\ 2,\ \ldots,\ n\},$$

tj. niech $\mathcal{L}_n^{(-)}$ oznacza podzbiór próby $\mathcal{L}_n$ złożony z obiektów klasyfikowanych błędnie przez klasyfikator łączony dany wzorem (9.7). Wówczas jego błąd, estymowany przy użyciu inicjującego algorytm wektora wag $\boldsymbol{w}^{(1)}$, jest równy

$$\hat{e}_R = \sum_{\{j\colon\, \boldsymbol{z}_j \in \mathcal{L}_n^{(-)}\}} w_j^{(1)}. \tag{9.13}$$

Ponieważ w każdej iteracji suma wag $w_j^{(b)}$ jest równa 1, więc korzystając z równości (9.12), mamy

$$1 = \sum_{j=1}^{n} w_j^{(B+1)} \geq \sum_{\{j\colon\, \boldsymbol{z}_j \in \mathcal{L}_n^{(-)}\}} w_j^{(B+1)} =$$

$$= \sum_{\{j\colon\, \boldsymbol{z}_j \in \mathcal{L}_n^{(-)}\}} w_j^{(1)} \prod_{b=1}^{B} \frac{\beta_b^{1-l_j^{(b)}}}{D_b}. \tag{9.14}$$

Zwróćmy uwagę, że klasyfikator $\hat{d}_{AdaB}$ błędnie zaklasyfikuje obserwację $\boldsymbol{x}_j$, jeżeli suma wag tych spośród klasyfikatorów składowych $\hat{d}_b$, które błędnie zaklasyfikują tę obserwację, jest nie mniejsza od sumy wag klasyfikatorów składowych klasyfikujących ją poprawnie. Zatem

$$\sum_{b=1}^{B} l_j^{(b)} \ln\left(\frac{1}{\beta_b}\right) \geq \sum_{b=1}^{B} (1 - l_j^{(b)}) \ln\left(\frac{1}{\beta_b}\right). \tag{9.15}$$

Potęgując przy podstawie e,

$$\prod_{b=1}^{B} \beta_b^{-l_j^{(b)}} \geq \prod_{b=1}^{B} \beta_b^{-(1-l_j^{(b)})}. \tag{9.16}$$

Mnożąc obustronnie przez $\prod_{b=1}^{B} \beta_b$ ($\beta_b > 0$),

$$\prod_{b=1}^{B} \beta_b^{1-l_j^{(b)}} \geq \prod_{b=1}^{B} \beta_b \prod_{b=1}^{B} \beta_b^{-(1-l_j^{(b)})}, \tag{9.17}$$

a stąd

$$\prod_{b=1}^{B} \beta_b^{2(1-l_j^{(b)})} \geq \prod_{b=1}^{B} \beta_b. \tag{9.18}$$

Pierwiastkując, otrzymujemy

$$\prod_{b=1}^{B} \beta_b^{1-l_j^{(b)}} \geq \prod_{b=1}^{B} \sqrt{\beta_b}. \tag{9.19}$$

Biorąc teraz ponownie zależność (9.14) i korzystając kolejno z równości (9.19) i (9.13), otrzymujemy

$$1 \geq \sum_{\{j:\, \boldsymbol{z}_j \in \mathcal{L}_n^{(-)}\}} w_j^{(1)} \prod_{b=1}^{B} \frac{\beta_b^{1-l_j^{(b)}}}{D_b} \geq \left(\sum_{\{j:\, \boldsymbol{z}_j \in \mathcal{L}_n^{(-)}\}} w_j^{(1)} \right) \prod_{b=1}^{B} \frac{\sqrt{\beta_b}}{D_b} = \\ = \hat{e} \prod_{b=1}^{B} \frac{\sqrt{\beta_b}}{D_b}. \tag{9.20}$$

Stąd

$$\hat{e}_R \leq \prod_{b=1}^{B} \frac{D_b}{\sqrt{\beta_b}}. \tag{9.21}$$

Z lematu (9.9) mamy

$$D_b = \sum_{j=1}^{n} w_j^{(b)} \beta_b^{1-l_j^{(b)}} \leq \sum_{j=1}^{n} w_j^{(b)} \left(1 - (1-\beta_b)(1-l_j^{(b)})\right) =$$

$$= \sum_{j=1}^{n} w_j^{(b)} (\beta_b + l_j^{(b)} - \beta_b l_j^{(b)}) =$$

$$= \beta_b \sum_{j=1}^{n} w_j^{(b)} + \sum_{j=1}^{n} w_j^{(b)} l_j^{(b)} - \beta_b \sum_{j=1}^{n} w_j^{(b)} l_j^{(b)} =$$

$$= \beta_b + \hat{e}_b - \beta_b \hat{e}_b = 1 - (1-\beta_b)(1-\hat{e}_b). \tag{9.22}$$

Stosując powyższą nierówność w zależności (9.21), otrzymujemy ostatecznie

$$\hat{e}_R \leq \prod_{b=1}^{B} \frac{1-(1-\beta_b)(1-\hat{e}_b)}{\sqrt{\beta_b}}. \tag{9.23}$$

Znajdziemy teraz wielkości $\beta_b = \beta_b(\hat{e}_b)$ minimalizujące ograniczenie górne występujące w otrzymanej zależności. Oznaczmy prawą stronę nierówności (9.23) przez U. Mamy

$$\frac{\partial U}{\partial \beta_b} = \frac{\beta_b(1-\hat{e}_b) - \hat{e}_b}{2\beta_b\sqrt{\beta_b}} \times A_b, \tag{9.24}$$

gdzie A_b jest pewną stałą dodatnią ($b = 1, 2, \ldots, B$).

Przyrównując pochodną do zera, otrzymujemy

$$\beta_b = \frac{\hat{e}_b}{1-\hat{e}_b}, \tag{9.25}$$

a więc zależność (9.5) algorytmu.

Ponadto

$$\frac{\partial^2 U}{\partial \beta_b^2}\left(\frac{\hat{e}_b}{1-\hat{e}_b}\right) = (1-\hat{e}_b)\left(\frac{1-\hat{e}_b}{\hat{e}_b}\right)^{3/2} \times \frac{A_b}{2}, \tag{9.26}$$

jest dodatnie, zatem U osiąga minimum w punkcie $\beta_b = \dfrac{\hat{e}_b}{1-\hat{e}_b}$.

Podstawiając (9.25) do (9.23), otrzymujemy tezę twierdzenia. ▪

Okazuje się, że teza powyższego twierdzenia jest prawdziwa również dla przypadku K klas ($K > 2$), jeżeli tylko ważone błędy klasyfikatorów składowych będą mniejsze od $1/2$[10].

TWIERDZENIE 9.2.

Niech będzie dane zagadnienie dyskryminacji K klas. Niech ponadto $\hat{e}_R$ oznacza estymator błędu klasyfikatora łączonego (9.7) uzyskany metodą ponownego podstawienia. Jeżeli $\hat{e}_b < 1/2$ dla $b = 1, 2, \ldots, B$, wówczas

$$\hat{e}_R < 2^B \prod_{b=1}^{B} \sqrt{\hat{e}_b(1 - \hat{e}_b)}, \tag{9.27}$$

gdzie $\hat{e}_b$ są ważonymi estymatorami błędu klasyfikatorów składowych danymi wzorem (9.4).

Dowód. Różnica między dowodem bieżącego twierdzenia a tw. 9.1 pojawia się przy zależności (9.15). Oznaczmy przez $\mathcal{D}$ zbiór wszystkich klasyfikatorów składowych algorytmu AdaBoost, tj.

$$\mathcal{D} = \{\hat{d}_b\colon b = 1, 2, \ldots, B\}.$$

Niech $\boldsymbol{z}_j \in \mathcal{L}_n$ będzie obserwacją próby zaklasyfikowaną błędnie przez klasyfikator łączony. Dokonajmy podziału zbioru $\mathcal{D}$ na 3 podzbiory

$$\begin{aligned}
\mathcal{D}^w &= \{\hat{d}_b \in \mathcal{D}\colon \hat{d}_b(\boldsymbol{x}_j) = \hat{d}_{AdaB}(\boldsymbol{x}_j) \neq y_j\} \\
\mathcal{D}^+ &= \{\hat{d}_b \in \mathcal{D}\colon \hat{d}_b(\boldsymbol{x}_j) = y_j\} \\
\mathcal{D}^- &= \{\hat{d}_b \in \mathcal{D}\colon \hat{d}_b(\boldsymbol{x}_j) \neq y_j \wedge \hat{d}_b(\boldsymbol{x}_j) \neq \hat{d}_{AdaB}(\boldsymbol{x}_j)\},
\end{aligned}$$

to znaczy na podzbiory: klasyfikatorów dających w wyniku odpowiednio: decyzję błędną, ale wygrywającą (tj. zgodną z decyzją klasyfikatora łączonego), decyzję poprawną oraz decyzję błędną, ale różną od decyzji klasyfikatora łączonego.

Zachodzi zależność

$$\sum_{\hat{d}_b \in \mathcal{D}^w} \ln\left(\frac{1}{\beta_b}\right) \geq \sum_{\hat{d}_b \in \mathcal{D}^+} \ln\left(\frac{1}{\beta_b}\right). \tag{9.28}$$

[10] Freund i Schapire argumentowali, iż dla słabych klasyfikatorów warunek $\hat{e}_b < 1/2$ w problemie wielu klas jest zbyt restrykcyjny. Zaproponowali więc inną wersję algorytmu *AdaBoost*, zwaną *AdaBoost.M2* (w odróżnieniu do opisanej tutaj wersji *AdaBoost.M1*). Zainteresowanych czytelników odsyłamy np. do pracy (Freund, Schapire, 1996).

Dodając do obu stron nierówności (dodatnie) wyrażenie $\sum_{\hat{d}_b \in \mathcal{D}^w} \ln(1/\beta_b) + \sum_{\hat{d}_b \in \mathcal{D}^-} \ln(1/\beta_b)$, otrzymujemy

$$2 \sum_{\hat{d}_b \in \mathcal{D}^w} \ln\left(\frac{1}{\beta_b}\right) + \sum_{\hat{d}_b \in \mathcal{D}^-} \ln\left(\frac{1}{\beta_b}\right) \geq \sum_{b=1}^{B} \ln\left(\frac{1}{\beta_b}\right). \tag{9.29}$$

Kolejno, chcemy dodać wyrażenie $\sum_{\hat{d}_b \in \mathcal{D}^-} \ln(1/\beta_b)$ bez zmiany zwrotu nierówności. Zagwarantujemy to, zakładając dodatkowo, że wszystkie składniki rozpatrywanej sumy są dodatnie. Równoważnie, że

$$\beta_b < 1, \qquad b = 1,\ 2,\ \ldots,\ B \tag{9.30}$$

(ponieważ docelowo $\boldsymbol{z}_j$ ma być dowolnie obranym elementem próby, prawdziwość powyższej nierówności zakładamy dla wszystkich wartości b – nie tylko dla tych będących wskaźnikami klasyfikatorów z $\mathcal{D}^-$).

W ten sposób lewa strona nierówności (9.29) jest równa podwojonej sumie wag wszystkich klasyfikatorów składowych dających w wyniku decyzję błędną. Zatem

$$2 \sum_{b=1}^{B} l_j^{(b)} \ln\left(\frac{1}{\beta_b}\right) \geq \sum_{b=1}^{B} \ln\left(\frac{1}{\beta_b}\right). \tag{9.31}$$

Zależność ta jest równoważna zależności (9.15). Dalej dowód przebiega analogicznie do dowodu tw. 9.1 (z nierówności $\beta_b = \hat{e}_b/(1-\hat{e}_b)$ i $\beta_b < 1$ mamy $\hat{e}_b < 1/2$). ■

Można pokazać (Freund, Schapire, 1997), że

$$\hat{e}_R < 2^B \prod_{b=1}^{B} \sqrt{\hat{e}_b(1-\hat{e}_b)} = \prod_{b=1}^{B} \sqrt{1 - 4\gamma_b^2} \leq \exp\left(-2 \sum_{b=1}^{B} \gamma_b^2\right), \tag{9.32}$$

gdzie $\gamma_b = 1/2 - \hat{e}_b$, tzn. estymator błędu $\hat{e}_R$ zbiega wykładniczo do zera.

Interesującym jest fakt, iż liczne eksperymenty wskazały, że poziom błędu estymowany za pomocą niezależnie pobranej próby testowej często maleje nawet wówczas, gdy ocena błędu szacowana na próbie uczącej osiągnęła już wartość zero!

Algorytm *AdaBoost* może być interpretowany jako iteracyjna procedura estymacji w ogólnym modelu logistycznym. W tym celu rozpatrzmy alternatywny (i równoważny) zapis przedstawionego uprzednio algorytmu *AdaBoost* dla przypadku dwóch klas, w której klasyfikatory $\hat{d}_b$ przyjmują wartości w zbiorze $\{-1,\ 1\}$. Ważone estymatory błędów (9.4) zapiszmy w postaci

$$\hat{e}_b = \mathrm{E}_{\boldsymbol{w}^{(b)}} I(\hat{d}_b(\boldsymbol{x}) \neq y), \tag{9.33}$$

gdzie $\mathrm{E}_{\boldsymbol{w}^{(b)}}$ jest operatorem wartości oczekiwanej względem próby uczącej $\mathcal{L}_n$ z wektorem wag $\boldsymbol{w}^{(b)}$, tj.

$$\hat{e}_b = \sum_{j=1}^{n} w_j^{(b)} I(\hat{d}_b(\boldsymbol{x}_j) \neq y_j), \tag{9.34}$$

natomiast proces aktualizacji wag (przed normalizacją) w postaci

$$w_j^{(b+1)} = w_j^{(b)} \exp\left[\tilde{\beta}_b I(\hat{d}_b(\boldsymbol{x}_j) \neq y_j)\right], \qquad j = 1,\ 2,\ \ldots,\ n, \tag{9.35}$$

gdzie

$$\tilde{\beta}_b = \ln\left(\frac{1-\hat{e}_b}{\hat{e}_b}\right). \tag{9.36}$$

Klasyfikator łączony wyraża się wzorem

$$\hat{d}_{AdaB}(\boldsymbol{x}) = \operatorname{sgn}\left(\sum_{b=1}^{B} \tilde{\beta}_b \hat{d}_b(\boldsymbol{x})\right). \tag{9.37}$$

Przypomnijmy, że w modelu logistycznym (patrz zależność (2.13))

$$p(y=1|\boldsymbol{x}) = \frac{e^{F(\boldsymbol{x})}}{1+e^{F(\boldsymbol{x})}} \tag{9.38}$$

zakładaliśmy, iż funkcja F (równoważnie odwzorowanie $p(1|\boldsymbol{x}) \mapsto \ln \dfrac{p(1|\boldsymbol{x})}{1-p(1|\boldsymbol{x})}$) jest funkcją liniową wektora obserwacji $\boldsymbol{x}$. W tym miejscu zrezygnujemy z tego założenia, stawiając zadanie estymacji funkcji F w pełnej ogólności.

Lemat 9.2. *Niech*

$$J(F) = \mathrm{E}\left(e^{-YF(\boldsymbol{X})}\right), \tag{9.39}$$

gdzie F jest pewną funkcją rzeczywistą, zaś E *operatorem wartości oczekiwanej względem łącznego rozkładu wektora* $(\boldsymbol{X},\ Y)$.
Wówczas J osiąga minimum dla funkcji

$$F(\boldsymbol{x}) = \frac{1}{2}\ln\frac{p(1|\boldsymbol{x})}{p(-1|\boldsymbol{x})}. \tag{9.40}$$

Dowód. Wystarczy znaleźć minimum funkcji

$$\mathrm{E}\left(e^{-YF(\boldsymbol{X})}|\boldsymbol{X} = \boldsymbol{x}\right) = p(1|\boldsymbol{x})e^{-F(\boldsymbol{x})} + p(-1|\boldsymbol{x})e^{F(\boldsymbol{x})}. \tag{9.41}$$

Mamy

$$\frac{\partial \mathrm{E}\left(e^{-YF(\boldsymbol{X})}|\boldsymbol{X} = \boldsymbol{x}\right)}{\partial F(\boldsymbol{x})} = -p(1|\boldsymbol{x})e^{-F(\boldsymbol{x})} + p(-1|\boldsymbol{x})e^{F(\boldsymbol{x})}. \tag{9.42}$$

Przyrównując pochodną do zera, otrzymujemy zależność (9.40).
Ponadto

$$\frac{\partial^2 \mathrm{E}\left(e^{-YF(\boldsymbol{X})}|\boldsymbol{X} = \boldsymbol{x}\right)}{\partial F^2(\boldsymbol{x})} = \mathrm{E}\left(e^{-YF(\boldsymbol{X})}|\boldsymbol{X} = \boldsymbol{x}\right) > 0, \tag{9.43}$$

co kończy dowód. ■

Po rozwikłaniu równania (9.40) otrzymujemy

$$p(1|\boldsymbol{x}) = \frac{e^{2F(\boldsymbol{x})}}{1 + e^{2F(\boldsymbol{x})}} = \frac{e^{F(\boldsymbol{x})}}{e^{-F(\boldsymbol{x})} + e^{F(\boldsymbol{x})}} \tag{9.44}$$

$$p(-1|\boldsymbol{x}) = \frac{1}{1 + e^{2F(\boldsymbol{x})}} = \frac{e^{-F(\boldsymbol{x})}}{e^{-F(\boldsymbol{x})} + e^{F(\boldsymbol{x})}}. \tag{9.45}$$

Jak widzimy – otrzymujemy model logistyczny (2.13) z dokładnością do stałej równej 2.

Załóżmy, że w pewnej iteracji zadania minimalizacji odwzorowania J danego wzorem (9.39) uzyskaliśmy przybliżenie rozwiązania równe $F(\boldsymbol{x})$. Załóżmy ponadto, że kolejne przybliżenie ma postać $F(\boldsymbol{x}) + cd(\boldsymbol{x})$, gdzie funkcja d przyjmuje wartości w zbiorze $\{-1, 1\}$ (odpowiednik klasyfikatorów $\hat{d}_b$ w kolejnych iteracjach algorytmu). Dla ustalonego c (i wektora $\boldsymbol{x}$)

dokonajmy rozwinięcia Taylora $J(F(\boldsymbol{x})+cd(\boldsymbol{x}))$ do rzędu drugiego w punkcie $d(\boldsymbol{x})=0$. Mamy

$$
\begin{aligned}
J(F(\boldsymbol{x})+cd(\boldsymbol{x})) &= \mathrm{E}\left(e^{-Y(F(\boldsymbol{X})+cd(\boldsymbol{X}))}|\boldsymbol{X}=\boldsymbol{x}\right) \approx \\
&\approx \mathrm{E}\left(e^{-YF(\boldsymbol{X})}(1-cYd(\boldsymbol{X})+c^2Y^2d^2(\boldsymbol{X})/2)|\boldsymbol{X}=\boldsymbol{x}\right) = \\
&= \mathrm{E}\left(e^{-YF(\boldsymbol{X})}(1-cYd(\boldsymbol{X})+c^2/2)|\boldsymbol{X}=\boldsymbol{x}\right), \qquad (9.46)
\end{aligned}
$$

gdzie ostatnia równość wynika z faktu, iż $y^2=1$ oraz $d^2(\boldsymbol{x})=1$.

Minimalizując powyższe wyrażenie względem $d(\boldsymbol{x})$, otrzymujemy

$$d(\boldsymbol{x}) = \arg\min_d \mathrm{E}_{\boldsymbol{w}}\left(1-cYd(\boldsymbol{X})+c^2/2|\boldsymbol{X}=\boldsymbol{x}\right), \qquad (9.47)$$

gdzie $\boldsymbol{w}=\boldsymbol{w}(\boldsymbol{x},\,y)=e^{-yF(\boldsymbol{x})}$, natomiast

$$\mathrm{E}_{\boldsymbol{w}}\left(g(\boldsymbol{X},\,Y)|\boldsymbol{X}=\boldsymbol{x}\right) = \frac{\mathrm{E}\left(\boldsymbol{w}(\boldsymbol{X},\,Y)g(\boldsymbol{X},\,Y)|\boldsymbol{X}=\boldsymbol{x}\right)}{\mathrm{E}\left(\boldsymbol{w}(\boldsymbol{X},\,Y)|\boldsymbol{X}=\boldsymbol{x}\right)}. \qquad (9.48)$$

Dla ustalonej wartości $c>0$ zadanie minimalizacji (9.47) jest równoważne zadaniu maksymalizacji

$$d(\boldsymbol{x}) = \arg\max_d \mathrm{E}_{\boldsymbol{w}}\left(Yd(\boldsymbol{X})|\boldsymbol{X}=\boldsymbol{x}\right), \qquad (9.49)$$

którego rozwiązaniem jest funkcja

$$d(\boldsymbol{x}) = \begin{cases} 1, & \text{gdy} \quad \mathrm{E}_{\boldsymbol{w}}\left(Y|\boldsymbol{X}=\boldsymbol{x}\right) = p_{\boldsymbol{w}}(1|\boldsymbol{x}) - p_{\boldsymbol{w}}(-1|\boldsymbol{x}) > 0 \\ -1, & \text{w przeciwnym przypadku.} \end{cases} \qquad (9.50)$$

Teraz możemy już wyznaczyć tę dodatnią wartość c, dla której $J(F(\boldsymbol{x})+cd(\boldsymbol{x}))$ osiąga minimum

$$c_0 = \arg\min_c \mathrm{E}_{\boldsymbol{w}}\left(e^{-cYd(\boldsymbol{X})}|\boldsymbol{X}=\boldsymbol{x}\right) = \frac{1}{2}\ln\left(\frac{1-e}{e}\right), \qquad (9.51)$$

gdzie

$$e = \mathrm{E}_{\boldsymbol{w}}\left(I(d(\boldsymbol{X})\neq Y)|\boldsymbol{X}=\boldsymbol{x}\right) \qquad (9.52)$$

(łatwo zauważyć, że warunek $c>0$ implikuje założenie poczynione w kroku 2d algorytmu, a mianowicie $e<1/2$).

Kolejne przybliżenie funkcji F w iterowanym procesie minimalizacji odwzorowania J danego wzorem (9.39) ma zatem postać

$$F(\boldsymbol{x}) \leftarrow F(\boldsymbol{x}) + \frac{1}{2} \ln \left(\frac{1-e}{e} \right) d(\boldsymbol{x}), \tag{9.53}$$

co jest równoważne (z dokładnością do stałej równej 2) dodaniu kolejnego składnika sumy w zależności (9.37).

Ponadto, w kolejnej iteracji składnik $cd(\boldsymbol{x})$ ma udział przy aktualizacji wektora wag

$$\boldsymbol{w}(\boldsymbol{x},\ y) \leftarrow \boldsymbol{w}(\boldsymbol{x},\ y) e^{-cyd(\boldsymbol{x})}. \tag{9.54}$$

Ponieważ $-yd(\boldsymbol{x}) = 2I(d(\boldsymbol{x}) \neq y) - 1$, otrzymujemy

$$\boldsymbol{w}(\boldsymbol{x},\ y) \leftarrow \boldsymbol{w}(\boldsymbol{x},\ y) \exp \left[\ln \left(\frac{1-e}{e} \right) I(d(\boldsymbol{x}) \neq y) \right], \tag{9.55}$$

a więc zależność (9.35) aktualizacji wag algorytmu *AdaBoost*.

Oczywiście powyższe rozważania teoretyczne są poczynione dla populacji i należy je odnieść do sytuacji, jaka występuje w praktyce. Z uwagi na posiadanie przez nas skończonej liczby obserwacji próby, algorytm *AdaBoost* jest tylko przybliżonym odpowiednikiem opisanej powyżej iteracyjnej minimalizacji funkcji (9.39).

Podamy teraz uogólnioną wersję algorytmu *AdaBoost* (dla zagadnienia 2 klas), tzw. *rzeczywisty AdaBoost*, zaproponowaną w przypadku ogólnym w pracy Friedmana, Hastiego i Tibshiraniego (2000). Tym razem wynikami klasyfikatorów składowych są nie etykiety klas, ale estymatory prawdopodobieństw a posteriori $\hat{p}(1|\boldsymbol{x})$, $\hat{p}(-1|\boldsymbol{x}) = 1 - \hat{p}(1|\boldsymbol{x})$.

Algorytm AdaBoost (rzeczywisty)

1. Ustal wektor wag $\boldsymbol{w}^{(1)} = (w_1^{(1)},\ w_2^{(1)},\ \ldots,\ w_n^{(1)})'$, przy czym

$$w_j^{(1)} \in [0,\ 1], \quad \sum_{j=1}^{n} w_j^{(1)} = 1.$$

(Zazwyczaj: $w_j^{(1)} = \frac{1}{n}$).

2. Dla $b = 1, 2, \ldots, B$:

2a) pobierz próbę bootstrapową $\mathcal{L}_n^{*b}$ z próby $\mathcal{L}_n$ według rozkładu

$$P(z_j \in \mathcal{L}_n^{*b}) = w_j^{(b)} \quad j = 1, 2, \ldots, n,$$

2b) konstruuj klasyfikator $\hat{d}_b$ dla próby bootstrapowej $\mathcal{L}_n^{*b}$

$$\hat{d}_b(\boldsymbol{x}) = \hat{p}^{(b)}(1|\boldsymbol{x})$$

Uwaga: Estymator prawdopodobieństwa a posteriori jest wyznaczany względem rozkładu wag $w_j^{(b)}$,

2c) oblicz

$$f_b(\boldsymbol{x}_j) = \frac{1}{2} \ln \frac{\hat{p}^{(b)}(1|\boldsymbol{x}_j)}{1 - \hat{p}^{(b)}(1|\boldsymbol{x}_j)}, \quad j = 1, 2, \ldots, n, \tag{9.56}$$

2d) aktualizuj wagi

$$w_j^{(b+1)} = \frac{w_j^{(b)} \exp(y_j f_b(\boldsymbol{x}_j))}{\sum_{i=1}^{n} w_i^{(b)} \exp(y_i f_b(\boldsymbol{x}_i))}, \quad j = 1, 2, \ldots, n. \tag{9.57}$$

3. Klasyfikuj obserwację $\boldsymbol{x}$ wg reguły

$$\hat{d}_{rAdaB}(\boldsymbol{x}) = \operatorname{sgn}\left(\sum_{b=1}^{B} f_b(\boldsymbol{x})\right). \tag{9.58}$$

Friedman, Hastie i Tibshirani (2000) pokazali, że powyższy algorytm – podobnie jak jego opisana wcześniej wersja dyskretna – jest iteracyjną metodą rozwiązania zadania minimalizacji funkcji (9.39). Tym samym, można go interpretować w terminach ogólnego modelu logistycznego.

Breiman (1998), prowadząc badania porównawcze nad baggingiem i boostingiem, klasę algorytmów typu boosting nazwał algorytmami *ARCing*[11], zaś sam algorytm *AdaBoost* – algorytmem *ARC-fs* (od pierwszych liter nazwisk jego twórców). W celu zbadania, czy sukces, jaki odniósł algorytm *ARC-fs*, ma korzenie w szczegółach technicznych, czy w samej metodzie adaptacyjnego repróbkowania, zaproponował on nowy algorytm oparty na boostingu – *ARC-x4*.

[11] Akronim słów: ang. **A**daptive **R**esample and **C**ombin**ing**.

Algorytm ARC-x4

1. Ustal wektor wag $\boldsymbol{w}^{(1)} = (w_1^{(1)},\ w_2^{(1)},\ \ldots,\ w_n^{(1)})'$, przy czym

$$w_j^{(1)} \in [0,\ 1], \quad \sum_{j=1}^{n} w_j^{(1)} = 1.$$

(Zazwyczaj: $w_j^{(1)} = \frac{1}{n}$).

2. Dla $b = 1,\ 2,\ \ldots,\ B$:

2a) pobierz próbę bootstrapową $\mathcal{L}_n^{*b}$ z próby $\mathcal{L}_n$ według rozkładu

$$\mathrm{P}(\boldsymbol{z}_j \in \mathcal{L}_n^{*b}) = w_j^{(b)}, \quad j = 1,\ 2,\ \ldots,\ n,$$

2b) konstruuj klasyfikator $\hat{d}_b$ dla próby bootstrapowej $\mathcal{L}_n^{*b}$,

2c) oblicz

$$m_j = \frac{1}{b}\sum_{i=1}^{b} I(\hat{d}_i(\boldsymbol{x}_j) \neq y_j), \qquad j = 1,\ 2,\ \ldots,\ n, \tag{9.59}$$

2d) aktualizuj wagi

$$w_j^{(b+1)} = \frac{1 + m_j^4}{\sum\limits_{k=1}^{n}(1 + m_k^4)} \qquad j = 1,\ 2,\ \ldots,\ n. \tag{9.60}$$

3. Klasyfikuj obserwację $\boldsymbol{x}$ wg reguły

$$\hat{d}_{ARCx4}(\boldsymbol{x}) = \arg \max_{1 \le k \le K} \sum_{b=1}^{B} I(\hat{d}_b(\boldsymbol{x}) = k). \tag{9.61}$$

Jak widzimy, waga obiektu $\boldsymbol{z}_j$ w $(b+1)$-kroku jest wyznaczana na bazie wielkości m_j będącej frakcją tych spośród b dotychczasowo wyuczonych klasyfikatorów, które zaklasyfikowały ją błędnie. Ponadto sam proces decyzyjny klasyfikatora łączonego odbywa się w tym przypadku na drodze głosowania przez zwykłą większość. Wartość 4 została wybrana eksperymentalnie, dając najlepsze rezultaty spośród wartości: 1, 2 i 4.

Co ciekawe, mimo że – jak przyznał sam Breiman (1998) – algorytm *ARC-x4* powstał *ad hoc*, w przeprowadzonych przez niego badaniach okazał się on równie dobry jak *AdaBoost*.

Breiman porównywał też zmiany wag w poszczególnych iteracjach. Okazało się, że w algorytmie *AdaBoost* są one zdecydowanie gwałtowniejsze niż w *ARC-x4*, gdzie zmiany te są bardziej regularne. Za miarę Breiman obrał odchylenie standardowe wagi odpowiadającej ustalonemu punktowi próby. Odchylenia te były znacznie większe w przypadku algorytmu Freunda i Schapire'a.

Przykład 9.1.
W celu ilustracji efektywności metod wzmacniania klasyfikatorów wykorzystamy rzeczywiste dane doświadczalne Wine[12]. Próba obejmuje 178 wyprodukowanych we Włoszech win, które następnie poddano analizie chemicznej. Rozpatrywano 13 cech typu ciągłego. Wyróżniono 3 klasy: klasa 1 (59 obserwacji), klasa 2 (71 obserwacji), klasa 3 (48 obserwacji).
Zastosowano metodę konstrukcji drzewa klasyfikacyjnego bez przycinania z użyciem indeksu Giniego (patrz podrozdz. 6.2). Aktualny poziom błędu tego klasyfikatora estymowany metodą sprawdzania krzyżowego (leave-one-out) wyniósł 0,1236. Kolejno zastosowano algorytmy Bagging (zarówno w wersji pierwotnej, jak i ważonej) oraz algorytm AdaBoost (wersja dyskretna). Wyniki eksperymentu dla początkowych iteracji rozpatrywanych algorytmów zawarte są w tab. 9.1–9.3. Kolumny 3–5 zawierają liczbę błędnych zaklasyfikowań obiektów danej klasy. Poziom błędnej klasyfikacji oznaczono przez $\hat{e}_R$, natomiast symbol r oznacza liczbę pojawiających się niejednoznaczności klasyfikatora $\hat{d}_{Bag1}$ danego równaniem (9.1) (Uwaga: W tym przypadku wielkość $\hat{e}_R$ jest wyznaczona dla sytuacji, w której wszystkie „sytuacje remisowe" zostały rozstrzygnięte w sposób błędny).

Tab. 9.1. Przebieg algorytmu Bagging (wersja pierwotna)

B	$\hat{e}_R$	klasa 1	klasa 2	klasa 3	r
1	0.0674	2	3	7	–
2	0.1067	–	2	1	16
3	0.0225	–	2	2	–
4	0.0112	–	–	–	2
5	0	–	–	–	–
6	0	–	–	–	–
7	0.0112	–	–	–	2
8	0	–	–	–	–

[12]Dane pochodzą ze składnicy danych Uniwersytetu Kalifornijskiego w Irvine. http://www.ics.uci.edu/~mlearn/MLRepository.html.

Tab. 9.2. Przebieg algorytmu Bagging (wersja ważona)

B	$\hat{e}_R$	klasa 1	klasa 2	klasa 3
1	0.0674	2	3	7
2	0.0337	1	3	2
3	0.0112	–	2	–
4	0.0056	–	–	1
5	0	–	–	–

Tab. 9.3. Przebieg algorytmu AdaBoost (wersja dyskretna)

B	$\hat{e}_R$	klasa 1	klasa 2	klasa 3
1	0.0674	2	3	7
2	0.0337	0	4	2
3	0.0056	–	1	–
4	0	–	–	–

Widzimy, że każda z metod prowadzi do bezbłędnej predykcji i to nie później niż po zaledwie 8 krokach (tj. przy 8-krotnym generowaniu próby bootstrapowej). Co więcej – analiza eksperymentu (przeprowadzonego w istocie dla $B = 50$) pokazuje, że w krokach późniejszych w żadnej z metod nie pojawiają się błędne zaklasyfikowania czy też niejednoznaczności w procesie predykcji (dla klasyfikatora (9.1)). Fakt ten wskazuje, że w rozpatrywanym przypadku – mimo stosunkowo wysokiego poziomu błędu (ponad 12% błędnych zaklasyfikowań), jaki otrzymamy, stosując pojedyncze drzewo klasyfikacyjne – algorytmy wzmacniania klasyfikatorów prowadzą do wyraźnej i bardzo szybkiej poprawy dokładności procesu decyzyjnego.

W ciągu paru kolejnych lat pojawiło się wiele odmian algorytmów opartych na idei boostingu. Wydaje się, że podejście to jest najintensywniej badane i rozwijane ze wszystkich metod łączonych w dyskryminacji (Kuncheva, 2004). Wymienić tu należy algorytm *LogitBoost* addytywnego modelu regresji logistycznej (Friedman, Hastie, Tibschirani, 2000) czy procedury powstałe wyłącznie z badań empirycznych: *MultiBoosting* (Webb, 2000), *DOOM* (Mason, Bartlet, Baxter, 2000), *AveBoost* (Oza, 2003).

10

Procedury kombinowane

W poprzednich rozdziałach, do rozwiązania zagadnienia klasyfikacji obiektu do jednej z K klas, używaliśmy wielu różnych metod (klasyfikatorów). Każda z nich była oparta na innej koncepcji. Może się zdarzyć, że żadna z rozpatrywanych metod indywidualnych nie jest jednoznacznie lepsza od pozostałych oraz że każda z metod może błędnie klasyfikować inne obserwacje. Wynika z tego, że uwzględniając w rozwiązaniu zagadnienia klasyfikacyjnego cały zbiór metod, możemy uzyskać pełną (a przynajmniej większą) informację dotyczącą klasyfikacji obiektu, chociaż żadna metoda z osobna nie ma tej własności. Zilustrujemy to na prostym przykładzie zaczerpniętym z monografii Webba (2003).

Przykład 10.1.
Na rysunku 10.1 są przedstawione dwie metody liniowe działające w następujący sposób: jeśli obserwacja znajduje się na lewo od B, to pierwszy klasyfikator

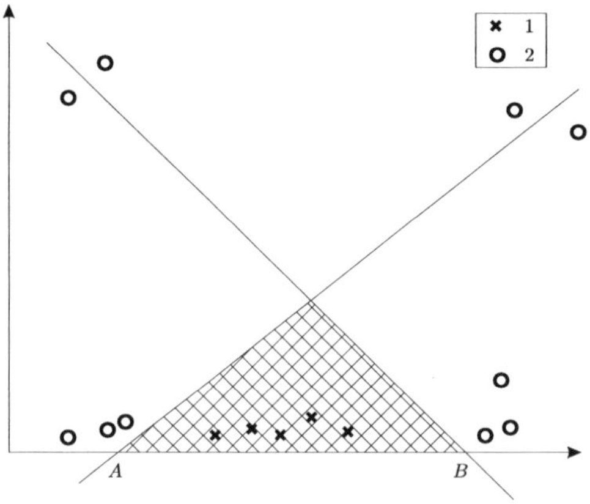

Rys. 10.1. Dwie metody liniowe

zalicza ją do pierwszej klasy, w przeciwnym razie do klasy drugiej, jeśli natomiast obserwacja znajduje się na prawo od A, to drugi klasyfikator zalicza ją do klasy pierwszej, a jeśli znajduje się na lewo od A, to do klasy drugiej. Jak widzimy obu klasyfikatorom wiele brakuje do osiągnięcia 100% skuteczności klasyfikacji. Skonstruujmy więc na bazie tych dwóch metod nowy klasyfikator łączony, działający następująco: jeśli oba klasyfikatory zgadzają się co do decyzji, że należy obserwację przypisać do klasy pierwszej, to do niej zalicza ją też klasyfikator łączony (obszar zakreskowany na rysunku), w przeciwnym razie obserwacja jest zaliczana do klasy drugiej. Tak skonstruowana metoda ma już dokładność klasyfikacji równą 100%.

Załóżmy, że dysponujemy c różnymi klasyfikatorami $\hat{d}_1, \hat{d}_2, \ldots, \hat{d}_c$ wyznaczonymi na bazie próby uczącej $\mathcal{L}_n = \{(\boldsymbol{x}_1, y_1), \ldots, (\boldsymbol{x}_n, y_n)\}$, gdzie $\boldsymbol{x}_i$ jest wektorem obserwowanych cech, a y_i etykietą klasy ($i = 1, 2, \ldots, n$). W dalszym ciągu będziemy poszukiwać procedury „kombinującej" informacje uzyskane z tych klasyfikatorów, dającej mniejszy błąd klasyfikacji.

10.1. Metoda selekcji

Podstawową ideą tej metody jest zastosowanie do klasyfikacji obiektu tego z c klasyfikatorów, który w pewnym otoczeniu punktu $\boldsymbol{x}$ daje możliwie najlepsze rezultaty. Czyli wybieramy lokalnie najlepszy klasyfikator. Otoczenie punktu $\boldsymbol{x}$ jest wyznaczone przez J jego najbliższych sąsiadów (patrz rozdz. 5). Zwyczajowo do ich wyznaczenia stosujemy zwykłą odległość euklidesową. Oznaczmy przez $NN(\boldsymbol{x})$ zbiór J najbliższych sąsiadów dla obserwacji $\boldsymbol{x}$.

Woods, Kegelmeyer i Bowyer (1997) zaproponowali, wykorzystując powyższą ideę, metodę CLA[1]. Za lokalnie najlepszy klasyfikator $\hat{d}_{CLA}$ wybieramy ten z klasyfikatorów $\hat{d}_j$, $j = 1, 2, \ldots, c$, którego dokładność (prawdopodobieństwo poprawnej klasyfikacji), w otoczeniu punktu $\boldsymbol{x}$ jest maksymalna.

Prawdopodobieństwo to szacujemy, przyjmując

$$\hat{p}^{(j)}(\boldsymbol{x}) = \frac{1}{J} \sum_{\boldsymbol{x}_i \in NN(\boldsymbol{x})} \sum_{k=1}^{K} I(\hat{d}_j(\boldsymbol{x}_i) = k) I(y_i = k), \qquad j = 1, 2, \ldots, c \tag{10.1}$$

[1] Po ang. Classifier's Local Accuracy (CLA).

lub bardziej restrykcyjnie, uwzględniając jedynie te obserwacje z otoczenia $NN(\boldsymbol{x})$, które należą do klasy $\hat{d}_j(\boldsymbol{x})$. Wzór (10.1) przyjmuje wtedy postać

$$\hat{p}^{(j)}(\boldsymbol{x}) = \frac{1}{J} \sum_{\boldsymbol{x}_i \in NN(\boldsymbol{x})} I(\hat{d}_j(\boldsymbol{x}_i) = \hat{d}_j(\boldsymbol{x}))I(y_i = \hat{d}_j(\boldsymbol{x})), \quad j = 1, 2, \ldots, c. \tag{10.2}$$

W zaproponowanej przez Giacinta i Roliego (2001) metodzie MCB[2] ograniczono dodatkowo otoczenie punktu $\boldsymbol{x}$ przez wprowadzenie miary podobieństwa obiektów.

Definicja 10.1.
Niech $\boldsymbol{x}_r$ i $\boldsymbol{x}_s$ będą wektorami obserwacji r-tego i s-tego obiektu. Miarę podobieństwa S tych obiektów wyznaczoną przez zbiór klasyfikatorów $\{d_1, d_2, \ldots, d_c\}$ określamy jako

$$S(\boldsymbol{x}_r, \boldsymbol{x}_s) = \frac{1}{c} \sum_{j=1}^{c} I(d_j(\boldsymbol{x}_r) = d_j(\boldsymbol{x}_s)). \tag{10.3}$$

Ograniczmy zatem otoczenie $NN(\boldsymbol{x})$ punktu $\boldsymbol{x}$ jedynie do tych punktów z próby uczącej, które są odpowiednio „mocno” do niego podobne, tzn. określmy otoczenie $NNS(\boldsymbol{x})$ punktu $\boldsymbol{x}$ w następujący sposób:

$$NNS(\boldsymbol{x}) = \{\boldsymbol{x}_i \in NN(\boldsymbol{x}) : \; S(\boldsymbol{x}, \boldsymbol{x}_i) > \gamma\}, \tag{10.4}$$

gdzie $0 < \gamma < 1$ jest ustaloną wartością progową.

Analogicznie jak w metodzie CLA, za lokalnie najlepszy klasyfikator $\hat{d}_{MCB}$ wybieramy ten z klasyfikatorów $\hat{d}_j$, $j = 1, 2, \ldots, c$, którego dokładność, w otoczeniu $NNS(\boldsymbol{x})$ punktu $\boldsymbol{x}$ jest maksymalna. Wzory (10.1) i (10.2) przyjmują postać

$$\hat{p}^{(j)}(\boldsymbol{x}) = \frac{1}{J} \sum_{\boldsymbol{x}_i \in NNS(\boldsymbol{x})} \sum_{k=1}^{K} I(\hat{d}_j(\boldsymbol{x}_i) = k)I(y_i = k), \quad j = 1, 2, \ldots, c \tag{10.5}$$

[2] Po ang. Multiple Classifier Behavior (MCB).

oraz

$$\hat{p}^{(j)}(\boldsymbol{x}) = \frac{1}{J} \sum_{\boldsymbol{x}_i \in NNS(\boldsymbol{x})} I(\hat{d}_j(\boldsymbol{x}_i) = \hat{d}_j(\boldsymbol{x}))I(y_i = \hat{d}_j(\boldsymbol{x})),$$
$$j = 1,\ 2,\ \ldots,\ c. \tag{10.6}$$

Widać wyraźnie, że jedyna różnica między metodami CLA i MCB tkwi w sposobie definiowania otoczenia punktu $\boldsymbol{x}$.

10.2. Łączenie klasyfikatorów

W poprzednim podrozdziale decyzja klasyfikacyjna oparta na zbiorze klasyfikatorów $\{\hat{d}_1,\ \hat{d}_2,\ \ldots,\ \hat{d}_c\}$ polegała na lokalnym wybieraniu najlepszego z nich. W podrozdziale tym omówimy metody, w których ostateczna decyzja klasyfikacyjna będzie oparta na klasyfikatorze łączonym, wykorzystującym informacje ze wszystkich c klasyfikatorów.

Zasadnicza różnica między tymi metodami jest związana ze sposobem, w jaki jest uzyskiwana informacja z klasyfikatorów składowych. Minimalna uzyskana z klasyfikatora składowego informacja to etykieta klasy przyporządkowana klasyfikowanej informacji. W tej sytuacji użyć możemy jedynie metod łączonych polegających na zliczaniu tzw. głosów. Dużo większe możliwości daje sytuacja, w której dla poszczególnych klasyfikatorów dysponujemy ocenami prawdopodobieństw a posteriori przynależności klasyfikowanego obiektu do każdej z klas. Przykładami metod łączonych opartych na prawdopodobieństwach a posteriori są różne odmiany procedur sumacyjnych i iloczynowych oraz metoda stosowa.

10.2.1. Metoda głosowania

Najprostszą metodą łączenia klasyfikatorów jest *metoda głosowania*[3]. Polega ona na tym, że klasyfikator łączony wskazuje na klasę, na którą wskazuje najwięcej klasyfikatorów indywidualnych, które jakby „oddają na nią głos". Formalnie

$$\hat{d}_V(\boldsymbol{x}) = \arg \max_{1 \le k \le K} \sum_{j=1}^{c} I(\hat{d}_j(\boldsymbol{x}) = k). \tag{10.7}$$

Największą zaletą tej metody jest fakt, że może być użyta dla dowolnych metod klasyfikacji. Jej największa wada tkwi w bardzo częstych sytuacjach

[3]Po ang. majority vote.

remisowych, występujących wtedy, gdy na dwie lub więcej klas wskazuje taka sama liczba metod, co zdarza się szczególnie często przy małej liczbie metod składowych.

Pewnym sposobem wyjścia z tej sytuacji jest wprowadzenie wag uzależnionych od jakości klasyfikatorów składowych. Ważona postać klasyfikatora (10.7) ma postać

$$\hat{d}_V(\boldsymbol{x}) = \arg \max_{1 \le k \le K} \sum_{j=1}^{c} w_{jk} I(\hat{d}_j(\boldsymbol{x}) = k), \tag{10.8}$$

gdzie

$$w_{jk} = \frac{\sum_{i=1}^{n} I(\hat{d}_j(\boldsymbol{x}_i) = k,\ y_i = k)}{\sum_{i=1}^{n} I(y_i = k)}.$$

Inną ciekawą metodą wykorzystującą jedynie „głosy" poszczególnych klasyfikatorów jest zaproponowana przez Huanga i Suena (1995) metoda BKS[4].

Klasyfikator łączony ma postać

$$\hat{d}_{BKS}(\boldsymbol{x}) = \arg \max_{1 \le k \le K} \sum_{i=1}^{n} I\Big(\sum_{j=1}^{c} I(\hat{d}_j(\boldsymbol{x}) = \hat{d}_j(\boldsymbol{x}_i)) = c\Big) I(y_i = k). \tag{10.9}$$

Występujący w (10.9) warunek zgodności decyzji dla wszystkich klasyfikatorów składowych postaci $\sum_{j=1}^{c} I(\hat{d}_j(\boldsymbol{x}) = d_j(\boldsymbol{x}_i)) = c$ jest bardzo restrykcyjny. Możemy go osłabić, żądając zgodności decyzji jedynie dla co najmniej m z c klasyfikatorów. Otrzymany w ten sposób klasyfikator ma postać

$$\hat{d}_{BKS}(\boldsymbol{x}) = \arg \max_{1 \le k \le K} \sum_{i=1}^{n} I\Big(\sum_{j=1}^{c} I(\hat{d}_j(\boldsymbol{x}) = \hat{d}_j(\boldsymbol{x}_i)) \ge m\Big) I(y_i = k), \tag{10.10}$$

gdzie liczba naturalna $1 \le m \le c$ jest ustaloną wartością progową.

[4] Po ang. Behavior-Knowledge Space (BKS).

10.2.2. Elementarne metody łączenia klasyfikatorów

Załóżmy teraz, że dla klasyfikatorów indywidualnych dysponujemy ocenami prawdopodobieństw a posteriori przynależności klasyfikowanego obiektu do każdej z klas. Oznaczmy je przez $\hat{p}^{(j)}(k|\boldsymbol{x})$, $j = 1, 2, \ldots, c$, $k = 1, 2, \ldots, K$.

W metodzie sumacyjnej, klasyfikacja bazuje na wykorzystaniu sum

$$s_k(\boldsymbol{x}) = \sum_{j=1}^{c} \hat{p}^{(j)}(k|\boldsymbol{x}), \qquad k = 1, \ldots, K. \tag{10.11}$$

Klasyfikowany obiekt zaliczamy do tej klasy, dla której suma (10.11) przyjmuje wartość maksymalną, tj.

$$\hat{d}_{SUM}(\boldsymbol{x}) = \arg\max_k s_k(\boldsymbol{x}).$$

W metodzie sumacyjnej oceną prawdopodobieństwa a posteriori jest

$$\hat{p}(k|\boldsymbol{x}) = \frac{s_k(\boldsymbol{x})}{\sum\limits_{i=1}^{K} s_i(\boldsymbol{x})}.$$

Słabością tej reguły jest jednakowy udział wszystkich wartości $\hat{p}^{(1)}(k|\boldsymbol{x}), \ldots, \hat{p}^{(c)}(k|\boldsymbol{x})$ w budowie klasyfikatora łącznego nawet wtedy, gdy istnieją znaczne różnice między jakością klasyfikatorów składowych. W literaturze często zamiast zwykłej sumy rozpatruje się średnią arytmetyczną z ocen prawdopodobieństw a posteriori. Nic nie stoi na przeszkodzie, aby stworzyć wersję ważoną tej metody, tzn. każdy klasyfikator otrzymałby wagę zależną od błędu klasyfikacji. Problemem pozostaje wybór odpowiedniego systemu wag. Pewne propozycje na ten temat możemy znaleźć w pracach Woodsa, Kegelmeyera i Bowyera (1997) oraz Góreckiego (2005b). W pierwszej z tych prac jest proponowany tzw. *dynamiczny wybór klasyfikatora*[5], polegający na podziale całej przestrzeni na rozłączne obszary, w taki sposób, że w każdym z nich wybierany jest tylko jeden klasyfikator (czyli dostaje on wagę 1, a pozostałe 0). Do podziału przestrzeni została wykorzystana metoda najbliższych sąsiadów. W pracy Góreckiego poszukiwanie optymalnych wag odbywało się na pewnej z góry zadanej siatce na zbiorze uczącym, następnie tak znalezione wagi były aplikowane do zbioru testowego.

[5] Po ang. dynamic classifier selection.

Formalna różnica między metodą iloczynową a sumacyjną polega jedynie na wykorzystaniu iloczynów prawdopodobieństw a posteriori zamiast ich sum. Bazując na iloczynach

$$r_k(\boldsymbol{x}) = \prod_{j=1}^{c} \hat{p}^{(j)}(k|\boldsymbol{x}), \qquad k = 1, \ldots, K, \tag{10.12}$$

klasyfikowany obiekt zaliczamy do tej klasy, dla której iloczyn (10.12) przyjmuje wartość maksymalną, tj.

$$\hat{d}_{PROD}(\boldsymbol{x}) = \arg\max_k r_k(\boldsymbol{x}).$$

Estymatorem prawdopodobieństwa a posteriori przynależności obserwacji $\boldsymbol{x}$ do klasy k w metodzie iloczynowej jest

$$\hat{p}(k|\boldsymbol{x}) = \frac{r_k(\boldsymbol{x})}{\sum_{i=1}^{K} r_i(\boldsymbol{x})}.$$

Reguła ta narzuca wzajemną zgodność między klasyfikatorami indywidualnymi w tym sensie, że jeżeli chociaż jeden z nich wskazuje, że przynależność obserwacji $\boldsymbol{x}$ do klasy k jest mało prawdopodobna, to jest mało prawdopodobne, by klasyfikator łączny zaliczył ją do tej grupy. Własność ta jest konsekwencją faktu, że iloczyn liczb z przedziału $(0,\ 1)$ jest mniejszy od jego najmniejszego czynnika.

Kolejna grupa metod, to metody bazujące na statystykach pozycyjnych. Do tej grupy należą:

- metoda minimum,
- metoda medianowa,
- metoda maksimum.

W tym przypadku klasyfikowany obiekt zaliczamy odpowiednio do klasy:

$$\begin{aligned}
\hat{d}_{MIN}(\boldsymbol{x}) &= \arg\max_k \min_j \hat{p}^{(j)}(k|\boldsymbol{x}), \\
\hat{d}_{ME}(\boldsymbol{x}) &= \arg\max_k \mathrm{Me}\ \hat{p}^{(j)}(k|\boldsymbol{x}), \\
\hat{d}_{MAX}(\boldsymbol{x}) &= \arg\max_k \max_j \hat{p}^{(j)}(k|\boldsymbol{x}).
\end{aligned}$$

Metoda minimum jest najbardziej pesymistyczna (wystarczy, że jeden klasyfikator jest zdecydowanie „na nie" i cała metoda będzie „na nie"), a metoda maksimum najbardziej optymistyczna (wystarczy, że jeden klasyfikator jest zdecydowanie „na tak" i cała metoda będzie „na tak"). Kompromis pomiędzy tymi metodami stanowi metoda medianowa.

Przykład 10.2.
Załóżmy, że dysponujemy pięcioma metodami klasyfikacji oraz problem klasyfikacyjny dotyczy trzech grup. Odpowiednia macierz prawdopodobieństwa a posteriori dla pewnej obserwacji $\boldsymbol{x}$ ma postać

$$\boldsymbol{P} = \begin{bmatrix} 0.1 & 0.5 & 0.4 \\ 0.0 & 0.0 & 1.0 \\ 0.4 & 0.3 & 0.4 \\ 0.2 & 0.7 & 0.1 \\ 0.1 & 0.8 & 0.1 \end{bmatrix}.$$

Teraz odpowiednie czynniki przynależności mają postać:

- metoda sumacyjna – (0.8 2.3 2.0),
- metoda iloczynowa – (0.0 0.0 0.0016),
- metoda minimum – (0.0 0.0 0.1),
- metoda maksimum – (0.4 0.8 1.0),
- metoda medianowa – (0.1 0.5 0.4).

Widzimy zatem, że metoda sumacyjna oraz medianowa zaliczają obiekt do drugiej klasy, a pozostałe do klasy trzeciej.

Okazuje się, że przy pewnych założeniach omawiane wcześniej metody łączenia klasyfikatorów są równoważne (w takim sensie, że wskazują na tę samą klasę).

TWIERDZENIE 10.1.

Dla dowolnej liczby klasyfikatorów, w przypadku rozpoznawania obiektów należących jedynie do dwóch klas, metody minimum i maksimum są sobie równoważne.

TWIERDZENIE 10.2.

Dla nieparzystej liczby klasyfikatorów, w przypadku rozpoznawania obiektów należących jedynie do dwóch klas, metody medianowa i głosowania są sobie równoważne.

Elementarne dowody obu twierdzeń można znaleźć np. w monografii Kunchevej (2004).

10.2.3. Regresja stosowa

Wolpert (1992) przedstawił ciekawą ideę łączenia klasyfikatorów, znaną jako *stacked generalization*. Zaproponował mianowicie zamiast poszukiwania najlepszego klasyfikatora w pewnym zbiorze klasyfikatorów, skonstruowanie nowego łączącego informacje ze wszystkich metod i będący ich liniową kombinacją. Ponieważ każda z metod klasyfikacji ma pewne zalety, ich połączenie wydaje się sensowne. Propozycja Wolperta została przetłumaczona na język statystyki matematycznej przez Breimana (1996a); autor nazwał ją *regresją stosową*[6]. Następnie LeBlanc i Tibshirani (1996) pokazali, jak można tę metodę wykorzystać w analizie dyskryminacyjnej. Klasyfikator połączony otrzymany za pomocą tej metody jest liniową kombinacją estymatorów prawdopodobieństw a posteriori otrzymanych z metod indywidualnych. Współczynniki tej kombinacji liniowej znajdujemy za pomocą metody najmniejszych kwadratów w odpowiednim modelu regresji. Regresja stosowa jest uogólnieniem metody sumacyjnej konstrukcji klasyfikatora połączonego.

Dysponujemy c klasyfikatorami oraz K klasami, mamy więc Kc estymatorów prawdopodobieństw a posteriori, które układamy w wektor

$$\hat{\boldsymbol{p}}(\boldsymbol{x}) = (\hat{p}^{(1)}(1|\boldsymbol{x}), \ \ldots, \ \hat{p}^{(1)}(K|\boldsymbol{x}), \ \ldots, \ \hat{p}^{(c)}(1|\boldsymbol{x}), \ \ldots, \ \hat{p}^{(c)}(K|\boldsymbol{x}))'.$$

Wektory ocen $\hat{\boldsymbol{p}}(\boldsymbol{x}_i)$ są układane w stos jako wiersze macierzy $\boldsymbol{P}$. Niech $\boldsymbol{u}_k$ będzie wektorem identyfikującym przynależność elementów zbioru uczącego do klasy k, tzn. i-ty element tego wektora ma postać

$$u_{ik} = \begin{cases} 1, & \text{jeśli } y_i = k, \\ 0, & \text{jeśli } y_i \neq k. \end{cases}$$

Model regresji stosowej ma postać

$$\boldsymbol{u}_k = \boldsymbol{P}\boldsymbol{\beta}_k + \varepsilon_k,$$

gdzie $\boldsymbol{\beta}_k$ jest Kc-wymiarowym wektorem nieznanych parametrów, ε_k jest n-wymiarowym wektorem błędów, o którym zakładamy, że ma wielowymiarowy rozkład normalny z wektorem wartości oczekiwanych $\mathbf{0}$ oraz macierzą

[6] Po ang. stacked regression.

kowariancji $e\boldsymbol{I}$. Estymator najmniejszych kwadratów $\hat{\boldsymbol{\beta}}_k$ otrzymujemy, rozwiązując układ równań normalnych

$$\boldsymbol{P}'\boldsymbol{P}\boldsymbol{\beta}_k = \boldsymbol{P}'\boldsymbol{u}_k$$

względem $\boldsymbol{\beta_k}$.

Estymatory prawdopodobieństw a posteriori otrzymane z klasyfikatorów składowych sumują się do jedności, czyli dla każdego $j = 1,\ 2,\ \ldots,\ c$ oraz dla każdego $i = 1,\ 2,\ \ldots,\ n$

$$\sum_{k=1}^{K} \hat{p}^{(j)}(k|\boldsymbol{x}_i) = 1.$$

Stąd kolumny macierzy $\boldsymbol{P}$ podlegają c ograniczeniom liniowym, macierz $\boldsymbol{P}$ nie jest zatem pełnego rzędu kolumnowego i $\boldsymbol{P}'\boldsymbol{P}$ jest macierzą osobliwą. Musimy zatem użyć jednej z wielu uogólnionych odwrotności, najczęściej używa się uogólnionej odwrotności Moore'a–Penrose'a macierzy $\boldsymbol{P}'\boldsymbol{P}$, którą oznaczamy przez $(\boldsymbol{P}'\boldsymbol{P})^{+}$. W ten sposób otrzymujemy estymator najmniejszych kwadratów nieznanych parametrów regresji stosowej postaci

$$\hat{\boldsymbol{\beta}}_k = (\boldsymbol{P}'\boldsymbol{P})^{+}\boldsymbol{P}'\boldsymbol{u}_k.$$

Mając dane oceny $\hat{\boldsymbol{\beta}}_1,\ \ldots,\ \hat{\boldsymbol{\beta}}_K$, obserwację $\boldsymbol{x}$ klasyfikujemy, wykorzystując iloczyn skalarny

$$\hat{u}_k = \hat{\boldsymbol{p}}'(\boldsymbol{x})\hat{\boldsymbol{\beta}}_k.$$

Każdy iloczyn skalarny jest sumą ważoną składowych wektora $\hat{\boldsymbol{p}}'(\boldsymbol{x})$, gdzie wagi są elementami wektora $\hat{\boldsymbol{\beta}}_k$. Ostatecznie obserwację $\boldsymbol{x}$ klasyfikujemy do klasy z największą wartością $\hat{u}_k$, tj.

$$\hat{d}_{stos}(\boldsymbol{x}) = \arg\max_{k} \hat{u}_k.$$

Te iloczyny skalarne są nazywane *czynnikami przynależności* i nie sumują się do jedności, a co gorsza mogą przyjmować wartości ujemne.

Cały problem powstawania ujemnych wartości czynników przynależności powstaje z powodu metody estymacji parametrów $\boldsymbol{\beta}_k$. Te współczynniki mogą być ujemne, zatem czynniki przynależności również. W wielu przypadkach jest to kłopotliwe, chociażby wtedy, gdy chcemy otrzymać z metody regresji stosowej prawdopodobieństwa a posteriori, aby móc je dalej wykorzystać w następnej metodzie łączonej. Najprościej jest zastąpić ujemne wartości czynników przynależności zerami i obliczyć oceny prawdopodobieństw a posteriori według wzoru

$$\tilde{u}_k = \frac{\hat{u}_k}{\sum\limits_{j=1}^{K} \hat{u}_j}, \qquad k = 1,\ 2,\ \ldots,\ K.$$

Wartości $\tilde{u}_k$ są już prawdopodobieństwami a posteriori. Metoda ta nie jest jednak w żaden sposób uzasadniona i jej stosowanie może być uzasadnione jedynie brakiem innych pomysłów. Zdecydowanie lepszym rozwiązaniem jest zastosowanie regresji z ograniczeniami już na poziomie estymacji współczynników $\boldsymbol{\beta}_k$. Zostało to już zauważone przez Breimana (1996a) oraz przez LeBlanca i Tibshiraniego (1996). W tej drugiej pracy został użyty zmodyfikowany nieco algorytm zaproponowany przez Lawsona i Hansona (1974). Efektywniejsze jest jednak użycie modelu przedstawionego przez Toutenburga i Roedera (1978). Model ten jest ogólniejszy od modelu wykorzystanego przez LeBlanca i Tibshiraniego (1996).

Toutenburg i Roeder (1978) rozważali następujący model regresji:

$$\boldsymbol{y} = \boldsymbol{X}\boldsymbol{\beta} + \boldsymbol{\varepsilon}, \tag{10.13}$$

gdzie: $\boldsymbol{y}$ jest T-wymiarowym wektorem obserwacji zmiennej zależnej, $\boldsymbol{X}$ jest $T \times h$-wymiarową macierzą obserwacji h zmiennych niezależnych (jest ona pełnego rzędu kolumnowego), $\boldsymbol{\beta}$ jest h-wymiarowym wektorem współczynników regresji oraz $\boldsymbol{\varepsilon}$ jest T-wymiarowym wektorem błędów, o którym zakładamy, że ma rozkład $N(\mathbf{0}, \sigma^2\boldsymbol{W})$ (macierz $\boldsymbol{W}$ jest znana, natomiast σ^2 jest nieznane). Dodatkowo mamy pewne informacje podane a priori dotyczące ograniczeń, jakie spełniają współczynniki regresji w tym modelu. Spełniają one mianowicie następujące ograniczenia:

$$r_{L_i} \leq \beta_i \leq r_{U_i}, \qquad i = 1,\ 2,\ \ldots,\ h,$$

gdzie: r_{L_i} oraz r_{U_i} są odpowiednio dolnym i górnym ograniczeniem dla parametru β_i, oraz

$$P(r_{L_i} \leq \beta_i \leq r_{U_i}) = 1 - \frac{1}{s^2}, \qquad i = 1,\ 2,\ \ldots,\ h$$

dla $s \geq 1$. Estymacja parametrów jest wykonywana za pomocą *regresji mieszanej*[7] – Theil i Goldberger (1961). To znaczy ograniczenia zostały wyrażone w postaci następujących liniowych ograniczeń stochastycznych:

$$\boldsymbol{r} = \boldsymbol{\beta} + \boldsymbol{v},$$

[7] Po ang. mixed regression.

gdzie $\boldsymbol{v}$ jest wektorem losowym z wektorem średnich $\mathbf{0}$ oraz macierzą kowariancji $(4p^2\boldsymbol{H})^{-1}$. $\boldsymbol{H}$ jest $h \times h$-wymiarową macierzą diagonalną, gdzie na przekątnej leżą elementy $(r_{U_i} - r_{L_i})^{-2}$, $\boldsymbol{r}$ jest h-wymiarowym wektorem o elementach $r_i = (r_{L_i} + r_{U_i})/2$. Prowadzi to do estymatora regresji mieszanej postaci

$$\hat{\boldsymbol{\beta}} = (\boldsymbol{X}^T\boldsymbol{W}^{-1}\boldsymbol{X} + 4s^2\sigma^2\boldsymbol{H})^{-1}(\boldsymbol{X}^T\boldsymbol{W}^{-1}\boldsymbol{y} + 4s^2\sigma^2\boldsymbol{H}\boldsymbol{r}). \tag{10.14}$$

Również Toutenburg i Roeder (1978) pokazali, że $\hat{\boldsymbol{\beta}}$ jest estymatorem nieobciążonym.

Zastosujmy teraz tę technikę do regresji stosowej. Wiemy, że macierz $\boldsymbol{P}$ nie jest pełnego rzędu kolumnowego, nie spełnia zatem założeń modelowych. Można oczywiście i tak użyć tego modelu, ponieważ nigdzie we wzorze (10.14) nie mamy do czynienia z odwracaniem tej macierzy (tutaj jest to macierz $\boldsymbol{X}$). Lepiej jednak, aby wszystkie założenia modelowe były spełnione. Na szczęście w tym przypadku wystarczy tylko niewielka modyfikacja, aby macierz $\boldsymbol{P}$ była odpowiednia. Ponieważ podczas jej konstrukcji wykorzystujemy wszystkie prawdopodobieństwa a posteriori, które oczywiście sumują się do jedności, zatem powstaje problem niepełnego rzędu. Nie musimy jednak wykorzystywać ich wszystkich, ponieważ ostatnie dla każdej metody jest liniową kombinacją poprzednich. W ten sposób możemy zredukować macierz $\boldsymbol{P}$ do $N \times (Kc - c)$-wymiarowej macierzy $\boldsymbol{P}_1$. Ta ostatnia jest już macierzą pełnego rzędu kolumnowego i nic nie stoi na przeszkodzie, aby została użyta.

Sformułujmy zatem model regresji stosowej z ograniczeniami

$$\boldsymbol{u}_k = \boldsymbol{P}_1\boldsymbol{\beta}_k + \boldsymbol{\varepsilon}_k, \tag{10.15}$$

gdzie: $\boldsymbol{u}_k$ jest N-wymiarowym wektorem obserwacji zmiennej zależnej, $\boldsymbol{P}_1$ jest $N \times c(K-1)$-wymiarową macierzą obserwacji zmiennych niezależnych (jest pełnego rzędu kolumnowego), $\boldsymbol{\beta}_k$ jest $c(K-1)$-wymiarowym wektorem współczynników regresji stosowej oraz $\boldsymbol{\varepsilon}_k$ jest N-wymiarowym wektorem błędów, o którym zakładamy, że ma rozkład $N(\mathbf{0}, \sigma_k^2\boldsymbol{W})$ (macierz $\boldsymbol{W}$ jest znana, σ_k^2 jest nieznana). Przy naszych założeniach możemy sformułować ograniczenia w przedziale $[0,\ a]$. Stąd

$$\hat{\boldsymbol{\beta}}_k = (\boldsymbol{P}_1^T\boldsymbol{W}^{-1}\boldsymbol{P}_1 + 4s^2\sigma_k^2a^{-2}\boldsymbol{I})^{-1}(\boldsymbol{P}_1^T\boldsymbol{W}^{-1}\boldsymbol{u}_k + 2s^2\sigma_k^2a^{-1}\mathbf{1}). \tag{10.16}$$

Chaturvedi i Wan (1999) zaproponowali użycie następującego estymatora:

$$\hat{\sigma}_k^2 = \frac{\boldsymbol{u}_h^T(\boldsymbol{W}^{-1} - \boldsymbol{W}^{-1}\boldsymbol{P}_1(\boldsymbol{P}_1^T\boldsymbol{W}^{-1}\boldsymbol{P}_1)^{-1}\boldsymbol{P}_1^T\boldsymbol{W}^{-1})\boldsymbol{u}_k}{N - c(K-1)} \tag{10.17}$$

parametru σ_k^2, $k = 1, 2, \ldots, K$. W modelu zakładamy znajomość macierzy kowariancji błędu, dana jest ona zatem a priori. Ponieważ jednak nie mamy żadnych wcześniejszych obserwacji, zakłada się w większości przypadków praktycznych, że błędy są nieskorelowane, czyli

$$\boldsymbol{W} = \boldsymbol{I}.$$

Przy powyższych założeniach otrzymujemy ostatecznie następujący estymator parametrów:

$$\hat{\boldsymbol{\beta}}_k = (\boldsymbol{P}_1^T \boldsymbol{P}_1 + 4s^2\hat{\sigma}_k^2 \boldsymbol{I})^{-1}(\boldsymbol{P}_1^T \boldsymbol{u}_k + 2s^2\hat{\sigma}_k^2 \mathbf{1}), \tag{10.18}$$

gdzie

$$\hat{\sigma}_k^2 = \frac{\boldsymbol{u}_k^T(\boldsymbol{I} - \boldsymbol{P}_1(\boldsymbol{P}_1^T\boldsymbol{P}_1)^{-1}\boldsymbol{P}_1^T)\boldsymbol{u}_k}{N - c(K-1)}.$$

W praktyce okazuje się, że powstaje jeszcze jeden problem. Mianowicie w obliczeniach numerycznych macierz $\boldsymbol{P}_1^T\boldsymbol{P}_1$ nie daje się najczęściej odwrócić. Z problemem tym spotykamy się jedynie przy estymacji parametru $\hat{\sigma}_k^2$. Gdy zatem nie estymujemy $\hat{\sigma}_k^2$, nie musimy używać uogólnionej odwrotności, natomiast w pozostałych przypadkach jest to, niestety, konieczne. W tym też tkwi odpowiedź na nasuwające się od początku pytanie: „Dlaczego nie usuwa się kolumn w oryginalnej regresji stosowej?". Usunięcie tych kolumn niczego nie zmienia w zagadnieniu odwracalności macierzy, a tylko wprowadza dodatkowe zbędne operacje.

Wyniki eksperymentalne (Górecki (2005a)) pokazują, że regresja stosowa daje lepsze rezultaty od zwykłej regresji stosowej. Oprócz zatem otrzymywania nieujemnych wartości czynników przynależności poprawiamy efektywność klasyfikacji. Dodatkowo nie musimy używać uogólnionej odwrotności Moore'a–Penrose'a w przypadku nieestymowania parametru $\hat{\sigma}_k^2$ (można przyjąć, że wszystkie błędy mają jednakową wariancję równą 1).

Interesujące jest, jak szybko zmniejsza się procent ujemnych wartości współczynników $\hat{\beta}_{ki}$ i czynników przynależności u_{ki}. Wynik doświadczeń zaprezentowano na rys. 10.2 (szczegóły w pracy Góreckiego (2005a)).

Zgodnie z oczekiwaniami procent ujemnych wartości $\hat{\beta}_{ki}$ zmniejsza się bardzo szybko i dla $s = 5$ mamy już poniżej 1% wartości ujemnych. Zauważmy dodatkowo, że praktycznie dla $s = 2$ nie mamy już ujemnych czynników przynależności. Jak widać najsensowniejszym wyborem wydaje się być wartość s wynosząca 1 lub 2. Właśnie dla tych wartości parametru osiągamy najmniejszy współczynnik błędu klasyfikacji, a równocześnie już dla $s = 1$

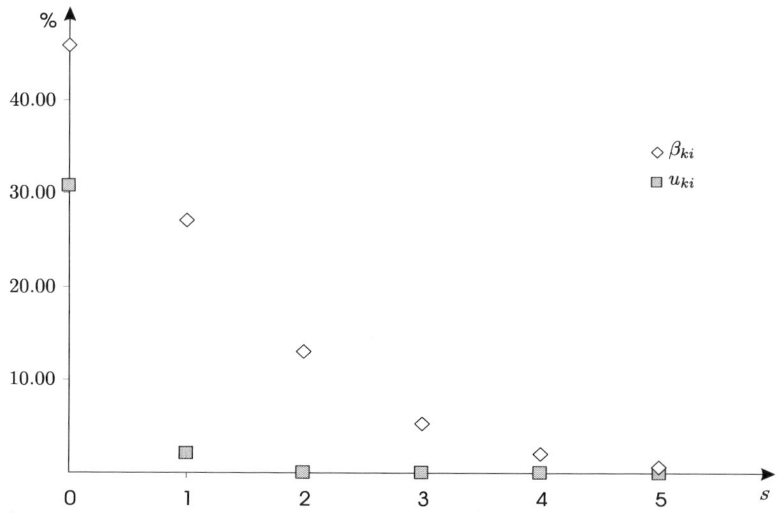

Rys. 10.2. Zależność średniej ilości ujemnych parametrów $\hat{\beta}_{ki}$ oraz u_{ki} od wartości parametru s

około 2% czynników przynależności jest ujemnych (oczywiście te ujemne musimy zastępować zerami, jeśli chcemy mieć nieujemne prawdopodobieństwa a posteriori).

Omawiane w punktach 10.2.2 i 10.2.3 procedury łączenia klasyfikatorów mają ten sam schemat (patrz rys. 10.3).

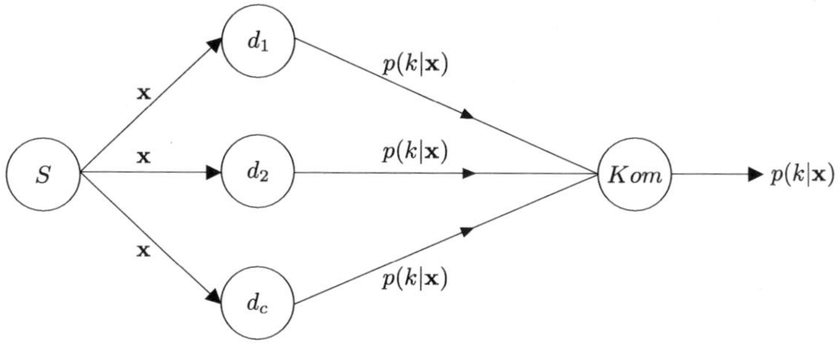

Rys. 10.3. Architektura klasyfikatora łącznego bazująca na wspólnej przestrzeni cech

Klasyfikujemy obiekt opisany zespołem cech S za pomocą c różnych klasyfikatorów (działających na tym samym zespole cech). Otrzymana z każdego klasyfikatora informacja (zazwyczaj są to prawdopodobieństwa a posteriori przynależności obiektu do każdej z klas) zostaje połączona tak, aby uzyskać jedną końcową informację o przynależności obiektu do poszczególnych klas. Łączenie informacji ma na celu zmniejszenie prawdopodobieństwa błędnej klasyfikacji. Najważniejszy w tym przypadku jest wybór metody łączenia informacji.

Na rysunku 10.4 przedstawiono inną sytuację, w której również zachodzi potrzeba łączenia informacji uzyskanych z wielu klasyfikatorów.

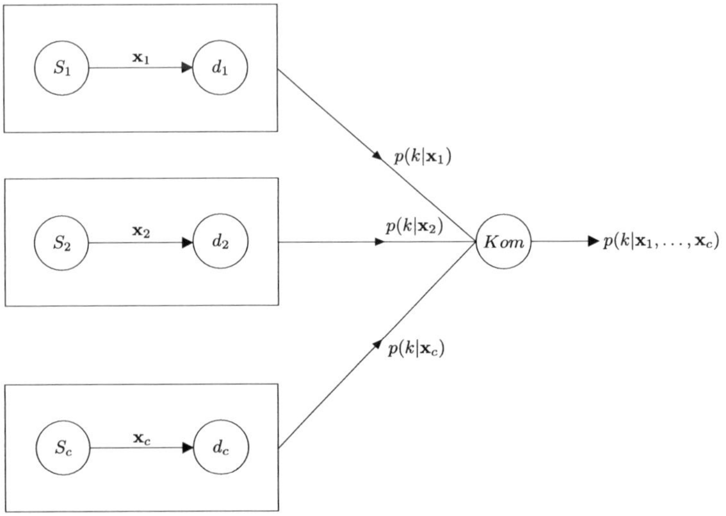

Rys. 10.4. Architektura klasyfikatora łącznego bazująca na różnych przestrzeniach cech

W tym przypadku każda metoda klasyfikacji jest konstruowana na bazie innego zbioru uczącego. Dokładnie zbiory te różnią się wartościami cech, za pomocą których są charakteryzowane klasyfikowane obiekty. Za przykład może tu służyć weryfikacja osoby. Cechami tymi mogą być: skan siatkówki, obraz twarzy, pismo odręczne itd. Do każdego zbioru cech używamy innej metody klasyfikacji. Dysponujemy zatem c zbiorami cech (S_1, S_2, ..., S_c) i dla każdego z nich istnieje pewna obserwacja $\boldsymbol{x}_1$, $\boldsymbol{x}_2$, ..., $\boldsymbol{x}_c$. Z każdym z nich jest związany również jeden klasyfikator (d_1, d_2, ..., d_c). Z każdego klasyfikatora otrzymujemy informację o prawdopodobieństwach a posteriori $p(k|\boldsymbol{x}_i)$ przynależności obiektu do klasy k dla i-tego podzbioru jego cech S_i. Następnie łączymy uzyskane w ten sposób informacje, uzyskując oszacowania prawdopodobieństw a posteriori $p(k|\boldsymbol{x}_1, \ldots, \boldsymbol{x}_c)$ przynależności do klas bazujące na obserwacjach wszystkich cech. W metodzie tej kluczowe znaczenie ma wybór podzbiorów cech, na których działają poszczególne klasyfikatory.

Wiele dalszych informacji dotyczących metod łączenia klasyfikatorów jest zawartych w monografii Kunchevej (2004).

CZĘŚĆ II

Statystyczne systemy uczące się bez nadzoru

Analiza skupień, redukcja wymiaru

11

Analiza składowych głównych

Analiza składowych głównych jest techniką redukcji wymiaru. *Składowe główne*[1] zostały po raz pierwszy zaproponowane przez Pearsona (1901), a następnie rozwinięte przez Hotellinga (1933). Analiza składowych głównych jest zaliczana do systemów uczących się bez nadzoru, a więc każdy element zbioru uczącego składa się jedynie z wektora cech. Zadaniem systemu uczącego się bez nadzoru jest opisanie obserwowanych danych (danych wejściowych) na podstawie wyłącznie nich samych. Można je określić jako zadanie wykrycia wewnętrznej struktury zbioru danych lub współzależności między tymi danymi.

Celem badacza może być redukcja danych, a dokładniej – liczby zmiennych. Polega ona na poszukiwaniu takiego zbioru zmiennych, mniej licznego od zbioru zmiennych oryginalnych, na których podstawie można z pewnym, ale możliwie najmniejszym błędem, odtworzyć wartości zmiennych oryginalnych. Aby taka redukcja była możliwa, między zmiennymi oryginalnymi muszą zachodzić zależności statystyczne. Nowe zmienne mogą być funkcjami zmiennych oryginalnych, bądź też zmienne oryginalne mogą być traktowane jako funkcje nowych zmiennych. W pierwszym przypadku nowe zmienne – składowe główne – są liniowymi funkcjami zmiennych oryginalnych. W drugim przypadku nowe zmienne – czynniki ukryte – są zmiennymi nieobserwowalnymi. Zakłada się, że zmienne obserwowalne są ich liniowymi funkcjami. Takie założenie przyjmuje się w *eksploracyjnej analizie czynnikowej*. Zajmiemy się teraz metodą składowych głównych.

Metoda składowych głównych ma głównie charakter eksploracyjny i umożliwia redukcję danych w przypadku zbioru skorelowanych ze sobą zmiennych. Zmienne te są traktowane w jednakowy sposób, tj. nie są one

---

[1] Po ang. principal components.

dzielone – tak jak w przypadku analizy regresji – na zmienne zależne i niezależne. Metoda ta przekształca oryginalne, skorelowane zmienne w nowe, nieskorelowane zmienne, tzw. składowe główne, które wyjaśniają w maksymalnym stopniu całkowitą wariancję z próby p zmiennych pierwotnych $x_1, \ldots, x_p$, tj. wielkość $\sum_{j=1}^{p} s_j^2 = \mathrm{tr}(\boldsymbol{S})$, gdzie $\boldsymbol{S}$ jest macierzą kowariancji z próby, a s_j^2 jest wariancją zmiennej x_j, $j = 1, \ldots, p$.

Każda nowa zmienna jest liniową funkcją oryginalnych zmiennych. Składowe główne są uporządkowane według udziału w redukcji wspólnego zróżnicowania oryginalnych zmiennych (wielkości całkowitej wariancji). Pierwsza składowa główna redukuje największą część tego zróżnicowania. Druga – kolejną największą część tego zróżnicowania, którego nie redukowała pierwsza składowa główna, itd. (Formalnie, wariancje składowych głównych są wartościami własnymi macierzy kowariancji oryginalnych zmiennych). Badacz może więc zredukować liczbę zmiennych – ograniczając się do kilku pierwszych składowych głównych – z możliwie małą stratą informacji. Oceną ograniczenia się tylko do kilku składowych głównych jest udział zredukowanej przez nie wariancji w wielkości całkowitej wariancji. Wyznaczanie wartości składowych głównych dla badanych obiektów nie przedstawia żadnej trudności i nie wymaga przyjmowania dodatkowych założeń. W sytuacji gdy oryginalne zmienne nie są skorelowane, zastosowanie metody składowych głównych nie zapewnia możliwości redukcji danych przy ograniczonej stracie informacji.

Metoda składowych głównych została po raz pierwszy zastosowana przez Hotellinga do analizy testów osiągnięć szkolnych. Wykazał on wówczas, że istnieją dwie składowe główne wyników testów: zdolności słowne i matematyczne.

Obecnie metoda ta jest często stosowana. Przekształcenie licznego zbioru skorelowanych zmiennych oryginalnych w kilka nieskorelowanych składowych głównych stanowi na ogół pierwszy etap dla zastosowania innych metod wielowymiarowej analizy statystycznej: analizy skupień czy też analizy dyskryminacyjnej.

Przegląd tematyki związanej ze składowymi głównymi można znaleźć w następujących monografiach: Jolliffe (1986), Jackson (1991) oraz Basilevsky (1994).

11.1. Definicja składowych głównych

Pierwsza składowa główna jest definiowana jako unormowana kombinacja liniowa mająca maksymalną wariancję z próby spośród wszystkich unormowanych kombinacji liniowych zmiennych pierwotnych $x_1,\ x_2,\ \ldots,\ x_p$. Dokładniej, dla wektora obserwacji $\boldsymbol{x} = (x_1,\ x_2,\ \ldots,\ x_p)'$ w próbie poszukujemy kombinacji liniowej

$$z_1 = a_{11}x_1 + a_{12}x_2 + \cdots + a_{1p}x_p = \boldsymbol{a}_1'\boldsymbol{x}, \tag{11.1}$$

której wariancja z próby

$$s_{z_1}^2 = \boldsymbol{a}_1'\boldsymbol{S}\boldsymbol{a}_1 \tag{11.2}$$

jest maksymalna, gdzie $\boldsymbol{S}$ jest macierzą kowariancji z próby $\boldsymbol{x}_1,\ \boldsymbol{x}_2,\ \ldots,\ \boldsymbol{x}_n$, natomiast wektor $\boldsymbol{a}_1$ spełnia warunek $\boldsymbol{a}_1'\boldsymbol{a}_1 = 1$, tj. kwadrat jego długości jest równy jeden.

Warunek $\boldsymbol{a}_1'\boldsymbol{a}_1 = 1$ jest wprowadzony po to, by zapewnić jednoznaczność (z wyjątkiem znaku) składowej głównej.

Wektor $\boldsymbol{a}_1$, który maksymalizuje wariancję (11.2), przy dodatkowym warunku $\boldsymbol{a}_1'\boldsymbol{a}_1 = 1$, jest wektorem charakterystycznym odpowiadającym największej wartości własnej λ_1 macierzy $\boldsymbol{S}$ lub inaczej największemu pierwiastkowi równania wyznacznikowego

$$|\boldsymbol{S} - \lambda\boldsymbol{I}| = 0. \tag{11.3}$$

Największa wariancja składowej głównej z_1 jest zatem największym pierwiastkiem równania (11.3).

W celu wyznaczenia drugiej składowej głównej, konstruujemy kombinację liniową

$$z_2 = \boldsymbol{a}_2'\boldsymbol{x} \tag{11.4}$$

taką, że jest ona nieskorelowana z z_1, ma maksymalną wariancję i spełnia warunek $\boldsymbol{a}_2'\boldsymbol{a}_2 = 1$.

Aby składowa z_2 była nieskorelowana ze składową z_1, musi zajść równość $\mathrm{Cov}(z_2,\ z_1) = 0$. Mamy

$$\mathrm{Cov}(z_2,\ z_1) = \boldsymbol{a}_2'\boldsymbol{S}\boldsymbol{a}_1. \tag{11.5}$$

Jednakże $\boldsymbol{S}\boldsymbol{a}_1 = \lambda_1\boldsymbol{a}_1$ i stąd

$$\mathrm{Cov}(z_2,\ z_1) = \lambda_1\boldsymbol{a}_2'\boldsymbol{a}_1 = 0,$$

co implikuje, że $\boldsymbol{a}_2'\boldsymbol{a}_1 = 0$.

Wariancja z próby kombinacji liniowej (11.4) jest równa

$$s_{z_2}^2 = \boldsymbol{a}_2'\boldsymbol{S}\boldsymbol{a}_2. \tag{11.6}$$

Stąd poszukujemy wektora $\boldsymbol{a}_2$ maksymalizującego (11.6) przy dodatkowych warunkach $\boldsymbol{a}_2'\boldsymbol{a}_2 = 1$ i $\boldsymbol{a}_2'\boldsymbol{a}_1 = 0$.

Wektor $\boldsymbol{a}_2$ jest wektorem własnym macierzy $\boldsymbol{S}$ odpowiadającym drugiej wartości własnej $\lambda_2 < \lambda_1$ ortogonalnym do wektora $\boldsymbol{a}_1$ i unormowanym tak, by kwadrat jego długości był równy jedności ($\boldsymbol{a}_2'\boldsymbol{a}_2 = 1$).

Ponieważ macierz $\boldsymbol{S}$ ma p wartości własnych, więc otrzymujemy p składowych głównych:

$$\begin{aligned} z_1 &= \boldsymbol{a}_1'\boldsymbol{x}, \\ z_2 &= \boldsymbol{a}_2'\boldsymbol{x}, \\ &\cdots\cdots\cdots \\ z_p &= \boldsymbol{a}_p'\boldsymbol{x}. \end{aligned} \tag{11.7}$$

Składowe główne $z_1, z_2, \ldots, z_p$ można zapisać w postaci

$$\boldsymbol{z} = \boldsymbol{A}\boldsymbol{x}, \tag{11.8}$$

gdzie

$$\boldsymbol{z} = \begin{bmatrix} z_1 \\ z_2 \\ \vdots \\ z_p \end{bmatrix}, \qquad \boldsymbol{A} = \begin{bmatrix} \boldsymbol{a}_1' \\ \boldsymbol{a}_2' \\ \vdots \\ \boldsymbol{a}_p' \end{bmatrix}.$$

11.2. Własności składowych głównych

Omówimy teraz szereg własności składowych głównych.

WŁASNOŚĆ 1.

Jeżeli wektor własny $\boldsymbol{a}_1$ macierzy kowariancji z próby $\boldsymbol{S}$ jest wyskalowany tak, by $\boldsymbol{a}_1'\boldsymbol{a}_1 = 1$, to wariancja z próby pierwszej składowej głównej z_1 jest równa

$$s_{z_1}^2 = \boldsymbol{a}_1'\boldsymbol{S}\boldsymbol{a}_1 = \lambda_1. \tag{11.9}$$

Stąd wartość własna λ_1 macierzy $\boldsymbol{S}$ jest równa wariancji z próby pierwszej składowej głównej $z_1 = \boldsymbol{a}'_1\boldsymbol{x}$.

Podobnie, wariancja z próby każdej innej składowej głównej jest równa odpowiedniej wartości własnej

$$s^2_{z_j} = \boldsymbol{a}'_j\boldsymbol{S}\boldsymbol{a}_j = \lambda_j, \qquad j = 2,\ 3,\ \ldots,\ p. \tag{11.10}$$

Składowa główna z_1 ma maksymalną wariancję λ_1, natomiast składowa główna z_p ma najmniejszą wariancję λ_p, gdzie $\lambda_1 > \lambda_2 > \cdots > \lambda_p$ są wartościami własnymi macierzy kowariancji z próby $\boldsymbol{S}$.

WŁASNOŚĆ 2.

Składowe główne są wzajemnie ortogonalne, tj. $\boldsymbol{a}'_j\boldsymbol{a}_k = 0$, *dla wszystkich* $j \neq k$, *ponieważ są wektorami własnymi symetrycznej macierzy* $\boldsymbol{S}$.

Ortogonalność składowych głównych pociąga za sobą własność ich nieskorelowania z próby.

Współczynnik korelacji z próby między składowymi głównymi z_j oraz z_k jest równy

$$r_{z_j z_k} = \frac{\boldsymbol{a}'_j\boldsymbol{S}\boldsymbol{a}_k}{\sqrt{\lambda_j\lambda_k}}.$$

Ponieważ $\boldsymbol{S}\boldsymbol{a}_k = \lambda_k\boldsymbol{a}_k$, więc $\boldsymbol{a}'_j\boldsymbol{S}\boldsymbol{a}_k = \lambda_k\boldsymbol{a}'_j\boldsymbol{a}_k$.
Stąd, jeśli $\boldsymbol{a}'_j\boldsymbol{a}_k = 0$, to $r_{z_j z_k} = 0$, dla wszystkich $j \neq k$.

WŁASNOŚĆ 3.

Jako kombinacje liniowe zmiennych pierwotnych, składowe główne reprezentują rzuty. Pierwsza składowa główna może być rozpatrywana jako prosta, na którą są rzutowane ortogonalnie (prostopadle) punkty $\boldsymbol{x}_1,\ \boldsymbol{x}_2,\ \ldots,\ \boldsymbol{x}_n$. *Prosta ta jest „prostą najlepszą" w tym sensie, że ma ona taki kierunek, dla którego suma kwadratów rzutów prostopadłych* $z_{1i} = \boldsymbol{a}'_1\boldsymbol{x}_i$, $i = 1,\ 2,\ \ldots,\ n$, *jest maksymalna. Ponadto, suma kwadratów prostopadłych odległości punktów od tej prostej jest minimalna. Pierwsza i druga składowa główna tworzą „najlepiej" dopasowaną płaszczyznę w podobnym sensie: suma kwadratów odległości prostopadłych punktów* $\boldsymbol{x}_1,\ \boldsymbol{x}_2,\ \ldots,\ \boldsymbol{x}_n$ *od tej płaszczyzny jest minimalna.*

Zatem n punktów $\boldsymbol{x}_1, \boldsymbol{x}_2, \ldots, \boldsymbol{x}_n$ z przestrzeni p-wymiarowej możemy przedstawić na płaszczyźnie dwóch pierwszych składowych głównych. Będą to punkty o współrzędnych ($z_{1i} = \boldsymbol{a}_1'\boldsymbol{x}_i$, $z_{2i} = \boldsymbol{a}_2'\boldsymbol{x}_i$), $i = 1, 2, \ldots, n$.

Uzyskany rysunek może być pomocny w wykryciu skupień punktów, obserwacji odstających czy też badaniu normalności rozkładów rozpatrywanych danych.

Ponieważ pierwsza składowa główna jest kombinacją liniową mającą maksymalną wariancję oraz n obserwacji wektorowych jest rzutowanych ortogonalnie na tę prostą, więc rzuty $z_{1i} = \boldsymbol{a}_1'\boldsymbol{x}_i$, $i = 1, 2, \ldots, n$, można wykorzystać do rangowania obserwacji.

WŁASNOŚĆ 4.

Suma wariancji z próby składowych głównych jest równa sumie wariancji z próby zmiennych pierwotnych

$$\sum_{j=1}^{p} s_{z_j}^2 = \sum_{j=1}^{p} \lambda_j = \operatorname{tr}(\boldsymbol{S}). \tag{11.11}$$

W analizie składowych głównych oczekujemy, że dla pewnego małego k, suma $\lambda_1 + \lambda_2 + \cdots + \lambda_k$ będzie bliska $\operatorname{tr}(\boldsymbol{S}) = \lambda_1 + \lambda_2 + \cdots + \lambda_p$. Jeśli tak jest, to k pierwszych składowych głównych wyjaśnia dobrze zmienność wektora $\boldsymbol{x} = (x_1, x_2, \ldots, x_p)'$ i pozostałe $p - k$ składowe główne wnoszą niewiele, ponieważ mają one małe wariancje z próby.

Wskaźnik

$$\frac{\lambda_1 + \cdots + \lambda_k}{\lambda_1 + \cdots + \lambda_p} \, 100\% \tag{11.12}$$

jest procentową miarą wyjaśniania zmienności wektora $\boldsymbol{x}$ przez pierwszych k składowych głównych.

WŁASNOŚĆ 5.

Macierz $\boldsymbol{S}$ można przedstawić w postaci rozkładu spektralnego

$$\boldsymbol{S} = \lambda_1\boldsymbol{a}_1\boldsymbol{a}_1' + \lambda_2\boldsymbol{a}_2\boldsymbol{a}_2' + \cdots + \lambda_p\boldsymbol{a}_p\boldsymbol{a}_p'. \tag{11.13}$$

Pierwszych k wyrazów sumy (11.13) jest przybliżeniem macierzy $\boldsymbol{S}$

$$\bar{\boldsymbol{S}} = \lambda_1 \boldsymbol{a}_1 \boldsymbol{a}_1' + \cdots + \lambda_k \boldsymbol{a}_k \boldsymbol{a}_k'. \tag{11.14}$$

Jest to najlepsze przybliżenie rzędu k w tym sensie, że suma kwadratów elementów $\boldsymbol{S} - \bar{\boldsymbol{S}}$ jest minimalna (Rao, 1964).

Zatem pierwszych k składowych głównych nie tylko wyjaśnia część wariancji całkowitej $\mathrm{tr}(\boldsymbol{S})$, ale również wyjaśnia nie wprost kowariancje.

WŁASNOŚĆ 6.

Składowe główne nie są niezmiennicze względem zmiany skali zmiennych pierwotnych.

Aby to pokazać, załóżmy że wektor $\boldsymbol{x}$ skalujemy za pomocą przekształcenia $\boldsymbol{u} = \boldsymbol{C}\boldsymbol{x}$, gdzie $\boldsymbol{C}$ jest macierzą nieosobliwą.

Wówczas macierz kowariancji z próby wektora $\boldsymbol{u}$ jest równa $\boldsymbol{CSC}'$. Składowe główne zmiennych przekształconych będą postaci $\boldsymbol{b}_j'\boldsymbol{u} = \boldsymbol{b}_j'\boldsymbol{C}\boldsymbol{x}$, gdzie $\boldsymbol{b}_j$ jest wektorem własnym macierzy $\boldsymbol{CSC}'$. Ale $\boldsymbol{b}_j'\boldsymbol{C}\boldsymbol{x}$ nie równa się $\boldsymbol{a}_j'\boldsymbol{x}$, składowej głównej uzyskanej z macierzy $\boldsymbol{S}$, ponieważ $\boldsymbol{b}_j = (\boldsymbol{C}')^{-1}\boldsymbol{a}_j$ nie jest wektorem własnym macierzy $\boldsymbol{CSC}'$.

Jeżeli

$$\boldsymbol{C} = \mathrm{diag}(s_{x_1}^{-1},\ s_{x_2}^{-1},\ \ldots,\ s_{x_p}^{-1}), \tag{11.15}$$

gdzie s_{x_i} jest odchyleniem standardowym z próby zmiennej pierwotnej x_i, to $\boldsymbol{CSC}' = \boldsymbol{R}$, gdzie $\boldsymbol{R}$ jest macierzą korelacji zmiennych pierwotnych $x_1,\ x_2,\ \ldots,\ x_p$. Tak więc składowe główne uzyskane z macierzy kowariancji z próby $\boldsymbol{S}$ i z macierzy korelacji z próby $\boldsymbol{R}$ są różne.

Często można spotkać następujące zalecenie: jeżeli wariancje w macierzy $\boldsymbol{S}$ są bardzo zróżnicowane, to chcąc uniknąć dominacji zmiennej o bardzo dużej wariancji powinniśmy posłużyć się macierzą korelacji $\boldsymbol{R}$.

W celu zilustrowania, jak zmieniają się wartości własne i wektory własne po przejściu z $\boldsymbol{S}$ do $\boldsymbol{R}$, wykorzystamy prosty przykład dwuwymiarowy, w którym jedna wariancja jest istotnie większa od drugiej. Załóżmy, że $\boldsymbol{S}$ i $\boldsymbol{R}$ mają następujące wartości:

$$\boldsymbol{S} = \begin{bmatrix} 1 & 4 \\ 4 & 25 \end{bmatrix}, \qquad \boldsymbol{R} = \begin{bmatrix} 1 & 0.8 \\ 0.8 & 1 \end{bmatrix}.$$

Dla macierzy $\boldsymbol{S}$ mamy: $\lambda_1 = 25.65$, $\lambda_2 = 0.35$, $\boldsymbol{a}_1 = [0.160,\ 0.987]'$, $\boldsymbol{a}_2 = [0.987,\ -0.160]'$, $\lambda_1/(\lambda_1+\lambda_2)\cdot 100\% = 98.65\%$.

Dla macierzy $\boldsymbol{R}$ mamy: $\lambda_1 = 1.8$, $\lambda_2 = 0.3$, $\boldsymbol{a}_1 = [0.707,\ 0.707]'$, $\boldsymbol{a}_2 = [0.707,\ -0.707]'$, $\lambda_1/(\lambda_1+\lambda_2)\cdot 100\% = 90.00\%$.

Porównując w tym przykładzie składowe główne uzyskane na bazie $\boldsymbol{S}$ i $\boldsymbol{R}$, stwierdzamy, że:

1. Procent wariancji z próby wyjaśniony przez pierwszą składową główną z_1 jest inny dla $\boldsymbol{S}$ i $\boldsymbol{R}$.
2. Współczynniki składowych głównych różnią się.
3. Współczynniki składowych głównych uzyskane z $\boldsymbol{R}$ nie są wyznaczone jednoznacznie. Na przykład wartości własne macierzy $\boldsymbol{R} = \begin{bmatrix} 1 & r \\ r & 1 \end{bmatrix}$ są równe $\lambda_1 = 1 + r$, $\lambda_2 = 1 - r$, ale składowe główne pozostają niezmiennicze dla wszystkich wartości r

$$z_1 = 0.707\frac{x_1}{s_{x_1}} + 0.707\frac{x_2}{s_{x_2}},$$

$$z_2 = 0.707\frac{x_1}{s_{x_1}} - 0.707\frac{x_2}{s_{x_2}}.$$

Składowe te nie zależą od r. Są one identyczne, dla $r = 0.01$ i $r = 0.99$. Dla $r = 0.01$ procent wariancji wyjaśniony przez z_1 wynosi $\lambda_1/(\lambda_1+\lambda_2)\cdot 100\% = 50.5\%$. Dla $r = 0.99$ otrzymujemy 99.5%. Stąd np. stwierdzenie, że pierwsza składowa główna uzyskana z macierzy korelacji z próby $\boldsymbol{R}$ wyjaśnia 90% zmienności, niewiele znaczy.

W ogólności, dla $p > 2$, składowe główne uzyskane z $\boldsymbol{R}$ zależą jedynie od ilorazów współczynników korelacji, a nie od ich rzeczywistych wartości i stąd składowe główne dla danej macierzy $\boldsymbol{R}$ będą służyć dla innych macierzy $\boldsymbol{R}$.

Powinniśmy zatem preferować składowe główne uzyskane z macierzy $\boldsymbol{S}$.

WŁASNOŚĆ 7.

Składowe główne jako osie obrotu. Jeżeli dokonamy przekształcenia $\boldsymbol{z}_i = \boldsymbol{A}\boldsymbol{x}_i$, *dla* $i = 1, 2, \ldots, n$, *gdzie* p *wierszy macierzy* $\boldsymbol{A}$ *jest równych* p *wektorom własnym macierzy kowariancji z próby* $\boldsymbol{S}$, *to macierz* $\boldsymbol{A}$ *jest ortogonalna oraz* $\boldsymbol{z}_i'\boldsymbol{z}_i = \boldsymbol{x}_i'\boldsymbol{A}'\boldsymbol{A}\boldsymbol{x}_i = \boldsymbol{x}_i'\boldsymbol{x}_i$.

Zatem odległości w przestrzeni zmiennych pierwotnych są równe odległościom w przestrzeni składowych głównych i te ostatnie reprezentują osie obrotu. Fakt ten zilustrujemy przykładem dwuwymiarowym.

Przykład 11.1.
Dokonano pomiarów długości i szerokości głowy u pierwszego dorosłego syna w 25 rodzinach. Wyniki pomiarów są zawarte w monografii Renchera (1995), s. 91.

Wektor wartości średnich i macierz kowariancji z próby są równe

$$\bar{\boldsymbol{x}} = \begin{bmatrix} 185.7 \\ 151.1 \end{bmatrix}, \qquad \boldsymbol{S} = \begin{bmatrix} 95.29 & 52.87 \\ 52.87 & 54.36 \end{bmatrix}.$$

Wartości własne i wektory własne macierzy $\boldsymbol{S}$ mają postać: $\lambda_1 = 131.52$, $\lambda_2 = 18.14$, $\boldsymbol{a}_1 = (a_{11},\ a_{12})' = (0.825,\ 0.565)'$, $\boldsymbol{a}_2 = (a_{21},\ a_{22})' = (-0.565,\ 0.825)'$.
Zachodzą związki: $\boldsymbol{a}_1'\boldsymbol{a}_1 = 1$, $\boldsymbol{a}_2'\boldsymbol{a}_2 = 1$, $\boldsymbol{a}_1'\boldsymbol{a}_2 = a_{11}a_{21} + a_{12}a_{22} = 0$.
Przesuwamy układ $x_1 0 x_2$ do punktu $(\bar{x}_1,\ \bar{x}_2)$ i obracamy o kąt α, którego $\operatorname{tg}\alpha = a_{12}/a_{11}$.
Pierwsza składowa główna $z_1 = \boldsymbol{a}_1'\boldsymbol{x}$ jest osią przechodzącą przez punkt (185.7, 151.1) i nachyloną do osi x_1 pod kątem $\operatorname{tg}\alpha = a_{12}/a_{11} = 0.565/0.825 = 0.6848$. Stąd $\alpha = 34°25'$.
Druga składowa główna $z_2 = \boldsymbol{a}_2'\boldsymbol{x}$ jest osią prostopadłą do osi z_1 i również przechodzi przez punkt (185.7, 151.1).
Alternatywnie, równanie składowej głównej z_2 ma postać: $z_2 = 0$, $z_2 = 0 = a_{21}(x_1 - \bar{x}_1) + a_{22}(x_2 - \bar{x}_2) = -0.565(x_1 - 185.7) + 0.825(x_2 - 151.1)$.
Układy $x_1 0 x_2$ oraz $z_1 0 z_2$ są pokazane na rys. 11.1. Zauważmy, że składowa główna z_1 może być rozpatrywana jako specyficzna prosta regresji. Jest ona

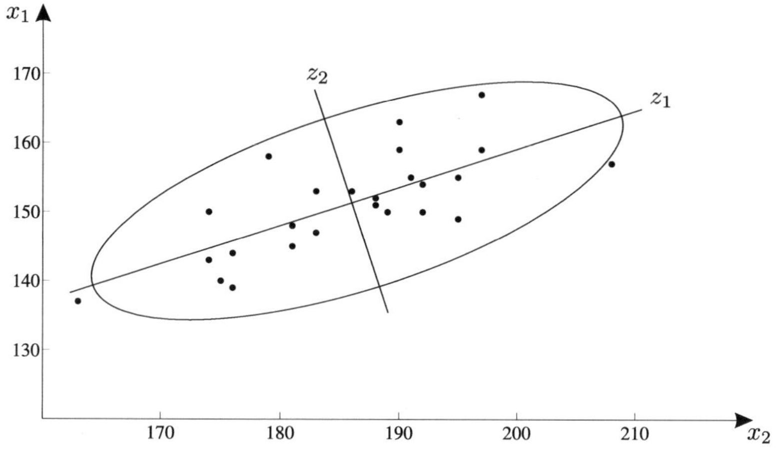

Rys. 11.1. Składowe główne jako osie obrotu

dopasowana do n punktów tak, aby suma kwadratów odległości prostopadłych punktów od tej prostej była minimalna. Odległość prostopadła każdego punktu od składowej głównej z_1 jest równa z_2. Stąd suma kwadratów odległości prostopadłych n punktów od składowej głównej z_1 jest równa

$$\sum_{i=1}^{n} z_{2i}^2 = \sum_{i=1}^{n} [\boldsymbol{a}_2'(\boldsymbol{x}_i - \bar{\boldsymbol{x}})]^2,$$

gdzie $\boldsymbol{a}_2$ jest drugim wektorem charakterystycznym macierzy $\boldsymbol{S}$. Wykorzystujemy różnice $\boldsymbol{x}_i - \bar{\boldsymbol{x}}$, ponieważ początek układu $x_1 0 x_2$ został przesunięty do punktu $(\bar{x}_1, \bar{x}_2)$.
Ponieważ $\boldsymbol{a}_2'(\boldsymbol{x}_i - \bar{\boldsymbol{x}}) = (\boldsymbol{x}_i - \bar{\boldsymbol{x}})'\boldsymbol{a}_2$, więc

$$\begin{aligned}\sum_{i=1}^{n} z_{2i}^2 &= \sum_{i=1}^{n} \boldsymbol{a}_2'(\boldsymbol{x}_i - \bar{\boldsymbol{x}})(\boldsymbol{x}_i - \bar{\boldsymbol{x}})'\boldsymbol{a}_2 = \\ &= \boldsymbol{a}_2' \sum_{i=1}^{n} (\boldsymbol{x}_i - \bar{\boldsymbol{x}})(\boldsymbol{x}_i - \bar{\boldsymbol{x}})'\boldsymbol{a}_2 = \boldsymbol{a}_2'\boldsymbol{A}\boldsymbol{a}_2 = \\ &= (n-1)\boldsymbol{a}_2'\boldsymbol{S}\boldsymbol{a}_2 = (n-1)\lambda_2.\end{aligned}$$

Dla pary zmiennych (x_1, x_2) zwykła regresja liniowa x_2 względem x_1 minimalizuje sumę kwadratów pionowych odległości punktów od prostej. Podobnie regresja x_1 względem x_2 minimalizuje sumę kwadratów poziomych odległości od prostej. Te trzy linie regresji są porównane na rys. 11.2 dla danych związanych z pomiarem długości i szerokości głowy pierwszego syna u 25 rodzin. Pierwsza składowa główna leży między tymi dwiema liniami regresji.

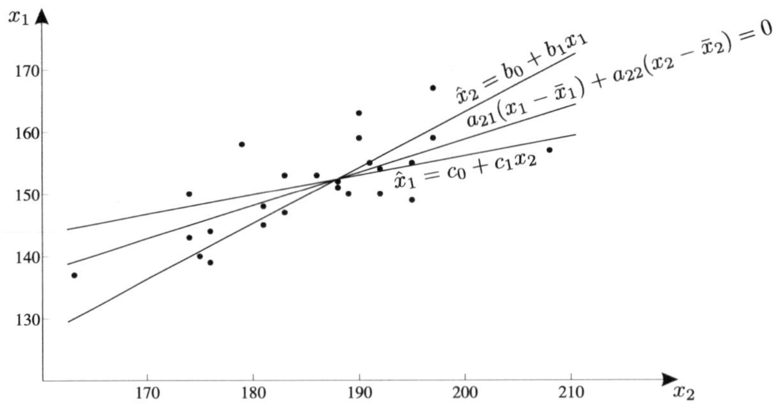

Rys. 11.2. Porównanie pierwszej składowej głównej z prostymi regresji

WŁASNOŚĆ 8.

Wkład zmiennych pierwotnych w budowę składowej głównej. Maksymalna wartość modułu współczynnika stojącego przy danej zmiennej pierwotnej wskazuje na maksymalny wkład tej zmiennej w budowę danej składowej głównej (z uwzględnieniem pozostałych zmiennych pierwotnych).

Do interpretacji składowych głównych wielu autorów rekomenduje wykorzystanie współczynników korelacji między zmiennymi pierwotnymi a daną składową główną, a nie wykorzystywanie współczynników danej składowej głównej. Oznaczmy przez $r_{x_i z_j}$ współczynnik korelacji między i-tą zmienną pierwotną x_i oraz j-tą składową główną z_j i przez $R_{x_i|z_1,\dots,z_t}$ współczynnik korelacji wielokrotnej między zmienną x_i oraz pierwszymi składowymi $z_1, \dots, z_t$. Wówczas

$$r^2_{x_i z_1} + r^2_{x_i z_2} + \dots + r^2_{x_i z_t} = R^2_{x_i|z_1,\ \dots,\ z_t}.$$

Jeżeli użyjemy p składowych, to $R^2_{x_i|z_1,\ \dots,\ z_p} = 1$.

Zatem $r^2_{x_i z_j}$ jest częścią podziału wartości $R^2_{x_i|z_1,\ \dots,\ z_t}$ oraz jest jednowymiarową miarą pokazującą, ile sama zmienna x_i wnosi do danej składowej, a nie, ile x_i wnosi w obecności innych x-ów. Stąd współczynniki korelacji $r_{x_i z_j}$ nie dostarczają wielowymiarowej informacji odnośnie łącznego wkładu zmiennych $x_1, \dots, x_p$ do danej składowej głównej (patrz Rencher (1998), s. 361).

Zatem nie rekomendujemy posługiwania się współczynnikami korelacji do interpretacji składowych głównych. Zalecamy natomiast posłużenie się modułami współczynników a_{ji} w j-tej składowej głównej

$$z_j = a_{j1}x_1 = a_{j2}x_2 + \dots + a_{jp}x_p = \boldsymbol{a}_j'\boldsymbol{x}$$

uzyskanej z macierzy kowariancji z próby $\boldsymbol{S}$.

Wektor $\boldsymbol{a}_j$ jest nazywany często wektorem ładunków j-tej składowej.

11.3. Metody pomijania składowych głównych

Posługując się składowymi głównymi, badacz musi zdecydować, ile z nich jest istotnych dla opisu konkretnych danych. Składowe nieistotne należy pominąć. Będą to składowe związane z najmniejszymi wartościami własnymi macierzy kowariancji z próby $\boldsymbol{S}$. Możemy posłużyć się następującymi kryteriami.

1. **Procent wariancji.** Jeśli wskaźnik (11.12) jest większy lub równy ustalonej przez nas wartości, np. 80%, to pozostałe $p - k$ składowe główne pomijamy.

2. **Wartość średnia z wartości własnych.** Pomijamy te składowe główne, których wartości własne są mniejsze od średniej

$$\bar{\lambda} = \frac{1}{p} \sum_{j=1}^{p} \lambda_j. \tag{11.16}$$

Jest to zarazem średnia wariancja zmiennych pierwotnych, ponieważ $\sum_{j=1}^{p} \lambda_j = \mathrm{tr}(\boldsymbol{S})$.

3. **Wykres piargowy (osypisko).** Wartości własne numerujemy w porządku malejącym. Na osi odciętych zaznaczamy numer wartości własnych, na osi rzędnych zaznaczamy wielkości wartości własnych i wielkości te łączymy odcinkami. Uzyskany w ten sposób wykres jest nazywany *wykresem piargowym* lub *osypiskiem*[2]. Termin ten wprowadził Cattell (1966). Przykładowy wykres piargowy jest pokazany na rys. 11.3. Pierwsze dwie wartości własne istotnie różnią się od pozostałych wykazujących trend liniowy o małym spadku. Składowe główne, poza dwiema pierwszymi, pomijamy.

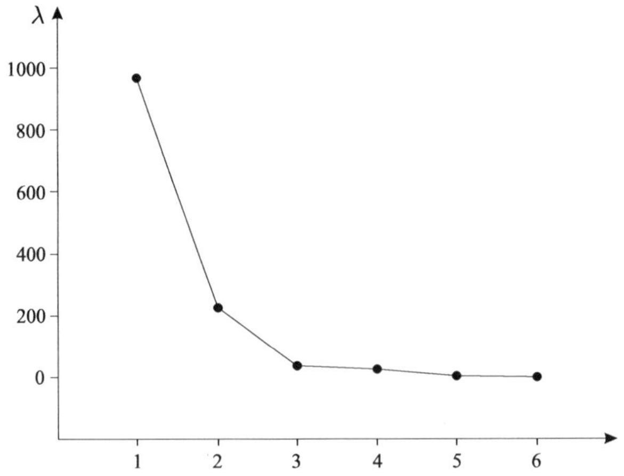

Rys. 11.3. Wykres piargowy

[2]Po ang. scree graph.

4. **Testy istotności.** Będziemy zakładać, że n-elementowa próba pochodzi z p-wymiarowego rozkładu normalnego z dodatnio określoną macierzą kowariancji $\boldsymbol{\Sigma}$ oraz że $n > p$ (przy tym założeniu macierz kowariancji z próby $\boldsymbol{S}$ jest z prawdopodobieństwem jeden również macierzą dodatnio określoną).

Chcemy sekwencyjnie testować hipotezy

$$H_k: \ \gamma_{k+1} = \cdots = \gamma_p = \gamma, \qquad k = 0,\ 1,\ \ldots,\ p-2, \tag{11.17}$$

gdzie $\gamma_1 \geq \cdots \geq \gamma_p$ są wartościami własnymi macierzy $\boldsymbol{\Sigma}$ oraz γ jest nieznaną wartością.

Test ilorazu wiarogodności hipotezy H_k postaci (11.17) jest opisany w następującym twierdzeniu Lawleya (1956) zawierającym również dodatkową informację dotyczącą dokładności aproksymacji rozkładem chi-kwadrat, która pochodzi od Jamesa (1969).

Twierdzenie 11.1.

Jeżeli hipoteza H_k jest prawdziwa i $n \to \infty$, to granicznym rozkładem prawdopodobieństwa statystyki

$$P_k = -\left(n - 1 - k - \frac{2q^2 + q + 2}{6q} + \sum_{i=1}^{k} \frac{\bar{\lambda}_q^2}{(\lambda_i - \bar{\lambda}_q)^2}\right) \ln V_k \tag{11.18}$$

jest rozkład chi-kwadrat z $(q+2)(q-1)/2$ stopniami swobody, gdzie $q = p-k$,

$$V_k = \left(\prod_{i=k+1}^{p} \lambda_i\right) \Big/ \left(\frac{1}{p-k} \sum_{i=k+1}^{p} \lambda_i\right)^{p-k}, \tag{11.19}$$

$$\bar{\lambda}_q = \frac{1}{q} \sum_{i=k+1}^{p} \lambda_i \tag{11.20}$$

oraz

$$\mathrm{E}(P_k) = \frac{1}{2}(q+2)(q-1) + O((n-1)^{-2}). \tag{11.21}$$

Z twierdzenia 11.1 wynika, że jeżeli liczebność próby n jest duża, to przybliżony test rozmiaru α hipotezy H_k odrzuca tę hipotezę wówczas, gdy

$$P_k > \chi^2(1-\alpha,\ (q+2)(q-1)/2), \tag{11.22}$$

gdzie P_k jest zdefiniowane wzorem (11.18), $q = p - k$ oraz $\chi^2(1-\alpha, (q+2)\cdot(q-1)/2)$ jest kwantylem rzędu $1-\alpha$ rozkładu chi-kwadrat z $(q+2)\cdot(q-1)/2$ stopniami swobody.

Załóżmy, że hipotezy H_k, $k = 0, 1, \ldots, p-2$ są testowane sekwencyjnie i że dla pewnego k hipoteza H_k po raz pierwszy nie została odrzucona. Oznacza to, że $q = p - k$ najmniejszych wartości własnych macierzy $\boldsymbol{\Sigma}$ ma tę samą wartość γ. Jeżeli wartość γ jest mało znacząca w porównaniu z pozostałymi wartościami własnymi, to możemy zdecydować o pominięciu ostatnich q składowych głównych i zająć się tylko k pierwszymi, uzyskując redukcję wymiarowości.

Przykład 11.2.
Dane pokrywają się z danymi z przykładu 1.4. Macierz kowariancji z próby jest równa

$$\boldsymbol{S} = \begin{bmatrix} 865.1 & 6.6 & -7.7 & -102.0 & 49.2 & -240.5 \\ 6.6 & 71.9 & 15.7 & 49.3 & -2.2 & 59.2 \\ -7.7 & 15.7 & 7.6 & 16.9 & -1.8 & 20.3 \\ -102.0 & 49.3 & 16.9 & 107.1 & -5.6 & 116.2 \\ 49.2 & -2.2 & -1.8 & -5.6 & 4.6 & -14.7 \\ -240.5 & 59.2 & 20.3 & 116.2 & -14.7 & 204.6 \end{bmatrix}.$$

W tabeli 11.1 podano wartości własne $\lambda_1, \ldots, \lambda_6$ macierzy $\boldsymbol{S}$, procent wariancji rozumianej jako $(\lambda_i / \sum_{i=1}^{6} \lambda_i)100\%$ oraz skumulowany procent wariancji, natomiast w tab. 11.2 podano wektory własne macierzy $\boldsymbol{S}$.

Tab. 11.1. Wartości własne macierzy $\boldsymbol{S}$ z przykładu 11.2

Wartości własne	% wariancji	Skumulowany % wariancji
967.7	76.7	76.7
226.6	18.0	94.7
35.7	2.8	97.5
25.8	2.0	99.6
3.9	0.3	99.9
1.2	0.1	100.0

Tab. 11.2. Wektory własne macierzy $\boldsymbol{S}$ z przykładu 11.2

a_1	a_2	a_3	a_4	a_5	a_6
-0.932	0.320	0.142	-0.068	0.022	0.057
0.023	0.439	-0.872	-0.121	-0.178	0.006
0.018	0.120	-0.130	0.027	0.906	-0.383
0.156	0.539	0.174	0.805	-0.047	0.063
-0.054	0.010	0.070	0.039	-0.381	-0.920
0.321	0.632	0.410	-0.574	-0.014	0.001

Z tabeli 11.1 wynika, że pierwsze dwie składowe główne wyjaśniają 94.7% wariancji całkowitej. Procent ten jest na tyle duży, że pozostałe 4 składowe główne możemy pominąć. Decyzję taką potwierdza postać wykresu piargowego (patrz rys. 11.3).
Wartość średnia z sześciu wartości własnych jest równa $\hat{\lambda} = 210.158$. Tylko dwie pierwsze wartości własne są większe od tej średniej, co oznacza, że dwie pierwsze składowe w stopniu wystarczającym wyjaśniają zmienność całkowitą sześciu cech.
Weźmy teraz pod uwagę test weryfikujący sekwencyjnie hipotezy (11.17). Wyniki testu opartego na statystyce (11.18) przedstawiono w tab. 11.3.

Tab. 11.3. Wyniki testu zastosowanego do danych z przykładu 11.2

k	P_k	Stopnie swobody	War. kryt.
0	808.34	20	31.41
1	460.29	14	23.69
2	200.07	9	16.92
3	151.77	5	11.07
4	22.33	2	5.99

Z tabeli 11.3 wynika, że wszystkie hipotezy o równości wartości własnych zostały odrzucone, co oznacza, że nie można pominąć żadnej ze składowych głównych.
Wynik ten przeczy wszystkim poprzednim kryteriom. Przyczyny należy poszukiwać w małej mocy użytego testu. Punkty zaznaczone na rys. 11.4 reprezentują 74 okazy chrząszczy skaczących w układzie dwóch pierwszych składowych głównych.
Z tabeli 11.2 wynika, że największy udział w budowie pierwszej składowej głównej ma zmienna pierwsza, ponieważ moduł współczynnika

stojącego przy tej zmiennej wynosi 0.932 i jest maksymalny wśród innych modułów współczynników. W budowie drugiej składowej głównej największy udział ma zmienna szósta.

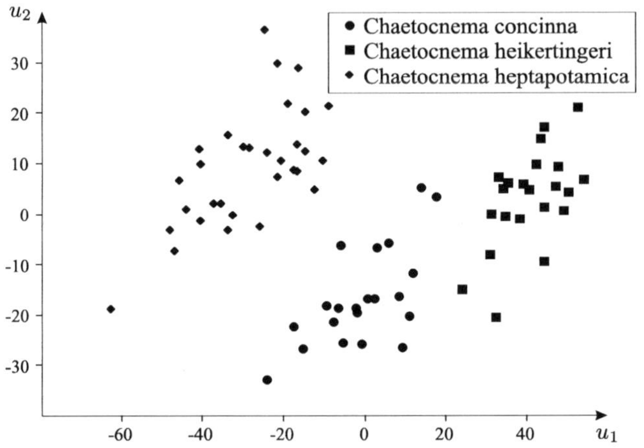

Rys. 11.4. Punkty reprezentujące 74 okazy chrząszczy skaczących w układzie dwóch pierwszych składowych głównych

12

Analiza skupień

Analiza skupień[1] jest narzędziem analizy danych służącym do grupowania n obiektów, opisanych za pomocą wektora p cech, w K niepustych, rozłącznych i możliwie „jednorodnych" grup – skupień. Obiekty należące do danego skupienia powinny być „podobne" do siebie (będziemy w tym celu używali różnych miar podobieństwa, a w zasadzie niepodobieństwa obiektów), a obiekty należące do różnych skupień powinny być z kolei możliwie mocno „niepodobne" do siebie. Głównym celem tej analizy jest wykrycie z zbiorze danych, tzw. „naturalnych" skupień, czyli skupień, które dają się w sensowny sposób interpretować.

Termin „analiza skupień" został wprowadzony przez Tryona (1939). Szczegółowe omówienie technik analizy skupień znaleźć można w wielu monografiach, między innymi Hartigana (1975), Kaufmana i Rousseuwa (1990) oraz Mirkina (2005).

Zwróćmy uwagę, że pod terminem tym kryje się szereg różnych algorytmów. Koncepcyjnie, najprostszym byłby następujący algorytm. Ustalamy liczbę skupień K oraz kryterium optymalnego podziału obiektów. Przeszukujemy wszystkie możliwe podziały n obiektów na K skupień, wybierając najlepszy podział ze względu na przyjęte kryterium optymalności. Bezpośrednie sprawdzenie wszystkich możliwych podziałów jest jednak, nawet przy niewielkim n, praktycznie niemożliwe. Ich liczba bowiem jest równa

$$\frac{1}{K!}\sum_{k=1}^{K}(-1)^{K-k}\binom{K}{k}k^n$$

i np. dla 100 obiektów i czterech skupień jest rzędu 10^{58}.

[1]Po ang. cluster analysis.

Dodatkowym problemem jest wybór końcowej liczby skupień. Często bardzo pomocne są w tym przypadku metody wizualizacji danych. W sytuacji gdy liczba cech jest większa niż trzy, zmuszeni jesteśmy dodatkowo do redukcji wymiaru danych. W tym celu korzystamy zazwyczaj z techniki analizy składowych głównych omówionej w rozdz. 11.

Przykład 12.1.
Na rysunku 12.1 pokazano, w układzie dwóch pierwszych składowych głównych, sześciowymiarowe dane dotyczące chrząszczy skaczących (patrz przykład 1.5).

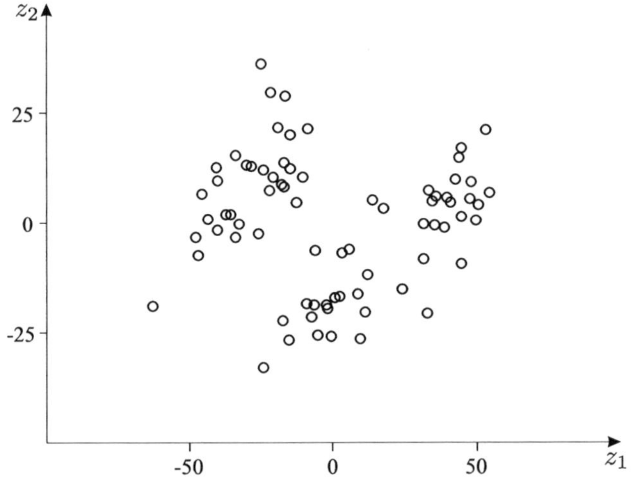

Rys. 12.1. Wykres prezentujący sześciowymiarowe dane w układzie dwóch pierwszych składowych głównych, wskazujący na trzy „naturalne" skupienia

Widać, że analizowane obiekty grupują się w trzy „naturalne" skupienia (odpowiadające trzem gatunkom chrząszczy skaczących).

W dalszym ciągu podamy również inne, bardziej automatyczne, kryteria wyboru końcowej liczby skupień.

12.1. Algorytmy hierarchiczne

Najprostszą i zarazem najczęściej używaną metodą analizy skupień jest *metoda hierarchiczna*. Wspólną cechą krokowych algorytmów tej metody jest wyznaczanie skupień przez łączenie (aglomerację) powstałych, w poprzednich krokach algorytmu, mniejszych skupień. Inne wersje tej metody zamiast idei łączenia skupień, bazują na pomyśle ich dzielenia.

Podstawą wszystkich algorytmów tej metody jest odpowiednie określenie miary niepodobieństwa obiektów. Miary takie zostały szczegółowo omówione w podrozdz. 5.1. Przypomnijmy, że miary niepodobieństwa to semimetryki (a często również metryki) na przestrzeni próby $\mathcal{X}$.

Omówmy teraz dokładnie algorytm aglomeracyjny. W jego pierwszym kroku każdy z obiektów tworzy oddzielne skupienie. Zatem skupień tych jest n. W kroku drugim w jedno skupienie połączone zostają dwa najbardziej podobne do siebie obiekty – w sensie wybranej miary niepodobieństwa obiektów. Otrzymujemy zatem $n-1$ skupień. Postępując analogicznie, tzn. łącząc (wiążąc) ze sobą skupienia złożone z najbardziej podobnych do siebie obiektów, w każdym następnym kroku, liczba skupień maleje o jeden. Obliczenia prowadzimy do momentu uzyskania zadeklarowanej, końcowej liczby skupień K lub do połączenia wszystkich obiektów w jedno skupienie. Graficzną ilustracją przebiegu aglomeracji jest wykres zwany *dendrogramem* (patrz rys. 12.2). Jest to (binarne) drzewo, którego węzły reprezentują skupienia, a liście pojedyncze obiekty. Liście są umieszczone na poziomie zerowym, pozostałe węzły drzewa są umieszczone na wysokości odpowiadającej mierze niepodobieństwa między skupieniami reprezentowanymi przez węzły potomki.

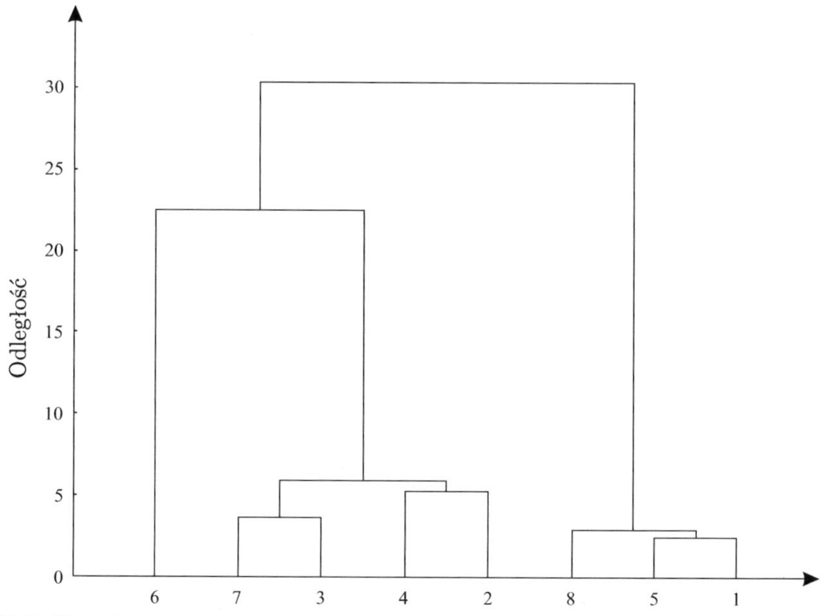

Rys. 12.2. Dendrogram

Algorytm ten wykorzystuje nie tylko miary niepodobieństwa między obiektami, są potrzebne nam również metody wiązania skupień. W jaki sposób możemy je zdefiniować?

Niech R i S oznaczają skupienia (dokładnie są to zbiory numerów obiektów je tworzących), a $\rho(R, S)$ oznacza miarę niepodobieństwa między nimi. Poniżej podano trzy najczęściej wykorzystywane sposoby jej określenia.

1. *Metoda pojedynczego wiązania (najbliższego sąsiedztwa)*[2]. Miara niepodobieństwa między dwoma skupieniami jest określona jako najmniejsza miara niepodobieństwa między dwoma obiektami należącymi do różnych skupień, tzn.

$$\rho(R,\ S) = \min_{i \in R,\ j \in S} \rho(\boldsymbol{x}_i,\ \boldsymbol{x}_j).$$

Zastosowanie tego typu odległości prowadzi do tworzenia wydłużonych skupień, tzw. łańcuchów.

2. *Metoda pełnego wiązania (najdalszego sąsiedztwa)*[3]. Miara niepodobieństwa między dwoma skupieniami jest określona jako największa miara niepodobieństwa między dwoma obiektami należącymi do różnych skupień, tzn.

$$\rho(R,\ S) = \max_{i \in R,\ j \in S} \rho(\boldsymbol{x}_i,\ \boldsymbol{x}_j).$$

Metoda ta jest przeciwieństwem metody pojedynczego wiązania. Jej zastosowanie prowadzi do tworzenia zwartych skupień o małej średnicy.

3. *Metoda średniego wiązania*[4]. Miara niepodobieństwa między dwoma skupieniami jest określona jako średnia miara niepodobieństwa między wszystkimi parami obiektów należących do różnych skupień, tzn.

$$\rho(R,\ S) = \frac{1}{n_R n_S} \sum_{i \in R} \sum_{j \in S} \rho(\boldsymbol{x}_i,\ \boldsymbol{x}_j),$$

gdzie: n_R i n_S są liczbami obiektów wchodzących w skład skupień R i S, odpowiednio.

Metoda ta jest swoistym kompromisem między metodami pojedynczego i pełnego wiązania. Ma ona jednak zasadniczą wadę. W odróżnieniu od dwóch poprzednich wykorzystywana w niej miara niepodobieństwa nie jest

[2]Po ang. single linkage.
[3]Po ang. complete linkage.
[4]Po ang. average linkage.

niezmiennicza ze względu na monotoniczne przekształcenia miar niepodobieństwa między obiektami. Różnice między tymi metodami wiązania skupień wyraźnie widać na dendrogramach z rys. 12.3. Szczególnie widoczny jest „efekt łańcucha" związany z wyborem metody pojedynczego wiązania.

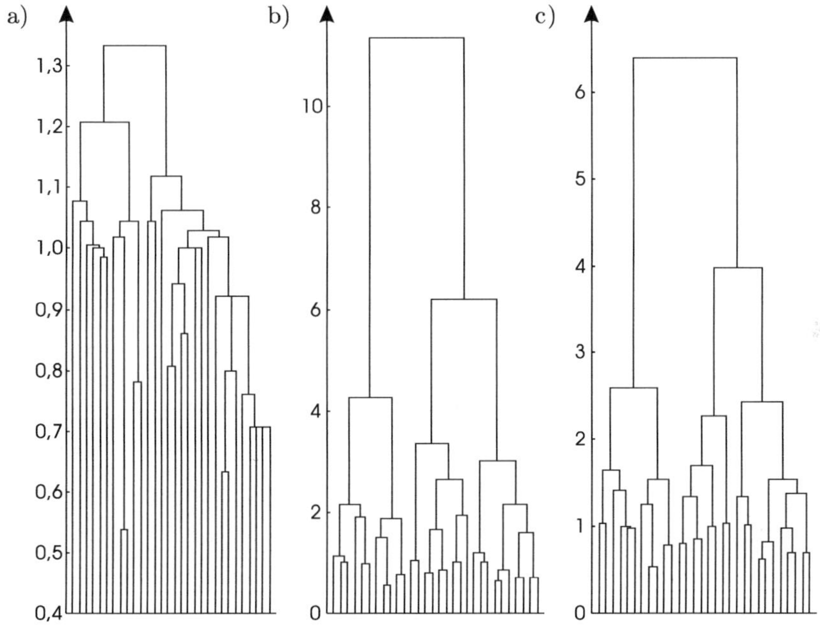

Rys. 12.3. Dendrogramy wyznaczone przy użyciu różnych metod wiązania skupień: a) metoda pojedynczego wiązania, b) metoda pełnego wiązania, c) metoda średniego wiązania

W każdym kroku procedury aglomeracyjnej powstaje nowe skupienie. Za każdym razem musimy obliczyć miary niepodobieństwa między tym nowym skupieniem a pozostałymi, wcześniej utworzonymi, skupieniami. Obliczenia te można wykonać rekurencyjnie. Załóżmy, że zostały połączone skupienia S i S'. W zależności od wybranej metody wiązania skupień, miary niepodobieństwa między skupieniami R i $S \cup S'$ wyrażają się poniższymi wzorami.

1. Dla metody pojedynczego wiązania

$$\rho(R,\ S \cup S') = \min\{\rho(R,\ S),\ \rho(R,\ S')\} =$$
$$= \frac{1}{2}(\rho(R,\ S) + \rho(R,\ S') - |\rho(R,\ S) - \rho(R,\ S')|).$$

2. Dla metody pełnego wiązania

$$\rho(R,\ S\cup S') = \max\{\rho(R,\ S),\ \rho(R,\ S')\} = \\ = \frac{1}{2}(\rho(R,\ S)+\rho(R,\ S')+|\rho(R,\ S)-\rho(R,\ S')|).$$

3. Dla metody średniego wiązania

$$\rho(R,\ S\cup S') = \frac{n_S}{n_S+n_{S'}}\rho(R,\ S)+\frac{n_{S'}}{n_S+n_{S'}}\rho(R,\ S').$$

Omówione dotychczas metody wiązania skupień, choć najczęściej stosowane, nie są jedyne. W przypadku gdy liczebności skupień są zdecydowanie różne, zamiast metodą średniego wiązania możemy posługiwać się jej ważonym odpowiednikiem. Wagami są wtedy liczebności poszczególnych skupień. Inna popularna metoda wiązania skupień pochodzi od Warda (1963). Do obliczania miary niepodobieństwa między skupieniami wykorzystuje on podejście analizy wariancji. W metodzie tej miara niepodobieństwa między skupieniami R i $S \cup S'$ jest dana wzorem

$$\rho(R,\ S\cup S') = \frac{n_R+n_S}{n_R+n_S+n_{S'}}\rho(R,\ S)+\frac{n_R+n_{S'}}{n_R+n_S+n_{S'}}\rho(R,\ S') - \\ - \frac{n_R}{n_R+n_S+n_{S'}}\rho(S,\ S').$$

Jak widać algorytm aglomeracyjny jest bardzo szybki i uniwersalny w tym sensie, że może być on stosowany zarówno do danych ilościowych, jak i jakościowych. Wykorzystuje on jedynie miary niepodobieństwa między obiektami oraz między skupieniami. Kończąc rozważania dotyczące metody aglomeracyjnej, podkreślić trzeba zasadniczy wpływ wybranej miary niepodobieństwa na uzyskane w końcowym efekcie skupienia. Do ustalenia końcowej liczby skupień możemy wykorzystać wykresy rozrzutu (przy wielu wymiarach w układzie dwóch pierwszych składowych głównych) wykonywane przy różnej liczbie skupień. Pomocny może być także dendrogram. Ustalamy wtedy progową wartość miary niepodobieństwa między skupieniami, po przekroczeniu której zostaje zatrzymany proces ich dalszego łączenia.

Innym podobnie uniwersalnym algorytmem jest tzw. *taksonomia wrocławska*. Jej idea polega na rozpięciu na zbiorze n obiektów *najkrótszego dendrytu* (drzewa spinającego), zbudowanego na bazie wybranej odległości (miary niepodobieństwa) między obiektami, a następnie na wydzieleniu skupień przez usunięcie najdłuższych jego krawędzi[5]. Najkrótszym dendrytem

[5] Algorytm taksonomii wrocławskiej jest równoważny algorytmowi aglomeracyjnemu opartemu na metodzie pojedynczego wiązania.

nazywamy drzewo, dla którego suma wag przy krawędziach jest najmniejsza. W naszym przypadku wagami są odległości między obiektami reprezentowanymi przez wierzchołki drzewa. W przypadku wielowymiarowych danych ilościowych wykres przedstawiający minimalny dendryt wygodnie jest przedstawić w układzie dwóch pierwszych składowych głównych (patrz rys. 12.4). Istnieje prosta procedura konstrukcji najkrótszego dendrytu zwana *algorytmem Kruskala*.

Algorytm Kruskala

1. Wybieramy krawędź o minimalnej wadze.
2. Z pozostałych krawędzi wybieramy tę o najmniejszej wadze, która nie prowadzi do cyklu (z krawędzi o jednakowych wagach wybieramy dowolną).
3. Powtarzamy krok drugi, aż do uzyskania najkrótszego dendrytu.

Jeżeli w najkrótszym dendrycie usuniemy l krawędzi o maksymalnych wagach, to rozpadnie się on na $K = l + 1$ części, odpowiadających K skupieniom. Końcowa liczba skupień wyznaczona jest przez progową wartość odległości między obiektami. Do jej wyznaczenia zastosować można poniższe kryterium.

Niech ρ_i oznacza wagę i-tej krawędzi dendrytu ($i = 1, 2, \ldots, n-1$). Obliczamy średnią $\bar{\rho}$ i odchylenie standardowe s_ρ wag wszystkich jego krawędzi, a następnie usuwamy te z nich, dla których

$$\rho_i > \bar{\rho} + cs_\rho, \qquad (12.1)$$

przy czym za stałą c przyjmujemy zazwyczaj 1 lub 2.

Przykład 12.2.
Załóżmy, że danych jest 8 obiektów. Miarę niepodobieństwa między nimi zapisano w postaci macierzy $\boldsymbol{D}$. Jako miarę niepodobieństwa obiektów przyjęto odległość euklidesową

$$\boldsymbol{D} = \begin{bmatrix} 0.00 & 27.75 & 24.65 & 29.00 & 2.50 & 46.86 & 22.87 & 2.65 \\ 27.75 & 0.00 & 4.37 & 5.32 & 26.33 & 20.88 & 6.58 & 29.22 \\ 24.65 & 4.37 & 0.00 & 5.93 & 23.32 & 23.73 & 3.63 & 26.25 \\ 29.00 & 5.32 & 5.93 & 0.00 & 27.86 & 19.47 & 6.85 & 30.81 \\ 2.50 & 26.33 & 23.32 & 27.86 & 0.00 & 45.37 & 21.76 & 3.21 \\ 46.86 & 20.88 & 23.73 & 19.47 & 45.37 & 0.00 & 25.98 & 48.36 \\ 22.87 & 6.58 & 3.63 & 6.85 & 21.76 & 25.98 & 0.00 & 24.68 \\ 2.65 & 29.22 & 26.25 & 30.81 & 3.21 & 48.36 & 24.68 & 0.00 \end{bmatrix}.$$

Wyniki grupowania za pomocą algorytmu aglomeracyjnego przedstawiono na rys. 12.2. Przy łączeniu skupień wykorzystano metodę średniego wiązania. Przyjmując wartość progową odległości równą 10, obiekty zostały rozbite na trzy skupienia: $\{1, 5, 8\}$, $\{2, 3, 4, 7\}$ i $\{6\}$.

Te same obiekty pogrupowano również za pomocą algorytmu opartego na najkrótszym dendrycie (patrz rys. 12.4). Przyjmując we wzorze (12.1) stałą c równą jeden, progowa wartość odległości między obiektami wynosi 16.87. Daje ona w konsekwencji, analogiczny jak w metodzie aglomeracyjnej, podział obiektów na trzy skupienia.

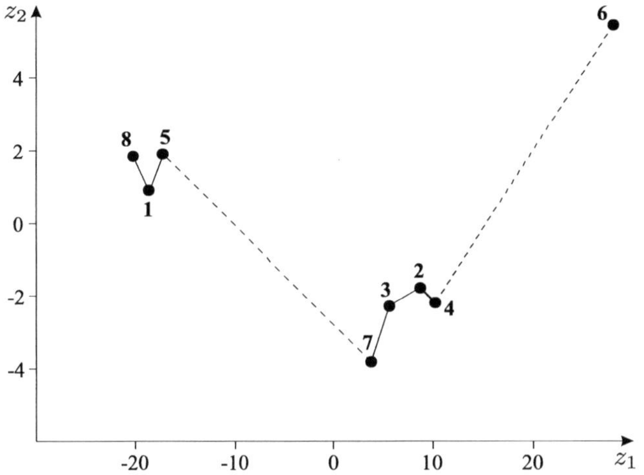

Rys. 12.4. Skupienia wyznaczone za pomocą algorytmu opartego na najkrótszym dendrycie

12.2. Metoda *K*-średnich

Zapewne najbardziej popularnym, niehierarchicznym algorytmem analizy skupień jest *algorytm K-średnich*. Przyporządkowanie n obiektów do zadanej liczby skupień K, odbywa się niezależnie dla każdej wartości K – nie bazując na wyznaczonych wcześniej mniejszych lub większych skupieniach.

Niech C_K oznacza funkcję, która każdemu obiektowi (dokładnie jego numerowi) przyporządkowuje numer skupienia, do którego jest on przyporządkowany (przy podziale na K skupień).

Zakładamy, że wszystkie cechy są ilościowe o wartościach rzeczywistych (przestrzeń próby to $\mathbb{R}^p$).

Główną ideą metody K-średnich jest taka alokacja obiektów, która minimalizuje zmienność wewnątrz powstałych skupień, a co za tym idzie maksymalizuje zmienność między skupieniami.

Dla ustalonej funkcji C_K, przez $W(C_K)$ i $B(C_K)$ oznaczmy macierze zmienności odpowiednio wewnątrz i między skupieniami.

Mamy

$$W(C_K) = \sum_{k=1}^{K} \sum_{C_K(i)=k} (\boldsymbol{x}_i - \bar{\boldsymbol{x}}_k)(\boldsymbol{x}_i - \bar{\boldsymbol{x}}_k)',$$

gdzie $\bar{\boldsymbol{x}}_k$ oznacza wektor średnich k-tego skupienia, tzn.

$$\bar{\boldsymbol{x}}_k = \frac{1}{n_k} \sum_{C_K(i)=k} \boldsymbol{x}_i,$$

a n_k jest liczebnością k-tego skupienia.

Macierz $B(C_K)$ ma postać

$$B(C_K) = \sum_{k=1}^{K} n_k(\bar{\boldsymbol{x}}_k - \bar{\boldsymbol{x}})(\bar{\boldsymbol{x}}_k - \bar{\boldsymbol{x}})',$$

gdzie $\bar{\boldsymbol{x}}$ oznacza średnią ogólną, tzn.

$$\bar{\boldsymbol{x}} = \frac{1}{n} \sum_{i=1}^{n} \boldsymbol{x}_i.$$

Poniższa, znana z analizy wariancji, zależność opisuje związek między tymi macierzami

$$T = W(C_K) + B(C_K),$$

gdzie, niezależna od dokonanego podziału na skupienia macierz zmienności całkowitej T jest dana wzorem

$$T = \sum_{i=1}^{n} (\boldsymbol{x}_i - \bar{\boldsymbol{x}})(\boldsymbol{x}_i - \bar{\boldsymbol{x}})'.$$

Powszechnie stosowane algorytmy metody K-średnich minimalizują ślad macierzy $W(C_K)$. Spotkać można również algorytmy minimalizujące jej wyznacznik lub ślad macierzy $B(C_K)W(C_K)^{-1}$.

Oznaczmy przez C_K^* funkcję realizującą optymalny podział n obiektów na K skupień.

Wtedy

$$C_K^* = \min_{C_K} \operatorname{tr}[W(C_K)] = \min_{C_K} \sum_{k=1}^{K} \sum_{C_K(i)=k} \rho_2(\boldsymbol{x}_i,\ \bar{\boldsymbol{x}}_k), \tag{12.2}$$

gdzie ρ_2 oznacza kwadrat odległości euklidesowej danej wzorem (5.3).

Poniżej prezentujemy podstawowe kroki algorytmu K-średnich.

Algorytm K-średnich

1. W losowy sposób rozmieszczamy n obiektów w K skupieniach. Niech funkcja $C_K^{(1)}$ opisuje to rozmieszczenie.
2. Dla każdego z K skupień obliczamy wektory średnich $\bar{\boldsymbol{x}}_k$, $(k = 1, 2, \ldots, K)$.
3. Rozmieszczamy ponownie obiekty w K skupieniach w taki sposób, że
$$C_K^{(l)}(i) = \arg \min_{1 \le k \le K} \rho_2(\boldsymbol{x}_i, \bar{\boldsymbol{x}}_k).$$
4. Powtarzamy kroki drugi i trzeci aż do momentu, gdy przyporządkowanie obiektów do skupień pozostanie niezmienione, tzn. aż do momentu, gdy $C_K^{(l)} = C_K^{(l-1)}$.

Istnieje wiele modyfikacji powyższego algorytmu. Przykładowo, losowe rozmieszczenie elementów w skupieniach – krok pierwszy algorytmu, zostaje zastąpione narzuconym podziałem, mającym na celu szybsze ustabilizowanie się algorytmu. Inna modyfikacja polega na przeliczeniu średnich skupień natychmiast po przesunięciu między skupieniami choćby jednego elementu.

Wszystkie wersje algorytmu K-średnich są zbieżne. Nie gwarantują one jednak zbieżności do optymalnego rozwiązania C_K^*. Niestety, w zależności od początkowego podziału, algorytm zbiega do zazwyczaj różnych lokalnie optymalnych rozwiązań (minimów lokalnych sumy (12.2)). W związku z tym, aby uzyskać najlepszy podział, zaleca się często wielokrotne stosowanie tego algorytmu z różnymi, wstępnymi rozmieszczeniami obiektów.

Wynik działania metody K-średnich dla danych z przykładu 12.1 zaprezentowano na rys. 12.5. Przedstawiono na nim sytuację, gdy $K = 3$ oraz $K = 4$.

Jak zatem dobrać końcową wartość K? Algorytm metody K-średnich bazuje na minimalizacji zmienności wewnątrz powstałych skupień, wyrażoną przez $\mathrm{tr}(W(C_K))$. Zwróćmy uwagę, że zmienność ta maleje wraz ze wzrostem liczby skupień (dla $K = n$ jest wręcz zerowa). Na rysunku 12.6 przedstawiono zależność między liczbą skupień K a wartością $W_K = \log(\mathrm{tr}(W(C_K)))$ dla analizowanych przez nas danych z przykładu 12.1.

Wykres ten przypomina wykres piargowy (patrz rys. 11.3) stosowany w analizie składowych głównych. Analizując różnice między W_K i W_{K+1},

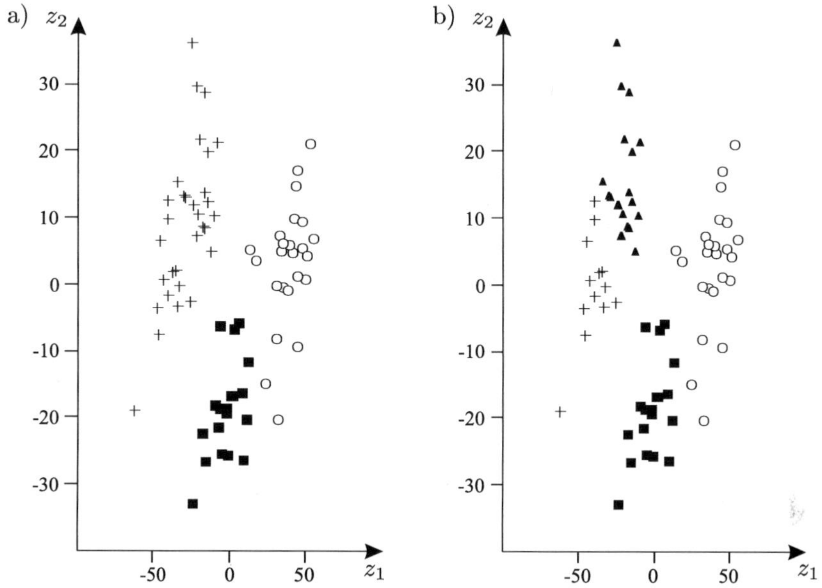

Rys. 12.5. Skupienia wyznaczone metodą K-średnich: a) $K = 3$, b) $K = 4$

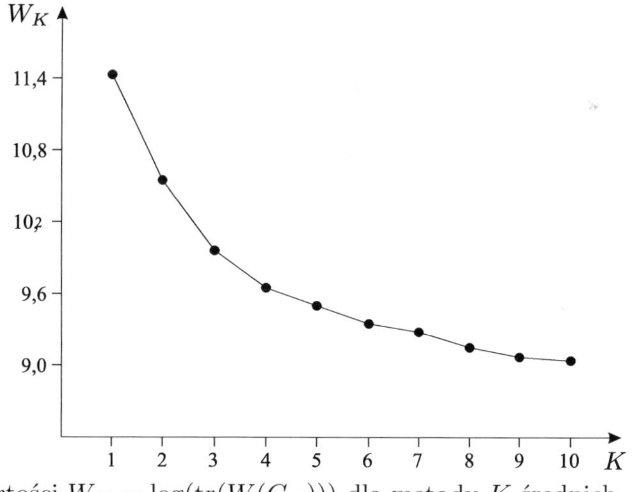

Rys. 12.6. Wartości $W_K = \log(\mathrm{tr}(W(C_K)))$ dla metody K-średnich

widzimy wyraźnie, że wartości $W_1 - W_2$ i $W_2 - W_3$ (być może również $W_3 - W_4$) są zdecydowanie większe od pozostałych różnic. Sugeruje to podział na trzy (lub cztery) skupienia. Trudno jest jednak precyzyjnie określić, którą z różnic uznać za istotnie małą.

W literaturze znaleźć można wiele pomysłów na automatyczne wyznaczania końcowej liczby skupień. Dwa z nich zasługują na szczególną uwagę.

Caliński i Harabasz (1974) zaproponowali, aby końcową liczbę skupień wybierać na podstawie wartości *pseudostatystyki* F[6] postaci

$$CH(K) = \frac{\mathrm{tr}(B(C_K))/(K-1)}{\mathrm{tr}(W(C_K))/(n-K)}. \tag{12.3}$$

Optymalną wartość K dobieramy tak, aby zmaksymalizować (12.3).

Inna, ciekawa koncepcja pochodzi od Hastiego, Tibshiraniego i Walthera (2001). Metoda ta bazuje na maksymalizacji tzw. *statystyki odstępu*[7] postaci

$$Gap(K) = W_K^* - W_K, \tag{12.4}$$

gdzie W_K^* jest obliczone na podstawie danych wygenerowanych z rozkładu jednostajnego na odpowiednio dobranej p-wymiarowej kostce.

Dokładnie, w celu uzyskania wartości statystyki (12.4), dla każdej z p-zmiennych generujemy n obserwacji z rozkładu jednostajnego na przedziale wyznaczonym przez zakres wartości tej zmiennej w pierwotnym zbiorze danych. Symulację tę powtarzamy B razy (zazwyczaj $B = 20$) i dla tak wyznaczonego, sztucznego zbioru danych obliczamy wartości W_K^b ($b = 1, 2, \ldots, B$).

Niech W_K^* i s_K^* oznaczają średnią i odchylenie standardowe obliczone na podstawie wartości $W_K^1, \ldots, W_K^B$. Ponadto, niech $s_K = \sqrt{1-(1/B)}s_K^*$.

Hastie, Tibshirani i Walther proponują, aby za końcową liczbę skupień przyjąć najmniejsze K, dla którego

$$Gap(K) \geq Gap(K+1) - s_{K+1}. \tag{12.5}$$

Jak widać, we wzorze (12.5) wykorzystano tzw. regułę „1SE". Reguła ta była już między innymi wykorzystywana w algorytmie przycinania drzew klasyfikacyjnych (patrz podrozdz. 6.3).

Rachunkowo, metoda oparta na statystyce odstępu, jest dużo bardziej skomplikowana od metody opartej na indeksie Calińskiego–Harabasza – co niewątpliwie jest jej wadą. Zaletą natomiast jest możliwość wykorzystania tej metody dla $K = 1$, tzn. wykorzystania jej w sytuacji braku jakichkolwiek podstaw do dzielenia zbioru danych.

Na rysunku 12.7 są przedstawione wartości indeksu Calińskiego–Harabasza i statystyki odstępu obliczone na podstawie danych z przykładu 12.1.

[6] Statystyka ta nazywana jest często *indeksem Calińskiego–Harabasza*.

[7] Po ang. gap statistic.

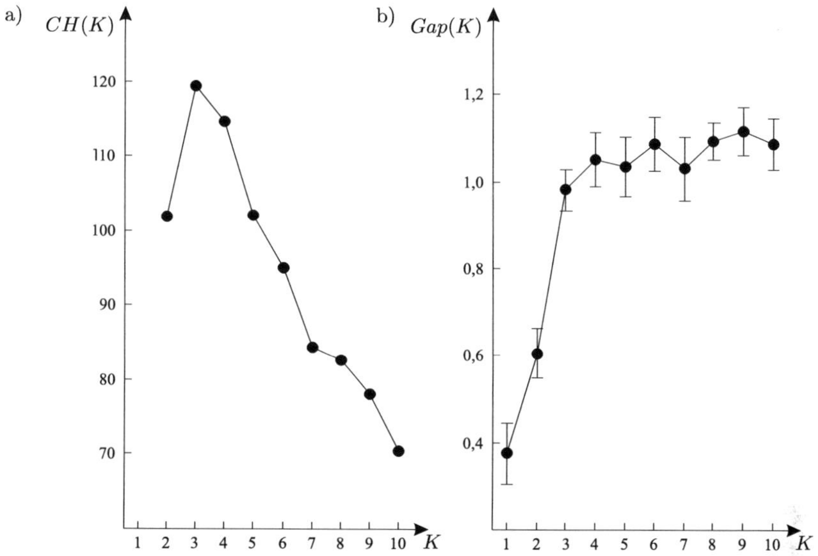

Rys. 12.7. a) Indeks Calińskiego–Harabasza, b) statystyka odstępu (wraz z odchyleniami standardowymi) dla danych z przykładu 12.1

Pierwsza z nich wskazuje na trzy, a druga na cztery „naturalne" skupienia.

W omawianej dotychczas metodzie K-średnich zakładaliśmy, że wszystkie zmienne są ilościowe. Umożliwiało to wykorzystanie do budowy algorytmu odległości euklidesowej między obiektami oraz przyjęcie za „środki" skupień wektorów średnich.

Modyfikacja metody K-średnich, dostosowana do danych jakościowych, nosi nazwę metody *K-median* (lub *K-medoid*). W metodzie tej kwadrat odległości euklidesowej zastępujemy inną, dostosowaną do danych, odległością między obiektami. Ponadto eliminujemy z algorytmu wektory średnich, zastępując je innymi „środkami" skupień.

Dla danych ilościowych oraz kwadratu odległości euklidesowej mamy

$$\bar{\boldsymbol{x}}_k = \frac{1}{n_k} \sum_{C_K(i)=k} \boldsymbol{x}_i = \arg \min_{\boldsymbol{y} \in \mathbb{R}^p} \sum_{C_K(i)=k} \rho_2(\boldsymbol{x}_i, \ \boldsymbol{y}). \tag{12.6}$$

Załóżmy teraz, że badane cechy są jakościowe i przyjmują wartości w zbiorze V. Wzorując się na równości (12.6), zdefiniujmy środki skupień w następujący sposób:

$$\boldsymbol{m}_k = \arg \min_{\boldsymbol{y} \in V^p} \sum_{C_K(i)=k} \rho(\boldsymbol{x}_i, \ \boldsymbol{y}), \tag{12.7}$$

gdzie ρ oznacza wybraną odległość (miarę niepodobieństwa) między obiektami.

Łatwo zauważyć, że środkiem skupienia może być jedynie wektor opisujący obiekt należący do tego skupienia. Równanie (12.7) przyjmuje zatem postać

$$\boldsymbol{m}_k = \arg \min_{\boldsymbol{x}_i:\ C_K(i)=k} \sum_{C_K(j)=k} \rho(\boldsymbol{x}_i,\ \boldsymbol{x}_j).$$

Algorytm K-median

1. W losowy sposób rozmieszczamy n obiektów w K skupieniach. Niech funkcja $C_K^{(1)}$ opisuje to rozmieszczenie.
2. Dla każdego z K skupień obliczamy środki skupień $\boldsymbol{m}_k$, $(k = 1, 2, \ldots, K)$, korzystając ze wzoru (12.7).
3. Rozmieszczamy ponownie obiekty w K skupieniach w taki sposób, że

$$C_K^{(l)}(i) = \arg \min_{1 \le k \le K} \rho(\boldsymbol{x}_i,\ \boldsymbol{m}_k).$$

4. Powtarzamy kroki drugi i trzeci aż do momentu, gdy przyporządkowanie obiektów do skupień pozostanie niezmienione, tzn. aż do momentu, gdy $C_K^{(l)} = C_K^{(l-1)}$.

12.3. Inne algorytmy analizy skupień

Prezentowane w poprzednich rozdziałach algorytmy analizy skupień nie wymagały szczegółowych założeń dotyczących rozkładów prawdopodobieństwa obserwowanych zmiennych. W omówionym poniżej algorytmie takie właśnie założenia odgrywają zasadniczą rolę.

Załóżmy zatem, że rozkład prawdopodobieństwa analizowanych cech daje się opisać za pomocą rozkładu prawdopodobieństwa będącego mieszaniną K rozkładów odpowiadających podziałowi na K skupień.

Dokładnie, niech rozkład ten opisuje gęstość

$$f(\boldsymbol{x}) = \sum_{k=1}^{K} \pi_k f(\boldsymbol{x};\ \theta_k),$$

gdzie $f(\boldsymbol{x};\ \theta_k)$ są gęstościami odpowiadającymi rozkładom poszczególnych skupień (znanymi z dokładnością do parametru θ_k), a π_k są parametrami tej mieszaniny $(\sum_{k=1}^{K} \pi_k = 1,\ \pi_1,\ \ldots,\ \pi_k \ge 0)$. Przykładowo, gdy $f(\boldsymbol{x};\ \theta_k)$

jest gęstością p-wymiarowego rozkładu normalnego, to $\theta_k = (\boldsymbol{\mu}_k, \boldsymbol{\Sigma}_k)$, gdzie $\boldsymbol{\mu}_k$ jest wektorem wartości oczekiwanych, a $\boldsymbol{\Sigma}_k$ macierzą kowariancji.

Oznaczmy przez $\Psi = \{\pi_k,\ \theta_k\colon\ k = 1, \ldots, K\}$ zbiór parametrów tej mieszaniny rozkładów. Do ich estymacji użyjemy metody największej wiarogodności. Przez L oznaczmy logarytm z funkcji wiarogodności postaci

$$L(\Psi;\ \boldsymbol{x}_1, \ldots, \boldsymbol{x}_n) = \log[\prod_{i=1}^{n}\sum_{k=1}^{K}\pi_k f(\boldsymbol{x}_i;\ \theta_k)]. \tag{12.8}$$

Bezpośrednia maksymalizacja powyższej funkcji nie jest możliwa. W tej sytuacji zazwyczaj jest używany *algorytm EM*[8].

Zapiszmy równanie (12.8) w następującej równoważnej postaci:

$$\begin{aligned} L(\Psi;\ \boldsymbol{x}_1, \ldots, \boldsymbol{x}_n) = & \sum_{i=1}^{n}\sum_{k=1}^{K}\psi_{ik}\log\pi_k + \sum_{i=1}^{n}\sum_{k=1}^{K}\psi_{ik}\log f(\boldsymbol{x}_i;\ \theta_k) - \\ & - \sum_{i=1}^{n}\sum_{k=1}^{K}\psi_{ik}\log\psi_{ik}, \end{aligned} \tag{12.9}$$

gdzie

$$\psi_{ik} = \frac{\pi_k f(\boldsymbol{x}_i;\ \theta_k)}{\sum_{k=1}^{K}\pi_k f(\boldsymbol{x}_i;\ \theta_k)}. \tag{12.10}$$

Widać zatem, że funkcje L możemy rozważać jako funkcje dwóch grup argumentów: π_k, θ_k oraz ψ_{ik}. Poniżej jest przedstawiony ogólny schemat algorytmu EM.

Algorytm EM

1. Ustalamy początkowe wartości parametrów $\pi_1, \ldots, \pi_K$ oraz $\theta_1, \ldots, \theta_K$.
2. (Krok E) Korzystając ze wzoru (12.10), obliczamy wartości ψ_{ik}.
3. (Krok M) Dla ustalonych wartości ψ_{ik} (z kroku drugiego) znajdujemy wartości parametrów π_k, θ_k $(k = 1, \ldots, K)$ maksymalizujące (12.9).
4. Powtarzamy kroki drugi i trzeci aż do momentu, gdy wartość funkcji (12.9) nie zmieni się istotnie.

[8] Nazwa algorytmu pochodzi od pierwszych liter podstawowych jego kroków: E – expectation, M – maximalization.

Podobnie jak w przypadku algorytmu K średnich, na końcowy rezultat duży wpływ ma wybór początkowych wartości parametrów. Również i ten algorytm nie gwarantuje osiągnięcia globalnego maksimum funkcji wiarogodności, zbiegając często do maksimów lokalnych.

Do ostatecznego przyporządkowania obserwacji do poszczególnych skupień wykorzystujemy wartości ψ_{ik}, przy czym obserwacja będzie należeć do k-tego skupienia, gdy wartość ψ_{ik}/ψ_k jest maksymalna, gdzie $\psi_k = \sum_{i=1}^{n} \psi_{ik}$.

Wróćmy jeszcze do szczególnie ważnej z punktu widzenia zastosowań sytuacji mieszaniny rozkładów normalnych. Można pokazać (patrz np. Heijden, Duin, Ridder i Tax, 2004), że wartości parametrów π_k, $\boldsymbol{\mu}_k$ i $\boldsymbol{\Sigma}_k$ $(k = 1, \ldots, K)$ maksymalizujące (12.9) mają postacie

$$\pi_k = \frac{1}{n} \sum_{i=1}^{n} \psi_{ik},$$

$$\boldsymbol{\mu}_k = \frac{1}{n\pi_k} \sum_{i=1}^{n} \psi_{ik} \boldsymbol{x}_i,$$

$$\boldsymbol{\Sigma}_k = \frac{1}{n\pi_k} \sum_{i=1}^{n} \psi_{ik} (\boldsymbol{x}_i - \boldsymbol{\mu}_k)(\boldsymbol{x}_i - \boldsymbol{\mu}_k)'.$$

Szczególnie dobre rezultaty zastosowania mieszanin rozkładów normalnych uzyskujemy w przypadku skupień o charakterze eliptycznym. Zilustrowano to wyraźnie na rys. 12.8, gdzie jedynie zastosowanie mieszaniny rozkładów normalnych doprowadziło do poprawnego podziału danych (rys. 12.8a). Zarówno rezultat działania algorytmu K średnich (rys. 12.8b), jak i rezultaty zastosowania algorytmów aglomeracyjnych (rys. 12.8c, 12.8d) są zdecydowanie niezadowalające.

Pozostaje jeszcze zagadnienie wyboru końcowej liczby skupień. Wydaje się naturalne, że wybór ten powinien być oparty na wartości funkcji wiarogodności. Wybór ten powinien być swoistego rodzaju kompromisem między większą liczbą skupień i co za tym idzie większą liczbą estymowanych parametrów, a prostszym modelem o mniejszej liczbie skupień i parametrów o mniejszej wartości funkcji wiarogodności.

Stąd w przypadku tej procedury sensowne wydaje się wykorzystanie reguły bazującej na tzw. *bayesowskim kryterium informacyjnym*[9] (BIC) postaci

[9] Po ang. Bayesian Information Criterion (BIC).

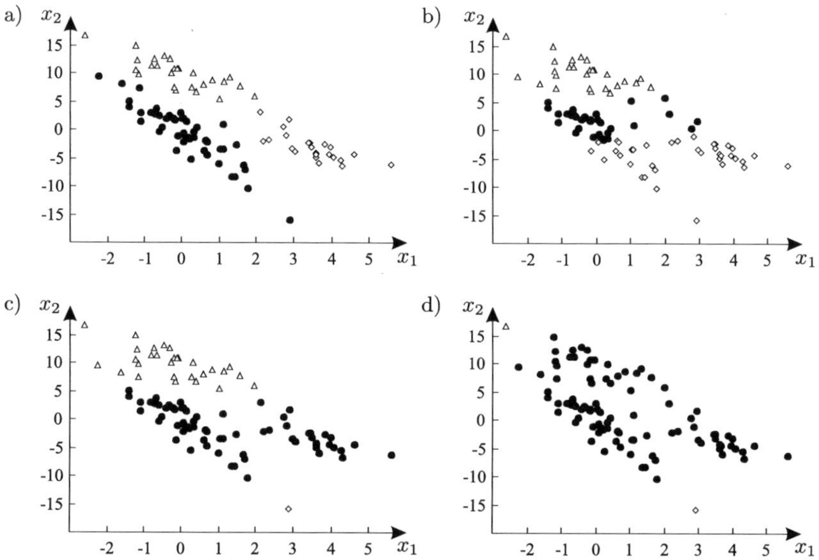

Rys. 12.8. Skupienia wyznaczone: a) metodą opartą na mieszaninie rozkładów normalnych, b) metodą K-średnich, c) metodą aglomeracyjną przy zastosowaniu metody średniego wiązania między skupieniami, d) metodą aglomeracyjną przy zastosowaniu metody pojedynczego wiązania między skupieniami

$$\text{BIC} = 2L(\Psi;\ \boldsymbol{x}_1,\ \ldots,\ \boldsymbol{x}_n) - \eta_k \log n,$$

gdzie η_k jest liczbą parametrów modelu. Wielkość $\eta_k \log n$ spełnia w tym kryterium funkcję kary za zbytnie skomplikowanie modelu.

Rozważane dotychczas algorytmy dotyczyły skupiania obiektów. W analogiczny sposób można je również wykorzystać do podziału zmiennych. Poza techniczną zamianą ról obiektów i zmiennych musimy jeszcze odpowiednio wybrać sposób określania miary niepodobieństwa zmiennych. Często do jej określenia używa się współczynników korelacji Pearsona lub Spearmana. Miara niepodobieństwa między i-tą i j-tą zmienną $(i,\ j = 1,\ \ldots,\ p)$ ma wtedy postać

$$\rho_{ij} = 1 - |r_{ij}|,$$

gdzie r_{ij} jest współczynnikiem korelacji między zmiennymi.

Jeszcze innym zagadnieniem jest jednoczesne skupianie zarówno obiektów, jak i zmiennych. Mówimy wtedy o *blokowej analizie skupień*. Przegląd tego typu algorytmów, stosowanych często w naukach społecznych i medycznych, można znaleźć w pracy Mechelena, Bocka i Boecka (2004).

13

Skalowanie wielowymiarowe

W rozdziale 11 została omówiona analiza składowych głównych, czyli metoda zrzutowania wielowymiarowego zbioru danych na przestrzeń, w której dane będą miały maksymalny rozrzut. Metoda taka umożliwia nam wizualne zbadanie danych przy minimalnej utracie INFORMACJI. Metoda ta jest, niestety, efektywna tylko wtedy, gdy dane leżą w dwuwymiarowej podprzestrzeni liniowej rozpiętej przez zmierzone zmienne. Co jednak z takim przypadkiem, gdy ta dwuwymiarowa podprzestrzeń nie jest „płaska" tylko w jakiś sposób zaburzona (tutaj doskonale pasuje wstęga Möbiusa, która sama w sobie jest oczywiście dwuwymiarowa, ale leży w przestrzeni trójwymiarowej). W takim przypadku analiza składowych głównych może nie wykryć istniejących zależności. Pomocne w tej sytuacji okazuje się *skalowanie wielowymiarowe*[1].

Jest to pewna technika redukcji danych, gdyż jej celem jest znalezienie takiego zbioru punktów w przestrzeni o niewielkiej liczbie wymiarów, na ogół w przestrzeni dwu- lub trójwymiarowej, który dobrze reprezentuje konfigurację badanych obiektów lub zmiennych w przestrzeni wielowymiarowej. W tej nowej przestrzeni możliwe jest wyjaśnienie podobieństw lub różnic między badanymi obiektami. Istnieje wiele typów skalowania wielowymiarowego. We wszystkich jednak zakładamy, że dysponujemy pewną macierzą niepodobieństw między obiektami oraz pewną miarą odległości. Celem jest natomiast znalezienie takiej konfiguracji punktów w przestrzeni $\mathbb{R}^e$, gdzie $e < p$, aby odległości między punktami były jak najbliższe w sensie podobieństwa. Wszystkie metody znajdują współrzędne punktów w tej nowej przestrzeni oraz jej wymiar e. Dwa główne typy skalowania wielowymiarowego to:

[1] Po ang. Multidimensional Scaling (MDS).

- *metryczne*[2] – zakładamy, że dane są ilościowe oraz że zależność między odległościami między punktami oraz danymi niepodobieństwami jest funkcyjna. Ten typ skalowania wielowymiarowego został zaproponowany przez Torgersona (1952);
- *niemetryczne*[3] lub *porządkowe*[4] – w tym przypadku zakładamy, że dane są jakościowe, ale istnieje między nimi pewien porządek. Metoda ta została zaproponowana przez Sheparda (1962) oraz Kruskala (1964).

13.1. Klasyczne skalowanie

W przypadku *klasycznego skalowania*[5] (zwanego również analizą współrzędnych głównych[6]) będziemy zakładać, że odległości między punktami są odległościami euklidesowymi. Mamy dany zbiór n obserwacji $\boldsymbol{x}_i$. Na podstawie macierzy $\boldsymbol{T}$ odległości euklidesowych między punktami chcemy znaleźć współrzędne punktów w przestrzeni $\mathbb{R}^e$ oraz wymiar e tej przestrzeni. Macierz odległości może być wyrażona następująco:

$$\boldsymbol{T} = \boldsymbol{X}\boldsymbol{X}', \tag{13.1}$$

gdzie $\boldsymbol{X} = (\boldsymbol{x}_1, \boldsymbol{x}_2, \ldots, \boldsymbol{x}_n)'$ jest $n \times p$-wymiarową macierzą obserwacji. Odległość między dwoma obserwacjami i oraz j może być zapisana w postaci

$$d_{ij}^2 = T_{ii} + T_{jj} - 2T_{ij}, \tag{13.2}$$

gdzie

$$T_{ij} = \sum_{k=1}^{p} x_{ik}x_{jk}.$$

Jeśli narzucimy założenie, że środek ciężkości zbioru punktów $\{\boldsymbol{x}_i,\ i = 1, 2, \ldots, n\}$ znajduje się w początku układu współrzędnych, wtedy równanie (13.2) może zostać odwrócone w celu przedstawienia elementów macierzy $\boldsymbol{T}$ w terminach macierzy niepodobieństwa. Otrzymujemy

$$T_{ij} = -\frac{1}{2}\left[d_{ij}^2 - d_{i\cdot}^2 - d_{\cdot j}^2 + d_{\cdot\cdot}^2\right], \tag{13.3}$$

[2]Po ang. metric.
[3]Po ang. non-metric.
[4]Po ang. ordinal.
[5]Po ang. classical scaling.
[6]Po ang. principal coordinates analysis.

gdzie

$$d_{i\cdot}^2 = \frac{1}{n}\sum_{j=1}^{n} d_{ij}^2,$$

$$d_{\cdot j}^2 = \frac{1}{n}\sum_{i=1}^{n} d_{ij}^2,$$

$$d_{\cdot\cdot}^2 = \frac{1}{n}\sum_{i=1}^{n}\sum_{j=1}^{n} d_{ij}^2.$$

Równanie (13.3) umożliwia zbudowanie macierzy $\boldsymbol{T}$ na podstawie danej macierzy niepodobieństw $\boldsymbol{D}$ (przy założeniu, że niepodobieństwa są odległościami euklidesowymi). Jedyne co teraz należy zrobić to dokonać faktoryzacji macierzy $\boldsymbol{T}$. Ponieważ jest to rzeczywista macierz symetryczna, więc może być zapisana w postaci

$$\boldsymbol{T} = \boldsymbol{U}\boldsymbol{\Lambda}\boldsymbol{U}', \tag{13.4}$$

gdzie: kolumny macierzy $\boldsymbol{U}$ są wektorami własnymi macierzy $\boldsymbol{T}$, a $\boldsymbol{\Lambda}$ jest macierzą diagonalną złożoną z wartości własnych macierzy $\boldsymbol{T}$. Stąd jako macierz współrzędnych bierzemy

$$\boldsymbol{X} = \boldsymbol{U}\boldsymbol{\Lambda}^{\frac{1}{2}}. \tag{13.5}$$

Wartości własne możemy uporządkować następująco:

$$\lambda_1 \geq \lambda_2 \ldots \geq \lambda_n.$$

Jeśli szukamy reprezentacji danych w przestrzeni o niższej liczbie wymiarów, to musimy użyć tylko największych wartości własnych. Metody ich wyboru zostały omówione w rozdz. 11 dotyczącym analizy składowych głównych. Jako nowe współrzędne wybieramy zatem

$$\boldsymbol{X} = \boldsymbol{U}_e\boldsymbol{\Lambda}_e^{\frac{1}{2}}. \tag{13.6}$$

Jeśli niepodobieństwa nie są odległościami euklidesowymi, część wartości własnych może być ujemna. Wciąż możemy użyć klasycznego skalowania, ale najmniejsza wzięta do reprezentacji wartość własna powinna być dodatnia oraz większa co do wartości bezwzględnej od największej ujemnej. W przeciwnym razie uzyskana reprezentacja może być niepoprawna.

Jeśli dysponujemy oryginalnym zbiorem danych, a nie macierzą niepodobieństw, klasyczna metoda skalowania wielowymiarowego jest tożsama z analizą składowych głównych.

13.2. Skalowanie metryczne

Klasyczne skalowanie wielowymiarowe jest szczególnym przypadkiem *skalowania metrycznego*. Skalowanie wielowymiarowe jest nie tyle ścisłą procedurą, ile raczej sposobem „zmiany rozmieszczenia" obiektów w sposób na tyle efektywny, aby otrzymać konfigurację, która jest najlepszym przybliżeniem obserwowanych odległości. Powstaje zatem pytanie, jak ocenić dobroć dopasowania. Funkcję oceniającą rozbieżność między danymi niepodobieństwami δ_{ij} a obliczonymi d_{ij} w przestrzeni $\mathbb{R}^e$ nazywamy *funkcją stresu*[7]. Naszym celem jest oczywiście minimalizacja funkcji stresu. Najczęściej stosuje się ważoną funkcję stresu postaci

$$S = \sum_{i=1}^{n} \sum_{j=1}^{n} a_{ij}(\delta_{ij} - d_{ij})^2. \tag{13.7}$$

Kruskal (1964) proponuje za wagi przyjąć

$$a_{ij} = \left(\sum_{i=1}^{n} \sum_{j=1}^{n} d_{ij}^2 \right)^{-1}. \tag{13.8}$$

Różne propozycje wag można znaleźć np. w pracach Sammona (1969), Kontza i Fukunagi (1972), Niemanna i Weissa (1979). Do najczęściej używanych należą

$$a_{ij} = \left(\sum_{i=1}^{n} \sum_{j=1}^{n} d_{ij} \right)^{-1} d_{ij}^{-1}, \tag{13.9}$$

$$a_{ij} = d_{ij}^{-2}. \tag{13.10}$$

Funkcja stresu jest niezmiennicza ze względu na przesunięcia, obroty, odbicia oraz jednostajne rozciąganie i zwężanie danych. W praktyce za najlepsze uważa się wagi postaci (13.9), które są pewnym kompromisem między wagami postaci (13.8) oraz postaci (13.10). Rozwiązanie problemu minimalizacji uzyskuje się za pomocą algorytmu największego spadku. Wyboru punktów startowych dokonuje się w taki sposób, aby miały one jak największą wariancję w przestrzeni docelowej $\mathbb{R}^e$.

W praktyce często jest używana bardziej ogólna postać wzoru (13.7)

$$S = \sum_{i=1}^{n} \sum_{j=1}^{n} a_{ij}(\phi(\delta_{ij}) - d_{ij})^2, \tag{13.11}$$

[7]Po ang. stress function.

gdzie funkcja ϕ należy do pewnej predefiniowanej klasy funkcji. Jeżeli np. należy do klasy funkcji liniowych, to otrzymujemy

$$S = \sum_{i=1}^{n} \sum_{j=1}^{n} a_{ij}(a + b\delta_{ij} - d_{ij})^2. \tag{13.12}$$

Niestety, w ogólności nie istnieje analityczne rozwiązanie problemu (13.11). Minimalizacja (13.11) odbywa się za pomocą iteracyjnej metody najmniejszych kwadratów[8]. W przypadku regresji liniowej (13.12), najpierw minimalizujemy ze względu na a i b, dla danego zbioru współrzędnych początkowych. Następnie przyjmujemy a i b za znane i minimalizujemy ze względu na współrzędne punktów d_{ij}. Proces ten powtarzamy aż do uzyskania zbieżności.

Przykład 13.1.
W przykładzie tym dysponujemy odległościami między 21 miastami Europy Zachodniej. Chcemy dokonać wizualizacji tych danych na płaszczyźnie. Jako miarę odległości przyjęto odległość euklidesową. W wyniku działania procedury uzyskano 11 niezerowych wartości własnych, przy czym 2 pierwsze wyjaśniają 86.79% zmienności. Dane w układzie dwóch pierwszych zmiennych zostały przedstawione na rys. 13.1.

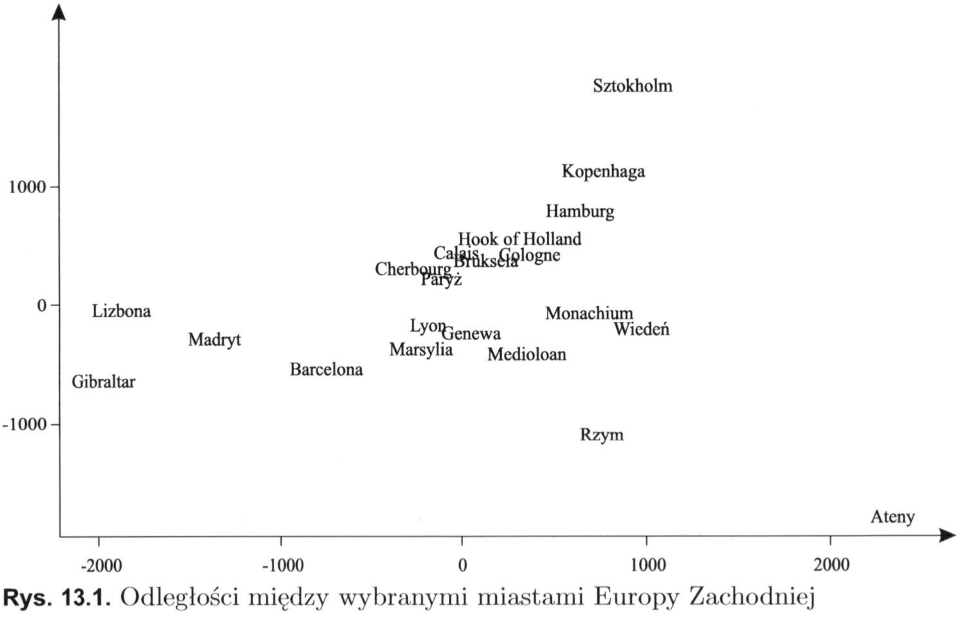

Rys. 13.1. Odległości między wybranymi miastami Europy Zachodniej

[8]Po ang. alternating least square method.

13.3. Skalowanie porządkowe

W tym przypadku nie dysponujemy klasyczną macierzą niepodobieństw między obiektami. Mamy jedynie pewne zmienne porządkowe, które różnicują obserwacje. Chcemy znaleźć taką konfigurację nowych punktów, aby ten porządek był możliwie jak najlepiej zachowany. W przeciwieństwie do klasycznego skalowania nie istnieje analityczne rozwiązanie tego zagadnienia. Co gorsza procedura poszukiwania rozwiązania jest iteracyjna i wymaga początkowej konfiguracji punktów. Najczęściej testuje się kilka takich początkowych konfiguracji. Żądanie, aby porządek odległości w nowej konfiguracji był taki sam jak w macierzy niepodobieństw jest tożsame z warunkiem, aby odległości były monotoniczną funkcją niepodobieństw. W przypadku skalowania porządkowego, współrzędne punktów w przestrzeni reprezentacji są poszukiwane w taki sposób, aby minimalizowały funkcję kosztu, która jest miarą stopnia odchylenia od monotoniczności relacji między d_{ij} oraz δ_{ij}. Oczywiście może nie być możliwe osiągnięcie rozwiązania, które jest doskonale monotoniczne, ale ostatecznie porządek d_{ij} powinien być tak blisko porządku δ_{ij}, jak to tylko możliwe. Można wykreślić odtworzone odległości dla danej liczby wymiarów względem obserwowanych danych wejściowych (odległości). Taki wykres rozrzutu jest znany jako *diagram Sheparda*. Wykres ten przedstawia odtworzone odległości wykreślone na osi pionowej względem pierwotnych niepodobieństw wykreślonych na osi poziomej. Wykres pokazuje także funkcję krokową. Linia ta przedstawia wartości $\hat{d}_{ij}$, to znaczy wynik transformacji monotonicznej danych wejściowych δ_{ij}. Jeśli wszystkie odtworzone odległości znajdowałyby się na linii krokowej, to porządek rangowy odległości (lub niepodobieństw) byłby dokładnie odtworzony. Odchylenia od linii krokowej wskazują na brak dopasowania. Przykładowy diagram Sheparda przedstawiono na rys. 13.2.

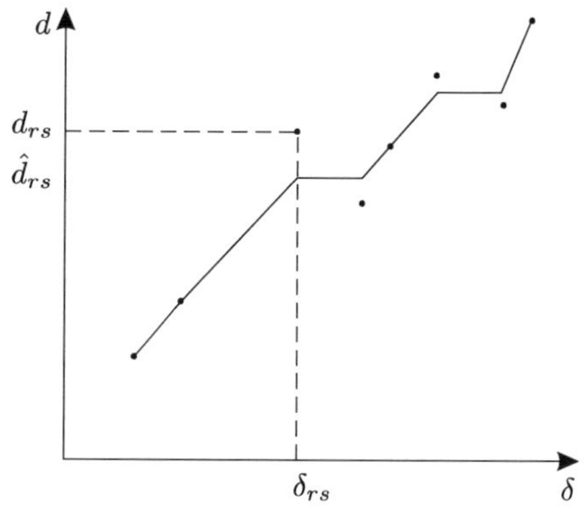

Rys. 13.2. Diagram Sheparda

Aby znaleźć odpowiednią konfigurację punktów, musimy wpierw precyzyjnie zdefiniować pojęcie monotoniczności. Istnieją dwie podstawowe definicje.

Definicja 13.1.
Podstawowy warunek monotoniczności[9]

$$\delta_{rs} < \delta_{ij} \Rightarrow \hat{d}_{rs} \leq \hat{d}_{ij} \tag{13.13}$$

oraz *drugorzędny warunek monotoniczności*[10]

$$\delta_{rs} \leq \delta_{ij} \Rightarrow \hat{d}_{rs} \leq \hat{d}_{ij}, \tag{13.14}$$

gdzie $\hat{d}_{rs}$ jest punktem na linii dopasowania odpowiadającym δ_{rs}. Wartości $\hat{d}_{rs}$ są nazywane *różnicami*[11] lub *pseudoodległościami*[12].

Różnica między tymi dwoma warunkami polega jedynie na innym rozwiązywaniu sytuacji remisowych, gdy $\delta_{rs} = \delta_{ij}$. W pierwszym przypadku $\hat{d}_{rs}$ i $\hat{d}_{ij}$ mogą się różnić, natomiast w przypadku drugiej definicji nie. Drugorzędny warunek monotoniczności jest uważany za zbyt restrykcyjny i często prowadzi do problemów ze zbieżnością. Możemy teraz zdefiniować funkcję stresu

$$S = \sum_{i<j} (d_{ij} - \hat{d}_{ij})^2. \tag{13.15}$$

Minimalizując funkcję (13.15), otrzymujemy monotoniczną linię regresji otrzymaną metodą najmniejszych kwadratów[13] (w rzeczywistości nie jest to linia, bo określona jest jedynie w punktach δ_{ij}). Funkcja (13.15) nie jest niezmiennicza ze względu na jednostajne rozszerzanie. Ta niedogodność zostaje usunięta przez odpowiednią normalizację stresu

$$S = \sqrt{\frac{\sum_{i<j} (d_{ij} - \hat{d}_{ij})^2.}{\sum_{i<j} d_{ij}^2}}. \tag{13.16}$$

[9] Po ang. primary monotone condition.
[10] Po ang. secondary monotone condition.
[11] Po ang. disparities.
[12] Po ang. pseudo-distances.
[13] Po ang. least square monotone regression line.

W celu optymalizacji funkcji (13.16) używa się nieliniowego schematu optymalizacji[14] (zobacz Press i inni (1992)), który wymaga początkowej konfiguracji punktów. Ta konfiguracja jest wybierana w sposób losowy lub jako wynik działania klasycznej metody skalowania. Wymiar przestrzeni e musi być również zadany, dlatego testuje się różne wymiary i wybiera optymalny.

Przykład 13.2.
W tym przykładzie dysponujemy danymi o 30 gatunkach traw porastających holenderskie wydmy. Dane pobierane były na 20 stanowiskach badawczych. Obserwowane były następujące cechy: nawożenie (5 poziomów), wilgotność (4 poziomy), wykorzystanie powierzchni (3 poziomy), typ uprawy (4 poziomy), miąższość (cecha ciągła). Dane pochodzą z monografii Jongmana, ter Braaka i van Tongerena (1987). Jako miara podobieństwa między obiektami została wykorzystana odległość Jaccarda. Po 3 krokach procedury iteracyjnej uzyskano minimalną wartość funkcji stresu równą 11.97273. Dane w przestrzeni dwóch pierwszych zmiennych zostały przedstawione na rys. 13.3. Natomiast na

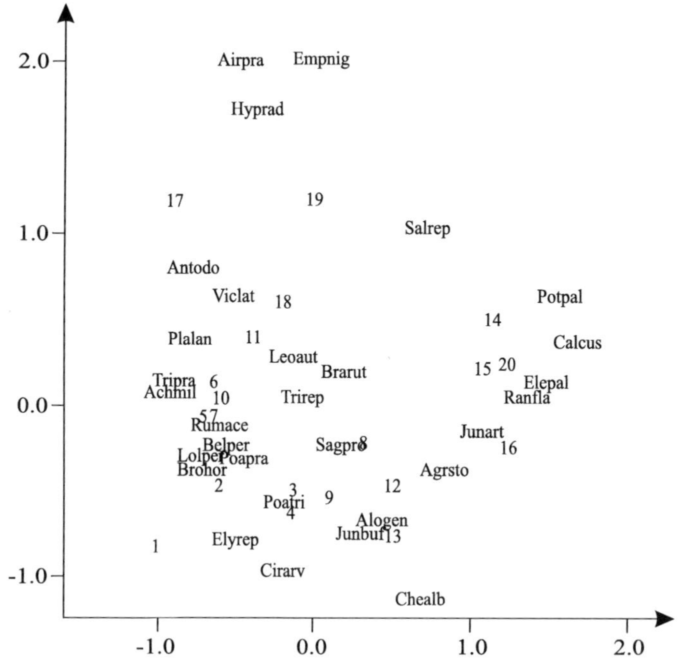

Rys. 13.3. Gatunki oraz stanowiska badawcze traw na wydmach holenderskich. Numery przedstawiają kolejne stanowiska badawcze, a nazwy to różne gatunki traw

[14] Po ang. nonlinear optimization schema.

rys. 13.4 został przedstawiony diagram Sheparda. Otrzymano, że dopasowanie funkcji stresu do linii regresji krokowej wynosi 98.6%, natomiast dopasowanie danych wynosi 92.5%.

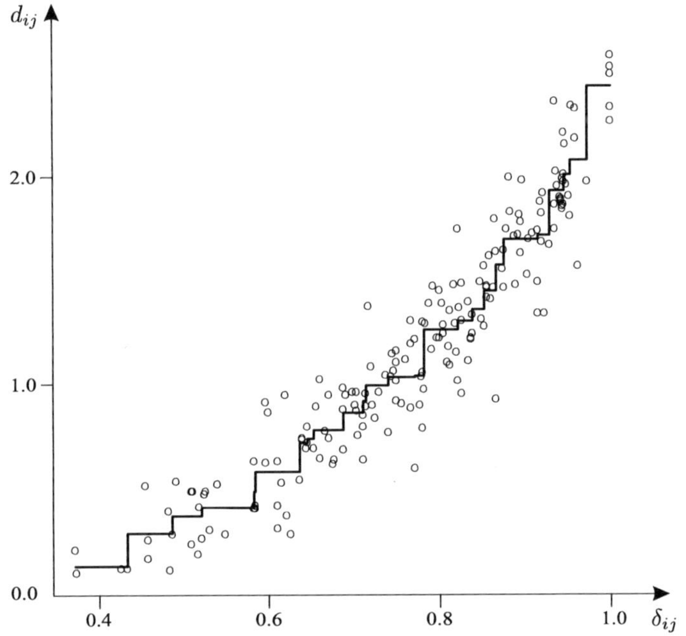

Rys. 13.4. Diagram Sheparda do przykładu 13.2

Zaleta skalowania wielowymiarowego polega na tym, że możemy analizować dowolny rodzaj macierzy odległości lub niepodobieństwa. Niepodobieństwa te mogą reprezentować oceny niepodobieństwa obiektów dokonane przez respondentów, procentową niezgodność między sędziami, liczbę przypadków, gdy badany nie umiał rozróżnić bodźców itd. Metody MDS są bardzo popularne w badaniach psychologicznych nad percepcją oraz w badaniach marketingowych, gdzie są stosowane w celu wykrycia liczby i natury wymiarów kryjących się za postrzeganiem różnych marek lub produktów. Skalowanie wielowymiarowe jest stosowane do poznania preferencji klientów, np. sposobu postrzegania przez nich marek lub poznania układu cech charakteryzujących pewne grupy klientów, np. segmentacji klientów według marek, z których korzystają i poziomu zamożności. Mówiąc ogólnie, metody MDS umożliwiają badaczowi zadawanie względnie neutralnych pytań („na ile marka A jest podobna do marki B").

Uwaga 13.1. Głównymi problemami praktycznymi, które pojawiają się w zastosowaniach metody MDS, są:

- procedura iteracyjna zakończy się po osiągnięciu minimum lokalnego, a nie minimum globalnego,
- wybór właściwej liczby wymiarów poszukiwanej przestrzeni,
- poprawna interpretacja uzyskanego rozwiązania.

Uwaga 13.2. Skalowanie wielowymiarowe może być rozważane jako alternatywa dla analizy czynnikowej[15], która to technika została precyzyjnie opisane np. w książce Koronackiego i Ćwika (2005). W analizie czynnikowej podobieństwa między obiektami są wyrażone w postaci macierzy korelacji. Za pomocą metody MDS, oprócz macierzy korelacji, można analizować dowolny rodzaj macierzy podobieństwa lub niepodobieństwa. Metody skalowania wielowymiarowego i analizy czynnikowej, mimo podobieństw, to zasadniczo różne metody. Analiza czynnikowa wymaga, aby związki między cechami były liniowe. Metoda MDS nie narzuca takich ograniczeń i może być stosowana pod warunkiem, że porządek rangowy odległości (lub podobieństw) jest sensowny. Mówiąc o różnicach, analiza czynnikowa zmierza do wyodrębnienia większej liczby czynników (wymiarów) niż MDS, przy czym MDS często dostarcza bardziej czytelnych, łatwiejszych do interpretacji rozwiązań.

[15]Po ang. factor analysis.

14

Analiza korespondencji

Analiza korespondencji[1] to technika, która umożliwia graficzne przedstawienie w niskowymiarowej przestrzeni numerycznych danych zawartych w tablicy wielodzielczej. Technika ta należy zatem – obok skalowania wielowymiarowego, analizy składowych głównych oraz analizy czynnikowej – do technik redukcji danych. Jest to technika eksploracyjnej analizy danych, której celem jest odkrywanie struktur i wzorców w zbieranych danych. Metody analizy korespondencji zostały rozwinięte głównie we Francji we wczesnych latach siedemdziesiątych przez lingwistę Benzécriego (1973). Analiza korespondencji jest używana najczęściej do analizy tzw. tablic kontyngencji, czyli rozkładów łącznych dwu zmiennych mierzonych zazwyczaj na skalach nominalnych. Skala nominalna uznawana jest za najsłabszą ze skal. Umożliwia ona jedynie klasyfikację ze względu na mierzoną własność. Nie ma możliwości uporządkowania obserwacji.

14.1. Algorytm analizy korespondencji

Technika ta dotyczy cech o charakterze jakościowym. Będziemy zakładać, że cecha pierwsza ma r kategorii, a cecha druga c. Tablica kontyngencji będzie oznaczana przez $\boldsymbol{N}$, liczebności poszczególnych pól tej tablicy wynoszą n_{ij}, $i = 1, 2, \ldots, r$; $j = 1, 2, \ldots, c$. Liczebności brzegowe (odpowiednio wierszowe i kolumnowe) będziemy oznaczać przez $n_{i\cdot} = \sum_{j=1}^{c} n_{ij}$ oraz $n_{\cdot j} = \sum_{i=1}^{r} n_{ij}$.

[1] Po ang. correspondence analysis.

Definicja 14.1.
Macierzą korespondencji $\boldsymbol{P}$ nazywamy macierz częstości zaobserwowanych

$$p_{ij} = \frac{n_{ij}}{n}, \qquad i = 1, 2, \ldots, r; \quad j = 1, 2, \ldots, c. \tag{14.1}$$

Częstości brzegowe będziemy oznaczać przez $p_{i\cdot} = \frac{n_{i\cdot}}{n}$ oraz $p_{\cdot j} = \frac{n_{\cdot j}}{n}$. Elementy te tworzą wektory częstości brzegowych, odpowiednio $\boldsymbol{r}$ oraz $\boldsymbol{c}$. Macierze częstości wierszowych wyznaczamy jako $\boldsymbol{D}_r = \mathrm{diag}(\boldsymbol{r})$, a macierze częstości kolumnowych jako $\boldsymbol{D}_c = \mathrm{diag}(\boldsymbol{c})$.

Definicja 14.2.
Częstości brzegowe wierszy są nazywane *masami wierszowymi*[2], natomiast częstości brzegowe kolumn *masami kolumnowymi*[3].

Definicja 14.3.
Profile wierszowe $\boldsymbol{R}$ obliczamy następująco:

$$\boldsymbol{R} = \boldsymbol{D}_r^{-1}\boldsymbol{P},$$

natomiast *profile kolumnowe* $\boldsymbol{C}$

$$\boldsymbol{C} = \boldsymbol{D}_c^{-1}\boldsymbol{P}.$$

Aby móc stosować analizę korespondencji, musi oczywiście wystąpić zależność między badanymi cechami. Musimy zatem zweryfikować klasyczną hipotezę o niezależności cech. Do sprawdzenia tej hipotezy służy statystyka

$$\chi^2 = \sum_{i=1}^{r}\sum_{j=1}^{c}\frac{(n_{ij} - np_{i\cdot}p_{\cdot j})^2}{np_{i\cdot}p_{\cdot j}},$$

która przy prawdziwości hipotezy zerowej ma asymptotyczny rozkład χ^2 z $(r-1)(c-1)$ stopniami swobody. Jeżeli wartość statystyki testowej χ^2 jest większa od wartości krytycznej $\chi^2_{(r-1)(c-1)}(1-\alpha)$, to mamy podstawy, aby sądzić, że cechy są zależne. W przeciwnym razie nie ma podstaw do odrzucenia hipotezy mówiącej o niezależności.

---
[2] Po ang. row masses.
[3] Po ang. column masses.

Jeśli stwierdzimy zależność badanych cech, możemy przejść do właściwej analizy korespondencji. Chcemy dokonać rzutowania oryginalnych danych na przestrzeń o jak najmniejszym wymiarze, przy czym w tej nowej przestrzeni powinna być zachowana możliwie najlepiej odległość χ^2 między profilami. Wymiar ten wynosi $K = \min(r-1,\ c-1)$. Aby to uzyskać, należy dokonać dekompozycji według wartości osobliwych pewnej macierzy[4]. Wybór macierzy może zostać wykonany na kilka sposobów. Tutaj przedstawimy jedynie najpopularniejsze podejście zaproponowane przez Greenacre'a (1984). W tym przypadku rozkład opiera się na macierzy $(\boldsymbol{P} - \boldsymbol{r}\boldsymbol{c}')$. W celu jednoczesnej graficznej prezentacji wystąpień kategorii obydwu cech, należy wyznaczyć macierz rezyduów standaryzowanych

$$\boldsymbol{A} = \boldsymbol{D}_r^{-\frac{1}{2}}(\boldsymbol{P} - \boldsymbol{r}\boldsymbol{c}')\boldsymbol{D}_c^{-\frac{1}{2}} \tag{14.2}$$

z elementami

$$a_{ij} = \frac{p_{ij} - r_i c_j}{\sqrt{r_i c_j}}.$$

Macierz rezyduów standaryzowanych podlega rozkładowi według wartości osobliwych

$$\boldsymbol{A} = \boldsymbol{U}\boldsymbol{\Gamma}\boldsymbol{V}'.$$

Definicja 14.4.
Wektory macierzy $\boldsymbol{U}$ są nazywane *osiami głównymi kolumn*[5], a wektory macierzy $\boldsymbol{V}$ *osiami głównymi wierszy*[6].

Macierz $\boldsymbol{\Gamma}$ jest macierzą diagonalną o niezerowych wartościach osobliwych γ_k macierzy $\boldsymbol{A}$ ułożonych w porządku nierosnącym.

Definicja 14.5.
Wartościami osobliwymi macierzy $\boldsymbol{A}$ nazywamy pierwiastki kwadratowe z wartości własnych macierzy $\boldsymbol{A}'\boldsymbol{A}$ oraz $\boldsymbol{A}\boldsymbol{A}'$.

Współrzędne kategorii obydwu cech są obliczane następująco:

$$\boldsymbol{F} = \boldsymbol{D}_r^{-\frac{1}{2}}\boldsymbol{U}\boldsymbol{\Gamma}, \quad \text{dla wierszy,}$$

$$\boldsymbol{G} = \boldsymbol{D}_c^{-\frac{1}{2}}\boldsymbol{V}\boldsymbol{\Gamma}, \quad \text{dla kolumn.}$$

---

[4] Po ang. Singular Value Decomposition (SVD).

[5] Po ang. principal axes for columns.

[6] Po ang. principal axes for rows.

Pierwsze kolumny macierzy $\boldsymbol{F}$ oraz $\boldsymbol{G}$ zawierają, odpowiednio, współrzędne wierszy i kolumn dla pierwszej osi głównej. Drugie kolumny zawierają współrzędne dla drugiej osi głównej itd. W przypadku prezentacji dwuwymiarowej wykorzystuje się dwie pierwsze kolumny macierzy $\boldsymbol{F}$ (dla wierszy) oraz dwie pierwsze kolumny macierzy $\boldsymbol{G}$ (dla kolumn).

Definicja 14.6.
Inercją całkowitą[7] λ nazywamy sumę kwadratów wszystkich wartości osobliwych macierzy $\boldsymbol{A}$. Używa się również określenia bezwładność.

Z powyższej definicji wynika, że

$$\lambda = \sum_{k=1}^{K} \gamma_k^2 = \sum_{k=1}^{K} \lambda_k = \mathrm{tr}\boldsymbol{A}'\boldsymbol{A} = \mathrm{tr}\boldsymbol{A}\boldsymbol{A}'.$$

Wskazane jest, aby inercja była możliwie największa. Jeśli wyrazimy w procentach inercję dla każdej nowej zmiennej, to użyte przez nas do prezentacji zmienne powinny zawierać jak najwięcej inercji. Istotne jest, że inercja ma swoją interpretację geometryczną jako miary rozproszenia profili w wielowymiarowej przestrzeni. Im większa jest inercja układu, tym punkty w tej przestrzeni są bardziej rozproszone. Kiedy wartość inercji wynosi 0, punkty profili skupiają się w jednym punkcie przestrzeni.

Uwaga 14.1. W klasycznej terminologii statystycznej procent inercji odzwierciedlałby, jaka część wariancji jest wyjaśniana przez daną zmienną w nowym układzie współrzędnych. Natomiast całkowita inercja jest miarą zróżnicowania elementów zawartych w tablicy kontyngencji.

Przykład 14.1.
Przykład dotyczy intensywności palenia papierosów wśród pracowników pewnej firmy. Zbiór danych został zaczerpnięty z książki Greenacre'a (1984). W celu zapisania informacji o liczbie jednoczesnych wystąpień dwóch kategorii z różnych cech wykorzystano tablicę kontyngencji (tab. 14.1).
Zgodnie z przyjętymi oznaczeniami liczba kategorii pierwszej cechy to $r = 5$, a drugiej $c = 4$. W tabeli 14.2 zamieszczono obliczoną zgodnie ze wzorem (14.1) macierz korespondencji.
Wartości zapisane w ostatniej kolumnie i ostatnim wierszu tab. 14.2 tworzą odpowiednio wektor $\boldsymbol{r}$ oraz $\boldsymbol{c}$. Są to również elementy przekątniowe, odpowiednio, macierzy $\boldsymbol{D}_r$ i $\boldsymbol{D}_c$. W celu wyznaczenia współrzędnych kategorii cech dokonano dekompozycji macierzy $\boldsymbol{A}$ według wartości osobliwych zgodnie ze wzorem

[7] Po ang. total inertia.

Tab. 14.1. Tabela kontyngencji oraz liczebności brzegowe

Grupa pracownicza	Intensywność palenia				$n_{i\cdot}$
	Nie pali	Mało	Średnio	Dużo	
Starsi kierownicy	4	2	3	2	11
Młodsi kierownicy	4	3	7	4	18
Starszy personel	25	10	12	4	51
Młodszy personel	18	24	33	13	88
Sekretarki	10	6	7	2	25
$n_{\cdot j}$	61	45	62	25	193

Tab. 14.2. Macierz korespondencji $\boldsymbol{P}$ oraz zaobserwowane częstości brzegowe

Grupa pracownicza	Intensywność palenia				$p_{i\cdot}$
	Nie pali	Mało	Średnio	Dużo	
Starsi kierownicy	0.0207	0.0104	0.0155	0.0104	0.0570
Młodsi kierownicy	0.0207	0.0155	0.0363	0.0207	0.0933
Starszy personel	0.1295	0.0518	0.0622	0.0207	0.2642
Młodszy personel	0.0933	0.1244	0.1710	0.0674	0.4560
Sekretarki	0.0518	0.0311	0.0363	0.0104	0.1295
$p_{\cdot j}$	0.3161	0.2332	0.3212	0.1295	

(14.2). Rzeczywisty wymiar przestrzeni cech wynosi $K = \min(r-1, c-1) = 3$. Macierz wartości osobliwych ma postać

$$\boldsymbol{\Gamma} = \begin{bmatrix} 0.2734 & 0 & 0 \\ 0 & 0.1001 & 0 \\ 0 & 0 & 0.0203 \end{bmatrix}.$$

Odpowiadające macierzy $\boldsymbol{\Gamma}$ macierze lewych $\boldsymbol{U}$ i prawych $\boldsymbol{V}$ wektorów osobliwych mają postać

$$\boldsymbol{U} = \begin{bmatrix} -0.0574 & 0.4621 & -0.8333 \\ 0.2892 & 0.7424 & 0.5061 \\ -0.7155 & 0.0548 & 0.1303 \\ 0.5753 & -0.3896 & -0.1098 \\ -0.2647 & -0.2838 & 0.1430 \end{bmatrix}, \quad \boldsymbol{V} = \begin{bmatrix} -0.8087 & 0.1713 & 0.0246 \\ 0.1756 & -0.6806 & -0.5223 \\ 0.4070 & -0.0417 & 0.7151 \\ 0.3867 & 0.7112 & -0.4639 \end{bmatrix}.$$

Współrzędne kategorii wierszowych $\boldsymbol{F}$ oraz kolumnowych $\boldsymbol{G}$ mają postać

$$\boldsymbol{F} = \begin{bmatrix} -0.0658 & 0.1937 & -0.0710 \\ 0.2590 & 0.2433 & 0.0337 \\ -0.3806 & 0.0107 & 0.0052 \\ 0.2330 & -0.0577 & -0.0033 \\ -0.2011 & -0.0789 & 0.0081 \end{bmatrix}, \quad \boldsymbol{G} = \begin{bmatrix} -0.3933 & 0.0305 & 0.0009 \\ 0.0995 & -0.1411 & -0.0220 \\ 0.1963 & -0.0074 & 0.0257 \\ 0.2938 & 0.1978 & -0.0262 \end{bmatrix}.$$

Na rysunku 14.1 przedstawiono kategorie analizowanych cech w przestrzeni dwuwymiarowej (dwie pierwsze zmienne wyjaśniają aż 99.51% wariancji). Kropki czarne oznaczają intensywność palenia, a kropki szare grupę pracowniczą. Wielkość kropek jest uzależniona od udziału danej kategorii w całości. Z rysunku możemy w łatwy sposób odczytać wiele zależności. Zauważmy, że młodszy personel pali średnio lub mało, starszy personel raczej nie pali, podobnie jak sekretarki. Najwięcej palą młodsi kierownicy (być może najbardziej narażeni na stres). Wszystko to możemy odczytać z wykresu, bez konieczności wczytywania się w tabele danych. Zysk z takiej prezentacji jest tym większy, im więcej jest kategorii cech.

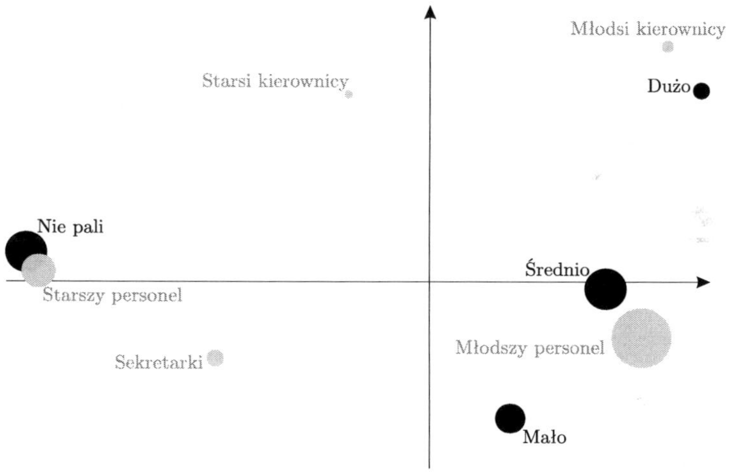

Rys. 14.1. Graficzna prezentacja zależności między grupą pracowniczą a intensywnością palenia

Przykład 14.2.
Przykład ten dotyczy pierwszej tury wyborów prezydenckich w Polsce w roku 2005. Dane do przykładu zostały zaczerpnięte ze strony Państwowej Komisji Wyborczej[8] i znajdują się w tab. 14.3.

[8]http://www.wybory2005.pkw.gov.pl/.

Tab. 14.3. Wyniki pierwszej tury wyborów prezydenckich w Polsce w roku 2005

Województwo	Borowski Marek	Kaczyński Lech	Kalinowski Jarosław	Lepper Andrzej	Tusk Donald	Pozostali
dolnośląskie	128230	331940	12368	142739	460623	39777
kujawsko–pomorskie	106619	213881	12979	159898	252363	24533
lubelskie	65399	310240	30684	221299	194709	27104
lubuskie	49044	89835	4265	56003	146695	13599
łódzkie	124107	320966	23853	206546	309721	39436
małopolskie	91530	575255	18742	135237	469978	39438
mazowieckie	209242	805153	51325	270194	827925	75435
opolskie	31731	88037	4281	50641	153243	12172
podkarpackie	52650	399540	19711	130089	186114	21156
podlaskie	36786	172417	10449	88929	112080	12637
pomorskie	78936	267749	7183	87830	444964	25979
śląskie	215751	607133	16103	172585	765204	65579
świętokrzyskie	51172	140570	15349	122378	120715	13666
warmińsko–mazurskie	55264	131218	8531	89667	191730	17599
wielkopolskie	166143	345120	26767	226752	527742	44703
zachodniopomorskie	82038	148873	6726	98307	265860	23231

Graficzne przedstawienie wyników analizy korespondencji znajduje się na rys. 14.2. Czarne kropki oznaczają kandydatów, natomiast szare województwa. Dodatkowo wielkość jest proporcjonalna w przypadku kandydata do liczby oddanych na niego głosów, a w przypadku województw do liczby głosów ogółem oddanych w tym województwie (małe kropki oznaczają więc małe znaczenie województwa w ogólnych wynikach wyborów). Z mapy korespondencji można teraz z łatwością odczytać np., że Jarosław Kalinowski największe poparcie ma w województwie lubelskim (oczywiście procentowo, bo w innych województwach, w których głosowało więcej osób, zyskał więcej głosów). Z drugiej strony województwo mazowieckie znalazło się praktycznie dokładnie między dwoma najpoważniejszymi kandydatami: Donaldem Tuskiem i Lechem Kaczyńskim. Pierwsza zmienna wyjaśnia 55.78% wariancji, a druga 37.22%, tak więc prezentując te dane na płaszczyźnie, tracimy jedynie 5% informacji o zmienności.

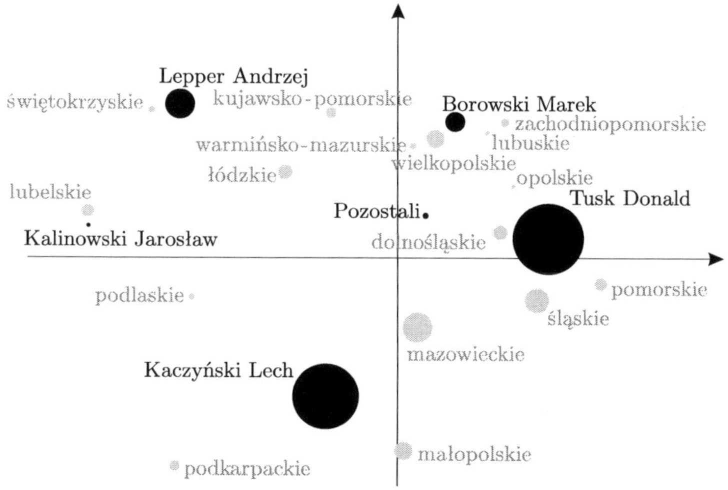

Rys. 14.2. Graficzna prezentacja wyników pierwszej tury wyborów prezydenckich w roku 2005

14.2. Wieloczynnikowa analiza korespondencji

Wieloczynnikowa analiza korespondencji (MCA[9]) może być rozpatrywana jako rozszerzenie prostej analizy korespondencji na zagadnienia o liczbie zmiennych większej od dwóch. Wieloczynnikowa analiza korespondencji to prosta analiza korespondencji prowadzona na macierzy kodów (zwana też złożoną macierzą znaczników), gdzie poszczególne wiersze odpowiadają kolejnym obserwacjom, a kolumny – wariantom zmiennych. W rzeczywistości, zazwyczaj analizuje się iloczyn wewnętrzny takiej macierzy, nazywany w wieloczynnikowej analizie korespondencji *macierzą Burta*.

Przykład 14.3.
Wróćmy do przykładu 14.1. W macierzy kodów każda ze 193 obserwacji jest reprezentowana przez jeden wiersz. Dla każdej obserwacji zaznaczamy 1 w kategorii, do której dany przypadek należy, natomiast w pozostałych kategoriach stawiamy 0. Wycinek macierzy kodów znajduje się w tab. 14.4. Tak utworzona macierz kodów jest traktowana jak tablica kontyngencji w klasycznej analizie korespondencji.

[9]Po ang. Multiple Correspondence Analysis.

Tab. 14.4. Macierz kodów

	Grupa pracownicza					Intensywność palenia			
Nr obserwacji	Starsi kier.	Młodsi kier.	Starszy personel	Młodszy personel	Sekretarki	Nie pali	Mało	Średnio	Dużo
1	1	0	0	0	0	1	0	0	0
2	1	0	0	0	0	1	0	0	0
3	1	0	0	0	0	1	0	0	0
4	1	0	0	0	0	1	0	0	0
5	1	0	0	0	0	0	1	0	0
...	...	...	...	...	...	...	...	...	...
191	0	0	0	0	1	0	0	1	0
192	0	0	0	0	1	0	0	0	1
193	0	0	0	0	1	0	0	0	1

Macierz Burta

Wykorzystywanie macierzy kodów nie jest zbyt wygodne, gdyż najczęściej jest bardzo dużo wierszy, przez co otrzymany wykres jest nieczytelny. Dodatkowo sama interpretacja uzyskanych wyników jest trudna. Najczęściej dokonuje się wieloczynnikowej analizy korespondencji na podstawie macierzy Burta

$$\boldsymbol{B} = \boldsymbol{Z}'\boldsymbol{Z},$$

gdzie przez $\boldsymbol{Z}$ oznaczamy wcześniej omówioną macierz kodów. Macierz Burta ma jasno określoną strukturę. Jest to symetryczna macierz blokowa, w której na przekątnej są umieszczone macierze diagonalne, zawierające liczebności wystąpień kategorii, a poza przekątną znajdują się tablice kontyngencji dla każdej pary badanych zmiennych. Sumy elementów każdej podmacierzy są zawsze takie same i równe całkowitej liczbie badanych obiektów. Całkowita liczebność macierzy Burta wynosi nQ^2, gdzie Q oznacza liczbę cech. Ze względu na symetrię macierzy $\boldsymbol{B}$ częstości brzegowe wierszy i kolumn są sobie równe i wynoszą

$$p_{i\cdot} = \frac{\boldsymbol{Z}'\boldsymbol{1}}{nQ}.$$

Są one elementami wektora $\boldsymbol{r}$. Macierz korespondencji obliczamy ze wzoru

$$\boldsymbol{P} = \frac{\boldsymbol{B}}{nQ^2}.$$

Ponownie ze względu na symetrię macierzy $\boldsymbol{B}$ wykonujemy nieco zmienioną dekompozycję. Mianowicie dekompozycji podlega macierz

$$\boldsymbol{A} = \boldsymbol{D}_r^{-\frac{1}{2}}(\boldsymbol{P} - \boldsymbol{r}\boldsymbol{r}')\boldsymbol{D}_r^{-\frac{1}{2}} = \boldsymbol{U}\boldsymbol{\Gamma}_B^2\boldsymbol{U}',$$

gdzie $\boldsymbol{U}$ jest macierzą wektorów własnych macierzy $\boldsymbol{A}$, a $\boldsymbol{\Gamma}_B^2$ jest macierzą diagonalą zawierającą wartości własne macierzy $\boldsymbol{A}$. Współrzędne kategorii są zawarte w macierzy

$$\boldsymbol{F} = \boldsymbol{D}_r^{-\frac{1}{2}}\boldsymbol{U}\boldsymbol{\Gamma}_B.$$

Przykład 14.4.
Dla danych z przykładu 14.1 macierz Burta została przedstawiona w tab. 14.5.

Tab. 14.5. Macierz Burta

	Starsi kier.	Młodsi kier.	Starszy person.	Młodszy person.	Sekret.	Nie pali	Mało	Średnio	Dużo
Starsi kier.	11	0	0	0	0	4	2	3	2
Młodsi kier.	0	18	0	0	0	4	3	7	4
Starszy personel	0	0	51	0	0	25	10	12	4
Młodszy personel	0	0	0	88	0	18	24	33	13
Sekretarki	0	0	0	0	25	10	6	7	2
Nie pali	4	4	25	18	10	61	0	0	0
Mało	2	3	10	24	6	0	45	0	0
Średnio	3	7	12	33	7	0	0	62	0
Dużo	2	4	4	13	2	0	0	0	25

Uwaga 14.2. Można się również spotkać w przypadku wieloczynnikowej analizy korespondencji z wykorzystaniem zamiast macierzy kodów lub macierzy Burta wielowymiarowej tablicy kontyngencji[10] lub łączonej tablicy kontyngencji[11]. Zainteresowanych odsyłamy do książek Greenacre'a (1993) oraz Stanimir (2005).

Uwaga 14.3. Klasyczna analiza korespondencji jest właściwie metodą metrycznego skalowania wielowymiarowego z odległością χ^2 jako miarą niepodobieństwa. Dokonanie analizy korespondencji odpowiada rozwiązaniu pewnego zadania jednoczesnego skalowania dla obiektów i zmiennych.

[10]Po ang. multiway contingency table.
[11]Po ang. concatenated table.

Literatura

[1] Agre P. E. (1997), *Computation and human experience*, Cambridge University Press. [192][1]

[2] Alpaydin E. (2004), *Introduction to machine learning*, MIT Press. [11]

[3] Anderson J. A. (1982), Logistic discrimination, In: Krishnaiah P. R., Kanal L. N. (eds), *Handbook of Statistics. Vol. II, Classification, Pattern Recognition, and Reduction of Dimension*. North-Holland, s. 169–191. [73]

[4] Anderson T. W., Bahadur R. R. (1962), Classification into two multivariate normal distributions with different covariance matrices, *Ann. Math. Statist.* **33**, s. 420–431. [55]

[5] Bartlett M. S. (1947), Multivariate analysis (with discussion), *J. Roy. Statist. Soc. Suppl.* **9(B)**, s. 176–197. [50]

[6] Basilevsky A. (1994), *Statistical factor analysis and related methods*, Wiley. [330]

[7] Benzécri J. P. (1973), *L'Analyse des donées. Tome 2: L'Analyse des correspondances*, Dunod. [372]

[8] Bishop Ch. (1996), *Neural networks for pattern recognition*, Clarendon Press. [221]

[9] Bishop C. M. (2006), *Pattern recognition and machine learning*, Springer. [11, 19]

[10] Bose R. C., Ray-Chaudhuri D. K. (1960), On a class of error-correcting binary group codes, *Information and Control*, **3**, s. 68–79. [285]

[11] Bradley A. P. (1997), The use of the area under the ROC curve in the evaluation of machine learning algorithms, *Pattern Recognition*, **30(7)**, s. 1145–1159. [103]

[12] Breiman L. (1996a), Stacked regression, *Machine Learning* **24**, s. 49–64. [320, 322]

[13] Breiman L. (1996b), Bagging predictors, *Machine Learning* **26**, s. 123–140. [292]

[1]Liczby w nawiasach kwadratowych podane na końcu każdej pozycji literatury oznaczają numery stron w niniejszej książce, na których dana pozycja jest cytowana.

[14] Breiman L. (1998), Arcing classifiers, *Annals of Statistics* **26**, s. 801–849. [308, 309]

[15] Breiman L. (2001), Random forest, *Machine Learning* **45**, s. 5–32. [186]

[16] Breiman L., Friedman J. H., Olshen R. A., Stone C. J. (1984), *Classification and regression trees*, Chapman and Hall. [155, 178]

[17] Burges C. (1998), A tutorial on support vector machines for pattern recognition, *Knowledge Discovery and Data Mining* **2(2)**, s. 121–167. [116, 119, 126]

[18] Caliński T., Harabasz J. (1974), A dendrite method for cluster analysis, *Communications in Statistics* **3**, s. 1–27. [356]

[19] Caliński T., Lejeune M. (1998), Dimensionality in MANOVA tested by a closed testing procedure, *J. Multivariate Anal.* 65, s. 181–194. [50]

[20] Carpenter G. A., Grossberg S. (1987a), A massively parallel architecture for a self-organizing neural pattern recognition machine, *Computer Vision, Graphics, and Image Processing*, **37**, s. 54–115. [263]

[21] Carpenter G. A., Grossberg S. (1987b), ART 2: Self-organization of stable category recognition codes for analog input patterns, *Applied Optics*, **26**, s. 4919–4930. [263]

[22] Carpenter G. A., Grossberg S. (1988), The ART of adaptive pattern recognition by a self-organizing neural network, *IEEE Computer*, **21(3)**, s. 77–88. [263]

[23] Cattell R. B. (1966), The scree test for the number of factors, *Multivariate Behavioral Research* **1**, s. 245–276. [340]

[24] Chaturvedi A., Wan A. (1999), Estimation of regression coefficients subject to interval constraints in models with non-spherical errors, *Sankhyā*, **61 Series B**, s. 433–442. [323]

[25] Cichosz P. (2000), *Systemy uczące się*, WNT. [10]

[26] Cover T., Hart P. (1967), Nearest neighbor pattern classification, *IEEE Trans. Information Theory*, **13(1)**, s. 21–27. [135, 141]

[27] Cox D. R. (1966), Some procedures associated with the logistic qualitative response curve, In: David F. (Ed.), *Research Papers on Statistics: Festschrift for J. Neyman*, Wiley, s. 55–71. [73]

[28] Day N. E., Kerridge D. F. (1967), A general maximum likelihood discriminant, *Biometrics*, **23**, s. 313–324. [73]

[29] Devroye L., Györfi L., Lugosi G. (1996), *A probabilistic theory of pattern recognition*, Springer-Verlag. [215]

[30] Dietterich T. G., Bakiri G. (1991), Error-correcting output codes: A general method for improving multiclass inductive learning programs, In: *Proceedings of the Ninth International Conference on Artificial Intelligence*, AAAI Press/MIT Press, s. 572–577. [283]

[31] Dietterich T. G., Bakiri G. (1995), Solving multiclass learning problems via error-correcting output codes, *Journal of Artificial Intelligence Research*, **2**, s. 263–286. [283]

[32] Dillon W. R., Goldstein M. (1984), *Multivariate analysis*, Wiley. [135]

[33] Duda R., Hart P., Stork D. (2001), *Pattern classification*, Wiley. [11, 141, 204]

[34] Efron B. (1983), Estimating the error rate of a prediction rule: Improvement on cross-validation, *J. Amer. Statist. Assoc.* **78**, s. 316–331. [67]

[35] Efron B., Tibshirani R. (1997), Improvement on cross-validation: The .632+ bootstrap method, *J. Amer. Statist. Assoc.* **92**, s. 548–560. [67]

[36] Fahlman S. E. (1988), Faster-learning variations on back-propagation: An empirical study, *Proc. of the 1988 Connectionist Models Summer School*, s. 38–51. [230]

[37] Fahlman S. E., Lebiere C. (1989), The cascade-correlation learning architecture, *Advances in Neural Information Processing Systems*, **2**, s. 524–532. [278]

[38] Fisher R. A. (1936), The use of multiple measurements in taxonomic problems, *Ann. Eugen.* **7**, s. 179–188. [42]

[39] Fix E., Hodges J. (1951), Discriminatory analysis. Nonparametric discrimination: Consistency properties, *Technical Raport*, **21-49-004**, *USAF School of Aviation Medicine, Randolph Field, TX.* [135]

[40] Fletcher R., Reeves C. M. (1964), Function minimization by conjugate gradients, *Computer Journal*, **7**, s. 149–154. [234]

[41] Freund Y. (1995), Boosting a weak learning algorithm by majority, *Information and Computation* **121**, s. 256–285. [295]

[42] Freund Y., Schapire R. E. (1996), Experiments with a new boosting algorithm, In: *Machine Learning: Proceedings of the 13th International Conference*, Morgan Kaufman, San Francisco, s. 148–156. [302]

[43] Freund Y., Schapire R. E. (1997), A decision-theoretic generalization of online learning and an application to boosting, *Journal of Computer and System Sciences* **55**, s. 119–139. [292, 295, 297, 303]

[44] Friedman J. H. (1989), Regularized discriminant analysis, *J. Amer. Statist. Assoc.* **84**, s. 165–175. [38, 40]

[45] Friedman J. (1997), On bias, variance, 0/1-loss and the curse-of-dimensionality, *Data Mining and Knowledge Discovery*, **1**, s. 55–77. [151]

[46] Friedman J., Hastie T., Tibhirani R. (2000), Additive logistic regression: A statistical view of boosting, *Annals of Statistics* **28**, s. 337–374. [307, 311]

[47] Fujikoshi Y. (1977), Asymptotic expansion for the distributions of some multivariate tests, In: Krishnaiah P. R. (eds) *Multivariate Analysis. Vol. IV*, North-Holland, s. 55–71. [50]

[48] Fukunaga K. (1990), *Introduction to statistical pattern recognition*, Academic Press. [11, 104]

[49] Gatnar E. (1998), *Symboliczne metody klasyfikacji danych*, Wydawnictwo Naukowe PWN. [10]

[50] Gatnar E. (2000), *Nieparametryczna metoda dyskryminacji i regresji*, Wydawnictwo Naukowe PWN. [10, 161]

[51] Giacinto G., Roli F. (2001), Dynamic classifier selection based on multiple classifier behaviour, *Pattern Recognition* **34**, s. 1879–1881. [314]

[52] Górecki T. (2005a), Stacked regression with restrictions, *Discussiones Mathematicae Probability and Statistics*, **25(I)**, s. 103–113. [324]

[53] Górecki T. (2005b), Effect of choice of dissimilarity measure on classification efficiency with nearest neighbor method, *Discussiones Mathematicae Probability and Statistics*, **25(II)**, s. 217–239. [144, 317]

[54] Górecki T. (2005c), *Sekwencyjna analiza dyskryminacyjna*, Praca doktorska, Uniwersytet im. Adama Mickiewicza w Poznaniu. [34]

[55] Greenacre M. (1984), *Theory and applications of correspondance analysis*, Academic Press. [374, 375]

[56] Greenacre M. (1993), *Correspondance analysis in practice*, Academic Press. [381]

[57] Grossberg S. (1982), *Studies of mind and brain: Neural principles of learning, perception, development, cognition, and motor control*, Reidel Press. [252]

[58] Hand D. J., Tiell R. J. (2001), A simple generalisation of the area under the ROC curve for multiple class classification problems, *Machine Learning* **45**, s. 171–186. [105]

[59] Hart P. (1968), The condensed nearest neighbor rule, *IEEE Transactions on Information Theory*, **16**, s. 515–516. [146]

[60] Hartigan J. A. (1975), *Clustering algorithms*, Wiley. [345]

[61] Hassibi B., Stork D. G., Wolf G. (1993), Optimal brain surgeon and general network pruning, *IEEE International Conference on Neural Networks*, s. 293–300. [277]

[62] Hastie T., Tibshirani R. (1998), Classification by pairwise coupling, *The Annals of Statistics* **26**, s. 451–471. [286]

[63] Hastie T., Tibshirani R., Friedman J.R. (2001), *The elements of statistical learning: Data mining, inference, and prediction*, Springer. [11]

[64] Hastie T., Tibshirani R., Walther G. (2001), Estimating the number of clusters in a dataset via the Gap statistic, *J. Roy. Statist. Soc.* **B 63**, s. 411–423. [356]

[65] Haykin S. (1994), *Neural networks, a comprehensive foundation*, Macmilian College Publising Company. [268]

[66] Hebb D. O. (1949), *The organization of behavior*, Wiley. [187, 209]

[67] van der Heijden F., Duin R. P. W., de Ridder D., Tax D. M. J. (2004), *Classification, parameter estimation and state estimation: An engineering approach using matlab*, Wiley. [360]

[68] Highleyman W. H. (1963), Data for character recognition studies, *IRE Trans. Electr. Computers*, **EC-12**, s. 135–136. [78]

[69] Hocquenghem A. (1959), Codes corecteurs d'erreurs, *Chiffres*, **2**, s. 147–156. [285]

[70] Hopfield J. J. (1982), Neural networks and physical systems with emergent computational abilities, *Proc. Natl. Acad. Sci. USA*, **79**, s. 2554–2558. [245]

[71] Hotelling H. (1933), Analysis of a complex of statistical variables into principal components, *J. Educ. Psychol.* **24**, s. 417–441. [329]

[72] Huang Y. S., Suen C. Y. (1995), A method of combining multiple experts for the recognition of unconstrained handwritten numerals, *IEEE Transactions on Pattern Analysis and Machine Intelligence* **17**, s. 90–93. [316]

[73] Jackson J. E. (1991), *A user's guide to principal components*, Wiley. [330]

[74] Jacobs R. A. (1988), Increased rates of convergence through learning rate adaptation, *Neural Networks*, **1(4)**, s. 295–307. [229]

[75] James A. T. (1969), Test of equality of the latent roots of the covariance matrix, In: Krishnaiah P. R. (eds) *Multivariate Analysis. Vol. II*, Academic Press, s. 205–218. [341]

[76] Jelonek J., Stefanowski J. (1998), Experiments on solving multiclass learning problems by n^2-classifier, In: *Proceedings of the 10th European Conference on Machine Learning (ECML-98)*, Chemnitz, Germany, s. 172–177. [290]

[77] Johnson R. A., Wichern D. W. (1982), *Applied multivariate statistical analysis*, Englewood Cliffs. [40]

[78] Jolliffe I. T. (1986), *Principal component analysis*, Springer. [330]

[79] Jongman R. H. G., ter Braak C. J. F., van Tongeren O. F. R. (1987), *Data analysis in community and landscape ecology*, Wageningen: Pudoc. [369]

[80] Karush W. (1939), Minima of functions of several variables with inequalities as side constraints, *Master's thesis, Department of Mathematics, University of Chicago.* [110]

[81] Kass G. V. (1980), An exploratory technique for investigating large quantities of categorical data, *Applied Statistics* **29**, s. 119–127. [155]

[82] Kaufman L., Rousseeuw P. (1990), *Finding groups in data: An introduction to cluster analysis*, Wiley. [136, 345]

[83] Kecman V. (2005), Basics of machine learning by support vector machines, In: Negoita M. Gh., Reusch B. (eds) *Real World Applications of Computational Intelligence*, Springer, s. 49–103. [107]

[84] Kim H., Loh W.-Y. (2001), Classification trees with unbiased multiway splits, *J. Amer. Statist. Assoc.* **96**, s. 598–604. [186]

[85] Kohavi R. (1995), *Wrappers for performance enhancement and oblivious decision graphs*, praca doktorska, Stanford University. [151]

[86] Kohonen T. (1995), *Self-organising maps*, Springer. [260]

[87] Kołmogorow A. (1957), On the representation of continuous functions of many variables by superposition of continuous functions of one variable and addition, *Doklady Akademii Nauk USSR*, **114**, s. 953–956. [224]

[88] Kong E. B., Dietterich T. G. (1995), Error-correcting output coding corrects bias and variance, In: *12th International Conference on Machine Learning*, Morgan Kaufmann, s. 313–321. [285]

[89] Koontz W. L. G., Fukunaga K. (1972), A nonlinear feature extraction algorithm using distance information, *IEEE Transactions on Computers*, **21(1)**, s. 56–63. [365]

[90] Koronacki J., Ćwik J. (2005), *Statystyczne systemy uczące się*, WNT. [10, 19, 371]

[91] Kosko B. (1988), Adaptive bi-directional associative memories, *IEEE Transactions on Systems Man and Cybernetics*, **18(1)**, s. 49–60. [252]

[92] Krawiec K., Stefanowski J. (2003), *Uczenie maszynowe i sieci neuronowe*, Wydawnictwo Politechniki Poznańskiej. [10]

[93] Kruskal J. B. (1964a), Multidimensional scaling by optinizing goodness of fit to a nonmetric hypothesis, *Psychometrika*, **29**, s. 1–28. [363, 365]

[94] Kruskal J. B. (1964b), Nonmetric multidimensional scaling: A numerical method, *Psychometrika*, **29**, s. 115–129. [363, 365]

[95] Krzanowski W. J., Marriott F. H. C. (1994), *Multivariate analysis. Volume 1 – Distributions, ordination and inference*, Edward Arnold. [136]

[96] Krzanowski W. J., Marriott F. H. C. (1995), *Multivariate analysis. Volume 2 – Classification, covariance structures and repeated measurements*, Edward Arnold. [136]

[97] Krzyśko M. (1990), *Analiza dyskryminacyjna*, WNT. [10]

[98] Krzyśko M. (2000), *Wielowymiarowa analiza statystyczna*, UAM-Poznań. [10, 20, 33, 38, 51]

[99] Kuhn H. W., Tucker A. W. (1951), Nonlinear programming, In: *Proceedings of the 2nd Berkely Symposium on Mathematical Statistics and Probabilities*, University of California Press, s. 481–492. [110]

[100] Kuncheva L. I. (2004), *Combining pattern classifiers: methods and algorithms*, Wiley. [292, 297, 311, 319, 326]

[101] Lange W. (1975), Cell number and cell density in the cerebellar cortex of man and other mammals, *Cell Tissue Res.*, **157(1)**, s. 115–124. [190]

[102] Larose D. T. (2006), *Odkrywanie wiedzy z danych*, Wydawnictwo Naukowe PWN. [10, 279]

[103] Lawley D. N. (1956), Test of significance for the latent roots of covariance and correlation matrices, *Biometrika* **43**, s. 128–136. [341]

[104] Lawley D. N. (1959), Test of significance in canonical analysis, *Biometrika* **46**, s. 59–66. [50]

[105] Lawson J., Hanson R. (1974), *Solving least squares problems*, Prentice-Hall. [322]

[106] LeBlanc M., Tibshirani R. (1996), Combining estimates in regression and classification, *J. Amer. Statist. Assoc.*, **91**, s. 1641–1650. [320, 322]

[107] LeCun Y., Denker J. S., Solla S. A. (1990), Optimal brain damage, *Advances in Neural Information Processing Systems*, **2**, s. 598–605. [277]

[108] Levenberg K. (1944), A method for the solution of certain non-linear problems in least squares, *Quarterly Journal of Applied Mathematics*, **II(2)**, s. 164–168. [238]

[109] Lippmann R. P. (1987), An introduction to computing with neural nets, *IEEE ASSP Magazine*, **5(2)**, s. 4–22. [249]

[110] Loh W.-Y., Shih Y.-S. (1997), Split selection methods for classification trees, *Statistica Sinica* **7**, s. 815–840. [155, 178]

[111] Lorentz G. (1976), The thirteenth problem of Hilbert, *Proceedings of Symposia in Pure Mathematics*, **28**, s. 419–430. [224]

[112] Lubischew A. A. (1962), On the use of discriminant functions in taxonomy, *Biometrics* **18**, s. 455–477. [52]

[113] MacWilliams F. J., Sloane N. J. A. (2006), *The theory of error-correcting codes*, Springer. [285]

[114] Maimon O., Rokach L. (2005), *The data mining and knowledge discovery handbook*, North-Holland. [11]

[115] Marcus R., Peritz E., Gabriel K. R. (1976), On closed testing procedures with special reference to ordered analysis of variance, *Biometrika* **63**, s. 655–660. [51]

[116] Marquardt D. W. (1963), An algorithm for least-squares estimation of non-linear parameters, *Journal of the Society of Industrial and Applied Mathematics*, **11(2)**, s. 431–441. [238]

[117] Mason L., Bartlet P. L., Baxter J. (2000), Improved generalization through explicit optimization of margins, *Machine Learning* **38**, s. 243–255. [311]

[118] McCulloch W. S., Pitts W. H. (1943), A logical calculus of the ideas immanent in nervous activity, *Bulletin of Mathematical Biophysics*, **5**, s. 115–133. [187]

[119] McCullagh P., Nelder J. A. (1989), *Generalized linear models*, 2nd ed., Chapman and Hall. [77]

[120] McEliece R., Posner E., Rodemich E., Venkatesh S.(1987), The capacity of the Hopfield associative memory, *IEEE Transactions on Information Theory*, **33(4)**, s. 461–482. [252]

[121] McKeon J. J. (1974), F approximation to the distribution of Hotelling's T_0^2, *Biometrika*, **61**, s. 381–383. [51]

[122] McLachlan G. J. (1992), *Discriminant analysis and statistical pattern recognition*, Wiley. [11, 64, 66]

[123] van Mechelen I., Bock H. H., de Boeck P. (2004), Two-mode clustering: A structured overview, *Statistical Methods in Medical Research* **13**, s. 363–394. [361]

[124] Mercer J. (1909), Functions of positive and negative type and their connection with the theory of integral equations, *Philosophical Transactions of the Royal Society of London* **209**, s. 415–446. [115]

[125] Merz C. J., Murphy P. M. (1998), UCI repository of machine learning databases. Machine readable data repository http://www.ics.uci.edu/~mlearn/MLRepository.html. Irvine, CA: University of California, Department of Information and Computer Science. [129, 176, 289, 310]

[126] Metropolis N., Rosenbluth A., Rosenbluth M., Teller A., Teller E. (1953), Equation of state calculation by fast computing machines, *J. Chem. Phys.*, **21**, s. 1087–1091. [241]

[127] Micó M. L., Oncina J., Vidal E. (1994), A new version of the nearest-neighbour approximating and eliminating search algorithm (AESA) with linear preprocessing time and memory requirements, *Pattern Recognition*, **15**, s. 9–17. [146]

[128] Mirkin B. (2005), *Clustering for data mining: A data recovery approach*, Chapman and Hall. [345]

[129] Niemann H., Weiss J. (1979), A fast-converging algorithm for nonlinear mapping of highdimensional data to a plane, *IEEE Transactions on Computers*, **28(2)**, s. 142–147. [365]

[130] Novikoff A.B. (1962), On convergence proofs on perceptrons, In: *Proceedings of the Symposium on the Mathematical Theory of Automata*, Polytechnic Institute of Brooklyn, s. 615–622. [196]

[131] Oja E. (1982), Simplified neuron model as a principal component analyzer, *J. Math. Biology*, **15**, s. 267–273. [210]

[132] Osuna E., Freund R., Girosi F. (1997), An improved training algorithm for support vector machines, In: Principe J., Giles L., Morgan N., Wison E. (eds) *Neural Networks for Signal Processing VII–Proceedings of the 1997 IEEE Workshop*, IEEE, New York: s. 276–285. [128]

[133] Oza N. C. (2003), Boosting with averaged weight vectors, In: Windeatt T., Roli F. (eds) *Lecture Notes in Computer Science LNCS*, Vol. 2709 of Proc. 4th International Workshop on Multiple Classifier Systems (MCS 2003), Springer, Guildford, s. 15–24. [311]

[134] Pearson K. (1901), On lines and planes of closest fit to systems of points in space, *Philos. Mag. Ser. 6* **2**, s. 559–572. [329]

[135] Pfister M., Rojas R. (1994), Qrprop – a hybrid learning algorithm which adaptively includes second order information, *Proc. 4th Dortmund Fuzzy Days*, s. 55–62. [231]

[136] Powell M. J. D. (1977), Restart procedures for the conjugate gradient method, *Mathematical Programming*, **12(1)**, s. 241–254. [236]

[137] Press W. H., Flannery B. P., Teukolsky S. A., Vetterling W. T. (1992), *Numerical Recipes. The art of Scientific Computing*, Cambridge University Press. [369]

[138] Quinlan J. R. (1993), *C4.5: Programs for machine learning*, Morgan Kaufmann. [155]

[139] Rao C. R. (1964), The use and interpretation of principal components in applied research, *Sankhya* **A 26**, s. 329–358. [335]

[140] Reed R. D., Marks R. J. (1999), *Neural smithing: Supervised learning in feedforward artifical neural networks*, MIT Press. [229]

[141] Rencher A. C. (1995), *Methods of multivariate analysis*, Wiley. [337]

[142] Rencher A. C. (1998), *Multivariate statistical inference and applications*, Wiley. [48, 339]

[143] Riedmiller M., Braun H. (1993), A direct adaptive method for faster backpropagation learning: The RPROP algorithm, *Proc. of the IEEE International Conference on Neural Networks*, s. 586–591. [230]

[144] Rumelhart D., Hinton G., Williams, R. (1986), *Learning internal representations by error propagation. In Parallel Distributed Processing, chapter 8*, MIT Press. [226]

[145] Rutkowska D., Piliński M., Rutkowski L. (1999), *Sieci neurnowe, algorytmy genetyczne i systemy rozmyte*, Wydawnictwo Naukowe PWN. [244]

[146] Rutkowski L. (2005), *Metody i techniki sztucznej inteligencji*, WNT. [205, 209]

[147] Sammon J. W. (1969), A Nonlinear Mapping for Data Structure Analysis, *IEEE Transactions on Computers*, **18(5)**, s. 401–409. [365]

[148] Schapire R. E. (1990), The strength of weak learnability, *Machine Learning* **5**, s. 197–227. [292, 295]

[149] Schölkopf B., Sung K., Burges C., Girosi F., Niyogi P., Poggio T., Vapnik V. (1997), Comparing support vector machines with gaussian kernels to radial basis function classifier, *IEEE Trans. Sign. Processing*, **45**, s. 2758–2765. [272]

[150] Sheather S. J., Jones M. C. (1991), A reliable data-based bandwidth selection method for kernel density estimation, *J. Roy. Statist. Soc. B* **53**, s. 683–690. [84]

[151] Shepard R. N. (1962a), The analysis of proximities: Multidimensional scaling with an unknown distance function. I, *Psychometrika*, **27**, s. 125–140. [363]

[152] Shepard R. N. (1962b), The analysis of proximities: Multidimensional scaling with an unknown distance function. II, *Psychometrika*, **27**, s. 219–246. [363]

[153] Silva F. M., Almeida L. B. (1990), Speeding up backpropagation, In: Eckmiller R. (eds.) *Advanced Neural Computers*, s. 151–158. [229]

[154] Silverman B. W. (1986), *Density estimation for statistics and data analysis*, Chapman and Hall. [86]

[155] Srivastava M. S. (2002), *Methods of multivariate statistics*, Wiley. [52]

[156] Stanimir A. (2005), *Analiza korespondencji jako narzędzie do badania zjawisk ekonomicznych*, Wydawnictwo AE we Wrocławiu. [381]

[157] Stone C. (1977), Consistent nonparametric regression, *Annals of Statistics* **5**, s. 595–645. [84]

[158] Tadeusiewicz R. (1993), *Sieci neuronowe*, Akademicka Oficyna Wydawnicza RM. [203, 204]

[159] Theil H., Goldberger A. S. (1961), On pure and mixed statistical estimation in economics, *International Economics Review*, **2**, s. 65–78. [322]

[160] Torgerson W. S. (1952), Multidimensional scaling: I. Theory and method, *Psychometrika*, **17**, s. 401–419. [363]

[161] Toutenburg H., Roeder B. (1978), Minimax-linear and theil estimator for restricted regression coefficients, *Statistics*, **9**, s. 499–505. [322, 323]

[162] Tryon R. C. (1939), *Cluster analysis*, Edwards Brothers. [345]

[163] Ultsch A. (1993), Self-organizing neural networks for visualization and classification, In: Opitz O., Lausen B., Klar R. (eds) *Information and classification*, Springer, s. 307–313. [261]

[164] Vapnik V. V. (1995), *The nature of statistical learning theory*, Springer. [11, 115, 123]

[165] Vapnik V. V., Chervonenkis A. Y. (1971), On the uniform convergence of relative frequencies of events to their probabilities, *Theory of Probability and Its Applications* **16**, s. 264–280. [107]

[166] Vapnik V. V., Chervonenkis A. Y. (1974), *Theory of pattern recognition (po rosyjsku)*, Nauka. [107]

[167] Vilenkin N. Ya. (1979), *Method of successive approximations*, Mir Publishers. [62]

[168] Ward J. H. (1963), Hierarchical grouping to optimize an objective function, *J. Amer. Statist. Assoc.* **58**, s. 236–244. [350]

[169] Webb G. I. (2000), Multiboosting: A technique for combining boosting and bagging, *Machine Learning* **40**, s. 159–196. [311]

[170] Webb A. (2003), *Statistical pattern recognition*, Wiley. [11, 19, 65, 215, 312]

[171] Wehberg S., Schumacher M. (2004), A comparison of nonparametric error rate estimation methods in classification problems, *Biometrical Journal* **46**, s. 35–47. [66]

[172] Werbos P. J. (1974), *Beyond regression: New tools for prediction and analysis in the behavioral sciences*, praca doktorska, Harvard University. [226]

[173] Werbos P. J. (1988), Backpropagation: Past and future, *Proc. of the IEEE International Conference on Neural Networks*, s. 343–353. [229]

[174] Widrow B., Hoff M. E. (1960), Adaptive switching circuits, *IRE WESCON Convention Record*, **4**, s. 96–104. [202, 219]

[175] Wilson D. (1972), Asymptotic properties of nearest neighbor rules using edited data, *IEEE Transactions on Systems, Man, and Cybernetics*, **SMC-2**, s. 408–421. [146]

[176] Wolpert D. (1992), Stacked generalization, *Neural Networks*, **5**, s. 241–259. [320]

[177] Woods K., Kegelmeyer W. P., Bowyer K. (1997), Combination of multiple classifiers using local accuracy estimates, *IEEE Transactions on Pattern Analysis and Machine Intelligence*, **19(4)**, s. 405–410. [313, 317]

Polski skorowidz pojęć

J

K

L

M

N

O

P

R

S

T

U

W

Z

Angielski skorowidz pojęć

T

V

W

WNT. Warszawa 2008. Wyd. I
Ark. wyd. 25,5. Ark. druk. 25,0
Symbol MF/84454/WNT